Cold and Ultracold Molecules

Durham University, UK
15–17 April 2009

FARADAY DISCUSSIONS
Volume 142, 2009

RSC Publishing

The Faraday Division of the Royal Society of Chemistry, previously the Faraday Society, founded in 1903 to promote the study of sciences lying between Chemistry, Physics and Biology.

EDITORIAL STAFF

Editor
Philip Earis

Assistant editor
Madelaine Chapman

Publishing assistant
Kate McCallum

Team leader, informatics
Gisela Scott

Technical editor
Joanna Pugh

Publisher
Janet Dean

Faraday Discussions (Print ISSN 1359-6640, Electronic ISSN 1364-5498) is published 4 times a year by the Royal Society of Chemistry, Thomas Graham House, Science Park, Milton Road, Cambridge, UK CB4 0WF. Volume 142 ISBN-13: 978 1 84755 8374

2009 annual subscription price: print+electronic £592, US $1,160; electronic only £533, US $1045. Customers in Canada will be subject to a surcharge to cover GST. Customers in the EU subscribing to the electronic version only will be charged VAT. All orders, with cheques made payable to the Royal Society of Chemistry, should be sent to RSC Distribution Services, c/o Portland Customer Services, Commerce Way, Colchester, Essex, UK CO2 8HP.
Tel +44 (0) 1206 226050;
E-mail sales@rscdistribution.org

If you take an institutional subscription to any RSC journal you are entitled to free, site-wide web access to that journal. You can arrange access *via* Internet Protocol (IP) address at www.rsc.org/ip. Customers should make payments by cheque in sterling payable on a UK clearing bank or in US dollars payable on a US clearing bank. Periodicals postage is paid at Rahway, NJ and at additional mailing offices. Airfreight and mailing in the USA by Mercury Airfreight International Ltd., 365 Blair Road, Avenel, NJ 07001, USA.

US Postmaster: send address changes to *Faraday Discussions*, c/o Mercury Airfreight International Ltd., 365 Blair Road, Avenel, NJ 07001. All despatches outside the UK by Consolidated Airfreight.

PRINTED IN THE UK

Faraday Discussions documents a long-established series of *Faraday Discussion* meetings which provide a unique international forum for the exchange of views and newly acquired results in developing areas of physical chemistry, biophysical chemistry and chemical physics.

ORGANISING COMMITTEE, Volume 142

Chair
Jeremy Hutson (Durham, UK)

Editor
Ed Hinds (Imperial College London, UK)
Rudolf Grimm (University of Innsbruck, Austria)
Gerard Meijer (Fritz-Haber-Institut, Berlin, Germany)
Tim Softley (Oxford, UK)
William Stwalley (University of Connecticut, USA)

FARADAY STANDING COMMITTEE ON CONFERENCES

Chair
D E Heard (Leeds, UK)

W A Brown (UCL, UK)
H M Colquhoun (Bath, UK)
G Jackson (Imperial, UK)
A Rodger (Warwick, UK)

© The Royal Society of Chemistry 2009. Apart from fair dealing for the purposes of research or private study, or criticism or review, as permitted under the Copyright, Designs and Patents Act 1988 and Related Rights Regulations 2003, this publication may only be reproduced, stored or transmitted, in any form or by any means, with the prior permission in writing of the Publishers or in the case of reprographic reproduction in accordance with the terms of licences issued by the Copyright Licensing Agency in the UK. US copyright law applicable to users in the USA. The Royal Society of Chemistry takes reasonable care in the preparation of this publication but does not accept liability for the consequences of any errors or omissions.

Royal Society of Chemistry: Registered Charity No. 207890.

∞The paper used in this publication meets the requirements of ANSI/NISO Z39.48-1992 (Permanence of Paper).

Cold and Ultracold Molecules

Faraday Discussions

www.rsc.org/faraday_d

A General Discussion on Cold and Ultracold Molecules was held at Durham University, UK on 15th, 16th and 17th April 2009.

RSC Publishing is a not-for-profit publisher and a division of the Royal Society of Chemistry. Any surplus made is used to support charitable activities aimed at advancing the chemical sciences. Full details are available from www.rsc.org

CONTENTS

ISSN 1359-6640; ISBN 978-1-84755-837-4

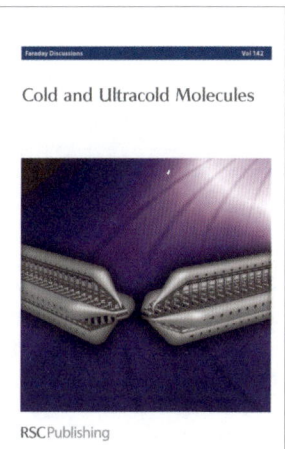

Cover
See Sebastiaan van de Meerakker, *Faraday Discuss.*, 2009, **142**, 113–126.

Artist's impression of a molecular beam scattering machine consisting of two Stark decelerators under a 90 degree crossing angle (design: Henrik Haak).

Image reproduced by permission of Dr Sebastiaan van de Meerakker, from *Faraday Discuss.*, 2009, **142**, 113.

INTRODUCTORY LECTURE

9   **Molecular collisions, from warm to ultracold**
Dudley Herschbach

PAPERS AND DISCUSSIONS

25   **Testing the time-invariance of fundamental constants using microwave spectroscopy on cold diatomic radicals**
Hendrick L. Bethlem and Wim Ubachs

37   **Prospects for measuring the electric dipole moment of the electron using electrically trapped polar molecules**
M. R. Tarbutt, J. J. Hudson, B. E. Sauer and E. A. Hinds

57   **Buffer gas cooling of polyatomic ions in rf multi-electrode traps**
D. Gerlich and G. Borodi

73   **Ion-molecule chemistry at very low temperatures: cold chemical reactions between Coulomb-crystallized ions and velocity-selected neutral molecules**
Martin T. Bell, Alexander D. Gingell, James M. Oldham, Timothy P. Softley and Stefan Willitsch

Enjoyed this discussion?
How about some others...

Faraday Discussions provide a unique discussion forum for original research in physical chemistry, chemical physics and biophysical chemistry.

Previous volumes of interest include:

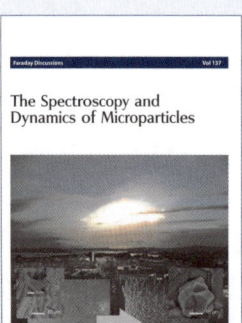

The Spectroscopy and Dynamics of Microparticles
Faraday Discussion 137
This volume focuses on micron sized particles dispersed in a gaseous or liquid medium. Work is presented in a range of areas including the optical spectroscopy, chemical dynamics, physical dynamics and new techniques for manipulating and characterising microparticles.

Chemical Concepts from Quantum Mechanics
Faraday Discussion 135
This volume brings together the communities of scientists working in the areas of high-resolution crystallography, Valence Bond (VB) theory, Molecular Orbital theory, conceptual Density Functional Theory (DFT) and Quantum Chemical Topology (QCT), driven by the importance of bridging the expanding gap between widely used chemical concepts and modern accurate physical and computational data on molecules.

Chemical Evolution of the Universe
Faraday Discussion 133
This volume focuses on recent astronomical observations of molecules in various regions and eras of the Universe, and on describing the processes that determine the chemistry. These processes include gas phase neutral exchanges and ion-molecule reactions, surface reactions on dust grains, and the radiative processing of dirty ices.

To view the full contents and, where available, purchase an article or a volume visit the website.

RSCPublishing   www.rsc.org/faraday

Registered Charity Number 207890

93	**General discussion**
113	**Collision experiments with Stark-decelerated beams** Sebastiaan Y. T. van de Meerakker and Gerard Meijer
127	**Dynamics of OH($^2\Pi$)–He collisions in combined electric and magnetic fields** Timur V. Tscherbul, Gerrit C. Groenenboom, Roman V. Krems and Alex Dalgarno
143	**Production of cold ND_3 by kinematic cooling** Jeffrey J. Kay, Sebastiaan Y. T. van de Meerakker, Kevin E. Strecker and David W. Chandler
155	**Manipulating the motion of large neutral molecules** Jochen Küpper, Frank Filsinger and Gerard Meijer
175	**Sympathetic cooling by collisions with ultracold rare gas atoms, and recent progress in optical Stark deceleration** P. F. Barker, S. M. Purcell, P. Douglas, P. Barletta, N. Coppendale, C. Maher-McWilliams and J. Tennyson
191	**Prospects for sympathetic cooling of polar molecules: NH with alkali-metal and alkaline-earth atoms – a new hope** Pavel Soldán, Piotr S. Żuchowski and Jeremy M. Hutson
203	**Continuous guided beams of slow and internally cold polar molecules** Christian Sommer, Laurens D. van Buuren, Michael Motsch, Sebastian Pohle, Josef Bayerl, Pepijn W. H. Pinkse and Gerhard Rempe
221	**General discussion**
257	**Broadband lasers to detect and cool the vibration of cold molecules** Matthieu Viteau, Amodsen Chotia, Dimitris Sofikitis, Maria Allegrini, Nadia Bouloufa, Olivier Dulieu, Daniel Comparat and Pierre Pillet
271	**Dark state experiments with ultracold, deeply-bound triplet molecules** Florian Lang, Christoph Strauss, Klaus Winkler, Tetsu Takekoshi, Rudolf Grimm and Johannes Hecker Denschlag
283	**Precision molecular spectroscopy for ground state transfer of molecular quantum gases** Johann G. Danzl, Manfred J. Mark, Elmar Haller, Mattias Gustavsson, Nadia Bouloufa, Olivier Dulieu, Helmut Ritsch, Russell Hart and Hanns-Christoph Nägerl
297	**Rotational spectroscopy of single carbonyl sulfide molecules embedded in superfluid helium nanodroplets** Rudolf Lehnig, Paul L. Raston and Wolfgang Jäger
311	**Self-organisation and cooling of a large ensemble of particles in optical cavities** Yongkai Zhao, Weiping Lu, P. F. Barker and Guangjiong Dong
319	**General discussion**
335	**Formation of ultracold dipolar molecules in the lowest vibrational levels by photoassociation** J. Deiglmayr, M. Repp, A. Grochola, K. Mörtlbauer, C. Glück, O. Dulieu, J. Lange, R. Wester and M. Weidemüller
351	**Ultracold polar molecules near quantum degeneracy** S. Ospelkaus, K.-K. Ni, M. H. G. de Miranda, B. Neyenhuis, D. Wang, S. Kotochigova, P. S. Julienne, D. S. Jin and J. Ye

361 **Ultracold molecules from ultracold atoms: a case study with the KRb molecule**
Paul S. Julienne

389 **Two-photon coherent control of femtosecond photoassociation**
Christiane P. Koch, Mamadou Ndong and Ronnie Kosloff

403 **A pump–probe study of the photoassociation of cold rubidium molecules**
Jovana Petrovic, David McCabe, Duncan England, Hugo Martay, Melissa Friedman, Alexander Dicks, Emiliya Dimova and Ian Walmsley

415 **Fano profiles in two-photon photoassociation spectra**
Maximilien Portier, Michèle Leduc and Claude Cohen-Tannoudji

429 **General discussion**

CONCLUDING REMARKS

463 **Concluding remarks: achievements and challenges in cold and ultracold molecules**
F. A. Gianturco and M. Tacconi

ADDITIONAL INFORMATION

479 **Poster titles**
483 **List of participants**
487 **Index of contributors**

Molecular collisions, from warm to ultracold

Dudley Herschbach*

Received 21st May 2009, Accepted 27th May 2009
First published as an Advance Article on the web 22nd July 2009
DOI: 10.1039/b910118g

This introductory article contrasts molecular collisions, particularly reactive collisions, in the familiar "warm" domain with the ultracold regime where the relative deBroglie wavelengths become long compared with the range of interaction of the collision partners. Ultracold collisions have much greater sensitivity to entrance channel interactions, so offer the prospect of tuning by external fields to control onset of reaction. However, for ultracold collisions, kinematic constraints impose severe limitations on the observable dynamical properties. In the exit channel for appreciably exoergic reactions, the deBroglie wavelengths become short, so the exit dynamics are much like those for warm collisions. Reactions of alkali dimers, halides, and monoxide molecules are discussed that seem especially congenial for cold collision studies.

I. Introduction

The Organizing Committee gave me a pleasant assignment: to "link recent progress in cold molecules with former developments in molecular beams." The advent of these fields as compelling frontiers in molecular physics is about four decades apart in time. Yet they are akin in enterprising spirit, buoyed by a sense of historical imperative. For cold molecules, new recruits and curious visitors will find excellent surveys of remarkable experimental and theoretical methods, prospects and challenges, in several current reviews[1-8] and a forthcoming evangelical book.[9] For molecular beams, both historical and new developments have been recently reviewed[10,11] and a classic canonical text[12] has been updated.[13]

Research with cold molecules, as with molecular beams, is very broad in scope, as exemplified admirably in this volume of *Faraday Discussions*. My paper considers almost solely one major area, probing the dynamics of collisions, especially reactive collisions. For anticipatory and adroit discussions focused on ultracold collisions, I particularly recommend a succinct account by Stwalley[1] and a comprehensive exposition by Bell and Softley.[8] After a prelude meant to supply complementary background, I contrast basic aspects of molecular collisions in the familiar "warm" regime with the cold and ultracold regimes, emphasizing observable properties and the role of kinematic constraints. In response to the blossoming cold molecule "alkali age", next I discuss some A + BC atom exchange reactions that seem especially amenable for cold collision studies. These involve three kinds of alkali molecules: dimers, halides, and monoxides. In closing, I invoke a benedictory comment by Otto Stern and confess to becoming captivated by long deBroglie matter waves.

Department of Chemistry and Chemical Biology, Harvard University, 12 Oxford St., Cambridge, Massachusetts, 02138, U.S.A. E-mail: dherschbach@yahoo.com

II. Prelude: antecedents and outlook

A. Cold molecules

Much of the appeal of exploring molecular interactions in the "cold" (<1 K) and "ultracold" (<1 mK) regimes does not stem from attaining low temperatures as such. Rather, a chief aim is to gain access to dramatic quantum phenomena that become prominent when deBroglie wavelengths become comparable to, or longer than, the size or the separation of molecules. This aspect has a direct historical antecedent in nuclear physics during the 1930s. For collisions of protons or neutrons at a kinetic energy of $\sim$1 MeV (equivalent to $\sim 10^{10}$ K), the deBroglie wavelength ($\sim$40 $\times$ 10^{-13} cm) is considerably longer than the range of nuclear forces. Consequently, s-wave scattering, resonances, and tunneling became major features in probing nuclear collisions. In molecular physics, it has proved much more difficult to reach the corresponding long wavelength realm. In contrast to the nuclear case, however, both intramolecular and intermolecular interactions can be strongly influenced by applying external electric or magnetic fields. Coherent excitation processes induced by laser light provide another powerful tool. Moreover, at least for prototype systems, molecular theory and electronic structure calculations can often offer guidance in the design and interpretation of experiments.

Attaining the ultracold molecular regime requires liberation from thermodynamics. That aspect has its motivating antecedent in Bose–Einstein condensation (BEC), long presumed to be attainable only for liquid helium. It was expected that, before the temperature and density required to form a degenerate quantum gas could be reached, thermodynamics would impose the mundane process of ordinary condensation. Now BEC has been obtained, in the nanokelvin range, for many atomic and a few molecular vapors. That required finding pathways along which the kinetics become much too slow to reach equilibrium states. In fact, this principle has long governed much of conventional chemical synthesis. Hordes of organic molecules, many crucial for biology, are thermodynamically unstable but are made by exploiting kinetically dominated pathways. That also must be why sizable organic molecules appear in the interstellar medium, despite the low cosmic abundance of carbon.[14] Perhaps the exotic example of ultracold gas phase molecules may embolden even pragmatic chemists to seek more extreme means to evade thermodyamics.

B. Molecular beam collisions

Like ultracold atomic and molecular research, the era of "warm" collision experiments with molecular beams began with an "alkali age." A key antecedent was surface ionization on a hot-wire, found in the 1920s in Otto Stern's laboratory to provide an extremely sensitive and specific detector for alkali species. Also in the 1920s, Michael Polanyi had shown that gas phase alkali atoms reacted rapidly with a wide variety of halogen-containing molecules.[15] Crossed-beam collision experiments of such reactions did not come until the mid-1950s, but over the next 20 years hundreds of beam studies of alkali atom reactions (as well as inelastic and elastic scattering) were conducted. Much variety, largely unanticipated, was found in the dynamics for reactions with different target molecules. Two chief categories became evident in the form of product angular distributions, $I(\theta)$, in the center-of-mass frame. For impulsive or *brisk* reactions $I(\theta)$ exhibits marked forward-backward asymmetry with respect to the incident alkali atom beam, indicating that the products separate very rapidly (typically in $\sim 10^{-13}$ s or less). For *lingering* reactions, $I(\theta)$ has forward-backward symmetry, indicating that an intermediate complex forms that lives long enough to execute at least a few full rotations (typically 5×10^{-12} or longer).

The brisk reactions, usually quite exoergic, involve electron transfer from the alkali atom to disrupt a target covalent bond and form an alkali halide with a considerably stronger, highly ionic bond. The electron transfer (termed "harpooning" by

Michael Polanyi) becomes energetically allowed within a curve-crossing radius R_C, where the ionic potential energy drops below the covalent potential; this can be estimated from the difference in ionization energy of the alkali atom and electron affinity of the halogen. Striking trends with R_C were found in the size of the reaction cross section, in the form of $I(\theta)$, which ranged from sharply forward ("stripping") to backward ("rebound"), and in the disposal of reaction exoergicity in product translation, vibration, and rotation. The lingering reactions, usually only modestly exoergic, have an attractive potential well that fosters transient formation of the intermediate complex. Often, rather than the "ski-run" exit typical for brisk reactions, the decay of the lingering complex is slowed by an exit centrifugal barrier.

From the late 1960s onward, molecular beam collision studies moved far beyond the alkali age, exploiting supersonic beam sources, mass spectrometric and ion imaging detection, and many varieties of chemiluminescence and laser spectroscopy. A host of atom exchange reactions employing beams of hydrogen, halogen, or oxygen atoms were studied; target molecules included many organic molecules, as well as favorite diatomics such as H_2, hydrogen halides, and halogens. The ability to control or analyze reactant and product dynamical properties was also much extended. Among them are means of spatially orienting reactant molecules prior to collision as well as observing the distribution of product rotational angular momentum and its orientation. In particular, these make accessible angular correlations between three (or more) vectors, such as the relative velocities of reactants and products plus the product molecule rotational orientation. Such stereodynamic correlations deserve special attention; as again shown in nuclear physics, they enable the recovery of information that is otherwise lost by averaging over intrinsically uncontrollable random initial conditions.[16,17]

For physical chemists the great appeal of molecular beams has been the ability to study reactions under single-collision conditions and thereby elucidate intimate reaction dynamics. However, the deeper mission is to be able to predict the rate and outcome of reactions. That requires us to relate the reaction dynamics to electronic structure and how it changes in the transition from reactants to products. In that task, crossed-beam collision experiments are handicapped because the preparation of reactants and observation of products are done only in asymptotic regions, where the collision partners are not interacting. This has been overcome by the development of femtochemistry, led by Ahmed Zewail,[18] which provides incisive means to follow dynamics between a laser excited transition state and its progression to the asymptotic separation of products.

Yet even experiments limited to asymptotic states have provided instances where the relation of dynamics to electronic structure became gratifyingly clear, at least qualitatively. A favorite example (pertinent to later discussion): the H + Cl_2 and K + ICH_3 reactions, both brisk rebound reactions, display essentially identical product angle-velocity contour maps[19] (aside from scale factors involving mass and exoergicity). That is remarkable, since the K reaction involves harpooning and the H reaction does not. A simple molecular orbital rationale was found using the "frontier orbital" concept of Fukui. For H–Cl–Cl, the axial node in that orbital lies midway between the Cl atoms, inducing strong repulsion between them; the repulsive energy release is quantitatively the same as that in photodissociation of Cl_2. The same holds for the K + ICH_3 reaction, where the harpooning electron enters the same orbital that governs photodissociation of ICH_3. For several related reactions, dynamical properties proved to be consistent with kindred molecular orbital arguments. Overall, much learned from alkali reactions turned out to presage the dynamics and its relation to electronic structure found in exploring the wider chemical realm.

III. Contrasting warm *vs* ultracold collisions

In these contrasts, for brevity I focus on the ultracold limit since properties for less cold collisions may be inferred by interpolation. The most evident contrast is in the deBroglie wavelengths, λ_T, for translational motion. In the warm regime, λ_T is usually so short that classical mechanics can provide a reliable description of most aspects of reaction dynamics. In the ultracold realm, where λ_T is much larger than the typical range of chemical interactions, quantum mechanics becomes essential. The situation is actually equivocal, however, because λ_T pertains just to the asymptotic approach and accelerations during collision can markedly shorten the local wavelength.

A. Elastic scattering

The contrast is huge for elastic scattering. In the warm realm, elastic collisions typically occur with very large orbital angular momentum (hundreds of $\hbar$ units). Ford and Wheeler provided, 50 years ago, a semiclassical treatment that treats the angular momentum as a continuous variable, proportional to the classical impact parameter.[20] This allows the quantum scattering amplitude to be accurately approximated by assembling quantities computed using purely classical mechanics. Different branches of the classical deflection function, corresponding to distinct ranges of impact parameters, are simply combined with suitable phase factors. The semiclassical results agree nicely with full-scale quantum partial wave calculations, but bring out much more clearly the origin of characteristic features (rainbow and glory structure, fast and slow interference oscillations). Moreover, the semiclassical analysis aids inversion of experimental data to determine atom–atom intermolecular potential functions.

In the ultracold limit, collisions involve only the s-wave phase shift (except for indistinguishable fermions, but then the p-wave is suppressed by the $l=1$ centrifugal barrier). Since the angular distribution is isotropic, the only observable is the total cross section, given by the scattering length. For a known intermolecular potential, $V(R)$, the scattering length can usually be determined from a simple analytical semiclassical approximation.[21] That is applicable, despite the extremely long asymptotic wavelength, because typically the intermolecular potential has an attractive well, van der Waals or chemical, deep enough to shrink the local wavelength to semiclassical size within the well. The scattering length is very sensitive to the shape of the potential, however, in the well region as well as at long-range. If $V(R)$ is not already accurately known, little about it can be inferred from an experimental measurement of the scattering length.

B. Kinematic properties

The conservation laws for energy, linear momentum, and angular momentum in collisions are of course the same, regardless of the initial conditions. The ultracold limit imposes important further constraints. These are illustrated in Fig. 1–3 for an A + BC → AB + C reactive collision. The captions define the notation and main points. As evident in Fig. 1, for the ultracold limit inelastic or reactive collisions can only occur if exoergic ($\Delta D_0 > 0$). Even a modest energy release into relative translation of the products E_T' can make the associated exit deBroglie wavelength much shorter, thereby converting "ultracold in" to "warm out." Sufficiently large E_T' would effectively convert "quantum waves in" to "semiclassical particles out."

In Fig. 2, the velocity vector diagram for the warm domain is of the kind introduced at a 1962 Faraday Discussion.[22] It was named a "Newton Diagram," because classical mechanics suffices as the vectors involved pertain to asymptotic states with the collision partners far apart. Since the center-of-mass vector **C** is a constant of motion, the proper frame for viewing collisional interactions should travel with **C**, so the diagram serves to display the transformation between the lab and cm frames. For any specified exit translational energy E_T', the magnitude of the velocity vector in the cm frame for the AB product, $\mathbf{u}_{AB}$, is fixed. Kinematics allows $\mathbf{u}_{AB}$ to have any direction, although its distribution must have azimuthal symmetry about the initial

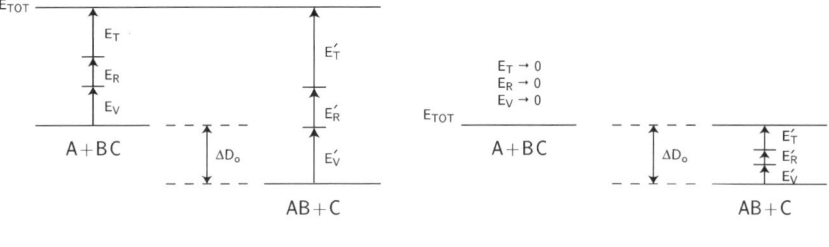

Fig. 1 Energy conservation for an A + BC → AB + C reaction. In the "warm" domain, the total energy available to the products is the sum of the asymptotic relative translational energy of A + BC, plus any initial rotational and/or vibrational excitation of the BC molecule, plus $\Delta D_0 = D_0(AB) - D_0(BC)$, the difference in dissociation energies of product and reactant molecules (relative to their lowest vibrational levels). In the ultracold limit, only ΔD_0 is available to be distributed among relative translation of AB + C and rotation, and/or vibration of AB. If present, electronic excitation or other energetic modes would enter in an analogous fashion.

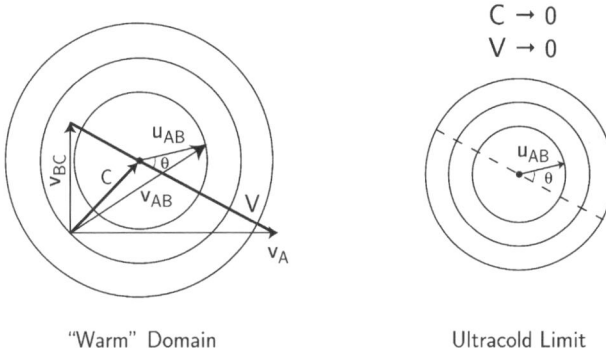

Fig. 2 Velocity vector diagrams for A + BC reaction ("Newton diagrams"). In the "warm" domain, **C** represents the center-of-mass velocity, which remains constant throughout the collision; **V** the relative velocity of A + BC. Spheres indicate the range of product AB velocity vector $\mathbf{u}_{AB}$ relative to **C** (*i.e.*, in center-of-mass frame) allowed by kinematics; $\mathbf{u}_{AB}$ can have any direction but its magnitude $u_{AB} = (m_C/m_{ABC})V'$ with $V' = (2E_T'/\mu')^{1/2}$ the product relative velocity, is determined (*cf.* Fig. 1) by the product reduced mass μ' and the asymptotic relative translational energy, $E_T' = E_{TOT} - E_R' - E_V'$. Hence the outermost sphere designates maximum u_{AB}, the others (only two shown) correspond to excitation of various rotational and vibrational states of the AB product, which reduces the energy appearing in translation. The product velocity vector in the lab frame is $\mathbf{v}_{AB} = \mathbf{C} + \mathbf{u}_{AB}$. Diagram for the atom C is determined by $m_C\mathbf{u}_C = -m_{AB}\mathbf{u}_{AB}$. In the ultracold limit, the lab frame becomes coincident with the center-of-mass frame and the radii of spheres shrink because E_{TOT} is reduced to just ΔD_0.

relative velocity vector **V**, by virtue of the "dart-board" random distribution of impact parameters. The actual angular distribution $I(\theta)$ of $\mathbf{u}_{AB}$ with respect to **V** is determined by dynamical forces.

In the ultracold limit, the lab and cm frames become coincident. Since the initial velocity vector vanishes, the $\mathbf{u}_{AB}$ vector no longer has a defined reference axis (the dashed line merely indicates the "ghost" of **V**). Thus, no matter what dynamical forces operate as the products exit, the observable product angular distribution becomes isotropic.

In Fig. 3, for the warm domain, conservation of the total angular momentum vector **J** requires the vector sum of collisional orbital momentum ***l*** and molecular rotation ***j*** for the reactants to equal that for the products (again distinguished by primes). The orbital momentum is perpendicular to the relative velocity vector,

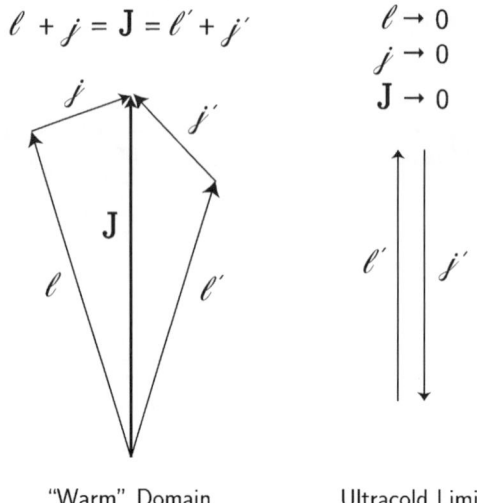

Fig. 3 Angular momentum conservation for A + BC reaction (without consideration of any electronic or nuclear spins). Total angular momentum **J** remains constant throughout the reactive collision; it is the sum of initial orbital angular momentum vector *l* for the approach of A + BC and rotational momentum vector *j* of the BC molecule, equal to the sum of final orbital momentum vector *l'* for the retreat of AB + C and rotational momentum vector *j'* for the AB molecule. In the ultracold limit, since $J \to 0$, the product orbital and rotational momentum vectors are constrained to be equal in magnitude but antiparallel, $l' = -j'$.

V; in classical mechanics its magnitude is given by $l = (2\mu E_T)^{1/2} b$, in terms of the reduced mass, relative translational energy, and impact parameter. The molecular rotational momentum may have any spatial orientation; its classical magnitude is $j = (2IE_R)^{1/2}$, in terms of the molecular moment of inertia and rotational energy. Various special cases give rise to distinctive "form factors" for the product angular distribution and other properties; two that often occur are

(I) $l \gg j$ with $l' \ll j'$ and (II) $l \gg j$ with $l' \gg j'$

These cases have proved especially useful in developing diagnostic models for reactive scattering; *e.g.*, for (I) see ref. 23, for (II) see ref. 24. Also very informative are angular correlations involving combinations of the angular momentum vectors and the initial and final relative velocity vectors, **V** and **V'**. Particularly for reactions governed by a statistical collision complex, these are readily evaluated by appropriate averaging over unobserved angles.[25]

In the ultracold limit, only the product vectors survive and $l' = -j'$ is a strict constraint, rendering these vectors equivalent, with both perpendicular to the exit relative velocity vector, **V'**. Unfortunately, that directional correlation is not observable, nor is any other in this limit. As seen in Fig. 2, the angular distribution of **V'** is isotropic, and therefore so is that for *j'*. However, from measurements of the distribution of the magnitude of *j'* (or of E_R') information can be extracted about the distribution of *b'*, the exit impact parameter. In the warm realm, a variety of reactions have exhibited an approximate antiparallel correlation, $l' \approx -j'$. These were particularly featured in a 1987 Faraday Discussion;[26] in combination with other quantities observable in the warm realm, the antiparallel propensity proved especially diagnostic.

Stwalley suggested that the ultra constraint, $l' = -j'$, might result in suppressing rotational states of the product molecule by an exit centrifugal barrier. Our

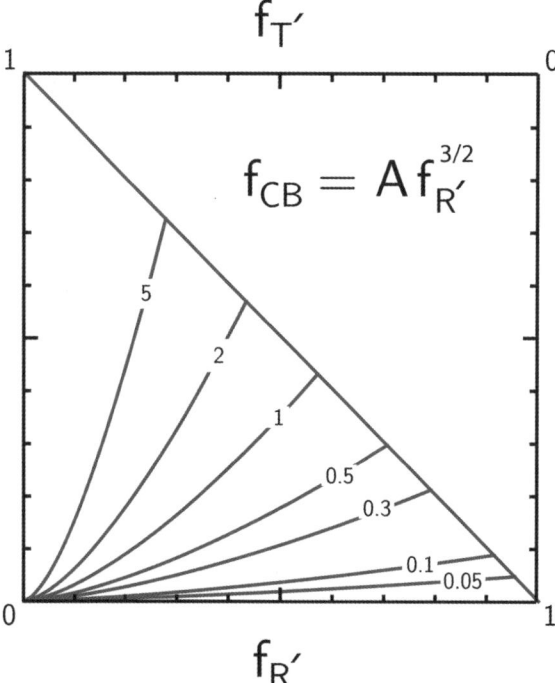

Fig. 4 Reduced variable plot for analysis of constraint arising from exit centrifugal barriers, for a long-range R^{-6} potential, that is imposed in the ultracold limit because the magnitudes of product orbital angular momentum l' and rotational momentum j' are then equal (cf. Fig. 3). For a specified product vibrational state v', the exoergicity $\Delta E = \Delta D_0 - E_{v'} = E_T' + E_R'$ is distributed between product translation and rotation (cf. Fig. 1). The ratios $f_{T'} = E_T'/\Delta E$ and $f_{R'} = E_R'/\Delta E$ are related to that specifying the exit centrifugal barrier height, $f_{CB} = V_{CB}(l')/\Delta E$, via $f_{T'} = 1 - f_{R'} > f_{CB} = A f_{R'}^{3/2}$ where the parameter A is given in eqn (2). The plot displays (for $A = 0.05, 0.1, ...$) the allowed ranges of $f_{R'}$ and $f_{T'}$ (upper abscissa axis); e.g., for $A = 0.3$, the maximum product rotational excitation is about $f_{R'} = 0.8$, and minimum product exit relative translation about $f_{T'} = 0.2$. The plot does not take into account any tunneling through the centrifugal barrier.

comments on this, particularly pertaining to alkali atom + dimer reactions, are among those compiled in this Discussion.[27] To provide a more general perspective, Fig. 4 presents a reduced variable plot. The AB + C product relative translational energy, AB rotational excitation energy, and centrifugal barrier height for a given AB vibrational state v' are all scaled by ΔE, the reaction exoergicity to form AB $(v', j' = 0)$. These scaled quantities are related by

$$f_{T'} = 1 - f_{R'} > f_{CB} = A f_{R'}^{3/2} \qquad (1)$$

with

$$A = 0.00555(\mu_{AB}/\mu_{AB,C})^{3/2}(\Delta E r_{v'}^6/C_6)^{1/2} \qquad (2)$$

where the reduced masses are in amu, ΔE in cm^{-1}, internuclear distance $r_{v'}$ of AB(v') in Angstroms and van der Waals coefficient C_6(AB − C) in atomic units. Usually only rough estimates are available for the exit C_6 coefficient. Typically $A < 0.1$ unless the product atom is hydrogen. In a 1973 Faraday Discussion, an analogous plot is given for the effect on $f_{T'}$ of exit centrifugal barriers from a loose complex subject to $l \approx l'$.[28]

C. Example: the F + H₂ reaction

At present the F + H$_2$ reaction offers the only opportunity to compare directly reactive scattering in the warm domain with the ultracold limit. Under warm conditions, this reaction and its isotopic variants have been extensively studied in molecular beam[11,29] and infrared chemiluminescence[30] experiments. There are also many theoretical treatments, with much attention to a variety of resonances.[11] For ultracold chemistry, F + H$_2$ has come to be regarded as a beacon. Full quantum dynamics calculations by Dalgarno and coworkers,[4,31,32] employing the best available potential energy surface,[33] found the low temperature limit of the rate constant is remarkably large, despite a barrier equivalent to ~400 K. This is attributed to enhancement of tunneling *via* the "presence of a virtual state associated with the van der Waals well in the entrance channel."

Fig. 5 exhibits the huge increase in the ultracold rate constant (lower panel) predicted to occur if the energy of a virtual or barely bound state in the entrance van der Waals well could be tuned to become a zero energy resonance (*i.e.*, coincident with the asymptotic F + H$_2$ energy). Here the tuning is done by adjusting the hydrogen atom mass. The scattering length (upper panel) becomes singular when the zero energy resonance is reached, at $m_H = 1.12$ amu. Note that no resonance appears

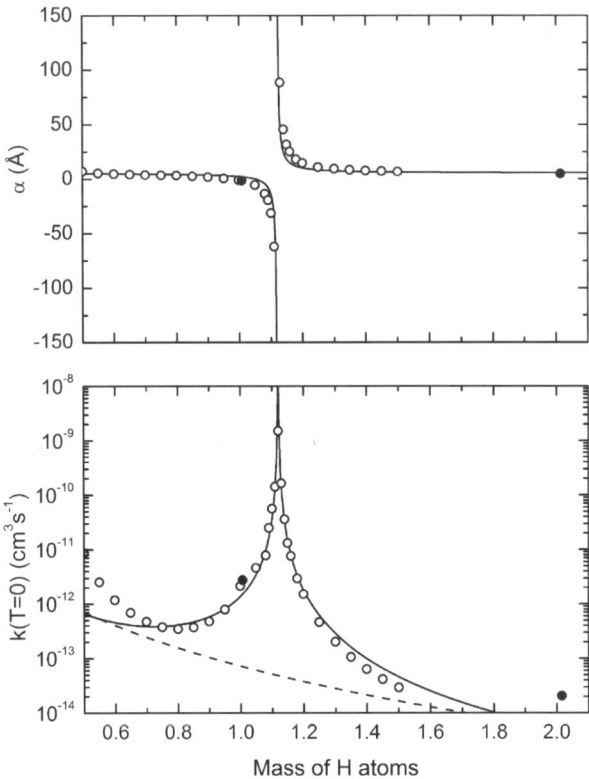

Fig. 5 For F + H$_2$ reaction at the near limit of zero kinetic energy, the upper panel shows the real part of the scattering length and the lower panel the corresponding reaction rate constant, as functions of the hydrogen atom mass. Points were obtained from quantum dynamics calculations of Dalgarno and coworkers[32] using an accurate potential energy surface;[33] black points identify the results for actual H and D atom masses. Curves are from a 1-D model using a square well preceding a square barrier, with parameters adjusted to fit the quantum dynamics results. The dashed curve in the lower panel shows the result of omitting the square well. Model calculations were done by Kelly Higgins at Harvard (unpublished).

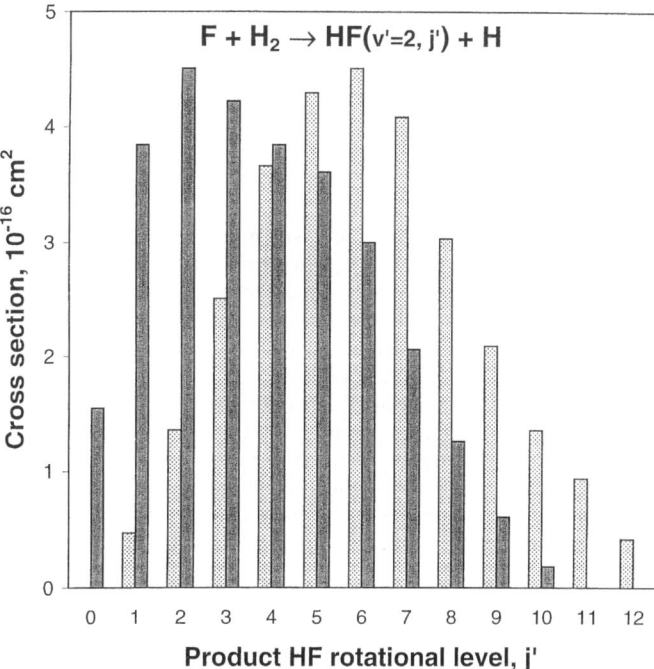

Fig. 6 Comparison of rotational angular momentum distributions for $v' = 2$ vibrational state of HF from F + H$_2$($v = 0, j = 0$) reaction: solid bars obtained from quantum dynamics calculations of Balakrishnan and Dalgarno[31] for reaction approaching the ultracold regime (7.3 mK); speckled bars from infrared chemiluminescence experiment of Polanyi and coworkers[30] at $\sim$300 K (data for $j' = 0$ are lacking because that state does not luminesce). Ordinate scale for experimental data is normalized to the theoretical cross sections.

near $m_H \sim 2$ amu and the rate constant for F + D$_2$ is 100-fold lower than for F + H$_2$. The points shown are from full 3-D quantum scattering matrix calculations of Dalgarno and coworkers. Curves are from a rudimentary 1-D model, simply using a square well preceding a square barrier.

Fig. 6 illustrates that in the exit channel the warm and ultracold regimes yield similar results for the distribution of rotational states of the product HF molecule. The downward shift of the j' distribution for the ultracold reaction is consistent with the lessened energy and angular momentum available, as the warm reaction has an "extra" $E_T \sim 200$ cm^{-1} and $l \sim 3$ or 4.

For both the warm and cold reactions, the $v' = 2$ vibrational state of HF is the most populated, exceeding others tenfold or more. These results for product energy disposal are compatible with classical trajectory calculations. Such comparisons suggest that, at least for a strongly exoergic reaction like F + H$_2$, in which the exit potential valley is like a "ski-run," the process is essentially "warm particles out" whether or not launched by "warm particles in" or by "cold waves in."

D. Electronic structure governance of dynamics

A chief pillar of molecular dynamics is the Born–Oppenheimer approximation. Within the wide domain of its suitability, the much more rapid motion of the electrons can be treated separately from the motion of the nuclei. Computing by quantum mechanics the electronic eigenenergies and wavefunctions for fixed configurations of the nuclei then enables construction of a potential energy surface that governs the motion of the nuclei. Although the separability of electronic motions breaks down when potential surfaces "cross," computations for fixed nuclei still provide basis functions for the general solution. In this venerable context, undertaking study of cold or ultracold collisions might seem to involve merely slowing down the initial translational and rotational motions of nuclei. The electrons just come along for the ride, and a slower ride should merely improve the separability approximation. The potential surface would not change. Accordingly, benchmark quantum dynamics calculations such as those done for $F + H_2$ use the same surface whether treating warm or ultracold collisions.

Yet physical chemists naturally seek heuristic models to relate reaction dynamics to a potential surface and the underlying electronic structures (as in much of the work alluded to in my prelude). Whether interpreting results of full quantum scattering calculations or devising models, for ultraslow collisions it is necessary to take into account that during the leisurely approach a reactant molecule may execute many vibrations and undergo large amplitude tumbling or oscillatory rotations. Thus, in slow collisions an effective adiabatic potential, averaged over such motions, becomes appropriate. For instance, in the $F + H_2$ case, the effective or "dressed" potential surface differs appreciably from the "bare" surface obtained from electronic structure calculations. For the effective potential, the depth of the entrance van der Waals well is less and decreases further as j for H_2 is increased.[34] The effective potential also has a considerably narrower potential barrier,[35] which enhances tunneling.

Another basic aspect of ultraslow collisions has to enter thinking about the relation of dynamics to a potential surface. When the incident deBroglie wavelength becomes long, it samples a large region of the potential surface, obscuring altogether our usual myopic view. In particular, this severely affects stereodynamic notions. For instance, early molecular beam experiments on exchange reactions of $X + ICl$ (with X = H, O, halogen, or CH_3) found that attack at the "I-end" was strongly favored; indeed the predominant product was always X–I although that bond is much less strong than X–Cl. That led to recognition of a broadly applicable molecular orbital criterion.[19] However, in an ultraslow collision, the long incident wave would engulf the I–Cl molecule, sampling both "ends" at once. The electronic structure would, of course, insist on the same preference for forming X–I bonds. But in slow, engulfing collisions the familiar way of visualizing how that comes about becomes problematic.

E. Assessment

The experimentally observable properties of chemical reaction dynamics that can be explored in warm collisions and in ultracold collisions differ greatly in character. Warm collisions offer much more scope for determining dynamical properties that can provide insights about how electronic structure governs reactions. Ultracold collisions have much greater sensitivity to entrance channel interactions and thereby offer the prospect of tuning these by external fields (electric, magnetic, laser) to control the onset of reaction and hence the overall rate.[3,6,8]

At the rudimentary level I've considered, going to or near the ultracold limit may seem to impoverish collision experiments because directional properties become unobservable. However, gaining tunable control in the entrance channel would be a satisfying reward for attaining ultracold conditions. That is already exemplified by magnetic tuning of Feshbach resonances to induce atoms to combine. Now, as reported by Ye at this Discussion, by means of STIRAP the resulting molecules can be coherently shuffled into a desired vibronic state.[36] Striving to generalize such uncanny manipulations is certainly a worthy goal. Use of external fields may also restore directional properties by providing a reference axis, akin to resurrecting in Fig. 2 the dashed "ghost" line. Yet there remains a nagging concern: whether means for control achieved for ultracold reactions could evolve into effective techniques for the wider world of warm chemistry.

Since reactive atoms and molecules often are open-shell species, with electronic spin and/or orbital angular momenta, entrance channel interactions can become complicated and even inaccessible in cold collisions. For instance, recently Ruedenberg and colleagues have carried out highly accurate *ab initio* calculations to determine the potential curve for the ground $^1\Sigma_g^+$ state of the fluorine molecule in the long-range region.[37] This shows that the approach of the atoms (each ^2P) is hindered by a substantial potential hump of height $\sim$12 K, located at a separation distance of $\sim$4 Å, resulting from repulsive interaction between the quadrupoles of the atoms (proportional to R^{-5}), contending with the attractive dispersion potential (proportional to R^{-6}). Moreover, in the long-range region two $^3\Pi_u$ states, which have attractive quadrupole interactions but covalent repulsion, drop below the $^1\Sigma_g^+$ state. That produces curve crossings near $\sim$3 Å involving reorientation of the quadrupole moments and possibly nonadiabatic couplings. To elucidate such complications will likely require resort to warm collision conditions.

IV. Alkali molecule candidates for cold reactions

Alkali atoms and dimers seem likely to long remain the ultraspecies in the ultracold realm. Further species notorious in the warm domain, such as OH radicals, have arrived in the cold realm, and the rapid development of Zeeman deceleration and other slowing techniques now promise to soon deliver others, such as hydrogen, halogen, and oxygen atoms as well as metastable rare gas atoms. There await a host of reactions, identified in warm collision experiments, that go on barrierless potential surfaces. Here I note just a few, involving three varieties of alkali molecules that may prove especially suitable for cold collision experiments.

A. Alkali dimers

Alkali dimer molecules are intriguing reagents for studies of chemical dynamics because the dimer bonds are abnormally weak. Many processes which at ordinary collision energies would be endoergic for alkali atoms (such as gas phase reactions with water) become exoergic for the dimer molecules. The dimer molecules can be regarded as chemical analogs of the deuteron, which in nuclear physics proved to be a uniquely useful projectile because the proton–neutron pair is weakly bound. Thus, reactions of alkali dimers might be expected to provide examples of chemical reaction dynamics analogous to the two prototype modes found in deuteron scattering, direct reaction *via* a stripping process or resonance reaction *via* compound nucleus formation.

This motivated early molecular beam studies of reactions of alkali dimers, readily produced in supersonic nozzle expansions and filtered out by magnetically deflecting away the accompanying atoms. Reactions were studied with hydrogen atoms,[38] with alkali atoms,[39] and with halogen atoms.[40] Much more could have been done, were it not for the siren call to go beyond the alkali age.

The halogen atom + alkali dimer reactions seem congenial candidates for cold collision experiments. The exoergicity is so large that these reactions produce intense chemiluminescence from electronically excited alkali atoms. This should be advantageous in cold and ultracold collision experiments, where the chief (or only) observable property is the total reaction cross section, which the atomic emission provides an extremely sensitive and specific means to monitor. Ian Lane has also emphasized this aspect; he proposes an elegant experiment: insert halogen atoms into a cloud of ultracold atoms, then sweep through a Feshbach resonance with a magnetic field to produce dimers that would promptly react with the halogen atoms to generate chemiluminescence.[41] Hydroxyl radicals could be used in the same way.

In the classic diffusion flame experiments of Michael Polanyi, Na and Cl_2 were diffused into opposite ends of a tube containing argon. Where the joint concentration peaked, about midway down the tube, reaction produced a deposit of NaCl. From the width of the salt deposit and the known diffusion rates, Polanyi could determine how rapidly the reaction occurred. Perhaps something akin to that could now be done, sending a decelerated halogen atom beam, perhaps further slowed by a buffer gas, into an alkali condensate. Polanyi flames were so named because he saw bright D-line emission from excited Na atoms, appearing over a wider region than the salt deposit. He was unable to establish whether the excitation came *directly* from $Cl + Na_2 \rightarrow NaCl + Na^*$ or *indirectly* from subsequent collisions of vibrationally excited NaCl with Na atoms. This question was much debated in the early days of molecular quantum theory; then there was much concern about the separability of electronic and nuclear motions, and Polanyi flames were intertwined with the Born–Oppenheimer and Franck–Condon concepts. Forty years later, the individual steps were studied under the single-collision conditions of beam experiments and both the direct and indirect processes were shown to occur readily.[40]

B. Alkali halides

In beam experiments exploring the Polanyi flame mechanisms, the vibrationally excited alkali halides used were generated from a prior crossed-beam reaction of alkali atoms with halogen molecules.[42] Such a source may be feasible to supply alkali halide molecules as input to a Stark decelerator. It could be enhanced by means of kinematic slowing[43,44] or by guiding devices[45] that exploit the very large dipole moments. In the prior reaction, the extent of vibrational excitation produced is inversely related to the product translational energy. Thus, selecting the input velocity of the alkali halide will also define fairly well its vibrational excitation. That would enable a great variety of cold collision experiments, including many that would produce conversion to electronic excitation and so facilitate detection.

A prosaic input source for alkali halides (with merely modest thermal energy) also appears suitable, as intense supersonic beams have been made by simply capturing vapor from molten salt in a flow of inert gas. This should be adaptable for preslowing by inverse seeding in xenon, as is now regularly used. Atom exchange reactions of alkali halides with alkali atoms (not involving high vibrational or electronic excitation) are of special interest. In warm collisions, these proved to be the prototype for reaction involving lingering complexes,[24] so have been extensively studied.[12] Under cold conditions, sending a beam of alkali halide A^+X^- into a cloud of alkali atoms B would liberate A atoms, readily detectable by laser excitation to monitor the reaction.

C. Alkali monoxides

A prior crossed-beam source can produce strong beams of alkali monoxides, using the reactions of alkali atoms with NO_2[46] or O_3.[47] A special feature is that the AO molecules can be obtained in either a $^2\Sigma$ or $^2\Pi$ ground state, depending on the choice of alkali atom and precursor molecule. Again, kinematic cooling and dipole guiding would be applicable. To my knowledge, warm collisions of alkali monoxides with alkali atoms have never been studied. These would be of particular interest because the monoxide bonding is highly ionic, A^+O^-, like that of alkali halides except it is open-shell because one less electron is present. Reactions with alkali atoms would again be expected to involve lingering complexes and could also be monitored by laser excitation of the liberated alkali atom. The particular interest for cold collisions is that the entrance channel interactions would involve couplings to electronic angular momentum (not available with alkali halides). As cold or ultracold collisions are best suited to probing the entrance interactions, these alkali oxide reactions seem inviting candidates to elucidate such electronic couplings.

If and when cold collisions with oxygen atoms can be studied, there awaits a link to an intriguing saga of atmospheric chemistry. It has to do with the sodium glow and meteor trails in the night sky. As the saga is traced elsewhere,[48] here I'll mention only a couple of aspects. In warm beam experiments the $Na + O_3$ reaction was found to produce NaO essentially only in its $A^2\Sigma^+$ electronic state, although formation of its ground $X^2\Pi$ state would be more exoergic.[47] This is striking because, considering the ionic bonding, in the $^2\Sigma$ state the hole in the outer p-shell of the O^- anion is in a σ orbital pointing towards the Na^+ cation, whereas in the $^2\Pi$ state that σ orbital has two electrons and the hole is in a π orbital transverse to the bond axis. Moreover, $O + NaO \rightarrow Na^* + O_2$, the reaction that produces the nightglow and meteor trails, only goes when NaO is in its $^2\Sigma$ state and not the $^2\Pi$ state. This preference has been rationalized by a correlation diagram argument,[48] but interesting questions of electronic stereospecificity remain. Cold collision experiments might prove capable of elucidating the role of electronic couplings, here involving the open-shell $O(^3P)$ atom as well.

V. Closing remarks

Otto Stern liked to speak of "that beauty and peculiar charm which so firmly captivates workers in the field of molecular beams." That applies just as much to the new frontier field of cold and ultracold molecules. It is a joy to witness the intrepid spirit and exploratory zest displayed at this Discussion. Also to admire the new tools, experimental and conceptual, that are being developed. Those tools will surely impact many other civilized fields, some seemingly remote from this frontier. For instance, just the ability to slow down molecules by a factor of ten, which in most cases does not even get down to the official border of the "cold" realm (<1 K), enhances a hundred-fold the deflecting power of a Stern–Gerlach field. That is enough to make it feasible to do high resolution rotational spectroscopy of nonpolar molecules.[49]

My first thinking about molecular collisions started in high school, when I became intrigued with figuring out how high up in the sky a molecule would have to be to have a mean free path between collisions that was longer than an orbit around the earth. Later, working with molecular beams, I was concerned with the mean free path of molecules within the beam and the distribution of their distance apart. As everyone knows, in a molecular beam the molecules are supposed to be sufficiently separated so they travel independently. Now, decades later, I find myself intrigued with deBroglie wavelengths. That led to a fantasy: what about a beam of molecules with deBroglie wavelengths longer than their distance apart? Wouldn't it attain quantum degeneracy, so if comprised of Bosons, condense into a single coherent, macroscopic quantum state? Nothing like the beams we've had ever since Otto Stern! When I confessed this fantasy to Gerard Meijer, he kindly informed me of work underway in Paris to make just such a source of continuous and coherent matter waves.[50]

There is another captivating experiment, also not represented here, that I must mention because it has such uncanny "beauty and peculiar charm." The experiment[51] was done in Lene Hau's lab at Harvard. She uses resonant laser fields interacting with ultracold, dense atom clouds. Her description: "The light fields interact strongly with electrons of the atoms, and couple directly to external atomic motion through recoil momenta imparted when photons are absorbed and emitted…Here we demonstrate that a slow light pulse can be stopped and stored in one Bose–Einstein condensate and subsequently revived from a totally different condensate…information is transferred through conversion of the optical pulse into a traveling matter wave." In conversation, she describes this as "sending off an atom, leaving its electron orbital behind." Hard for a fan of molecular orbitals like me to grasp! Surely Hau has opened up a brave new world for chemical physics.

Acknowledgements

I am grateful for the opportunity to take part in this meeting. It has awakened echoes extending back to the first Faraday Discussion meeting I attended, held in 1962 at Trinity College, Cambridge. I thank Kelly Higgins and Sam Lipoff for preparing the Figures for my lecture.

References

1. W. C. Stwalley, *Can. J. Chem.*, 2004, **82**, 709.
2. J. Doyle, B. Friedrich, R. V. Krems and F. Masnou-Seeuws, *Eur. Phys. J. D*, 2004, **31**, 149.
3. R. V. Krems, *Int. Rev. Phys. Chem.*, 2005, **24**, 99.
4. P. F. Weck and N. Balakrishnan, *Int. Rev. Phys. Chem.*, 2006, **25**, 283.
5. J. M. Hutson and P. Soldán, *Int. Rev. Phys. Chem.*, 2007, **26**, 1.
6. R. V. Krems, *Phys. Chem. Chem. Phys.*, 2008, **10**, 4079.
7. I. W. M. Smith (ed.) *Low temperatures and cold molecules*, World Scientific Publishing, ISBN 978-1-84816-209-9, 2008.
8. M. T. Bell and T. P. Softley, *Mol. Phys.*, 2009, **107**, 99.
9. R. V. Krems, W. C. Stwalley, and B. Friedrich (ed.), *Cold molecules: theory, experiment, applications*, CRC Press, Taylor and Francis Group, London, 2009.
10. R. Campargue (ed.) *Atomic and Molecular Beams: The State of the Art 2000*, Springer, Berlin, 2001.
11. Special Issue, Chemical Dynamics, *J. Chem. Phys.*, 2006, **125**, 132101, (in honor of Y.T. Lee).
12. R. Levine and R. B. Bernstein, *Molecular reaction dynamics and chemical reactivity*, Oxford University Press, New York, 1987.
13. R. Levine, *Molecular reaction dynamics*, Cambridge University Press, London, 2004.
14. W. Klemperer, *Annu. Rev. Phys. Chem.*, 1995, **46**, 1.
15. M. Polanyi, *Atomic Reactions*, Williams and Norgate, London, 1932.
16. R. B. Bernstein, D. R. Herschbach and R. D. Levine, *J. Phys. Chem.*, 1987, **91**, 5365.
17. D. Herschbach, *Eur. Phys. J. D*, 2006, **38**, 1.
18. A. H. Zewail, *J. Phys. Chem.*, 2000, **104**, 5660.
19. D. R. Herschbach, *Angew. Chem., Int. Ed. Engl.*, 1987, **26**, 1221.
20. K. W. Ford and J. A. Wheeler, *Ann. Phys.*, 1959, **7**(259), 287.
21. G. F. Gribakin and V. V. Flambaum, *Phys. Rev. A*, 1993, **48**, 546; V. V. Flambaum, G. F. Gribakin and C. Harabati, *Phys. Rev. A*, 1999, **59**, 1998.
22. D. R. Herschbach, *Discuss. Faraday Soc.*, 1962, **33**, 149.
23. A. J. McCaffery, *J. Chem. Phys.*, 2008, **129**, 224303.
24. W. B. Miller, S. A. Safron and D. R. Herschbach, *Discuss. Faraday Soc.*, 1967, **44**, 108.
25. J. D. Barnwell, J. G. Loeser and D. R. Herschbach, *J. Phys. Chem.*, 1983, **87**, 2781.
26. S. K. Kim and D. R. Herschbach, *Faraday Discuss. Chem. Soc.*, 1987, **84**, 159 and subsequent discussion pages 182 and 188.
27. W. C. Stwalley and D. Herschbach, *Faraday Discuss.*, 2009, **142**, 87, contributions to discussion section.
28. D. R. Herschbach, *Faraday Discuss. Chem. Soc.*, 1973, **55**, 233.
29. Y. T. Lee, *Science*, 1987, **236**, 793.
30. J. C. Polanyi, *Science*, 1987, **236**, 680.
31. N. Balakrishnan and A. Dalgarno, *Chem. Phys. Lett.*, 2001, **341**, 652.
32. E. Bodo, F. A. Gianturco, N. Balakrishnan and A. Dalgarno, *J. Phys. B*, 2004, **37**, 3641.
33. K. Stark and H.-J. Werner, *J. Chem. Phys.*, 1996, **104**, 6515.
34. T. Takayanagi and Y. Kurosaki, *J. Chem. Phys.*, 1998, **109**, 8929.
35. E. Rosenman, S. Hochman-Kowal, A. Persky and M. Baer, *Chem. Phys. Lett.*, 1998, **257**, 421.
36. S. Ospelkaus, K. K. Ni, M. H. G. de Miranda, B. Neyenhuis, D. Wang, S. Kotochigova, P. S. Julienne, D. S. Jin and J. Ye, *Faraday Discuss.*, 2009, **142**, DOI: 10.1039/b821298h.
37. L. Bytautas and K. Ruedenberg, *J. Chem. Phys.*, 2009, **130**, 204101.
38. Y. T. Lee, R. J. Gordon and D. R. Herschbach, *J. Chem. Phys.*, 1971, **54**, 2410.
39. D. J. Mascord, H. W. Cruse and R. Grice, *Mol. Phys.*, 1976, **32**, 131.
40. W. S. Struve, J. R. Krenos, D. L. McFadden and D. R. Herschbach, *J. Chem. Phys.*, 1975, **62**, 404.
41. I. Lane, *Faraday Discuss.*, 2009, **142**, 431, comment in discussion section.
42. M. C. Moulton and D. R. Herschbach, *J. Chem. Phys.*, 1966, **44**, 3010.
43. M. S. Elioff, J. J. Valentini and D. W. Chandler, *Science*, 2003, **302**, 1940; K. E. Strecker and D. W. Chandler, *Phys. Rev. A*, 2008, **78**, 063406.

44 N.-N. Liu and H. Loesch, *Phys. Rev. Lett.*, 2007, **98**, 103002.
45 H. J. Loesch and B. Scheel, *Phys. Rev. Lett.*, 2000, **85**, 2709.
46 R. R. Herm and D. R. Herschbach, *J. Chem. Phys.*, 1970, **52**, 1317.
47 X. Shi, D. R. Herschbach, D. R. Worsnop and C. E. Kolb, *J. Phys. Chem.*, 1993, **97**, 2113.
48 D. R. Herschbach, in R. Campargue (ed.) Atomic and Molecular Beams: The State of the Art 2000, Springer, Berlin, 2001, p. 12.
49 T. J. McCarthy, M. T. Timko and D. R. Herschbach, *J. Chem. Phys.*, 2006, **125**, 133501.
50 T. Lahaye, J. M. Vogels, K. J. Gunter, Z. Wang, J. Dalibard and D. Guery-Odelin, *Phys. Rev. Lett.*, 2004, **93**, 093003.
51 N. S. Ginsberg, S. R. Garner and L. V. Hau, *Nature*, 2007, **445**, 623.

Testing the time-invariance of fundamental constants using microwave spectroscopy on cold diatomic radicals

Hendrick L. Bethlem* and Wim Ubachs

Received 28th October 2008, Accepted 5th January 2009
First published as an Advance Article on the web 27th May 2009
DOI: 10.1039/b819099b

The recently demonstrated methods to cool and manipulate neutral molecules offer new possibilities for precision tests of fundamental physics theories. We here discuss the possibility of testing the time-invariance of fundamental constants using near degeneracies between rotational levels in the fine structure ladders of molecular radicals. We show that such a degeneracy occurs between the $J = 6$, $\Omega = 1$ and the $J = 8$, $\Omega = 0$ levels of the various natural isotopomers of carbon monoxide in its a$^3\Pi$ state. As a result, the 2-photon transition that connects these states is 300 times more sensitive to a variation of m_p/m_e than a pure rotational transition. We present a molecular beam apparatus that might be used to measure these transitions with a fractional accuracy of 10^{-12}. Ultimately, the precision of an experiment on metastable CO will be limited by the lifetime of the a$^3\Pi$ state. We will discuss other possible molecules that have a suitable level structure and can be cooled using one of the existing methods.

Introduction

The equivalence principle of general relativity postulates that the outcome of any non-gravitational experiment is independent of position and time. In theories that attempt to unify gravity with other fundamental forces, on the other hand, violation of the equivalence principle may occur and can be consistently described. The Kaluza–Klein theories from the 1920s as well as modern string theories, for instance, introduce additional compactified dimensions, and the size of these—yet unobserved— dimensions determines the strength of the fundamental forces. If the size of these dimensions should happen to change over time, the strength of the forces in four-dimensional space-time would change as well. Such a change would manifest itself as a change of the coupling constants and particle masses.[1]

From an experimental perspective, it is most practical to search for variation of dimensionless quantities. The fine structure constant, α, representing the strength of the electro-weak force, and the proton–electron mass ratio, $\mu = m_p/m_e$, which is a measure for the strength of the strong force, are the prime targets of modern research in this area. Possible variations can be detected from a wide variety of physical phenomena, but the extreme accuracy that can be obtained in the determination of frequencies or wavelengths of spectral lines in atoms and molecules makes spectroscopy the ideal testing ground for searches of varying α and μ.[2] The proton–electron mass ratio is of particular interest as theoretical models predict that the variation of μ could be significantly larger than the variation of α. Calmet and

Laser Centre Vrije Universiteit, de Boelelaan 1081, NL-1081 HV Amsterdam, The Netherlands

Fritsch, for instance, predict that the variation of μ is 38 times larger than the variation of α.[3]

Recent astrophysical data suggest that the fine structure constant, α, has increased over cosmological time. The combined analysis over more than 100 quasar systems has produced a value of a relative change of $\Delta\alpha/\alpha = -0.57 \pm 0.10 \times 10^{-5}$, which is at the 5$\sigma$ significance level.[4] Based on the spectral lines of molecular hydrogen in two quasar systems at redshifts of $z = 2.6$ and $z = 3$, an indication was found of a relative decrease of the proton–electron mass ratio, μ, of two parts in 10^5 over cosmological time.[5] If one assumes that the constants change linearly over time, this implies a fractional change on the order of 10^{-15} per year. To test the time-variation of fundamental constants in the current epoch, frequency standards based on different atomic and molecular transitions are being compared as a function of time. As these standards have in general a different dependence on α and μ, a possible time-variation of α and/or μ will lead to a frequency shift. In contrast to astrophysical observations, which measure the constants' value over a significant fraction of the age of the universe, laboratory tests cover only a short time span. Their advantage, however, is their great accuracy, reproducibility and unequivocal interpretation.

Currently, laboratory experiments have found no indications for the time-variation of any fundamental constant. The most stringent limit is set by a comparison between an optical mercury ion clock and a caesium fountain clock over 6 years. Assuming invariance of other constants, this results in a limit for the variation of $\Delta\alpha/\alpha < 1.3 \times 10^{-16}$ yr^{-1}.[6] The most stringent *independent* test of the time-variation of μ is set by comparing vibrational transitions in SF$_6$ with a caesium fountain over 2 years, which has resulted in a limit for the variation of $\Delta\mu/\mu < 5.6 \times 10^{-14}$ yr^{-1}.[7]

The sensitivity of any experiment looking for a frequency shift, $\Delta\nu$, due to the time-variation of a fundamental constant, X, depends both on the size of the shift, *i.e.*, the inherent sensitivity of the atomic or molecular transition, and on the ability to measure this shift. As a measure for the inherent sensitivity of a transition, we introduce a sensitivity coefficient, K, via:

$$\frac{\Delta\nu}{\nu} = K_X \frac{\Delta X}{X}$$

Generally, two strategies can be followed: (i) one uses a system that is naturally suitable for precision measurements—*i.e.*, a system that can be well controlled by *e.g.*, laser cooling techniques, offers a high Q-factor, a high signal to noise, *etc.*— but has typically a sensitivity on the order of unity. (ii) one uses a system that has a large sensitivity, but that is not necessarily ideal for precision measurements. At present, most currently running and proposed[8] experiments take the first approach. A notable exception is the experiment of Cingöz *et al.* on atomic dysprosium.[9]

Atomic dysprosium has a unique property of an accidental near degeneracy between two high lying energy levels that have a different symmetry and move in opposite directions if α varies. As a result, the rf transition between the two levels has a fractional sensitivity to a variation in α of about 10^6. Note that, given a certain change of α, the resulting absolute frequency shift of the energy levels in dysprosium is comparable to that in, for instance, the mercury ion. However, whereas in the mercury ion this shift needs to be measured on an optical frequency, in dysprosium this shift can be measured on an rf frequency. The latter is usually much simpler. For example, the fractional frequency uncertainty of the current best optical standard— the mercury ion clock—is on the order of 10^{-17}, which corresponds to an absolute frequency uncertainty of 10 mHz.[6] The current best microwave standard—the caesium fountain clock—has a fractional frequency uncertainty of 10^{-16}, which corresponds to an absolute frequency uncertainty of 1 µHz.[10]

Whereas in atoms a near degeneracy between levels of different symmetry is very rare, in molecules this happens rather frequently. DeMille *et al.*[11] proposed a test based on a near degeneracy between the $a^3\Sigma_u^+(v = 37)$ and the $X^1\Sigma_g^+(v = 138)$ in Cs_2. Flambaum[12] proposed to use diatomic molecules with unpaired electrons, such as LaS, LaO, LuS, LuO, YbF and similar ions, that have a cancellation between their hyperfine and rotational interval. Kozlov and Flambaum[13] proposed to use diatomic molecules that have a cancellation between their fine structure and vibrational interval and identified Cl_2^+, CuS, IrC, SiBr and HfF$^+$ as suitable candidates.[14] These cancellations occur in heavy molecules which, however, do not lend themselves well to the recent developed cooling and manipulation techniques based on time-varying electric or magnetic fields.[15]

In this paper we propose another system that is suitable for detecting μ-variation in a laboratory experiment, in this case because of a near degeneracy between rotational levels in the fine structure ladders of a $^2\Pi$ or $^3\Pi$ state. Such a degeneracy is shown to occur in the various natural isotopomers of carbon monoxide in its $a^3\Pi$ state, which is metastable. We discuss the precision that might be obtained using metastable CO and discuss other (including ground-state) molecules that have a suitable level structure.

The sensitivity of molecular transitions to a variation of α or μ

In the Born–Oppenheimer approximation, the motion of the molecule is separated into an electronic and nuclear part. The physical basis of this separation is that the nuclei are much heavier than the electrons and hence move at much slower speeds. The nuclear motion can be further separated into a vibrational part and a rotational part. The energy difference between electronic states is much larger than the energy difference between vibrational states within an electronic state, and the energy difference between vibrational states is correspondingly larger than the energy difference between rotational states. We thus have a hierarchy of states which reveals itself in electronic, vibrational and rotational spectra of molecules. The energy of a closed shell diatomic molecule can be written as;

$$E_{v,J} = T_e + \omega_e(v + 1/2) + \omega_e x_e(v + 1/2)^2 + \ldots + B_e J(J + 1) - D_e J^2(J + 1)^2 + \ldots - \alpha_e(v + 1/2)J(J + 1) + \ldots$$

with T_e being the electronic energy, ω_e and $\omega_e x_e$ being the harmonic an anharmonic vibrational constant, B_e and D_e being the rotational constant and the centrifugal distortion constant, and v and J are the vibrational and rotational quantum numbers, respectively. α_e is the lowest order coupling term between the vibrational and rotational motion in the molecule.[16,17]

In order to determine the sensitivity of a certain molecular transition to a possible change of the proton to electron mass ratio, μ, we must determine the dependence of the molecular constants to μ. If we assume that the neutron to electron mass ratio changes at the same rate as the proton to electron mass ratio, we find that:

$$\frac{\Delta \mu}{\mu} = \frac{\Delta \mu_N}{\mu_N}$$

with μ_N being the reduced nuclear mass of the molecule. Thus, the dependence of the molecular constants on μ can be found by using the well known scaling relations upon isotopic substitution—the favorite method of the spectroscopist to aid the assignment of molecular spectra. The following relations hold[16]

$$T_e \propto \mu^0, \quad \omega_e \propto 1/\sqrt{\mu}, \quad \omega_e x_e \propto 1/\mu$$
$$B_e \propto 1/\mu, \quad D_e \propto 1/\mu^2, \quad \alpha_e \propto 1/\mu^{1.5}$$

Thus we see that pure electronic transitions do not depend on μ (but do depend on α). Transitions between low lying vibrational levels have $K_\mu = \frac{1}{2}$, while K_μ is below

½ for transitions between higher vibrational levels due to anharmonicity. Transitions between low lying rotational states have $K_\mu = 1$, while K_μ is below 1 for transitions between higher lying rotational states where centrifugal distortion becomes significant. Note that in more complex molecules, additional degrees of freedom and intramolecular motions play a role which may lead to an enhancement of the sensitivity to a variation in μ. The inversion frequency in the ground state of ammonia, for instance, has a sensitivity equal to $K_\mu = 4.6$.[18,19]

Let us now turn to diatomics with open shell electronic states—so called free radicals. The interaction between the electronic orbital angular momentum, L, the spin angular momentum, S, and the rotational angular momentum, R, give rise to additional structure as compared to a closed shell molecule. Two cases can be distinguished:[16,17] (i) Hund's case (a); L is strongly coupled to the internuclear axis and S couples to the electronic orbital angular momentum vector *via* spin-orbit interaction. (ii) Hund's case (b); the S is decoupled from the internuclear axis, the orbital angular momentum couples to R to form N, which then couples to S.

We will be interested in molecular states that are well described by Hund's case (a) coupling scheme for reasons that will become obvious. In Hund's case (a), molecular levels are labeled by the total angular momentum, J, by the projection of L on the internuclear axis, Λ, by the projection of S on the internuclear axis, Σ, and by the projections of J on the internuclear axis, Ω, and on the space fixed axis, M. There are $2S + 1$ fine structure levels, characterized by their Ω values. These levels have energies of $A_e \Lambda \Sigma$, with A_e being the spin–orbit constant. Each fine-structure state has a pattern of rotational levels with energies given by $B_e J(J + 1) - D_e J^2 (J + 1)^2$..., with $J = \Omega$ being the lowest level. In light molecules, A_e is comparable to B_e, whereas in heavy molecules A_e becomes comparable to ω_e. In Hund's case (a), the electronic motion can in first order be separated from the nuclear motion; that is, the energy differences between the different Ω-manifolds depend on α but not on μ. This has as interesting consequence that when two rotational levels of different Ω-manifolds have a near degeneracy, the sensitivity of a transition connecting these two levels to a variation of α and μ is enhanced by a factor A_e/ν, with ν being the frequency of the transition in energy units. As an example we will study such near degeneracies in metastable CO.

Metastable CO

The metastable $a^3\Pi$ state in CO is one of the most extensively studied triplet states of any molecule. It has been studied using rf,[20] microwave[21] and UV-spectroscopy.[22] The lifetime of the $a^3\Pi$ state depends strongly on the rotational level and ranges from 2.6 ms in the $v = 0$, $J = 1$, $\Omega = 1$ level to 140 ms in the $v = 0$, $J = 2$, $\Omega = 2$ level. Rotational levels in the $v = 0$ and $v = 1$ of the $a^3\Pi$ state can be directly excited using laser radiation around 206 and 199 nm, respectively.[23] Due to its favorable properties, metastable CO was used in the first demonstration of Stark deceleration.[24] Recently, metastable CO was trapped in an electrostatic trap and its lifetime directly measured by monitoring the trap-decay.[25]

We have calculated the excitation energy of the rotational levels of the $a^3\Pi$ state using the model of Field *et al.*[22] In this model, the wavefunctions are a superposition of pure Hund's case (a) wavefunctions.[26] The energies of the lower rotational levels are shown in Fig. 1. The component of the total electronic angular momentum along the internuclear axis Ω takes on the values $\Lambda + \Sigma = 0,1,2$, resulting in three Ω-ladders. Each J-level is split by the Λ-doubling into two components with opposite parity. The Λ-doubling in the $\Omega = 0$ state is large and relatively independent of J. The Λ-doubling in the $\Omega = 1$ state is much smaller and in the $\Omega = 2$ state even smaller still.

For pedagogic reasons, we will first examine 1-photon transitions between the different Ω-manifolds. We are looking for transitions with as small a frequency as

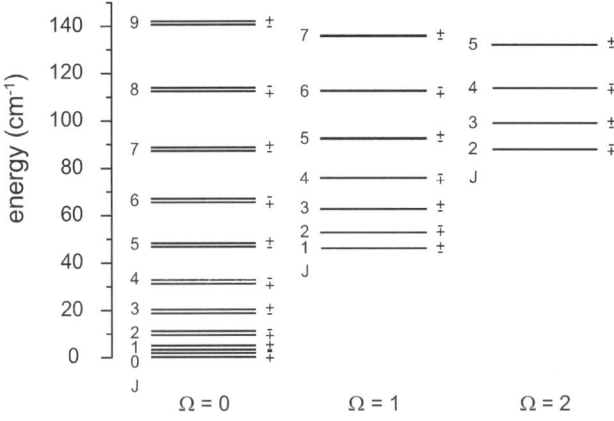

Fig. 1 Energy of the lowest rotational levels of the a³Π (v = 0) state of ¹²C¹⁶O.

possible, hence we look at transitions $J-1, \Omega=2 \rightarrow J, \Omega=1$ and $J, \Omega=1 \rightarrow J+1$, $\Omega=0$. As the total parity changes sign in a 1-photon transition, transitions only connect upper Λ-doublet levels to upper Λ-doublet levels and lower Λ-doublet levels to lower Λ-doublet levels. Hence, for every J, we have a total of 4 transitions. These 4 transitions are shown in the upper panel of Fig. 2 as a function of J. Naively, we expected a minimum transition frequency to occur when $A_e \approx 2B_eJ$. For the a³Π state of ¹²C¹⁶O, $A_e = 41.45$ cm⁻¹ and $B_e = 1.68$ cm⁻¹,[22] thus, we expected a minimum transition frequency when $J \approx 12$. From the figure, we see that although the energy difference initially becomes rapidly smaller as J increases—as is obvious from the tangent at low J shown as dashed lines—the energy difference converges to a constant. We can understand this as a transition from a Hund's case (a) to a Hund's case (b) coupling scheme. At higher J, the spin becomes uncoupled from the internuclear axis and the electronic and nuclear motion can no longer be separated. As a consequence the sensitivity to a μ-variation for transition between the different Ω-manifolds at high J is similar to pure rotational transitions.

Let us now turn to 2-photon transitions, *i.e.*, transitions from $J-2, \Omega=2 \rightarrow J$, $\Omega=1$ and $J, \Omega=1 \rightarrow J+2, \Omega=0$. Again transitions only connect upper Λ-doublet levels to upper Λ-doublet levels and lower Λ-doublet levels to lower Λ-doublet levels because parity is conserved in a 2-photon transition. Hence, we have 4 transitions for every J. These 4 transitions are shown in the lower panel of Fig. 2 as a function of J. A minimum energy difference is now expected when $A_e \approx 4B_eJ$; *i.e.*, when $J \approx 6$, as is indeed observed.

In Fig. 3, the transitions from $J=4, \Omega=2 \rightarrow J=6, \Omega=1$ and $J=6, \Omega=1 \rightarrow J=8$, $\Omega=0$ are shown for the most common isotopomers of CO. Here the effects of nuclear spin and hyperfine structure have been neglected. The molecular constants are obtained *via* isotope scaling of the constants of ¹²C¹⁶O *via* the relations discussed earlier. The large isotope shift of the transitions is indicative of a large sensitivity to a variation of μ. The solid lines plotted in the figure, follow the relation E = constant + $\mu_N \times 72.5$ [GHz amu⁻¹], with μ_N being the reduced nuclear mass. From this we can determine the sensitivity, $K_\mu = 2\mu_N \times 72.5$ [GHz amu⁻¹]/ΔE, with ΔE being the energy difference between the two near degenerate levels. Similar calculations have been performed for the $v = 1$ state (not shown). Due to the smaller B constant in the $v = 1$, the energy differences of the different transitions are slightly larger (more positive). As a result, in the $v = 1$ state the $J = 6, \Omega = 1$ lies slightly above the $J = 8, \Omega = 0$, whereas in the $v = 0$ state the $J = 6, \Omega = 1$ lies slightly below the $J = 8, \Omega = 0$.

All transitions with frequencies below 25 GHz are listed in Table 1 together with their sensitivity to a variation of μ. Particularly interesting are the two transitions in

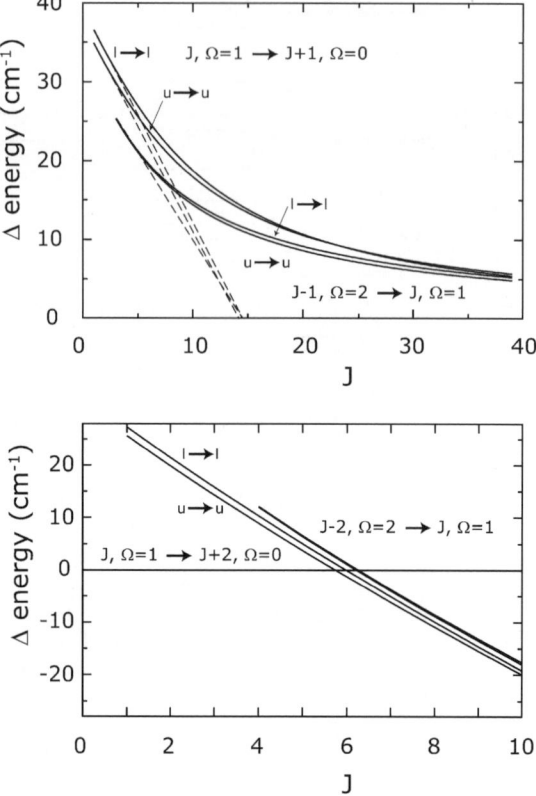

Fig. 2 Energy difference between $J-1$, $\Omega=2$ and J, $\Omega=1$ levels and between $J+1$, $\Omega=0$ and J, $\Omega=1$ levels (1-photon transitions) (upper panel) and energy difference between $J-2$, $\Omega=2$ and J, $\Omega=1$ levels and between J, $\Omega=1$ and $J+2$, $\Omega=0$ levels (2-photon transitions) (lower panel) in the $a^3\Pi$ ($v=0$) state of $^{12}C^{16}O$.

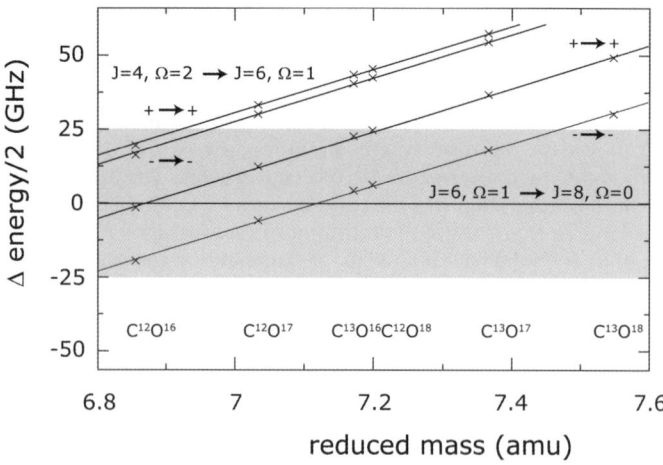

Fig. 3 Energy of selected 2-photon transitions in the $a^3\Pi$ ($v=0$) state of (different isotopomers of) CO. The sensitivity for a change of μ is apparent from the large isotope shift of the transition.

Table 1 Frequencies of selected 2-photon transitions in the a$^3\Pi$ state of (different isotopomers) of CO together with their sensitivity to a μ-variation

	ν/GHz	Isotope	Transition	Sensitivity, K_μ
1	1.648	$C^{12}O^{16}$	$v=0, J=6, \Omega=1, + \rightarrow v=0, J=8, \Omega=0, +$	-302
2	1.9	$C^{12}O^{17}$	$v=1, J=6, \Omega=1, - \rightarrow v=1, J=8, \Omega=0, -$	-264
3	2.459	$C^{12}O^{16}$	$v=1, J=6, \Omega=1, + \rightarrow v=1, J=8, \Omega=0, +$	202
4	4.1	$C^{13}O^{16}$	$v=0, J=6, \Omega=1, - \rightarrow v=0, J=8, \Omega=0, -$	126
5	6.0	$C^{12}O^{17}$	$v=0, J=6, \Omega=1, - \rightarrow v=0, J=8, \Omega=1, -$	-86
6	6.1	$C^{12}O^{18}$	$v=0, J=6, \Omega=1, - \rightarrow v=0, J=8, \Omega=0, -$	86
7	8.0	$C^{13}O^{16}$	$v=1, J=6, \Omega=1, - \rightarrow v=1, J=8, \Omega=0, -$	65
8	10.0	$C^{12}O^{18}$	$v=1, J=6, \Omega=1, - \rightarrow v=1, J=8, \Omega=0, -$	52
9	12.2	$C^{12}O^{17}$	$v=0, J=6, \Omega=1, + \rightarrow v=0, J=8, \Omega=0, +$	42
10	16.040	$C^{12}O^{16}$	$v=0, J=6, \Omega=1, - \rightarrow v=0, J=4, \Omega=2, -$	31
11	16.1	$C^{12}O^{17}$	$v=1, J=6, \Omega=1, + \rightarrow v=1, J=8, \Omega=0, +$	32
12	17.8	$C^{13}O^{17}$	$v=0, J=6, \Omega=1, - \rightarrow v=0, J=8, \Omega=0, -$	30
13	19.253	$C^{12}O^{16}$	$v=0, J=6, \Omega=1, + \rightarrow v=0, J=4, \Omega=2, +$	26
14	19.470	$C^{12}O^{16}$	$v=0, J=6, \Omega=1, - \rightarrow v=0, J=8, \Omega=0, -$	-26
15	19.917	$C^{12}O^{16}$	$v=1, J=6, \Omega=1, - \rightarrow v=1, J=4, \Omega=2, -$	25

$^{12}C^{16}O$ at 1.648 GHz and 2.459 GHz (#1 and #3 of Table 1). When μ changes, the frequencies of these two transitions will change in opposite directions, *i.e.*, when μ becomes larger, the transition frequency in the $v = 0$ will decrease while the transition frequency in the $v = 1$ will increase. Combined these two transitions have a sensitivity which is 500 times larger than an ordinary rotational transition. It is instructive to compare the sensitivity to what one would expect in a pure Hund's case (a). For the transition at 1.648 GHz we expect a sensitivity of $-A/\nu = -41.45 \times 29.979/1.648 = -754$ which is 2.5 times larger than calculated using the model that includes the coupling between the different Ω-ladders. This is a warning that neglecting relevant couplings leads to an overestimation of the sensitivity.

Transitions between levels of different Ω-ladders are sensitive to variations of both μ and α. As the spin orbit energy scales as α^2,[12,13] in a pure Hund's case (a) the sensitivity of the 2-photon transitions to a variation of α is given by 2 A/ν. Mixing between different Ω-ladders will decrease the sensitivity to a variation of α. To first order, K_α will decrease in the same way as K_μ, hence, for the different transitions listed in Table 1, K_α is estimated to be two times larger than K_μ.

A proposed experiment in metastable CO

In this section we will present a molecular beam machine that is currently under construction at the Laser Centre Vrije Universiteit to measure 2-photon microwave transitions in metastable CO, and estimate the accuracy that may be achieved. We will focus on the $J = 6, \Omega = 1, + \rightarrow J = 8, \Omega = 0, +$ transitions in the $v = 0$ and $v = 1$ of $^{12}C^{16}O$ at 1.648 GHz and the 2.459 GHz, respectively, as these offer the highest sensitivity and are located in a convenient frequency range. Furthermore, $C^{12}O^{16}$ is the most abundant isotope (99%) and both C^{12} and O^{16} have a nuclear spin equal to zero. As both transitions differ only in vibrational energy (and sensitivity coefficient), systematic shifts due to stray magnetic or electric fields will be similar, and hence will cancel to first order. The other transitions listed in Table 1, as well as Λ-doublet transitions that are located in the same frequency range might be used to check for various systematic errors.

As mentioned, astronomical observations suggest that $\Delta\mu/\mu$ might change by 2 × 10^{-15} yr^{-1}.[5] The most stringent laboratory test of μ-variation is set by comparing vibrational transitions in SF$_6$ with a caesium fountain over 2 years, which has

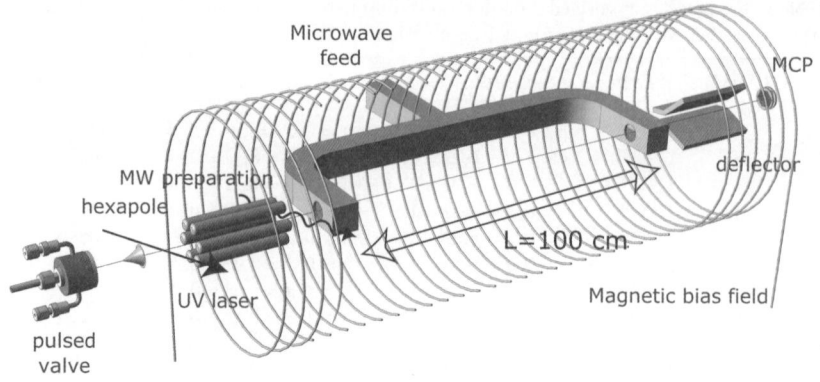

Fig. 4 Schematic view of a molecular beam apparatus to measure 2-photon microwave transitions in metastable CO.

resulted in a limit for the variation of $\Delta\mu/\mu < 5.6 \times 10^{-14}$ yr^{-1}.[7] In order to achieve a sensitivity in the order of 10^{-15} yr^{-1} the 2-photon transitions should be measured with a fractional accuracy of 10^{-12}. Although demanding, this seems possible.

Fig. 4 shows a schematic of a molecular beam apparatus that will be used to measure 2-photon microwave transitions in metastable CO. The machine is largely similar to the familiar magnetically deflected cesium beam clock[27]—clocks that have routinely reached fractional accuracies below 10^{-13}—the main difference being that the inhomogeneous magnetic fields are replaced by inhomogeneous electric fields. Due to its large vapor pressure and small polarisability, it is straightforward to produce an intense beam of CO molecules. Using pulsed laser light around 206 or 199 nm, the spin forbidden $a^3\Pi_1(v = 0, J = 6, -) \leftarrow X^1\Sigma^+(J = 6)$ or $a^3\Pi_1(v = 1, J = 6, -) \leftarrow X^1\Sigma^+(J = 6)$ is driven. These states are the upper components of the $J = 6, \Omega = 1$ state and have a positive Stark shift. Hence molecules in these states can be focused into the microwave cavity using either a quadrupole or hexapole or a combination of the two. Before the molecules enter the cavity they are driven to the lower component of the $J = 6, \Omega = 1$ state by inducing the Λ-doubling transition using pulsed microwave radiation around 6.5 GHz.

The 2-photon transition is measured using Ramsey's separated oscillatory fields method.[28] In the first cavity a coherent superposition of the lower component of the $J = 6, \Omega = 1$ and $J = 8, \Omega = 0$ state is created. During free flight between the two cavities, the phase of this coherent superposition will evolve. If there is no phase difference between the wavefunction and the microwave field—the microwave field is exactly at resonance—all molecules will be driven to the $J = 8, \Omega = 0$ state. If there is a π difference between the phase of the microwave field and the wavefunction, molecules will be driven to $J = 6, \Omega = 1$ state. Thus, if the frequency of the microwave field is scanned, an interference pattern is observed. The width of the fringes is determined by the time it takes for the molecules to pass the distance between the two cavities. If the distance between the cavities is taken to be a meter and the molecules have a velocity of about 500 m s^{-1}, the fringes will have a width of about 250 Hz. A homogeneous magnetic field is applied throughout the beam tube, to shift all resonances out of resonance, except for the $M = 0 \rightarrow M = 0$. Note that, as this magnetic field is perpendicular to the electric field, the magnetic field does not influence the pre and post state-selection. After the microwave cavity the molecules pass an inhomogeneous electric field that will deflect molecules in the $J = 6, \Omega = 1$ state while it will not influence the trajectories of molecules in the $J = 8, \Omega = 0$ state. Metastable CO molecules are detected by letting them impinge onto a multichannel plate (MCP) detector. The 6 eV internal energy of the metastable CO molecules is sufficient to free electrons at the surface of the MCP detector. These electrons are multiplied and accelerated towards a phosphor screen. Subsequently, a digital camera is used to record the image on the phosphor screen.[29] The possibility to record the number of molecules in the $J = 6$, $\Omega = 1$ and $J = 8$, $\Omega = 0$ state *simultaneously* makes it possible to normalize the beam on a shot to shot basis which increases the signal to noise ratio. Note that any molecules remaining in the upper component of the $J = 6$, $\Omega = 1$ will be deflected in the opposite direction to molecules in the lower component and hence will not cause any background signal.

A frequency measurement is characterized by its statistical uncertainty and its accuracy—how well the measured frequency agrees with the unperturbed molecular frequency. A measure of the statistical instability as a function of measurement time is provided by the Allen variance,[27]

$$\sigma_y(\tau) = \frac{1}{Q} \frac{1}{\sqrt{N\tau/\tau_c}}$$

where Q is the quality factor of the transition given by the transition frequency over the measured linewidth, $\nu/\Delta\nu$, N is the number of metastable molecules detected per cycle, τ is the measurement time and τ_c is the duration of a single cycle. As mentioned, if we assume that the two cavities are separated by a distance of 1 m, and the velocity of the beam is on the order of 500 m s^{-1}, the interaction time is 2 milliseconds and the measured linewidth will be about 250 Hz. This corresponds to a Q-factor of 1×10^7. We estimate that it should be possible to detect 1000 or more molecules per cycle. At a repetition frequency of 10 Hz, we find $\sigma_y = 1 \times 10^{-9}\, \tau^{-\frac{1}{2}}$. This implies that a measurement time of about 1 h is required to reach an accuracy of 2×10^{-11}, and 300 h to reach an accuracy of 1×10^{-12}.

Ultimately, the precision will be limited by systematic shifts. With a careful design it should be feasible to keep errors due to the second order Zeeman and Stark shifts and blackbody radiation below 10^{-12}. Note that in caesium fountain clocks[10] these systematic effects are compensated at the level of 10^{-16}. Moreover, as the two transitions that will be compared differ only in vibration, we expect that most systematic errors will—to a very large extent—cancel. An open question is if the high power, necessary to induce the 2-photon transitions, leads to unacceptably large light shifts. Again, the shifts in both transitions are expected to be similar and they will partly cancel.

Other molecules

The precision that will be obtained on measuring the proposed 2-photon transitions in metastable CO will be limited by the lifetime of the metastable CO, which is on the order of 3 milliseconds for the $J = 6$, $\Omega = 1$ and $J = 8$, $\Omega = 0$ state. In this section we will discuss a number of other molecules where a cancellation of the fine structure and rotational interval might be expected to occur. We are particularly interested in molecules that can be decelerated using time-varying electric fields, such that the transition can be measured in a slow molecular beam[17,30] or in a molecular fountain.[18] In Bethlem *et al.*[31] a list of candidate molecules for Stark deceleration was presented. Most molecules on that list have by now indeed been decelerated. If we discard all polyatomic molecules—which are interesting in their own right—we are left with CH, CF, CO (a$^3\Pi$), LiH, NH (a$^1\Delta$), NO, OH, OD, SH. In what follows, we will discuss the suitability of these molecules to detect μ-variation using near degenerate rotational levels in different Ω-manifolds.

CH

The ground state of CH is $^2\Pi$. The spin orbit constant and rotational constant are $A = 27.95$ cm^{-1} and $B = 14.457$ cm^{-1}, respectively.[32] Due to the small ratio of A to B

the molecule is best described by a Hund's case (b) coupling scheme even for low J, and no degeneracies will occur. For CD, the spin orbit constant and rotational constant are $A = 27.95$ cm^{-1}, and $B = 7.8$ cm^{-1}. Although the ratio of A to B is increased by a factor of 2 compared to CH, CD is still very much case (b) even in the lowest J.

CF

The ground state of CF is $^2\Pi$. The spin orbit constant and rotational constant are $A = 77.12$ cm^{-1} and $B = 1.417$ cm^{-1}, respectively.[32] Due to the large ratio of A to B, CF is well described by Hund's case (a). The two Ω-manifolds are expected to have a near-degeneracy at $J \approx 13$. More detailed calculations are needed to determine the frequency of the 2-photon transition and its sensitivity. However, the fact that the degeneracy occurs at high J, makes CF less suitable.

LiH

The ground state of LiH is $^1\Sigma$. As $\Lambda = 0$, it follows a Hund's case (b) coupling scheme.

NH

NH can be decelerated in the long lived a$^1\Delta$ state. This state is best described in Hund's case (b).

NO

The ground state of NO is $^2\Pi$. The spin orbit constant and rotational constant are $A = 123.26$ cm^{-1} and $B = 1.672$ cm^{-1}, respectively.[32] Due to the large ratio of A to B, NO is well described by Hund's case (a). For ^{14}N^{16}O the two Ω-manifolds have near-degeneracies at $J = 18.5$ and $J = 19.5$. The smallest frequencies are around 70 GHz, which results in an enhancement of about 60. Smaller frequencies and correspondingly larger enhancement might be obtained with other isotopes of NO, however, the high J at which the degeneracy occurs is a disadvantage.

OH

The ground state of OH is $^2\Pi$. The spin orbit constant and rotational constant are $A = -139.21$ cm^{-1} and $B = 84.881$ cm^{-1}, respectively.[32] Due to the negative A constant the $^2\Pi_{3/2}$ lies below the $^2\Pi_{1/2}$ state. If this would not have been the case, no near-degeneracies would have occurred due to the small A to B ratio. However, as a result of spin–orbit inversion, a near-degeneracy occurs between the $J = 7/2$, $\Omega = 3/2$ and $J = 3/2$, $\Omega = 1/2$ states with a 2-photon transition frequency of 220 GHz and an estimated enhancement to a μ-variation of about 10. The situation might be more favorable in OD.

SH

The ground state of SH is $^2\Pi$. The spin orbit constant and rotational constant are $A = -376.9$ cm^{-1} and $B = 9.461$ cm^{-1}, respectively.[32] Due to the large ratio of A to B, SH is well described by Hund's case (a). The two Ω-manifolds are expected to have a near-degeneracy at $J \approx 12$. More detailed calculations are needed to determine the frequency of the 2-photon transition and its sensitivity.

Conclusions

In this paper we have shown that the sensitivity to a variation of the proton–electron mass ratio, μ, and the fine-structure constant, α, is enhanced due to near degeneracy between rotational levels in the fine structure ladders of molecular radicals. This is an extension to the work of Flambaum[12] and Flambaum and Kozlov,[13] who showed that the relative sensitivity to a variation of μ and α is enhanced due to cancellation of hyperfine intervals with rotational intervals and cancellation of fine structure intervals with vibrational intervals. Whereas the cancellations discussed in ref. [12] and [13] occur in heavy molecules, the cancellations discussed in this paper occur in light molecules. As techniques to manipulate and cool molecules are more adapted to light molecules, a higher precision might be expected for these molecules.

We have presented a detailed calculation, including all relevant couplings, that show that a degeneracy occurs in the various natural isotopomers of carbon monoxide in its a$^3\Pi$ state. The most suitable transitions are the $J = 6$, $\Omega = 1$, + → $J = 8$, $\Omega = 0$, + transitions in the $v = 0$ and $v = 1$ of $^{12}C^{16}O$ at 1.648 GHz and the 2.459 GHz, respectively, which have a combined sensitivity to a μ-variation on the order of 500, and an estimated combined sensitivity to an α-variation of 1000. It seems possible to measure the 2-photon transitions with a fractional precision on the order of 10^{-12}, which would result in a limit for μ-variation on the order of 10^{-15} yr^{-1}.

The enhanced sensitivity due to the near-degenerate fine-structure levels should occur in many other light molecules as well. It is our hope to identify a molecule that has a suitable level structure and that can be cooled using one of the existing cooling techniques. We have performed a check for a number of seemingly promising molecules. But it turned out that these molecules all have either a too small A/B ratio—such that the molecule is no longer well described in a Hund's case (a) coupling scheme— or a too large A/B ratio—in which case the near degeneracy occurs at high J. We encourage the attendants of the Faraday meeting to come forward with suggestions for suitable molecules.

Acknowledgements

We thank Giel Berden and Gerard Meijer for many stimulating discussions on metastable CO and Sam Meek for useful discussions on the Zeeman effect. We thank Ruud van Putten for help with the calculations. H. L. B acknowledges financial support from the Netherlands Organisation for Scientific Research (NWO) *via* a VIDI-grant and from the European Research Council *via* a Starting Grant.

References

1 See for instance J.-P. Uzan, *Rev. Mod. Phys.*, 2003, **75**, 403.
2 S. G. Karshenboim and E. Peik, *Eur. Phys. J., Special Topics*, 2008, **163**, 1, and references therein.
3 X. Calmet and H. Fritzsch, *Eur. Phys. J. C*, 2002, **24**, 693.
4 M. T. Murphy, J. K. Webb and V. V. Flambaum, *Mon. Not. Roy. Astron. Soc.*, 2003, **345**, 609.
5 E. Reinhold, R. Buning, U. Hollenstein, A. Ivanchik, P. Petitjean and W. Ubachs, *Phys. Rev. Lett.*, 2006, **96**, 151101.
6 T. M. Fortier and et al., *Phys. Rev. Lett.*, 2007, **98**, 070801.
7 A. Shelkovnikov, R. J. Butcher, C. Chardonnet and A. Amy-Klein, *Phys. Rev. Lett.*, 2008, **100**, 150801.
8 T. Zelevinsky, S. Kotochigova and J. Ye, *Phys. Rev. Lett.*, 2008, **100**, 043201.
9 A. Cingöz, A. Lapierre, A.-T. Nguyen, N. Leefer, D. Budker, S. K. Lamoreaux and J. R. Torgerson, *Phys. Rev. Lett.*, 2007, **98**, 040801.
10 S. Bize and et al., *J. Phys. B.: Atom., Mol. Opt. Phys.*, 2005, **38**, S449.
11 D. DeMille, S. Sainis, J. Sage, T. Bergeman, S. Kotochigova and E. Tiesinga, *Phys. Rev. Lett.*, 2008, **100**, 043202.
12 V. V. Flambaum, *Phys. Rev. A*, 2006, **73**, 034101.
13 V. V. Flambaum and M. G. Kozlov, *Phys. Rev. Lett.*, 2007, **99**, 150801.
14 We have performed calculations on CuS using molecular constants given by J. M. Thompsen and L. M. Ziurys, *Chem. Phys. Lett.*, 2001, **344**, 75. We indeed found a near degeneracy between the X $^2\Pi_{1/2}$ ($v = 0$, $J = 75½$) and the X $^2\Pi_{3/2}$ ($v = 1$, $J = 76½$).

15 S. Y. T. van de Meerakker, H. L. Bethlem and G. Meijer, *Nat. Phys.*, 2008, **4**, 595.
16 G. Herzberg, *Spectra of Diatomic Molecules*, D. van Nostrand Company, Inc., Princeton, 1950.
17 J. Brown and A. Carrington, *Rotational Spectroscopy of Diatomic Molecules*, Cambridge University Press, Cambridge, 2003.
18 J. van Veldhoven, J. Küpper, H. L. Bethlem, B. Sartakov, A. J. A. van Roij and G. Meijer, *Eur. Phys. J. D*, 2004, **31**, 337.
19 H. L. Bethlem, M. Kajita, B. Sartakov, G. Meijer and W. Ubachs, *Eur. Phys. J. Special Topics*, 2008, **163**, 55.
20 B. G. Wicke, R. W. Field and W. Klemperer, *J. Chem. Phys.*, 1972, **56**, 5758.
21 R. J. Saykally, K. M. Evenson, E. R. Comben and J. M. Brown, *Mol. Phys.*, 1986, **58**, 735.
22 R. W. Field, S. G. Tilford, R. A. Howard and J. D. Simmons, *J. Mol. Spectrosc.*, 1972, **44**, 347.
23 M. Drabbels, S. Stolte and G. Meijer, *Chem. Phys. Lett.*, 1992, **200**, 108.
24 H. L. Bethlem, G. Berden and G. Meijer, *Phys. Rev. Lett.*, 1999, **83**, 1558.
25 J. J. Gilijamse, S. Hoekstra, S. A. Meek, M. Metsälä, S. Y. T. van de Meerakker, G. Meijer and G. C. Groenenboom, *J. Chem. Phys.*, 2007, **127**, 221102.
26 The coefficients of these wave functions are listed in R. T. Jongma, Molecular Beam Experiments and Scattering Studies with State-selected Metastable CO, PhD Thesis, University of Nijmegen, The Netherlands, 1997.
27 See, for instance, J. Vanier and C. Audoin, *The Quantum Physics of Atomic Frequency Standards*, IOP Publishing, Bristol, 1989.
28 N. F. Ramsey, *Rev. Mod. Phys.*, 1990, **62**, 541.
29 R. T. Jongma, Th. Rasing and G. Meijer, *J. Chem. Phys.*, 1995, **102**, 1925.
30 E. R. Hudson, H. J. Lewandowski, B. C. Sawyer and J. Ye, *Phys. Rev. Lett.*, 2006, **96**, 143004.
31 H. L. Bethlem, F. M. H. Crompvoets, R. T. Jongma, S. Y. T. van de Meerakker and G. Meijer, *Phys. Rev. A*, 2002, **65**, 053416.
32 K. P. Huber and G. Herzberg, *Constants of Diatomic Molecules*, van Nostrand Reinhold Company, New York, 1979.

Prospects for measuring the electric dipole moment of the electron using electrically trapped polar molecules

M. R. Tarbutt,* J. J. Hudson, B. E. Sauer and E. A. Hinds

Received 18th November 2008, Accepted 15th January 2009
First published as an Advance Article on the web 21st May 2009
DOI: 10.1039/b820625b

Heavy polar molecules can be used to measure the electric dipole moment of the electron, which is a sensitive probe of physics beyond the Standard Model. The value is determined by measuring the precession of the molecule's spin in a plane perpendicular to an applied electric field. The longer this precession evolves coherently, the higher the precision of the measurement. For molecules in a trap, this coherence time could be very long indeed. We evaluate the sensitivity of an experiment where neutral molecules are trapped electrically, and compare this to an equivalent measurement in a molecular beam. We consider the use of a Stark decelerator to load the trap from a supersonic source, and calculate the deceleration efficiency for YbF molecules in both strong-field seeking and weak-field seeking states. With a 1 s holding time in the trap, the statistical sensitivity could be ten times higher than it is in the beam experiment, and this could improve by a further factor of five if the trap can be loaded from a source of larger emittance. We study some effects due to field inhomogeneity in the trap and find that rotation of the electric field direction, leading to an inhomogeneous geometric phase shift, is the primary obstacle to a sensitive trap-based measurement.

1 Introduction

The permanent electric dipole moment (edm) of the electron, or other fundamental particle, can only be non-zero if both parity (P) and time-reversal (T) invariance are violated.[1,2] The weak interaction violates P, while T violation is equivalent to CP violation provided CPT invariance is accepted. The CP violation of the Standard Model generates edms that are far too small to detect.[3] It also fails to account for the predominance of matter over antimatter in the universe.[4] Many of the proposed extensions of the Standard Model contain new sources of CP violation and result in vastly larger edms;[5,6] indeed, some models predict values for the electron edm that are very close to the current experimental limit. A measurement of a non-zero edm would be firm evidence for new physics.

Heavy, paramagnetic atoms and molecules offer high sensitivity to the electron edm. In an applied electric field, E_a, the change in energy of such an atom or molecule resulting from the electron edm, d_e, is $d_e P(E_a) E_{int}$. Here, E_{int} is a structure-dependent effective electric field whose magnitude scales as the cube of the nuclear charge,[7] and $P(E_a)$ is the degree of polarization of the atom or molecule in the applied field. The most sensitive measurement to date, $d_e = (6.9 \pm 7.4) \times 10^{-28}$ e cm, was made using a beam of Tl atoms,[8] for which E_{int} is large but P is small

Centre for Cold Matter, Blackett Laboratory, Imperial College London, Prince Consort Road, London, United Kingdom SW7 2AZ

for realistic laboratory fields. Because polar molecules can have much larger values of P they offer even higher sensitivity.[9–11] Using a continuous, thermal beam of YbF, we made the first measurement of the electron edm with a molecular system.[12] By upgrading this experiment to a cold, pulsed source, its statistical sensitivity has been greatly improved, and a new measurement is underway.[13,14] Using the most recent advances in the production and manipulation of cold molecules, it seems likely that still higher precision can be obtained. The ability to trap polar molecules electrically[15] is particularly attractive for precision measurements such as the electron edm, because of the very long interaction times that are then available.

The shot-noise limit on the statistical error in a molecule-based measurement of the edm, δd_e, in the usual units of e cm, is given by

$$\delta d_e = \frac{\hbar}{e} \frac{1}{|P| E_{int} \tau \sqrt{N}}. \tag{1}$$

Here, E_{int} is measured in V cm^{-1}, τ is the amount of time each molecule spends inside the apparatus (the coherence time), and N is the total number of molecules that participate in the experiment. The inefficiencies of state preparation and readout, and any intervening state manipulations, are incorporated in our definition of N. In our current edm experiment, pulses of YbF molecules created at 4 K in a supersonic source travel at 600 m s^{-1} through a 60 cm long interaction region where the applied electric field is 20 kV cm^{-1}. There are 25 pulses per second, approximately 2500 molecules are detected in each pulse, and 50% of the running time is devoted to data-taking. For YbF, $E_{int} = 26$ GV cm^{-1}, and the polarization factor at 20 kV cm^{-1} is $|P| = 0.7$. Using these values in eqn (1), we estimate our shot-noise limited sensitivity to be 7×10^{-28} e cm/$\sqrt{day}$. As is often the case, the experiment operates slightly above the shot-noise limit, magnetic field noise and source intensity fluctuations being the dominant sources of additional noise. At present, our statistical error is about 1.3 times the shot-noise limit.

To increase the sensitivity, we have to increase the product $\sqrt{N}\tau$. There are, of course, many possible ways to do this. An increase in N might be obtained by improving the detection efficiency or by increasing the time-averaged flux of molecules from the source. These would benefit both the beam experiment and the trap experiments considered here. Another strategy is to increase the fraction of molecules transmitted from source to detector, which might be achieved using molecular optics near the source to collimate the beam. For that to yield an improvement, the transverse phase-space acceptance of the optics must exceed that of the machine without the optics. We will discuss beam optics below in the context of Stark decelerators, and we will see that the acceptance of our current machine is already comparable to that of realistic beam optics. An increase in τ could be achieved without slowing or trapping the molecules, simply by making the machine longer. For a freely diverging molecular beam however, there is no advantage to increasing the interaction length L since the coherence time is proportional to L and the number of molecules reaching the detector scales as L^{-2}, leaving $\tau\sqrt{N}$ constant. If optics are used near the source to collimate or focus the molecular beam, then an increase in L can be beneficial.

In this paper, motivated by the prospect of an enormous increase in coherence time, we consider the possibility of measuring the electron edm using electrically trapped molecules. Given the difficulties involved in producing trappable molecules, we can expect the increase in τ to be accompanied by a reduction in N. We evaluate the efficiency of Stark deceleration and so obtain an estimate for the statistical sensitivity that could be obtained. We then consider a few of the very many systematic effects that may scupper the measurement. We focus mainly on a measurement using YbF molecules, though our considerations are easily modified to other relevant molecular systems.

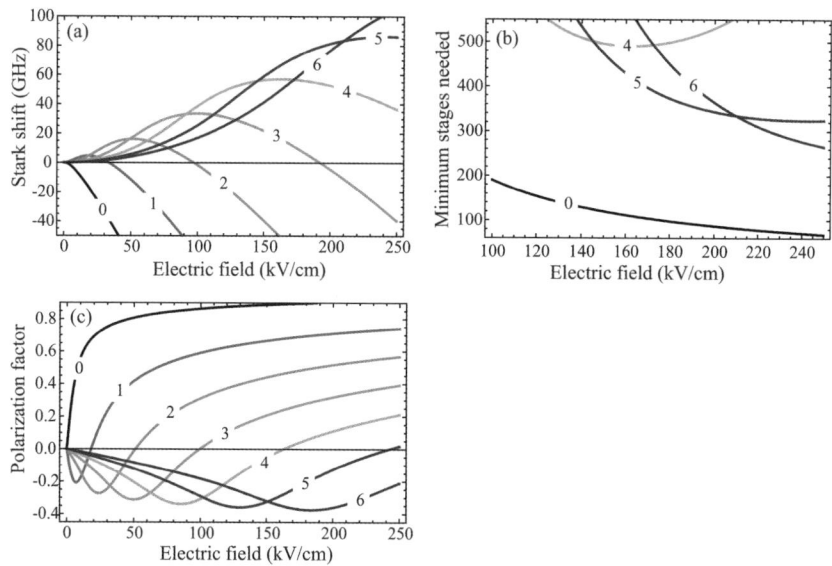

Fig. 1 Electric field dependence of (a) the Stark shifts of ($N = 0$–6, $M = 0$) states of YbF, (b) the minimum number of deceleration stages needed to stop 340 m s^{-1} YbF molecules in a selection of these states, and (c) the polarization factor of each state. Each curve is labelled by the value of the rotational quantum number, N.

2. Deceleration

If the molecules are to be trapped, they must have a low velocity. We consider the prospects of bringing to rest the YbF molecules produced by our current supersonic source.[16] We take 340 m s^{-1} and 4 K as the starting speed and temperature, these being typical when using room temperature xenon as the carrier gas. We have obtained speeds as low as 290 m s^{-1} by cooling the xenon, but only at the expense of molecular flux. It seems possible that a buffer gas source[17,18] will provide slower YbF beams with higher intensity, but this has not yet been demonstrated.

In designing a suitable decelerator, we can choose which rotational state of the molecule to use. Fig. 1(a) shows the Stark shifts of the (N, $M = 0$) rotational states of YbF in fields up to 250 kV cm^{-1}. Here, N and M are the quantum numbers of the rotational angular momentum and its projection onto the electric field axis. In the ground state the molecule has a negative Stark shift and can only be decelerated using the alternating gradient (AG) method.[19–21] The other states shown have positive Stark shifts at low field, and negative Stark shifts at high field. Molecules prepared in these excited rotational states could be slowed down with a "conventional" Stark decelerator for weak-field seekers,[22] which we will refer to as a WF decelerator. It would have to be operated at fields below the turning point of the state. Fig. 1(b) shows the minimum number of deceleration stages needed to decelerate YbF molecules from 340 m s^{-1} to zero. Deceleration in the ground state (Section 2.2) is obviously advantageous in terms of stages needed, but the difficulty of implementing the AG method is a disadvantage. Deceleration in the weak-field seeking states (Section 2.1) is easier to implement but requires more stages. The best choice of state depends on the operating field. At 200 kV cm^{-1}† the number of WF deceleration stages is minimized for $N = 5$. Lower rotational states are not as efficient because their turning points are reached at lower fields, as illustrated by the curve for $N = 4$ in the figure. Higher rotational

† We have operated decelerators at this field. Higher fields are very difficult to sustain.

states are also less efficient because their Stark shifts are smaller at this field, as illustrated by the curve for $N = 6$.

We have looked at the rotational state dependence of deceleration, but what about the edm experiment itself? Its sensitivity depends on the rotational state and the applied electric field through the polarization factor. This is given in terms of the Stark shift, W, the electric field, E, and the dipole moment, μ, by $P = -(1/\mu)dW/dE$. Fig. 1(c) shows this polarization factor as a function of electric field for the various rotational states with $M = 0$. At any given field the ground state offers the largest polarization factor, which is one reason why our current experiment uses this state. An experiment that used a weak-field seeking state would have a smaller, though still useful, value of $|P|$. For example, the $N = 5$ state discussed above has $|P| = 0.36$ at $E = 130$ kV cm^{-1}.

2.1 Deceleration in a rotationally excited state

In this section we consider a practical decelerator for molecules in the $(N, M) = (5, 0)$ state. We consider a design with a 4 mm square aperture and a periodicity of $2L = 24$ mm, similar to the one used in Berlin to decelerate OH.[23]‡ Using voltages of ± 40 kV, and a synchronous phase angle of 66°, 444 stages are required to decelerate from 340 m s^{-1} to zero. Each unit cell of the decelerator has two deceleration stages, one that focusses along x, and the other along y, so the overall length is 5.328 m. These are the parameters we use in the rest of this section.

To estimate how many molecules we can expect to decelerate, we need to calculate the phase-space acceptance of the decelerator. We start with the longitudinal acceptance. Since for our case the Stark shift is not at all linear (see Fig. 1), we will need to take a short diversion to extend the usual treatment of the longitudinal acceptance e.g. refs 24 and 25. The change in kinetic energy of the synchronous molecule in each deceleration stage is usually well approximated by $\Delta K = -\Delta K_{max}\sin\phi_0$, with ΔK_{max} being the maximum possible energy change per stage (a positive number), and ϕ_0 the synchronous phase angle which goes from 0 to 2π over the period $2L$. For our case, this is a poor approximation, and we find it necessary to retain an extra term in the Fourier expansion of ΔK. We shall derive a general formula for the longitudinal acceptance, and then apply it to our case. We write the change in kinetic energy using the expansion

$$\Delta K = \Delta K_{max} \sum_{n \text{ odd}} (-1)^{(n+1)/2} a_n \sin(n\phi_0). \quad (2)$$

Here, the a_n are coefficients that depend on the details of the decelerator and Stark shift of the molecule, and satisfy $\sum_{n \text{ odd}} a_n = 1$. Next, we use the approximation of a constantly acting force to write down the equation of motion of a non-synchronous molecule in terms of its relative phase, $\tilde{\phi}$, and its relative velocity $\tilde{v}$,

$$m\pi\tilde{v}\frac{d\tilde{v}}{d\tilde{\phi}} = \Delta K_{max} \sum_{n \text{ odd}} (-1)^{(n+1)/2} a_n \left[\sin(n\phi_0 + n\tilde{\phi}) - \sin(n\phi_0)\right], \quad (3)$$

where m is the mass. Integrating this equation of motion we obtain the contours of constant (relative) energy,

$$\left(\frac{\tilde{v}}{v_i}\right)^2 + \frac{1}{\pi N_{min}} \sum_{n \text{ odd}} (-1)^{(n+1)/2} a_n \left[\frac{1}{n}\cos(n\phi_0 + n\tilde{\phi}) + \tilde{\phi}\sin(n\phi_0)\right] = \text{constant}, \quad (4)$$

where v_i is the initial velocity and N_{min} is the minimum number of deceleration stages needed to bring the molecules to rest. The potential well represented by the

---

‡ The Berlin decelerator had $2L = 22$ mm.

second term in eqn (4) has a maximum at $\tilde{\phi} = \pi - 2\phi_0$, irrespective of the values of the a_n, for any sensible decelerator configuration. This is easily verified by inspecting the derivative of this potential with respect to $\tilde{\phi}$. A molecule that reaches this maximum with zero relative velocity is moving on the separatrix between bound and unbound motion. The equation of the separatrix is thus found to be

$$\left(\frac{\tilde{v}}{v_i}\right)^2 + \frac{1}{\pi N_{min}} \sum_{n \text{ odd}} (-1)^{(n+1)/2} a_n \left[\frac{2}{n} \cos\left(n\phi_0 + \frac{1}{2}n\tilde{\phi}\right) \cos\left(\frac{1}{2}n\tilde{\phi}\right) \right.$$
$$\left. + (2\phi_0 + \tilde{\phi} - \pi) \sin(n\phi_0) \right] = 0. \tag{5}$$

For our case, the first two terms in this sum give a sufficiently accurate result. Using the electric fields obtained from a finite element model, and the Stark shift shown in Fig. 1(a), we can determine ΔK numerically. Fitting this to the first two terms of eqn (2), with $a_3 = 1 - a_1$, we obtain $a_1 = 0.825$. The bold line in Fig. 2(a,c) shows the longitudinal acceptance area calculated using the first two terms of eqn (5), and this value of a_1. Integrating the area inside this separatrix gives a longitudinal phase-space acceptance of $A_z = 23.4$ mm m s^{-1}. Note that the calculated acceptance area is smaller, 15.2 mm m s^{-1}, when only the first term in eqn (2) is used (i.e. $a_1 = 1$).

Next, we estimate the transverse acceptance, following the procedure outlined in e.g. refs 19 and 26. The restoring force in the transverse direction is approximately linear in the transverse coordinate, $F_x \simeq -k_x(z)x$, but with a spring constant that is a strongly-varying function of the longitudinal coordinate, with period $2L$. The wavelength of the transverse oscillation is very much longer than $2L$, except when the molecules have been decelerated to very low speeds. We calculate a mean spring

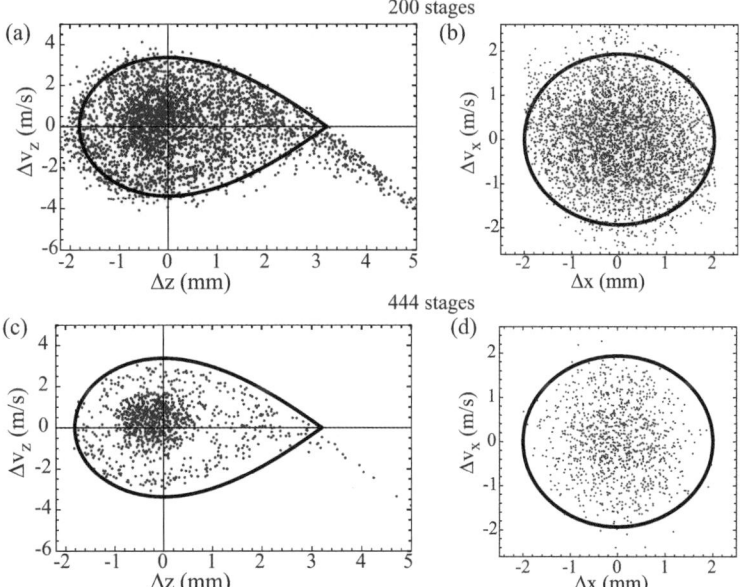

Fig. 2 Phase-space acceptance of a Stark decelerator for YbF in the (5, 0) state. Dots show the initial positions in phase-space of those molecules transmitted and decelerated, as determined from a three dimensional numerical simulation. Bold lines show the acceptances calculated analytically as discussed in the text. (a, c) Longitudinal acceptance after 200, 444 stages. (b, d) Transverse acceptance after 200, 444 stages.

constant, $\bar{k}_x$, by integrating $k_x(\phi)$ from $\phi_0 - \pi$ to ϕ_0 in the first switch state, and then from ϕ_0 to $\phi_0 + \pi$ in the second switch state, finally dividing by 2π. In this way, we find the mean transverse angular oscillation frequency for our molecules of mass m to be $\omega_x = \sqrt{\bar{k}_x/m} = 2\pi \times 154$ Hz when $\phi_0 = 66°$. The corresponding phase-space acceptance in the x-direction is $A_x = \pi \omega_x r_0^2$, where $r_0 = 2$ mm is half the size of the decelerator's aperture. Thus, this approximate calculation gives us $A_x = 12.2$ mm m s^{-1}. The bold line in Fig. 2(b,d) shows the transverse acceptance obtained using this approximate method. The total 6D acceptance is $A = A_x A_y A_z = A_x^2 A_z = 3487$ mm^3 (m s^{-1})^3.

The acceptance of a real Stark decelerator tends to be less than the idealized estimates given above. Coupling of the longitudinal and transverse motions can lead to unstable regions in phase-space.[26] Further loss occurs at very low speeds as the wavelength of transverse oscillations becomes comparable to the decelerator periodicity and the molecules are thrown out of the decelerator in regions where the focussing is weak.[27] To calculate the true phase-space acceptance, we simulated the motion of molecules through the decelerator. Fig. 2 shows the results obtained for 200 stages, where the final speed is 252 m s^{-1}, and for 444 stages where the final speed is 14 m s^{-1}, slow enough that one final stage – the trap – will bring the molecules to rest. To produce these plots, the trajectories of molecules chosen at random from a larger phase-space volume were simulated through the decelerator. The initial coordinates of successful trajectories, those that reached the end *and* were decelerated, are marked by the dots. The acceptance is the volume of the initial distribution multiplied by the successful fraction. After 200 stages, the region of phase-space that contributes to the slowed beam is very similar to that predicted by the simple theory outlined above. In the longitudinal direction, part (a), some molecules are accepted from regions that are slightly outside the bold line. This is because off-axis molecules that travel closer to the rods of the decelerator experience a slightly deeper longitudinal potential than those on the axis. The region inside the bold line is not uniformly filled, the area closer to the origin having a higher density because it is less susceptible to coupling of the transverse and longitudinal motions. In the transverse direction, part (b), we see again that a few molecules outside the region given by the simple theory contribute to the acceptance because the transverse confinement is greater for some of the non-synchronous molecules. Once again, the central region has a higher density of accepted particles. The total 6D phase-space acceptance for 200 stages is 1168 mm^3 (m s^{-1})^3, which is a factor of 3 smaller than the idealized prediction. An increase in the number of stages to 444, so that the molecules can be trapped, reduces the acceptance to 352 mm^3 (m s^{-1})^3. In the longitudinal direction, Fig. 2(c), the coupling of transverse and longitudinal motions results in an unstable region of phase-space within the separatrix, as investigated in ref. 26. The accepted molecules come from a region close to the origin, along with the 'halo' region around this. In the transverse direction, part (d), molecules coming from regions close to the rods of the decelerator tend to be focussed too strongly at low speed and so are lost.

In our supersonic beam, the rotational temperature is typically about 4 K, and the population in the (5, 0) state is only a few percent of the ground state population. To increase the flux of slow molecules, we will need to transfer population into our chosen state. This could be done by driving the microwave transition from the ground state in the presence of a static electric field. Without the static field it is impossible to drive this transition, since the total angular momentum quantum number has to change by 5 units. In the presence of a field, states of the same M but different N are mixed and the selection rule on N is no longer a good one. In the limit of very strong fields the molecules are strongly polarized along the field axis and the eigenstates, known as pendular states, are those of a two-dimensional angular oscillator.[28] They are labelled by M and by a vibrational quantum number v_p, which only changes by one unit in an electric dipole transition. The state that correlates to the field-free (5, 0) state is the one with $(v_p, M) = (10, 0)$, while the

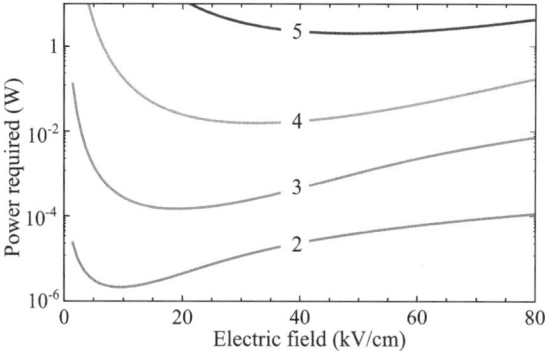

Fig. 3 Driving rotational transitions in a static electric field. Molecules enter the field in the $(N, M) = (0, 0)$ state, and exit in the $(N, 0)$ state. The power needed for a π-pulse is plotted as a function of the static electric field strength, for the experimental parameters given in the text. Curves are labelled by their value of N. Note the logarithmic scale.

ground state has $(v_p, M) = (0, 0)$. It follows that the required transition is forbidden in the strong-field limit as well as in the weak-field limit. At intermediate fields, it may be possible to drive the transition.

We suppose that ground state YbF molecules, travelling at 340 m s^{-1}, cross a resonant microwave beam at right angles. The microwaves have a cylindrical focus, 20 mm long in the direction of the molecular beam, and 4 mm in the perpendicular direction. Fig. 3 shows the power required to drive a π-pulse to the state that correlates to the field-free $(N, M = 0)$ state, as a function of the applied, uniform, static electric field strength.§ As anticipated by the considerations above, the required power is high at both low and high fields. The power requirement increases rapidly with N, and the electric field that minimizes this power also increases with N. For the $(5, 0)$ state, the power is minimized in a static field of 49 kV cm^{-1}, and is then 2 W at 286 GHz. This seems unfeasible. Furthermore, the power requirement increases if the transition is inhomogenously broadened, and it will be unless the static field is uniform to 1 part in 10^7. A better route takes the molecule *via* the $(3, 0)$ state. At 31 kV cm^{-1} the transitions correlating to $(0,0) \rightarrow (3,0)$ and $(3,0) \rightarrow (5,0)$ have the same frequency, 126 GHz. At this field strength, the power requirements are 232 μW and 7 μW respectively. Inhomogeneous broadening is severe for the first transition whose Stark shift is about 2GHz per kV cm^{-1} at this field. If the field in the interaction region is non-uniform at the 10^{-4} level, the power needed increases to about 200 mW, a demanding, but not unfeasible, requirement.

2.2 Alternating gradient deceleration in the ground state

Whereas weak-field seeking molecules are naturally focussed through a Stark decelerator, strong-field seekers are naturally defocussed. The alternating gradient (AG) decelerator provides a solution to this focussing problem, as explained in detail in refs 21 and 31. The molecules move through an alternating sequence of focussing and defocussing lenses, always passing closer to the axis in the defocussing lenses than in the focussing lenses. Because the transverse force is proportional to the off-axis displacement, these molecules are on stable trajectories. In the ideal case, each lens focusses the molecules in one transverse direction, and defocusses them in the other, the force components being $F_x = kx$ and $F_y = -ky$. The two directions alternate in successive lenses. The stability of the trajectories, and the transverse

---

§ To apply this plot to a different rigid rotor molecule, X, multiply the power by $(\mu_{YbF}/\mu_X)^2$ and multiply the electric field by $(B_X/B_{YbF})(\mu_{YbF}/\mu_X)$.

phase-space acceptance, are determined by the product $\Omega\tau$, where $\Omega = \sqrt{k/m}$ is the angular oscillation frequency inside a focussing lens, and τ is the time spent in each lens. When $\Omega\tau \ll 1$, the focussing by one lens is almost exactly counteracted in the next and the net effect is very weak focussing. The trajectories are stable but have large amplitude, so the acceptance is small. As $\Omega\tau$ increases, the transverse extent of the beam modulates, always being larger in the focussing lenses than in the defocussing lenses. Molecules with greater transverse speeds can now be transmitted, and so the acceptance grows. At some critical value of $\Omega\tau$, the focussing becomes so strong that the molecules are overfocussed and all trajectories become unstable.

AG deceleration of ground state YbF has been demonstrated experimentally[20] using a short prototype machine. Other strong-field seeking molecules have also been decelerated using similar machines.[19,29,30,] In these experiments, the degree of deceleration has been limited and the transmission rather low. The AG decelerator is particularly sensitive to lens aberrations which tend to reduce its transmission enormously. Here, we consider a particular design where we aim to minimize the detrimental effects of aberrations and provide a very large reduction in the velocity. The design is illustrated in Fig. 4(a). Each lens is formed by a pair of rods, radius R, their axes parallel and separated by $2(R + r_0)$, similar to the geometry used in all AG experiments to date. Specifically, we take $r_0 = 2$ mm and $R = 9.3$ mm. At the exit end, each rod is terminated by a semi-ellipsoid of radii 9.3 mm along x and y, and 13.3 mm along z. At the entrance, the edge of the rod is rounded with a 2.7 mm radius of curvature. The total length of each electrode is 29.3 mm, and the distance from the entrance of one lens to that of the next is 34.7 mm.

The choice of R/r_0 is determined by a trade-off between the strength of focussing and the size of the non-linear contributions to the force.[32] When R/r_0 is large, the focussing will be weak and the acceptance will be small. When it is small the beneficial effect of the enhanced focussing is outweighed by the detrimental effect of large

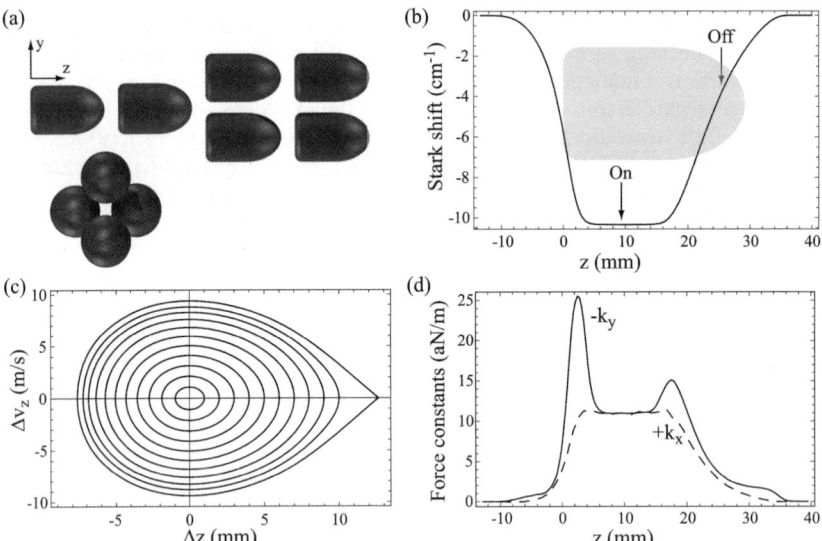

Fig. 4 Alternating gradient deceleration of ground state YbF. (a) Projections of the decelerator onto the yz- and xy-planes. (b) The z-dependence of the Stark shift, with the position of the central electrode indicated in grey. The entrance end of this electrode is positioned at $z = 0$. (c) Longitudinal phase-space acceptance calculated from the potential and the turn-on and -off positions shown in (b). (d) The focussing (dashed line) and defocussing (solid line) force constants, $k_x = -\partial F_x/\partial x$ and $k_y = -\partial F_y/\partial y$. For ease of comparison, $-k_y$ is plotted along with $+k_x$.

nonlinear forces. Our choice, $R/r_0 = 14/3$, is close to optimum for a decelerator where the length of the lenses is approximately equal to the length of the gaps between lenses, and where $\Omega\tau$ has its best value of about 0.7. As shown in Fig. 11 of ref. 32, the transverse acceptance is then $0.012 \times 2\mu_{\text{eff}}E_0 r_0^2/m$. For our parameters, this is 70 mm^2 (m s^{-1})^2, a factor of 2 smaller than the transverse acceptance estimated for the WF decelerator above. A higher transverse acceptance can be achieved using a more complicated electrode structure to form each lens, but only by a factor of about 2.5.[32]

The deceleration regions are the ellipsoidal ends of the rods. Fig. 4(b) shows the z-dependence of the Stark shift for our lens structure, along with a cross-section of an electrode so that its position relative to the map is clear. The lens is charged to ± 40 kV, and its neighbours are grounded. To make a first estimate of the longitudinal acceptance, we consider only those molecules travelling along the axis of the machine, ignoring any coupling between transverse and longitudinal motions. Then, the position $\tilde{z}$ of a molecule relative to that of the synchronous molecule is given by the mean (relative) acceleration[31]

$$\frac{d^2\tilde{z}}{dt^2} = \frac{W(z_{\text{on}} + \tilde{z}) - W(z_{\text{on}}) - W(z_{\text{off}} + \tilde{z}) + W(z_{\text{off}})}{mD}, \quad (6)$$

where $W(z)$ is the z-dependent Stark shift shown in Fig. 4(b), z_{on} and z_{off} are the positions of the synchronous molecule when the fields of the lens are turned on and off, and D is the lens-to-lens distance. By solving this equation of motion using the turn-on and turn-off positions indicated, we obtain the phase-space trajectories of non-synchronous molecules shown in Fig. 4(c). The outer trajectory is the separatrix between stable and unstable motion. Using this model, we obtain a longitudinal acceptance of 270 mm m s^{-1}, almost 12 times as large as that for deceleration in the (5,0) state (Fig. 2). In the Δz direction, the accepted region is about four times larger and is a direct result of the larger spacing between deceleration stages. In the Δv_z direction, the accepted region is about three times larger, because of the larger Stark shift in the ground state. Within the model of uncoupled longitudinal and transverse motion, the total 6D acceptance of the AG decelerator is 18900 mm^3 (m s^{-1})^3.

In reality, the transverse and longitudinal motions are rather intimately coupled. We will arrange the lenses so that $\Omega\tau$ is close to the optimum value for the synchronous molecule. The non-synchronous molecules will have different values of this parameter. Referring to Fig. 4(b), a molecule that is ahead will spend less time in the lens and so will not be focussed as effectively, while a molecule that is behind will spend more time in the lens and may be overfocussed and become unstable. Furthermore, deceleration is always accompanied by transverse defocusing. This effect is shown in Fig. 4(d), where the transverse force constants $k_x = -\partial F_x/\partial x$, $k_y = -\partial F_y/\partial y$ are plotted as a function of z. Inside the lens, where there is no longitudinal electric field gradient, these two force constants are equal and opposite, $k_x = -k_y = 11$ aN m^{-1}. Moving into the fringe fields of the lens, the focussing force constant decreases along with the electric field, whereas the defocusing force constant increases to a value that depends on the longitudinal curvature of the electrodes. The smaller the radius of curvature, the larger this defocusing peak becomes. We have chosen a fairly gradual termination of the electrodes at the exit end, where the deceleration occurs, and the magnitude of k_y peaks at 15 aN m^{-1}. At the entrance end, the electrodes begin suddenly, and so the defocusing constant rises to a much larger value of 25 aN m^{-1}. However, molecules that stay close to the synchronous phase never see this excess defocusing, because the voltages are not yet on when they pass through this region. Since the focussing is most effective when the gaps between successive lenses are kept short, it is good to keep this entrance region short; that is why we have opted for an asymmetry between the entrance and exit ends. When the turn-on and turn-off positions are as indicated in part (b) of

Fig. 4, the mean value of k_y seen by the synchronous molecule exceeds that of k_x by 19%. This makes the trajectories unstable if $\Omega\tau$ is too small, making it even more important to keep this parameter close to its optimum value. Provided that is done, the excess defocussing has a rather small effect. We note that molecules lagging behind the synchronous one see the very large defocussing constant at the lens entrance, and we can expect these to be lost from the decelerator.

If the turn-on and turn-off positions are kept constant, the value of τ increases as the molecules slow down. Eventually, the limit of stability is reached and the molecules are lost. To avoid that, the lens length seen by the molecules needs to be reduced in harmony with the deceleration: the lenses should be long at the beginning and short at the end. Since deceleration occurs at the end of every lens, it is better to use n short focussing lenses, followed by n short defocussing lenses, instead of using long lenses and alternating each time. By using a larger value of n at the beginning than at the end, efficient deceleration can be combined with stability of focussing. We have simulated the motion of ground state YbF molecules through an AG decelerator having the electrode structure described above. The decelerator was divided into four sections, with $n = 4, 3, 2, 1$ and with 56, 36, 32, 15 deceleration stages in these sections respectively. The first section reduces the speed from 340 m s^{-1} to 266 m s^{-1}, and the subsequent sections reduce this further to 205, 128 and 66 m s^{-1} respectively. In order to avoid a mismatch between the phase-space distribution exiting one section and the acceptance of the next, the turn-on position was gradually adjusted within each section so that the value of $\Omega\tau$ remained approximately the same throughout the deceleration process.

Fig. 5 shows the accepted phase-space volume of the first section, projected onto the (x, v_x), (y, v_y) and (z, v_z) planes. The plots were obtained as described above for the WF decelerator. In the (x, v_x) plane the accepted molecules form a diverging beam because the first four lenses of the decelerator focus in this direction, while in (y, v_y) they form a converging beam because those lenses defocus in that direction. In the (z, v_z) plane, the shape of the accepted area is similar to the calculated one shown in Fig. 4(c), though a little smaller. In each plane, there is a dense central core of accepted molecules, with a less dense region near the edges. These regions of lower density are due to the coupling between the two transverse motions, and between the transverse and longitudinal motions. The acceptance of this first 56-stage section is 2750 mm^3 (m s^{-1})^3, more than twice that found for the first 200 stages of the WF decelerator, where the final speed was nearly the same. Taken alone, the acceptances of the other three sections are similar, but when we concatenate them, the net acceptance falls. We calculate 850 mm^3 (m s^{-1})^3 for the first two sections, 340 mm^3 (m s^{-1})^3 for three sections and 254 mm^3 (m s^{-1})^3 for all four. There are several causes for this additional loss. Firstly, those molecules near the edges of the acceptance region in Fig. 5 are on metastable trajectories and will be lost in subsequent sections. Secondly, there remains some mismatch between one section and the next, despite our efforts to control this. Thirdly, as the turn-on position is moved towards the turn-off position the excess defocussing at the end of each lens

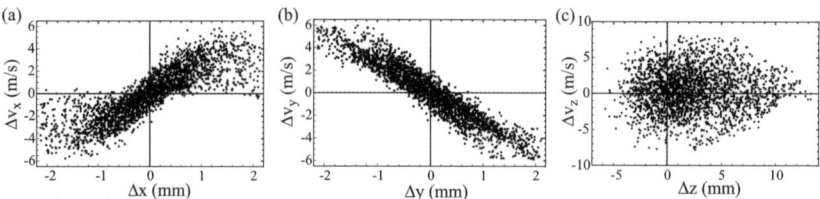

Fig. 5 Phase-space acceptance of the first section of AG decelerator described in the text, projected onto the (a) (x, v_x), (b) (y, v_y) and (c) (z, v_z) planes. The section has 56 stages and reduces the speed from 340 m s^{-1} to 266 m s^{-1}.

plays an increasingly important role and can result in additional loss. Nevertheless, the acceptance of the AG design studied here is similar to that of the WF decelerator and it is likely that further design iterations will improve on this.

This 139-stage AG decelerator removes 96% of the initial kinetic energy, but the molecules are still not slow enough to be trapped. Five more stages are needed. These final stages are particularly difficult to design because of the large fractional changes in speed, and they would need to be designed in conjunction with the trap to ensure that the molecules remain focussed and are coupled efficiently into the trapping region. An alternative is to use a combination of AG and WF deceleration. In this scheme, the AG decelerator would be used to remove the majority of the energy, and then the molecules would be switched into the weak-field seeking state for the final reduction of the velocity.

3. Trapping

The electric field of the trap is also the field that interacts with the edm being measured. The edm signal is proportional to the integral over time of the polarization factor, P, which in turn depends on the electric field as shown in Fig. 1(c). The trap should have a bias field so that $|P|$ remains large and approximately constant, *i.e.* close to its turning point for a weak-field seeking state, or close to unity for a strong-field seeking state.

It is not difficult to design a very deep, suitably-biased, electrostatic trap for weak-field seeking YbF molecules in the (5, 0) state. As an example, we consider using the "chain-link" trap discussed in ref. 33. This has a minimum field at its centre, which we choose to be 130 kV cm^{-1} in order to maximize the value of $|P|$ and minimize its variation. For modest applied voltages, the phase-space acceptance of such a biased Stark trap is of the order 10^4 mm^3 (m s^{-1})^3, very much larger than the phase-space volume occupied by the molecules coming from the Stark decelerator. So, with a suitably careful coupling of the molecules into the trap, all the available slow molecules could be trapped and used in an edm measurement.

If we are instead to trap molecules in the strong-field seeking ground state, an ac trap is required. For this, we consider a cylindrical ac trap of the kind first proposed by Peik[34] and recently used to trap ammonia molecules.[35,36] The centre of the trap is a saddle point, and the confining and deconfining directions alternate so that the molecules are trapped dynamically – the same principle that is used to guide molecules through the AG decelerator. An electric field of 100 kV cm^{-1} at the centre of the trap would gives a polarization factor of 0.86. For the ammonia experiments, the acceptance of this trap was found to be 270 mm^3 (m s^{-1})^3. Scaling to the YbF case, with its larger mass and larger dipole moment, we obtain an acceptance of about 50 mm^3 (m s^{-1})^3. This small acceptance makes trapping in the ground state rather unattractive for an edm measurement. Additionally, the large currents associated with the switching of the high voltages are likely to be incompatible with the extremely high level of magnetic field control needed for the experiment.

4. Statistical sensitivity

We now have all the information needed to estimate the statistical sensitivity of a trap experiment relative to that of our beam experiment, using eqn (1). It is convenient to write the total number of participating molecules as $N = \rho V r T$, where ρ is the phase-space density, V is the phase-space acceptance of the experiment, r is the mean number of shots per second and T is the total integration time. The first and last factors will be common to both beam and trap experiments and so factor out in the comparison. We introduce a figure of merit, which is the ratio of the sensitivity in a trap experiment to that in our current beam experiment,

$$S = \frac{\delta d_{e,\text{beam}}}{\delta d_{e,\text{trap}}} = \frac{|P_{\text{trap}}|}{|P_{\text{beam}}|} \frac{\tau_{\text{trap}}}{\tau_{\text{beam}}} \sqrt{\frac{V_{\text{trap}}}{V_{\text{beam}}}} \sqrt{\frac{r_{\text{trap}}}{r_{\text{beam}}}}. \quad (7)$$

In the beam, the volume of phase-space occupied by the participating molecules is $V_{\text{beam}} \approx 6 \times 10^4$ mm^3 (m s^{-1})^3. The other parameters are $r_{\text{beam}} = 12.5$ s^{-1}, $|P_{\text{beam}}| = 0.7$ and $\tau_{\text{beam}} = 10^{-3}$ s.

Consider first an experiment where the number of available molecules is limited by the acceptance of the WF Stark decelerator modelled above, which we found to be 350 mm^3 (m s^{-1})^3. We take a coherence time in the trap of 1 s, and we suppose that the trap is reloaded every 1.2 s. With a polarization factor of 0.36 we obtain a figure of merit of $S \approx 10$. This would be a very significant improvement in the sensitivity of the experiment, offering the possibility of reaching well below 10^{-28} e cm in statistical uncertainty. Note that the repetition rate of the proposed trap experiment is limited by the coherence time, not the pulse rate of the source, and so its sensitivity could be improved even further by loading multiple traps, though at a cost of further experimental complexity.

Finally, let us consider an experiment that is limited by the acceptance of the trap itself rather than the method used to fill the trap. This may become possible by loading the trap from a buffer gas source, or from an improved decelerator design. In this case $V_{\text{trap}} \approx 10^4$ mm^3 (m s^{-1})^3 and we find $S \approx 50$. All other factors being equal, in particular the phase-space density produced at the source, the statistical sensitivity would then be about 2×10^{-29} e cm after $T = 24$ hours of integration. To realize such exceptional sensitivity we would, of course, need to reduce all other sources of noise to this level. We see no fundamental reason why that should not be possible.

5. Physics in the trap

So far we have considered how the statistical sensitivity of an edm experiment might be improved by decelerating YbF molecules and loading them into a trap. Now we consider the impact of the trap environment on such an experiment. Our present edm experiment measures the precession of the molecular spin in a combination of nominally parallel electric and magnetic fields. We prepare the spin in a direction perpendicular to the fields and, after a fixed time, we measure the accumulated precession angle, which depends on the magnetic moment and the electron edm. In the trap, the electric field is necessarily inhomogeneous – that is how it exerts a force. Therefore, as molecules move around within the trap, they experience a substantial variation in the magnitude and direction of the field and this fluctuating environment varies from one molecule to another. It is therefore essential to consider what happens to the spin precession angle when the direction of the local electric field at the position of the molecule is changing with time. In the following, we consider two consequences of this: an effective magnetic field inhomogeneity, and a geometric phase.

5.1 Effective magnetic field inhomogeneity

Because of the large tensor Stark shift induced by the electric field of the experiment, only the magnetic field component parallel to the local electric field direction contributes to the precession angle.[12] The accumulated phase angle due to a magnetic field $\vec{B}$ is

$$\phi_B = \frac{g\mu_B}{\hbar} \int \vec{B}(\vec{r}(t)) \cdot \vec{\varepsilon}(\vec{r}(t)) \, dt, \quad (8)$$

where g and μ_B are the g-factor and Bohr magneton, $\vec{\varepsilon}$ is a unit vector along the local electric field direction, and the integral is taken over the trajectory of the molecule, $\vec{r}(t)$.

In a biased electric trap (Section 3), the trapping field is superimposed on a uniform bias electric field. We define $\hat{z} = \vec{\varepsilon}(0)$, the electric field direction at the centre of the trap. For those molecules that stay close to the centre, the local electric field direction is always close to $\hat{z}$, whereas molecules that reach the outer regions of the trap will see $\vec{\varepsilon}$ making larger angles with $\hat{z}$. It follows that the integral in eqn (8) changes from one trajectory to the next *even if the magnetic field is perfectly uniform*. This spread in the accumulated phase is a source of noise in the experiment and may degrade the experimental sensitivity if it is too large. The size of the phase spread depends on the temperature of the molecules relative to the Stark shift at trap centre (expressed in temperature units). If the temperature is high the molecules will explore a large range of electric field directions and the spread will be high. If cold, the molecules remain close to the trap centre where the electric field direction is close to $\hat{z}$ everywhere.

To explore this, we simulated the motion of molecules in the chain-link trap discussed in Section 3, and evaluated ϕ_B for each trajectory using eqn (8). The molecules were drawn from an initial phase-space volume of (4 mm × 4 m s^{-1})^3. As discussed above, the molecules are most sensitive to fields along $\hat{z}$ and so we applied a uniform magnetic field, B_z, in this direction. We find that both the mean and the standard deviation of the accumulated phases increase linearly with the coherence time, τ. Specifically we find a mean of $\langle \phi_B \rangle \approx 0.9\phi_0$ and a standard deviation of $\Delta\phi_B \approx 0.05\phi_0$, where $\phi_0 = g\mu_B B_z \tau/\hbar$ is the phase accumulated by a molecule that remains at the centre of the trap throughout. This phase spread should be compared to the error in a shot-noise limited measurement of the edm phase, which is $1/\sqrt{N}$ rad (see eqn (1)). Therefore, to ensure that the magnetic field does not increase the noise and so degrade the sensitivity, we require $\Delta\phi_B < 1$. For $\tau = 1$ s, the corresponding requirement on magnetic field is $B_z < 200$ pT. Magnetic field gradients will also contribute noise as different molecules sample different regions of the trap. Our simulations show that, for a coherence time of 1 s, these gradients need to be smaller than 50 nT m^{-1} if they are not to degrade the sensitivity. These field requirements are not too demanding; they can be satisfied using a multi-layer magnetic shield with an overall shielding factor of 10^6, *e.g.* ref. 37.

5.2 Geometric phase

We show below that, for a sufficiently strong electric field, the motion of the spin follows the electric field direction adiabatically, and that the spin polarization will always lie in the plane perpendicular to the local electric field. In this adiabatic limit, one might guess that the accumulated phase of the spin would be the time-integral of the instantaneous precession frequency. However, in the case of a rotating magnetic field, it is known from the work of Berry[38] that the accumulated phase angle has a second component, known as the geometric phase. This depends on the history of the field direction but not on the precession dynamics. We might expect that a similar geometric phase will appear in the spin precession of a molecule when the electric field changes direction. If we are to obtain the electron edm from the precession angle, we must be able to control any such geometric phase with high precision.

While the geometric phase of a spin precessing in a rotating magnetic field has been studied in great detail,[38] the case of a rotating electric field is less well-studied and the few treatments that do exist tend to be phrased in the somewhat arcane language of differential geometry.[39] In the interest of clarity, therefore, we first present a simple quantum mechanical treatment of the geometric phase in a rotating electric field, before going on to consider the impact of this phase on a potential edm experiment.

In the limit of large electric field, which is well-fulfilled in our experiments, the magnetic field makes no significant contribution to the geometric phase evolution. This is because the large tensor Stark shift strongly suppresses the interaction with magnetic field components perpendicular to the electric field.[12] Therefore we neglect the applied magnetic field in the following treatment, even though it is possible in principle to include it.[40]

5.2.1 Spin-one system in a rotating electric field.

Let us consider a spin-one system, corresponding to the $F = 1$ state of the YbF molecule used in our edm experiment. The Stark interaction mixes many states of the molecule, but for our present purpose it is adequately described by an effective Hamiltonian restricted to the $F = 1$ manifold, consisting of a scalar and a rank-2 tensor. With the electric field directed along the z-axis this takes the form

$$\hat{H}_{E_z} = \delta(E)\hat{T}_0^{(0)} + \varepsilon(E)\hat{T}_0^{(2)} \tag{9}$$

where the coefficients $\delta(E)$ and $\varepsilon(E)$ are phenomenological parameters that depend on the electric field strength E. If desired, they can be derived from a full calculation of the Stark shift including all of the molecular levels. $\hat{T}_0^{(n)}$ is the z-component of the rank-n irreducible spherical tensor operator. The matrix elements of this operator in the $|F = 1, m_F\rangle$ basis can be written, in order of ascending m_F, as

$$\hat{H}_{E_z} = \begin{pmatrix} 0 & 0 & 0 \\ 0 & \Delta & 0 \\ 0 & 0 & 0 \end{pmatrix}, \tag{10}$$

where we have simplified the parametrization by redefining the zero of energy and introducing $\Delta = \sqrt{1/3}\,\delta(E) - \sqrt{2/15}\,\varepsilon(E)$. We have dropped the explicit functional dependence of the parameter Δ on the electric field in order to streamline the notation.

When the electric field rotates to a new direction given by the Euler angles $\{\alpha(t), \beta(t), \gamma(t)\}$, the Hamiltonian becomes

$$\hat{H}(t) = R(t)^{-1}\hat{H}_{E_z}R(t) \tag{11}$$

where $R(t)$ is the rotation operator. We adopt the Euler angle convention of Weissbluth:[41] a rotation about the z-axis by γ, a rotation about the y'-axis by β, and finally a rotation about the z''-axis by α. The rotation matrix acting on states and operators (rather than on coordinates) is then given by

$$R(t) = \begin{pmatrix} e^{i(\alpha+\gamma)}\cos^2\left(\frac{\beta}{2}\right) & \frac{1}{\sqrt{2}}e^{i\alpha}\sin\beta & e^{i(\alpha-\gamma)}\sin^2\left(\frac{\beta}{2}\right) \\ -\frac{1}{\sqrt{2}}e^{i\gamma}\sin\beta & \cos\beta & \frac{1}{\sqrt{2}}e^{-i\gamma}\sin\beta \\ e^{i(\gamma-\alpha)}\sin^2\left(\frac{\beta}{2}\right) & -\frac{1}{\sqrt{2}}e^{-i\alpha}\sin\beta & e^{-i(\alpha+\gamma)}\cos^2\left(\frac{\beta}{2}\right) \end{pmatrix}. \tag{12}$$

We note that the cylindrical symmetry of $\hat{H}_{E_z}$ around z also makes $\hat{H}(t)$ symmetric around z'' and therefore independent of α.¶ The Schrödinger equation takes the form

$$R(t)^{-1}\hat{H}_{E_z}R(t)C(t) = i\hbar\frac{\partial}{\partial t}C(t), \tag{13}$$

¶ Explicitly, $\hat{H}(t) = \dfrac{1}{2\sqrt{2}}\begin{pmatrix} \sqrt{2}\Delta\sin^2\beta & -e^{-i\gamma}\Delta\sin 2\beta & -\sqrt{2}e^{-2i\gamma}\Delta\sin^2\beta \\ -e^{i\gamma}\Delta\sin 2\beta & 2\sqrt{2}\Delta\cos^2\beta & e^{-i\gamma}\Delta\sin 2\beta \\ -\sqrt{2}e^{2i\gamma}\Delta\sin^2\beta & e^{i\gamma}\Delta\sin 2\beta & \sqrt{2}\Delta\sin^2\beta \end{pmatrix}$.

where $C(t)$ is the column matrix of coefficients c_{m_F} from the expansion $|\psi\rangle = \sum_{m_F} c_{m_F}(t)|m_F\rangle$. The same state $|\psi\rangle$ can be expanded on the basis that rotates with the electric field and has z'' as its quantization axis. This has expansion coefficients $C_R(t) = R(t)C(t)$. In terms of these, eqn (13) becomes

$$R(t)^{-1} \hat{H}_{E_z} R(t) R(t)^{-1} C_R(t) = i\hbar \frac{\partial}{\partial t} R(t)^{-1} C_R(t), \qquad (14)$$

which simplifies to

$$\left(\hat{H}_{E_z} + \hbar \hat{G}(t)\right) C_R(t) = i\hbar \frac{\partial}{\partial t} C_R(t), \qquad (15)$$

where

$$\hat{G}(t) = -iR(t) \frac{\partial}{\partial t} R(t)^{-1}. \qquad (16)$$

We see that the evolution of the state expressed in the rotating z'' basis is governed by an effective Hamiltonian made up of the static $\hat{H}_{E_z}$, plus a term $\hbar \hat{G}(t)$ involving the time-dependence of the rotation. This second term is much like a fictitious force in classical mechanics. As we shall see, it generates a geometric phase.

We can write $\hat{G}(t)$ explicitly, using (16) and (12):

$$\hat{G}(t) = \frac{1}{\sqrt{2}} \begin{pmatrix} -\sqrt{2}(\dot{\alpha} + \dot{\gamma}\cos\beta) & e^{i\alpha}(\dot{\gamma}\sin\beta + i\dot{\beta}) & 0 \\ e^{-i\alpha}(\dot{\gamma}\sin\beta - i\dot{\beta}) & 0 & e^{i\alpha}(\dot{\gamma}\sin\beta + i\dot{\beta}) \\ 0 & e^{-i\alpha}(\dot{\gamma}\sin\beta - i\dot{\beta}) & \sqrt{2}(\dot{\alpha} + \dot{\gamma}\cos\beta) \end{pmatrix}, \qquad (17)$$

where dotted quantities are derivatives with respect to time. We wish to consider the adiabatic limit, where the tensor Stark splitting Δ is always much larger than the rotation rate. In this limit, it is a good approximation to replace $\hat{G}(t)$ by

$$\hat{G}_a(t) = \begin{pmatrix} -\dot{\alpha} - \dot{\gamma}\cos\beta & 0 & 0 \\ 0 & 0 & 0 \\ 0 & 0 & \dot{\alpha} + \dot{\gamma}\cos\beta \end{pmatrix}. \qquad (18)$$

The subscript a indicates that this is the effective operator in the adiabatic limit. This approximation follows from the formal property of matrices that the eigenvalues and eigenvectors of $\begin{pmatrix} a & b & 0 \\ b^* & \Delta & b \\ 0 & b^* & -a \end{pmatrix}$ approach those of the diagonal matrix $\begin{pmatrix} a & 0 & 0 \\ 0 & \Delta & 0 \\ 0 & 0 & -a \end{pmatrix}$ in the limit of large Δ. It is also familiar from time-independent perturbation theory, which applies in the limit of slow rotations, where the effects of the off-diagonal elements G_{ij} are smaller than those of the diagonal elements G_{ii} by a factor of order $\frac{G_{ij}}{\Delta}$.

The operator $\hat{G}_a(t)$ is the central result of this section: in the adiabatic limit, the dynamics of a spin-one system in a rotating electric field are the same as those in the static field plus a time-dependent fictitious magnetic field directed along the electric field. This breaking of degeneracy between the $m_F = \pm 1$ sub-levels has nothing to do with the static interaction. On the contrary, it is a vector property caused by the

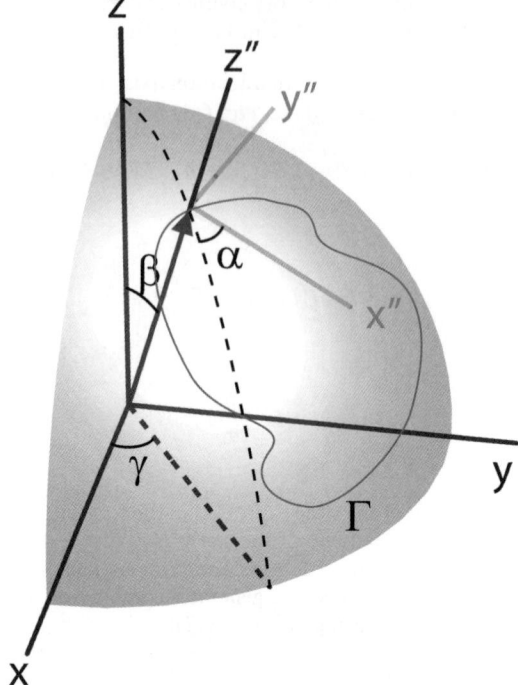

Fig. 6 Geometry of the problem. The z'' axis is along the electric field direction defined by (β, γ). When the field rotates, a unit vector along z'' follows the path Γ (thin solid line) on the surface of a unit sphere. The x'' axis is tangent to the sphere and lies at an angle α to the dashed great circle. This angle changes under parallel transport around the loop, as given by eqn (19), and does not in general return to its original value α_0 at the end of the loop. The change in α is equal to the solid angle subtended at the origin by the curve Γ.

vector nature of rotation.‖ We also note that the adiabatic condition only requires the splitting Δ between the $m_F = 0$ and $m_F = \pm 1$ sub-levels to be large compared with the rotation rates. The splitting between $m_F = \pm 1$ sub-levels is unimportant as the angular momentum operator does not couple these states.

In order to calculate $\hat{G}_a$ from eqn (18), we need the Euler angles as a function of time. The electric field rotation fixes $\beta(t)$ and $\gamma(t)$ unambiguously, but we have complete freedom to choose $\alpha(t)$ because of the cylindrical symmetry of the Stark interaction around z''. It is convenient to pick $\alpha(t)$ such that $\hat{G}_a$ vanishes, which requires

$$\alpha(t) = \alpha_0 - \int_0^t \dot{\gamma}(\tau)\cos(\beta(\tau))\, d\tau, \qquad (19)$$

where α_0 is a constant of integration. In this particular rotated basis, the adiabatic Hamiltonian is just that of the molecule in a static electric field. However, when the field makes some excursion before returning to its original direction, the final value of α, given by (19), will not be the same as the starting value α_0. This is the essence of the geometric phase.

‖ Rotation is generated by angular momentum, a vector quantity.

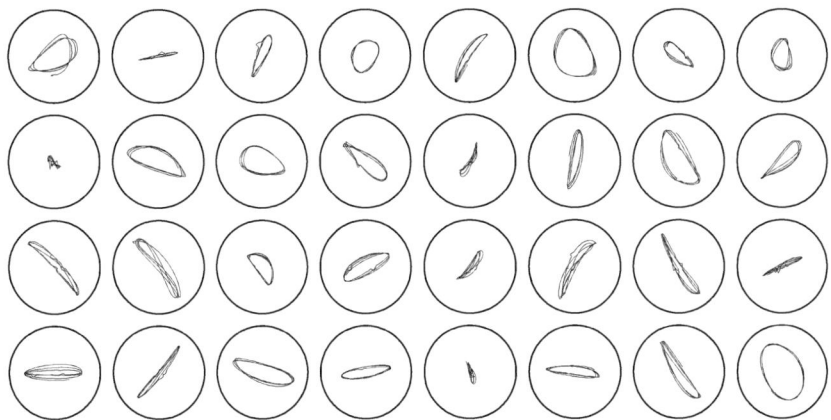

Fig. 7 x-y projection of the normalised local electric field at the positions of 32 molecules evolving in the trap for 10 ms. Each circle shows the curve swept out on the sphere by the tip of the electric field direction vector for a single molecule, as viewed from the north pole.

Fig. 6 illustrates the geometry of our problem. A unit vector along z'' follows the rotation of the electric field, mapping out the path Γ on the surface of a sphere. At any given point on the path, with coordinates (β, γ), the x'' axis lies in the tangent plane, and makes an angle α to the great circle of constant γ. Following an initial choice of α_0, the subsequent evolution of α under eqn (19) is given by *parallel-transport* of the x''-y'' plane along the path Γ. This means that for each infinitesimal displacement along the path, the new basis is the one most nearly parallel to the old basis. In the interest of brevity, we simply state this here and do not prove it.[42] At the end of a rotation around a closed path Γ, it is easy to verify that $\alpha - \alpha_0$, as given by eqn (19), is equal to the solid angle Ω subtended at the origin by the curve.** This is an example of a more general result from differential geometry, that the so-called holonomy angle introduced by parallel-transport around a curve is equal to the integral of the Gaussian curvature over the region bounded by the curve. For a sphere, which has constant Gaussian curvature $1/r^2$, this integral is equal to the solid angle.

This phase shift has a mechanical consequence. Consider a spin prepared at $t = 0$ along $x = x''(0)$. Under adiabatic rotation of the electric field, the spin undergoes parallel transport into the new x'' direction. At the end of a closed loop, although the spin has returned to the initial $x - y$ plane, its direction $x''(t)$ is now rotated through an angle $\alpha(t) - \alpha_0 = \Omega$, equal to the solid angle subtended by the path Γ. It is this spin rotation that poses the main challenge for measuring the electron edm using trapped YbF.

5.2.2 Geometric phase in the trap. We have made a numerical simulation of thirty-two molecular trajectories in the chain-link trap of Section 3 in order to evaluate the effect of the geometric phase. This calculation was made as realistic as possible by using finite-element-analysis to determine the field. The molecules were loaded into the trap with a uniform distribution over ± 2 mm in all three spatial directions and over ± 2 m s^{-1} in all three velocity components. Fig. 7 shows x-y projections of the unit vector along the electric field at the position of each molecule. These views, seen along the the z-axis, show the path Γ (illustrated schematically in Fig. 6) that determines the geometric phase. It is immediately clear without further

---

** More correctly, it is $\Omega - 2\pi$, but the 2π comes from our definition of the Euler angles and has no physical consequence

calculation that the accumulated geometric phases are large and different for each molecule. The solid angle for each orbit of the trap can approach a sizeable fraction of 2π. These simulations last for 10 ms, during which time the molecules execute 3 or 4 oscillations. The result is a large inhomogeneous spread of the spin direction in the x-y plane that completely depolarizes the molecular ensemble. Therefore the spin decoherence time is only a few ms, not the 1 s proposed in Section 4. The geometric phase is clearly the primary obstacle to making an edm measurement, or indeed any other spin precession measurement, in an electrostatic trap.

6. Conclusions and outlook

The use of trapped polar molecules to measure the electron edm offers a very substantial increase in statistical sensitivity. By loading an electrostatic trap using a Stark decelerator a sensitivity gain of 10 is feasible. Deceleration is possible in either the ground state or a rotationally excited state with comparable efficiencies. The efficiency of AG deceleration is particularly sensitive to the electrode geometry and timing sequence and it seems likely that its losses could be reduced further through refinement of these parameters. A buffer gas source might be able to deliver more molecules to the trap and so increase the statistical sensitivity even further.

However, the trap is a difficult environment for an edm experiment. The spatial variation in the direction of the electric field which provides the trapping force has two deleterious effects. The first is an inhomogeneous broadening of the magnetically-induced spin-precession angle, which can be kept under control by making the applied magnetic field small enough. The second is an inhomogeneous geometric phase. The long coherence time that seemed to be possible in the trap is precluded by this geometric phase, which scrambles the spin precession signal in just a few milliseconds.

One possibility for overcoming this problem might be to use a molecule with Ω-doublet structure, for example ThO.[43] In these molecules there are states with equal m_F but opposite sensitivity to the edm. A superposition of two such states prepared within the same molecule would experience a common geometric phase shift but a differential edm phase shift. This might provide the basis for a reduced sensitivity to the geometric phase, but it would be vulnerable to differential Stark shifts between the two Ω-doublet levels resulting from the variation in electric field strength. Further analysis is required to determine if this could be used as the basis for a workable edm measurement. A molecule with a pure spin 1/2 ground state, and therefore no tensor Stark splitting, would not adiabatically follow the electric field and so would be immune to the geometric phase arising from the rotating electric field. However, the molecule would lose its anisotropic response to magnetic fields which is a key feature for suppressing both the magnetic geometric phase and the troublesome $v \times E$ effect.[8] Another possibility is to use the decelerator as a source for a molecular fountain where it would seem to be much easier to control the electric and magnetic fields and thereby control the systematic effects. The coherence time approaches 1 s in a 1 m-high fountain. Note that the molecular beam exiting the decelerator would need to be expanded to ≈ 10 cm diameter to obtain the required degree of collimation.

The difficulties associated with the trap environment occur because the temperature of the molecules is similar to the trap depth and so they explore most of the trap. If the molecules were far colder they would remain near the trap centre where the field is very uniform. This might be achieved by loading from a source with much higher phase-space density, so that we could afford to discard all but the coldest fraction, or by actively cooling the molecules to ultralow temperatures. There has been rapid progress in the formation of deeply-bound ultracold polar molecules using photoassociation, Feschbach resonance, and coherent transfer techniques,[44–46] and these methods might be extended to molecules suitable for edm measurement. Several research groups are now exploring the direct application of laser cooling techniques to molecules.[47] In this context we note that the vibrational Franck–Condon structure of YbF[48] is favourable for cycling transitions. Other routes to ultracold polar molecules include sympathetic cooling[49] and cavity-assisted cooling.[50] If a scheme can be devised to cool a suitable heavy polar molecule to low temperature, edm measurements would not just be improved but revolutionized.

Acknowledgements

We are grateful to the Royal Society and to the STFC for supporting us.

References

1 E. M. Purcell and N. F. Ramsey, *Phys. Rev.*, 1950, **78**, 807.
2 L. D. Landau, *Zh. Eksp. Teor. Fiz*, 1957, **32**, 405; L. D. Landau, *Sov. Phys. JETP*, 1957, **5**, 405.
3 M. Pospelov and I. B. Khriplovich, *Sov. J. Nucl. Phys.*, 1991, **53**, 638.
4 S. Barr, G. Segrè and H. A. Weldon, *Phys. Rev. D*, 1979, **20**, 2494.
5 W. Bernreuther and M. Suzuk, *Rev. Mod. Phys.*, 1991, **63**, 313; W. Bernreuther and M. Suzuk, *Rev. Mod. Phys.*, 1992, **66**, 633, errata.
6 E. D. Commins, *Adv. At. Mol. Opt. Phys.*, 1999, **40**, 1.
7 E. A. Hinds, *Phys. Scr.*, 1997, **T70**, 34.
8 B. C. Regan, E. D. Commins, C. J. Schmidt and D. DeMille, *Phys. Rev. Lett.*, 2002, **88**, 071805.
9 P. G. H. Sandars, *Phys. Rev. Lett.*, 1967, **19**, 1396.
10 P. G. H. Sandars, in *Atomic Physics 4*, ed. G. zu Pulitz, Plenum, New York, 1975, p. 71.
11 O. P. Sushkov and V. V. Flambaum, *Zh. Eksp. Theo. Fiz*, 1978, **75**, 1208; O. P. Sushkov and V. V. Flambaum, *Sov. Phys. JETP*, 1978, **48**, 608.
12 J. J. Hudson, B. E. Sauer, M. R. Tarbutt and E. A. Hinds, *Phys. Rev. Lett.*, 2002, **89**, 023003.
13 J. J. Hudson, P. C. Condylis, H. T. Ashworth, M. R. Tarbutt, B. E. Sauer and E. A. Hinds, in *Laser Spectroscopy 17*, World Scientific, Singapore, pp. 219–226, 2005.
14 B. E. Sauer, H. T. Ashworth, J. J. Hudson, M. R. Tarbutt and E. A. Hinds, in *Atomic Physics 20*, Plenum, New York, pp. 44–52, 2006.
15 H. L. Bethlem, G. Berden, F. M. H. Crompvoets, R. T. Jongma, A. J. A. van Roij and G. Meijer, *Nature (London)*, 2000, **406**, 491.
16 M. R. Tarbutt, J. J. Hudson, B. E. Sauer, E. A. Hinds, V. A. Ryzhov, V. L. Ryabov and V. F. Ezhov, *J. Phys. B*, 2002, **35**, 5013.
17 S. E. Maxwell, N. Brahms, R. deCarvalho, D. R. Glenn, J. S. Helton, S. V. Nguyen, D. Patterson, J. Petricka, D. DeMille and J. M. Doyle, *Phys. Rev. Lett.*, 2005, **95**, 173201.
18 D. Patterson and J. M. Doyle, *J. Chem. Phys.*, 2007, **126**, 154307.
19 H. L. Bethlem, A. J. A van Roij, R. T. Jongma and G. Meijer, *Phys. Rev. Lett.*, 2002, **88**, 133003.
20 M. R. Tarbutt, H. L. Bethlem, J. J. Hudson, V. L. Ryabov, V. A. Ryzhov, B. E. Sauer, G. Meijer and E. A. Hinds, *Phys. Rev. Lett.*, 2004, **92**, 173002.
21 H. L. Bethlem, M. R. Tarbutt, J. Küpper, D. Carty, K. Wohlfart, E. A. Hinds and G. Meijer, *J. Phys. B*, 2006, **39**, R236.
22 H. L. Bethlem, G. Berden and G. Meijer, *Phys. Rev. Lett.*, 1999, **83**, 1558.
23 S. Y. T. van de Meerakker, P. H. M. Smeets, N. Vanhaecke, R. T. Jongma and G. Meijer, *Phys. Rev. Lett.*, 2005, **94**, 023004.
24 H. L. Bethlem, G. Berden, A. J. A. van Roij, F. M. H. Crompvoets and G. Meijer, *Phys. Rev. Lett.*, 2000, **84**, 5744.
25 S. Y. T. van de Meerakker, N. Vanhaecke, H. L. Bethlem and G. Meijer, *Phys. Rev. A*, 2005, **71**, 053409.
26 S. Y. T. van de Meerakker, N. Vanhaecke, H. L. Bethlem and G. Meijer, *Phys. Rev. A*, 2006, **73**, 023401.
27 B. C. Sawyer, B. K. Stuhl, B. L. Lev, E. R. Hudson and J. Ye, *Eur. Phys. J. D*, 2008, **48**, 197.
28 J. M. Rost, J. C. Griffin, B. Friedrich and D. R. Herschbach, *Phys. Rev. Lett.*, 1992, **68**, 1299.
29 K. Wohlfart, F. Grätz, F. Filsinger, H. Haak, G. Meijer and J. Küpper, *Phys. Rev. A*, 2008, **77**, 031404R.
30 K. Wohlfart, F. Filsinger, F. Grätz, J. Küpper and G. Meijer, *Phys. Rev. A*, 2008, **78**, 033421.

31 M. R. Tarbutt, J. J. Hudson, B. E. Sauer and E. A. Hinds, arXiv:0803.0967, 2008.
32 M. R. Tarbutt and E. A. Hinds, *New J. Phys.*, 2008, **10**, 073011.
33 N. E. Shafer-Ray, K. A. Milton, B. R. Furneaux, E. R. I. Abraham and G. R. Kalbfleisch, *Phys. Rev. A*, 2003, **67**, 045401.
34 E. Peik, *Eur. Phys. J. D*, 1999, **6**, 179.
35 J. van Veldhoven, H. L. Bethlem and G. Meijer, *Phys. Rev. Lett.*, 2005, **94**, 083001.
36 H. L. Bethlem, J. van Veldhoven, M. Schnell and G. Meijer, *Phys. Rev. A*, 2006, **74**, 063403.
37 D. Budker, V. Yashchuk and M. Zolotorev, *Phys. Rev. Lett.*, 1998, **81**, 5788.
38 M. V. Berry, *Proc. R. Soc. London, Ser. A*, 1984, **392**, 45.
39 F. Wilczek and A. Zee, *Phys. Rev. Lett.*, 1984, **52**, 2111.
40 P. Hoodbhoy, *Phys. Rev. A*, 1988, **38**, 3766.
41 M. Weissbluth, *Atoms and Molecules*, Academic Press, New York, 1978.
42 This result can be proved using elementary geometry and calculus. At each point on the curve Γ, one needs to consider the cone whose apex lies on the z axis and whose surface is tangent to the sphere at that point. One then uses the facts that the cone is developable, and that lines of constant γ are great circles.
43 A. C. Vutha, O. K. Baker, W. C. Campbell, J. M. Doyle, G. Gabrielse, Y. V. Gurevich, P. Hamilton, N. Hutzler, M. A. H. M. Jansen and D. DeMille, *Search for the electron's electric dipole moment with cold ThO molecules*, Poster TU2, XXI International Conference on Atomic Physics, Storrs CT, 2008.
44 J. M. Sage, S. Sainis, T. Bergeman and D. DeMille, *Phys. Rev. Lett.*, 2005, **94**, 203001.
45 S. Ospelkaus, A. Pe'er, K. -K. Ni, J. J. Zirbel, B. Neyenhuis, S. Kotochigova, P. S. Julienne, J. Ye and D. S. Jin, *Nat. Phys.*, 2008, **4**, 622.
46 J. Deiglmayr, A. Grochola, M. Repp, K. Mörtlbauer, C. Glück, J. Lange, O. Dulieu and M. Weidemüller, *Phys. Rev. Lett.*, 2008, **101**, 133004.
47 M. D. Di Rosa, *Eur. Phys. J. D*, 2004, **31**, 395–402; M. Viteau, A. Chotia, M. Allegrini, N. Bouloufa, O. Dulieu, D. Comparat and P. Pillet, *Science*, 2008, **321**, 232; B. K. Stuhl, B. C. Sawyer, D. Wang and J. Ye, *Phys. Rev. Lett.*, 2008, **101**, 243002.
48 K. L. Dunfield, C. Linton, T. E. Clarke, J. McBride, A. G. Adam and J. R. D. Peers, *J. Mol. Spectrosc.*, 1995, **174**, 433.
49 M. Lara, J. L. Bohn, D. Potter, P. Sold'an and J. M. Hutson, *Phys. Rev. Lett.*, 2006, **97**, 183201.
50 P. Domokos and H. Ritsch, *Phys. Rev. Lett.*, 2002, **89**, 253003.

PAPER

Buffer gas cooling of polyatomic ions in rf multi-electrode traps

D. Gerlich* and G. Borodi

Received 24th November 2008, Accepted 5th January 2009
First published as an Advance Article on the web 28th May 2009
DOI: 10.1039/b820977d

Cooling all degrees of freedom of a molecule, a cluster, or even a nanoparticle which is suspended in a vacuum, is an experimental challenge. Without suitable schemes, cold or ultracold chemical reactions are not feasible. Methods such as laser based preparation of *very slow* atoms, decelerating molecules to low velocities with electric fields or freezing molecular ions into Coulomb crystals, are generally not suitable to cool the vibrational or rotational motion of molecules. This contribution describes a new method in which a beam of slow atoms or molecules (H, He, H_2, or D_2) is used for cooling charged particles confined in a multi-electrode rf trap. For reaching sub-K temperatures, the fast part of a cold effusive beam is removed with a shutter before the slow remaining neutrals interact with the ion cloud. The development of a pulsed cold beam source is discussed as well as suitable methods for determining the ion temperature. A challenging application is to prepare internally cold CH_5^+ for spectroscopy or chemistry. New experimental results for hydrogen abstraction in collisions with slow H atoms are reported at energies of a few meV. For evaluating these measurements and for predicting effective rate coefficients at lower energies, the kinematic conditions of the slow neutral beam–ion trap arrangement have been analyzed in detail. The potential of cooling ions such as protonated methane or H_3^+ with slow energy selected H atoms is briefly mentioned. An interesting process is the formation of weakly bound ions such as H_4^+ or CH_6^+ *via* radiative or ternary association. Such ions are ideal candidates for preparing the corresponding collision complexes very close (μeV) to the dissociation continuum using infrared transitions.

1 Introduction

Several new instruments are currently emerging to study cold molecular ions and low-energy ion–molecule reactions. The majority of them are based on inhomogeneous rf fields either using 22-pole traps[1,2,3,4] mounted onto a cold head or linear quadrupoles in combination with laser cooled atomic ions and sympathetically cooled molecular ions.[5,6,7] The relevant methods have been summarized recently in the book, entitled *Low temperatures and cold molecules*.[8] This interesting collection contains a variety of remarks concerning the meaning of cold, ultracold, slow *etc*. Generally accepted definitions of the *temperature T* such as "*T* is a measure of the average translational kinetic energy associated with the disordered microscopic motion of atoms and molecules" need to be reconsidered. In many innovative experiments the ensemble average is replaced by a time average. One also has to be aware that the language used in various subfields can lead to erroneous conclusions. For

Faculty of Natural Science, Technical University, 09107 Chemnitz, Germany. E-mail: gerlich@physik.tu-chemnitz.de

example, laser cooled neutral atoms are really moving very slowly, *i.e.*, they are translationally ultracold, while the mK or μK ions in coulomb crystals oscillate in phase with the frequency of the storing rf field, usually with rather high kinetic energies. Another example is related to chemical reactions such as $Ca^+ + CH_3F \rightarrow CaF^+ + CH_3$, which has been studied with laser cooled Ca^+ ions and a few K slow but internally 300 K *hot* neutrals.[7] Since, especially at low collision energies, a collision complex usually utilizes the energy from all degrees of freedom, one should only call a bimolecular reaction *cold* if the *total energy* is a few meV or below. Another often overlooked fact is that the laser cooled ions or atoms spend quite some time in the electronically excited state. In order to differentiate between different aspects, it is useful to distinguish between *cold collisions*[9] and the production and study of *cold molecules*.

The complexity of states and optical transitions does not allow one to directly laser cool molecules. One scheme used is laser stimulated association of two ultracold atoms. This is a very interesting process in itself; however, it is restricted to very few specific molecules. The effort to create these diatoms in the vibrational–rotational ground state is huge. Another strategy to produce ultracold neutral molecules is to slow them down in Stark decelerators.[11] Although this method has become the subject of intense studies in recent years its applications are restricted to a small number of specific molecules. An advantage of the method is that the cold molecules are in selected states. In the case of molecular ions, a variety of single- and multi-photoionization schemes has been developed to create them in specific states. Well-controlled but also complex is the creation of ground state ions using methods such as ZEKE (zero kinetic energy electrons), PFI (pulsed field ionization), or similar schemes. However, controlling the velocity of these ground state ions simultaneously is hard to achieve.

In this contribution to the Faraday Discussion on *Cold and Ultracold Molecules* we discuss recent progress in cooling polyatomic ions with buffer gas. For many years our main motivation was to investigate experimentally ion–molecule reactions relevant to interstellar chemistry as discussed several years ago[12] and summarized recently.[9,13] Therefore, most experiments have been performed at 10 K or above. However, the experimental scheme we use can be extended to relax ions to temperatures below 1 K, *i.e.*, to move towards the field of *ultracold ion chemistry*. The new method is based on the combination of ion trapping in suitable rf fields with very slow beams of neutrals.

2 Experimental

High frequency multi electrode rf traps

Since most methods of preparing cold molecules are rather inefficient, experimental studies usually rely on suitable traps where the slow objects can be collected and confined. Long storage times and a well-controlled environment provide ideal conditions for studying the trapped ensemble and its interaction with electromagnetic radiation, coolants, or reactants. For cooling ions and for studying low energy collisions it is necessary to understand the cooling process in detail and the impairment of the energy by the forces which are used for confinement. In general, random relaxation *via* multiple collisions works only in traps with real three-dimensional potential minima. Therefore, *dynamic* storage devices such as magnetic traps, where the Lorentz force plays a role, or electrostatic traps, in which centrifugal forces lead to quasi-periodic orbits, are not suitable for slowing particles down to zero velocity.

Our solution is based on rf ion traps with effective potential minima where the ions finally relax to. The method, which has been documented thoroughly in ref. 14 is based on inhomogeneous, electrical fields $\mathbf{E}_o(\mathbf{r},t)$, which vary in space and time. In most applications, the time dependence used is a harmonic oscillation with the frequency Ω,

$$E_0(\mathbf{r},t) = E_0(r)\cos(\Omega t) \tag{1}$$

Although it seems to be contradictory to the aim of ion cooling, high frequency rf trapping makes use of iterative modulation of the kinetic energy. The success of our applications is to use, instead of quadrupoles, a wide nearly field free trapping volume created by multi-electrode arrangements such as a stack of rings or many parallel rods. In such structures the region where the kinetic energy is modulated, is very small. Of special importance for sub-K buffer gas cooling is the experimental fact that we use an atomic or molecular beam instead of filling the whole trap with neutrals. Restriction of the beam diameter passing through the ion cloud, already confines the relaxing collisions to regions of low field.

For semi-quantitative estimates of ion heating and cooling one can make use of the fact that the momentary energy never goes above 3 times the initial value and that, in time average, the energy is conserved. Mathematically, *i.e.*, under ideal experimental boundary conditions, there is no limitation towards zero motion. In order to understand this, one can study numerically the equation of motion of a particle under the influence of such a field under inclusion of collisions.[14] For heavy ions cooled with a light buffer gas it is sufficient to introduce a *friction* term. An analytical treatment of ion cooling in an rf field is based on the adiabatic approximation. As outlined in ref. 14 trapping ions without rf heating requires that the dimensionless parameter η, defined by

$$\eta(\mathbf{r}) = \frac{2q|\nabla E_0|}{m\Omega^2}, \tag{2}$$

must be very small in those regions where the ions are moving. If this condition is fulfilled, the time-averaged influence of the oscillatory force can be described by the effective potential

$$V^*(\mathbf{r}) = \frac{q^2 E_0^2}{4m\Omega^2}. \tag{3}$$

In these two equations m and q are the particles' mass and charge, respectively. Usually it is sufficient, for practical applications, to postulate $\eta < 0.3$. For a mathematical treatment of ion cooling one makes use of the fact that, for many geometries, both $V^*(\mathbf{r})$ and $\eta(\mathbf{r})$ are functions of the coordinate $\mathbf{r}$ and both go to 0 for $|\mathbf{r}| \to 0$. In other words, the adiabatic approximation becomes better with decreasing kinetic energy E_k. In the case of a linear multipole trap ($2n$ pole) one can derive the relation

$$\eta \sim E_k^{(n-2)/(2n-2)} \tag{4}$$

In practice, several technical problems restrict the efficiency of ion cooling in rf traps. One is due to so-called "patch effects", *i.e.* potential distortions on the surfaces of electrodes. They can lead to local electrostatic fields, which pull slow ions into regions of strong rf fields. Superconducting electrodes may eliminate or reduce such surface effects. Another problem is caused by parasitic low frequency voltages, which can heat the motion of the ions. A high quality resonance circuit helps significantly to create a very pure sinusoidal field as described by eqn (1).

Cooling in traps, buffer gas

There are quite general schemes to cool a finite ensemble of charged particles in a trap. Examples include evaporation, optical pumping or chemically removing energetic ions. Interesting methods are resistive cooling or active-feedback cooling,[15] which make use of the currents induced by the motions of the charges in suitable electrodes. If molecules are not exposed to any other interaction, long enough trapping finally equilibrates the internal excitation with the black body radiation penetrating the ion cloud. Typical examples are the rotational temperature of polar molecules such as CH^+ in dense interstellar clouds, which is given by the 2.7 K of the cosmic background radiation or the so-called ultracold HD^+ ions in a Coulomb crystal which have a population of rotational states determined by the 300 K environment.[6] This means that innovative instruments, which aim to reach temperatures of 1 K or below, either need an efficient cooling mechanism or the environment of the trap has to be operated at low enough temperatures.

Our strategy for contributing to the new interdisciplinary field of molecular matter at very low temperatures ($T < 1$ K) uses buffer gas cooling, *i.e.*, elastic and inelastic collisions with very slow, velocity selected beams of non-reactive neutrals. It is based on the idea to take only the slow tail from a pulsed thermal beam of cold gas. Depending on the specific cross sections, collisions cool more or less all degrees of freedom efficiently, including complex ions. One of the restrictions of buffer gas cooling is that the ions must be capable of surviving multiple collisions. For example, clustering with the neutrals used for cooling should be avoided. In our experiment this process is significantly suppressed by operating at low enough densities that only radiative association plays a role. One favorable fact of cooling internal degrees of freedom of heavy ions (mass m_2) *via* collisions with light buffer gas (m_1) is that the internal temperature of the ions is determined by the collision temperature. Transformation into the center-of-mass system reveals that the ion motion contributes only with the fraction $m_2/(m_1 + m_2)$. This means, for example, that very slow H atoms are well suited to get very cold CH_5^+ (see below).

The CEB-22PT

The apparatus developed for sub-K cooling trapped ions with a beam of neutrals and for studying inelastic and reactive collisions is shown schematically in Fig. 1. The main parts (storage ion source, mass filter, temperature variable ion trap, and external detector) are standard and have been described in detail elsewhere (see ref. 9,10 and references therein). Briefly, externally created ions are injected into the 22PT *via* the electrostatic quadrupole bender (QP). There they are confined in a cylindrical volume (~8 mm diameter, 25 mm length) by the effective potential (eqn 3) and small dc voltages applied to the entrance and exit electrodes. Voltages of some 10 mV are sufficient for closing the trap if the ions are cold. For mass

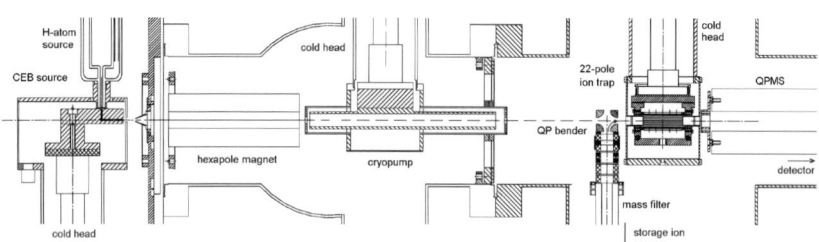

Fig. 1 Schematic diagram of the Cold Effusive Beam 22-Pole Trap Apparatus (CEB-22PT). This instrument has been developed for exposing trapped ions to an effusive beam of very slow atoms or molecules (H, He, H_2 or D_2). Primary ions are produced in a storage ion source (not shown), mass filtered, and injected into the ion trap *via* an electrostatic quadrupole bender (QP). For analysis, the ions are extracted, mass analyzed in a quadrupole mass spectrometer (QPMS) and detected. Large differential pumping capacity is provided using several turbomolecular pumps and a specially designed cryopump. For reaching the sub-K range, the neutral beam is pulsed such that only the slow part of the Maxwellian distribution is allowed to pass through the 22-pole trap. Hydrogen atoms are velocity selected using the transmission features of magnetic fields.

analysis and detection, the ions are extracted (QPMS). A first description of this beam-trap arrangement and some preliminary results have been reported in ref. 16. Additional information and temperature dependent rate coefficients (above 10 K) for the interaction of cold CO_2^+ ions with slow hydrogen molecules and atoms can be found in ref. 17. More details concerning the preparation of a beam of cold H-atoms and their reactions with trapped hydrocarbon ions are published in ref. 18.

In this contribution, we discuss special features and modifications of this instrument needed for cooling trapped ions to sub-K temperatures. One module which is still under development is a special beam source for creating a cold effusive beam. As shown in Fig. 2, it is composed of a copper cylinder mounted onto a cryocooler (Sumitomo SRDK-101E). For producing a continuous beam, the inner cylinder is operated at such a temperature that the vapor pressure of the gas (usually He or H_2) is still high enough for getting a sufficiently dense cold effusive beam penetrating the trapped ion cloud. For producing intense cold gas pulses, the inner cylinder is operated at the lowest temperature (3.6 K) and the helium or hydrogen gas condenses on the walls. A short voltage pulse is used for locally raising the temperature of a thin substrate evaporating some gas (see left side in Fig. 2). Since this process takes place inside the cold tube, evaporation and cryopumping leads to a well directed pulsed beam emerging from the exit hole and to a very low overall gas consumption and gas load. For reducing the "temperature" of the gas pulse, time of flight selection is used, *i.e.*, the fast part of the effusive beam is rejected and only the very slow neutrals are allowed to enter the trap. For this a fast UHV compatible shutter is needed, which is capable of being operated with ms switching time and at repetition frequencies of hundreds of Hz. Presently we perform tests with a rotating wheel and a tuning fork chopper.

Although molecular beams are well-described in the literature, most applications concentrate on high Mach numbers and high densities and almost nothing can be found on the slow velocity tail of an effusive beam. In order to test the idea, velocity

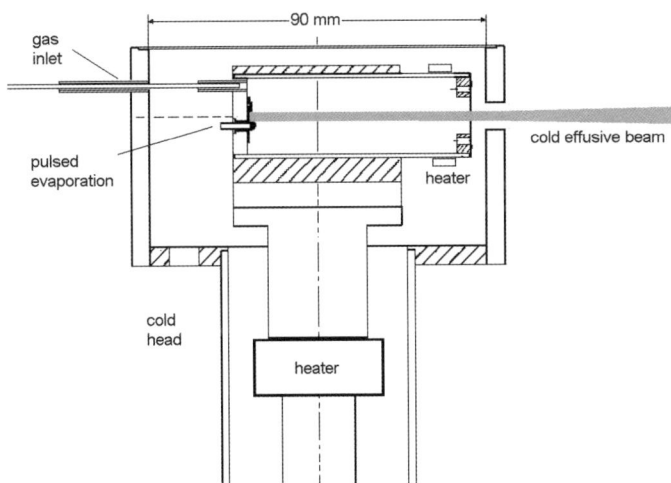

Fig. 2 Source for forming cold effusive beams (CEB) of helium or hydrogen molecules. The inner cylinder which is mounted on a cold head (SRDK-101E, lowest temperature 3.6 K) is operated at a temperature where the vapor pressure is sufficiently high that enough gas flows through the exit hole (2–5 mm). Alternatively, tests are underway to work at lower temperatures and to use a heating pulse for evaporating a short gas pulse on the left side of the cylinder. Cryopumping and an adequate temperature gradient for gas recycling significantly reduces the background gas load.

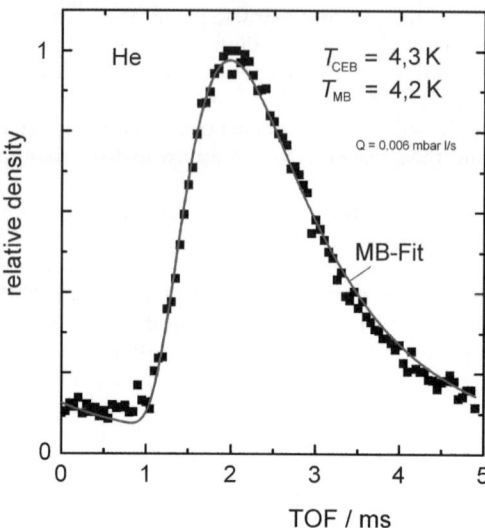

Fig. 3 Measured time of flight distribution of an effusive He beam at $T_{CEB} = 4.3$ K. The data have been recorded in the arrangement shown in Fig. 1 with the H-atom source replaced by the CEB source shown in Fig. 2. The flight path was 37.5 cm. The overall agreement with a Maxwell–Boltzmann distribution (solid line) is satisfying; however, a detailed comparison reveals that, despite a low flow of only 0.006 mbar l s^{-1}, some acceleration of the slow molecules towards the mean velocity already occurs.

and density distributions of He atoms and D_2 molecules emerging from the source have been determined using time of flight measurements. A chopper wheel with 2 slits (2 mm wide) interrupts the beam on a 37 mm radius. For a chopping frequency of 200 Hz, one gets a time resolution of ~20 μs. The neutrals are detected at the location of the ion trap using an electron bombardment ionizer. The flight path is 37.5 cm. The data shown in Fig. 3. have been recorded at $T_{CEB} = 4.3$ K for He atoms. Comparison with a Maxwell–Boltzmann distribution, calculated for an effusive beam with $T_{MB} = 4.2$ K, reveals good overall agreement. Cutting off the first half of the beam intensity, leads to a velocity distribution the mean velocity of which already corresponds to 1 K.

Using a calibrated ionizer, such measurements show in accordance with theoretical estimates that one can reach time averaged number densities of sub-K atoms of some 10^8 cm^{-3} for the geometry shown in Fig. 1. Such densities require cooling times of several seconds, in specific cases several minutes. In principle, this is no problem with the rf trapping technique; however, one has to make sure that the background pressure in the trap is sufficiently low. In order to achieve this, a special cryopump has been constructed which is shown in Fig. 1. It consists of a 15 cm long porous carbon tube mounted onto a cold head (Sumitomo SRDK-101E, reaching 3.6 K). The geometry has been chosen such that only the directed beam coming from the beam source can pass this region without hitting the cold surface. In addition it is planned to use shutters on the differential wall for improving the separation of the vacuum chambers. Most important is finally to operate the ion trap itself at such low temperatures that the surrounding box adsorbs all gas with high efficiency, of course with the exception of the directed beam, which passes the ion cloud without hitting a surface. Presently, the ion trap is mounted onto a cold head which reaches only 10 K (Leybold RGD 210).

We are also exploring the possibility of using velocity selected hydrogen atoms not only for studying chemical reactions but also for cooling trapped ions. With the

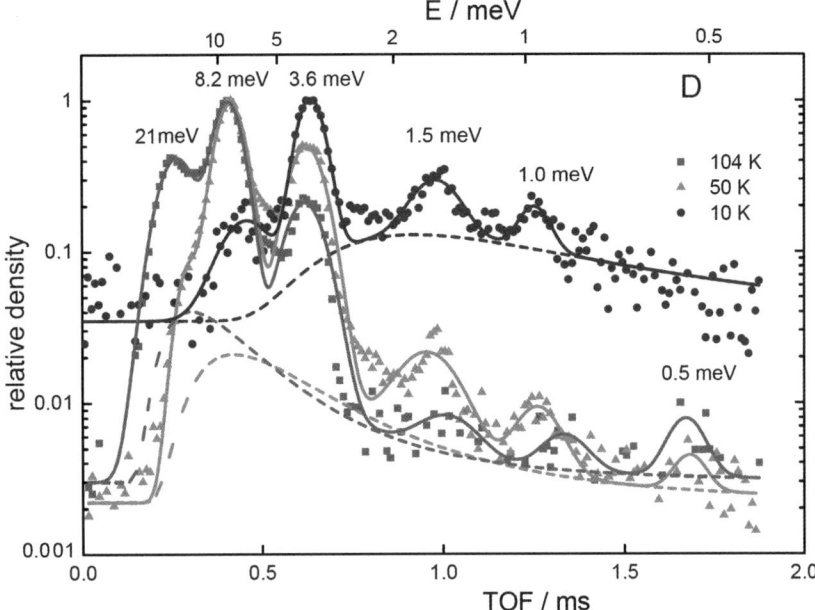

Fig. 4 Relative time of flight distributions of D atoms measured at three different temperatures of the accommodator. Without the hexapole magnets, thermal distributions result in TOF distributions indicated by the dashed lines (not at scale). The transmission features of two hexapoles favour mainly three groups of hydrogen atom with kinetic energies which are indicated. Maximal transmission is obtained for D atoms with (8.2 ± 3.3) meV. For favouring slower atoms weaker magnetic fields are required.

setup shown in Fig. 1, mean H-atom number densities of some 10^9 cm^{-3} have been reached at 50 K. Fig. 4 shows a few velocity distributions measured for a beam of D atoms recorded at different temperatures of the accommodator. As explained elsewhere,[17,18] the distributions are structured due to the transmission properties of the two guiding hexapole magnets used in this experiment. With this arrangement, the highest transmission obtained for D (and also H) atoms was at a kinetic energy of (8.2 ± 3.3) meV. Cooling the accommodator to 10 K leads to a beam in which the 3.6 meV part dominates with a half width smaller than 1 meV. Additional peaks appear at 1.5 and 1 meV. Comparison with the 10 K Maxwell–Boltzmann distributions (dashed line) reveals that weaker magnets should be used for favoring 1 meV atoms or slower ones. For velocity selection, it is also conceivable to change to other geometries such as deflecting magnets.

In order to test the cooling efficiency of the cold effusive beam and for determining the actual energy content ("temperature") of the trapped ensemble, several methods are under development. Rather direct "thermometers" are based on *in situ* determination of rotational populations and translational distributions (Doppler profile). Related tests have been performed with diatomic ions such as the homonuclear N_2^+.[19] Infrared active ions such as CH^+ and CO^+ are also interesting. N_2^+ is an ideal candidate because it can be excited with cheap but narrow bandwidth laser-diodes. Photoabsorption is detected *via* the slightly endothermic (179 meV) laser induced charge transfer $N_2^+ + Ar \rightarrow Ar^+ + N_2$. So far this process has been used successfully by filling the ion trap with some Ar target gas; however, due to condensation, this limits the temperature range to above 35 K. In order to apply the method at lower temperatures, a skimmed Ar beam has to be integrated into the machine. Corresponding tests are ongoing; however, operating two beams, one for cooling and the other one for chemical probing, has not yet been achieved.

Other methods for probing the temperature of an ultracold ion ensemble are based on the temperature dependence of collision processes such as weakly endothermic reactions (see below) or radiative association. One strategy is to follow the production of He–N_2^+, a well-studied, weakly bound (12.5 meV) van der Waals molecule.[20] If they are formed in sufficient numbers (conversion of a few %) laser induced fragmentation of the van derWaals bond can be used to determine rotational temperatures. Completely new methods are to combine sub-K trapping with the sensitivity of modern GHz and THz techniques. Ideal test cases are rotational transitions in cold molecular ions such as CO^+, deuterated variants of H_3^+, or CH_5^+. These ions may be studied *via* suitable reactions or even *via* absorption spectroscopy.

Kinematics

In order to assess the capabilities and limitations of the present arrangement for cooling ions in elastic or inelastic collisions to temperatures below 1 K or for studying reactive collisions at total energies of a few meV or below, a detailed analysis of kinematic averaging is required. Some special aspects, *e.g.* how to reach very low relative velocities and how to cool efficiently the internal energy of molecular ions, have been discussed recently in ref. 9 and 10 respectively. Averaging over quantities like velocity or angular spread of the colliding partners is a quite general problem in gas phase experiments and many investigations have been devoted to such problems;[14,21,22,23] however, the combination of ion traps and a molecular beam needs some special attention. An example is an experiment reported recently.[7] In this arrangement, ions, confined in a Coulomb crystal, collide with a slow beam of neutrals. For large crystals, the collision energy distribution is dominated by the oscillatory motion of the ions imposed by the rf field. For ions sitting close to the center line of the trap, the distribution of the collision energy has to be calculated by integration over the velocity distributions of both reactants.

The situation in an ion trap is well-defined if the stored ions and the neutral reactants are in thermal equilibrium with the surrounding walls at a common temperature T_{22PT}. Adding non-reactive buffer gas for cooling does not change the situation as long as the densities are low enough to make ternary processes very scarce. In this case one extracts reaction rate coefficients from the measured temporal change of the number of ions, $N_i(t)$ by fitting the parameters of an appropriate rate equation system. In the case of a simple two-channel process, the decay of the number of primary ions is described by

$$\dot{N}_1 = dN_1/dt = -kn_2N_1 \qquad (5)$$

where n_2 is the number density of the neutrals and $k = k(T_{22PT})$ is the *thermal* rate coefficient (by definition). $N_1(t)$ decreases exponentially with the time constant $\tau = (kn_2)^{-1}$, while the number of product ions, $N_2(t)$, augments complementarily. For slow reactions or low number densities (*i.e.*, $t \ll \tau$), the linear approximation $N_2 = N_1(0)~t/\tau$ is often sufficient.

The situation is more complicated if one performs experiments with a beam–beam, beam–gas cell or beam ion–cloud arrangement. For deriving the relationships between measured quantities and rate coefficients (or cross sections), one usually starts with the ideal case of two well-collimated monochromatic beams with velocities $\mathbf{v}_1$ and $\mathbf{v}_2$, interacting in a scattering volume $d\tau$. The number of products (or collisions) per unit time is given by[23]

$$d\dot{N}_2 = g\sigma(g)n_1n_2d\tau \qquad (6)$$

Here $g = |\mathbf{g}| = |\mathbf{v}_1 - \mathbf{v}_2|$ is the relative velocity, n_1 and n_2 are the projectile and target densities, respectively. $\sigma(g)$ is the *elementary* (or *intrinsic*) integral cross section. The corresponding elementary rate coefficient is obviously given by

$$k = g\sigma(g). \quad (7)$$

In contrast to the ideal case, the conditions of a realistic experiment are less well defined. Therefore, the product rate actually obtained, $d\dot{N}_2$, is an average over the velocity distributions of the reactants. Instead of eqn 6 one obtains the six-dimensional integral

$$d\dot{N}_2 = \int_{v_1} d\mathbf{v}_1 \int_{v_2} d\mathbf{v}_2 \, g\sigma(g) n_1(\mathbf{r}, \mathbf{v}_1) n_2(\mathbf{r}, \mathbf{v}_2) d\tau \quad (8)$$

Assuming that the velocity distributions are independent of the spatial coordinate $\mathbf{r}$, the density functions $n_i(\mathbf{r},\mathbf{v}_i)$ can be factored using normalized probability functions $f_i(\mathbf{v}_i)$ ($i = 1,2$)

$$n_i(\mathbf{r},\mathbf{v}_1) = n_i(\mathbf{r}) f_i(\mathbf{v}_i) \quad (9)$$

With $f_1(\mathbf{v}_1)$ and $f_2(\mathbf{v}_2)$ we define $f(g)$, the distribution of the relative velocity,

$$f(g) \, dg = \iint_{(v1 \, v2)^*} d\mathbf{v}_1 \, d\mathbf{v}_2 \, f_1(\mathbf{v}_1) f_2(\mathbf{v}_2). \quad (10)$$

The asterisk indicates symbolically that the integration must be restricted to that subspace $(\mathbf{v}_1, \mathbf{v}_2)$ where $|\mathbf{v}_1 - \mathbf{v}_2| \in [g, g + dg]$. $f(g)dg$ denotes the probability that the relative velocity lies in the interval $[g, g + dg]$. With the mean relative velocity, given by

$$\langle g \rangle = \int_0^\infty g f(g) dg \quad (11)$$

we define the effective cross section and the effective rate coefficient as

$$\sigma_{\mathit{eff}}(\langle g \rangle) = \int_0^\infty \frac{g}{\langle g \rangle} \sigma(g) f(g) \, dg, \quad (12)$$

$$k_{\mathit{eff}} = \int_0^\infty g\sigma(g) f(g) dg. \quad (13)$$

With this, we obtain a result that is very similar to eqn (8),

$$d\dot{N}_2 = \langle g \rangle \sigma_{\mathit{eff}}(\langle g \rangle) n_1(\mathbf{r}) n_2(\mathbf{r}) d\tau \quad (14)$$

For evaluating specific experimental results, the velocity and spatial distributions functions must be known. For the present experiment, we assume here for simplicity that the beam of hydrogen atoms is mono-energetic and that the ions are thermalized to the temperature of the trap. For this situation, eqn 10 can be evaluated analytically, resulting in the generalized Maxwell–Boltzmann distribution.[14,21]

$$f(g) = (m_2/2\pi kT_2)^{1/2} \times \frac{g}{v_1}\left[\exp\left(-\frac{m_2}{2kT_2}(g - v_1)^2\right) - \exp\left(-\frac{m_2}{2kT_2}(g + v_1)^2\right)\right]. \quad (15)$$

In order to be consistent with the literature, we have used in this formula v_1 for the laboratory velocity of the H atoms and T_2 and m_2 for the ions, which play the role of the target in this treatment. An approximation for the half width (FWHM) of this distribution is

$$\Delta E = \sqrt{11.1\ m_1/m_2\ E_1 k T_2}, \quad (16)$$

where m_1 and E_1 are the mass and the laboratory energy of the H atoms, respectively. The following example illustrates the importance of this result. For the H + CH_5^+ system at $T_2 = 10$ K and $E_1 = 3.6$ meV, one obtains $\Delta E = 1.4$ meV, while the half width of the H atom peak (see Fig. 7) is only 0.75 meV. At low H atom velocities, the generalized Maxwell–Boltzmann distribution approaches a normal Maxwellian $f_M(g;T_C)$ with a reduced temperature $T_C = m_1/(m_1 + m_2)T_2$. This has the interesting consequence that, in the case of H + CH_5^+, one can reach sub-K collision temperatures with 10 K ions, provided the H-atom beam is slow enough.

Inspection of Fig. 4 reveals that our H-atom beam is not yet ideal and the assumption of a mono-energetic beam cannot be made. Therefore, further analysis requires to integrate the generalized Maxwell–Boltzmann distribution numerically over the velocity distribution of the H atom beam. An analytical approximation, describing the hydrogen atoms with the three Gaussians leads to rather good results.

3 Results

Preparing cold ions *via* photoionization

Photoionization has become a widely used technique for the preparation of state selected ions in the study of ion–molecule reactions in the gas phase. We have tested a variety of schemes to prepare internally cold and state selected ions. An example is the special photoionization source for mass-, energy-, and state-selected ions

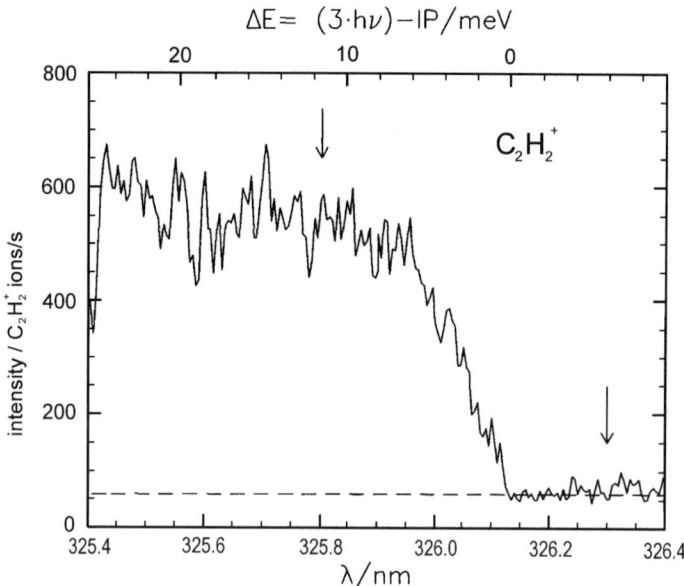

Fig. 5 Multi- or multiple photoionization of acetylene for producing cold $C_2H_2^+$ ions. The laser has been scanned in the wavelength region where the energy of three photons is equal to the ionization potential of $C_2H_2^+$. It has been operated with 1 mJ per pulse corresponding to 2.5×10^9 W cm^{-2}. The upper scale indicates the maximum possible internal energy of the product ion, if one assumes absorption of three photons. Using chemical probing, the internal excitation has been determined at the wavelengths marked with arrows.

described in ref. 24. Low temperature applications of state-selective preparation of diatomic ions *via* resonance enhanced multi-photon ionization have been summarized recently.[9] Another laser based strategy for getting ions in specific states is to use optical pumping schemes, *e.g.* to remove ions in unwanted states from the trap by using selective photo-fragmentation. In a well-isolated finite ensemble of trapped ions, this process (hole burning) can be used to tailor the population.

One of our attempts at creating a cold ensemble of $C_2H_2^+$ ions *via* 3-photon ionization is shown in Fig. 5. The ionization energy (IP) of acetylene is 11.40 eV. Ionization with three photons of the same wavelength is energetically possible below 326.15 nm. As can be seen, the ion signal increases significantly above the threshold for 3-photon ionization. Energy conservation leads to a maximal internal energy of a few meV (see upper scale) provided no additional photon is absorbed. In this energy range no resonance enhanced multiphoton ionization has been observed. An investigation of the dependence of the ion intensity on the laser power did not show a significant change from below and above threshold, where 4 or 3 photons are needed, respectively. Most probably, there are some near resonant or predissociating states involved.

In order to get information on the internal energy of the product ions, chemical probing with H_2 has been used. As shown in Fig. 3.23 of ref. 9, the reaction $C_2H_2^+ + H_2 \rightarrow C_2H_3^+ + H$ strongly depends on the ion temperature. The rate coefficient with the ions produced at 326.3 nm (see arrow, four photons are needed for ionization) indicates a significant internal energy. For evaluating the reactivity of the ion ensemble at 325.8 nm (left arrow), it is assumed that the number of ions, created *via* 4-photon processes does not change in this range (extrapolated dotted line). The number of $C_2H_3^+$ product ions revealed that, within statistical error, all products come from this part. So most probably the additional ions created just above the threshold, are internally cold; however, the multi photon scheme is not very useful since the portion of exited ions (~18%) is too large for most applications. A general conclusion from our photoionization studies is, that in almost all cases, collisional thermalization is superior if one works with traps.

CH_5^+, a polyatomic test case

One of our motivations to develop new cooling methods and to extend our energy range towards the sub-meV range is the interesting cation CH_5^+ and its interaction with H atoms. Many high quality calculations have shown in recent years (see ref. 25 and references therein) that the potential energy surface of protonated methane is very flat near the various minima. The hydrogen atoms can change their location on a sub-ns time scale. This leads to very complex spectra especially if one does not cool the highly fluxional ion to very low temperatures. This ion has been formed and cooled in a supersonic jet and high-resolution absorption spectra have been recorded in the 3000 cm^{-1} range.[25] Also a temperature variable 22-pole ion trap has been used successfully in combination with a free electron laser to record infrared spectra of CH_5^+.[26] In this experiment, the applied chemical probing scheme was based on laser induced proton transfer to carbon dioxide requiring the trap to be operated at and above 110 K. Both experiments have provided important spectral information; however, it is far from being sufficient to predict high resolution rotational transitions of CH_5^+ with the accuracy required for detecting this important ion in space.

One possible strategy to get more direct spectral information on the CH_5^+ ground state is to detect the $J = 1 \leftarrow J = 0$ absorption spectrum, which is predicted to be in the 220–235 GHz region.[27] Spectra in the mm-wave range will contain a lot of information on the large amplitude motion of the H atoms. One possible basis for such an experiment is to cool the ions to a few K. This can be achieved in the present experiment using the continuous or pulsed beam of He. Another very attractive method is to use a slow beam of H-atoms. The scheme is illustrated in Fig. 6, which shows a part of the potential energy surface relevant for H–CH_5^+ collisions. The data

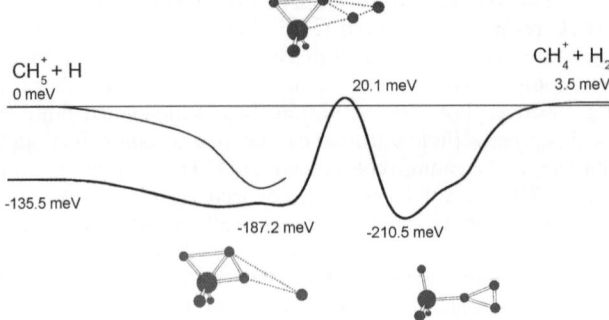

Fig. 6 Potential energy surface for describing low energy collisions of CH_5^+ with H. According to the calculations in ref. 28 (energy profile shown as heavy line) formation of $CH_4 + H_2$ is 12.6 ± 5.2 kJ mol^{-1} endothermic. In order to explain the experimental results shown in Fig. 7, the reaction must be nearly thermoneutral (thin line).

(heavy solid line) have been taken from a recent high level *ab initio* calculation.[28] Based on these results the reaction

$$CH_5^+ + H \rightarrow CH_4^+ + H_2 \quad (17)$$

is 12.7 ± 5.2 kJ mol^{-1} endothermic at 0 K. As a consequence there should be no competition between inelastic cooling with slow H atoms and the hydrogen abstraction reaction, provided one starts with cold enough ions. However, first experimental studies performed with the ion trapping apparatus, resulted in rate

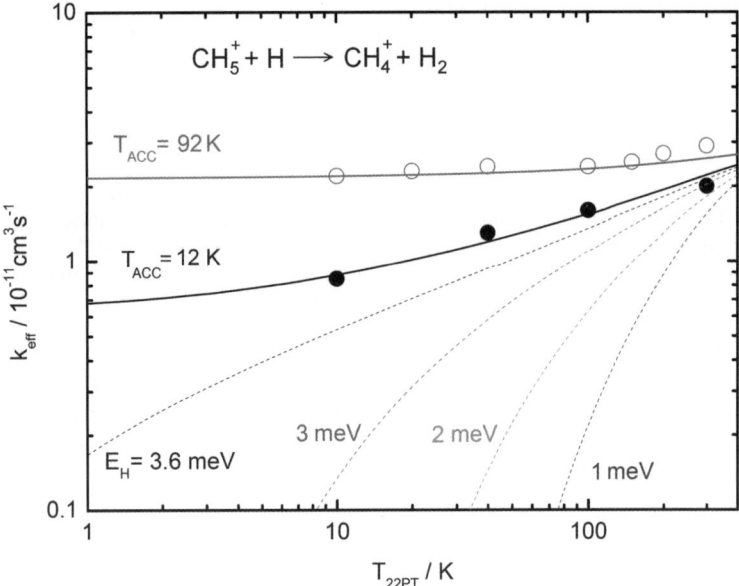

Fig. 7 The rate coefficients for the reaction $CH_5^+ + H \rightarrow CH_4 + H_2$ are only weakly dependent on T_{22PT}, *i.e.*, on the internal excitation of the ions. The ions are cooled in more than 10^4 collisions with He buffer gas. Lowering T_{ACC} from 92 K (open circles) to 12 K (filled circles) leads to smaller rate coefficients; however the observed reactivity is not at all in accordance with the known energetics (see Fig. 6). The solid lines are the result of a detailed analysis of the data, using an analytical *ansatz* for the rate coefficient and accounting for the kinematic broadening. The dashed lines are predictions from the model assuming a monoenergetic beam of H atoms with the indicated kinetic energies E_H.

coefficients which are in contradiction to this.[16] In these studies, the ions have been pre-cooled with an intense pulse of He to the temperature of the walls surrounding the trap, 10 K. These results ($T_{ACC} = 92$ K) are shown together with new measurements ($T_{ACC} = 12$ K) in Fig. 7. It is obvious that the reaction rate decreases if one uses slower H atoms; however, the measured data are in contradiction to the calculated endothermicity.

In order to account for the kinematic conditions, the data have been evaluated using the measured velocity distributions shown in Fig. 4 and accounting in detail for the thermal motion of the ions as discussed above. Some assumptions have been made concerning the elementary rate coefficients in order to calculate the effective rate coefficients from eqn (13). The threshold for reaction (17) has been used as a free parameter, E_0. Various assumptions have been made concerning the influence of the internal excitation of the CH_5^+ ions. After several tests, the experimental results have been reproduced successfully using the following *ansatz* for the elementary rate coefficient

$$k(E_t) = k_0 \left((1 - E_0/(E_t + \alpha k T_{22PT}))\right)^{1/2}. \tag{18}$$

E_t is the translational energy of the relative motion. The internal energy of the stored ions is described just by the temperature T_{22PT}. The factor α weights the internal energy of the CH_5^+ ion relative to the translational energy in helping to overcome the endothermicity. In a more sophisticated evaluation, and if more data are available, we will start with elementary state to state rate coefficients $k_{if} = g\sigma_{if}(g)$, accounting at least for the rotational population. Fitting the experimental data (solid lines in Fig. 7) results in a very low value for the threshold, $E_0 = 3.5 \pm 0.1$ meV. This is in obvious contradiction to the established energetics of reaction (17). Very surprising is also the fact, that internal energy is very inefficient, which can be seen from the value $\alpha = 0.1 \pm 0.1$. The scaling factor $k_0 = (3.2 \pm 0.2) \times 10^{-11}$ cm^3 s^{-1} is not so critical and has been adjusted for overall agreement. An obvious conclusion from these results is that we need colder H atoms in order to really probe the threshold onset of reaction (17). Predictions for monoenergetic H atom beams with 1, 2, or 3 meV are shown in Fig. 7 as dashed lines. In principle the preparation of a colder beam of H-atoms by operating the accommodator at the limit of H-atom condensation and by using a weaker hexapole magnet is straightforward. First velocity distributions of a thermal 8 K H-atom beam have been obtained recently.

Although very high number densities of He buffer gas have been used (1.5×10^{13} cm^{-3}) and storage times up to 1 s, we cannot completely exclude the possibility that there are some metastable excited ions in the trap which react. Therefore, new measurements will use higher He number densities and extend the storage time up to 100 s. This would also allow us to remove possible exited ions chemically towards $CH_4^+ + H_2$. For such studies and for cooling CH_5^+ directly with H atoms, our corrected potential (guessed, thin line in Fig. 6) has several attractive features. For example, it would allow us to synthesize cold CH_5^+ ions just by a collision of cold CH_4^+ with H_2. The very small exoergicity will result in a cold ensemble of CH_5^+ since the internally exited ions react away to CH_4^+. Even more attractive is the possibility that rotationally exited CH_5^+ ions react faster with H atoms than non-rotating ions. Such behavior would allow us to record directly rotational spectra by using cold H atoms not only as cooling atoms but also for chemical probing.

In addition to the methane cation[29] and the protonated methane[30] the CH_6^+ collision complex belongs to the class of fluxional molecules. Its synthesis *via* radiative or ternary association followed by collisional cooling is also within the reach of the present experiment. Threshold IR photofragmentation of this ion is a real challenge for preparing the ultracold $CH_5^+ + H$ collision complex (or also $CH_4^+ + H_2$) in the vicinity of the dissociation limit. Finally there are many possibilities to perform experiments with partial deuteration or with the bosonic system $CD_5^+ + D$.

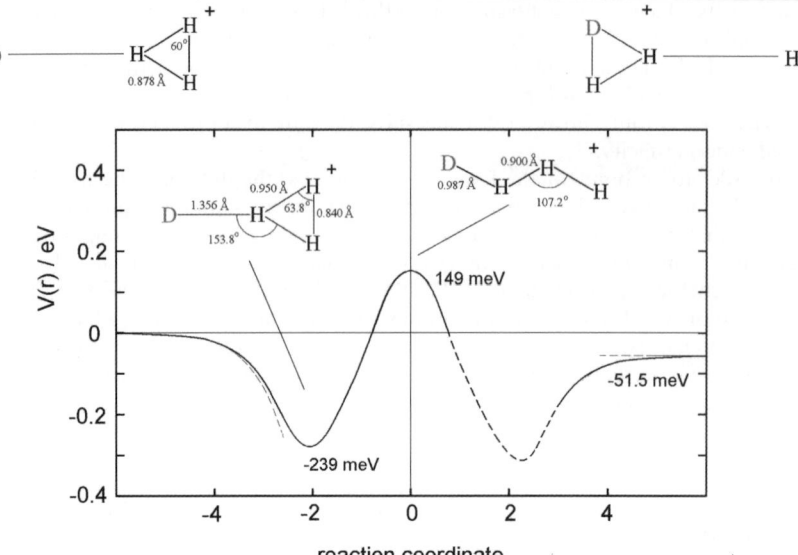

Fig. 8 Schematic illustration of the H_4^+ potential energy surface in the $H_3^+ + H$ region. The indicated well depth and the barrier height have been taken from ref. 33. The much higher lying $H_2^+ + H_2$ part (not shown here) is well characterized; however, in order to understand scrambling in low energy $H_3^+ + H$ collisions such as *ortho–para* transitions or H–D exchange in partly deuterated combinations, the energies in the vicinity of the H_3D^+ transition state need to be characterized with very high accuracy, a challenge for quantum chemistry.

Ultracold H_3^+ and H_4^+

H_3^+ ions and deuterated variants play an important role in interstellar chemistry. Various aspects of the importance of these ions at low temperatures have been discussed recently.[3,31,32] There are many motivations for studying this simplest triatomic ion at low temperatures including ultrahigh resolution spectroscopy, the role of nuclear spin in collisions, fermionic and bosonic behavior in chemical reactions, or the relevance of the deuterated ion for dense interstellar clouds.

Very challenging are studies of the ultracold H_4^+ ion or the collision complex with various H–D ratios. The relevant potential energy surface is shown schematically in Fig. 8. The indicated energies (well depth and barrier height) have been taken from ref. 33. For studying the cold H_4^+ ion itself in more detail, the potential energy surface needs to be known with spectroscopic accuracy, a challenge for quantum chemistry. For predicting *ortho–para* transitions in H–H exchange or deuteration in H–D exchange, the energies in the vicinity of the H_4^+ transition state need to be characterized in more detail.

One of the present projects with the cold H-atom beam is to obtain an ensemble of H_3^+ ions in a real thermal equilibrium *via* a sequence of exchange reactions

$$H_3^+ + H \leftrightarrow H_3^+ + H. \tag{19}$$

Already at a temperature of 5 K, the (1,1) ground state of H_3^+ (see Fig. 2 of ref. 32) would be populated with a probability of 99.7%. Lowering the temperature to 4 K reduces the population of the competing, 22.9 cm^{-1} higher lying (1,0) state to 5×10^{-4}. However, nuclear spin restrictions and traces of *ortho*-H_2 make it very hard to get to these limits with cold molecular hydrogen as buffer gas. Collisions with He do not change the total nuclear spin. Therefore reaction (19) is used for cooling H_3^+. The actual state population can be determined using laser probing as already shown in previous experiments.[34,35] Such detailed experiments will allow us to study the influence of nuclear spin restrictions in great detail. Looking at the barrier, it may take many collisions to really get into thermal equilibrium with the cold H-atom beam. It is an interesting question whether the lifetimes of the collision complex become long enough at low energies that tunneling leads to full scrambling.

4 Conclusions

For many years, and many applications, buffer gas cooling in rf traps has been based on cryogenically cooling the trap and the surrounding walls. In order to overcome the limitation imposed by condensation of neutral species, our trapping method has now been combined successfully with cold effusive beams. First tests with a 4.3 K He beam have been performed. The temperature regime below 1 K for ion chemistry is within our reach. Chopped very slow beams of He as coolant appear particularly promising at the frontier of cooling all degrees of freedom in the laboratory frame. Another interesting method is based on slow H-atoms. Attractive and technically possible is the combination of our ion traps with a trap for ultracold H atoms; however, there has not yet been any attempt to develop such an instrument. The use of superconducting electrodes for creating both the rf and the magnetic field may allow one to construct a rather simple setup. As already mentioned above, the use of superconducting materials can also reduce potential distortions on surfaces. It certainly will create less heat and improve the quality of the resonance circuit needed for getting an rf field without parasitic oscillations.

An alternative strategy in ultracold ion chemistry which has gained a lot of attention in the last decade, is to work with Coulomb crystals in rf ion traps. Especially attractive are linear strings of ions localized along the axis of a linear multipole trap. For working with cold molecular ions the process of sympathetic cooling is of significant importance. In addition one has to use specific methods such as photoionization for preparing internally cold or state selected ions.[7] A unique feature of this method is that single specific ions can be addressed and manipulated in the crystalline arrangements.

All these methods are still far from producing an ionic ensemble in the quantum degenerate regime. However, many recent results, such as spectra for large molecules of biological and astrochemical relevance[10] or the examples discussed in this contribution, show that there is a lot to do between room temperature and the sub-K regime. Applying, in addition to buffer gas cooling, optical manipulation and detection schemes to the stored molecular ions will allow us to make unique contributions to the emerging field of cold and maybe also ultracold chemistry.

References

1 A. Dzhonson, D. Gerlich, E. J. Bieske and J. P. Maier, Apparatus for the study of electronic spectra of collisionally cooled cations: *para*-dichlorobenzene, *J. Mol. Struct.*, 2006, **795**, 93.
2 O. V. Boyarkin, S. R. Mercier, A. Kamariotis and T. R. Rizzo, Electronic spectroscopy of cold, protonated tryptophan and tyrosine, *J. Am. Chem. Soc*, 2006, **128**, 2816.
3 O. Asvany, E. Hugo, F. Müller, F. Kühnemann, S. Schiller, J. Tennyson and S. Schlemmer, Overtone spectroscopy of H_2D^+ and D_2H^+ using laser induced reactions, *J. Chem. Phys.*, 2007, **127**, 154317–1.
4 R. Otto, J. Mikosch, S. Trippel, M. Weidemüller and R. Wester, Nonstandard behavior of a negative ion reaction at very low temperatures, *Phys. Rev. Lett.*, 2008, **101**, 0632011.
5 M. Drewsen, Cooling, identification and spectroscopy of super-heavy element ions, *Eur. Phys. J. D*, 2007, **45**, 125.
6 J. C. J. Koelemeij, B. Roth, A. Wicht, I. Ernsting and S. Schiller, Vibrational spectroscopy of HD^+ with 2-ppb accuracy, *Phys. Rev. Lett.*, 2007, **98**, 1730021.
7 S. Willitsch, M. T. Bell, A. D. Gingell, S. R. Procter and T. P. Softley, Cold reactive collisions between laser-cooled ions and velocity-selected neutral molecules, *Phys. Rev. Lett.*, 2008, **100**, 0432031.
8 I W M Smith (ed.), *Low temperatures and cold molecules*. World Scientific Publishing, ISBN 978-1-84816-209-9, 2008.

9 D. Gerlich, The study of cold collisions using ion guides and traps. in: *Low temperatures and cold molecules*, World Scientific Publishing, ed. Ian W M Smith, ISBN 978-1-84816-209-9, 2008, p. 121.
10 D. Gerlich, The production and study of ultra-cold molecular ions. in: *Low temperatures and cold molecules*, World Scientific Publishing, ed. Ian W M Smith, ISBN 978-1-84816-209-9, 2008, p. 295.
11 S. Y. T. van de Meerakker, N. Vanhaecke and G. Meijer, Stark deceleration and trapping of OH radicals, *Annu. Rev. Phys. Chem.*, 2006, **57**, 159.
12 D. Gerlich, Experimental investigations of ion molecule reactions relevant to interstellar chemistry, *J. Chem. Soc., Faraday Trans.*, 1993, **89**, 2199.
13 D. Gerlich and M. Smith, Laboratory astrochemistry: studying molecules under inter- and circumstellar conditions, *Phys. Scr.*, 2006, **73**, C25.
14 D. Gerlich, Inhomogeneous electrical radio frequency fields: a versatile tool for the study of processes with slow ions, *Adv. Chem. Phys.*, 1992, **82**, 1.
15 H. G. Dehmelt, Radiofrequency spectroscopy of stored ions I: storage, *Adv. At. Mol. Phys.*, 1967, **3**, 53.
16 A. Luca, G. Borodi, D. Gerlich, *Interactions of ions with Hydrogen atoms. Progress report in XXIV ICPEAC 2005, Rosario, Argentina, July 20–26, 2005*, ed. F. D. Colavecchia, P. D. Fainstein, J. Fiol, M. A. P. Lima, J. E. Miraglia, E. C. Montenegro, and R. D. Rivarola, 2005, p. 20.
17 G. Borodi, A. Luca and D. Gerlich, Reactions of CO_2^+ with H, H_2 and deuterated analogues, *Int. J. Mass. Spectrom.*, 2009, **280**, 218.
18 D. Gerlich, G. Borodi, A. Luca and M. Smith, *Reactions between cold hydrocarbon ions and slow hydrogen atoms*, 2009, submitted to *J. Phys. Chem.*
19 S. Schlemmer, T. Kuhn, E. Lescop and D. Gerlich, Laser excited N_2^+ in a 22-pole trap, experimental studies of rotational relaxation processes, *Int. J. Mass. Spectrom.*, 1999, **185**, 589.
20 E. J. Bieske, A. M. Soliva, A. Friedmann and J. P. Maier, Electronic spectra of N_2^+–$(He)_n$ ($n = 1, 2, 3$), *J. Chem. Phys.*, 1992, **96**, 28.
21 P. J. Chantry, Doppler broadening in beam experiments, *J. Chem. Phys.*, 1971, **55**, 2746.
22 R. D. Levine, R. B. Bernstein, Molecular Reaction Dynamics. in *Molecular Reaction Dynamics*, Oxford University Press, Oxford, 1974.
23 D. Gerlich, Kinematic averaging effects in thermal and low energy ion-molecule collisions: influence on product ion kinetic energy distributions, *J. Chem. Phys.*, 1989, **90**, 127.
24 S. Mark, T. Glenewinkel-Meyer and D. Gerlich, REMPI in a focusing rf-quadrupole: a new source for mass and state selected ions, *Int. Rev. Phys. Chem.*, 1996, **15**, 283.
25 X. Huang, A. B. McCoy, J. M. Bowman, L. M. Johnson, C. Savage, F. Dong and D. J. Nesbitt, Quantum deconstruction of the infrared spectrum of CH_5^+, *Science*, 2006, **311**, 60.
26 O. Asvany, P. Kumar, B. Redlich, I. Hegemann, S. Schlemmer and D. Marx, Understanding the infrared spectrum of bare CH_5^+, *Science*, 2005, **309**, 1219.
27 P. R. Bunker, B. Ostojic and S. Yurchenko, A theoretical study of the millimeterwave spectrum of CH_5^+, *J. Mol. Struct.*, 2004, **695–696**, 253.
28 B. Wang and Hua Hou, Computational study of the ion–molecule reactions involving fluxional cations: $CH_4^+ + H_2$ to $CH_5^+ + H$ and isotope effect, *J. Chem. Phys.*, 2005, **109**, 8537.
29 H. J. Wörner, X. Qian and F. Merkt, Jahn–Teller effect in tetrahedral symmetry: large-amplitude tunneling motion and rovibronic structure of CH_4^+ and CD_4^+, *J. Chem. Phys.*, 2007, **126**, 1443051.
30 A. B. McCoy, B. J. Braams, A. Brown, X. Huang, Z. Jin and J. M. Bowman, *Ab initio* diffusion Monte Carlo calculations of the quantum behavior of CH_5^+ in full dimensionality, *J. Phys. Chem. A*, 2004, **108**, 4991.
31 T. Oka, Introductory remarks, *Philos. Trans. R. Soc. London, Ser. A*, 2006, **364**, 2847.
32 D. Gerlich, F. Windisch, P. Hlavenka, R. Plašil and J. Glosik, Dynamical constraints and nuclear spin caused restrictions in HmD_n^+ collision systems and deuterated variants, *Philos. Trans. R. Soc. London, Ser. A*, 2006, **364**, 3007.
33 G. E. Moyano, D. Pearson and M. A. Collins, Interpolated potential energy surface and dynamics for atom exchange between H and H_3^+, and D and H_3^+, *J. Chem. Phys.*, 2004, **121**, 12396.
34 J. Mikosch, H. Kreckel, R. Wester, R. Plasil, J. Glosik, D. Gerlich, D. Schwalm and A. Wolf, Action spectroscopy and temperature diagnostics of H_3^+ by chemical probing, *J. Chem. Phys.*, 2004, **121**, 11030.
35 J. Glosik, P. Hlavenka, R. Plašil, F. Windisch and D. Gerlich, Action spectroscopy of H_3^+ using overtone excitation, *Philos. Trans. R. Soc. London, Ser. A*, 2006, **364**, 2931.

Ion-molecule chemistry at very low temperatures: cold chemical reactions between Coulomb-crystallized ions and velocity-selected neutral molecules

Martin T. Bell,[a] Alexander D. Gingell,[a] James M. Oldham,[a] Timothy P. Softley*[a] and Stefan Willitsch†*[b]

Received 22nd October 2008, Accepted 27th November 2008
First published as an Advance Article on the web 12th May 2009
DOI: 10.1039/b818733a

The recent development of a range of techniques for producing cold atoms and molecules at very low translational temperatures $T \leq 1$ K has provided the opportunity to investigate collisional processes in a new physical regime. We have recently presented a new experimental method to study low-temperature reactive collisions between translationally cold ions and neutral molecules (S. Willitsch *et al.*, *Phys. Rev. Lett.* 2008, **100**, 043203). Our technique relies on the combination of a quadrupole-guide velocity selector for the generation of translationally cold neutral molecules with a facility to produce ordered structures of cold ions (Coulomb crystals) by laser cooling in a linear quadrupole ion trap. The strong localisation of the ions in the trap in combination with the high sensitivity of laser-induced-fluorescence detection enabled us to study chemical reactions on the single-particle level, down to temperatures of $T \approx 1$ K. In the current paper, we present a detailed characterisation of the scope and limitations of this method based on our study of the reaction between laser-cooled Ca^+ ions and velocity-selected CH_3F molecules. The properties of our cold-neutrals source and the dependence of the measured rate constant on the shape of the Coulomb crystals, trapping and laser-cooling parameters are discussed. An extension of our technique for the study of low-temperature reactions with sympathetically cooled molecular ions (translational temperature $T > 10$ mK) is presented and first results on the charge-transfer reaction between OCS^+ and ND_3 are discussed. Finally, perspectives for further developments of our method are explored.

1 Introduction

The experimental study of reactive collision dynamics in the cold ($T \leq 1$ K) and ultracold ($T \leq 1$ mK) regions offers a new physical regime in which to test quantum mechanical descriptions of chemical processes. The average molecular velocities at such temperatures may be of the order of m s^{-1} (cold), or even cm s^{-1} (ultracold) leading to several possible implications for studies of collision dynamics. First, the

---

[a] *Department of Chemistry, University of Oxford, Chemistry Research Laboratory, Mansfield Road, Oxford, OX1 3TA, United Kingdom. E-mail: tim.softley@chem.ox.ac.uk*
[b] *Department of Chemistry, University of Basel, Klingelbergstrasse 80, 4056 Basel, Switzerland. E-mail: stefan.willitsch@unibas.ch*

† Presenting author.

collisional angular momentum ($|l| = \mu v_{rel} b$) will be substantially reduced as the relative velocity v_{rel} decreases and ultimately the collision will be characterized by one or a few angular momentum states. Thus the normal averaging of the collision dynamics over these states is reduced, and sharp resonances in the energy-dependent cross section are more likely to be revealed experimentally. Second, the de Broglie wavelength associated with the translational motion will become large compared to the range of molecular interactions. The intrinsic wave properties of the collisional process may thus be enhanced and effects of tunnelling, barrier reflection and interference are expected to be important.[1,2] Third, the low kinetic energy makes the reaction dynamics highly sensitive to the long-range part of the potential energy surface, especially for processes with a negligible reaction barrier. Non-adiabatic effects may become important in open-shell systems where multiple potential energy surfaces may converge at long range to closely spaced limits. The study of ultracold collisions offers the possibility of new forms of control using electromagnetic fields[3] and it is of relevance to the rapidly developing field of ultracold matter physics with its links to condensed matter physics and quantum information processing.[4]

The recent and ongoing development of a range of cold and ultracold molecular sources in the gas phase[1] – *e.g.*, Stark, Zeeman and optical deceleration,[5–7] photoassociation and association by Feshbach resonance magnetic tuning with ultracold alkali metal atoms,[8–10] collisional cooling in crossed beams[11] or buffer gases,[12] velocity selection[13] and sympathetic cooling of molecular ions in laser-cooled atomic ion traps[14,15] – has stimulated efforts to study reactive and inelastic scattering processes in these temperature ranges.[16,17] The experimental challenges remain considerable however, for two reasons. First, reactive systems that are of particular interest from a theoretical viewpoint, for example the F + H_2 reaction,[2] are not necessarily the ones to which cooling/deceleration techniques can be applied; thus the quest to develop *general* methods for producing cold and ultracold molecules continues. Second, the number densities and the absolute numbers of cold and ultracold molecules produced by existing sources are typically several orders of magnitude lower than conventional supersonic molecular beam sources; hence, the rate of production of reaction products generated when two such sources are combined to study bimolecular collisions is likely to be significantly lower than needed for conventional detection methods.

In a recent paper[18] we reported preliminary results of a new experimental setup designed to study the reactions of ions with neutral molecules at translational temperatures of the order 1 K or below. The experiment combines a quadrupole velocity selector for neutral dipolar species with laser-cooled Ca^+ ions in a radiofrequency ion trap, which form ordered structures known as "Coulomb crystals". As we demonstrated in our preliminary report, and as reviewed elsewhere[14] the use of Coulomb crystals enables an extraordinary level of detection sensitivity, allowing reactions to be observed at rates as low as a few molecules per hour. The mass-spectrometric identification of CaF^+ ions produced from the reaction of Ca^+ with CH_3F was reported in reference 18.

In this paper we provide new results which illustrate the scope and limitations of this new experimental technique including extension to the reactions of molecular ions which can be trapped in the Coulomb crystals by sympathetic cooling.[14]

2 Experimental

2.1 Experimental overview

An overview of the experimental apparatus is shown in Fig. 1 (see also ref. 14 and 18). Ca^+ ions are produced by non-resonant multiphoton ionization of Ca atoms, effusing from a resistively heated oven, using the third harmonic of a nanosecond Nd:YAG laser at 355 nm. The ionization occurs in the centre of a linear quadrupole ion trap to which radiofrequency (RF) and DC fields are applied (RF amplitude 150

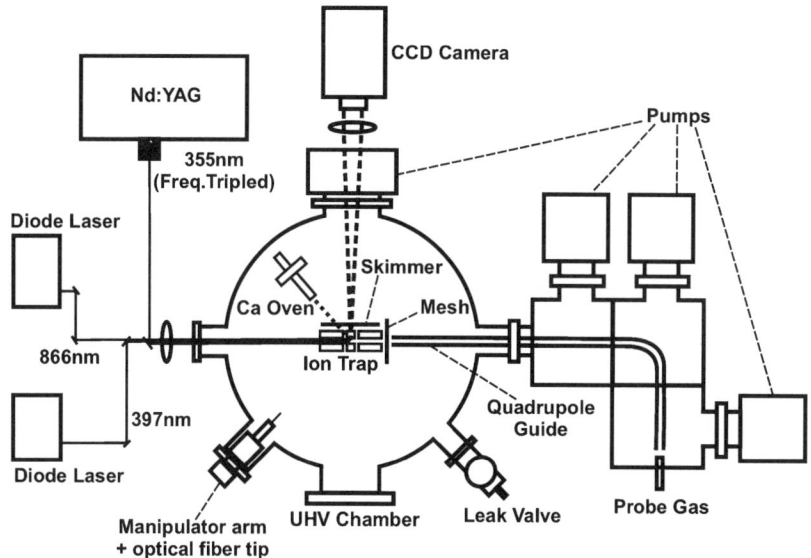

Fig. 1 Overview of the experimental setup, showing the ion trap in the centre of the figure combined with the quadrupole velocity selector to the right.

V, frequency $2\pi \times 3.8$ MHz, DC potentials 1–5 V). The Ca$^+$ ions are laser-cooled on the $4s(^2S_{1/2}) - 4p(^2P_{1/2})$ transition at 397 nm using an external cavity enhanced 1 MHz linewidth diode laser beam (1 mW Toptica DL100) directed along the axis of the trap. A second diode laser beam at 866 nm (5 mW Toptica DL100) is employed to repump the population that accumulates in the $3d(^2D_{3/2})$ level to the $4p(^2P_{1/2})$ level. Under the action of laser cooling the ions reach temperatures for which the potential energy exceeds the kinetic energy and undergo a phase transition to form a Coulomb crystal,[14] in which regularly ordered structures of ions are observed, with a typical ion spacing of 20 μm. According to the number of ions trapped and the field amplitudes applied, the crystals may be one-dimensional strings or two/three-dimensional structures (*e.g.*, see Fig. 6 and 7). The ions are detected through their fluorescence induced by the cooling laser, which continuously irradiates the ion cloud throughout the experiment. A microscope objective lens is used in conjunction with a CCD camera to observe a two-dimensional projection of the crystal, although for three-dimensional crystals the shallow depth of focus of the lens limits the observation range to the central two or three layers of the crystal. Individual ions are resolvable (see *e.g.*, Fig. 6(a)(i)). Near the central axis of the RF trap the effective temperature of the ions is of the order 10 mK (see Section 2.3 below for full details). Crystals can be maintained and observed in the trap at a background pressure of 5×10^{-10} mbar for periods of several hours.

The ion trap is combined with a quadrupole guide velocity selector for polar neutral molecules, following the original design of Rempe and coworkers.[13,19] The quadrupole guide is constructed from 2 mm diameter polished stainless steel cylindrical rods, rod spacing 1 mm, and incorporates a 90 degree bend with radius of curvature of 12.5 mm. The neutral molecules (CH$_3$F or ND$_3$ in the experiments described below) are injected effusively at room temperature into the guide through a 1.2 mm diameter capillary connected to a thermal reservoir *via* a leak valve. Static potentials up to ±5 kV are applied to the guide rods with alternating polarity, and low-field seeking molecules are confined by the quadrupole field to the centre of the guide. This field also provides the centripetal force required to guide molecules around the bend, but the force is insufficient to guide the majority of molecules in the 300 K sample: these high-velocity molecules overshoot the bend and are removed by the vacuum pump. Only a small percentage of the molecules, *i.e.*, those which have the lowest velocities in the tail of the Maxwell–Boltzmann distribution, are transmitted around the bend and then guided towards the ion trap. Three stages of differential pumping between the effusive source and the ion trap are employed to maintain the ultrahigh vacuum (5×10^{-10} mbar) in the ion trap chamber. The end of the quadrupole guide is located within 5 mm of the end of the ion trap, with a grounded mesh used to isolate the fields of the two devices. The effective translational temperature of the gas is of the order 1 K (see Section 2.2) and, as described below, the velocity distribution can be varied by changing the field applied to the quadrupole. The density of neutral molecules at the Coulomb crystal is estimated by placing a calibrated quadrupole mass spectrometer at the position of the ion trap centre, and is found to be 10^5–10^6 cm^{-3} (see Section 2.2).

In the reaction experiments the Coulomb crystal is first formed, and then the quadrupole guide is switched on to admit the flux of cold polar neutral molecules. Images of the crystals are recorded as a function of time, typically for periods of 15 minutes to 1 hour. The ion trap has a deep potential energy well for ions (≥ 1.2 eV under typical trapping conditions). Therefore the product ions from reactive processes are typically retained within the trap even for highly exothermic processes. Through the strong Coulombic interaction with the Ca$^+$ ions, the product ions are sympathetically cooled and are condensed into the Coulomb crystal. The dynamics of the ions in the radiofrequency fields are such that ions lighter than Ca$^+$ are located near the central axis of the crystal whereas heavier ions are located outside the Ca$^+$ shell. As shown in ref. 14 and 18 the presence of these non-fluorescing ions can be detected by scanning an axial radiofrequency field through the resonant motional frequencies of the crystal and observing the heating of the crystal at these resonances. The number of ions of each type as a function of time is determined by comparing simulated images based on molecular dynamics calculations of the crystal with the observed ones.[14] Room temperature reactions can also be studied by leaking a low partial pressure of the neutral reactant gas into the ion trap chamber.

2.2 Properties of the quadrupole velocity selector

In the experiments reported here, we have successfully produced continuous beams of translationally cold molecules using a variety of symmetric and asymmetric top polar molecules, with masses in the range 20–61 u and dipole moments between 1.5 and 3.92 D. The fluxes of transmitted molecules are measured by placing a quadrupole mass spectrometer (QMS) at the exit of the quadrupole velocity selector. The inset of Fig. 2 shows the QMS current at the appropriate parent-ion mass for various molecules as the guiding voltage (V_0) is varied. Increasing the voltage allows a greater fraction of molecules to be extracted from the initial velocity distribution, which leads to a greater flux detected by the QMS. For molecules in a single low-field seeking state with a linear Stark shift, an approximate theoretical upper bound for the guided flux can be shown to be given by[19]:

$$\phi_{\text{guide}} \approx \phi_0 \, \tilde{v}_{\text{max}}^2 \, \tilde{v}_{z,\text{max}}^2$$
$$\approx \phi_0 \left(\frac{R}{2r_0}\right)\left(\frac{W_{\text{max}}}{k_B T}\right)^2. \tag{1}$$

where $\tilde{v}_{\text{max}}$ and $\tilde{v}_{z,\text{max}}$ are the maximum transverse and longitudinal velocities that can be confined by the guide, expressed using the reduced velocity, $\tilde{v} = v/v_w$, where $v_w = \sqrt{2k_B T/m}$ is the most probable velocity for the Maxwell–Boltzmann distribution. Consequently, the flux of guided molecules depends on: ϕ_0, the flux of molecules entering the guide from an effusive source at a temperature T; R, the radius of the quadrupole bend; r_0, the radius of the quadrupole electrodes; and W_{max},

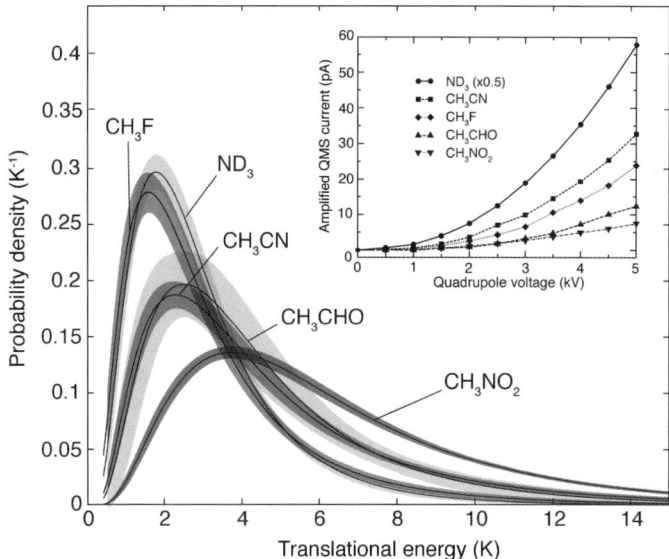

Fig. 2 Kinetic energy distributions for various guided molecules at a quadrupole voltage of $V_0 = \pm 4$ kV. Inset: transmitted flux (quadrupole mass spectrometer signal) through the quadrupole guide for the same molecules as a function of applied potential V_0. The shaded areas indicate the uncertainties (1σ) in the probability densities.

the depth of the potential that confines the molecules inside the straight section of the guide.

While the filtering efficiency does not depend directly on the molecular mass (the acceptance depends on kinetic energy, not velocity[13]), the flux of molecules entering the guide has an inverse dependence on the mass. The total flux of molecules exiting the guide, which is predicted to increase with the squared dipole moment, should therefore be highest for light, strongly polar molecules. For the symmetric top molecules ND_3, CH_3CN and CH_3F, which have a linear Stark shift over the range of electric fields present in the guide, a quadratic increase in the flux is observed as the quadrupole voltage is increased as expected from Eqn 1. Lower fluxes were obtained using the asymmetric top molecules CH_3CHO and CH_3NO_2, the latter being found to produce the least intense beam. As this molecule has an asymmetry parameter, $\kappa = 0.46$, which is far from the oblate symmetric top limit ($\kappa = +1$), only a weak Stark effect is expected for most of the rotational states populated in the thermal beam and, consequently, the velocity filtering is inefficient. Methanol molecules were also found to be weakly guided, but the detected signal was too weak for further characterization of the beam. Attempts were also made to guide the isotopomers of water: H_2O, HDO and D_2O, but the fluxes of molecules were so low that no increase in the mass spectrometer signal could be seen above the background, even after baking the chamber to remove traces of water. Higher fluxes are obtainable using a quadrupole with a larger bend radius, as has been shown recently by Motsch et al.[20]

Absolute fluxes of molecules exiting the guide can be determined by calibrating the mass spectrometer measurements with those of a separate hot-cathode ionization pressure gauge. The calibration gas is admitted via a leak valve and the ionization gauge readings are converted to absolute number densities using a gas-specific correction factor. These values are then linearly extrapolated to the range of channeltron-amplified currents measured by the mass spectrometer in the guiding experiments. In the case of CH_3F, the number densities measured at a distance of 28 mm from the guide exit were found to range from 1.2(5) × 10⁵ cm⁻³ for $V_0 = \pm 2.0$ kV to 1.1(5) × 10⁶ cm⁻³ for $V_0 = \pm 5.0$ kV.

2.2.1 Velocity distributions.

By pulsing the high-voltages applied to the guide electrodes, the distribution of arrival times at the mass spectrometer can be measured to give the velocity distribution of the guided molecules. A square waveform is used to modulate the electrode voltages and the resulting time-resolved signal from the mass spectrometer is averaged over several thousand cycles. Fig. 3(a) shows the leading edge of the mass spectrometer signal when ND_3 molecules are guided at a range of quadrupole voltages. As before, the amplitudes of the signal are seen to increase with quadrupole voltage and the rise times of the signals increase as the voltage is raised due to the broader range of velocities selected by the guide.

The data in Fig. 3(a) were found to be well modelled using a Gompertz function, $f(t) = \exp(-\exp(-k[t - t_c]))$, where the parameters k and t_c are determined by a non-linear least-squares fitting procedure and the velocity distributions, $f(v)$, are obtained from the derivative of the time-dependent mass-spectrometer signal, $S(t)$ via

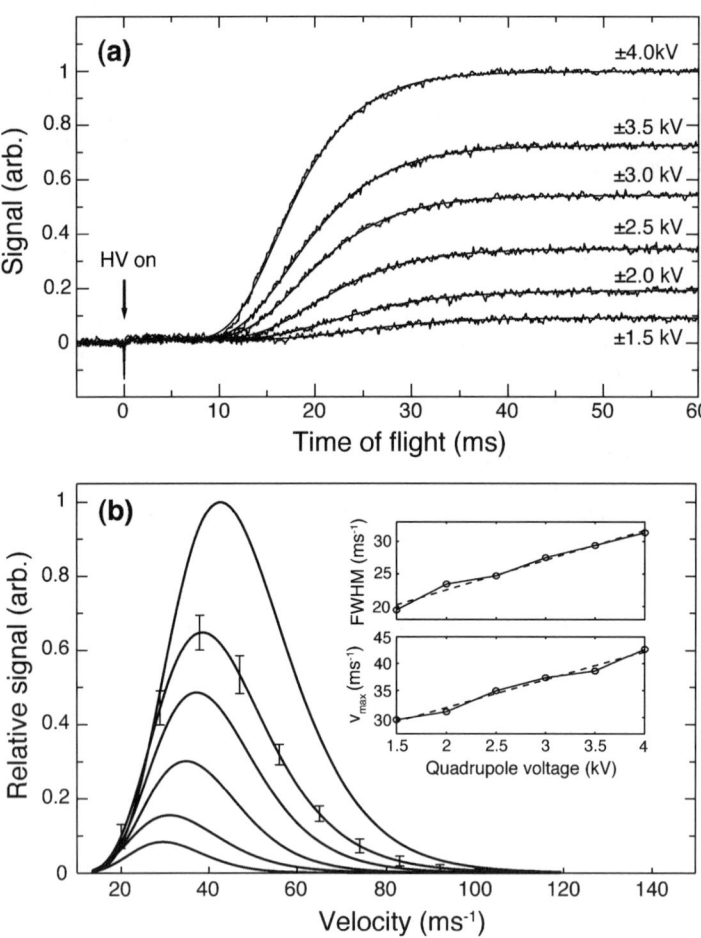

Fig. 3 (a) Transmitted ND_3 flux in an experiment in which a square-wave pulse is applied to the quadrupole guide for various voltages specified in the figure. (b) Velocity distribution for the ND_3 as a function of voltage, derived from the measurements in (a). The inset shows the widths and maximum velocities for each distribution.

$$f(v)\,dv \propto \frac{L}{v^2} \dot{S}(L/v)\,dv, \qquad (2)$$

where $L = 811$ mm is the path length from the quadrupole-guide entrance to the centre of the ionisation cage of the mass spectrometer.

Fig. 3(b) shows the measured velocity distributions for guiding ND_3 molecules at a range of different quadrupole voltages. As shown in the inset to the figure, the most probable velocity and widths of the guided flux distribution are seen to increase with the quadrupole voltage, which is consistent with a higher proportion of faster molecules being selected from the initial velocity distribution. The measured velocity distributions show that only very small numbers of molecules are produced with velocities lower than 20 ms^{-1}. Trajectory simulations show that this is due to acceleration of the molecules by the fringe fields at the end of the quadrupole. Upon exiting the guide, CH_3F molecules injected into the quadrupole bend with an initial velocity of 10 ms^{-1} are observed to be accelerated to around 20–25 ms^{-1} by the electric field gradient between the ends of the quadrupole and the grounded mesh. As the electric field is zero at the centre of the quadrupole, the amount of acceleration the molecules receive increases when their oscillatory motion propels them off axis, leading to a spread of different final velocities. While this effect is not expected to be so pronounced for molecules with a weak Stark effect, it does potentially limit the production of molecules at translational temperatures much below 1 K.

Translational energy distributions for a range of different molecules at a guiding voltage of ± 4.0 kV are shown in Fig. 2. The lowest energy beams are obtained for light molecules such as ND_3 and CH_3F, which have peak kinetic energies of 1.6 K and 1.8 K respectively. For heavier molecules, the distributions are shifted to higher kinetic energies and in the case of CH_3NO_2, a peak kinetic energy of 3.8 K was found. As the cut-off velocities for the various different molecules depend on the magnitude of the electric fields in the guide, beams of lower energy can be produced by lowering the quadrupole voltage, although this occurs at the expense of beam intensity.

2.2.2 Rotational state distributions.

The guiding efficiency of the quadrupole guide will show a strong dependence on rotational quantum state through the varying Stark shifts for these states. Thus the rotational state distribution of molecules exiting the guide will be changed from that of the source distribution. However, the continuous partial flux of molecules exiting the guide in each quantum state is very small and measurements of the rotational state populations were not attempted in the present study (although such measurements are possible using depletion spectroscopy, see ref. 21). Instead, numerical simulations of the guiding process have been performed to determine the different behaviour of the many Stark states populated in the thermal source.

Monte Carlo trajectory simulations of molecules in the quadrupole guide were carried out by numerical integration, using an adaptive Runge–Kutta method, of the equation of motion:

$$\mathbf{F} = m[\ddot{x} - (R+x)\dot{\theta}^2]\hat{\mathbf{x}} + m\ddot{y}\hat{\mathbf{y}} + m[2\dot{x}\dot{\theta} + (R+x)\ddot{\theta}]\hat{\mathbf{s}} \qquad (3)$$

where $\mathbf{F} = -\nabla W(x, y, \theta)$ is the three-dimensional force on the molecule, expressed in the curvilinear coordinate system in which $\hat{\mathbf{x}}$ and $\hat{\mathbf{y}}$ are unit vectors in the radial direction and the vertical direction (perpendicular to the plane of the bend) respectively and $\hat{\mathbf{s}}$ is the unit vector tangent to the circle of radius R which follows the centre-line of the quadrupole guide. The angle θ is defined with reference to the start of the bend. As the radius of the bend is much larger than the inner radius of the quadrupole, the electric field in the bent section of the guide can be locally approximated by the field in the straight sections. The starting positions of the trajectories were uniformly distributed over a 1.2 mm diameter disk located 0.5 mm from the quadrupole entrance, while initial velocities were obtained randomly from the calculated effusive capillary source flux distribution using a transformation sampling procedure. The electric field magnitude inside the quadrupole was calculated on a two-dimensional grid using a finite element method and the necessary electric field gradients were obtained from centred finite differences. A similar three-dimensional calculation for a small volume containing the end of the guide and the grounded mesh was also performed so that the fringe-field acceleration of the molecules could be included in the simulation. Calculations of the molecular Stark effect were performed by diagonalizing the combined rotational and Stark Hamiltonian matrices. The wavefunctions obtained were adiabatically matched over the range of electric fields in the guide and least-squares fitting of fifth-order polynomials was used to parameterize each of the Stark curves, thereby allowing rapid calculation of the field-dependent effective dipole moment function. The number of trajectories propagated for each of the low-field seeking Stark states was weighted by a Boltzmann-factor based on the zero-field rotational energies and including the degeneracies associated with nuclear spin statistics.

In Fig. 4 a simulated longitudinal kinetic energy distribution of CH_3F molecules guided at a voltage of $V_0 = \pm 4.0$ kV is presented in comparison with the experimental distribution. The peak position of the experimental distribution is well reproduced by the simulation, whilst the width appears to be slightly overestimated, with apparently fewer faster molecules being produced experimentally. However, the results of the simulation have not been corrected for the detection efficiency of the mass spectrometer. Two competing effects are expected: at low velocities the increased divergence of the beam leads to poor overlap with the detection region whilst, at higher velocities, the probability of the molecules being ionized by the detector is decreased. Nevertheless the agreement with experiment is considered sufficient to evaluate the rotational state distributions.

Fig. 5 shows the normalized rotational energy distributions obtained from the Monte Carlo simulations for velocity-selected beams of CH_3F, ND_3 and HDO. For symmetric top molecules, the first-order Stark effect is larger for molecules with high-K quantum numbers as these states have higher effective dipole moments. These high-K states are found to be the most strongly populated in the transmitted beam, and the total number of populated states remains high because they exist across the spectrum of rotational energies. In the case of HDO, which is an asymmetric top, the initial state distribution is relatively sparse as a consequence of the

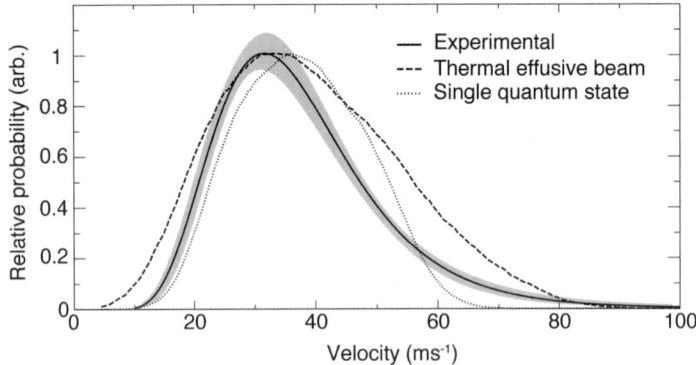

Fig. 4 Simulated CH_3F velocity distribution *versus* experimental at $V_0 = \pm 4$ kV. The simulations are presented for the thermal source distribution at 300 K and for a single quantum state $|J, MK> = |1, -1>$ in a low field seeking state. The distributions are normalised to the same peak height. The shaded area indicates the uncertainties (1σ) in the experimental probability densities.

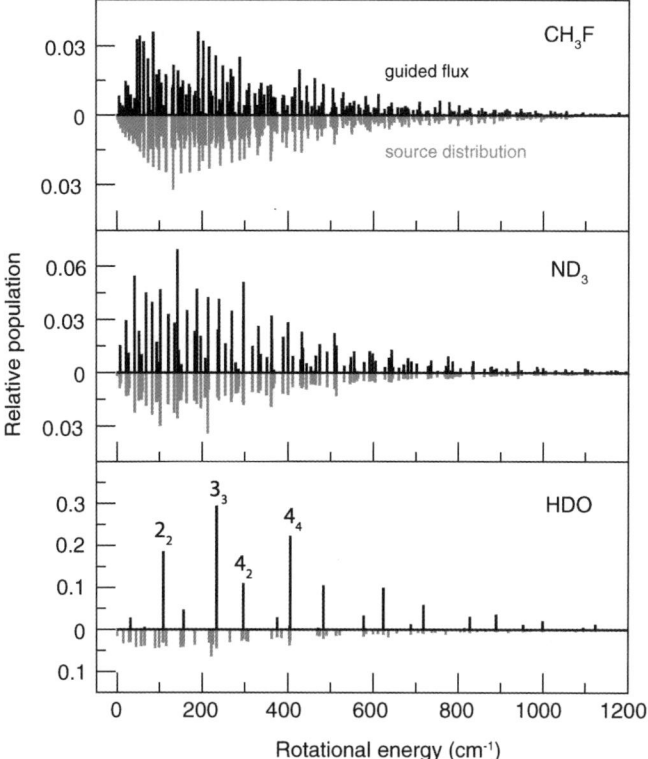

Fig. 5 Rotational energy distributions for (a) CH$_3$F, (b) ND$_3$ and (c) HDO obtained from the Monte Carlo simulations. The upper stick diagram in each case shows the transmitted flux in each rotational state whereas the lower diagram shows the distribution in the source at 300 K.

large rotational constants of the molecule, and a higher degree of selectivity is exhibited, with approximately 50% of the population spread over four rotational levels. For H$_2$O, as Motsch et al.[20] have discussed, the selectivity can be even higher, approaching 80% for the $|J_\tau M\rangle = |1_1 1\rangle$ state. This is a consequence of the second-order Stark effect: in the absence of any accidental near-degeneracies between rotational levels, the Stark effect becomes weaker with increasing rotational energy and so only molecules in low-lying rotational states survive the velocity-filtering process.

2.3 Structural and thermal properties of Coulomb crystals

Laser cooling is a well-established technique to prepare samples of atomic ions at temperatures in the millikelvin range, see, e.g., refs 14 and 15. The spontaneous emission of the ions generated in the laser-cooling process can be imaged with a CCD camera coupled to a microscope. The spacing between the laser-cooled ions is sufficiently large to spatially resolve individual particles. The fluorescence images serve as a useful diagnostic to monitor the laser cooling and to study the structural and dynamic properties of the laser-cooled ions. Ions in radiofrequency traps exhibit two types of motion: a slow thermal ("secular") motion in the effective potential of the trap superimposed with a fast oscillating motion ("micromotion") driven by the periodically changing RF fields.[22] Laser-cooling experiments are performed in an adiabatic regime in which the two types of motion can be separated. Only the secular motion of the ions is cooled, which is sufficient to cause Coulomb crystallization of the ions in the trap.[23,24] The micromotion, however, is constantly driven by the RF fields and therefore is not affected by laser cooling. Its amplitude is governed by the magnitude of the inhomogeneous electric fields in the trap and therefore depends on the position of the ion within the trapping region. Because of the quadrupolar trap symmetry, the amplitude of the micromotion is zero on the central axis and increases in the direction of the electrodes. Thus, the effective kinetic energy of the ions, which contains contributions from both the secular and the micromotion, is a function of the position of the ion inside the Coulomb crystal. The coldest ions are located on, or close to, the trap axis.

For reactive-scattering studies it is imperative to characterize the effective kinetic energies of the Coulomb-crystallized ions, as these energies contribute to the collision energies in the experiment. The average secular kinetic energy can be readily determined from an analysis of the thermal motion of the ions observed in the fluorescence images (see ref. 24). The amplitude of the micromotion, on the other hand, is generally too small to be observable with commonly used imaging systems. However, its contribution to the effective ion kinetic energy can be determined indirectly by simulating the fluorescence images using molecular dynamics (MD) calculations in the fully time-dependent trapping potential and computing the kinetic energies from the ion trajectories.

For the MD simulations we adopted the following model for the force acting on an ion i with charge Q_i:

$$F_{i,tot} = F_{i,trap} + F_{i,Coulomb} + F_{i,LC} + F_{i,heating} \qquad (4)$$

The first term on the right-hand side represents the force $F_{i, trap} = -Q_i \nabla_i V$ arising from the harmonic trapping potential

$$V = \frac{V_{RF}}{2R_0^2}(x^2 - y^2)\cos(\Omega_{RF} t) + \frac{\eta V_{end}}{z_0^2}(z^2 - (1/2)(x^2 + y^2)), \qquad (5)$$

where V_{RF} and Ω_{RF} stand for the RF amplitude and frequency, respectively, R_0 is the closest distance from the central trap axis to the surface of the electrodes, $2z_0$ denotes the distance between the endcaps of the trap, V_{end} is the voltage applied to the endcaps and η represents a geometric parameter.[14] The second term in Eqn 4 represents the Coulomb force between the ions

$$F_{i,Coulomb} = -\frac{1}{4\pi\varepsilon_0}\nabla_i \sum_{i \neq j} \frac{Q_i Q_j}{r_{ij}}, \qquad (6)$$

the third is the laser-cooling force along the z axis $F_i,LC = -\beta \dot{z}_i$ where β is a damping coefficient[25] and the fourth denotes a stochastic force which is introduced to effectively take into account the heating of the ions caused by collisions and by imperfections of the trap. The heating force is computationally implemented by imparting random velocity kicks of a defined magnitude to the ions at every time step in the simulation.

Fig. 6 shows a series of simulated and experimental fluorescence images of Coulomb crystals (a) and their corresponding effective energy distributions (b). The trapping parameters used were $V_{RF} = 125$ V and $\Omega_{RF} = 3.8$ MHz. The thermal (secular) energy of the ions was kept fixed at $E_{sec}/k_B = 17$ mK which is typical of that achieved in our experiments. Under these trapping conditions, a small sample of only 15 ions at low endcap voltage ($\eta V_{end} = 0.07$ V) crystallises as a string located on the central trap axis (Fig. 6 (a) (i)). The corresponding effective-kinetic-energy distribution of the ions (dashed-dotted trace in Fig. 6 (b)) is very narrow with an average kinetic energy of $\langle E_{kin,eff}\rangle/k_B = 22$ mK. A larger ensemble of ≈ 900 ions at $\eta V_{end} = 1.20$ V forms a "prolate" spheroid-shaped crystal consisting of shells of ions with the long axis oriented along the central trap axis (Fig. 6 (a) (ii)). Because most ions are displaced from the central trap axis, the contribution of the micromotion to the effective kinetic energy of the ions is significant and manifests itself in

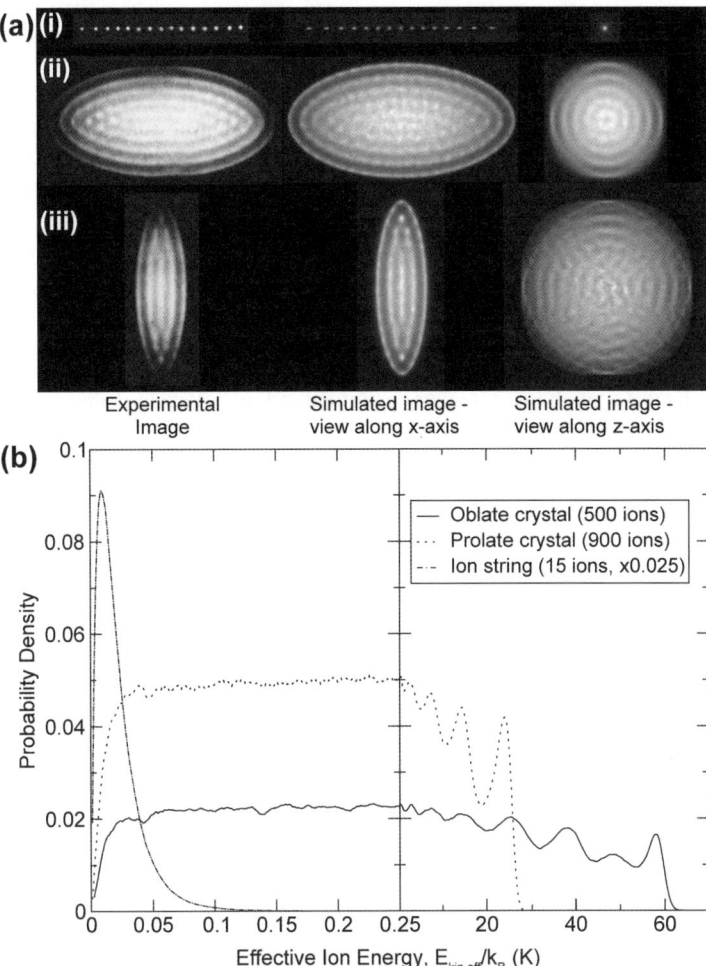

Fig. 6 (a) Experimental and simulated Coulomb crystal images for (i) a string of 15 ions, (ii) a prolate crystal of 900 ions (long axis of spheroid along the trap axis), and (iii) an oblate crystal of 500 ions (long axis perpendicular to the trap axis). (b) Calculated ion kinetic energy distributions for the crystals in (a), including both secular and micromotion. Note the change in the horizontal scale at $E_{kin,eff}/k_B = 0.25$ K.

a broad kinetic energy distribution echoing the shell structure of the Coulomb crystal (dashed trace in (b)). The average kinetic energy of the ions in this crystal is computed to be $\langle E_{kin,eff} \rangle /k_B = 11.9$ K.

At $\eta V_{end} = 2.76$ V, the same ensemble of ions assumes an "oblate" spheroidal form with the long axis oriented perpendicular to the trap axis (Fig. 6 (a) (iii)). In this configuration, a large proportion of the ions is located far from the trap axis where they experience the strong RF fields close to the electrodes. The corresponding kinetic-energy distribution (solid line in Fig. 6 (b)) is comparatively broad with the "hottest" ions exhibiting kinetic energies as high as $E_{kin,eff}/k_B \approx 60$ K. In this crystal the average kinetic energy of the ions amounts to $\langle E_{kin,eff} \rangle /k_B = 26.5$ K.

These results show that the effective kinetic energy of the ions and hence the collision energy in a reactive-scattering experiment depends critically on the geometric shape of Coulomb crystals and thereby on the trapping parameters. A significant change of shape during the course of a reaction, because of changing trapping parameters or a changing composition for example, would lead to a change in the collision energy. To determine the overall effective collision-energy distributions in our experiments, we convoluted the effective ion-kinetic-energy distributions extracted from the MD simulations with the experimental kinetic-energy distributions of the velocity-selected neutral molecules. The resulting collision-energy distribution allowed us to compute average collision energies $\langle E_{coll} \rangle$ for the relevant experiments.

3 Cold chemical reactions with laser- and sympathetically-cooled ions

3.1 Determination of rate constants

For the $Ca^+ + CH_3F \rightarrow CaF^+ + CH_3$ reaction, which serves as a test system for our new technique, a reaction experiment using a prolate-shaped medium sized Ca^+ Coulomb crystal of about 900 ions is presented in Fig. 7. Panel (a) shows the crystal before (left image) and after (right image) the reaction whereby the CH_3F molecules were velocity selected at a quadrupole voltage of $V_0 = 3.0$ kV. Over the course of the reaction, the number of Ca^+ ions decreases as a result of reactive collisions with CH_3F molecules. In addition, a flattening of the remaining Ca^+ crystal can be observed, which is attributed to the formation of shells of sympathetically cooled product ions surrounding the central Ca^+ core. The total energy release of the reaction (≈ 1.05 eV[18]) is smaller than the depth of our ion trap (typically ≥ 1.2 eV) so that a substantial part of the product ions remain trapped (in fact the mass ratio of the products is such that a maximum of only 20% of the energy can be released as kinetic energy of the CaF^+ product). The CaF^+ ions aggregate at the edges of the crystal owing to the shallower depth of the effective trapping potential for heavier species. This change of shape is indicative of the formation of bi-component Coulomb crystals,[26] which is also evidenced by the simulated images of the Ca^+/CaF^+ bi-component crystals shown in panel (b) of Fig. 7. Panel (c) shows the same simulated images in which the sympathetically cooled CaF^+ ions have been made visible. Because of the finite depth of our imaging system, the images only

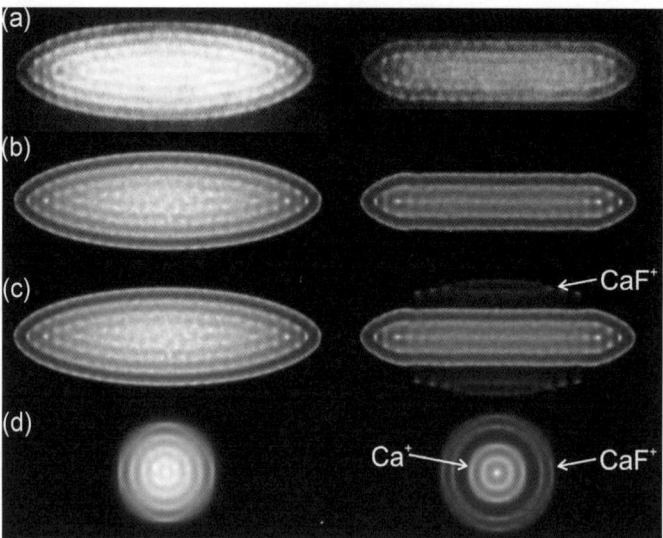

Fig. 7 (a) Experimental image of a 900-ion prolate crystal before (left) and after (right) the reaction with CH_3F. (b) Molecular dynamics simulation of the Coulomb crystals in (a). CaF^+ ions are not shown. (c) As (b) except that the CaF^+ ions are made visible. (d) End view of the simulated crystals before and after reaction.

show a slice through the three-dimensional Coulomb crystals and the shells of CaF^+ ions appear as "caps" enclosing the central Ca^+ core. Panel (d) shows a view of the simulated crystals along the trap axis in which the shell structure of the sympathetically cooled ions becomes apparent.

Rate constants for chemical reactions between laser-cooled ions and velocity selected neutrals are determined by monitoring the decrease of the number of Ca^+ ions as a function of reaction time as shown in our preliminary publication.[18] The number of Ca^+ ions is directly determined from the fluorescence images. However, because the ions cannot be individually counted in images of large crystals containing hundreds of ions as used in the present study, the number of ions was established by (a) a comparison with simulated images as demonstrated previously by Zhang et al.[24] and as illustrated in Fig. 7, or (b) a measurement of the apparent volume of the Ca^+ ion core by assuming a constant ion density (the "fluid model"[18,27]). The bimolecular rate constant is determined from a fit of the reaction-time dependent number of Ca^+ ions to a pseudo-first-order rate law as discussed in our preliminary report.[18]

3.2 Dynamics of the Ca^+ + CH_3F reaction

The rate constants obtained for the Ca^+ + CH_3F reaction using our new method with a 250-ion crystal are $k = 1.3(6) \times 10^{-9}$ cm^3 s^{-1} for $V_0 = 3.0$ kV (corresponding to $E_{coll}/k_B \approx 3$ K) and $k = 4.2(4) \times 10^{-10}$ cm^3 s^{-1} for $T = 300$ K as already reported in our preliminary publication.[18] In Fig. 8 a comparison of the experimental results with theoretical calculations is presented. The temperature-independent Langevin

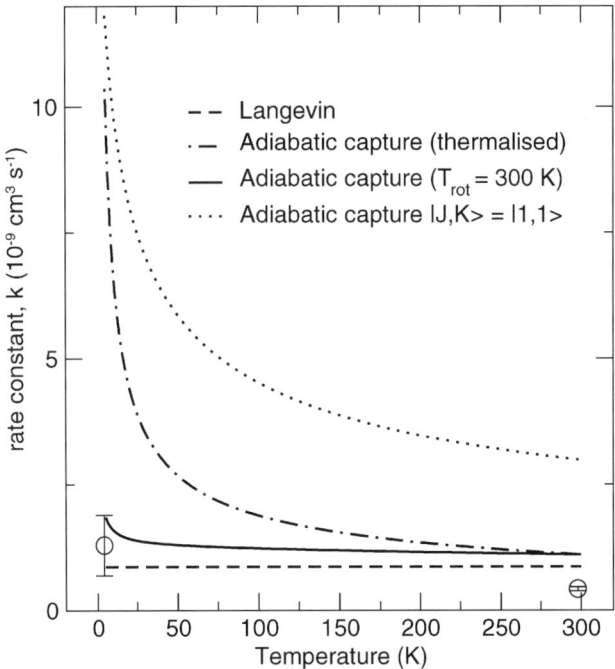

Fig. 8 Calculated and observed (circles with error bars) rate constants for the Ca^+ + CH_3F reaction. The calculated rates were obtained using Langevin theory (dashed) and adiabatic capture theory. The adiabatic capture rate constants are calculated assuming different rotational temperatures for the methyl fluoride reactant: dashed-dotted line, translational and rotational temperatures equal; solid line, 300 K rotational temperature; dotted line, state-selected $|J, K\rangle = |1, 1\rangle$.

rate constant is computed to be $k_L = 9.4 \times 10^{-10}$ cm^3 s^{-1} (dashed horizontal line). The other curves represent predictions of the temperature-dependent rate constant using the adiabatic-capture theory of Clary for the reaction of an ion with a strongly polar molecule.[28] These calculations were performed by diagonalising the Hamiltonian matrix for the charge–dipole and charge-induced-dipole interactions in a symmetric top basis over a grid of ion-molecule separations to obtain rotationally-adiabatic potential energy curves. Classical capture theory was then used to calculate state-selected cross sections for the full range of total angular momentum values, and these cross sections used to obtain thermally averaged rate constants.

The solid line represents the calculated temperature dependence of the rate constant whereby the neutral molecules exhibit a thermal rotational state distribution at $T_{rot} = 300$ K (which is the case relevant for the present study). For comparison, the dashed-dotted line shows the rate constant calculated for a sample with thermal internal state distribution at a temperature of $T_{rot} = T_{trans}$ and the dotted line represents k for a fully state-selected reaction with the CH$_3$F molecules in the $|J, K\rangle = |1, 1\rangle$ rotational state. The measured rate constants are systematically smaller than those predicted by adiabatic capture theory. Moreover, the room-temperature value is almost a factor of three smaller than the one predicted by Langevin theory which can be regarded as a lower limit for the rate constant within the framework of simple ion-molecule capture models.

The present results indicate that the reaction of calcium ions with fluoromethane is considerably slower than expected on the basis of simple ion-molecule capture models. The same conclusion was already reached previously by Harvey *et al.*, in a combined theoretical and experimental study of the reaction at room temperature.[29] Based on the results of their *ab initio* calculations Harvey *et al.* attributed the slow rate of the reaction to the presence of a submerged barrier along the reaction coordinate, which acts as a bottle neck for the reactive flux from the reactants to the products. However, as shown below the involvement of the excited electronic states of the calcium ion in the present experiments, for which a steady state population is produced in the laser-cooling process, must also be considered in interpreting the experimental results (see Section 3.3).

3.3 Dependence of the rate constant on experimental parameters

In general the rate of reaction is dependent on the quantum state and collision energy of the reaction partners. In our experiment the velocity-selected neutrals are translationally cold, but exhibit a broad distribution of rotational-state population in the vibrational ground state (see Section 2.2.2). The cold atomic ions, on the other hand, exhibit a substantial degree of electronic excitation owing to the photo-excitation during laser cooling. Thus, the measured rate constant represents an average over a range of rovibronic reaction channels and collision energies.

While we cannot change the internal-state distributions of the velocity-selected neutrals with our present experimental setup, we have studied the dependence of the rate constant on the electronic-state populations of the Ca$^+$ ions, which are dependent on the detuning of the 397 nm cooling laser from the 4s(^{2}S$_{1/2}$) → 4p(^{2}P$_{1/2}$) transition frequency. Fig. 9 shows how the rate of loss of reactant ions decreases as the laser detuning increases for a 350-ion crystal. In this experiment, the powers for the 397 nm and 866 nm laser beams were held fixed and the CH$_3$F molecules were velocity selected at $V_0 = 5.0$ kV. A similar decrease in rate constant was also observed for the room temperature reaction, in which the reactant CH$_3$F is leaked into the ion trap chamber. The measured rate constants are shown in the inset of the figure. Using tabulated Einstein coefficients for the 4s(^{2}S$_{1/2}$) → 4p(^{2}P$_{1/2}$) and 3d(^{2}D$_{3/2}$) → 4p(^{2}P$_{1/2}$) transitions[30] the relative populations in the excited ^{2}P$_{1/2}$ and ^{2}D$_{3/2}$ states are estimated to be 27% for $\Delta\nu = 19$ MHz and 10% for $\Delta\nu = 114$ MHz. These calculated populations can be used to derive rate constants for the ground state and the combined (3d and 4p) excited state reaction. At room

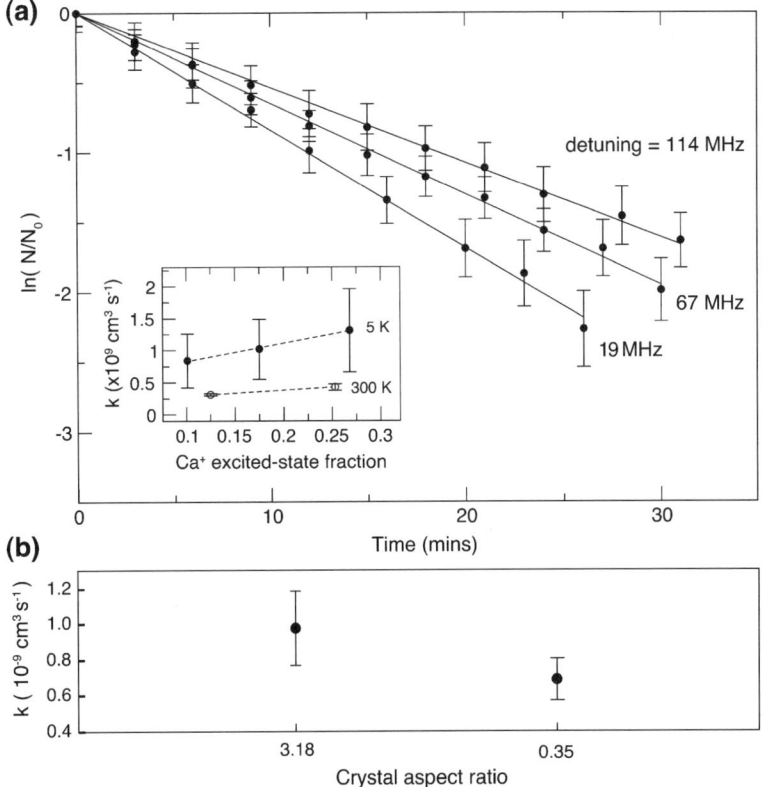

Fig. 9 (a) Experimentally measured rate plots for the Ca$^+$ + CH$_3$F reaction using a prolate crystal and quadrupole guide voltage of $V_0 = \pm 5$ kV for various cooling-laser detunings. N is the number of ions in the crystal and N_0 the initial number at the start of the reaction. The inset shows the derived rate constants as a function of Ca$^+$ excited state (4p and 3d) fraction. (b) Experimentally determined rate constants for a prolate crystal (aspect ratio 3.18) and an oblate crystal (0.35) with the neutral molecules guided using $V_0 = 3$ kV.

temperature the ground state reaction rate constant $k = 0.2 \times 10^{-9}$ cm^3 s^{-1} is found to be a factor of ≈ 6 smaller than the value predicted by adiabatic capture theory in agreement with the previous work of Harvey et al.[29] The rate constant derived for the excited states is $k = 1.1 \times 10^{-9}$ cm^{3}s^{-1}, which is close to the adiabatic capture theory prediction. The effect of the submerged barrier on the ground state reaction has been discussed above, but these measurements suggest that the excited state reaction is barrierless. For the low temperature ($T_\text{trans} \approx 3$ K) measurements the rate constants determined are $k = 0.5 \times 10^{-9}$ cm^3 s^{-1} and $k = 3.4 \times 10^{-9}$ cm^3 s^{-1} for the ground and excited states, respectively. The ratio of rate constants is similar to the 300 K measurements and the increase of the excited-state rate constant with decreasing temperature is in reasonable agreement with the approximately two-fold increase predicted by adiabatic capture theory. However it should be noted that, as indicated in Fig. 9, the error bars are somewhat larger at lower temperature because of the uncertainty in the neutral molecule flux calibration.

Fig. 9 (b) shows the rate constant obtained for different aspect ratios, defined as the ratio of the two principal axes of a spheroidal Coulomb crystal. Large aspect ratios indicate a prolate crystal with a narrow ion-kinetic-energy distribution, while low aspect ratios define oblate crystals with broad kinetic-energy distributions containing comparatively hot ions (see Fig. 6). In this experiment the trapping parameters, crystal sizes and crystal aspect ratios were chosen so that the collision energies were clearly dominated by the ion kinetic energies ($\langle E_{\text{coll}}/k_B \rangle = 14$ K and 29 K for the prolate and oblate crystals, respectively). The rate constants obtained for velocity-selector voltages $V_0 = 3.0$ kV and cooling-laser detunings $\Delta \nu = 20$ MHz agree within the uncertainty limits indicating only a weak dependence on the collision energy in the studied range.

3.4 OCS$^+$ + ND$_3$ reaction

In principle the setup described above can be extended to a wide range of ion–molecule reactions involving Ca$^+$ (or other alkaline earth ions) and a range of dipolar neutral species, such as those transmitted by the quadrupole in Fig. 2. Recent studies have indicated the potential importance of Ca$^+$ reactions in the upper atmosphere.[31] However the set up can also be used to study reactions of sympathetically cooled molecular ions, and we report here a preliminary study of the reaction of OCS$^+$ with ND$_3$ as an example. In these experiments, OCS$^+$ ions are generated in the centre of the ion trap by (2 + 1) Resonance-Enhanced Multiphoton Ionization (REMPI) at 280.15 nm[32] of a low partial pressure of OCS gas (10^{-8} mbar) leaked into the ion trap chamber. The weakly focused 280 nm beam, which is the frequency doubled output of nanosecond Nd:YAG-pumped dye laser, crosses the trap axis perpendicularly and is typically admitted for around 60 s with an energy of 0.8 mJ per pulse at 10 Hz. OCS$^+$ ions are sympathetically cooled into the crystal and form a layer around the Ca$^+$ ions, which is shown by the apparent flattening of the crystal shown in Fig. 10 (b). Some S$^+$ ions are also generated by photodissociation of the OCS$^+$ [32] and these form a dark core of the crystal, being of lower mass than the Ca$^+$ ions. The ions are exposed to a room temperature sample of ND$_3$ (with which the Ca$^+$ ions do not react) and a dark core expands as a consequence of the formation and sympathetic cooling of ND$_3$ ions via the charge transfer reaction OCS$^+$ + ND$_3$ → OCS + ND$_3^+$. Molecular dynamics simulations of the multicomponent crystals enable the determination of the number of ions of each type present (by comparing simulated and experimental images) as a function of time and hence the determination of a room temperature rate constant. The value obtained, $k = 1.8(2) \times 10^{-9}$ cm^3 s^{-1} is in agreement with the previous experimental value from ref. 33. The experiment was repeated using a sample of ND$_3$ guided by the velocity

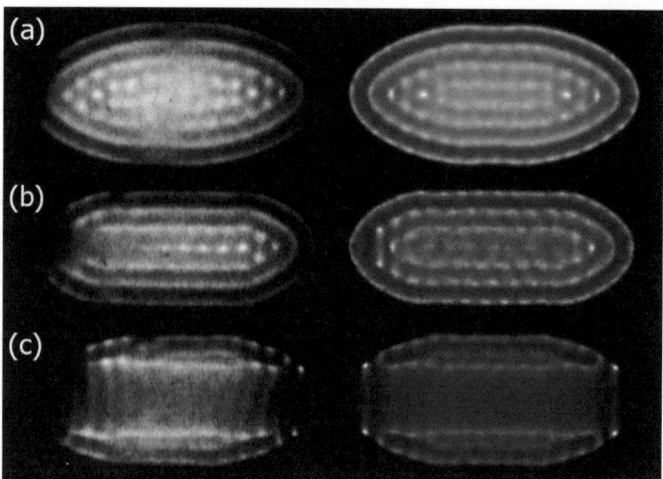

Fig. 10 Coulomb crystal images observed in the course of the room temperature OCS$^+$ + ND$_3$ reaction (left) and simulated images (right). (a) Ca$^+$ Coulomb crystal before production of OCS$^+$. (b) With sympathetically cooled OCS$^+$. (c) After reaction with ND$_3$ with ND$_3^+$ product ions in the core.

selector at $V_0 = 3$ kV. A rate constant of $k = 2.0(5) \times 10^{-9}$ cm^3 s^{-1} was obtained from these preliminary measurements, which correspond to an effective translational temperature of about 5 K. This work together with the previous experiments of Staanum et al.[34] and Roth et al.[35] provide a demonstration of the wide ranging potential of this technique.

4 Discussion and future outlook

In this paper we have presented new results that illustrate the potential of using Coulomb crystals in combination with a neutral molecule velocity selector for the study of cold chemical processes. The quadrupole velocity selector is a relatively simple device to implement and to operate and has the flexibility to guide a wide range of polar species as illustrated in this work. The experimental setup allows the reactions of the calcium ions to be studied directly with single-particle sensitivity as illustrated by the reaction with fluoromethane. One complication illustrated by the results for this reaction is the significant steady-state population of the excited electronic state of the ions which may contribute to the reaction. However such effects can be partly disentangled from the ground state reaction as illustrated here by systematically varying the laser cooling conditions (and hence varying the populations). The facile sympathetic cooling of molecular ions into these crystals, and the ability to simulate the well defined properties of the crystals, determined only by the masses of species present and the applied fields, allows the use of the Coulomb crystal as a framework within which other types of reaction can be studied, as illustrated for the OCS$^+$ + ND$_3$ reaction. As shown by Staanum et al.[34] the most precisely defined conditions for studying reactions could be obtained by reducing the crystal size to a single pair of ions, one calcium ion and one molecular ion, so that very precise mass analysis and temperature control is achievable.

It is seen that the definition of temperatures for these collisional experiments is not a straightforward matter. The effective ion temperature is seen to vary with the radial position of the ions, and therefore the best control and the lowest temperature would be obtained using long strings of ions. For larger crystals the average temperature could vary over the course of the reaction as the size of the reactive shell decreases. The quadrupole guide selects molecules within a range of kinetic energies but the vibration–rotation population is only partially selected. A general limitation of the velocity-selection technique is that the properties of the effusive source fundamentally limit the fluxes and densities of cold molecules that can be produced. Increasing the source pressure leads to "boosting" of the velocity distribution which depletes the beam of molecules with low kinetic energies, whilst cooling the gas to compress the Maxwell–Boltzmann distribution increases the guiding efficiency but is eventually limited by condensation of the molecules inside the effusive nozzle. A solution to these problems is to combine the effusive source with a buffer gas cooling cell containing helium at a temperature of around 5 K. Such a modification is particularly appealing as the internal degrees of freedom of the molecules can be cooled to enhance the quantum-state purity of the beam. Indeed, recent work by van Buuren et al.[36] has shown that genuinely (both internally and translationally) cold beams of formaldehyde can be produced using this approach. Depletion spectroscopy measurements demonstrated that up to 80% of the guided H$_2$CO molecules resided in the low-field seeking state of the $|J, K_{-1}, K_{+1}\rangle = |1, 1, 0\rangle$ level. By tuning the density of helium atoms near the nozzle exit, the population in the other low-lying rotational states could also be increased, which suggests the interesting possibility of controlling the state distributions for collisional studies. As the translational degrees of freedom are cooled by the buffer gas, the emittance of the initial beam can be better matched to the acceptance of the guide and these experiments showed fluxes approaching 10^{11} molecules s^{-1}, with densities inside the guide of around 10^9 molecules cm^{-3}.

Weaker two-dimensional ac trapping has also been demonstrated by applying an alternating dipolar voltage configuration to the quadrupole electrodes.[37] Although molecules in both low-field and high-field seeking states can be guided this way, the well-depths associated with this kind of electrodynamic trapping are only of the order of 20 mK and so very much lower fluxes of molecules are produced. The operation of the guide with ac voltages is reminiscent of a quadrupole mass filter for ions and the selection of molecules and atoms according to their electric- or magnetic-dipole moment-to-mass ratio closely resembles the mass-to-charge ratio selection of ions in mass spectrometry. Consequently, one can imagine filtering different conformations of molecules (particularly biomolecules) according to their individual dipole moments. Such an experiment has been recently demonstrated using the direct neutral-molecule analogue of a quadrupole mass filter to spatially separate two conformers of 3-aminophenol formed in a supersonic molecular beam.[38]

Possible future developments of this experiment include the replacement of the velocity selector with a Stark decelerator.[5] This would provide a comparable flux of translationally cold neutral molecules to the quadrupole velocity selector but with a much narrower and precisely controllable velocity distribution[39] and the selection of molecules in specific quantum states. It is likely that collision energies well into the millikelvin range would be achievable in combination with a string of cold ions.

State selection of the molecular ions is, in principle, achievable *via* appropriate REMPI processes suggesting the possibility to study completely state selected reactions.[40,41] However, in a room temperature ion trap, the rotational distribution of most molecular ion species will thermalise on a short timescale compared to the very long timescale of these experiments. This problem could be mitigated by cooling the trap environment to cryogenic temperatures thus reducing the intensity of the black-body radiation field, or by working with non-polar ions. Translationally colder ions could in principle be generated using sideband cooling of the motional states of the calcium ions within a trap. And a recent experiment has demonstrated the possibility to combine laser cooling in an ion trap with laser cooled atoms in a MOT to study atomic charge transfer process of Yb atoms and Yb ions.[42]

In conclusion, the present work illustrates that the extremely high sensitivity available in experiments with Coulomb crystals will enable a detailed experimental probing of chemical reaction dynamics in the cold, and possibly ultracold temperature ranges. Ion–molecule reactions have dynamical properties that may differ significantly from neutral–neutral reactions, but the predominance of ion–molecule chemistry in low-temperature astrophysical environments reinforces the importance of understanding the dynamics of this category of chemical reaction.

5 Acknowledgments

The authors are grateful to the EPSRC (Portfolio Partnership Grant GR/S71750) for financial support of this work.

References

1 M. T. Bell and T. P. Softley, *Mol. Phys.*, 2009, **107**, 99.
2 P. F. Weck and N. Balakrishnan, *Int. Rev. Phys. Chem.*, 2006, **25**, 284.
3 R. Krems, *Phys. Chem. Chem. Phys.*, 2008, **10**, 4079.
4 P. Rabl, D. DeMille, J. M. Doyle, M. D. Lukin, R. J. Schoelkopf and P. Zoller, *Phys. Rev. Lett.*, 2006, **97**, 033003.
5 S. Y. T. van de Meerakker, H. L. Bethlem and G. Meijer, *Nature Phys.*, 2008, **4**, 595.
6 N. Vanhaeke, U. Meier, M. Andrist, B. H. Meier and F. Merkt, *Phys. Rev. A*, 2007, **75**, 031402.
7 R. Fulton, A. I. Bishop and P. F. Barker, *Phys. Rev. Lett.*, 2004, **93**, 243004.

8 J. Deiglmayr, A. Grochola, M. Repp, K. Mörtlbauer, C. Glück, J. Lange, O. Dulieu, R. Wester and M. Weidemüller, *Phys. Rev. Lett.*, 2008, **101**, 133004.
9 F. Lang, K. Winkler, C. Strauss, R. Grimm and J. Hecker Denschlag, *Phys. Rev. Lett.*, 2008, **101**, 133005.
10 K.-K. Ni, S. Ospelkaus, M. H. G. de Miranda, A. Pe'er, B. Neyenhuis, J. J. Zirbel, S. Kotochigova, P. S. Julienne, D. S. Jin and J. Ye, *Science*, 2008, **322**, 321.
11 M. Elioff, J. Valentini and D. W. Chandler, *Science*, 2003, **302**, 1940.
12 W. C. Campbell, E. Tsikata, Hsin-I. Lu, L. D. van Buuren and J. M. Doyle, *Phys. Rev. Lett.*, 2007, **98**, 213001.
13 S. A. Rangwala, T. Junglen, T. Rieger, P. W. H. Pinkse and G. Rempe, *Phys. Rev. A*, 2003, **67**, 043406.
14 S. Willitsch, M. T. Bell, A. Gingell and T. P. Softley, *Phys. Chem. Chem. Phys.*, 2008, **10**, 7200.
15 M. Drewsen, I. Jensen, J. Lindballe, N. Nissen, R. Martinussen, A. Mortensen, P. Staanum and D. Voigt, *Int. J. Mass Spectrom.*, 2003, **229**, 83.
16 E. R. Hudson, C. Ticknor, B. C. Sawyer, C. A. Taatjes, J. J. Lewandowski, J. R. Bochinski, J. L. Bohn and J. Ye, *Phys. Rev. A*, 2006, **73**, 063404.
17 J. J. Gilijamse, S. Hoekstra, S. Y. T. van de Meerakker, G. C. Groenenboom and G. Meijer, *Science*, 2006, **313**, 1617.
18 S. Willitsch, M. T. Bell, A. Gingell, S. R. Procter and T. P. Softley, *Phys. Rev. Lett*, 2008, **100**, 043203.
19 T. Junglen, T. Rieger, S. A. Rangwala, P. W. H. Pinkse and G. Rempe, *Eur. Phys. J. D*, 2004, **31**, 365.
20 M. Motsch, L. D. van Buuren, C. Sommer, M. Zeppenfeld, G. Rempe and P. W. H. Pinkse, *Phys. Rev. A*, 2009, **79**, 013405.
21 M. Motsch, M. Schenk, L. D. van Buuren, M. Zeppenfeld, P. W. H. Pinkse and G. Rempe, *Phys. Rev. A*, 2007, **76**, 061402.
22 D. Gerlich, *Adv. Chem. Phys.*, 1992, **82**, 1.
23 J. P. Schiffer, *J. Phys. B: At. Mol. Opt. Phys.*, 2003, **36**, 511.
24 C. B. Zhang, D. Offenberg, B. Roth, M. A. Wilson and S. Schiller, *Phys. Rev. A*, 2007, **76**, 012719.
25 H. J. Metcalf and P. van der Straten, *Laser Cooling and Trapping*, Springer, New York, 1999.
26 L. Hornekær, N. Kjærgaard, A. M. Thommesen and M. Drewsen, *Phys. Rev. Lett.*, 2001, **86**, 1994.
27 U. Fröhlich, B. Roth and S. Schiller, *Phys. Plasmas*, 2005, **12**, 073506.
28 D. C. Clary, *Chem. Phys. Lett.*, 1995, **232**, 267.
29 J. N. Harvey, D. Schröder, W. Koch, D. Danovich, S. Shaik and H. Schwarz, *Chem. Phys. Lett.*, 1997, **273**, 164.
30 D. F. V. James, *Appl. Phys. B*, 1998, **66**, 181.
31 J. M. C. Plane, T. Vondrak, S. Broadley, B. Cosic, A. Ermline and A. Fontijn, *J. Phys. Chem.*, 2006, **110**, 7874.
32 R. A. Morgan, A. Orr-Ewing, D. Ascenzi, M. N. R. Ashfold, W. J. Buma, C. R. Scheper and C. A. de Lange, *J. Chem. Phys.*, 1996, **105**, 2141.
33 D. Smith, N. G. Adams and W. Lindinger, *J. Chem. Phys.*, 1981, **75**, 3365.
34 P. F. Staanum, K. Højbjerre, R. Wester and M. Drewsen, *Phys. Rev. Lett.*, 2008, **100**, 243003.
35 B. Roth, P. Blythe, H. Wenz, H. Daerr and S. Schiller, *Phys. Rev. A*, 2006, **73**, 042712.
36 L. D. van Buuren, C. Sommer, M. Motsch, S. Pohle, M. Schenk, J. Bayerl, P. W. H. Pinkse and G. Rempe, *Phys. Rev. Lett.*, 2009, **102**, 033001.
37 T. Junglen, T. Rieger, S. A. Rangwala, P. W. H. Pinkse and G. Rempe, *Phys. Rev. Lett.*, 2004, **92**, 223001.
38 F. Filsinger, U. Erlekam, G. von Helden, J. Küpper and G. Meijer, *Phys. Rev. Lett.*, 2008, **100**, 133003.
39 C. E. Heiner, H. L. Bethlem and G. Meijer, *Phys. Chem. Chem. Phys.*, 2006, **8**, 2666.
40 S. R. Mackenzie and T. P. Softley, *J. Chem. Phys.*, 1994, **101**, 10609.
41 S. Willitsch and F. Merkt, *Int. J. Mass Spectrom*, 2005, **245**, 14.
42 A. T. Grier, M. Cetina, F. Oručević, V. Vuletić, arXiv:0808.3620v1.

General discussion

Dr Lane opened the discussion of the paper by Professor Herschbach: Chemists are particularly interested in the transition state region of a reaction. What prospects do you see for ultracold molecules in this area of research? Would their usefulness depend on whether the reaction has an early or late barrier?

Professor Herschbach replied: The transition state of a reaction, as Dr Lane points out, is indeed a chief concern of chemists. It refers to a critical stage that the transient liaison of the reactants must reach in order to metamorphose into products. Most often, the transition state is considered to be at the crest of an activation energy barrier separating the entrance and exit valleys of the potential energy surface. The barrier is termed "early" if its crest lies in the entrance valley, "late" if in the exit valley. Traversing an early barrier requires reactant translational kinetic energy at least slightly in excess of the barrier height, whereas a similar level of reactant vibrational excitation is ineffective. In contrast, for crossing a late barrier reactant vibrational excitation is highly effective but translation is not. This pattern, and more nuanced aspects, was demonstrated particularly well by John Polanyi and his colleagues in extensive classical trajectory simulations of reactions on a variety of potential surfaces.[1]

In ultracold collisions, chemical reactions cannot be expected to occur at an appreciable rate unless an activation energy barrier is lacking (as in the reactions of open-shell atoms or free radicals mentioned in my paper) or tunneling through the barrier is unusually facile (as for F + H_2, enhanced by very small reduced mass, low and narrow barrier, and strong resonance). There would thus seem to be no practical prospect of ultracold probing of transition states of the sort usually considered, located at barrier crests. However, that may hold only if translational kinetic energy is required to reach the crest. The crest region could instead be accessed in reactions of ultracold reactant molecules prepared in a vibrationally excited state. (Now achieved for alkali dimers, but not pertinent here as their reactions are barrierless.) According to the Polanyi pattern, this would be expected to enable reactions governed by a late barrier but not by an early barrier. That pattern was demonstrated in the classical realm, so invites a theoretical study to see whether or not it holds also for ultracold collisions.

1 J. C. Polanyi, *Chem. Scr.*, 1987, **27**, 229; J. C. Polanyi, *Science*, 1987, **236**, 68.

Professor Stwalley addressed Professor Herschbach: I wanted to point out that in the reactions of ultracold molecules, the uppermost final rotational state populated can be limited because of angular momentum rather than energy constraints. For example,[1] in consideration of the reaction of translationally ultracold KRb($v = 0$, $j = 0$) with K, the exoergicity to form K_2($v' = 0$, $j' = 0$) is 225 cm^{-1}. If the exoergicity is put into rotation instead, K_2 molecules with j' up to 62 could be formed. However, since the initial collisional angular momentum l must be small (say $l < 4$) at ultracold temperatures to surmount the centrifugal barriers in the entrance channel and since $j = 0$, the total angular momentum (excluding spins) must be small; consequently, a final $j' = 62$ must be accompanied by a final l' of ~60 to conserve the small total angular momentum. However, the exit channel centrifugal barrier for $l' \sim 60$ is then likely to be sufficiently large to prevent the products [Rb + K_2($v' = 0$, $j' = 62$)] from separating.

1 D. Wang *et al.*, *Eur. Phys. J. D*, 2004, **31**, 165.

Professor Herschbach responded: For A + BC reactions of low exoergicity that proceed *via* an attractive short-range potential well, the product departure is often inhibited by centrifugal barriers in the exit valley. The inhibition introduces a characteristic form factor in the product kinetic energy distribution.[1] As Professor Stwalley points out, the exit angular momentum constraint, $l' = -j'$, imposed in the ultracold limit might be expected to suppress to some extent the population of rotational states of the AB product molecule.

The maximum value of the product rotational excitation $E_R'(j') = B_v'j'(j' + 1)$ and corresponding momentum allowed by energy conservation is given by $\Delta E \geq E_R'(j')$, where ΔE is the reaction exoergicity to form the product vibrational state v', with $j = 0$, and B_v' is the rotational constant of the AB molecule in the v' state. The question is whether a substantially lower limit on j' results from the requirement that the product translational kinetic energy, $E_T' = \Delta E - E_R'(j')$, exceeds the exit centrifugal barrier $V_{CB}(l')$, with the exit orbital momentum specified by $l' = j'$.

Pertinent evidence and estimates indicate that for ultracold reactions of alkali atoms with dimers, the allowed range of j' is generally not much affected by the centrifugal barrier. In a review of quantum dynamics calculations,[2] Hutson and Soldán display product rotational distributions resulting from inelastic/reactive vibrational quenching collisions of spin-polarized $M + M_2(v = 1, j = 0) \rightarrow M + M_2(v' = 0, j')$. These occur on quartet surfaces with deep attractive wells. Both for M = Na and K, the rotational distributions extend to the maximum j' allowed by energy conservation: 20 and 24, respectively. Subject to symmetry restrictions (only even j' for bosons, odd for fermions), all accessible levels are populated. These results are consistent with the usual formula for estimating $V_{CB}(l')$ for an R^{-6} exit potential,

$$V_{CB}(l') = 0.38[l'(l' + 1)/\mu']^{3/2}C_6^{-1/2},$$

where V_{CB} is in cm^{-1}, the AB + C reduced mass μ' in amu and the van der Waals coefficent C_6 in atomic units (hartree-bohr^6). For the Na case, $V_{CB}(20) = 0.70$ cm^{-1} (with $C_6 \sim 6000$ a.u.), whereas $\Delta E = 23.5$ cm^{-1}. For the K case, $V_{CB}(24) = 0.44$ cm^{-1} (with $C_6 = 9050$ a.u.), whereas $\Delta E \sim 16$ cm^{-1}.

The reaction considered by Stwalley,

$$^{39}K + {}^{39}K^{85}Rb(v = 0, j = 0) \rightarrow {}^{85}Rb + {}^{39}K_2(v' = 0, j'),$$

occurs on the ground singlet surface, also likely subject to a strongly attractive well. With $\Delta E = 225$ cm^{-1} and $B_0 = 0.05666$ cm^{-1}, energy conservation allows K$_2$ levels up to $j' = 62$. A value for C_6(Rb–K$_2$) is not available; however, it is surely larger than C_6(Rb–K) ~ 4200 a.u.[3] Thus the latter value can be used to obtain an upper bound: $V_{CB}(62) < 6$ cm^{-1}. Deducting the requisite $E_T' \sim 6$ cm^{-1} from ΔE would then at most reduce the maximum allowed rotational excitation of K$_2$ to $j' = 61$.

1 S. A. Safron, N. D. Weinstein, D. R. Herschbach and J. C. Tully, *Chem. Phys. Lett.*, 1972, **12**, 564.
2 J. M. Hutson and P. Soldán, *Mol. Phys.*, 2007, **26**, 1.
3 H. L. Kramer and D. R. Herschbach, *J. Chem. Phys.*, 1970, **53**, 2792.

Dr Hudson opened the discussion of Dr Bethlem's paper: I wonder about systematic errors in the proposed measurement. Consideration of statistical sensitivity is important for establishing a bound on the experiment's sensitivity. But it is essential to also carefully consider possible systematic effects to truly estimate an experiment's potential. In particular, it would be interesting to make a quantitative estimate of the effect of the Stark shift and the consequent level of electric field control required.

The variety of transitions on offer in a molecule might in this respect be advantageous. As you suggest, one can imagine finding transitions whose frequencies can be combined in a way that eliminates the Stark effect. It would be interesting, then, to expand upon Table 1 by showing each transition's relative Stark shift.

It might be particularly useful to consider systematic effects in the context of choosing an ideal molecule. The relative merits of candidate molecules could appear somewhat different in this light.

Dr Bethlem answered: The Stark shift decreases rapidly as a function of J. For the considered transitions it is on the order of 1 Hz per $(V/cm)^2$. Note that the Stark shift is similar for all transitions in the table as all these transitions involve the same rotational levels – although being in different vibrational states or in different isotopes. In order to monitor the residual electric field we plan to use (1-photon) Λ-doublet transitions in a low J level. Note that in the proposed experiment we compare transitions in the $v=0$ and $v=1$ state such that residual Stark shifts will in fact cancel to first order, as will many other systematic effects.

Professor Dr van der Avoird enquired: In the considerations in your paper you assume that the frequencies of the molecular vibrations and rotations are purely determined by the nuclear masses. It has been shown, however, both by theory[1] and by computations of Schwenke and Tennyson on H_2O, that due to non-Born–Oppenheimer effects these frequencies are not determined by the masses of the bare nuclei, but by atomic masses in which the mass of the electrons has to be included to some extent. What make things complicated is that the effective atomic masses turn out to be different for different vibrations and again different for the rotations. If the ratio of the atomic masses and the nuclear masses were constant, this would not affect your results, I presume, but if this ratio varies even by a small extent (due to a time variance of the electromagnetic interactions, for example) this is perhaps a serious problem in the determination of the proton/electron mass ratio. Have you considered this problem?

1 W. Kutzelnigg, *Mol. Phys.*, 2007, **105**, 2627.

Dr Bethlem responded: In principle such considerations should be taken into account. However one should realize that knowledge of the K-sensitivity coefficients beyond accuracies of a few percent is not necessary as long as no variation is detected. For the case of H_2 the influence of non-Born–Oppenheimer (rotational-electronic interactions between the $B^1\Sigma_u^+$ and $C^1\Pi_u$ states) effects was included in the derivation of sensitivity coefficients. In addition the influence of adiabatic (mass-dependent) effects in the H_2 molecule on the resulting values for the K-coefficients was explicitly calculated and found to be less than 1% of the actual K values.[1]

The representation of the level energies is a separate issue. In virtually all cases the experimental accuracy on level energies of molecules is better than their *ab initio* representations. One strategy for determining K values may start from a semi-empirical representation of level energies (*e.g.* in a Dunham expansion); the intrinsic problem is then how the coefficients of the model scale with varying proton to electron mass ratio. In the case of a Dunham model this is well-known, even for the mass-scaling of ro-vibrational coupling terms. An alternative strategy for determining K values is to compute level energies over a range of values for m, as was done by Meshkov *et al.*[2] for H_2, and by Ivanov *et al.*[3] for HD. Here the shortcoming is that the representation of level energies is not as good, specifically for hydrogen at the 1 cm^{-1} accuracy level. For the example of H_2 both methods[1,2] compare to the few percent accuracy level.

1 W. Ubachs, R. Buning, K. S. E. Eikema and E. Reinhold, *J. Mol. Spectrosc.*, 2007, **241**, 155.
2 V. V. Meshkov, A. V. Stolyarov, A. Ivanchik and D. A. Varshalovich, *JETP Lett.*, 2006, **83**, 303.
3 T. I. Ivanov, M. Roudjane, M. O. Vieitez, C. A. de Lange, W. U. L. Tchang-Brillet and W. Ubachs, *Phys. Rev. Lett.*, 2008, **100**, 093007.

Dr Küpper asked: In your paper you discuss the sensitivity of change of μ and suggest the use of transitions between different Ω-ladders in CO. Did you consider transitions between different nuclear spin ladders? I understand these transitions are weak, but you are looking for narrow transitions to reach high resolution anyway. What would be the sensitivity for microwave-transitions between rotational levels in different nuclear spin symmetry ladders, *i.e.*, in molecules like ammonia?

Dr Bethlem replied: The energy level schemes of symmetric top molecules look – at a first glance – very similar to those of molecular radicals. However, whereas in radicals the energy difference between the different rotational manifolds (Ω-ladders) is due to the motion of the electrons, in symmetric top molecules the energy difference between different rotational manifolds (K-ladders) is due to the rotation along the different symmetry axes of the molecule. The energy difference between different K-ladders and the energy difference between rotational levels within a single K-ladder have the same dependence on the fundamental constants. As a consequence, transitions between different K-ladders do not show an anomalously large isotope shift as was explicitly checked for different isotopes of ammonia.

Mr Lemeshko addressed Dr Bethlem: You assume that the neutron-to-electron mass ratio changes at the same rate as the proton-to-electron one. Is it also predicted by theories beyond the Standard Model?

Dr Bethlem said: The proton and neutron are much heavier than the quarks they are composed of; the main contribution to their mass comes, in fact, from the interaction between the quarks – the number of gluons present inside the proton and neutron, if you like. As a consequence, the proton and neutron masses are a measure of the strength of the strong force and are expected to change at the same rate, at least in first order. Note that the electron is not a composite particle, and its mass arises from the electroweak sector. Hence, the proton to electron mass ratio probes phenomena of the hadronic sector with respect to those of the electroweak sector.[1,2]

1 V. V. Flambaum, private communication.
2 V. V. Flambaum, D. B. Leinweber, A. W. Thomas, and R. D. Young, *Phys. Rev. D*, 2004, **69**, 115006.

Professor Meijer remarked: The metastable state of CO is close to the $v = 27$ and 28 of the electronic ground-state. Would it be possible to make a $X^1\Sigma^+ \leftarrow a^3\Pi$ transition, or are the Franck–Condon factors for this prohibitively low?

Dr Bethlem replied: As already mentioned, a near degeneracy between a highly excited vibrational level in the ground state and a vibrational level in an excited electronic state potentially leads to a very large enhancement of the sensitivity to a variation of the proton-to-electron mass ratio. The sensitivity of the transition mentioned is enhanced by a factor of 100 000. Moreover, as both levels have a relatively long lifetime, the linewidth of the transition may be quite narrow. Unfortunately however, the two electronic states have a minimum at nearly the same internuclear distance which results in a Frank–Condon factor for the $X^1\Sigma^+(0) \leftarrow a^3\Pi$ (27,28) below 10^{-20}, so it seems hardly feasible to drive this transition.

Dr Vanhaecke asked: Have you considered the possibility of using an additional source of photons to dress one of the states involved in the transition you want to measure? Could it constitute an additional parameter which could be used to better characterize the transition?

Dr Bethlem answered: We have not considered such opportunities yet, for the reason that it may further complicate the experiment. Particularly, if the final precision depends on the intensity or frequency of the additional source of photons, it will only shift the problem.

Professor Ye said: It would be interesting to explore a much larger energy range where one can compare two molecular levels with vastly different dependence on (me/mp). The maximal enhancement factor obtained is through the measurement of an entire molecular potential. Can one think of optical 2-photon transitions to explore much higher vibrational levels of one potential and compare it against the lower vibrational levels of a different potential?

Dr Bethlem responded: It is important to distinguish the *absolute* sensitivity of a transition, *i.e.*, the frequency shift, δv, due to a certain variation of a fundamental constant from the *relative* sensitivity, *i.e.*, the relative shift, $\delta v/v$, due to a possible variation of a fundamental constant. Ideally both are large, but most systems that are currently being discussed have either a large absolute sensitivity or a large relative sensitivity.

As you mention, a large absolute sensitivity can be obtained through the measurement of an entire molecular potential using optical 2-photon transitions – as is also discussed in you paper on Sr_2,[1] however, the relative sensitivity of optical transitions is rather small. For comparison, the absolute sensitivity of the optical transitions in Sr_2 is about 10 times larger, but the relative sensitivity is about 1000 times smaller than the microwave transitions in CO discussed in our paper.

1 T. Zelevinsky, S. Kotochigova, and J. Ye, *Phys. Rev. Lett.*, 2008, **100**, 043201.

Professor Ye asked: What is the required level of control of residual electric fields, considering that the two molecular levels have a differential Stark shift? A drifting electric field could limit the achievable sensitivity as an important source for systematic error.

Dr Bethlem replied: The Stark shift of the considered transitions is on the order of 1 Hz per $(V/cm)^2$. For a measurement with a fractional uncertainty below 10^{-12}, we need to keep residual electric fields below 50 mV cm^{-1}, which seems feasible. Moreover, we will compare transitions in the $v = 0$ and $v = 1$, which have near identical Stark shifts (the dipole moment in the $v = 1$ is only $5 \cdot 10^{-4}$ larger than in the $v = 0$), so these will cancel.

Dr Tarbutt commented: The sensitivity enhancement of these proposed experiments is based on the notion that it is easier to measure a given frequency shift using a low-frequency transition than a high-frequency transition. The underlying assumption here is that the precision of the experiment is limited by the relative precision of the frequency measurement so that you get a better absolute precision by measuring a smaller frequency. That will sometimes be the case, but certainly not always. I think you already make this point in your paper, but I want to emphasize it here because I think it's an important one. For example, an experiment may well be limited by a systematic error that has a certain absolute size, irrespective of whether the measurement is made on a large or a small interval. In such cases, the sensitivity enhancement does not improve the experiment at all. The characterization of certain systems by a "sensitivity factor" can be useful, but there is an inevitable temptation to think that bigger is always better, which would be very misleading indeed. In fact, a system with an enormous sensitivity factor may make a far worse experiment than one with a small factor, because it happens to suffer from a particularly severe systematic effect. A related point is that the absolute precision of a frequency measurement tends to bottom out as the frequency is lowered. In going from optical to microwave the precision improves, but once the transition is already in the microwave region, there is little to be gained from going to still lower frequencies.

Dr Bethlem responded: We fully agree. It is generally not easier to measure a certain frequency shift at 100 MHz than at 1 GHz. But, it is certainly easier to measure a certain frequency shift at 2 GHz than at 1.2 THz – which is the 'effective' energy difference that is being probed in our proposed experiment. On your remark on systematic errors; in our paper we propose to compare two transitions in the $v = 0$ and $v = 1$ of the $a^3\pi$ state in CO, which shift in opposite directions when the proton-to-electron mass ratio changes. This will greatly reduce systematic errors and may turn out to be more important than the increased fractional sensitivity.

Professor Tennyson addressed Dr Bethlem: I have two questions/comments.

First you compare the (prospective) observations of shifts in laboratory experiments recorded over less than a decade in the current epoch with astronomical results where the major sensitivity comes from observations of objects billions of years ago. Is this a like-for-like comparison? Do theories predict that any change in fundamental constants is linear with time or would one expect greater changes in the early universe?

Second I note microwave electronic spectra recorded some years ago by Carrington et al.[1] These workers observed transitions from the highest vibrational state of the $X^2\Sigma_g^+$ ground state of H_2 to the ground vibrational state of the $A^2\Sigma_u^+$ state. The two electronic state converge to the same asymptote; the excited state is basically repulsive but has a shallow minimum which supports a long-range vibrational state. This observation provides a means of observing electronic state in near degeneracy. The H_2^+ system has the advantage of being amenable to *ab initio* calculations which reach experimental precision.[2] This means that the effect of changing any fundamental constant can be thoroughly and accurately investigated *a priori*. To this end I made a quick calculation on the effect of changing the proton to electron mass ratio. Increasing the proton mass by 0.1% lowers the vibrational ground state by 0.6 cm^{-1} but only lowers the last bound vibrational state by 0.02 cm^{-1}.

While on this topic I note that many of the rare gas dimers being prepared at ultracold temperatures, such as Rb_2, are often prepared in the highest vibrational state of the system. Rb_2 is electronically similar to H_2 and therefore these high-lying vibrational states of the $X^1\Sigma_g^+$ ground state could undergo electronic transitions at very long wavelengths to low-lying states of the first excited $a^3\Sigma_u^+$ state, which again only has a shallow minimum. This is of course a singlet–triplet transition so may not be expected to be as strong as those observed in H_2^+. To achieve the equivalent spin-allowed transitions using alkali metal dimers one would have to prepare ions such as Rb_2^+

1 A. Carrington, I. R. McNab and C. A. Montgomerie, *Chem. Phys. Lett.*, 1989, **160**, 237.
2 C. A. Leach and R. E. Moss, *Ann. Rev. Phys. Chem.*, 1995, **46**, 55.

Dr Bethlem replied: To the first question, indeed there are theories which predict variation of fundamental constants occurs only during some cosmological episodes. In the model by Barrow et al.[1] the fundamental constants change only during the cosmological epoch of the matter-dominated universe; in the radiation era before $z \sim 1000$ there is no variation, nor is there any variation in the present epoch of accelerated expansion, which may have started at $z = 0.5$–1. In such a scenario no variations would occur on laboratory time scales. On the other hand, if the variation of the fundamental constants is somehow connected to the size of the universe they might be changing still. The fact of the matter is that laboratory measurements – needed to convince most of us – can only be performed now. It is noted that recent laboratory measurements already put tighter constraints on rates of change of the fine-structure constant[2,3] than observations on cosmological time scales do.[4]

In answer to the second question, a near degeneracy between a highly excited vibrational level in the ground state and a vibrational level in an excited electronic state potentially leads to a very large enhancement of the sensitivity to a variation of the proton-to-electron mass, as the enhancement scales with the energy difference between the two electronic states divided by the energy between the two near degenerate levels. However, for vibrational levels near the dissociation energy the sensitivity is much reduced due to the anharmonicity of the potential wells – as indeed follows from your calculations as well. A test based on a near degeneracy between the $a^3\Sigma_u^+(v = 37)$ and the $X^1\Sigma_g^+(v = 138)$ in Cs_2 was proposed by DeMille et al.[5] The enhancement factor in their proposal is on the order of 1000 – again lower than one would naively expect due to anharmonic terms in the potential well. The Rb_2 system is similar to Cs_2 and similar arguments hold.

1 J. D. Barrow, H. Sandvik, and J. Mageijo, *Phys. Rev. D*, 2002, **65**, 063504.
2 A. D. Ludlow et al., *Science*, 2008, **319**, 1805.
3 T. Rosenband et al., *Science*, 2008, **319**, 1808.
4 M. T. Murphy, J. K. Webb, V. V. Flambaum, *Month. Not. Roy. Astron. Soc.*, 345 (2003) 609.
5 D. DeMille, S. Sainis, J. Sage, T. Bergeman, S. Kotochigova, and E. Tiesinga, *Phys. Rev. Lett.*, 2008, **100**, 043202.

Professor Herschbach commented: In the analysis of rotational spectra, it is generally assumed that the molecule consists of neutral atoms, each represented as a point mass that includes all its electrons. That is unrealistic, although adequate for many purposes. However, in early precision measurements of isotope mass ratios in diatomic molecules by microwave spectroscopy, the cloud-like electron distribution and its distortion during rotation was found to appreciably affect the molecular moment of inertia.[1] For LiBr and LiI, it proved possible in effect to "weigh an electron," as the rotational spectra for the ^{6}Li and ^{7}Li isotopes showed that one electron had indeed been transferred from lithium to the halogen atom.[2] Effects of the electron distribution evidently are not considered to handicap testing the time-invariance of the proton-electron mass ratio. Is that so because the fractional change in the p-e mass ratio is taken to be equal to the change in the reduced nuclear mass of the molecule? As Dr Bethlem points out, that equality requires the neutron-electron mass ratio to change at the same rate as the p-e mass ratio.

1 C. H. Townes and A. L. Schawlow, *Microwave Spectroscopy*, McGraw-Hill Book Company, New York, 1955.
2 A. Honig, M. Mandel, M. L. Stitch and C. H. Townes, *Phys. Rev.*, 1954, **96**, 629.

Dr Bethlem responded: For testing variation of the proton-to-electron mass ratio in the case of molecules other than H_2 we simply take the scaling of the proton-to-electron mass ratio μ to be similar to that of the reduced mass of the molecule. This implies that we do not make a distinction between protons and neutrons (see above), nor do we care about the effects of electrons, since those are at the 10^{-3} level. The important issue at stake is that in experiments testing μ-variation the variation over time is measured on a relative scale $\Delta\mu/\mu$. It is not the aim of such experiments to measure a value of μ or to bring a theoretical model of a molecule (*ab initio* or otherwise) in agreement with an experimental determination.

We note that there exists an ongoing activity of high-precision molecular spectroscopy with the aim of determining μ.[1] This is pursued in the simplest existing molecular entity H_2^+. In this case of course all the subtleties of masses should be accounted for.

1 S. Schiller and V. Korobov, *Phys. Rev. A*, 2005, **71**, 032505.

Dr Pinkse opened the discussion of Dr Tarbutt's paper: Would a square box potential trap solve or reduce the depolarization caused by the geometric phase in a harmonic trap?

Dr Hudson responded: On the bottom of a square box potential one could hope to better control the geometric phase. The problem will come at the wall. Here one would expect to generate large geometric phases, or even drive non-adiabatic transitions between states. Fundamentally, the difficulty is that, at around a kelvin, the molecule's kinetic energy is comparable to the depth of a realistic trapping potential. Changing the shape of the potential can't avoid this fact.

Dr Bethlem asked: You show that inhomogeneous geometrical phase shifts are a problem for doing an EDM experiment in an electrostatic trap. Is there any hope that these problems are less severe in a 2D-guide. Particularly, do you think it will be possible to do a sensitive EDM experiment in a guided fountain?

Dr Hudson replied: In a 2D-guide the field could, ideally, be confined to lie in a plane. A field rotating in a plane, returning exactly to its starting point, introduces no geometric phase shift. This is essentially the idea behind controlling the geometric phase in the nascent trapped ion EDM measurement in Eric Cornell's group.

Consider though, to set the scale of the problem, that in our current EDM measurement we measure phase shifts of order μrad. This would set very stringent limits on both the mechanical tolerances of the guide and the level of field control. I am sceptical that it will prove possible to make a sensitive EDM measurement in a guided fountain, but it is certainly worth investigating further.

Mr Zeppenfeld enquired: In how far could the geometric phase which dephases the spin precession caused by the EDM for EDM measurements in an electric trap be compensated by using a second pair of states as a reference?

Dr Hudson responded: There are no such states in the electronic ground state of YbF, but such states do exist. Ω-doubling, found in molecular states with a non-zero projection of electronic angular momentum on to the internuclear axis, results in pairs of states that acquire equal geometric phases, but have oppositely signed EDM interactions. ThO, for instance, is a promising molecule for EDM measurement with an Ω-doubled metastable state, which is being investigated by the ACME collaboration.

Using these Ω-doublet states to suppress the effect of the geometric phase in a trap, however, would be difficult. The Ω-doublet states Stark shift oppositely in an electric field. As the geometric-phase induced dephasing is an inhomogeneous effect it would be necessary to prepare a superposition of Ω-doublet states. The superposition would be very rapidly decohered across the molecular ensemble by the spatially inhomogeneous trapping electric field.

Dr Küpper commented: In the Miller group it was observed that $(HCN)_3$ is a linear chain inside helium droplets, but nevertheless the field-free rotationally resolved ν_n-band shows a Q-branch.[1] This might be attributed to electric fields (anisotropic interactions) inside the droplets.[2]

Of course, in an alternative approach we have routinely applied external electric fields of 20–100 kV cm^{-1} to create pendular states and to observe the connecting 'pendular spectra'.

1 K. Nauta, PhD thesis, University of North Carolina, Chapel Hill, NC, USA, 2000.
2 K. Nauta, D. T. Moore and R. E. Miller, *Faraday Discuss.*, 1999, **113**, 261.

Dr Mudrich opened the discussion of Professor Jäger's paper: What motivates the assumption of a droplet-size dependent line width (besides the similarity of the line shape with log-normal distribution), given the fact that at all studied conditions the nanodroplets are much larger than any solvation layer surrounding the molecules?

What could be the origin of the observed fast rotational relaxation, given the superfluid nature of the He environment? Is a "critical rotational velocity" (in analogy with Landau's critical velocity) conceivable, which may lead to damping in dependence of the rotational excitation J?

Professor Jäger answered: We initially speculated that the observed line shapes are a result of a droplet size dependence of the B rotational constant. For smaller clusters with up to 70 helium atoms, such dependence was indeed observed.[1] It is unclear however, which physical mechanism could be responsible for such behaviour in the larger helium nanodroplets. Our studies with different droplet sizes (see Fig. 9 in our paper) also clearly indicate that a droplet-size dependence is not the main contributor to the overall line widths. In the microwave study of a pure inversion transition of ammonia embedded in helium nanodroplets, it was found that the overall line shape could be well reproduced by assuming a splitting of the molecular energy levels into energy level manifolds.[2] Such splitting could occur through, for example, coupling of the molecular rotation with a particle in a spherical box motion. In this case the droplet size determines the details of the energy level manifolds and influences the line width. There is also evidence that the angular momentum imparted to the droplet by the collision with the dopant molecule has significant effects on the spectroscopic signatures. This will clearly be droplet-size dependent.[3,4] The existence of a critical rotational velocity at the microscopic level is an interesting thought. Pitaevskii and Stringari have treated superfluid effects in rotating helium clusters.[5] They considered surface excitations of the clusters and determined a critical rotational frequency, in analogy to the critical velocity in bulk helium. Also in this case the droplet size would be important, as it determines the density of surface states. It is not clear, however, how the observed asymmetric line shapes could be rationalized with this model.

1 A. R. W. McKellar, Y. Xu, and W. Jäger, *Phys. Rev. Lett.*, 2006, **97**, 183401.
2 R. Lehnig, N. V. Blinov, and W. Jäger, *J. Chem. Phys.*, 2007, **127**, 241101.
3 N. Pörtner, A. F. Vilesov, and M. Havenith, *Chem. Phys. Lett.*, 2003, **368**, 458.
4 K. K. Lehmann and A. M. Dokter, *Phys. Rev. Lett.*, 2004, **92**, 173401.
5 L. Pitaevskii and S. Stringari, *Z. Phys. D*, 1990, **16**, 299.

Dr Küpper enquired of Professor Jäger: You describe the asymmetric line shapes and suggest they are due to droplet-size variations. From various experiments in the Miller group (and others) we know that surface effects can have a strong influence on the spectra of embedded molecules.

Is it feasible to suggest that the droplet-size dependence of your lines is due to the averaged or fractional interaction of the molecule with the surface? This would obviously scale such that the interaction is weaker for larger droplets.

Professor Jäger replied: That is an interesting point considering that both the rotation–translation coupling and impurity–ripplon interactions increase as the dopant molecule approaches the surface. Lehmann has considered both these contributions and could simulate the experimental line shape of the R(0) infrared transition of OCS.[1] However, it was not possible to reproduce the increase in line width with increasing J quantum number. While this initial calculation assumed a Boltzmann distribution of the translational states of OCS in a helium droplet, Dokter and Lehmann later showed that the angular momentum resulting from the pick-up process needs to be considered.[2] They concluded that high angular momentum translational states may have substantially larger populations than predicted from a Boltzmann distribution at the droplet temperature. We are not aware of any line shape simulations based on this model.

1 K. K. Lehmann, *Mol. Phys.*, 1999, **97**, 645.
2 K. K. Lehmann and A. M. Dokter, *Phys. Rev. Lett.*, 2004, **92**, 173401.

Dr Küpper asked: In your article you discuss the differences between vibrational and rotational relaxation of molecules in helium droplets. One of the first and most extreme cases where such differences of behaviour were observed, is in HF.[1]

Moreover, the rotational relaxation of HF ($v = 1$) leads to a line width of 0.43 cm^{-1} (~12.9 GHz), whereas your line widths are <1 GHz. How can this be rationalized?

Could you measure the pure $J = 1 \leftarrow 0$ rotational transition of HF to determine the rotational lifetime of $J = 1, v = 0$?

1 K. Nauta and R. E. Miller, *J. Chem. Phys.*, 2000, **113**, 9466.

Professor Jäger answered: It seems evident that there is not a single mechanism that is responsible for the observed widths and shapes of rotational and rovibrational transitions. In the case of HF, the rotational excitation is at about 40 cm^{-1} (1.2 THz) and at this energy there are more relaxation channels available, *i.e.* the bulk modes, than for the much heavier OCS molecule (only surface modes). The increased line width for HF is thus the result of faster relaxation rates.

We are quite interested in the measurement of pure rotational transitions of lighter molecules embedded in helium nanodroplets. We do have sources, so-called backward wave oscillators, which can produce radiation up to frequencies of about 1 THz and are in the process of implementing them into our instrument. The $J = 1 \leftarrow 0$ rotational transition of HF is currently out of reach.

Dr Küpper continued: In Fig. 4 of your paper you compare the sensitivity of mass spectrometric and bolometric detection. Already in ref. 1 we have argued that bolometric detection is the best choice for such experiments. What do you consider the advantage of a MS over a bolometer (besides running costs)?

1 J. Küpper and J. M. Merritt, Int. Rev. Phys. Chem., 2007, **26**, 249; J. Küpper and J. M. Merritt, Int. Rev. Phys. Chem., 2007, **26**, 288.

Professor Jäger responded: The reason why we first took spectra using the mass spectrometer was simply that it was implemented into our helium nanodroplet spectrometer and functional before the bolometer system. However, there are a number of reasons for using a mass spectrometer for detection. For example, the mass spectrometer can be used as diagnostics for the helium droplet beam and it makes it possible to verify that a specific molecular species has been picked up and transported by the helium droplet beam. With the mass spectrometer, it is possible to shine a laser beam coaxially and counter-propagating to the droplet beam. This increases the overlap between laser beam and droplet beam compared to a perpendicular intersection, although the use of a multipass cell can partially compensate for this. Roger Miller and his group have used optically selected mass spectrometry experiments to investigate electron ionization mechanisms and fragmentation patterns of dopants embedded in helium nanodroplets.[1]

1 W. K. Lewis, B. E. Applegate, J. Sztray, B. Sztray, T. Baer, R. J. Bemish, and R. E. Miller, *J. Am. Chem. Soc.*, 2004, **126**, 11283.

Professor Stwalley asked: Traditionally, cold molecules were first studied using matrix isolation spectroscopy and then using supersonic molecular beams. Your beautiful experiments illustrate helium cluster spectroscopy, a very powerful merging of the matrix isolation and molecular beam technologies and traditions. What do you see as possible future applications for helium cluster spectroscopy in cold and ultracold molecular spectroscopy?

Professor Jäger replied: There have already been a number of contributions from helium nanodroplet experiments to the field of cold atoms and molecules, apart from purely spectroscopic studies of molecules and clusters. Among the most exciting are studies of chemical reactions at low temperatures. Vilesov and coworkers, for example, have investigated the exothermic chemical reaction Ba + N_2O → BaO + N_2 as it occurs at the low temperature of 0.38 K in helium droplets.[1] The Freiburg group looked at reactions of alkali metal clusters with water molecules embedded in helium droplets and used femtosecond photoionization. They found that Rb and Cs react completely with water in the cold droplet environment.[2] The Drabbels research group has studied photodissociation reactions of CH_3I and CF_3I embedded in helium droplets.[3] It would also be of significant interest to observe a reaction where quantum effects dominate the outcome, such as is expected for F + HD → HF (DF) + D (H),[4] at helium droplet temperatures.

The ultracold temperature regime cannot be reached in helium nanodroplet experiments, since the droplets attain (on the time scale of the experiment) final temperatures of 0.38 K and 0.15 K for ^{4}He and ^{3}He, respectively. However, important spectroscopic information can be obtained about systems that are relevant to ultracold physics and chemistry, such as alkali metal dimers and trimers. Examples are the formation and spectroscopic investigation of alkali dimers[5] and mixed alkali dimers (*e.g.* LiCs).[6]

1 E. Lugovoj, J. P. Toennies, and A. Vilesov, *J. Chem. Phys.*, 2000, **112**, 8217.
2 S. Müller, S. Krapf, Th. Koslowski, M. Mudrich, and F. Stienkemeier, *Phys. Rev. Lett.*, 2009, **102**, 183401.
3 A. Braun and M. Drabbels, *J. Chem. Phys.*, 2007, **127**, 114305.
4 N. Balakrishnan and A. Dalgarno, *J. Phys. Chem. A*, 2003, **107**, 7101.
5 F. Stienkemeier, W. E. Ernst, J. Higgins, and G. Scoles, *J. Chem. Phys.*, 1995, **102**, 615.
6 M. Mudrich, O. Bünermann, F. Stienkemeier, O. Dulieu, and M. Weidemüller, *Eur. Phys. J. D*, 2004, **31**, 291.

Dr Chandler addressed Professor Jäger and Dr Mudrich : Can a collision of a helium droplet with a trapped but internally hot molecular ion be used to cool the ion to sub kelvin temperatures without losing the ion from the trap?

Dr Mudrich replied: The idea of cooling the internal degrees of freedom of molecular ions stored in ion traps by shooting a beam of He nanodroplets into the trap appears to be very interesting. In order to answer the question, we have to estimate the following quantities: the cooling capacity of the droplets, the kinetic energy of the droplets, which is transferred to the new ion–droplet complex *versus* the trap depth, and the stability of the ion–droplet complex.

The cooling capacity due to evaporation of single He atoms amounts to about 5 K per He atom for a small droplet containing 1000 atoms, which gives about 0.4 eV in total. This is sufficient to efficiently cool small molecular ions down to 0.4 K. The kinetic energy of such a droplet propagating at 300 m s^{-1} is 1.9 eV, which is in the range of the trap depth of typical ion traps. In a real experiment, however, the fact that the evaporating ion–droplet complexes have different masses that change with time raises the issue of which trap parameters, *i.e.* RF-frequencies, *etc.*, to use for maintaining stable trapping conditions. Besides, multiple collisions with He droplets will eventually kick the ions out of the trap unless cooling of the translational motion is applied. Concerning the stability of ion–droplet complexes, we know from experiments that no fragmentation occurs when electrostatically extracting ion-doped He droplets using electric fields of order 1000 V cm^{-1}. Thus, the entire droplet would contribute to evaporative cooling. However, it is probable that a few He atoms will remain strongly bound to the ionic core forming "snowball" complexes.

Professor Jäger answered: This probably depends predominantly on the depth of the trap and the linear momentum that is transferred from the droplet to the ion. There is some control over the droplet linear momentum, simply by adjusting nozzle temperature and backing pressure to change the average droplet size. Helium nanodroplets are very efficient heat baths and cooling rates for ions of 1016 K sec^{-1} have been reported.[1] Roger Miller and his group have reported the production of (untrapped) cooled ions by electron impact ionization of doped helium droplets.[1] The ionization process causes significant heating of the resulting ion. The rapid energy transfer to the helium environment leads to cooling, can close certain fragmentation channels,[2] and leads to evaporation of the entire helium droplet. However, it was noted that the ion cooling was far from complete.[1]

1 W. K. Lewis, B. E. Applegate, J. Sztray, B. Sztray, T. Baer, R. J. Bemish, and R. E. Miller, *J. Am. Chem. Soc.*, 2004, **126**, 11283.
2 W. K. Lewis, C. M. Lindsay, and R. E. Miller, *J. Chem. Phys.*, 2008, **129**, 201101.

Dr Küpper said: I believe that there is strong overlap between the two fields with respect to the reactive processes, which have already been discussed for cold and ultracold systems. Helium droplets, on the other hand, provide great opportunities for studying potential energy surfaces through high resolution spectroscopy of entrance and exit channel complexes. For example, we have stabilised and investigated multiple structural isomers of HF, HCN, and HCCCN with halogen atoms.[1,2,3] Moreover, you can even stabilise clusters containing two halogen atoms, providing chemical energy densities similar to TNT.[3] Generally, the investigation of atoms, molecules and clusters in helium droplets can provide a variety of accurate information for experiments with ultracold atoms and molecules.[4]

1 J. M. Merritt, J. Küpper and R. E. Miller, *Phys. Chem. Chem. Phys.*, 2005, **7**, 67.
2 J. M. Merritt, J. Küpper and R. E. Miller, *Phys. Chem. Chem. Phys.*, 2007, **9**, 401.
3 J. Küpper and J. M. Merritt, *Int. Rev. Phys. Chem.*, 2007, **26**, 249; J. Küpper and J. M. Merritt, *Int. Rev. Phys. Chem.*, 2007, **26**, 288.
4 M. Mudrich, O. Bünermann, F. Stienkemeier, O. Dulieu and M. Weidemüller, *Eur. J. Phys. D*, 2004, **31**, 291.

Professor Gianturco commented on Professor Dr Willitsch's paper: Mario Tacconi and I have looked at the possibility of creating an experimental setup in which a cloud of ultracold, laser cooled Rb atoms are introduced in a Coulomb crystal arrangement where MgH$^+$($X^1\Sigma^+$) and Mg$^+$(^{2}S) along with the laser-cooling populated excited species MgH$^+$(A^{1}Sigma$^+$) and Mg$^+$(^{2}P) are contained. Our computational findings suggest the most likely pathways for a series of "reactive" events coming from the interaction of Mg$^+$, MgH$^+$ and Rb mixtures within the trap. Details of the *ab initio* calculations could be found in ref. 1, where the low-lying electronic states of the MgHRb$^+$ ionic complex are described, while the MgRb$^+$ system, thought to be also present in the Coulomb crystal (CC), has been extensively discussed in ref. 2.

Fig. 1 reports the region of curve crossing and the final, lowest-lying state of the charge-exchange process. The corresponding details for the atomic cation and for the RbMg$^+$ complex are given in Fig. 2.

The global picture arising from our analysis of the bi-crystal processes is therefore:

* Within the trap environment we are considering here, the most likely way in which the MgH$^+$ molecular ion can react with neutral Rb is *via* the radiative charge transfer process from the upper levels (entrance channels) shown by Fig. 1

* The neutral, cold partner Rb can further directly interact with the crystallized Mg$^+$ ions: our calculations tell us that, in this case, the only relevant process available is spontaneous photoemission and that the Mg$^+$ can effectively compete with MgH$^+$ in the reaction with Rb atoms.

* Our computational analysis of the atomic and molecular competing processes predict that the reactive events which involve the molecular ion MgH$^+$ can have an

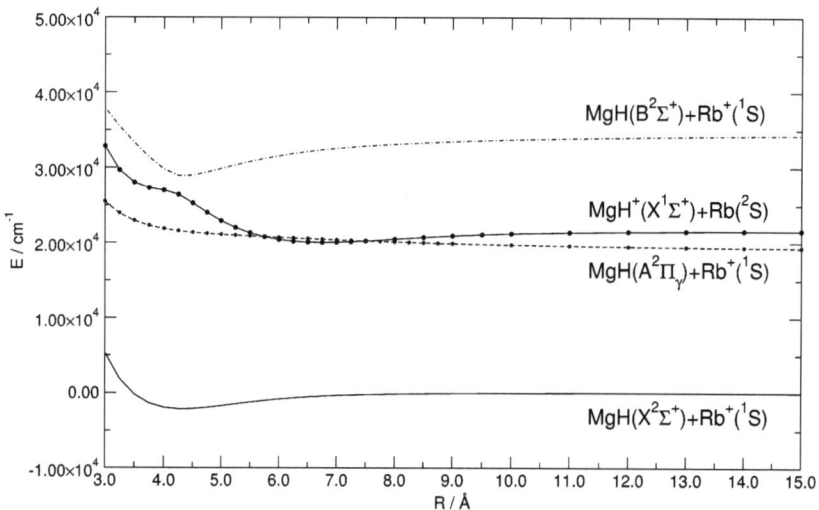

Fig. 1 Potential energy curves of MgH$^+$ interacting with Rb along the collinear geometry. Adapted from ref. 1.

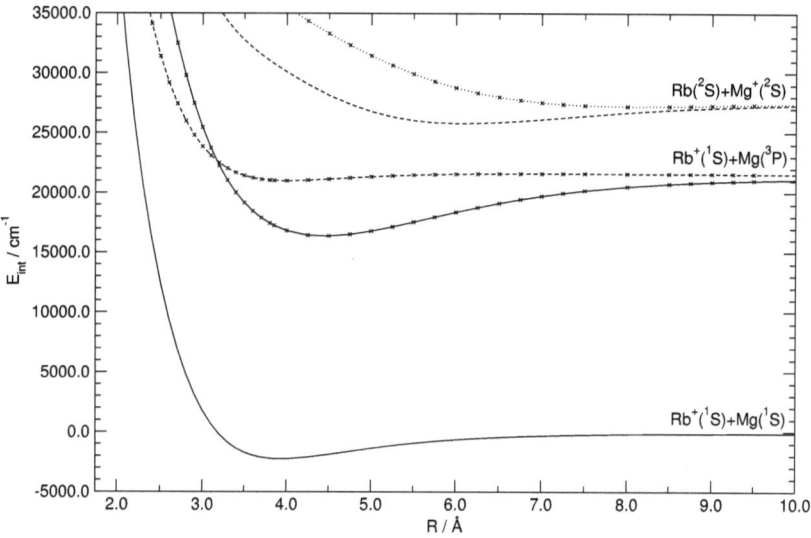

Fig. 2 Potential energy curves of Mg$^+$ interacting with Rb. Adapted from ref. 2.

efficiency which is comparable to that in which the ionic partner is the Mg$^+$ and that the Rb addition to the trap is very likely to produce sizable amounts of cold, trapped Rb$^+$ cations in the form of stable ionic complexes. We have also briefly considered the possibility of the Penning Ionization events whereby both couples of cations, *i.e.* MgH$^+$ plus Rb$^+$ and Mg$^+$ plus Rb$^+$ are produced. In our case we have tentatively provided an estimate of the expected rate coefficient K_P. Following ref. 3, one defines $K_P = f_W P_i K_C$, where f_W is the fraction of the collisions from which a continuum state can be reached without violating spin conservation (here $f_W = 1$ for both cases), P_i is the mean ionization probability per close collision, and K_C is the rate coefficient for the close collision. Thus, our estimates for the Penning Ionization process could be $10^{-12} \lesssim K_P \leqslant 2.87 \times 10^{-10}$ cm^3 s^{-1}, which is a fairly large value.

1 M. Tacconi and F. A. Gianturco., *Eur. Phys. J. D*, 2008, **46**, 443.
2 M. Tacconi and F. A. Gianturco, *Eur. Phys. J. D*, 2009, **54**, 31.
3 K. L. Bell, A. Dalgarno, and A. E. Kingston. *J. Phys. B*, 1968, **1**, 18.

Dr Wester opened the discussion of Professor Dr Gerlich's paper: Cryogenic radiofrequency multipole traps are a very useful tool to study negatively charged ions, their reactions and their interaction with light at temperatures between a few kelvin and room temperature. In particular the 22-pole ion trap[1] is employed, because it features a large trapping volume void of strong electric fields. The translational and rotational temperature of the trapped ions is controlled *via* buffer gas cooling with helium.

The dynamics of anion–molecule reactions are typically governed by a complex multi-well Born–Oppenheimer hypersurface. As a consequence, the reaction rate and the energy partitioning among the different translational and rovibrational degrees of freedom depends crucially on the complex few-body dynamics at short range.[2] Currently, open questions relate to the role of tunneling through the intermediate potential energy barrier and the importance of Feshbach scattering resonances.[3] In this respect anion–molecule reactions are qualitatively different from many cation–molecule reactions, which occur at the Langevin or capture rate and are thus not influenced substantially by the short-range dynamics. Cross sections for chemical reactions and photodetachment processes are also important to understand the role of anions in interstellar molecular clouds, where several negatively charged species have recently been detected.[4]

In a study of a negative ion reaction at the lowest currently achieved temperatures, reaction rate coefficients have been obtained for the proton transfer reaction NH$_2^-$ + H$_2$ → NH$_3$ + H$^-$.[5] Here the rate coefficient rises from 2 to 11% between room temperature and 20 K and decreases again for even lower temperatures down to 8 K. The increase may be explained by a classical complex-mediated reaction mechanism. The observed maximum at finite temperature, however, is currently unexplained. In this temperature regime only the lowest *para* and *ortho* angular momentum states of the two reactants are populated and only a small number of partial waves contribute to the reactive scattering. Furthermore the mean spacing of the Eigenstates of the reaction complex is larger than the temperature. Thus, quantum mechanics are assumed to dominate the reaction dynamics. Detailed scattering calculations are needed to clarify this and understand the low-temperature dynamics.

The most important light-matter interaction of negative ions is the photodetachment process. It is widely used in photoelectron spectroscopy to investigate the properties of both anions and neutrals. In plasmas the total cross section for photodetachment represents an important destruction mechanism for negative ions. In a 22-pole ion trap the absolute cross section can be measured with high systematic accuracy by observing the loss of ions in a radiation field.[6] In this way, the absolute detachment cross sections for O$^-$ and OH$^-$ were obtained at different wavelengths and as a function of the buffer gas temperature.[7] For O$^-$ an overall relative accuracy of 5% was achieved. This benchmark result challenges state-of-the-art theoretical calculations, which currently deviate by about 35%.[8] Photodetachment near threshold, thus using photon energies slightly larger than the electron affinity, has proven to be a useful tool to characterize the rotational state population of cold molecular anions in the ion trap. This will be explored in the future to study state-selected cold collisions, both inelastic and reactive.

Trapped negative chlorine anions, which are stable with respect to reactions with background gas and to black body-induced photodetachment, are useful to study the trapping conditions in multipole rf traps. Specifically, the stability conditions in a 22-pole ion trap could be experimentally accessed and an experimental estimate for the maximum stability parameter could be deduced.[9]

1 D. Gerlich, *Phys. Scr.*, 1995, **T59**, 256.
2 J. Mikosch, S. Trippel, C. Eichhorn, R. Otto, U. Lourderaj, J. X. Zhang, W. L. Hase, M. Weidemüller and R. Wester, *Science*, 2008, **319**, 183.
3 S. Schmatz, *ChemPhysChem*, 2004, **5**, 600.
4 M. C. McCarthy, C. A. Gottlieb, H. Gupta and P. Thaddeus, *Astrophys. J. Lett.*, 2006, **652**, L141.
5 R. Otto, J. Mikosch, S. Trippel, M. Weidemüller and R. Wester, *Phys. Rev. Lett.*, 2008, **101**, 063201.
6 S. Trippel, J. Mikosch, R. Berhane, R. Otto, M. Weidemüller and R. Wester, *Phys. Rev. Lett.*, 2006, **97**, 193003.
7 P. Hlavenka, R. Otto, S. Trippel, J. Mikosch, M. Weidemüller and R. Wester, *J. Chem. Phys.*, 2009, **130**, 061105.
8 O. Zatsarinny and K. Bartschat, *Phys. Rev. A*, 2006, **73**, 022714.
9 J. Mikosch, U. Frühling, S. Trippel, D. Schwalm, M. Weidemüller and R. Wester, *Phys. Rev. Lett.*, 2007, 98, 223001.

Professor Gianturco asked: We have recently studied the potential energy surfaces and the scattering behaviour at ultralow energies of an anionic species such as OH^- interacting with He and with Rb. In the latter case we have been able to compute not only the elastic cross sections but also the rotovibrational quenching cross sections at the nanokelvin regimes. As expected, the quenching cross sections for collisions with the Rb partner turned out to be much larger than in the case of He. They also appeared from calculations to markedly depend in size on the initial internal state of the anion. My question, therefore, is about the possibility of actually controlling the internal anionic temperatures in the experiments in order to reduce the importance of the quenching losses .Do you have any suggestion as to how this could be done and, at the same time, what sort of internal temperatures could be expected in the trapping of OH^-?

Dr Wester answered: In current experiments with negative ions in cryogenic ion traps[1] translational as well as rotational temperatures of a few kelvin are achieved using helium buffer gas cooling. For trapped OH^- anions at 10 kelvin a population of about 99% in the $J = 0$ rotational ground state is reached. Preliminary work in our ion trap shows that the population of the higher rotational states can be increased in a controlled way by increasing the temperature of the buffer gas. Cryogenic ion traps should therefore provide good starting conditions for experiments on inelastic collisions of small molecular anions.

1 R. Wester, *J. Phys. B*, 2009, in press; arXiv:0902.0475.

Professor Softley asked Professor Dr Gerlich: In your paper you have commented; "especially at low collision energies, a collision complex usually utilizes the energy from all degrees of freedom, one should only call a bimolecular reaction *cold* if the *total energy* is below a few meV." While we are grateful for this advice, we would argue that for many ion–molecule reactions it is valid to use the phrase "translationally cold" to describe reactions (as we have done in our paper) even where the reactants are internally warm. In the capture theory model of ion–dipole reactions, which has been shown to be valid for a number of reactions down to temperatures around 10 K, the translational and rotational effects on the reaction rate are physically quite distinct. Rotational effects are connected with the alignment of the dipole along the collision direction; as the rotational quantum number increases, this alignment becomes less pronounced leading to a possible decrease in the reaction rate. On the other hand the effect of translational cooling is attributable to the shift of the centrifugal barrier and how its height changes relative to the changing kinetic

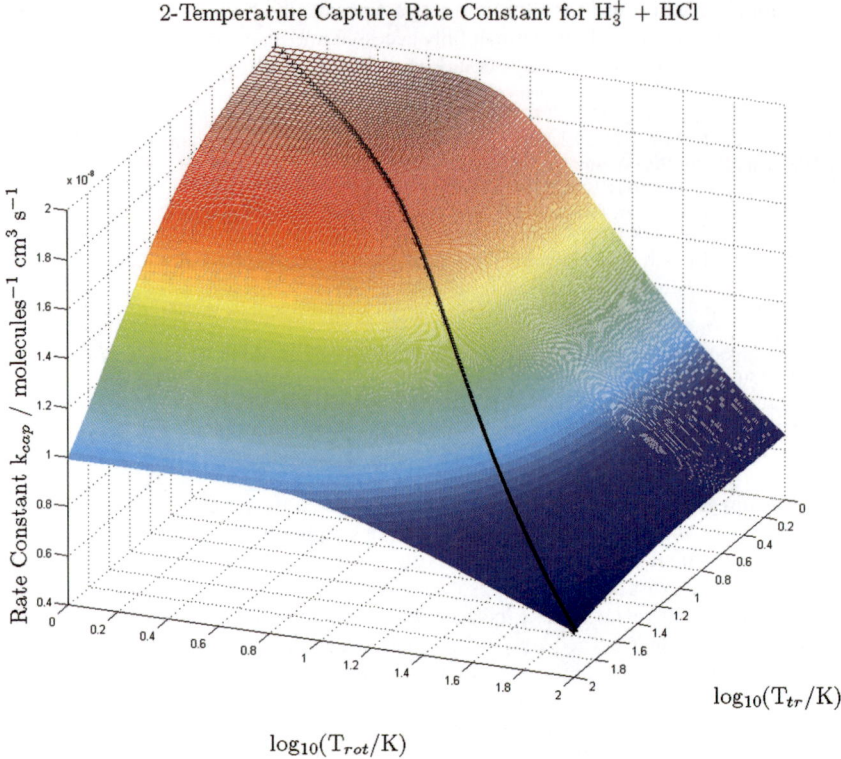

Fig. 3

energy. Fig. 3 shows an example rotationally adiabatic capture theory calculation[1] for the H_3^+ + HCl reaction showing how the reaction rate constant varies with rotational and translational temperatures separately. We have also considered the difference between translationally cold reactions which are either rotationally warm or rotationally cold in Fig. 8 of our paper. While the effects of decreasing rotational temperature or decreasing translational temperature are qualitatively similar, they are quantitatively different.

However it is interesting to consider what is the most useful kind of rate constant measurement in this cold or ultracold regime. In some senses this is a completely artificial world where there are no natural environments with which we can compare thermal rate constants k_T for $T < 1$ K. In fact if the interest is really in identifying the dominance of quantum effects in this regime then ideally one would try to measure internal-quantum-state-selected experiments, to limit the number of channels that contribute to the reaction and therefore enhance the probability of observing resonant effects in the cross section *etc.*

Would you comment on why you consider that the measurement of rate constants for thermally equilibrated systems is an important objective in the sub-kelvin range?

1 L Harper, M. T. Bell and T. P. Softley, Unpublished work.

Professor Dr Gerlich responded: Various discussions during the meeting have revealed that there is no general agreement concerning the meaning of cooling, ultracold molecules, cold chemistry, *etc.* My personal opinion has been summarized recently in the two chapters I contributed to Professor Smith's book.[1] Although

I do not intend to continue the discussion on this level, I am tempted to mention that the title of your paper contains misleading formulations such as "ion–molecule chemistry at very low temperatures" or "cold chemical reactions". I agree with your formulation that you can study translationally cold collisions in your very interesting experiment, provided that the ions are really localized on the symmetry axis of the rf driven quadrupole and that the cold effusive beam traverses the ion cloud without hitting any surface which would create a 300 K neutral background.

I am in accord with the statement that the relative velocity determines the very first step of an ion–molecule reaction. Capture models have become quite popular in the past decades (see for example ref. 2). However, I do not know any chemical reaction where the outcome of a collision is just determined by the capture and is completely independent of the internal energy of the reactants. Our experience shows that the formation of products depends on many details of the collision dynamics. For example, very long lifetimes of the intermediate complex often lead to radiative association, sometimes even in competition with an exothermic product channel such as in $C_3^+ + H_2$.[3] Many other effects influence the outcome of a reaction, *e.g.* zero point energies, tunneling through barriers, non-adiabatic coupling or restrictions due to symmetry selection rules in the case of identical atoms. All this makes the field of cold ion chemistry very interesting.

There is no final answer to your question on what the most useful experimental approach is for studying chemical reactions in the cold or ultracold regime. The various techniques including swarm methods, merged beams and various traps have been discussed recently.[4] There is no doubt that it would be ideal to measure state to state cross sections over a wide range of translational energies, maybe even as a function of relative orientation. Photoionization methods in combination with supersonic beams and the merged beam technique are a step in this direction.[5] On the other side there are many advantages to starting an experiment with a fully thermalized system since one knows precisely the population of the states. Temperature dependent studies allow one to record and assign, for example, unknown spectra (*e.g.* CH_5^+) or to determine state specific rate coefficients if only a few states are involved (*e.g.* H_3^+). Moreover, in a well-isolated finite ensemble of trapped ions, laser based strategies can be used to modify and tailor the thermal population, *e.g. via* hole burning.

Finally I want to comment on the statement that the sub-K regime is really an artificial world. Presently our universe is at 2.725 K; however, further expansion may lead to natural environments with $T < 1$ K. For predicting star formation in such regions, sub-K collisions are needed.

1 I. W. M. Smith (ed.), *Low temperatures and cold molecules*, World Scientific Publishing, ISBN 978-1-84816-209-9, 2008.
2 E. E. Nikitin and J. Troe, *Phys. Chem. Chem. Phys.*, 2005, **7**, 1540.
3 I. Savic and D. Gerlich, *Phys. Chem. Chem. Phys.*, 2005, **7**, 1026.
4 D. Gerlich, The study of cold collisions using ion guides and traps, in *Low temperatures and cold molecules*, World Scientific Publishing, ed. I. W. M. Smith, ISBN 978-1-84816-209-9, 2008, p. 121.
5 T. Glenewinkel-Meyer and D. Gerlich, *Isr. J. Chem.*, 1997, **37**, 343.

Dr Stoecklin commented: Concerning the possible use of the adiabatic capture theory for ion dipole reaction mentioned by Professor Softley, I would like to mention that the adiabatic capture theory cannot work in the very low collision energy range as it is based on a classical treatment of translation. However, an alternative approach was proposed by Dashevskaya *et al.*[1] where the Wigner regime is treated properly and the adiabatic capture approximation is used at higher energy.

1 E. I. Dashevskaya, A. I. Maergoiz, J. Troe, I. Litvin and E. E. Nikitin, *J. Chem. Phys.*, 2003, **118**, 7313.

Professor Sims addressed Professor Softley: In your comment, you asked why we should concern ourselves with the measurement of thermal rate coeffents when detailed measurements on single quantum states provide a more penetrating comparison with theory. While of course this would be true in an ideal world where it is possible to make such measurements of reactive or inelastic cross sections on an absolute basis, in practice this is extremely difficult to accomplish.[1] The advantage of thermal rate coefficient measurements at extremely low temperatures (indeed, at any temperature) is that the quantum state distributions of the colliding species are very well defined and known, and by the use of standard pseudo-first-order kinetics techniques, the absolute values of these rate coefficients can be reliably determined. This constitutes a significant advantage over many molecular beam techniques, and enables a more revealing comparison with theory than that possible with relative measurements alone. Of course, in the end, the two types of technique are highly complementary.

1 G. Scoles. In Atomic and Molecular Beam Methods; Scoles, G., Ed.; Oxford University Press: New York, Oxford, 1988; Vol. 1; p 4.

Dr Segal asked Professor Gianturco: Sympathetic cooling of molecular ions using laser cooled atomic ions is a well established technique. One well known limitation of this technique is that, while the long range Coulomb collisions are effective at equilibrating the translation temperatures of the co-trapped species, these collisions do not affect the internal degrees of freedom of the molecules. The molecular ions therefore remain rotationally hot and settle at a temperature close to that of the ambient black body field. One suggestion as a means of cooling the rotational motion is to co-locate cold atoms in a MOT (*e.g.* Rb) with a mixed cloud or Coulomb crystal of molecular ions and laser cooled atomic ions in an ion trap and then depend on relatively short range (r^{-4}) atom–ion collisions to cool the internal degrees of freedom of the molecules. In this context the chemical reactions described in your comment constitute loss channels *i.e.* the hope is that the rate of rotational cooling is sufficiently fast that it occurs on a timescale that is short in comparison with the timescale for the chemical reactions to take place. Are you able to calculate the rate of rotational cooling in order to be able to make this comparison?

Professor Gianturco replied: We have not done it as yet, although we have put together the relevant computational machinery to do so. However, one should also remember here that the "reaction" that entails a charge transfer process between MgH^+ and Rb is a strongly exothermic process as shown by our calculations and by the figures presented at the Discussion. Hence, what is important here is that the internally "hot" ion be able to cool its internal energy content with just a few collisions before charge exchange takes over. We already have some results for the rotational quenching of other ionic partners like OH^- and OH^+ but in collisions with He, which presents a less strong interaction with the molecules than Rb. However, the rotational relaxation rates for OH^+, also a polar cation like MgH^+, are already the largest exhibited by ultracold collisions with He and vary between 10^{-10} and 10^{-9} depending on the initial rotational state of OH^+.[1] Thus, it is reasonable to expect that the rotational cooling of MgH^+ in the presence of Rb might occur on a faster timescale than that of the reactions we are discussing in our computational work.

1 L. Gonzalez-Sanchez *et al.*, *Eur. Phys. J. D*, 2007, **44**, 65.

Professor Dr Gerlich addressed Professor Sims: Your comment that flow systems (such as CRESU) work in a real thermal equilibrium, sounds like a general advantage; however, one certainly gets more information if one can extract state specific (or state-to-state) rate coefficients from an experiment. Ion traps can be operated in various ways. With high buffer gas concentration (usually He), one gets near-thermal conditions. If one modifies the trapped cloud of ions (usually a few hundred) in a well-defined way, *e.g. via* laser induced processes, one can derive very specific information on elastic, inelastic or reactive collision processes.

Professor Dr Willitsch asked Professor Gianturco: The predominance of the Penning Ionisation channel over the direct-charge exchange reaction in excited Mg^+/MgH^+ + Rb strikes me as quite remarkable. Do you expect a similar behaviour in other systems *e.g.* Ca^+ + Rb, as well?

Professor Gianturco answered: We were also surprised by the behaviour of the MgH^+/Mg^+–Rb system in terms of the expected efficiency of the Penning Ionization process in the trap, although we must consider the results as coming from a fairly simplified treatment of the PI dynamics, which definitely require further exploration. In relation to that process being just as important in the heavier cations, my presumption at the moment is that perhaps it should be just as important, since we have more excited states available and closer to each other; level congestion is usually a good reason for increasing couplings to the nearby continuum, which is the main cause for the more likely occurrence of PI events.

Dr Küpper asked Professor Dr Gerlich: On page 3 of your paper you discuss technical issues with the applied voltages to your rf traps. Already around 1970 the equivalence of rectangular wave forms and sinosoidal wave forms for ion-guides and traps was shown. Could you comment on their respective advantages and disadvantages, also with special emphasis on parasitic frequencies, heating of ions, and technical implementation details?

Professor Dr Gerlich replied: The motion of a system under the influence of a force varying in time and space is a rather fundamental subject, well studied in theoretical and experimental physics. One of the mathematical treatments for understanding the influence of the time dependent electric field in our applications is based on a special adiabatic approximation which has been discussed in textbooks of classical mechanics (*e.g.* ref. 1). This calculation reveals that the effective potential is proportional to the time average of the square of the field, *e.g.* to $<|E(t)|^2>$. Based on this it is obvious that, in principle, one can use any alternating field, *i.e.* any symmetric wave form. However, one can get problems with low frequency components since they can lead to parasitic heating of the ion motion while the high frequency components of rectangular pulses do not contribute much to the guiding force ($V^* \sim 1/\Omega^2$). Nonetheless, one should avoid them since they cause electronic noise. An important practical reason for using pure sinusoidal oscillations with a single frequency Ω is that we usually connect the electrode arrangement (equivalent to a capacitance) with a coil, resulting in a resonance circuit. As discussed in my contribution, we presently test low temperature traps with superconducting circuits. This will lead to very high quality circuits, to a very pure frequency and to extremely low power dissipation.

1 L. D. Landau and E. M. Lifshitz, *Theoretical Physics*, Pergamon, Oxford, 1960, vol. 1, p. 93.

Collision experiments with Stark-decelerated beams

Sebastiaan Y. T. van de Meerakker* and Gerard Meijer

Received 5th November 2008, Accepted 16th December 2008
First published as an Advance Article on the web 29th April 2009
DOI: 10.1039/b819721k

The crossed molecular beam technique has been established as a mature and important experimental method for detailed studies of molecular interactions, and has contributed enormously to our present understanding of molecular reaction dynamics. The Stark deceleration technique yields unprecedented control over both the internal and external degrees of freedom of polar molecules in a molecular beam, offering new possibilities in scattering experiments. In particular, Stark-decelerated molecular beams allow detailed molecular scattering studies as a function of the collision energy, from low to high collision energies, and with a very high energy resolution. We discuss a variety of experimental geometries that exploit this new molecular beam technology for scattering experiments, ranging from crossed beam arrangements and molecular synchrotrons to surface scattering set-ups.

I. Introduction

Acquiring a detailed understanding of the interaction between individual molecules is a central theme in molecular physics, and is essential, for instance, to understand the dynamic behavior of larger systems. The study of collisions between neutral atoms and molecules in the gas-phase is a well-established approach to probe the potential energy surfaces (PES) that govern molecular interactions. The level of detail that can be reached in these experiments depends on the quality of the preparation of the collision partners prior to the collision event. "In the ultimate experiment one would determine the cross section and angular distribution of the products for a completely specified collision", as Raphael Levine and Richard Bernstein put it in their classic book "*Molecular Reaction Dynamics and Chemical Reactivity*".[1] The parameters that are to be specified include the quantum state of the reactants and products, the alignment or orientation of the reactants and products, the angular and velocity distribution of the products, and the energy of the collision. Experimentally, the crossed molecular beam technique is ideally suited to obtain detailed information of the PES.[2] Crossed molecular beams enable the study of molecular encounters under single collision conditions, and allow the realization of many of the desiderata listed above.[3] A rich variety of experimental geometries have been engineered to obtain control over the internal quantum states,[4] the orientation,[5] and the collision energy of the scatterers.[6,7] Sophisticated laser-based detection methods have been developed to analyze the state, angular, and velocity distributions of the molecules after the collision.[8–11] These techniques also allowed the determination of product pair correlations in bi-molecular reactions.[12]

Stark-decelerated molecular beams offer interesting new prospects for increased control over the molecular reactants in scattering experiments. The Stark decelerator

---

Fritz-Haber-Institut der Max-Planck-Gesellschaft, Faradayweg 4–6, 14195 Berlin, Germany. E-mail: basvdm@fhi-berlin.mpg.de

for neutral polar molecules is the equivalent of a linear accelerator (LINAC) for charged particles, and exploits the interaction of a polar molecule with inhomogeneous time-varying electric fields.[13] The deceleration (or acceleration) process can be seen as slicing a packet of molecules with a narrow velocity distribution out of the most intense part of the molecular beam pulse. This packet can then be decelerated or accelerated to any velocity, maintaining the narrow velocity distribution and the particle density in the packet. In a crossed beam configuration, this offers the revolutionary possibility to study fully state-selected (in)elastic and reactive scattering as a function of the continuously variable collision energy with a high intrinsic energy resolution. The computer-controlled velocity of the molecular beam(s) allows one to scan the collision energy in an otherwise fixed experimental geometry. The deceleration process is highly quantum-state specific, and the state purity of the bunches of selected molecules that emerge from the decelerator is measured to be close to 100%. Contamination of inelastic state-to-state scattering data by initial populations in different quantum states is therefore negligible. The high state purity of the beam also allows sensitive background-free detection of inelastically scattered products. The molecules that exit the decelerator are all naturally spatially oriented, allowing steric effects to be studied. Last but not least, the possibility to produce molecular beams with a low velocity gives access to scattering studies at low collision energies, an experimentally almost unexplored regime.

One of the interesting aspects of cold collisions involving molecules is the occurrence of scattering resonances. These resonances occur, for instance, when the collision energy is degenerate with a bound state of the collision complex.[14] These resonant states can also be formed near the energetic thresholds for inelastic scattering, where the collision deposits all available energy into internal molecular degrees of freedom.[15] Molecular collision studies as a function of the collision energy, and the careful mapping of scattering resonances, provide sensitive probes for the potential energy surfaces. Low collision energies also allow external control over the collision dynamics by electromagnetic fields.[16] At collision energies near one Kelvin, the perturbations due to the Zeeman and/or Stark effect become comparable to the translational energy. In this regime, molecular collisions are highly susceptible to externally applied fields, opening the possibility for controlled chemistry.[17]

These "tamed" molecular beams that have been produced using either electric, magnetic or optical fields,[18] can be used to advantage in a new generation of scattering experiments. It is the challenge, however, to design experiments that exploit these new possibilities while maintaining the already existing high level of control that has set the standard in the field of molecular scattering. In this paper we will present our view of how the Stark deceleration technique can be implemented in scattering experiments to probe the intermolecular interaction potential in detail. We will discuss the feasibility of these experiments, and we will describe the first bi-molecular study that is planned in our laboratory, $i.e.$, the scattering between Stark-decelerated beams of fully state-selected OH ($X\,^2\Pi_{3/2}$, $J = 3/2$, f) and NO ($X\,^2\Pi_{1/2}$, $J = 1/2$, f) radicals. Finally, we will discuss the potential of molecular storage rings for scattering experiments and we will describe a surface scattering experiment with slow, ground-state CO molecules that is currently being set up.

II. The density question

A critical parameter that determines the success of any crossed beam scattering experiment is the product of the molecular particle densities[50] of the colliding beams. The particle densities in the colliding beams should be sufficient to build up a detectable signal of the scattering products in the time the two beams intersect. Unfortunately, the relevant state-to-state inelastic collision rates for bi-molecular scattering are not accurately known and vary from system to system, but we can estimate the required particle densities by assuming a scattering rate k of 10^{-10}–10^{-11} cm^3 molecule^{-1} s^{-1}. These rates are typically found for diatom-rare gas atom systems at high collision energies, as well as for collisions between di-atoms at low collision energies.[19] The required particle densities n_A and n_B in each of the beams can now be estimated from

$$N_{\text{product}} = n_A n_B k V \Delta t \quad (1)$$

where N_{product} is the number of scattered molecules just after the beams have collided, and Δt is the time during which both beams interact. The scattering volume V in our experiments is approximately $3 \times 3 \times 3$ mm^3. The detection limit of the scattering products depends strongly on the molecular species of interest, on the detection technique that is used, on the spatial overlap of the detection laser with the scattering volume as well as on the background level in the experiment. In pulsed beam experiments, product molecular densities (N_{product}/V) typically between 10^3–10^4 cm^{-3} can be detected with a reasonable signal-to-noise ratio within a reasonable signal accumulation time. We will use this number density as a figure of merit to estimate the required particle densities in the colliding beams, and realize that for selected systems this estimate is under- or over-valued.

We assume that detectable scattering products are only produced during an interaction time of ~10 μs of the intersecting beams, leading to a required product density $n_A n_B$ of 10^{18}–10^{20} cm^{-6} to be able to experimentally observe scattering. Note that the particle density in each individual beam is in principle inconsequential, as the state purity of both beams is close to 100%. The inelastically scattered molecules can therefore be detected background free, independent of the particle density and rotational temperature of the parent molecular beams.

III. Molecular traps

It appears appealing to relax the requirements on the particle densities by increasing the interaction time between the molecules. In recent years a variety of traps, either electric,[20,21] magnetic, or a combination of both,[22] have been developed that allow the confinement of molecules for times up to several seconds.[23] Traps offer an interaction time that is typically five orders of magnitude larger than the interaction time in a crossed beam configuration. This advantage is deceptive, however, as the scattered products are, in general, not accumulated during this interaction time. In most trapping experiments, molecules are initially confined in a low-field seeking quantum state. If the collision induces a transition from the low-field seeking into a high-field seeking state, the molecules can no longer be confined to the trap. Inelastic channels are usually also available that populate low-field seeking states that can potentially be confined to traps. However, these collisions either populate hyperfine states that cannot be probed independently from the population in the initial quantum state, or provide the scattered molecules with a recoil energy that is larger than – or comparable to – the trap depth. Scattering between molecules in a trap can therefore only be inferred from the observed depletion of trapped molecules as a function of time. To be able to experimentally identify trap loss due to collisions of trapped molecules, this density dependent loss channel needs to be comparable in magnitude to the other loss mechanisms. Referring back to eqn 1, even a molecular density of 10^9 molecules cm^{-3} in the trap will only lead to a 10% reduction of the trapped molecules in the first second of trapping, if we assume a scattering rate of 10^{-10} cm^3 molecule^{-1} s^{-1}. Clearly, the advantage of the long interaction time afforded by the trap is undone by the inferior detection sensitivity of the scattering events.

Detailed molecular scattering studies are therefore better performed in a beam environment than in a trap. The study of molecular collisions in the *cold* (1 mK–1 K) temperature regime, *i.e.*, at temperatures that are routinely reached in traps, is also possible with molecular beams when two merging molecular beams with a well-defined velocity are used that slowly overtake each other. This experimental arrangement has the advantage that the collision energy can be varied, and that experimental parameters in the interaction region can easily be controlled. Moreover, the molecular densities that are reached in decelerated beams are generally higher than those that are obtained in traps. In section VII we discuss different experimental possibilities to achieve merged molecular beams.

The confinement of molecules in traps *is* required to study molecular collisions in the *ultracold* (<1 mK) regime that is inaccessible to crossed beam configurations. These collisions are dominated by *s* or *p*-wave scattering and are sensitive to the long-range part of the potential. In particular the dipole–dipole interaction for polar species gives rise to rich physics that is foreign to collisions at higher temperatures.[24,25] Recently, tremendous progress has been made in producing ground-state ultracold polar molecules *via* photoassociation spectroscopy,[26–28] bringing ultracold bi-molecular collision studies of alkali di-atoms within reach. For more chemically relevant molecules like ammonia or the hydroxyl radical that have only been trapped at tens of mKs thus far, further cooling schemes like evaporative or sympathetic cooling need to be developed.

IV. Scattering Stark-decelerated OH radicals with Xe atoms

Recently, we have carried out the first crossed beam scattering experiment using a Stark-decelerated beam. In this experiment, a Stark-decelerated beam of OH radicals collided with a conventional beam of Xe atoms.[29] The velocity tunability and the narrow velocity spread of the OH beam allowed us to tune the center-of-mass collision energy over the energetic thresholds for inelastic scattering into rotationally excited states. The velocity of the incoming OH radicals was varied from 33 m s^{-1} to 700 m s^{-1}, and the behavior of the cross sections for inelastic scattering around the energetic thresholds were accurately measured. In Fig. 1 the measured relative inelastic scattering cross sections are shown as a function of the collision energy for the scattering channels that are indicated in the figure.

A new *ab initio* potential energy surface for the OH–Xe system was calculated, and fully converged coupled-channels calculations were performed. The resulting

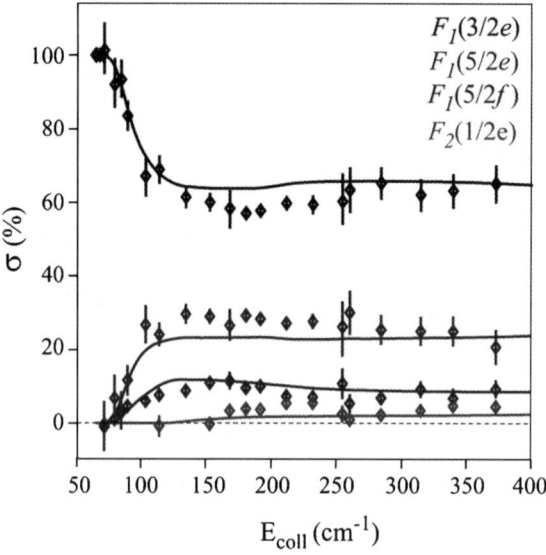

Fig. 1 Collision energy dependence of the relative inelastic collision cross section for scattering of OH ($X\ ^2\Pi_{3/2}$, $J = 3/2$, f) radicals with Xe atoms for the indicated scattering channels (see Fig. 3 for a rotational energy level diagram of the OH radical). The solid lines result from *ab initio* coupled-channels calculations.

relative inelastic cross-sections are shown as solid lines in Fig. 1 and agree excellently with the measured inelastic cross-sections throughout the range of collision energies that were probed.

In the OH–Xe experiment, the energy resolution of about 13 cm^{-1} that could be obtained was almost exclusively determined by the velocity spread in the Xe beam. The logical next step is to develop experimental methods to study bi-molecular collisions using two decelerated molecular beams under a 90 degree crossing angle. In such an experiment, molecular scattering between (combinations of) fully state-selected molecules like OH, NH, ND$_3$, CO, H$_2$CO, and SO$_2$ can be studied with a collision energy that can be varied from below 1 cm^{-1} to conventional collision energies (few hundred cm^{-1}). The energy resolution is ≤ 1 cm^{-1}, one order of magnitude better than in the OH–Xe experiment, and typically two orders of magnitude better than in conventional crossed molecular beam experiments. In these experiments, detailed studies of threshold phenomena, scattering resonances, and collisions at low temperatures are possible.

V. More efficient Stark decelerators

The particle densities that have been obtained thus far in Stark-decelerated molecular beams are, unfortunately, insufficient for scattering studies using *two* crossed Stark-decelerated beams. Depending on the molecular species and the final velocity of the packet of molecules, densities that are typically reached range from 10^6 to 10^9 molecules cm^{-3}. From Stark deceleration and trapping experiments in our laboratory, we know that the highest densities can be obtained for the molecules ND$_3$ and CO ($a\ ^3\Pi$). The density of decelerated OH beams, although often used in deceleration and trapping experiments, is already one to two orders of magnitude less, while decelerated beams of NH ($a\ ^1\Delta$) radicals are yet somewhat less intense.

During the last few years, an improved understanding of the motion of molecules in a Stark decelerator has been obtained. In particular, it was found that the

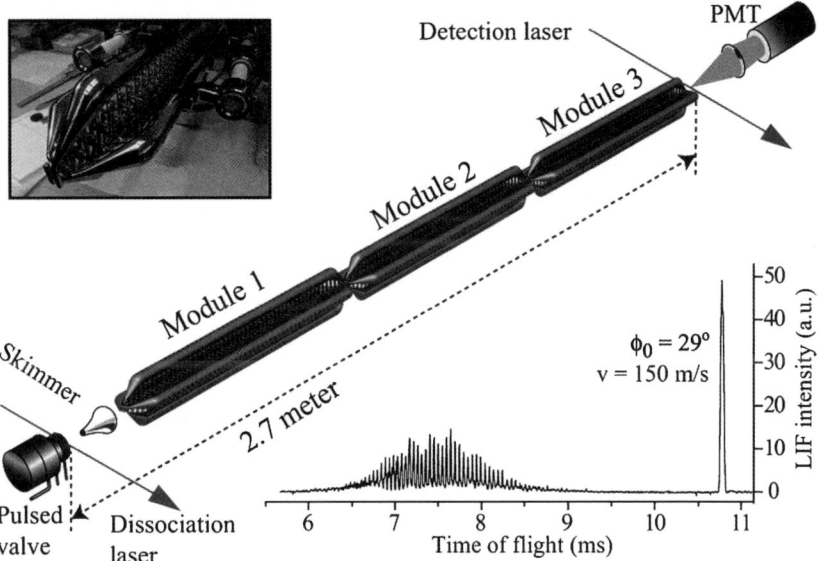

Fig. 2 A 2.6 meter long Stark decelerator that consists of 3 modules of about 100 electric field stages each. In the top left inset a photograph of a decelerator module is shown. In the bottom right corner an experimental time of flight profile of OH radicals is shown that is recorded at the exit of the decelerator. In this measurement, a packet of OH radicals is decelerated from 350 m s^{-1} to 150 m s^{-1}, using the decelerator in the $s = 3$ overtone mode at a phase angle of 29°.

coupling between the longitudinal and transverse motion of the molecules can substantially reduce the number of molecules that reach the exit of the decelerator.[30] In a Stark decelerator with the original electrode geometry,[13] this coupling between the longitudinal and transverse motion can be effectively switched off when the decelerator is operated in the so-called $s = 3$ overtone mode.[31] In this mode, only every third electric field stage is used for deceleration, while extra transverse focusing is provided by the intermediate stages. The performance of this scheme has been experimentally studied in our laboratory by passing a beam of OH radicals through a 2.6 meter long Stark decelerator consisting of 316 electric field stages.[32] A schematic representation of this Stark decelerator is shown in Fig. 2, together with a typical measured Time Of Flight (TOF) profile of a beam of OH radicals that exits the decelerator.

In this machine, the OH radicals can be detected either after 103 or after 316 electric field stages, enabling a direct comparison between the conventionally used mode of operation ($s = 1$) and the $s = 3$ overtone mode of operation under otherwise identical conditions, in particular using the same phase angle ϕ_0. We have demonstrated and quantified that indeed for many applications, the acceptance of the Stark decelerator in the $s = 3$ mode significantly exceeds that of a decelerator in the conventionally used mode of operation. A gain up to a factor of 4 is observed, and is found to depend strongly on the phase angle that is used. Simulations indicate that, compared to the decelerators that have been commonly used so far ($s = 1$, ≈ 100 stages, $\phi_0 = 50$–$60°$), a gain of a factor of 10 is realistic for a large range of final velocities if the decelerator were to be built even longer. It is important to note that this gain in number density is accompanied by a reduction in the longitudinal translational temperature by a factor of 3.[31] With this new deceleration scheme, scattering experiments between Stark-decelerated molecular beams have become feasible.

VI. Scattering between OH and NO radicals

Based on scientific relevance and feasibility, the first bi-molecular scattering experiment that is planned in our laboratory using two crossed Stark decelerators is the scattering between state-selected OH ($X\,^2\Pi_{3/2}$, $J = 3/2$) and NO ($X\,^2\Pi_{1/2}$, $J = 1/2$) radicals. Collisions involving radicals, and in particular the OH and the NO radical, play a key role in atmospheric and interstellar chemistry as well as in combustion processes.[33] Hence, the scattering of OH and NO with (mostly) rare-gas atoms has received much experimental and theoretical attention. For both species, ingenious experiments have been performed in which the radicals have been state-selected,[34,35] and in which steric effects have been studied by orienting the radicals prior to the collision by static electric fields.[36,37] Recently, reactive scattering has also been observed using state-selected beams of OH radicals.[38] Theoretically, high level *ab initio* PES's have been calculated, and the comparison with experimental data has yielded a wealth of information on the interaction between open shell $^2\Pi$ molecules and other atoms or molecules.

As a prototypical system for the interaction between *two* open shell $^2\Pi$ molecules, we plan to study the inelastic scattering between fully state-selected OH and NO molecules as a function of the collision energy. This will provide detailed information on the intermolecular interactions in this system,[39] and will probe the potential energy surface beyond the bound part of the potential that can be studied spectroscopically.[40] Although challenging, these systems are also still theoretically tractable,[41] so that experimental cross sections can be compared to theoretically calculated ones. These studies are not only relevant for atmospheric and combustion processes but are also relevant to predict the feasibility of evaporative cooling of molecules like OH in electric or magnetic traps, that solely relies on the collision properties of $^2\Pi$ molecules at low collision energies.

The rotational energy level schemes of the OH and NO radicals are shown in Fig. 3. The spin–orbit interaction splits the $X\,^2\Pi$ electronic state into two rotational

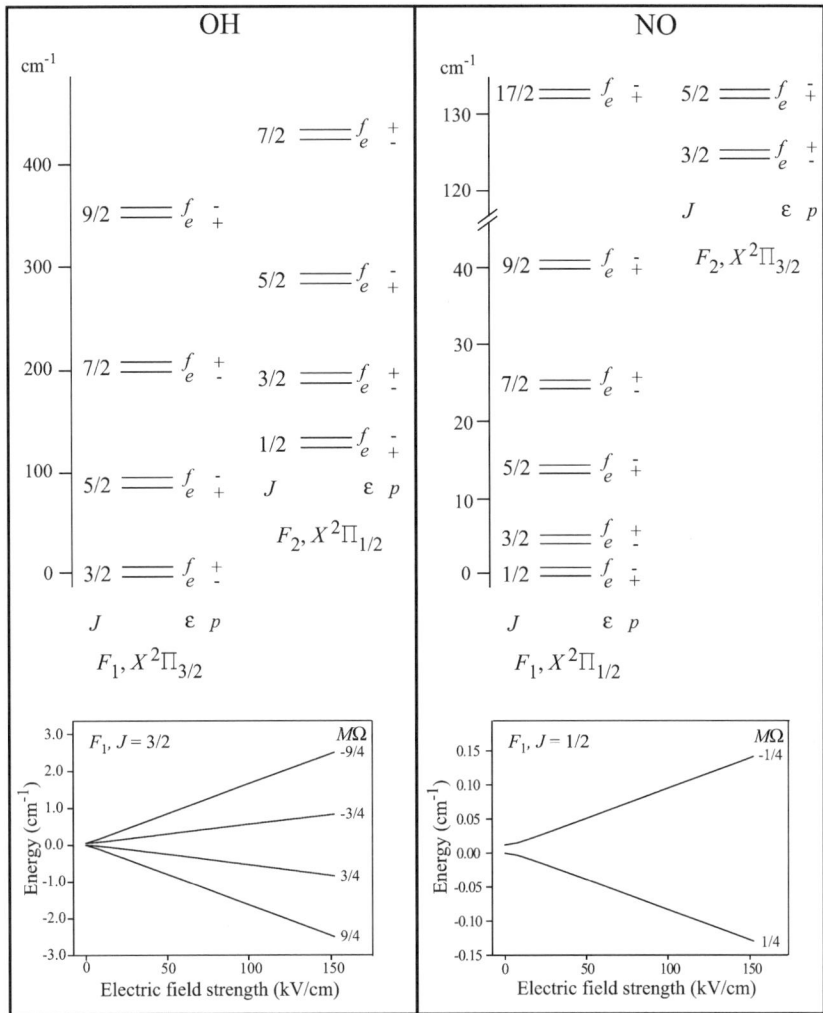

Fig. 3 Rotational energy level diagram of the electronic and vibrational ground states of the OH (left) and NO (right) radicals. The Λ-doublet splitting of the energy levels is greatly exaggerated for reasons of clarity. The parity and spectroscopic symmetry labels are indicated for each level. The Stark energy diagram for the rotational ground state is shown in the lower part for both species. Note the different energy scales that are used.

manifolds with $|\Omega| = 1/2$ and $|\Omega| = 3/2$, respectively. In the united atom picture, the NO radical has one electron in the 3pπ valence shell, and the $|\Omega| = 1/2$ manifold is lower in energy than the $|\Omega| = 3/2$ manifold. The OH radical has three electrons in the 2pπ shell, one electron short of filling the shell. Consequently, the energy level scheme of the OH radical is said to be inverted and the $|\Omega| = 3/2$ manifold is lower in energy than the $|\Omega| = 1/2$ one. The OH and NO radicals have rotational constants of 18.5 and 1.7 cm^{-1}, respectively, and the rotational energy level spacing is thus much larger for OH than for NO. Both molecules have a spin–orbit constant that is similar in magnitude (−139.7 cm^{-1} for OH and 123.1 cm^{-1} for NO).

A schematic representation of the crossed beam scattering machine that is being developed in our laboratory is shown in Fig. 4. The machine consists of two Stark decelerators under a 90 degree crossing angle, where each decelerator consists in

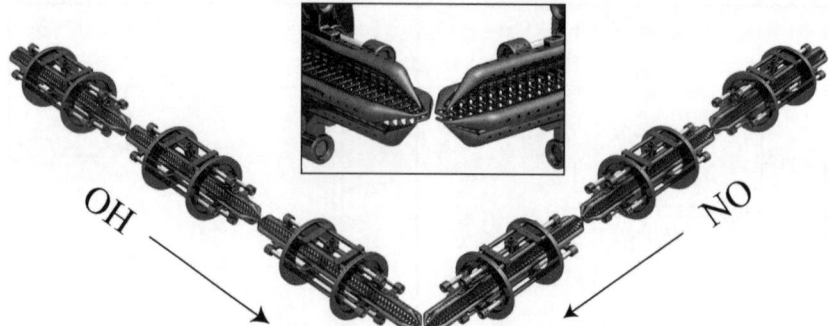

Fig. 4 Schematic representation of the crossed beam machine. Each decelerator has a modular design and consists in total of 316 electric field stages. The decelerators are designed such that the last electric field stages of either one of the decelerators are very close to the collision zone, while at the same time providing excellent optical access to detect the scattering products (Design: Henrik Haak).

total of 316 electric field stages and has a length of 2.6 meters. Both decelerators have the same design. The last electric field stages are mounted on conically shaped rods. This allows the exits of both decelerators to be very close to the collision zone, while providing excellent optical access to detect the scattering products. In the crossed beam experiment with OH and NO radicals, one decelerator will be used to accelerate or decelerate a beam of OH radicals as has been described in section V. The second decelerator will be used to produce cold packets of NO radicals by mere velocity filtering.

The NO radical possesses only a relatively small dipole moment of 0.16 D. Therefore, deceleration or acceleration of a beam of NO molecules in the $X\,^2\Pi_{1/2}$, $J = 1/2$ state over a large range of velocities is rendered impossible, even for the 300-stage long decelerators that we are constructing. NO radicals in the $X\,^2\Pi_{3/2}$, $J = 3/2$ state experience a significantly larger Stark shift, and the deceleration in this state would be possible. This state also possesses a magnetic moment of 1.2 Bohr magneton,

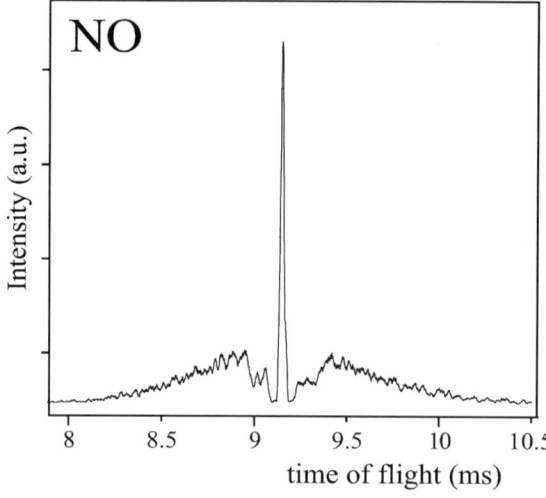

Fig. 5 Simulated time-of-flight profile of a beam of NO ($^2\Pi_{1/2}$, $J = 1/2$) radicals that emerges from the 316 stage Stark decelerator when the decelerator is operated at $\phi_0 = 0°$. The longitudinal temperature of the molecules in the central narrow peak is about 10 mK.

making it a good candidate for efficient deceleration with a Zeeman decelerator.[42] Although the $X\,^2\Pi_{3/2}$, $J = 3/2$ state will hardly be populated in a pulsed molecular beam, NO molecules can be prepared in this state from the rotational ground state by stimulated emission pumping *via* the $A^2\Sigma^+$ state. The dipole moment of NO molecules is sufficient, however, to *guide* $(\phi_0 = 0°)$[43] $X\,^2\Pi_{1/2}$, $J = 1/2$ state molecules through the decelerator, leaving the velocity of the selected part of the beam unchanged. In this mode, the decelerator acts as a velocity selector: only molecules within a limited initial velocity range that is determined by the Stark effect of the molecule are kept together, and arrive in the collision region as a compact bunch. The small dipole moment of NO results in a narrow velocity spread and thus a low temperature of the bunch. To produce a cold packet of NO radicals with a velocity that is as low as possible, a cooled pulsed valve will be employed and Xe will be used as a carrier gas. The TOF profile that can thus be obtained is shown in Fig. 5. In this simulation, a molecular beam of NO ($^2\Pi_{1/2}$, $J = 1/2$) molecules with a mean velocity of 280 m s^{-1} and with a 10% velocity spread is guided through the Stark decelerator. The intense and narrow peak in the center of the arrival time distribution contains the selected part of the beam. These cold NO molecules have a velocity spread of about 3 m s^{-1} (or 1% of the mean velocity).

The timing of the experiment can be chosen such that both the decelerated/accelerated OH beam and the cold part of the NO beam arrive simultaneously, but temporally separated from the remainder of the OH and NO beam pulses, in the collision zone. The Xe atoms from the carrier gas that is used to produce the NO molecular beam arrive simultaneously with the cold packet of NO radicals in the collision zone. These atoms, however, are not kept together by the Stark decelerator, and their particle density is reduced to a negligible number during the 2.7 meter flight path to the collision center.

When the velocity of the OH beam is varied between 50 m s^{-1} and 800 m s^{-1}, the center-of-mass collision energy is scanned from 35 cm^{-1} to 325 cm^{-1}. This energy range encompasses many energetic thresholds for inelastic scattering in both the OH radical and the NO radical, and many resonances in the inelastic cross sections can be expected. The energy resolution is about 1 cm^{-1} at the lowest collision energies, and scales with the square root of the collision energy.

In view of the small dipole moment of the NO radical, the Stark decelerator will only select a small fraction of the initial pulse of NO radicals. We can estimate this fraction from the calculated acceptance of the Stark decelerator. The acceptance for NO radicals is about one order of magnitude less than the acceptance for OH radicals under otherwise identical conditions. Molecular beams of NO radicals can be produced with particle densities that exceed the densities that are reached for OH radicals by orders of magnitude, however. The NO radical is one of the few open shell radicals that is stable enough that it can be stored at large pressures in the gas-phase, and can be directly seeded at large quantities in a carrier gas. From the acceptance of the Stark decelerator and from typical densities of molecular beam pulses, we expect a particle density of $\geq 10^{10}$ molecules cm^{-3} in the $X\,^2\Pi_{1/2}$, $J = 1/2$, f state in the cold packet. For the OH beam, a particle density of 10^9 cm^{-3} appears realistic with our long Stark decelerator, operating in the $s = 3$ overtone mode. With these particle densities we expect a scattering product density of about 10^4 molecules cm^{-3}. This is well within the detection limit for both the OH and NO radicals.

The OH radical can be sensitively detected using a Laser Induced Fluorescence (LIF) scheme on the strong dipole-allowed $A\,^2\Sigma^+ \leftarrow X\,^2\Pi$ transition in the ultraviolet part of the spectrum. The NO radical can be conveniently detected by a two-color (1 + 1′) Resonance Enhanced Multi Photon Ionization (REMPI) detection scheme. In this scheme, the first photon (226 nm) excites the NO radicals to the $A\,^2\Sigma^+$ state. The second photon (either 193 nm or 266 nm) ionizes the molecule, and the resulting ions can be detected using a standard time-of-flight mass spectrometer. Both transitions are very strong, and can be saturated with unfocussed laser beams.

The detection laser beams can therefore be matched in size to the intersection of the molecular beams, and the inelastically scattered NO radicals are then detected with almost unity detection efficiency. This REMPI detection scheme for NO radicals is at least two orders of magnitude more sensitive than LIF detection schemes, and allows the detection of particle densities as low as 10^2 molecules cm^{-3}.

The expected signal levels in the OH + NO scattering experiment can also be estimated from the signal levels that were obtained in the OH + Xe scattering experiment (see section IV). In that experiment, depending on the inelastic collision channel, 0.1%–0.5% of the incoming OH radicals were inelastically scattered. From this, the density of Xe atoms in the target beam was estimated to be 10^{12}–10^{13} atoms cm^{-3}. The expected density of NO radicals in the proposed OH + NO experiments is two to three orders of magnitude lower. This is compensated for by the at least two orders of magnitude superior detection sensitivity of NO radicals, and by the about one order of magnitude higher density of OH radicals. The signal-to-noise levels, therefore, are expected to be at least similar to those in the OH + Xe experiment.

VII. The molecular synchrotron

The combination of Stark-decelerated beams with molecular storage rings offers interesting alternatives to the crossed beam set-up for molecular scattering studies. In its simplest form a storage ring is a trap in which the particles – rather than having a minimum potential energy at a single location in space – have a minimum potential energy on a circle. As in traps, packets of molecules can be confined for times up to seconds, allowing for a large interaction time of the colliding particles. The advantage of a storage ring over a trap is that a storage ring accepts packets with a variable velocity, enabling scattering studies as a function of the collision energy. In addition, as molecules with a non-zero mean velocity can be confined in the ring, superior number densities of the molecular packets can be obtained. As in traps, the advantage of the long interaction time in storage rings is of limited use to study state-to-state inelastic collision processes. Inelastically scattered molecules in general do not experience the required potentials to stay confined in the ring. Molecular storage rings, however, are the ideal experimental arrangements to study the total (elastic + inelastic + reactive) scattering cross section, that can be inferred from the measured loss-rates of the stored molecules.

The first storage ring for neutral polar molecules was devised by bending a long electrostatic hexapole into a torus.[44] A Stark-decelerated beam of deuterated ammonia molecules was decelerated to approximately 100 m s^{-1} and injected into the ring, in which up to 50 round trips could be observed. As a result of the longitudinal velocity spread of the packet of molecules, however, the packet gradually spreads out along the ring on making successive round trips. In the second version of a molecular storage ring, bunching elements were added to the storage ring to counteract this spreading. This storage ring consists of two hexapole half-rings that are separated by a 2 mm gap. By appropriately switching the voltages as the molecules pass through the gaps between the two half-rings, molecules experience a force that keeps them together longitudinally in a compact bunch. This structure is the neutral analogue of a synchrotron for charged particles.[45] The synchrotron ensures that the density of stored molecules remains constant for each round trip. The broken symmetry of the synchrotron allows the injection of multiple packets into the ring, either collinear or counterpropagating, without affecting the packets that are already stored. While circling the ring, these particles can be made to interact repeatedly at well defined times and at distinct positions. The open electrode geometry allows control over experimental parameters in the interaction regions, and the introduction of electromagnetic fields and/or stationary targets in the beam path. The number of packets that can be stored simultaneously, and the efficiency of the bunching process, depend on the number of individual segments of the synchrotron. We are currently testing a molecular synchrotron that consists of 40 short hexapole segments.

In Fig. 6 a variety of possible experimental arrangements is shown that can be used to study collisions using a molecular synchrotron.

In Fig. 6(a) the simplest arrangement is shown to study molecular collisions in the synchrotron. The synchrotron can be loaded with multiple packets of molecules from adjacent pulses of a single injection beamline. The velocity of each packet is set by the Stark decelerator, allowing multiple trailing packets of molecules in the ring that overtake each other. Collisions between different species can be performed using the arrangements that are shown in Fig. 6(b) and (c). The open electrode geometry of the synchrotron allows the positioning of magnetic or optical traps for molecules or atoms in the beam path. For instance, the combination of a molecular synchrotron with a sample of ultracold atoms enables the study of collisions that are relevant for sympathetic cooling schemes in molecular traps. Collisions between different molecular species are most easily performed when two separate injection beam-lines are used. In Fig. 6(c) an arrangement is shown to load multiple counter-propagating packets of molecules in the ring.

The synchrotron technology can also be exploited for detailed studies of bi-molecular state-to-state inelastic scattering in the cold (1 K–10 K) temperature regime. A possible experimental arrangement for this is shown in Fig. 6(d). A Stark-decelerated packet of molecules with a forward velocity in the 100 to 150 m s^{-1} range is bent using a quarter segment of a synchrotron. The synchrotron technology ensures that the trajectory of the packet is manipulated while keeping the molecules together in a compact bunch. A second Stark-decelerated beam is merged with the first beam, and the molecules can be made to interact in free flight some distance downstream of the exit of the bend. The velocities of both beams can be made slightly different from each other, giving access to collision energies as low as 1 K. This experimental arrangement allows the controlled variation of the collision energy, and allows for

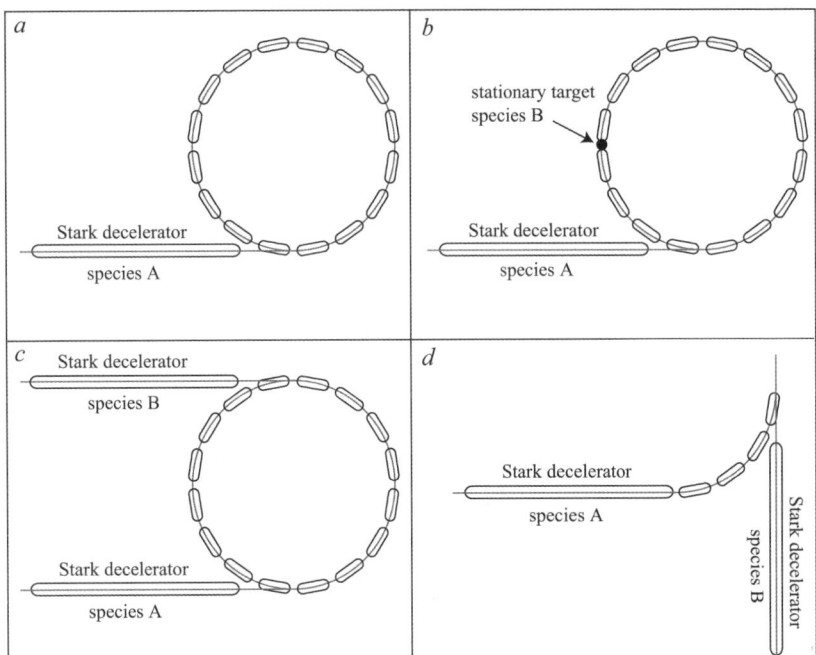

Fig. 6 Schematic presentation of four different experimental configurations that can be used to study molecular collisions, all making use of (parts of) a molecular synchrotron.

full flexibility to control experimental parameters in the interaction region. In addition, the low collision energies are obtained with molecular beams with a relatively high forward velocity and number density; for cold collisions an interesting alternative to traps and crossed beam geometries.

VIII. Molecule surface scattering

Molecular beams with a tunable velocity also offer new possibilities in surface scattering studies. The use of directed molecular beams to probe the interaction of molecules with well-defined single crystal surfaces, is a mature field of research. Beams of state-selected molecules, in particular ground-state CO and NO molecules, are routinely used for these studies.[46] After scattering from the surface, the internal ro-vibrational energy, the kinetic energy and the angular distribution of the scattered molecules are probed to obtain information on the molecule–surface interaction.[47,48] Although the velocity of the incoming beam of molecules can normally only be changed by changing the carrier gas or the temperature of the source, the component of the velocity perpendicular to the surface – which is assumed to be the only relevant one – can be changed by simply changing the angle of incidence of the beam with the surface. It is actually an interesting question whether indeed changing the perpendicular component of the velocity by rotating the sample, for instance, has the same effect as changing the speed of the molecules under normal incidence conditions. At very low velocities of the incoming molecules, this is most likely no longer true, and the Stark decelerator offers the possibility to experimentally address this issue.

In selecting a molecule for Stark deceleration and subsequent interaction with a surface, it is important to select a molecule/surface system that has already been well-characterized. The main new feature that the Stark-decelerated beams can contribute to these studies is that beams can be produced with very low velocities, and still with a very narrow velocity spread. Obviously, the molecules that can be most efficiently decelerated, are molecules with a rather large electric dipole moment. These are not the molecules one would like to use to study the rather subtle effects at low velocities in the molecule–surface interaction, however, as these molecules typically experience a rather strong interaction with the surface.

In collaboration with Alec M. Wodtke (UCSB, Santa Barbara, CA, USA) we are setting up an experiment to study the interaction of slow beams of ground-state CO ($X^1\Sigma^+$) molecules with an Au(111) surface. In the ground-state, CO only has a very small dipole moment, and the interaction with the Au(111) surface is rather weak. To be able to decelerate CO molecules, they are laser-prepared in the low-field seeking component of the metastable $a^3\Pi_1$, $v = 0$, $J = 1$ state prior to entering the Stark decelerator using pulsed laser radiation at 206 nm. In the metastable state, the CO molecules have a dipole moment of 1.5 Debye, similar to the one for OH and ND_3, and they live long enough (>2.6 ms) to make it through the decelerator without significant radiation losses. In the decelerator, the metastable CO molecules stay together in a compact bunch, enabling their efficient transfer back to selected ro-vibrational levels in the electronic ground-state *via* Stimulated Emission Pumping (SEP)[49] at the exit of the decelerator. In this way, a fully state-selected beam of CO ($X^1\Sigma^+$) molecules in any one of the vibrational levels $v'' = 1–6$, and always in either one of the lowest three rotational levels, can be prepared with a tunable velocity in the 10–200 m s^{-1} range. This slow beam will then scatter from a temperature controlled single crystal Au (111) surface, to gain information on the nature of energy transfer in molecule surface collisions in the limit of zero translational energy of incidence.

IX. Conclusions

The Stark deceleration technique offers a new degree of control over both the internal and external degrees of freedom of polar molecules in a molecular beam.

So far, this new molecular beam technology has been used mainly to decelerate packets of molecules to standstill, and to subsequently confine these molecules in a trap. These tamed molecular beams, however, also hold interesting prospects for scattering experiments. The exquisite level of control over molecules allows detailed molecular scattering studies as a function of the collision energy, from low to high collision energies, and with an unprecedented energy resolution. It is the challenge to design experiments that optimally exploit these new possibilities, expanding upon the rich variety of experimental methods that have been developed over decades to obtain detailed information on the molecular interaction potential. For this, we have discussed a variety of molecular systems and experimental arrangements that can be used for specific applications, ranging from crossed molecular beam set-ups to molecular synchrotrons and surface scattering machines.

The experiments that have been discussed are all challenging, but realistic. The first crossed beam experiment using a Stark-decelerated beam of OH radicals and a conventional beam of Xe atoms has already been successfully performed. New Stark decelerators with superior efficiency have been developed with which bimolecular scattering between two decelerated beam pulses becomes feasible. This has been discussed in more detail for the OH ($X\,^2\Pi_{3/2}$, $J = 3/2, f$) + NO ($X\,^2\Pi_{1/2}$, $J = 1/2, f$) system, but, for instance, the scattering of ND_3 with ND_3 would experimentally most certainly be feasible as well.

The use of tamed molecular beams is of advantage in all experiments in which the velocity and/or the internal state purity of the beam are important parameters. In scattering experiments, these beams bring us one step closer to Levine and Bernstein's "ultimate scattering experiment".

References

1. R. Levine and R. Bernstein, *Molecular reaction dynamics and chemical reactivity*, Oxford University Press, New York, 1987.
2. Y. T. Lee, *Angew. Chem.*, 1987, **26**, 939.
3. G. Scoles, ed., *Atomic and molecular beam methods*, vol. 1 & 2, Oxford University Press, New York, NY, USA, 1988 & 1992.
4. X. Yang, *Annu. Rev. Phys. Chem.*, 2007, **58**, 433.
5. S. Stolte, *Ber. Bunsen-Ges. Phys. Chem.*, 1982, **86**, 413.
6. D. Skouteris, D. E. Manolopoulos, W. Bian, H. J. Werner, L. H. Lai and K. Liu, *Science*, 1999, **286**, 1713.
7. R. T. Skodje, D. Skouteris, D. E. Manolopoulos, S.-H. Lee, F. Dong and K. Liu, *Phys. Rev. Lett.*, 2000, **85**, 1206.
8. A. T. J. B. Eppink and D. H. Parker, *Rev. Sci. Instrum.*, 1997, **68**, 3477.
9. K. T. Lorenz, D. W. Chandler, J. W. Barr, W. Chen, G. L. Barnes and J. I. Cline, *Science*, 2001, **293**, 2063.
10. X. Liu, J. J. Lin, S. Harich, G. C. Schatz and X. Yang, *Science*, 2000, **289**, 1536.
11. S. A. Harich, D. Dai, C. C. Wang, X. Yang, S. Der Chao and R. T. Skodje, *Nature*, 2002, **419**, 281.
12. J. J. Lin, J. Zhou, W. Shiu and K. Liu, *Science*, 2003, **300**, 966.
13. H. L. Bethlem, G. Berden and G. Meijer, *Phys. Rev. Lett.*, 1999, **83**, 1558.
14. J. L. Bohn, A. V. Avdeenkov and M. P. Deskevish, *Phys. Rev. Lett.*, 2002, **89**, 203202.
15. N. Balakrishnan, A. Dalgarno and R. C. Forrey, *J. Chem. Phys.*, 2000, **113**(2), 621.
16. R. V. Krems, *Int. Rev. Phys. Chem.*, 2005, **24**, 99.
17. R. V. Krems, *Phys. Chem. Chem. Phys.*, 2008, **10**, 4079.
18. S. Y. T. van de Meerakker, H. L. Bethlem and G. Meijer, *Nat. Phys.*, 2008, **4**, 595.
19. A. V. Avdeenkov and J. L. Bohn, *Phys. Rev.*, 2002, **A 66**(5), 052718.
20. H. L. Bethlem, G. Berden, F. M. H. Crompvoets, R. T. Jongma, A. J. A. van Roij and G. Meijer, *Nature*, 2000, **406**, 491.
21. S. Y. T. van de Meerakker, P. H. M. Smeets, N. Vanhaecke, R. T. Jongma and G. Meijer, *Phys. Rev. Lett.*, 2005, **94**, 023004.
22. B. C. Sawyer, B. L. Lev, E. R. Hudson, B. K. Stuhl, M. Lara, J. L. Bohn and J. Ye, *Phys. Rev. Lett.*, 2007, **98**, 253002.
23. S. Hoekstra, J. J. Gilijamse, B. Sartakov, N. Vanhaecke, L. Scharfenberg, S. Y. T. van de Meerakker and G. Meijer, *Phys. Rev. Lett.*, 2007, **98**(13), 133001.

24 L. Santos, G. V. Shlyapnikov, P. Zoller and M. Lewenstein, *Phys. Rev. Lett.*, 2000, **85**, 1791.
25 M. Baranov, L. Dobrek, K. Góral, L. Santos and M. Lewenstein, *Phys. Scr., T*, 2002, **102**, 74.
26 J. Deiglmayr, A. Grochola, M. Repp, K. Mörtlbauer, C. Glück, J. Lange, O. Dulieu, R. Wester and M. Weidemüller, *Phys. Rev. Lett.*, 2008, **101**, 133004.
27 F. Lang, K. Winkler, C. Strauss, R. Grimm and J. H. Denschlag, *Phys. Rev. Lett.*, 2008, **101**, 133005.
28 K. K. Ni, S. Ospelkaus, M. H. G. de Miranda, A. Pe'er, B. Neyenhuis, J. J. Zirbel, S. Kotochigova, P. S. Julienne, D. S. Jin and J. Ye, *Science*, 2008, **322**, 231.
29 J. J. Gilijamse, S. Hoekstra, S. Y. T. van de Meerakker, G. C. Groenenboom and G. Meijer, *Science*, 2006, **313**, 1617.
30 S. Y. T. van de Meerakker, N. Vanhaecke, H. L. Bethlem and G. Meijer, *Phys. Rev. A*, 2006, **73**(2), 023401.
31 S. Y. T. van de Meerakker, N. Vanhaecke, H. L. Bethlem and G. Meijer, *Phys. Rev. A*, 2005, **71**(5), 053409.
32 L. Scharfenberg, H. Haak, G. Meijer and S. Y. T. van de Meerakker, *Phys. Rev. A*, 2009, **79**, 023410.
33 J. A. Miller and R. J. Kee, *Annu. Rev. Phys. Chem.*, 1990, **41**, 345.
34 M. C. van Beek, J. J. ter Meulen and M. H. Alexander, *J. Chem. Phys.*, 2000, **113**, 628.
35 J. J. van Leuken, F. H. W. van Amerom, J. Bulthuis, J. G. Snijders and S. Stolte, *J. Phys. Chem.*, 1995, **99**, 15573.
36 M. C. van Beek, J. J. ter Meulen and M. H. Alexander, *J. Chem. Phys.*, 2000, **113**, 637.
37 J. J. van Leuken, J. Bulthuis, S. Stolte and J. G. Snijders, *Chem. Phys. Lett.*, 1996, **260**, 595.
38 D.-C. Che, T. Matsuo, Y. Yano, L. Bonnet and T. Kasai, *Phys. Chem. Chem. Phys.*, 2008, **10**, 1419.
39 K. Sharkey, I. R. Sims, I. W. M. Smith, P. Bocherel and B. R. Rowe, *J. Chem. Soc., Faraday Trans.*, 1994, **90**, 3609.
40 A. Dehayem-Kamadjeu, O. Pirali, J. Orphal, I. Kleiner and P. Flaud, *J. Mol. Spectrosc.*, 2005, **234**, 182.
41 M. T. Nguyen, R. Sumathi, D. Sengupta and J. Peeters, *Chem. Phys.*, 1998, **230**, 1.
42 N. Vanhaecke, U. Meier, M. Andrist, B. H. Meier and F. Merkt, *Phys. Rev. A*, 2007, **75**(3), 031402.
43 H. L. Bethlem, F. M. H. Crompvoets, R. T. Jongma, S. Y. T. van de Meerakker and G. Meijer, *Phys. Rev. A*, 2002, **65**(5), 053416.
44 F. M. H. Crompvoets, H. L. Bethlem, R. T. Jongma and G. Meijer, *Nature*, 2001, **411**, 174.
45 C. E. Heiner, D. Carty, G. Meijer and H. L. Bethlem, *Nat. Phys.*, 2007, **3**, 115.
46 J. A. Barker and D. J. Auerbach, *Surf. Sci. Rep.*, 1985, **4**, 1.
47 A. M. Wodtke, J. C. Tully and D. J. Auerbach, *Int. Rev. Phys. Chem.*, 2004, **23**, 513.
48 A. W. Kleyn, *Chem. Soc. Rev.*, 2003, **32**, 87.
49 M. Silva, R. Jongma, R. W. Field and A. M. Wodtke, *Annu. Rev. Phys. Chem.*, 2001, **52**(1), 811.
50 In scattering experiments, one usually refers to fluxes rather than densities. As in most pulsed laser based detection techniques the density of the scattered molecules at a given time is recorded; we use densities here throughout.

Dynamics of OH($^2\Pi$)–He collisions in combined electric and magnetic fields

Timur V. Tscherbul,[*ab] Gerrit C. Groenenboom,[c] Roman V. Krems[d] and Alex Dalgarno[ab]

Received 29th October 2008, Accepted 16th December 2008
First published as an Advance Article on the web 19th May 2009
DOI: 10.1039/b819198k

We use accurate quantum mechanical calculations to analyze the effects of parallel electric and magnetic fields on collision dynamics of OH($^2\Pi$) molecules. It is demonstrated that spin relaxation in ^{3}He–OH collisions at temperatures below 0.01 K can be effectively suppressed by moderate electric fields of order 10 kV cm^{-1}. We show that electric fields can be used to manipulate Feshbach resonances in collisions of cold molecules. Our theoretical results can be tested in experiments with OH molecules in Stark decelerated molecular beams and electromagnetic traps. PACS numbers: 33.20.-t, 33.80.Ps.

I. Introduction

The rapid progress in the research field of cold molecules holds great promise for new and important discoveries at the interface of physics and chemistry.[1–7] The remarkable properties of cold molecules allow for their application in quantum information processing, condensed-matter physics, physical chemistry, and precision spectroscopy. While some of these applications rely on specific properties of molecular ensembles, others exploit the diversity of molecular electronic states and energy level structures. For example, recently proposed schemes for quantum computing with polar $^2\Sigma$ molecules in optical lattices make use of intermolecular dipole–dipole interactions and molecule–field couplings to enable communication between quantum bits.[1] The experiments aiming to measure the electric dipole moment of the electron based on the spectroscopic study of cold YbF radicals exploit the relativistic distortion of molecular orbitals.[2] The fine and hyperfine perturbations in the spectra of $^2\Pi$ molecules allow for the development of new frequency standards and tests of physics beyond the Standard Model.[3,4]

Particularly noteworthy are the experiments probing inelastic energy transfer and chemical reactions of cold molecules.[5–7] The measurements of molecular collision properties at low temperatures are important for the development of buffer gas cooling techniques, which rely on collisional thermalization of molecules in a cell filled with cryogenic ^{3}He gas. This technique produces molecules at temperatures between 0.1 K and 0.5 K and allows for further evaporative cooling after removal of the buffer gas. In order to permit observations and evaporative cooling, molecules must remain trapped for a long time, which is possible if the ratio of elastic to

---

[a]*Institute for Theoretical Atomic, Molecular, and Optical Physics, Harvard-Smithsonian Center for Astrophysics, Cambridge, Massachusetts 02138, USA. E-mail: tshcherb@cfa.harvard.edu*
[b]*Harvard-MIT Center for Ultracold Atoms, Cambridge, Massachusetts 02138, USA*
[c]*Theoretical Chemistry, Institute for Molecules and Materials, Radboud University Nijmegen, Heyendaalseweg 135, 6525 AJ Nijmegen, The Netherlands. E-mail: gerritg@theochem.ru.nl*
[d]*Department of Chemistry, University of British Columbia, Vancouver, British Columbia V6T 1Z1, Canada*

inelastic spin-changing collisions exceeds 10^4.[8] Previous experimental[9,10] and theoretical[11] work has shown that this requirement is generally satisfied for Σ-state diatomic molecules with large rotational constants. However, it is unclear whether molecules in electronic states other than Σ may be sympathetically cooled in a buffer gas cell. The mechanism of spin relaxation in collisions of molecules in $^2Π$ electronic state remains unknown.

The OH radical is one of the first molecular species that were cooled and trapped at millikelvin temperatures.[5,6,12–14] The ground electronic state of OH is of $^2Π$ symmetry, and the energy levels of the molecule depend linearly on the strength of an applied electric field above ~5 kV cm^{-1}. As a result, the OH radicals can be efficiently decelerated, trapped, and manipulated using moderate static or time-varying electric fields available in the laboratory.[5,12–14] A variety of novel experiments with trapped OH molecules has been reported. The ability to fine tune the collision energy of a Stark-decelerated beam allowed Meijer and co-workers to study threshold behavior of Xe–OH rotationally inelastic scattering at collision energies as low as 50 cm^{-1}.[5] High-precision spectroscopy of cold trapped OH was used to study the time variation of the fine structure constant,[4] measure the lifetimes of vibrationally excited states,[13] and determine the rates of optical pumping due to blackbody radiation.[14] Sawyer *et al.* have recently reported measurements of cross sections for elastic and inelastic collisions of magnetically trapped OH molecules with He atoms and D_2 molecules at kinetic energies of 60 cm^{-1} and above.[6]

Theoretical studies of low-energy collisions of OH molecules have been reported by several groups.[5,15–18,20] Avdeenkov and Bohn studied ultracold collisions of OH molecules[15,16] and discovered weakly-bound dimer states supported by the dipole–dipole interaction forces in the presence of an external electric field.[16] Ticknor and Bohn found a significant suppression of inelastic relaxation rates for the low-field-seeking states of OH in a magnetic field.[17] Lara *et al.* analyzed the effects of non-adiabatic and hyperfine couplings on field-free Rb–OH collisions.[18] Their results indicated that sympathetic cooling of OH molecules by collisions with Rb atoms might be challenging due to large inelastic loss rates. González-Sánchez, Bodo, and Gianturco[20] considered field-free collisions of rotationally excited OH molecules with He atoms and found sharp propensity rules for rotational and Λ-doublet changing transitions at ultracold temperatures.

Here, we present a theoretical analysis of OH($^2Π$) collision dynamics in *combined* electric and magnetic fields. We have previously demonstrated that spin-changing collisions of Σ-state molecules can be efficiently manipulated by superimposed electric and magnetic fields.[21–23] Building on our previous work[21,22] and the results of Bohn and co-workers,[15–17] we develop a rigorous quantum theory of collisions between $^2Π$ molecules and structureless atoms in external fields and calculate the dependence of the cross sections for He–OH collisions on electric and magnetic fields. Our results suggest that collisions of OH molecules with He atoms can be efficiently manipulated with the external fields. In particular, we demonstrate an efficient mechanism for suppression of spin relaxation in $^2Π$ molecules with electric fields.

II. Theory

The quantum mechanical formalism for collisions of diatomic molecules in $^2Π$ electronic states in the absence of external fields has been presented by several authors (see, *e.g.*, ref. 20,24,25). Here, we focus on the theoretical aspects relevant for incorporating the effects of electromagnetic fields in scattering calculations. Section IIA presents the discussion of the influence of the electric and magnetic fields on the energy level structure of $^2Π$ molecules. In Section IIB, we discuss the Hamiltonian of the collision complex and the coupled-channel representation of the scattering wave function. A derivation of the matrix elements for the interaction potential operator in the basis of scattering states is presented in Section IIC.

A. The OH molecule in superimposed electric and magnetic fields

The Hamiltonian for a $^2\Pi$ molecule such as OH can be written as[26–28]

$$\hat{H}_{mol} = \hat{H}_{rot} + \hat{H}_{SO} + \hat{H}_\Lambda + \hat{H}_E + \hat{H}_B, \quad (1)$$

where $\hat{H}_{rot}$ is the angular part of the rotational kinetic energy, $\hat{H}_{SO}$ is the spin–orbit (SO) interaction, $\hat{H}_\Lambda$ is the Λ-doubling Hamiltonian, and the terms $\hat{H}_E$ and $\hat{H}_B$ describe the interactions with external electric and magnetic fields. The first term in eqn (1) can be written as[29]

$$\hat{H}_{rot} = B_e(\hat{J}^{SF} - \hat{L}^{SF} - \hat{S}^{SF})^2, \quad (2)$$

where B_e is the rotational constant, $\hat{J}^{SF} = \hat{N}^{SF} + \hat{L}^{SF} + \hat{S}^{SF}$ is the total angular momentum, $\hat{N}^{SF}$ is the rotational angular momentum of the nuclei, $\hat{L}^{SF}$ is the electronic orbital angular momentum, and $\hat{S}^{SF}$ is the electron spin. In eqn (1), we have neglected the hyperfine interaction due to the nuclear spin of H. The hyperfine interaction constant of OH is an order of magnitude smaller than the Λ-doublet splitting, and the hyperfine effects may alter collision dynamics at temperatures below 4 mK.[17]

The angular momentum operators in eqn (2) are defined in the space-fixed frame. However, the symmetry properties of the electronic wave functions are most conveniently exploited in the molecule-fixed frame, with the z-axis oriented along the OH bond. The row vector of molecule-fixed angular momentum operators can be defined as $\hat{J}^{SF} = \hat{J}^{MF}\mathbf{R}(\bar{\alpha},\bar{\beta},0)$, where $\mathbf{R}$ is the matrix of direction cosines and $(\bar{\alpha},\bar{\beta})$ are the Euler angles which specify the orientation of the diatomic molecule in the space-fixed coordinate system. Alternatively, one can define the column vector $\hat{J}^{SF} = \mathbf{R}(\bar{\alpha},\bar{\beta},0)\hat{J}^{MF}$.[29] The molecule-fixed angular momentum operators do not commute, and the choice of convention affects the products of operators that arise upon transforming eqn (2) to the molecule-fixed frame. It is easy to show that the frame transformation adopted in this work leads to $\hat{J}^{SF} \cdot \hat{S}^{SF} = \hat{J}^{MF} \cdot \hat{S}^{MF}$, whereas that of ref. 29 leads to $\hat{J}^{SF} \cdot \hat{S}^{SF} = \hat{S}^{MF} \cdot \hat{J}^{MF}$. The matrix elements of $\hat{H}_{rot}$ are independent of the convention. In the following, we will assume that the operators $\hat{J}, \hat{L},$ and $\hat{S}$ are defined in the MF frame, and the superscript "MF" will be omitted.

The SO interaction is given by

$$\hat{H}_{SO} = A\hat{L} \cdot \hat{S}, \quad (3)$$

where A is the SO interaction constant.

Explicit expressions for the remaining terms in eqn (1) are given below. The energy levels of a $^2\Pi$ molecule can be evaluated by diagonalizing the Hamiltonian (1) in Hund's case (a) basis

$$|JM\Omega\rangle|\Lambda\Sigma\rangle = \left(\frac{2J+1}{4\pi}\right)^{1/2} D^{J*}_{M\Omega}(\bar{\alpha},\bar{\beta},0)|\Lambda\Sigma\rangle, \quad (4)$$

where M and Ω are the projections of $\hat{J}$ onto the space-fixed and molecule-fixed quantization axes, $D^{J*}_{M\Omega}$ is the Wigner D-function, and $|\Lambda\Sigma\rangle$ is the electronic wave function. The molecule-fixed projections of $\hat{L}$ and $\hat{S}$ are denoted as Λ and Σ. For a $^2\Pi$ electronic state, they take the values $\Lambda = \pm 1$ and $\Sigma = \pm \frac{1}{2}$. The off-diagonal matrix elements of the SO interaction between the $^2\Pi$ state and the excited electronic states give rise to the Λ-doubling effect[26,27] described below. After neglecting the cross terms, eqn (3) may be written as

$$\hat{H}_{SO} = A\hat{L}_z\hat{S}_z, \quad (5)$$

where the subscript z refers to the molecule-fixed projections of the angular momentum operators. The Λ-doubling is described by the effective Hamiltonian[26,27]

$$\hat{H}_\Lambda = \frac{1}{2}e^{-2i\phi}\left[-q\hat{J}_+^2 + (p+2q)\hat{J}_+\hat{S}_+\right] + \frac{1}{2}e^{2i\phi}\left[-q\hat{J}_-^2 + (p+2q)\hat{J}_-\hat{S}_-\right], \quad (6)$$

where $\hat{J}_\pm = \hat{J}_x \mp i\hat{J}_y$ and $\hat{S}_\pm = \hat{S}_x \pm i\hat{S}_y$ are the ladder operators, ϕ is the azimuthal angle of the electron in the molecule-fixed frame, and p and q are the phenomenological Λ-doubling parameters. Using the phase convention for the electronic wave functions, $\langle \Lambda = \pm 1|e^{\pm 2i\phi}|\Lambda' = \mp 1\rangle = -1$, the matrix elements of the Hamiltonian (6) can be written as

$$\langle \Lambda|\hat{H}_\Lambda|\Lambda'\rangle = \frac{1}{2}\left(\delta_{\Lambda',-1}\delta_{\Lambda,1}\left[q\hat{J}_+^2 - (p+2q)\hat{J}_+\hat{S}_+\right]\right. \\ \left. + \delta_{\Lambda',1}\delta_{\Lambda,-1}\left[q\hat{J}_-^2 - (p+2q)\hat{J}_-\hat{S}_-\right]\right). \quad (7)$$

The interaction with the magnetic field of strength B has the form

$$\hat{H}_\text{B} = \mu_0 B(\hat{L} + 2\hat{S}) \cdot \hat{B}, \quad (8)$$

where $\hat{B}$ is the unit vector in the direction of the external magnetic field. To first order, the interaction of the molecule with the dc electric field can be written as

$$\hat{H}_\text{E} = -Ed\cos\chi, \quad (9)$$

where χ is the polar angle of the molecule in the space-fixed frame, E is the electric field strength, and d is the permanent electric dipole moment of the molecule. Here, we assume that both the electric and magnetic fields are oriented along the space-fixed z-axis. The more general case of crossed electric and magnetic fields is considered elsewhere.[30]

It is convenient to use parity-adapted Hund's case (a) basis functions

$$\left|JM\bar{\Omega}\varepsilon\right\rangle = \frac{1}{2}\left\{\left|JM\bar{\Omega}\right\rangle\left|\Lambda = 1, \Sigma = \bar{\Omega} - 1\right\rangle \\ + \varepsilon(-)^{J-1/2}\left|JM - \bar{\Omega}\right\rangle\left|\Lambda = -1, \Sigma = -\bar{\Omega} + 1\right\rangle\right\}, \quad (10)$$

where $\bar{\Omega} = |\Omega|$, and the parity index $\varepsilon = \pm 1$ characterizes the inversion symmetry of the basis functions [$\varepsilon(-)^{J-\frac{1}{2}} = 1$ for e-parity states and -1 for f-parity states]. Note that in the parity-adapted basis, $\bar{\Omega} > 0$ and the quantum number Λ does not have a definite value. Expanding the rotational kinetic energy in terms of the ladder operators and using eqn (5), we obtain

$$\hat{H}_\text{rot} + \hat{H}_\text{SO} = B_e[\hat{J}^2 - 2\hat{J}_z^2 - \hat{J}_+\hat{S}_- - \hat{J}_-\hat{S}_+] + (A + 2B_e)\hat{L}_z\hat{S}_z, \quad (11)$$

where we have omitted the terms $\hat{L}^2$ and $\hat{S}^2$ which would only result in an overall energy shift. The matrix elements of the rotational and spin–orbit Hamiltonians can now be evaluated in the parity-adapted basis (10). They have the form

$$\left\langle JM\bar{\Omega}\varepsilon|\hat{H}_\text{rot} + \hat{H}_\text{SO}|J'M'\bar{\Omega}'\varepsilon'\right\rangle = \delta_{\varepsilon\varepsilon'}\delta_{JJ'}\delta_{MM'}\left\{B_e\left[J(J+1) - 2\bar{\Omega}^2\right]\delta_{\bar{\Omega}\bar{\Omega}'} + (A+2B_e)\right. \\ \left. (\bar{\Omega}-1)\delta_{\bar{\Omega}\bar{\Omega}'} - B_e\left[\delta_{\bar{\Omega},\bar{\Omega}'-1}\alpha_-(J',\bar{\Omega})\times\alpha_-(S,\bar{\Omega}'-1) - \delta_{\bar{\Omega},\bar{\Omega}'+1}\alpha_+(J',\bar{\Omega})\alpha_+(S,\bar{\Omega}-1)\right]\right\}, \quad (12)$$

where $\alpha_{\pm}(J,\bar{\Omega}) = \sqrt{J(J+1) - \bar{\Omega}(\bar{\Omega}\pm 1)}$. Combining eqns (7) and (10), we obtain the following compact expression for the Λ-doubling matrix elements

$$\langle JM\bar{\Omega}\varepsilon|\hat{H}_\Lambda|J'M'\bar{\Omega}'\varepsilon'\rangle = \frac{1}{2}\delta_{\varepsilon\varepsilon'}\delta_{JJ'}\delta_{MM'}\varepsilon(-)^{J-1/2}\left[q\alpha_-\left(J',\bar{\Omega}'\right)\alpha_-\left(J',\bar{\Omega}'-1\right)\delta_{\bar{\Omega}',2-\bar{\Omega}'}\right.$$

$$\left. -(p+2q)\alpha_-\left(J',\bar{\Omega}'\right)\alpha_+\left(S,\bar{\Omega}'-1\right)\delta_{\bar{\Omega},1-\bar{\Omega}'}\right]. \quad (13)$$

In order to evaluate the matrix elements of the Zeeman Hamiltonian in the basis (10), it is necessary to transform the operator (8) to the molecule-fixed frame

$$\hat{H}_B = \mu_0 B \sum_{q=-1}^{1}\left(\hat{L}_q + 2\hat{S}_q\right)D_{0q}^{1*}(\bar{\alpha},\bar{\beta},0), \quad (14)$$

where only the $q=0$ molecule-fixed component of $\hat{L}$ survives on the right-hand side. Evaluating the integrals over the product of three Wigner D-functions, we find

$$\langle JM\bar{\Omega}\varepsilon|\hat{H}_B|J'M'\bar{\Omega}'\varepsilon'\rangle = \mu_0 B \delta_{\varepsilon\varepsilon'}\delta_{MM'}(-)^{M'-\bar{\Omega}'}$$

$$[(2J+1)(2J'+1)]^{1/2}\begin{pmatrix}J & 1 & J'\\M & 0 & -M'\end{pmatrix}\times\left[\sqrt{2}\alpha_+\left(S,\bar{\Omega}'-1\right)\begin{pmatrix}J & 1 & J'\\\bar{\Omega} & -1 & -\bar{\Omega}'\end{pmatrix}\right.$$

$$\left.-\sqrt{2}\alpha_-\left(S,\bar{\Omega}'-1\right)\begin{pmatrix}J & 1 & J'\\\bar{\Omega} & 1 & -\bar{\Omega}'\end{pmatrix} + (2\bar{\Omega}-1)\begin{pmatrix}J & 1 & J'\\\bar{\Omega} & 0 & -\bar{\Omega}'\end{pmatrix}\right]. \quad (15)$$

The matrix elements of the interaction with electric fields have a similar form

$$\langle JM\bar{\Omega}\varepsilon|\hat{H}_E|J'M'\bar{\Omega}'\varepsilon'\rangle = -Ed\delta_{\varepsilon,-\varepsilon'}\delta_{MM'}(-)^{M'-\bar{\Omega}'}[(2J+1)(2J'+1)]^{1/2}$$

$$\times\begin{pmatrix}J & 1 & J'\\M & 0 & -M'\end{pmatrix}\begin{pmatrix}J & 1 & J'\\\bar{\Omega} & 0 & -\bar{\Omega}'\end{pmatrix}. \quad (16)$$

This expression shows that electric fields couple the states of the opposite inversion parity.

B. Collision dynamics

The He–OH($^2\Pi$) collision complex can be described by the Jacobi vectors $\mathbf{R}$ – the separation of He from the center of mass of OH and $\mathbf{r}$ – the internuclear distance in OH. The angle between the vectors is denoted by θ. In the following, it will be convenient to use the unit vectors $\hat{R} = \mathbf{R}/R$ and $\hat{r} = \mathbf{r}/r$, where $R = |\mathbf{R}|$, $r = |\mathbf{r}|$. The Hamiltonian of the collision complex can be written in atomic units as[17,20,24]

$$\hat{H} = -\frac{1}{2\mu R}\frac{\partial^2}{\partial R^2}R + \frac{\hat{\ell}^2}{2\mu R^2} + \hat{V}(R,r,\theta) + \hat{H}_{\text{mol}}, \quad (17)$$

where μ is the reduced mass of the ^{3}He–OH system, $\hat{\ell}$ is the orbital angular momentum for the collision, $\hat{V}(R,r,\theta)$ is the electrostatic interaction potential, and $\hat{H}_{\text{mol}}$ is the Hamiltonian of the OH molecule in the presence of external electric and magnetic fields (see Section IIA). We assume that the internuclear distance of

OH is fixed at the equilibrium value of 1.226 Å. The wave function of the He–OH collision complex Ψ satisfies the Schrödinger equation at a total energy E, and can be expanded over the complete coupled-channel basis

$$\Psi = \frac{1}{R}\sum_{\beta} F_{\beta}(R)\psi_{\beta}(\hat{R},\hat{r}), \tag{18}$$

where $\psi_{\beta}(\hat{R},\hat{r})$ are the angular basis functions. The fully uncoupled angular basis set can be defined as a direct product of the parity-unadapted Hund's case (a) functions and spherical harmonics

$$|JM\Omega\rangle|\Lambda\Sigma\rangle|\ell m_{\ell}\rangle, \tag{19}$$

where the spherical harmonics $|\ell m_{\ell}\rangle = Y_{\ell m_{\ell}}(\hat{r})$ describe the orbital motion of the He atom around the OH fragment. To be consistent with spectroscopic nomenclature, it is convenient to use a slightly modified basis given by

$$|JM\bar{\Omega}\varepsilon\rangle|\ell m_{\ell}\rangle, \tag{20}$$

where $|JM\bar{\Omega}\varepsilon\rangle$ are Hund's case (a) basis functions of definite parity (10). The basis sets (19) and (20) are related by a unitary transformation, and are equivalent. We note that a more general basis set was used by Lara et al. in their theoretical study of Rb + OH collisions in the presence of non-adiabatic and hyperfine interactions.[18,19] These authors used the OH basis functions in the form $|(JI)\Omega Fm_F\rangle$, where the total angular momentum $\hat{F} = \hat{J} + \hat{I}$ is the vector sum of $\hat{J}$ and the nuclear spin of the molecule $\hat{I}$. These basis functions reduce to eqn (20) if the nuclear spin of OH is neglected. In order to describe the transitions in the Rb atom, Lara et al. used the standard hyperfine basis functions $|F_a m_{F_a}\rangle$. For collisions with structureless atoms, $|F_a m_{F_a}\rangle = |00\rangle$, and eqn (23) of ref. 19 reduces to our eqn (20).

Substituting the coupled-channel expansion (18) into the Schrödinger equation, we obtain a system of coupled second-order differential equations

$$\left[\frac{d^2}{dR^2} + 2\mu E\right]F_{\beta}(R) = 2\mu \sum_{\beta'}\langle\psi_{\beta}(\hat{R},\hat{r})|\hat{V}(R,\theta) + \frac{\hat{\ell}^2}{2\mu R^2} + \hat{H}_{\text{mol}}|\psi_{\beta'}(\hat{R},\hat{r})\rangle F_{\beta'}(R). \tag{21}$$

In order to solve these equations, it is necessary to evaluate the matrix elements on the right-hand side. In the uncoupled representation (20), the operator $\hat{\ell}^2$ is diagonal with matrix elements given by $\ell(\ell + 1)$.[11] The matrix elements of the asymptotic Hamiltonian $\hat{H}_{\text{mol}}$ in basis (20) are

$$\langle JM\bar{\Omega}\varepsilon|\langle\ell m_{\ell}|\hat{H}_{\text{mol}}|J'M'\bar{\Omega}'\varepsilon'\rangle|\ell' m'_{\ell}\rangle = \delta_{\ell\ell'}\delta_{m_{\ell}m'_{\ell}}\langle JM\bar{\Omega}\varepsilon|\hat{H}_{\text{mol}}|J'M'\bar{\Omega}'\varepsilon'\rangle, \tag{22}$$

where the expression on the right-hand side is evaluated in section IIA. All that remains to complete the definition of the system of coupled equations (21) is to evaluate the matrix elements of the interaction potential. This is described in the following section.

Once the coupled equations are solved, the asymptotic wave function is transformed to the field-dressed basis $|\gamma\rangle|\ell m_{\ell}\rangle$, which diagonalizes the asymptotic Hamiltonian $\hat{H}_{\text{mol}}$. The transformation can be written as

$$|\gamma\rangle|\ell m_{\ell}\rangle = |\ell m_{\ell}\rangle\sum_{JM\bar{\Omega}\varepsilon} C_{JM\bar{\Omega}\varepsilon,\gamma}|JM\bar{\Omega}\varepsilon\rangle, \tag{23}$$

where $C_{JM\bar{\Omega}\varepsilon,\gamma}$ are the components of the eigenvector of $\hat{H}_{\text{mol}}$ corresponding to the eigenstate γ with energy ε_γ. The matrix of the transformation (23) is diagonal in ℓ and m_ℓ. The S-matrix can be obtained from the transformed wave function using the standard asymptotic matching procedure.[36] The cross sections for transitions between the field-dressed states of OH can be expressed as

$$\sigma_{\gamma \to \gamma'} = \frac{\pi}{k_\gamma^2} \sum_{M_{\text{tot}}} \sum_{\ell, m_\ell} \sum_{\ell', m'_\ell} |\delta_{\gamma\gamma'} \delta_{\ell\ell'} \delta_{m_\ell, m'_\ell} - S^{M_{\text{tot}}}_{\gamma\ell m_\ell; \gamma'\ell'm'_\ell}|^2, \qquad (24)$$

where the wavevector $k^2_\gamma = 2\mu(E - \varepsilon_\gamma) = 2\mu E_{\text{coll}}$, E_{coll} is the collision energy, and the summation in eqn (24) is performed in a cycle over the total angular momentum projection.[11]

We used the following molecular constants of OH (in cm^{-1}): $B_e = 18.55$, $A = -139.273$, $p = 0.235608$, $q = -0.03877$.[17,26] The value of the permanent electric dipole moment of OH($^2\Pi$) $d = 1.68$ D was taken from ref. 17. The coupled-channel expansion (18) included all basis states with $J \leq \frac{11}{2}$ and $\ell \leq 5$, which resulted in a total of 622 coupled channels for $M_{\text{tot}} = \frac{1}{2}$. The close-coupled equations (21) were solved numerically using the improved log-derivative method[37] on a grid of R from 2 to 65 a_0 with a step size of 0.01 a_0. The resulting cross sections were converged to within 5%.

C. Matrix elements of the interaction potential

The matrix elements of the interaction potential between the states with definite Λ can be expanded in reduced Wigner D-functions

$$V_{\Lambda\Lambda'}(R, \theta) = \sum_\lambda D^{\lambda *}_{0, \Lambda' - \Lambda}(0, \theta, 0) V_{\lambda, \Lambda' - \Lambda}(R). \qquad (25)$$

Since in our case $|\Lambda| = |\Lambda'| = 1$, only the terms in eqn (25) with $\Lambda' - \Lambda = 0, \pm 2$ are different from zero. They can be obtained by expanding the half-sum and half-difference of the two ground-state potential energy surfaces of A' and A'' symmetry[24,25]

$$\frac{1}{2}(V_{A'} + V_{A''}) = \sum_{\lambda=0}^{\lambda_{\max}} P_\lambda(\cos\theta) V_{\lambda 0}(R) \qquad (26)$$

$$\frac{1}{2}(V_{A''} - V_{A'}) = \sum_{\lambda=2}^{\lambda_{\max}} d^\lambda_{02}(\cos\theta) V_{\lambda 2}(R), \qquad (27)$$

where $d^\lambda_{0\mu}(\cos\theta)$ are the reduced Wigner D-functions and $P_{\lambda\mu}(\cos\theta)$ are the associated Legendre polynomials, which are related through[31]

$$d^\lambda_{0\mu}(\cos\theta) = \left[\frac{(\lambda - \mu)!}{(\lambda + \mu)!}\right]^{1/2} P_{\lambda\mu}(\cos\theta). \qquad (28)$$

The interaction potential (25) can be evaluated in the parity-unadapted basis using Eqs. (4), (19) and the generalized spherical harmonics addition theorem[25,31]

$$d^\lambda_{0\mu}(\cos\theta) = (-)^{-\mu} d^\lambda_{0,-\mu}(\cos\theta) = (-)^\mu \sum_{m_\lambda} \left[\frac{4\pi}{2\lambda + 1}\right]^{1/2} D^\lambda_{m_\lambda \mu}(\bar{\alpha}, \bar{\beta}, 0) Y_{\lambda m_\lambda}(\hat{R}). \qquad (29)$$

Combining this expression with eqn (25) and evaluating the integrals over the products of three D-functions, we obtain

$$\langle JM\Omega|\langle\Lambda\Sigma|\langle\ell m_\ell|\hat{V}(R,\theta)|J'M'\Omega'\rangle|\Lambda'\Sigma'\rangle|\ell'm'_\ell\rangle =$$

$$\delta_{\Sigma\Sigma'}[(2J+1)(2J'+1)(2\ell+1)(2\ell'+1)]^{1/2}(-)^{m_\ell+M'-\Omega'}$$

$$\times \sum_{\lambda,m_\lambda} \begin{pmatrix} J & \lambda & J' \\ M & m_\lambda & -M' \end{pmatrix} \begin{pmatrix} J & \lambda & J' \\ \Omega & \Lambda'-\Lambda & -\Omega' \end{pmatrix} \begin{pmatrix} \ell & \lambda & \ell' \\ -m_\ell & m_\lambda & m'_\ell \end{pmatrix} \begin{pmatrix} \ell & \lambda & \ell' \\ 0 & 0 & 0 \end{pmatrix}$$

$$\times V_{\lambda,\Lambda'-\Lambda}(R).$$
(30)

The expansion coefficients have the property $V_{\lambda,\Lambda'-\Lambda}(R) = V_{\lambda,\Lambda-\Lambda'}(R)$.[25] We note that the 3-j symbols in eqn (30) vanish unless $m_\lambda = M' - M = m_\ell - m'_\ell$. Thus, the electrostatic interaction potential only couples the states with the same total angular momentum projection $M_{\text{tot}} = M + m_\ell = M' + m'_\ell$. A transformation of the interaction potential matrix elements (30) to the parity-adapted basis (10) using the symmetry properties of 3-j symbols[31] yields

$$\langle JM\bar{\Omega}\varepsilon|\langle\ell m_\ell|\hat{V}(R,\theta)|J'M'\bar{\Omega}'\varepsilon'\rangle|\ell'm'_\ell\rangle$$

$$= [(2J+1)(2J'+1)(2\ell+1)(2\ell'+1)]^{1/2}(-)^{m_\ell+M'-\bar{\Omega}'}$$

$$\times \sum_{\lambda,m_\lambda} \frac{1}{2}[1+\varepsilon\varepsilon'(-)^\lambda] \begin{pmatrix} J & \lambda & J' \\ M & m_\lambda & -M' \end{pmatrix} \begin{pmatrix} \ell & \lambda & \ell' \\ -m_\ell & m_\lambda & m'_\ell \end{pmatrix} \begin{pmatrix} \ell & \lambda & \ell' \\ 0 & 0 & 0 \end{pmatrix}$$

$$\times \left[\delta_{\bar{\Omega}\bar{\Omega}'}\begin{pmatrix} J & \lambda & J' \\ \bar{\Omega} & 0 & -\bar{\Omega}' \end{pmatrix} V_{\lambda 0}(R) - \varepsilon'(-)^{J'-1/2}\delta_{\bar{\Omega}',2-\bar{\Omega}}\begin{pmatrix} J & \lambda & J' \\ \bar{\Omega} & -2 & \bar{\Omega}' \end{pmatrix} V_{\lambda 2}(R)\right].$$
(31)

An analysis of the expression in square brackets shows that the rotational levels in the same SO manifold ($\bar{\Omega}' = \bar{\Omega}$) are coupled by the half-sum PES (26), whereas the levels belonging to different SO manifolds ($\bar{\Omega}' = 2 - \bar{\Omega}$) are coupled by the half-difference PES. Similarly, the factor $\frac{1}{2}[1 + \varepsilon\varepsilon'(-)^\lambda]$ ensures that the couplings between the states of the same parity ($\varepsilon = \pm 1 \leftrightarrow \varepsilon' = \pm 1$) are induced by the anisotropic terms with even λ, and those of the states of opposite parity ($\varepsilon = \pm 1 \leftrightarrow \varepsilon' = \mp 1$) are induced by the anisotropic terms with odd λ.

III. Results

The upper panel of Fig. 1 shows the energy levels of OH as functions of the magnetic field. At zero field, the absolute ground state of the molecule is a Λ-doublet with $J = \frac{3}{2}$. Magnetic fields further split the e and f components of the doublet into four Zeeman levels characterized by $M = -\frac{3}{2}, -\frac{1}{2}, \frac{1}{2}, \frac{3}{2}$ (in order of increasing energy). As mentioned above, M is rigorously conserved in parallel fields and we will use this label to classify the molecular states. The magnetic levels corresponding to different manifolds cross at $B \sim 0.1$ T, where the Zeeman shift becomes comparable to the Λ-doublet splitting. The matrix elements of the Zeeman Hamiltonian (15) are diagonal in ε and independent of its sign. Therefore, the e and f Zeeman manifolds are identical and the crossings between the levels of different manifolds are not avoided, as illustrated in the upper panel of Fig. 1. Electric fields couple the opposite parity states, leading to avoided crossings similar to those encountered in $^2\Sigma$ molecules.[21,30] An important difference is that crossings in $^2\Pi$ molecules occur at small magnetic fields $B \sim 0.1$ T corresponding to typical Λ-doublet splittings of tenths of cm^{-1}. In contrast, the electric-field induced crossings in $^2\Sigma$ molecules occur between different rotational levels, which requires magnetic fields on the order of several Tesla.[21,22]

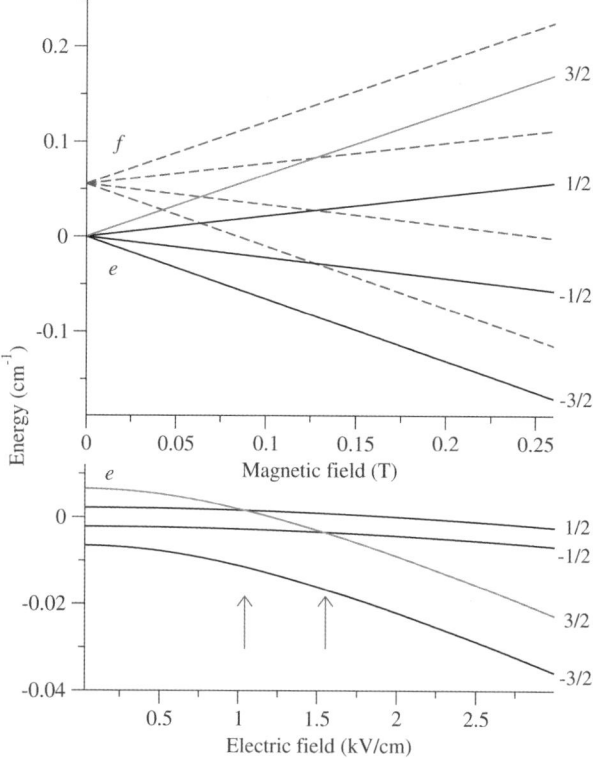

Fig. 1 Zeeman energy levels of OH. The initial state for scattering calculations (see text) is denoted by the red (light grey) solid line. The values of M for individual magnetic sublevels are shown to the right of the curves.

The lower panel of Fig. 1 shows the electric field dependence of the e-manifold energy levels in the presence of a static magnetic field of 0.01 T. The four e-states shift downwards with increasing the field, whereas the corresponding f-states (not shown) shift in the opposite direction. It follows from eqn (16) that the $|M| = 3/2$ states have larger g-factors, so their energy decreases faster than that of the $|M| = 1/2$ states. As a result, two crossings occur at electric fields of about 1 and 1.5 kV cm^{-1} shown by the vertical arrows in the lower panel of Fig. 1. Note that the location of the crossings depends on magnetic field strength, shifting to higher electric fields with increasing magnetic field. In the following, we will consider collisions of OH molecules initially in the state $|J = 3/2, M = 3/2, \varepsilon = -1\rangle$ denoted by the red (light grey) line in Fig. 1. The OH molecules selected in this state gain potential energy with increasing magnetic field and can be confined in a permanent magnetic trap. The expansion of the initial state in terms of Hund's case (a) basis functions (23) is

$$\left| J = \frac{3}{2}, M = \frac{3}{2}, \varepsilon = -1 \right\rangle = 0.985 \left| J = \frac{3}{2}, \bar{\Omega} = \frac{3}{2}, M = \frac{3}{2}, \varepsilon = -1 \right\rangle$$
$$+ 0.174 \left| J = \frac{3}{2}, \bar{\Omega} = \frac{1}{2}, M = \frac{3}{2}, \varepsilon = -1 \right\rangle. \quad (32)$$

This equation illustrates that OH is not a pure Hund's case (a) molecule: different $\bar{\Omega}$ components of the basis (10) are mixed by the cross terms $\hat{J}_+\hat{S}_-$ and $\hat{J}_-\hat{S}_+$ in eqn (11). Because the rotational constant of OH is large compared to Zeeman splittings, the mixing coefficients in eqn (32) are independent of M and magnetic field. In the presence of an electric field, the initial state (32) contains an admixture of basis functions of opposite parity ($\varepsilon = 1$), whose contribution increases linearly with the field strength. The energy of the state given by eqn (32) decreases with increasing electric field as shown in the lower panel of Fig. 1.

Fig. 2 shows the cross sections for elastic scattering and inelastic spin relaxation in ^{3}He–OH collisions as functions of collision energy. The behavior of the cross sections in the $E_{coll} \rightarrow 0$ limit is dictated by the Wigner threshold laws:[32] the inelastic cross section increases as $E_{coll}^{-\frac{1}{2}}$ and the elastic cross section is independent of E_{coll}. Magnetic fields modify the energy dependence of the cross sections. At small magnetic fields, the cross sections continue to decrease with E_{coll} down to 10^{-5} K and start to follow the threshold behavior as the energy is further decreased. The turnover point moves to higher energy with increasing magnetic field. At very large fields, the spin relaxation cross section always increases with decreasing collision energy. This behavior is qualitatively similar to that observed for collisions of molecules in Σ electronic states.[11] The conservation of the total angular momentum projection (see section IIB) implies that the spin relaxation transition $|M = \frac{3}{2}\rangle \rightarrow |M' = \frac{1}{2}\rangle$ should be accompanied by the transition $m_\ell \rightarrow m'_\ell = m_\ell + 1$ which leads to a centrifugal barrier in the outgoing collision channel. The centrifugal barrier suppresses inelastic processes as long as the energy defect between the initial and final Zeeman levels does not exceed the barrier height.[11,33] At higher magnetic fields (or collision energies), the centrifugal barrier is easily surmounted and spin relaxation rates increase dramatically, as illustrated in Fig. 2.

An interesting feature apparent in Fig. 2 is the rapid increase of spin relaxation as the collision energy is varied through the Λ-doubling threshold (0.06 cm^{-1}). This is caused by the opening of new relaxation channels in the higher-energy f-manifold (see the upper panel of Fig. 1). At collision energies above 0.1 K, both elastic and inelastic cross sections display a rich resonance structure. The resonance pattern is rather dense and features both shape and Feshbach resonances. This is in contrast

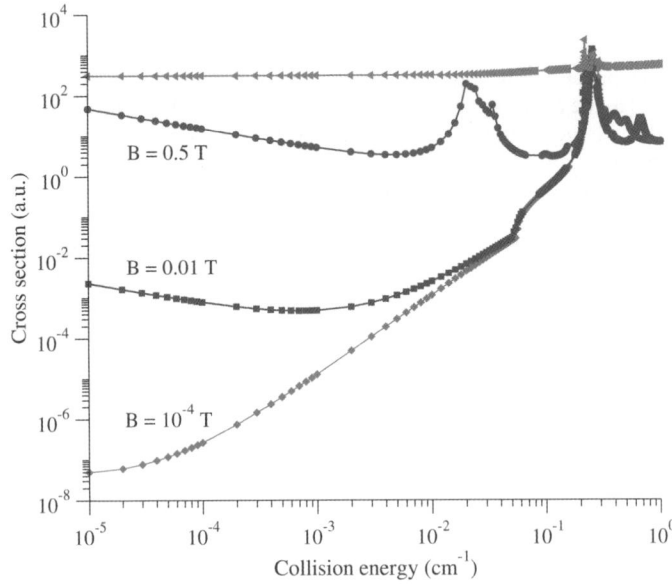

Fig. 2 Cross sections for elastic scattering (upper trace) and spin relaxation in ^{3}He–OH collisions as functions of collision energy at different magnetic fields indicated in the graph. The elastic cross section is shown for $B = 10^{-4}$ T.

with Σ-state molecules where a single (or at most several) shape resonances are typically present.[21,30] The resonances grow in number with increasing magnetic field. At $B = 0.5$ T, the spin relaxation cross section shows two distinct peaks. We attribute the peaks to shape resonances in the outgoing collision channels which are split by the magnetic field. A similar separation of shape resonances has been observed for NH($^3\Sigma^-$)–He collisions.[34] From Fig. 2, the ratio of the cross sections for elastic scattering and spin relaxation varies from 1 to 100 in the temperature interval 0.01–1 K. Therefore, cryogenic cooling and magnetic trapping of OH using ^{3}He buffer gas would be extremely challenging. The rate constant for spin relaxation is 1.2×10^{-11} cm^3 s^{-1} at $T = 0.1$ K and $B = 0.01$ T, which corresponds to the OH trapping lifetime of ~0.1 ms at the buffer gas density of 10^{15} cm^{-3}. Although spin relaxation is suppressed at collision energies below 10 mK and magnetic fields <0.01 T, this regime is far beyond the capability of modern cryogenic cooling techniques.

Equations (31) and (32) establish that different Zeeman levels of OH are directly coupled by the atom–molecule interaction potential. Therefore, spin relaxation in collisions of $^2\Pi$ molecules with $^1S_0$ atoms is a direct process, in which all Zeeman states get mixed up in a collision mediated by the atom–molecule interaction potential. The non-relativistic interaction potential for molecules in Σ electronic states is diagonal in spin states, and spin-changing transitions in $^2\Sigma$ molecules occur *via* a two-step mechanism through the coupling of the ground and the first excited rotational states and the spin–rotation interaction.[11] Collision-induced spin relaxation in $^3\Sigma$ molecules follows a similar mechanism involving the spin–spin interaction. In the case of OH, direct couplings of different magnetic sublevels arise due to the anisotropic terms ($\lambda > 0$) in the expansion of the interaction potential over the Wigner D-functions (25).

Fig. 3 shows the cross sections for spin relaxation as a function of collision energy at selected values of the electric field. Electric fields suppress spin relaxation in the ultracold regime. The origin of this effect is explained below. At collision energies larger than 0.1 cm^{-1}, the suppression is much less efficient. Another interesting effect is shifting and splitting of scattering resonances by electric fields. The suppression of

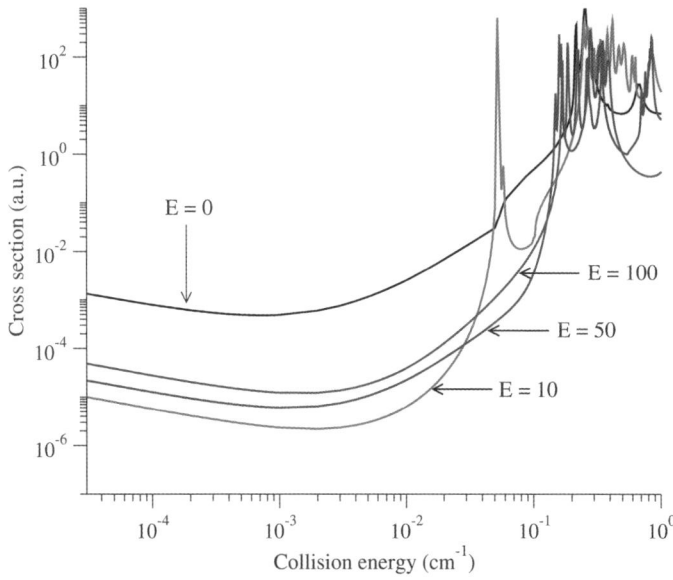

Fig. 3 Cross sections for spin relaxation in ^{3}He–OH collisions as functions of collision energy calculated for several electric field strengths (in kV cm^{-1}) indicated in the graph. The magnetic field is fixed at 0.01 T.

shape resonances is caused by the electric field-induced mixing of different partial waves.[23] The number of Feshbach resonances grows as the electric field is increased from zero to 10 kV cm^{-1}, and the resonances shift to lower energies. We attribute this to the electric field-induced couplings between the opposite parity states, which are uncoupled in the absence of the field, leading to additional avoided crossings and Feshbach resonances. The results shown in Fig. 3 suggest that collisional spin relaxation of OH at temperatures below 0.01 K can be suppressed by two orders of magnitude with electric fields of the order of 10 kV cm^{-1}. The suppression is most efficient at low collision energies. The He–OH spin relaxation rate in the absence of an electric field is 4.8×10^{-16} cm^3 s^{-1} at $T = 0.01$ K and $B = 0.01$ T. This value decreases to 4.4×10^{-18} in an electric field of 50 kV cm^{-1}. At 0.1 K, the rates are 1.1×10^{-11} and 3.4×10^{-12} cm^3 s^{-1}, respectively. Therefore, cryogenic cooling of magnetically trapped OH may be greatly facilitated in the presence of an electric field. A new electromagnetic trap for OH molecules[35] should be particularly suitable for experimental demonstration of the electric field-enhanced evaporative cooling.

In Fig. 4, we show the magnetic field dependence of spin relaxation cross sections at a collision energy of 10^{-3} cm^{-1} and zero electric field. The cross section shows sharp resonances superimposed on a smoothly varying background. The Feshbach resonances arise due to the coupling of the initial channel $|J = \frac{3}{2}, M = \frac{3}{2}, \varepsilon = -1\rangle$ (32) with the closed channels $|J' = \frac{3}{2}, M', \varepsilon = 1\rangle$ induced by the interaction potential. The closed channels are the bound states of the He$\cdots$OH van der Waals complex that correlate to the upper Zeeman manifold in Fig. 1 in the limit $R \rightarrow \infty$. The inset in Fig. 4 demonstrates that resonances may lead to a 100-fold enhancement of spin relaxation cross sections at certain magnetic fields. The elastic cross section (not shown in Fig. 4) is not affected by Feshbach resonances.

Fig. 5 displays the electric field dependence of spin relaxation cross sections at magnetic fields of 10^{-3} and 0.01 T and collision energy of 10^{-3} cm^{-1}. The cross sections increase monotonously at low electric fields but exhibit two sudden drops at $E > 0.5$ kV cm^{-1}. The location of the dips in Fig. 5 depends on the magnitude of the applied magnetic field. For $B = 0.01$ T, they occur at electric fields of about 1 and 1.5 kV cm^{-1}. These values coincide with the positions of the level crossings

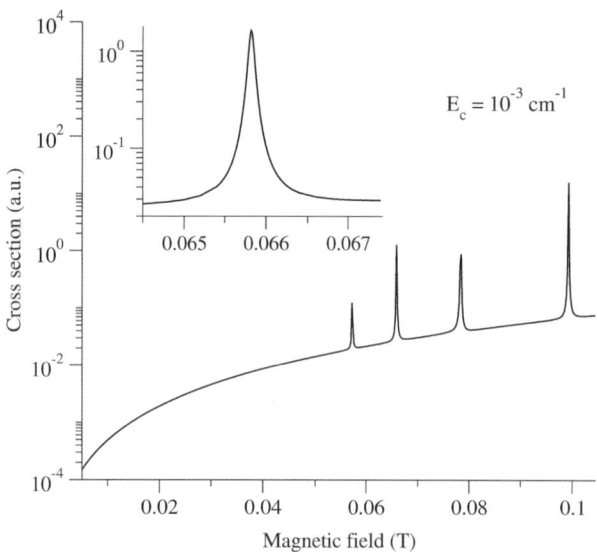

Fig. 4 Magnetic field dependence of spin relaxation cross sections at $E_{coll} = 10^{-3}$ cm^{-1} in the absence of an electric field.

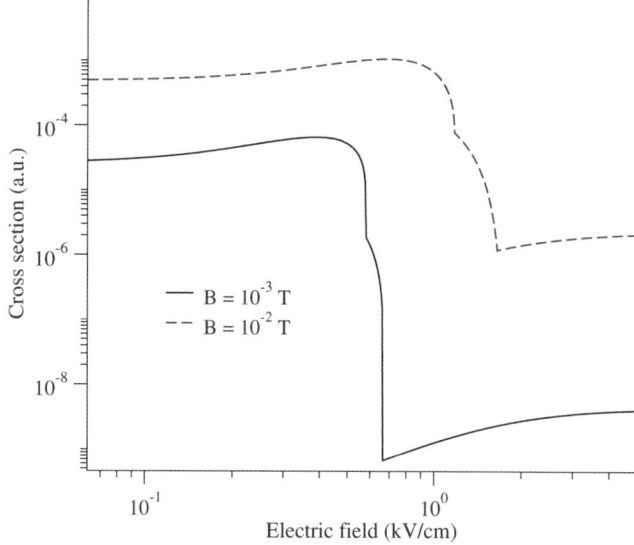

Fig. 5 Cross sections for spin relaxation as functions of electric field at $B = 10^{-3}$ T (lower curve), $B = 0.01$ T (upper curve). $E_{\text{coll}} = 1$ mK for both curves. The initial state for $B = 0.01$ T is shown by the red (light grey) line in the lower panel of Fig. 1.

marked by the vertical arrows in Fig. 1. As the electric field increases, the energy of the initial state $|J = \tfrac{3}{2}, M = \tfrac{3}{2}, \varepsilon = -1\rangle$ becomes lower than that of the two $|M| = \tfrac{1}{2}$ states. At electric fields larger than ~ 2 kV cm^{-1}, both of the $|M| = \tfrac{1}{2}$ channels become energetically forbidden, and spin relaxation can only occur via the $|J = \tfrac{3}{2}, M = \tfrac{3}{2}, \varepsilon = -1\rangle \rightarrow |J' = \tfrac{3}{2}, M' = -\tfrac{3}{2}, \varepsilon' = -1\rangle$ transition. An analysis of state-resolved cross sections shows that this transition is the least probable of all spin relaxation channels. As a consequence, the total inelastic cross section decreases by four to seven orders of magnitude depending on the magnetic field. The data shown in Fig. 2 and 5 demonstrate that spin relaxation of OH molecules in high electric field-seeking states can be completely suppressed by properly chosen combinations of electric and magnetic fields. The control is especially robust at low collision energies (of the order of 1 mK) and magnetic fields not exceeding 0.01 T. However, Fig. 1 shows that for any given value of magnetic field, it should be possible to chose an electric field at which the $|M| = \tfrac{1}{2}$ channels are closed. The necessary electric field can be determined from the positions of the crossings in the lower panel of Fig. 1.

IV. Summary

We have developed a rigorous quantum theory for collisions of molecules in $^2\Pi$ electronic states with structureless atoms in the presence of superimposed electric and magnetic fields. The matrix elements of the electrostatic potential and molecule–field interactions have been derived in the fully uncoupled representation of Hund's case (a) basis functions. The theory has been applied to elucidate the mechanisms of inelastic transitions in low-energy collisions of OH molecules with He atoms. Our results suggest that spin relaxation in collisions of OH molecules proceeds via direct coupling of different Zeeman states induced by the anisotropy of the atom–molecule interaction potential. The rate constants for spin relaxation at temperatures above 0.1 K are of the order of 10^{-12} cm^3 s^{-1}, leading to very short trapping lifetimes. We conclude that sympathetic cooling of OH molecules using cryogenic ^{3}He gas at densities $>10^{15}$ cm^{-3} and temperatures 0.1–1 K appears unfeasible.

We have demonstrated that spin relaxation of OH molecules at temperatures below 0.01 K can be efficiently manipulated by electric fields of moderate strength (~ 10 kV cm^{-1}) available in the laboratory. The mechanism of control is based on suppressing certain relaxation pathways *via* the Stark level crossings induced by superimposed electric and magnetic fields. The crossings are sensitive only to the strength of the applied fields, so the results shown in Fig. 5 are expected to be valid for collisions of OH with other atomic or molecular partners (changing the collision partner may, however, modify the details of the collision dynamics). This can be used to facilitate evaporative cooling of $^2\Pi$ molecules in electromagnetic traps.[35] We have found that electric fields modify the collision energy dependence of inelastic cross sections and may lead to the formation and splitting of Feshbach resonances. The magnetically tunable Feshbach resonances may be used to create weakly bound He$\cdots$OH complexes. It would be interesting to explore the effects of magnetic fields on spin–orbit, vibrational, and rotational predissociation of these complexes. The methods to control inelastic collisions presented in this work may be realized experimentally using Stark decelerated beams[5,12] and electromagnetic traps.[6,35]

Acknowledgements

We thank Hossein Sadeghpour for useful comments. This work was supported by the Chemical Science, Geoscience, and Bioscience Division of the Office of Basic Energy Science, Office of Science, U. S. Department of Energy and NSF grants to the Harvard-MIT Center for Ultracold Atoms and to the Institute for Theoretical Atomic, Molecular, and Optical Physics at Harvard University and Smithsonian Astrophysical Observatory.

References

1 P. Rabl, D. DeMille, J. M. Doyle, M. D. Lukin, R. J. Schoelkopf and P. Zoller, *Phys. Rev. Lett.*, 2006, **97**, 033003.
2 E. Hinds, *Phys. Scr., T*, 1997, **70**, 34.
3 B. L. Lev, E. R. Meyer, E. R. Hudson, B. C. Sawyer, J. L. Bohn and J. Ye, *Phys. Rev. A*, 2006, **74**, 061402R.
4 E. R. Hudson, H. J. Lewandowski, B. C. Sawyer and J. Ye, *Phys. Rev. Lett.*, 2006, **96**, 143004.
5 J. J. Gilijamse, S. Hoekstra, S. Y. T. van de Meerakker, G. C. Groenenboom and G. Meijer, *Science*, 2006, **313**, 1617.
6 B. C. Sawyer, B. K. Stuhl, D. Wang, M. Yeo and J. Ye, *Phys. Rev. Lett.*, 2008, **101**, 203203.
7 R. V. Krems, *Phys. Chem. Chem. Phys.*, 2008, **10**, 4079, and references therein.
8 W. Ketterle and N. J. van Druten, in *Advances in Atomic, Molecular, and Optical Physics*, vol. 37, p. 181, San Diego, Academic, 1996.
9 J. D. Weinstein, R. deCarvalho, T. Guillet, B. Friedrich and J. M. Doyle, *Nature (London)*, 1998, **395**, 148.
10 W. C. Campbell, E. Tsikata, H.-I. Lu, L. D. van Buuren and J. M. Doyle, *Phys. Rev. Lett.*, 2007, **98**, 213001.
11 R. V. Krems and A. Dalgarno, *J. Chem. Phys.*, 2004, **120**, 2296.
12 S. Y. T. van de Meerakker, N. Vanhaecke and G. Meijer, *Annu. Rev. Phys. Chem.*, 2006, **57**, 159.
13 S. Y. T. van de Meerakker, N. Vanhaecke, M. P. J. van der Loo, G. C. Groenenboom and G. Meijer, *Phys. Rev. Lett.*, 2005, **95**, 013003.
14 S. Hoekstra, J. J. Gilijamse, B. Sartakov, N. Vanhaecke, L. Scharfenberg, S. Y. T. van de Meerakker and G. Meijer, *Phys. Rev. Lett.*, 2007, **98**, 133001.
15 A. V. Avdeenkov and J. L. Bohn, *Phys. Rev. A*, 2002, **66**, 052718.
16 A. V. Avdeenkov and J. L. Bohn, *Phys. Rev. Lett.*, 2003, **90**, 043006.
17 C. T. Ticknor and J. L. Bohn, *Phys. Rev. A*, 2005, **71**, 022709.
18 M. Lara, J. L. Bohn, D. Potter, P. Soldán and J. M. Hutson, *Phys. Rev. Lett.*, 2006, **97**, 183201.
19 M. Lara, J. L. Bohn, D. Potter, P. Soldán and J. M. Hutson, *Phys. Rev. A*, 2006, **75**, 012704.
20 L. González-Sánchez, E. Bodo and F. Gianturco, *Phys. Rev. A*, 2006, **73**, 022703.
21 T. V. Tscherbul and R. V. Krems, *Phys. Rev. Lett.*, 2006, **97**, 083201.

22 T. V. Tscherbul and R. V. Krems, *J. Chem. Phys.*, 2006, **125**, 194311.
23 T. V. Tscherbul, *J. Chem. Phys.*, 2008, **128**, 244305.
24 M. H. Alexander, *J. Chem. Phys.*, 1982, **76**, 5974.
25 M. H. Alexander, *Chem. Phys.*, 1985, **92**, 337.
26 J. M. Brown, K. Kaise, C. M. L. Kerr and D. J. Milton, *Mol. Phys.*, 1978, **36**, 553.
27 J. M. Brown and A. J. Merer, *J. Mol. Spectrosc.*, 1979, **74**, 488.
28 M. P. J. van der Loo and G. C. Groenenboom, *J. Chem. Phys.*, 2007, **126**, 114314.
29 M. C. G. N. van Vroonhoven and G. C. Groenenboom, *J. Chem. Phys.*, 2002, **117**, 5240.
30 E. Abrahamsson, T. V. Tscherbul and R. V. Krems, *J. Chem. Phys.*, 2007, **127**, 044302.
31 R. N. Zare, *Angular Momentum*, Wiley, New York, 1988.
32 E. P. Wigner, *Phys. Rev.*, 1948, **73**, 1002.
33 A. Volpi and J. L. Bohn, *Phys. Rev. A*, 2002, **65**, 052712.
34 W. C. Campbell, T. V. Tscherbul, H.-I. Lu, E. Tsikata, R. V. Krems and J. M. Doyle, *Phys. Rev. Lett.*, 2008, **102**, 013003.
35 B. C. Sawyer, B. L. Lev, E. R. Hudson, B. K. Stuhl, M. Lara and J. L. Bohn, *Phys. Rev. Lett.*, 2007, **98**, 253002.
36 B. R. Johnson, *J. Comp. Phys.*, 1973, **13**, 445.
37 D. E. Manolopoulos, *J. Chem. Phys.*, 1986, **85**, 6425.

PAPER

Production of cold ND₃ by kinematic cooling

Jeffrey J. Kay,[a] Sebastiaan Y. T. van de Meerakker,[b] Kevin E. Strecker*[a] and David W. Chandler*[a]

Received 29th October 2008, Accepted 5th January 2009
First published as an Advance Article on the web 19th May 2009
DOI: 10.1039/b819256c

We have produced translationally cold ammonia (ND₃) molecules in various quantum states by kinematic cooling. In these experiments, ND₃ molecules are brought nearly to rest in the $(J, K) = (2,0), (2,1), (2,2), (3,1), (3,2),$ and $(3,3)$ rotational levels of the ground vibronic state by rotationally-inelastic collisions with Ne atoms. The cold molecules are produced in quantum-state-dependent velocity distributions whose laboratory frame velocities are measured to be between 21 m s⁻¹ (E_{trans}/k = 530 mk) and 32 m s⁻¹ (E_{trans}/k = 1.2 K), and are calculated to be between 7.5 m s⁻¹ (E_{trans}/k = 70 mK) and 27 m s⁻¹ (E_{trans}/k = 880 mK). Due to systematic experimental effects, the measured velocities are upper limits to the actual velocities. These temperatures are low enough that it should be possible to use electrostatic traps to confine cold molecules in many of these quantum states.

I. Introduction

Kinematic cooling is a general technique by which a vast array of molecules can be translationally cooled using crossed atomic and molecular beams.[1–4] In this technique, molecules are brought to rest by single collisions with an atomic partner of similar mass. The technique relies on velocity cancellation: through appropriate choice of velocities and masses of the collision partners, one can arrange for the molecule to be scattered in such a way that its post-collision velocity in the center-of-mass frame cancels the center-of-mass velocity of the collision pair, leaving it stationary in the laboratory frame. Kinematic cooling is a very generally-applicable technique because the efficiency with which molecules can be cooled is determined solely by the energetics of the collision, rather than by any particular physical property of the species comprising the collision pair (other than their relative masses). It is also quantum-state-selective, and in principle cold molecules can be prepared in any quantum state which can be populated by collisions. From a practical point of view, the technique relies on robust, mature, and relatively inexpensive crossed-beam scattering technology, which has been thoroughly developed over several decades, making it accessible to a wide number of potential workers in the field. Kinematic cooling generally cools molecules to mK temperatures, comparable to those achieved using Stark deceleration,[5,6] Zeeman deceleration,[7–10] and buffer-gas cooling,[11,12] but not yet to the much lower temperatures accessible by photoassociation[13–16] or Feschbach association.[17–19]

In our experiments we use velocity-mapped ion imaging[20,21] to directly measure the velocity distribution of the scattered products. The imaging technique, which associates the velocity of each scattered molecule with a unique spatial location

[a]*Sandia National Laboratories, Livermore, CA 94550, USA. E-mail: kstreck@sandia.gov; chand@sandia.gov*
[b]*Fritz-Haber-Institut der Max-Planck-Gesellschaft, Faradayweg 4-6, 14195 Berlin, Germany*

on a two-dimensional detector, is indispensable here: in addition to providing useful diagnostic information on the collision dynamics, it also allows one to discriminate the signal due to near-stationary cold molecules from that due to the other scattered products, purely by inspection of the raw data. Indeed, the success of a given experiment can often be judged within minutes of its inception.

Our laboratory has previously produced cold NO molecules at sub-Kelvin temperatures by kinematic cooling and performed a detailed analysis of the cooling process.[1-4] Our next major goal is to trap cold molecules produced by this method, using static fields. In this regard, ND_3 is a perfect candidate: neon atoms provide an ideal collision partner for efficient cooling, and the Stark structure of the ground state of ND_3 allows the construction of deep electrostatic traps using modest electric fields.[22-24] As a prelude to trapping, we have performed a series of experiments to identify conditions under which ND_3 can be cooled effectively and in quantum states that can be easily trapped. These are the results that we present here.

II. Experiment

Our experiments are performed in a 90° crossed-beam scattering apparatus in which the scattered products are detected using velocity-mapped ion imaging. The scattering apparatus is divided into several vacuum chambers: two "source" chambers, two "differential" chambers, and one "scattering" chamber, all of which are separately pumped. Pulsed atomic and molecular beams are produced in the two source chambers by home-built piezoelectrically-actuated pulsed valves containing either a 5% mixture of ND_3 in neon (at a total pressure of 40 psig) or pure neon (at a pressure of 80 psig). The pulsed valves operate at a repetition rate of 30 Hz and emit pulses approximately 150 μs in length. The typical rotational temperature of the molecular beam is approximately 5–10 K. Each beam passes through a 0.8 mm nickel skimmer (Beam Dynamics, Inc.) as it crosses from its respective source chamber into its respective differential chamber, and passes through a 0.8 mm collimating hole as it crosses into the scattering chamber. The two beams intersect in the scattering chamber, between two annular electrodes and along the axis of a time-of-flight imaging detector. All chambers are equipped with turbomolecular pumps and maintain typical base pressures of 10^{-8} Torr when the pulsed valves are off. The pressure in the scattering chamber is typically 10^{-7} Torr when the pulsed valves are operating.

The velocity distribution of the scattered ND_3 molecules is measured using velocity-mapped ion imaging. Molecules are state-selectively ionized by resonant absorption of two identical photons through the $\tilde{B}$ state. Discussion of the spectroscopic details of the $\tilde{B}$–$\tilde{X}$ transition in ND_3 can be found in[25,26] and references therein. In ND_3, all rovibrational levels of the $\tilde{X}$ ground state have nonzero spin-statistical weight and are doubled by inversion. The symmetry of the ground and excited electronic states and spin statistics of the three deuterium nuclei conveniently separate transitions originating from the two inversion components of the ground state into an alternating pattern: only transitions originating from the *lower* inversion component of the ground state appear in bands in which v_2 (the number of quanta of vibration in the "umbrella" mode of the excited state) is *even*, and only transitions from the *upper* inversion component appear in those bands in which v_2 is *odd*. To monitor the upper and lower inversion components we chose respectively the $\tilde{B}(5)$ and $\tilde{B}(4)$ bands, as these appear with reasonable intensity due to favorable Franck–Condon overlap between the upper and lower quantum states. Due to nuclear spin considerations and the low rotational temperature of the molecular beam, most ND_3 molecules are left in either the $J = 0$, $K = 0$ or $J = 1$, $K = 1$ states, with only a few percent of molecules occupying higher rotational levels. The molecules exist in two nuclear spin configurations, *ortho* and *para*, which do not interconvert by collisions or absorption of radiation, and are cooled independently of one another. The spin considerations consequently pose a restriction on the collision dynamics: molecules in the $(J, K) = (0, 0)$ state only scatter into rotational states with $K = 3n$ (0, 3, 6, ...), and molecules in the $(J, K) = (1, 1)$ state only scatter into rotational states with $K = 3n + 1$ or $K = 3n + 2$ (1, 2, 4, 5, 7, 8, ...).

Radiation of 315–325 nm wavelength for excitation of the two-photon transition is produced by frequency-doubling the output of a tunable pulsed dye laser (Lambda-Physik Scanmate, 5–7 ns pulse length, 0.1 cm^{-1} bandwidth) operating with DCM dye (Exciton, Inc.) using a KDP crystal (Inrad). The dye laser is pumped by the 532 nm output of a pulsed Nd:YAG laser (Coherent Infinity). The frequency-doubled output of the dye laser is focused by a spherical lens into the collision volume, coplanar with and bisecting the atomic and molecular beams. ND_3^+ ions produced by this laser pulse are then accelerated through the ion optics toward an imaging detector. The voltages applied to the electrodes comprising the ion optic assembly are set for velocity-mapping conditions, which maps each velocity component onto a unique spatial position on the detector. The detector itself consists of two microchannel plates arranged in a chevron configuration, placed before a phosphor screen. The voltage applied to the front plate of the detector is pulsed, producing a narrow mass gate which admits signal due to ND_3^+ and discriminates against undesired signals produced by other ions. Images of the phosphor screen are acquired by a thermoelectrically-cooled CCD camera (Princeton Instruments, 1024 × 1024 pixels, cooled to −50 °C), continuously throughout the experiment, typically 5 seconds per exposure, for a total of 225 exposures. An automated background subtraction program runs continuously throughout the experiment, and alternates the timing of the atomic beam such that on even-numbered exposures both the atomic and molecular beams are most intense when the laser fires, and on odd-numbered exposures the atomic beam arrives several milliseconds after the laser fires while the timing of the molecular beam remains constant. The odd-numbered exposures are then subtracted from the even-numbered exposures to reduce systematic noise and eliminate the portion of the signal due to rotationally-hot unscattered molecules in the molecular beam.

III. Results and discussion

A Newton diagram detailing the collision dynamics is shown in Fig. 1. The vectors representing the laboratory-frame velocities of the two beams, $\vec{v}_{ND_3}$ and $\vec{v}_{Ne}$, intersect at 90°, and since their magnitudes are approximately equal, the figure is nearly symmetric. The point of intersection of $\vec{v}_{ND_3}$ and $\vec{v}_{Ne}$ defines the origin in the laboratory frame, where the velocity of either species is exactly zero. The vector $\vec{v}_{cm}$ represents the velocity of the center-of-mass in the laboratory frame, and the vectors $\vec{u}_{ND_3}$ and $\vec{u}_{Ne}$ represent the pre-collision velocities of ND_3 and Ne in the center-of-mass frame. Upon collision, the particles can deflect one another, and the angle between the post-collision velocities $\vec{u}_{ND_3}'$ and $\vec{u}_{Ne}'$ and their pre-collision counterparts defines the center-of-mass scattering angle θ. Now, during an elastic collision, there is no interconversion of kinetic and internal energy, and therefore the velocities do not change magnitude, only direction. In this case the pre-collision and post-collision velocity vectors will lie along a circle of radius $|u_{ND_3}'| = |u_{Ne}'|$. This is the outer circle shown in Fig. 1. During an inelastic collision, however, internal excitation of the molecule occurs, and some kinetic energy is converted into rotational energy. The velocities of the scattered products are therefore lower, and both $\vec{u}_{ND_3}$ and $\vec{u}_{Ne}$ will lie along a circle of smaller radius. This is the inner circle shown in Fig. 1. The velocity-mapped images acquired in our experiments tell us exactly how the velocities of the scattered molecules are distributed along these rings.

What we are most concerned with here, however, is the identification of cold molecules and the measurement of their velocity in the laboratory frame. It is clear from Fig. 1 that, when the collision partners have identical masses, molecules will only truly be stationary in the laboratory frame if they are elastically scattered, since only the Newton sphere corresponding to elastic scattering passes through the

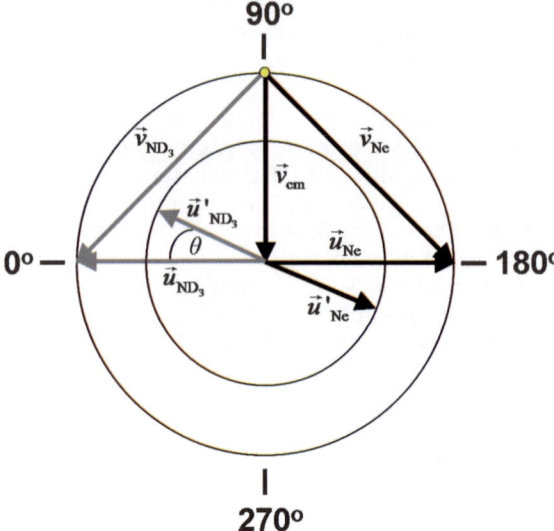

Fig. 1 Newton diagram for ND₃/Ne scattering. Unprimed vectors represent pre-collision quantities, primes represent post-collision quantities.

laboratory origin. Any transfer of energy into the internal degrees of freedom of the molecule will necessarily contract the radius of the Newton sphere, precluding any point on that sphere from having zero laboratory-frame velocity. However, in our case, when the amount of energy transferred into rotation is relatively low, some points on the post-collision Newton sphere will have velocities that are nearly zero, and indeed molecules can still be prepared in low rotational levels with low translational temperatures (and low enough velocity to be trapped with reasonable electric fields). Finally, it should be noted that for collision systems in which the mass of the atom is larger than that of the molecule, it is possible to create cold molecules with a velocity distribution exactly centered about zero. This is possible, for example, in the NO/Ar system.[1–4]

Our previous analysis[2] of the cooling process has led to a number of useful relationships between the relevant post-collision quantities and the directly-measurable pre-collision velocities of the collision partners, all of which may be derived from classical mechanics using vector algebra. The scattering angle θ_{cm} at which the scattered products have their lowest velocity is given by the relation

$$\cos(\theta_{cm}) = \frac{E_{Ne} - E_{ND_3}}{\sqrt{\left(E_{ND_3} + \frac{m_{ND_3}}{m_{Ne}} E_{Ne}\right)\left(E_{ND_3} + \frac{m_{Ne}}{m_{ND_3}} E_{Ne}\right)}}. \tag{1}$$

Here, E_{Ne} and E_{ND_3} are the pre-collision translational kinetic energies of the two species, and m_{Ne} and m_{ND_3} are their masses. The subscript "cm" indicates that for this particular scattering angle, the post-collision velocity $\vec{u}_{ND_3}{}'$ is perfectly anti-aligned with the velocity of the *center of mass* $\vec{v}_{cm}$. In our case, the masses of the collision partners are equal, so eqn (1) reduces to

$$\cos(\theta_{cm}) = \frac{E_{Ne} - E_{ND_3}}{E_{Ne} + E_{ND_3}}. \tag{2}$$

Were the velocities of the collision partners perfectly equal, molecules scattered at exactly 90 degrees would have the lowest attainable velocity. As we shall see, in our experiments the pre-collision velocity of the ND₃ molecules is slightly larger than the pre-collision velocity of the Ne atoms, and therefore the optimum scattering angle is slightly larger than 90 degrees. The post-collision velocity of the center of mass is given by

$$|v_{cm}| = \left(\frac{m_{Ne}^2}{M^2}|v_{Ne}|^2 + \frac{m_{ND_3}^2}{M^2}|v_{ND_3}|^2\right)^{1/2}, \quad (3)$$

where M represents the total mass $m_{Ne} + m_{ND_3}$. The post-collision velocity of the ND$_3$ molecules in the center-of-mass frame is given by

$$|u'_{ND_3}| = \left(\frac{m_{Ne}^2}{M^2}|v_{Ne}|^2 + \frac{m_{Ne}^2}{M^2}|v_{ND_3}|^2 - \frac{2m_{Ne}}{m_{ND_3}M}E'_{rot}\right)^{1/2}, \quad (4)$$

where E_{rot}' is the rotational energy of the molecule after the collision. The laboratory-frame velocity of a molecule scattered backward at θ_{cm} is then simply calculated as the difference

$$|v_{ND_3}'| = |v_{cm}| - |u_{ND_3}'|. \quad (5)$$

When the two vectors are of equal magnitude and perfectly anti-aligned, the anticipated vector cancellation occurs, and molecules are left perfectly stationary. However, as mentioned above, it will be seen from eqn (3) and (4) that in the case of equal masses, the vectors are only equal in length when E_{rot}' is zero, *i.e.* for elastic scattering. It should be noted that, because of the velocity spreads of each of the pulsed beams, each collision partner actually appears with a *range* of initial velocities, and therefore the preceding discussion applies to every conceivable *pair* of velocities of the collision partners. The cold molecules therefore appear with a finite velocity distribution, whose width may be estimated from eqn (3)–(5) using the half-widths of the velocity distribution of the parent atomic and molecular beams.

The above discussion provides us with all we need to analyze the images from our experiments. For visual clarity, we begin with an image of scattered molecules in the $(J,K) = (7,7)$ state, shown in Fig. 2(a). In this case E_{rot}' is quite high (188.59 cm^{-1}) and therefore $|u_{ND_3}'|$ is correspondingly quite low. The Newton spheres for elastic and inelastic scattering are therefore well-separated. There are four prominent features in the image: (i) a ring of radius $|u_{ND_3}'|$, whose intensity at any point is proportional to the number of molecules scattered at that particular angle, (ii) an irregularly-shaped "beam spot" at 0°, from unscattered ND$_3$ molecules in the (7,7) state that are present in the parent molecular beam, (iii) a second elliptical "beam spot" at 180°, which appears in our images due to residual ammonia in the

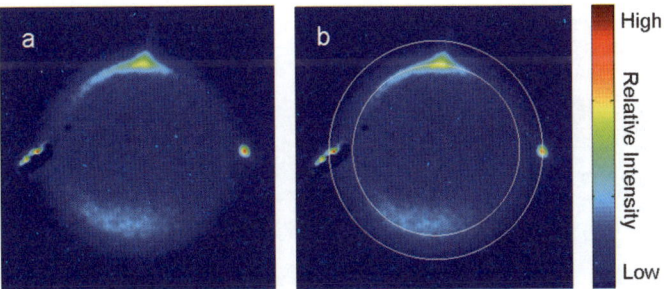

Fig. 2 Image of scattered products in $J = 7$, $K = 7$. Outer ring corresponds to the velocity of elastically-scattered products; inner ring corresponds to velocities of inelastically-scattered products. The left side of each image corresponds to 0° scattering, and the top corresponds to 90° scattering. Relatively slow molecules are found near 90°.

valve that normally emits pure neon, and (iv) an area of high intensity along the ring of inelastically-scattered products, near 90°, due to molecules with low laboratory-frame velocities, which preferentially accumulate in the scattering volume over time.

In Fig. 2(b), the same image is shown, but with two Newton spheres superimposed. The inner Newton sphere is fit to the scattering ring to determine the appropriate center and radius, and the outer sphere is concentric with the inner sphere and its radius is calculated using the centers of the velocity distributions associated with the two beams. The fit is quite good; the outer Newton sphere passes through the center of both "beam spots", and the inner one is centered perfectly on the ring of scattered products, just as demanded by the Newton diagram in Fig. 1. Given the measured pre-collision laboratory-frame velocities of ND_3 and Ne (790 m s^{-1} for Ne and 850 m s^{-1} for ND_3), eqn (1) predicts that the molecules with lowest velocity will lie at $\theta = 94.2°$; indeed, the most intense point on the ring is measured to occur at 93.8°. The difference in radius of the two spheres is measured to be 21 pixels; given the velocity calibration of our detector (5.13 m s^{-1} per pixel), this corresponds to a difference in velocity of $|v_{ND_3}'| = |v_{cm}| - |u_{ND_3}'| = 107.7$ m s^{-1}. Using eqn (3)–(5), we predict a difference of 102.0 m s^{-1}, or approximately 20 pixels. Given our measurement error in locating the laboratory-frame origin, which is expected to be ±1–2 pixels, this result is perfectly reasonable. The intensity, asymmetry, and distortion of the feature at 94° are a direct consequence of the preferential detection of slow molecules in our experiment: the slowest molecules tend to accumulate within the laser focal volume and therefore appear in the images as an intense "spot". The finite velocity spreads of the atomic and molecular beams also produce a distribution of velocities in the scattered products, and molecules with lower-than-average final velocities tending to remain in the focal volume longest. This gives rise to the apparent "pulling" of the intensity of the ring toward the laboratory-frame origin.

The images of the lower-lying rotational states, which are expected to yield the coldest molecules, are shown in Fig. 3. For these states, E_{rot}' is much lower

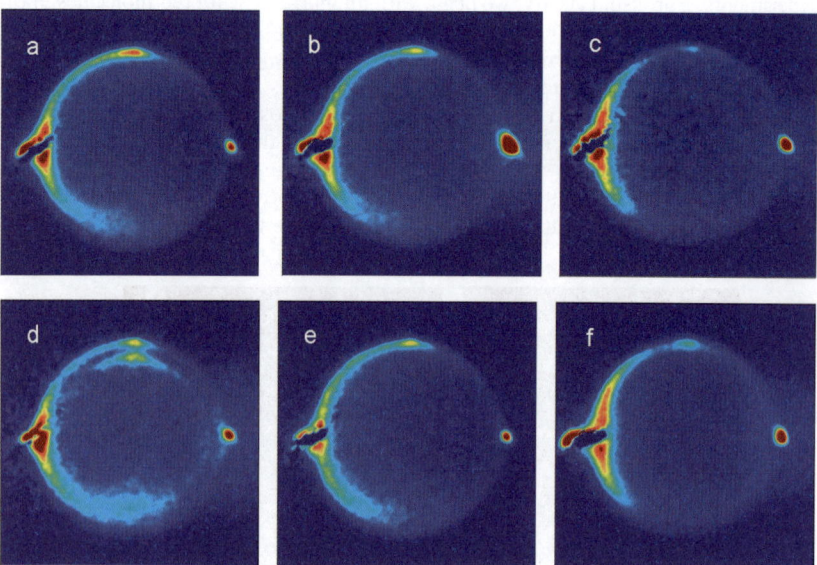

Fig. 3 Images of inelastically-scattered products in low-lying quantum states. (a) $J = 2, K = 0$; (b) $J = 2, K = 1$; (c) $J = 2, K = 2$; (d) $J = 3, K = 1$; (e) $J = 3, K = 2$; (f) $J = 3, K = 3$. A second ring, caused by scattering into a higher rotational level, is found in the interior of (d), and is observed due to a blended optical transition.

(14.48 cm^{-1} for the $J = 2$, $K = 2$ state, up to 51.42 cm^{-1} for the $J = 3$, $K = 1$ state), and consequently $|u_{ND_3}'|$ and $|v_{ND_3}'|$ are much lower as well. Although scattering tends toward relatively small angles for these low quantum states, in every case there is enough scattering around 90 degrees to produce a clear build-up of cold molecules. To determine the velocities of the scattered products, the data are fit in the same manner as above. A cross-section through the scattering ring at 95 degrees is then taken, and the velocity distribution is fit to a Gaussian to determine the center velocity and the velocity spread. The measured velocities, and those calculated using eqn (3)–(5), are tabulated in Table 1. As shown in the Table, the velocities of the centers of the distributions associated with the cold molecules are measured to be 20–30 m s^{-1} (with an uncertainty of ±10 m s^{-1}), and the velocity spreads of the cold molecules are typically found to be ±20 m s^{-1} FWHM. The calculations indicate that the molecules are even slower, and predict center velocities ranging from 7.5 m s^{-1} up to 22.8 m s^{-1} with velocity spreads of ±1 m s^{-1} or less.

The final velocities measured here represent a strict upper limit: known uncertainties associated with the experiment preclude accurate velocity measurements on such a fine scale. Several factors necessarily inflate both the measured velocity and the velocity spread. The image obtained in our experiment is in reality a two-dimensional projection of a three-dimensional shell of scattered molecules. It is well-known from previous work on velocity-mapped ion imaging[20,21] that this projection necessarily leads to the appearance of an asymmetric velocity distribution: in the absence of competing effects, the scattering ring would appear to have a sharp outer edge and gradually decreasing intensity toward the center of the image. The recoil due to the departing photoelectron during the photoionization step imparts a ~20 m s^{-1} velocity spread to the detected ND$_3^+$ ions, which gives rise to a 4-pixel blur on top of the asymmetry of the projection. Together, these effects give rise to a larger-than-expected velocity distribution. Finally, the resolution of the detection system affects our ability to reliably measure the apparent center velocity: one pixel is approximately 5 m s^{-1}, and the error in locating the laboratory origin is expected to be 1–2 pixels. This introduces an error of 5–10 m s^{-1} in the determination of the center velocity. However, precise velocity measurements were never the goal of this work: our main concern was to identify conditions under which significant numbers of relatively slow molecules could be produced, and from the images it is clear that significant numbers of cold ND$_3$ molecules can be produced in each of these quantum states. In previous work we have verified, for the NO/Ar scattering system, that these calculated temperatures can actually be experimentally obtained.[4]

Table 1 Measured and calculated velocities of ND$_3$ in low-lying quantum states

(J, K)	v_{ND_3}' (meas.)[a]	$\bar{E}_{trans}/k$ (meas.)[c]	$\Delta E_{trans}/k$ (meas.)[d]	v_{ND_3}' (calc.)[b]	$\bar{E}_{trans}/k$ (calc.)[c]	$\Delta E_{trans}/k$ (calc.)[d]
(2, 0)	26 ± 20 m s^{-1}	810 mK	460 mK	16 ± 0.6 m s^{-1}	310 mK	360 μK
(2, 1)	27 ± 18 m s^{-1}	880 mK	390 mK	11 ± 0.4 m s^{-1}	140 mK	170 μK
(2, 2)	23 ± 16 m s^{-1}	640 mK	320 mK	8 ± 0.3 m s^{-1}	70 mK	80 μK
(3, 1)	28 ± 16 m s^{-1}	940 mK	320 mK	27 ± 1.0 m s^{-1}	880 mK	1.1 mK
(3, 2)	21 ± 17 m s^{-1}	530 mK	340 mK	24 ± 0.8 m s^{-1}	680 mK	810 μK
(3, 3)	32 ± 16 m s^{-1}	1.2 K	790 mK	23 ± 0.8 m s^{-1}	630 mK	750 μK

[a] Indicates measured center velocity and velocity spread. Uncertainty in center velocity is ±10 m s^{-1}. [b] Indicates calculated center velocity and velocity spread. [c] Ratio of translational kinetic energy to k in the laboratory frame of reference, assuming stated center velocity. [d] Ratio of translational kinetic energy to k in the moving frame of reference, assuming stated velocity spread.

For benchmark comparison with other cooling techniques, we have associated "temperatures" with the measured velocities and velocity spreads, shown in Table 1. We define them for both the laboratory frame of reference and a frame of reference moving at the center velocity of the distribution. (Our definition of a temperature for a moving frame of reference is analogous to the definition of a translational temperature for a molecular beam: although the molecules are moving at very high velocities in the laboratory frame, the velocity distribution can be quite narrow, and one often speaks of its low "temperature" in the moving frame of reference of the beam.) For the laboratory frame velocity, we report the ratio of the mean kinetic energy $\bar{E}_{trans} = \frac{1}{2}mv^2_{center}$ to the Boltzmann constant k, which associates a temperature with the center of the velocity distribution of the cold molecules. For the moving frame we report the ratio of the kinetic energy associated with the velocity half-width $\Delta E_{trans} = \frac{1}{2}mv^2_{HWHM}$ to k, which removes the contribution of the laboratory frame velocity and associates a temperature with the velocity distribution. Defined in this way, the average speed of the distribution (in either frame) is then identical to the most-probable speed of a molecule in a true thermal system of the same temperature. As shown in the table, our measured velocities correspond to temperatures between 530 mK–1.2 K in the laboratory frame, and temperatures between 320–790 mK in the moving frame. The calculations predict much lower temperatures: 70–880 mK in the laboratory frame, and 80 μK–1.1 mK in the moving frame.

An independent confirmation that the molecules are indeed as cold as the calculation predicts may be had by acquiring delayed images. This is accomplished by introducing an adjustable time delay (here tens of microseconds) between the temporal center of the pulsed beams and the arrival of the laser pulse that ionizes the molecules. At the temporal center of the pulsed beams, scattering is most intense, and both fast- and slow-moving scattered particles within the laser focal volume will be detected. Images acquired during this time will therefore display intense scattering rings, with a significant accumulation of slow-moving molecules near $\theta = \theta_{cm}$. As the delay is increased, the densities of the gas pulses become progressively lower, scattering events become more rare, and fast particles scattered at earlier times preferentially move out of the focal volume of the laser. Slow-moving molecules, however, tend to remain within the detected volume for longer periods of time. Images acquired at later times will therefore display progressively weaker rings, with significant intensity near $\theta = \theta_{cm}$ but much weaker intensity at all other angles. Finally, at long delays, the gas pulses will have completely exited the collision volume, scattering will have ceased altogether, and all fast-moving particles scattered at earlier times will have moved out of the laser focal volume. At this point, only cold molecules remain, and the only remnant of the original scattering ring will be a cold "spot" near $\theta = \theta_{cm}$. Fig. 4 shows three images, monitoring the $J = 2$, $K = 2$ quantum

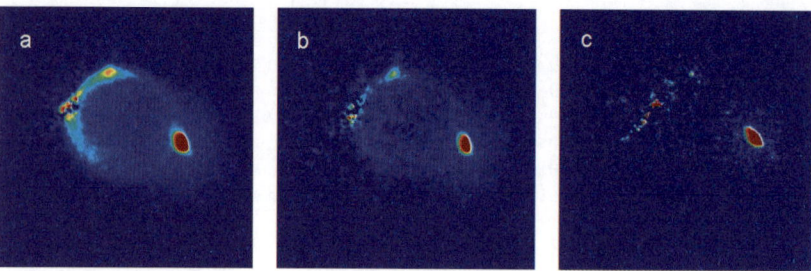

Fig. 4 Images of the $J = 2$, $K = 2$ scattered products, with variable delay introduced between pulsed valve and laser. (a) Image acquired on falling edge of molecular beam pulse. (b) Image acquired 30 μs after image shown in (a). (c) Image acquired 40 μs after image shown in (a). Cold molecules are observed to persist, while molecules with higher velocities escape the detection region.

state, with variable delay. The first image is acquired on the falling edge of the molecular beam pulse, where scattering is still clearly visible. The second image is acquired 30 microseconds later; here, the scattering ring is clearly significantly diminished in intensity, but the cold molecules are still clearly visible. The third image is acquired at a delay of 40 microseconds, at which point the scattering ring is completely absent while cold molecules persist.

For the low-lying quantum states discussed here, it is also possible to enhance the number of cold molecules produced by "tilting" the Newton sphere. This is accomplished by changing the velocity of either one, or both, of the collision partners, such that the laboratory-frame origin lies at a lower scattering angle where the cross section is higher. This is shown schematically in Fig. 5. Seeding the ND_3 in krypton, for example, reduces v_{ND_3}, and shifts the origin to a lower angle, in this case $\theta = 65°$. This was tacitly shown in Fig. 4, where images of $J = 2$, $K = 2$ were acquired with the ND_3 seeded in Krypton. The accumulation of cold molecules at the top of the scattering ring is indeed seen to lie at a scattering angle near 65°. The ratio of the intensity of cold molecules to forward-scattered molecules is clearly much higher than was the case for ND_3 seeded in neon; compare Fig. 3(c).

Finally, we wish to discuss prospects for trapping the cold ND_3 produced in these experiments. Cold samples of ND_3 molecules, produced by Stark deceleration and velocity filtering from effusive beam sources,[5,6,27] have already been confined in traps[22,24] and are of primary importance to the cold molecules community. The cold packets of ND_3 molecules that we have produced here can readily be confined in an electrostatic quadrupole trap that surrounds the collision volume. The fact that cold molecules can be produced inside the trapping potential is a great advantage; since there is no need to release the trap to admit additional cold molecules, the trap can in principle be continuously loaded over many cycles of the experiment without losing or heating molecules that have already been confined. Simultaneous trapping of multiple species, in selected and well-defined quantum states, is also possible and offers interesting prospects for the study of state-to-state collision and reaction dynamics at very low temperatures. The main disadvantage to our

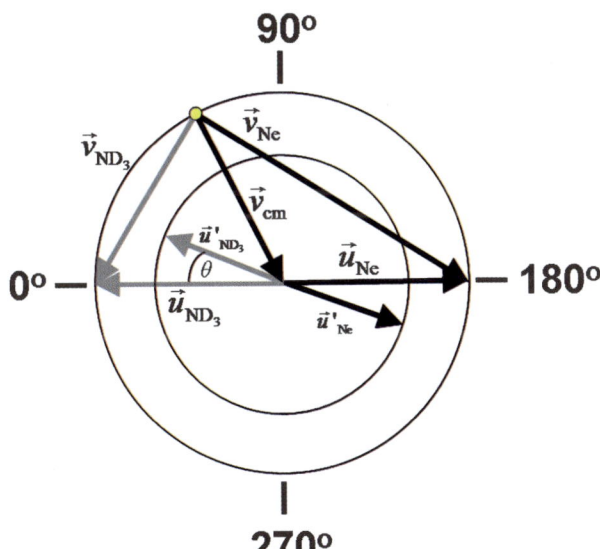

Fig. 5 Newton diagram for ND_3/Ne scattering, with ND_3 seeded in krypton. Unprimed vectors represent pre-collision quantities, primes represent post-collision quantities. Note that the origin is shifted to lower scattering angle θ as compared with the diagram shown in Fig. 1.

technique is that secondary collisions may affect the number of molecules which can be trapped. Since our cold molecules are, by necessity, produced at the busy intersection of two supersonic beams, re-heating by collisions with uncooled species in either beam can lead to loss of molecules from the trap. Since this loss process is difficult to quantify, it is not possible for us to estimate the total number of molecules we expect to trap or the phase space density of a trapped sample. However, since a substantial number of molecules are indeed left in the wake of the atomic and molecular beams, as evidenced by the images in Fig. 4, we are confident that it will be possible to trap significant numbers of cold molecules using our technique.

IV. Conclusions

We have produced cold ND_3 molecules in a variety of quantum states by kinematic cooling. We have measured the laboratory-frame velocity and velocity spread for the coldest inelastically scattered molecules the (J, K) = (2,0), (2,1), (2,2), (3,1), (3,2), and (3,3) rotational levels of the ground vibronic state. These have been compared with the expected velocity and velocity spreads. The measured distributions range in "temperature" from 530 mK to 1.2 K and represent an upper limit to the actual temperature. At these temperatures it should be possible to use electrostatic traps to confine cold molecules in many of these quantum states.

Acknowledgements

The authors acknowledge Mr Mark Jaska for valuable technical support. Funding for this work was provided by the U. S. Department of Energy, Office of Basic Energy Science. Sandia is a multiprogram laboratory operated by Sandia Corporation, a Lockheed Martin Company, for the U. S. Department of Energy.

References

1 M. S. Elioff, J. J. Valentini and D. W. Chandler, *Science*, 2003, **302**, 1940.
2 M. S. Elioff, J. J. Valentini and D. W. Chandler, *Eur. Phys. J. D*, 2004, **31**, 385.
3 K. E. Strecker and D. W. Chandler, in *Low Temperatures and Cold Molecules*, edited by I. W. Smith World Scientific, 2008.
4 K. E. Strecker and D. W. Chandler, *Phys. Rev. A*, 2008, **78**, 063406.
5 H. L. Bethlem, G. Berden and G. Meijer, *Phys. Rev. Lett.*, 1999, **83**, 1558.
6 S. Y. T. van de Meerakker, H. L. Bethlem and G. Meijer, *Nat. Phys.*, 2008, **4**, 595.
7 S. D. Hogan, D. Sprecher, M. Andrist, N. Vanhaecke and F. Merkt, *Phys. Rev. A*, 2007, **76**, 023412.
8 E. Narevicius, C. G. Parthey, A. Libson, J. Narevicius, I. Chavez, U. Even and M. G. Raizen, *New J. Phys.*, 2007, **9**, 358.
9 S. D. Hogan, A. W. Wiederkehr, M. Andrist, H. Schmutz and F. Merkt, *J. Phys. B*, 2008, **41**, 081005.
10 E. Narevicius, A. Libson, C. G. Parthey, I. Chavez, J. Narevicius, U. Even and M. G. Raizen, *Phys. Rev. Lett.*, 2008, **100**, 093003.
11 J. M. Doyle, B. Friedrich, J. Kim and D. Patterson, *Phys. Rev. A*, 1995, **52**, R2515.
12 J. Kim, B. Friedrich, D. P. Katz, D. Patterson, J. D. Weinstein, R. DeCarvalho and J. M. Doyle, *Phys. Rev. Lett.*, 1997, **78**, 3665.
13 H. R. Thorsheim, J. Weiner and P. S. Julienne, *Phys. Rev. Lett.*, 1987, **58**, 2420.
14 R. Wynar, R. S. Freeland, D. J. Han, C. Ryu and D. J. Heinzen, *Science*, 2000, **287**, 1016.
15 M. Viteau, A. Chotia, M. Allegrini, N. Bouloufa, O. Dulieu, D. Comparat and P. Pillet, *Science*, 2008, **321**, 232.
16 J. G. Danzl, E. Haller, M. Gustavsson, M. J. Mark, R. Hart, N. Bouloufa, O. Dulieu, H. Ritsch and H.-. Nägerl, *Science*, 2008, **321**, 1062.
17 E. A. Donley, M. R. Claussen, S. T. Thompson and C. E. Wieman, *Nature*, 2002, **417**, 6888.
18 C. A. Regal, C. Ticknor, J. L. Bohn and D. S. Jin, *Nature*, 2003, **424**, 6944.
19 K. E. Strecker, G. B. Partridge and R. G. Hulet, *Phys. Rev. Lett.*, 2003, **91**, 080406.
20 A. J. R. Heck and D. W. Chandler, *Ann. Rev. Phys. Chem.*, 1995, **46**, 335.
21 A. T. J. Eppink and D. H. Parker, *Rev. Sci. Inst.*, 1997, **68**, 3477.

22 H. L. Bethlem, G. Berden, F. M. H. Crompvoets, R. T. Jongma, A. J. A. van Roij and G. Meijer, *Nature*, 2000, **406**, 491.
23 S. Y. T. van de Meerakker, P. H. M. Smeets, N. Vanhaecke, R. T. Jongma and G. Meijer, *Phys. Rev. Lett.*, 2005, **94**, 023004.
24 T. Rieger, T. Junglen, S. A. Rangwala, P. W. H. Pinske and G. Rempe, *Phys. Rev. Lett.*, 2005, **95**, 173002.
25 M. N. R. Ashfold, R. N. Dixon, R. J. Stickland and C. M. Western, *Chem. Phys. Lett.*, 1987, **138**, 201.
26 M. N. R. Ashfold, R. N. Dixon, N. Little, R. J. Stickland and C. M. Western, *J. Chem. Phys.*, 1988, **89**, 1754.
27 T. Junglen, T. Rieger, S. A. Rangwala, P. W. H. Pinske and G. Rempe, *Eur. Phys. J. D*, 2004, **31**, 365–373.

Manipulating the motion of large neutral molecules

Jochen Küpper,* Frank Filsinger and Gerard Meijer

Received 10th November 2008, Accepted 5th January 2009
First published as an Advance Article on the web 26th May 2009
DOI: 10.1039/b820045a

Large molecules have complex potential-energy surfaces with many local minima. They exhibit multiple stereo-isomers, even at very low temperatures. In this paper we discuss the different approaches for the manipulation of the motion of large and complex molecules, like amino acids or peptides, and the prospects of state- and conformer-selected, focused, and slow beams of such molecules for studying their molecular properties and for fundamental physics studies.

I. Introduction

Over the last few years there have been tremendous advances in the preparation of cold and ultracold samples of small molecules, either by association of ultracold atoms, or by direct cooling methods. Using the association technique, ultracold heteronuclear ground-state dimers were recently produced.[1,2] Direct cooling methods allow the preparation of trapped samples of more complex molecules (*i. e.*, ammonia)[3] and they promise possibilities for their extension to large molecules like the "building blocks of life".[4] Recently, we have demonstrated the alternating gradient (AG) deceleration of the prototypical large molecule benzonitrile.[5] Once such molecules are decelerated to a quasi-standstill in the laboratory frame, they can be trapped in ac traps, which have been demonstrated for small molecules in high-field-seeking states.[6,7]

For many applications in physics and chemistry ensembles of large molecules all in one or in a few quantum states would be highly beneficial. For many years, small molecules have been state-selected and focused using static multipole fields.[8,9] For about ten years it has also been feasible to change the velocity of small molecules in low-field-seeking states using the Stark decelerator.[10] The samples of slow molecules produced can be trapped in static or dynamic fields, can be injected into storage rings, or can be used for various molecular physics applications.[11]

Larger molecules, however, have practically only high-field-seeking (hfs) states at the relevant electric field strengths. To illustrate this the Stark curves of some low rotational states of benzonitrile (C_7H_5N) are shown in Fig. 1. In order to focus molecules in these hfs states one would need to create a maximum of electric field in free space, which is not possible according to Maxwell's equations.[8,9] However, large molecules have been deflected using static fields[13,14] and their motion was manipulated in this way in matter–wave interferometry experiments.[15] Moreover, the rotational motion of large molecules has been restricted using brute-force orientation in dc electric fields,[16–18] using laser alignment,[19–21] or using mixed dc and laser fields.[14,22,23]

In order to confine the motion of large molecules, one has to use dynamic focusing in alternating-gradient (AG) setups.[24,25] We have demonstrated that alternating-gradient focusing can be used to focus and decelerate large molecules. In a prototype

Fritz-Haber-Institut der Max-Planck-Gesellschaft, Faradayweg 4–6, 14195 Berlin, Germany. E-mail: jochen@fhi-berlin.mpg.de

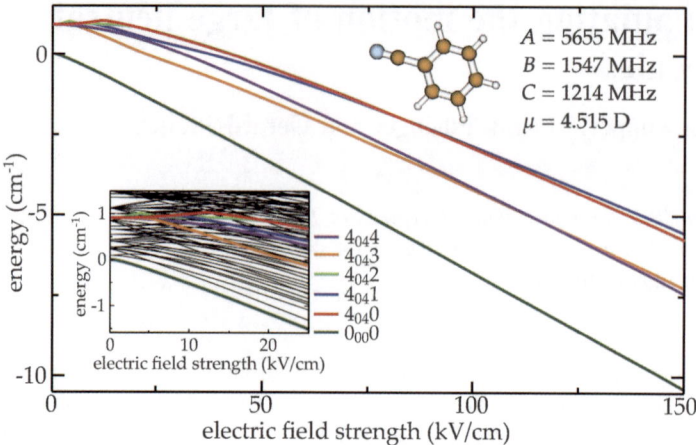

Fig. 1 Energy of selected low rotational states of benzonitrile as a function of electric field strength. In the upper inset the molecular structure of benzonitrile and its relevant molecular parameters are given.[12] In the lower inset, all states with a field-free energy below 1.2 cm^{-1} are shown at smaller field strengths.

experiment we have decelerated benzonitrile (C_7H_5N) molecules from a supersonic jet.[5] In similar experiments, we have demonstrated that the frequency characteristics for the dynamic focusing in an AG setup can be used to separate species with distinct mass-to-dipole moment ratios.[26] Equivalent to the m/q-selection of charged particles in a mass spectrometer, these experiments perform an m/μ-selection. We have demonstrated the selection of the *cis* and *trans* conformers of 3-aminophenol (C_6H_7NO) and are currently performing first selection experiments on the conformers of biomolecules. Slow, albeit warm, beams of thermally stable, large molecules can also emerge directly from an oven.[27]

The spectroscopic investigation of complex molecules isolated in the gas-phase has also seen big advances over the last decade.[28,29] These advances are largely due to the ability to create intense molecular beams of large molecules and to ingenious multi-resonance laser spectroscopy schemes that allow us to disentangle the signatures of different isomers[30] present even in cold molecular beams.[31] However, for many novel studies it would be very advantageous, or even necessary, to separate the individual isomers, to select quantum states, or to slow down these molecules. There is a large interest in performing coherent X-ray diffractive imaging of biomolecules[32] using novel free-electron-laser X-ray sources. It would be very advantageous to perform initial experiments on well-defined targets: ensembles of molecules all with the same structure (conformation) and all strongly oriented in space. Similar arguments can be made for high-harmonic generation[33] or tomographic orbital reconstruction[34] experiments using (large) molecules.

In this paper we will describe the different experimentally proven methods for the manipulation of the motion of large molecules. We will focus on the manipulation of translational motion, where we will present detailed descriptions and experimental results obtained in our laboratory. We place special emphasis on the ability to separate conformers with these experiments and will compare the approaches with one another.

II. Experimental approaches

Several complementary experimental approaches for the manipulation of the translation and for the state selection of large neutral molecules exist. Here we limit the discussion to the use of inhomogeneous electric fields, although similar experiments could be performed using magnetic[35–37] or laser fields.[38,39] All these methods rely on the strong cooling provided by the supersonic expansion, where rotational and translational temperatures in the moving frame of the molecular beam of the order of 1 K are routinely achieved.[40]

Whereas this article describes the manipulation of the translational motion of molecules, methods to manipulate the rotational motion have also been demonstrated. There are several electric field based methods that are applicable for large molecules, which are generally asymmetric rotors. The conceptually simplest method to confine the angular distribution of polar molecules is the interaction of the molecular dipole with a strong homogeneous electric field, as proposed independently by Loesch and Remscheid[16] and by Friedrich and Herschbach.[17] This approach has been experimentally demonstrated many times and is summarised elsewhere.[41,42] This "brute force orientation" of large molecules has been exploited, for example, to determine transition moment angles in the molecular frame.[43,44] Applying strong, non-resonant laser fields to the molecules also provokes angular confinement.[20] The crucial influence of the population of rotational states has been experimentally investigated.[21] Clearly, the state-selection of the lowest rotational states, performed by the experiments described below, allows for considerably stronger degrees of alignment.[14] Recently, strong alignment and orientation by mixed dc electric and laser fields[45,46] has been demonstrated for the large asymmetric top molecule iodobenzene.[14] For a more-in-depth discussion of alignment and orientation the reader is referred to the existing excellent reviews.[20,47]

II.A. Electric beam deflection

II.A.1. General description. The possibility of deflecting polar molecules from a molecular beam using inhomogeneous electric fields was first described by Kallmann and Reiche in 1921.[48,90] In 1926 Stern suggested that the technique could be used for the quantum state separation of small diatomic molecules at low temperatures.[49] The electric deflection of a molecular beam was experimentally demonstrated by Wrede – a doctoral student of Otto Stern – in 1927 and the dipole moment of KI was determined.[50] In 1939 Rabi *et al.* introduced the molecular beam resonance method, by using two deflection elements of oppositely directed gradients in succession to study the quantum structure of atoms and molecules.[51] By 1956 a number of different approaches to beam deflection were discussed in Ramsey's textbook.[8] Since then, electric beam deflection has been used extensively, for example, to determine polarisabilities and dipole moments of clusters and molecules.[13,52]

Commonly, the so-called two-wire-field geometry, depicted in Fig. 2, is used,[8] but more advanced electrode geometries have also been employed, for example, in recent matter–wave interferometry experiments.[15] Nevertheless, the classical two-wire field is widely employed, due to its experimental simplicity and the quite good results that can be obtained with it.

II.A.2. Deflection of large molecules. Using an experimental setup as shown in Fig. 2 we have, in collaboration with the group of Henrik Stapelfeldt (University of Aarhus, Denmark), deflected large molecules, *i. e.*, iodobenzene (C_6H_5I), and performed laser alignment and mixed-field orientation of the produced state-selected samples.[14] For these studies it was crucial to start with cold molecular beams as they are produced in high-stagnation-pressure expansions.[53] In Fig. 3a the calculated deflection profiles of benzonitrile from molecular beams of varying temperature is shown. For the lowest temperature (1 K) almost the complete beam can be deflected. With increasing temperature, the overall deflection significantly decreases and instead a considerable broadening of the beam is obtained. Due to the large number of quantum states involved we have not performed quantitative calculations for temperatures above 4 K, but approximate calculations suggest that at 10 K no significant deflection is obtained any more. In Fig. 3b the contribution of the lowest

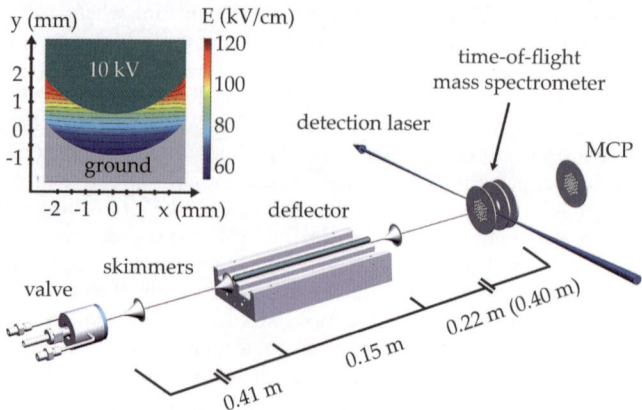

Fig. 2 Experimental setup of the electric m/μ deflector for quantum-state selection of large molecules. An internally cold molecular beam is produced by expanding a mixture of a few percent of the investigated molecule in several bar of rare gas. The resulting supersonic jet is skimmed a few centimetres downstream of the nozzle for differential pumping and again about 41 cm downstream of the nozzle by a 1 mm diameter skimmer for beam collimation. Then the molecular beam enters the electric deflector, where the inhomogeneous field shown enlarged in the inset provides a force on polar molecules in the vertical direction. Behind the deflector another 1.5 mm skimmer provides further differential pumping for the molecules entering the detection chamber, where the (vertical) deflection profiles can be measured by scanning a pulsed dye laser up and down in a resonance-enhanced multi-photon-ionisation time-of-flight mass-spectrometry detection scheme.

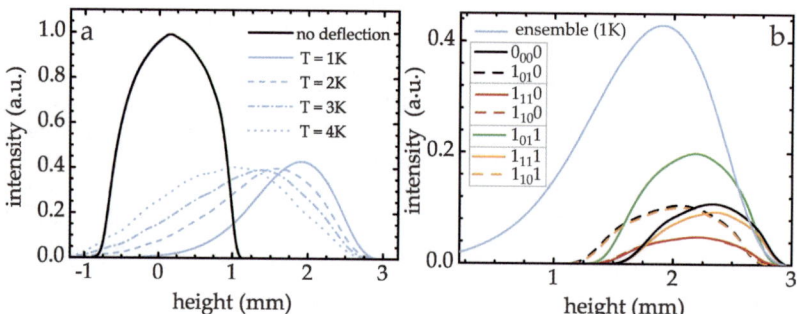

Fig. 3 Deflection profiles a) for molecular beams of benzonitrile at different rotational temperatures and b) for a molecular beam of benzonitrile at 1 K and individual profiles of the lowest energy rotational states. The intensities of all individual states are according to their population at 1 K and are scaled by a factor of 10. All calculations are made for the setup shown in Fig. 2 using a distance from deflector to detector of 22 cm.

individual rotational states, *i. e.*, all states with $J = 0$ or 1, to the deflection profiles is depicted. These states belong to four different Hamiltonian matrix blocks as shown in the figure legend (see appendix A for details) and the lowest energy states, shown as solid lines in Fig. 3b, are all very polar and behave very similarly under the influence of the electric field. The $J_{K_aK_c} = 0_{00}$ and 1_{11} states especially behave extremely similarly – they are the ground states of *para*- and *ortho*-benzonitrile, respectively – and differ practically only in intensity, which is due to the difference in population.

In any case, using a focused laser or a narrow slit one could perform experiments selectively with the molecular ensemble at a given height. Choosing, for example, a height of $z = 2.7$ mm at 1 K, where the density is 5% of the density at the peak of the free flight, only a few quantum states are populated, which are all very polar and have very similar effective dipole moments μ_{eff} (the negative gradient of the Stark curve). Therefore, such an ensemble can be strongly oriented using dc electric fields or mixed dc and laser fields.

For cold beams of small molecules with large rotational constants and, therefore, only a few rotational states populated in the molecular beam, this method would allow the preparation of samples of individual rotational states.[49] For low-field-seeking states this can also be achieved using electric multipoles, which also provide transverse focusing,[9,54] or, even cleaner, using the Stark decelerator.[10,55] The deflector, nevertheless, also allows us to individually address high-field-seeking states, *e.g.*, absolute ground states.

The large differences between the calculated deflection profiles for benzonitrile at different temperatures also demonstrate that deflection profiles can be used as a sensitive measure of the internal temperatures of molecular beams,[56] especially for low rotational temperatures, where the strongly polar quantum states are populated the most and where meaningful quantum-mechanical calculations can still be performed.

II.A.3. Conformer selection. For more complex molecules, typically multiple conformers are present in a molecular beam.[31] These conformers all have the same mass, but typically exhibit largely different dipole moments. Therefore, one can use the distinct forces exerted on the molecules by inhomogeneous electric fields to spatially separate the molecules based on their m/μ ratios. We have already demonstrated this for 3-aminophenol using dynamic focusing, see section IIB and reference 26. Here, we want to discuss the possibility of using an electric deflector for the same purpose.

In Fig. 4 the simulated deflection profiles of the *cis* and *trans* conformers of 3-aminophenol are shown for the experimental setup described above and the known molecular parameters.[57] For clarity we have assumed equal population of the two species in these simulations; the actual populations can be estimated to be 1 : 4 for *cis*- : *trans*-3-aminophenol from their intensities in electronic spectra. From the simulations at 1 K it is clear that a large fraction of the *cis*-3-aminophenol conformers can be deflected out of the original beam and even out of the distribution of the *trans* conformer. Moreover, even at 4 K one can foresee the performance of experiments with a pure sample of *cis*-3-aminophenol. On the other hand, no clean sample of

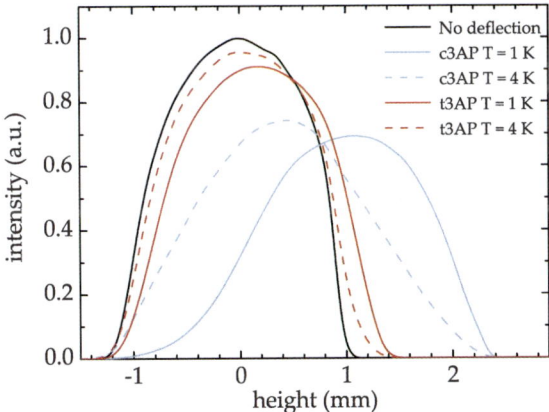

Fig. 4 Simulated deflection profiles of a beam of *cis*-3-aminophenol (c3AP) and *trans*-3-aminophenol (t3AP) using the setup described above (with a deflector-to-detector distance of 40 cm) at a voltage of 12 kV applied to the rod and rotational temperatures of 1 and 4 K. In these simulations the populations of *cis*- and *trans*-3-aminophenol in the original beam are assumed to be equal.

the less polar *trans*-3-aminophenol conformer can be produced this way. This is even more so, since typically the internal temperatures of the supersonic beams are not thermal[58] and a high-temperature fraction, corresponding to mostly unpolar quantum states, exists. Therefore, in general, only the most polar conformer, can be separated from the others. These results should be compared to the demonstrated selection of the conformers of 3-aminophenol using the AG focusing selector described in section IIB, where each conformer can be addressed individually.

II.B. Dynamic focusing selectors

II.B.1. General description. Whereas deflection experiments allow the spatial dispersion of quantum states, they do not provide any focusing. For small molecules in low-field-seeking states this issue could be resolved using multipole focusers with static electric fields, which were developed independently in Bonn[59,60] and in New York[61,62] in 1954/55. About ten years later, molecular samples in a single rotational state were used for state specific inelastic scattering experiments by the Bonn group[63] and, shortly thereafter, for reactive scattering.[64,65]

However, for large molecules all quantum states are practically hfs at the relevant electric field strengths. Therefore, focusing with static electric fields is not possible. Instead, one has to retreat to dynamic focusing schemes, which are also used in the operation of the LINAC, quadrupole ion guides, or Paul traps for charged particles. Dynamic focusing of neutral polar molecules was described by Auerbach *et al* in 1966.[24] Successively, it was experimentally demonstrated[66–69] and successfully applied in maser experiments.[70,71]

In these early experiments the switching frequency was defined by the beam velocity and the electrode geometry, requiring a setup with electric field lenses of alternating orientation. Nowadays, however, it is possible to electronically switch the necessary electric fields and one can use, for example, a four-wire setup, shown in Fig. 5, with varying voltages applied to create the necessary alternating gradient focusing at any frequency and always applying the maximum field strength and gradient, *i.e.*, always applying maximal forces.

II.B.2. Focusing of molecules in low- and high-field-seeking states. In order to characterise the operation of the AG focuser we have performed initial experiments using ammonia (NH_3) in its $J_K = 1_1$ rotational state. NH_3 in this state, the rotational ground state of *para* ammonia, exhibits a quadratic Stark effect at low and moderate electric field strengths that converges to a linear Stark shift once the Stark energy is comparable to the inversion splitting, as shown in Fig. 6. Ammonia molecules in the lfs $MK = -1$ component can be focused using a static quadrupole field. This was already exploited by Gordon, Zeiger and Townes in the original demonstration of the MASER[62] and was performed by us for initial optimisation of expansion conditions and laser detection. However, ammonia molecules in both polar quantum states ($MK = -1$ and $+1$) can be confined to the beam axis using AG focusing. This is demonstrated by the measurements in Fig. 7. Here, the transmissions of NH_3 in its $J_K = 1_1$, $MK = -1$ and $J_K = 1_1$, $MK = 1$ states are plotted as a function of the frequency used to switch between the two electric field configurations. The individual states are selectively detected by choosing appropriate 2 + 1-REMPI transitions. The transmission for the molecules in hfs quantum states (Fig. 7b) shows a frequency dependence as expected: At low switching frequencies the molecules are strongly defocused in one direction and lost from the focuser. At high frequency the time averaged potential is approximately flat and no focusing occurs, therefore, the transmission is also low. In between, at the appropriate switching frequency, AG focusing works and the transmission is high. The experimentally measured transmission is strongly modulated in that frequency range. This modulation is due to the overlap between the strongly focused detection laser ($w_L \approx 40$ μm) and the changing shape of the molecular packet. This shape strongly depends on the phase in the

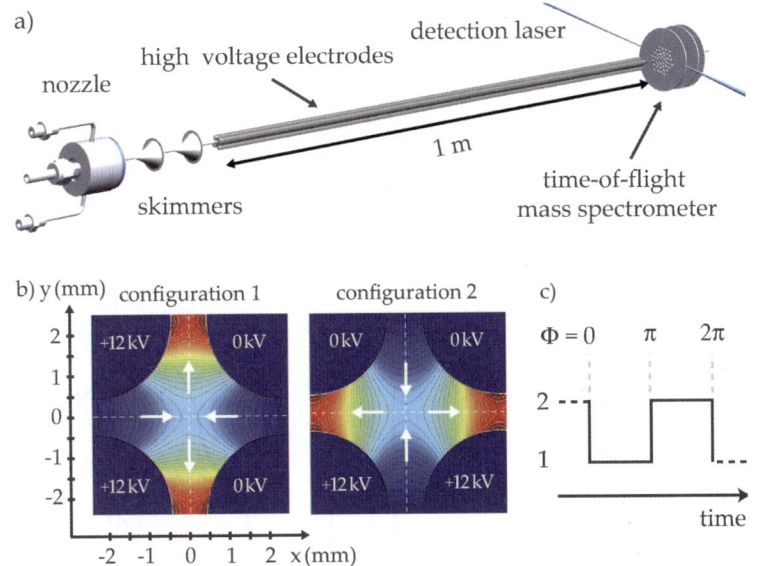

Fig. 5 a) Experimental setup of the dynamic-focusing electric m/μ selector for quantum-state selection of large molecules. An internally cold molecular beam is produced by expanding a mixture of a few percent of the investigated molecule in several bar of rare gas. The resulting supersonic jet is skimmed a few centimetres downstream of the nozzle for differential pumping and again about 18 cm downstream of the nozzle by a 1 mm diameter skimmer for beam collimation. Then the molecular beam enters the electric selector, where the switched inhomogeneous fields given in the inset b) provide a focusing force on polar molecules towards the molecular beam axis. The transmission through the selector is monitored in a resonance-enhanced multi-photon-ionisation time-of-flight mass-spectrometry detection scheme. c) Definition of the phase of the switching cycle, where configuration 1 corresponds to focusing in the x-direction and 2 to focusing in the y-direction for molecules in high-field-seeking states.

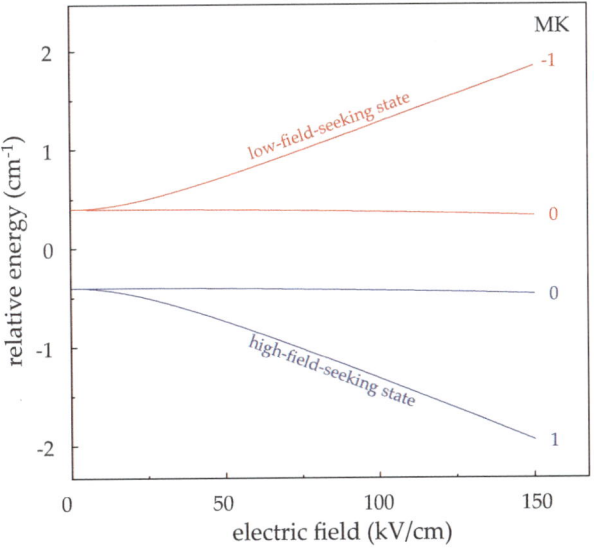

Fig. 6 The energy of ammonia (NH_3) in its $J_K = 1_1$ rotational state as a function of electric field strength (neglecting hyperfine structure).

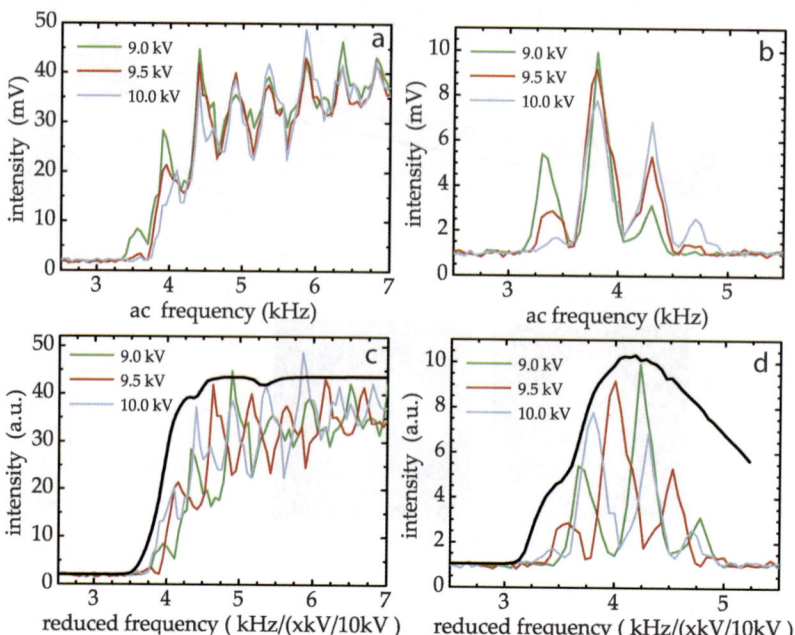

Fig. 7 Measured transmission through the m/μ selector for ammonia in its lfs ($MK = -1$, left graphs) and hfs ($MK = 1$, right graphs) states of the $J_K = 1_1$ rotational state. In the upper plots the experimental transmission as a function of the switching frequency between the two electric field configurations, shown in Fig. 5b, is plotted for different applied voltages. The observed strong modulation in all measurements is due to changes in the end-phase of the switching cycle and, therefore, the overlap between the focused detection laser and the molecular packet. The differences in the peak intensities are due to the changed frequency dependence according to the focusing strength at the different voltages. In the lower graphs the same data is plotted as a function of a reduced frequency $\tilde{f} = f/(U/10^4 \text{ V})$ with the applied switching frequency f and voltage U. The envelope of these data nicely represents the simulated overall transmission through the selector (solid black lines), independent of the detection (overlap) function.

switching cycle as depicted in Fig. 8. Due to the short focusing device – corresponding to a short residence time of the molecules in the device – the start phase, the end phase, and the switching frequency cannot be independently optimised. We chose to optimise the start phase for optimum transmission and then the end phase is determined from the applied switching frequency. In order to reduce these experimental artifacts, we measured the transmission curves for different applied high-voltages U. For comparison the frequencies f of these different measurements are converted to a reduced frequency $\tilde{f} = f/(U/10^4 \text{ V})$ taking into account the effective focusing

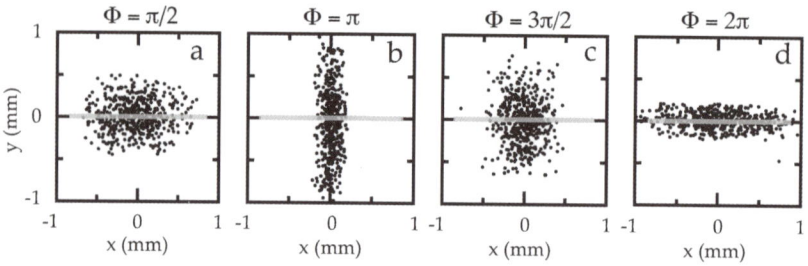

Fig. 8 Transverse phase-space distribution of the molecular packet in the detection region as a function of the end-phase Φ of the switching cycle depicted in Fig. 5c. The horizontal grey lines depict the area probed by the focused detection laser.

strength. The resulting transmission characteristics are displayed in Fig. 7d. The envelope of these measurements clearly represents the expected overall transmission curve, free of strong effects due to changes in the detection efficiency.

The measurements on ammonia in its lfs $J_K = 1_1$, $MK = -1$ states, shown in Fig. 7a and c, demonstrate the versatility of the device. Clear evidence of dynamic focusing is obtained, with similar characteristics as for the hfs states. Two differences in the frequency dependence of the transmission are obvious: at high frequencies the transmission practically does not decrease, and the low-frequency-onset of the transmission curve is shifted towards higher frequencies. The latter demonstrates the quantum state dependence of the process and that, at least in principle, one could use the device to separate the two quantum states. Moreover, since focusing and defocusing forces are interchanged between the lfs and hfs states, the phase effects shown in Fig. 8 are shifted by π, which can also be used to discriminate between the two states in laser experiments or by narrow slits in the beam path.

The large transmission at high frequencies is due to the fact that the time-averaged potential (the average of the two potentials in Fig. 5) is not really flat, but has a minimum of electric field on the molecular beam axis. Therefore, under these conditions molecules in lfs states are focused, whereas molecules in hfs states are defocused. This minimum has a depth of 7.5 kV cm^{-1}, corresponding to a 2-dimensional trap depth for ammonia in its lfs state of 0.11 K.[91]

II.B.3. Conformer selection. For given applied voltages the transmission through the selector is a function of the effective dipole moment μ_{eff} of the molecule's quantum state and of the switching frequency f:[24,25] the larger the dipole moment, the higher the optimal switching frequency. Consequently, for a given switching frequency f, the transmission depends on the molecule's μ_{eff}. This can, in principle, be exploited for the separation of individual quantum states. As the dipole moments of different conformers of large molecules often vary quite dramatically, one can use the selective focusing to separate the individual conformers from each other. This has been demonstrated for the *cis-* and *trans*-conformers of 3-aminophenol.[26]

In comparison to the separation by deflection proposed in section II.A.3, the AG focusing selector can provide enhanced transmission for any polar conformer. However, the focused molecules are confined to the molecular beam axis, where there is a background of molecules in unpolar quantum states and of atomic seed gas. In principle, the background could be removed by a beam stop on the molecular beam axis.[9] However, such a beam stop would take away a very large fraction of the beam intensity. Moreover, the central part of the beam is typically also the coldest internally, and, therefore, the most polar one. A similar effect can be obtained by tilting the focuser against the incoming molecular beam axis. We have done this in the selection experiments on 3-aminophenol and it did provide a somewhat improved contrast, although even then a considerable amount of background was observed.[26]

II.B.4. Resolution. In the conformer selection experiment described above, the focusing selector has been operated under conditions of maximum throughput, the equivalent of a quadrupole *ion guide*. Just as the m/q-resolution of quadrupole mass spectrometers can be improved at the cost of transmission by applying a dc offset, the m/μ-resolution of the focusing selector can be improved at the cost of transmission by changing the duty cycle of the applied high-voltage switching sequence. This effect is demonstrated by the simulations in Fig. 9, where the transmission as a function of frequency is shown for the ground states of 3-aminophenol for duty cycles $\tau_x/(\tau_x + \tau_y) = 0.5$, 0.46, 0.44, and 0.42, respectively. Here τ_x and τ_y are the fractions of a switching period for which *x*- and *y*-focusing are applied (with $\tau_x + \tau_y = 1$). From the calculations it is clear that the changed duty cycle improves the resolution, whereas the intensity decreases simultaneously. In principle, the effect is completely equivalent to the dc offset in a quadrupole guide for charged particles. However, it has to be taken into account that the charged particles experience

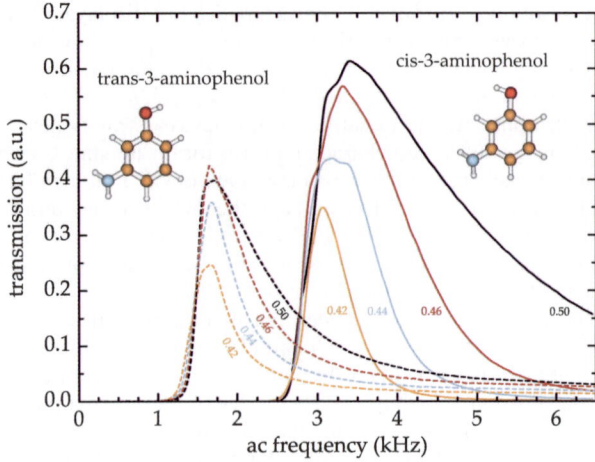

Fig. 9 Calculated transmission through the selector for *cis*- (full lines) and *trans*-3-aminophenol (dashed lines) in their rotational ground state for different duty cycles. The duty cycle for each individual simulation is given in the figure.

a harmonic potential inside the guide, whereas the large neutral molecules, discussed here, experience a strongly non-harmonic force. This is due to the often highly non-linear Stark effect (see Fig. 1) and the higher-order terms in the created electric fields.[25,72] Nevertheless, the effect can be used to separate species with smaller dipole moment differences or, for the same samples, to achieve stronger discrimination between different species. We are currently experimentally verifying these simulations.

It has also to be taken into account, that the molecules experience only a few electric field switching periods inside the current device. Together with the initial spatial spread of the molecular packet, this results in the switching frequency not being well-defined for the molecular ensemble.

In order to be able to routinely operate under conditions with higher resolution we plan to set up a focuser with an improved overall transmission, by scaling up the transverse phase-space acceptance, and a larger residence time, by making the device longer. This should enable us to separate individual conformers of many complex molecules.

II.C. Alternating gradient decelerator

The alternating gradient decelerator, depicted in Fig. 10, combines the dynamic focusing of the selector with deceleration equivalent to the Stark decelerator[11] and it has been described in detail elsewhere.[24,25] Generally, alternating-gradient deceleration is applicable to molecules in any polar quantum state, *i. e.*, it allows the deceleration of molecules in lfs states and in hfs states. This has been demonstrated for the deceleration of OH radicals in their lfs and hfs components of the $^2\Pi_{3/2}$, $v = 0$, $J = 3/2$ Λ-doublet.[73] Experiments on the alternating-gradient deceleration of diatomic molecules in hfs states have also been performed for the diatomic molecules CO*[74] and YbF.[75]

It has also been laid out that, in order to decelerate molecules in lfs states using the AG decelerator, the fields have to be switched on twice per electrode pair – once in between successive pairs in order to provide longitudinal bunching and deceleration, and once inside the electrode pair, in order to provide transverse focusing. In Fig. 11 a deceleration sequence of OH radicals in their lfs state from 305 to 200 m s^{-1} using 27 AG lenses is given. This is the lowest-velocity molecular packet obtained from an AG decelerator so far and it clearly demonstrates the versatility of the AG decelerator. It has to be taken into account, however, that the phase-space acceptance is at least one order of magnitude smaller than for deceleration of OH radicals in their lfs state using the normal Stark decelerator.[73]

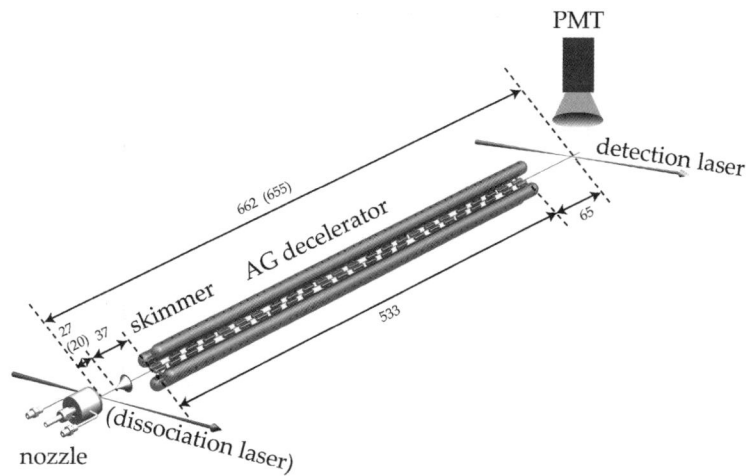

Fig. 10 Experimental setup of the alternating gradient decelerator. An internally cold molecular beam is produced by expanding a mixture of a few percent of the investigated molecule in approximately 1 bar of rare gas. In experiments with OH, an expansion of HNO_3 is irradiated by a pulsed laser inside the expansion region in order to photodissociate HNO_3 to form OH, which is successively cooled in the remaining expansion. The resulting supersonic jet is skimmed a few centimetres downstream from the nozzle for differential pumping and enters the 54 cm long decelerator. The time-resolved transmission of individual quantum states through the decelerator is monitored in a laser-induced-fluorescence detection scheme using a narrow-linewidth continuous-wave dye laser.[5,73]

For molecules in hfs states the fields are switched on when the molecules are inside the electrode pair (AG lens), where the molecules are focused in the transverse direction. When the molecules exit the lens, they are decelerated before the field is switched off again sometime before the molecules reach the minimum of the electric field, which is at the centre between two successive lenses. In principle, the focusing works the same way as described for the selector. However, in the decelerator the switching frequency is determined by the distance of successive electrode pairs and the velocity of the molecules and can, therefore, not be varied independently. Moreover, in order to obtain maximal fields on the molecular beam axis for the deceleration process, the fields are created by only two cylindrical electrodes and their geometric orientation defines the focusing and defocusing direction. However, one can change the overall focusing strength for the given geometry/directions by the duration the fields are switched on, or equivalently, by the distance f the molecules travel inside the electrodes. In principle the focusing strength can also be changed by changing the applied voltage. However, this would reduce the maximum field strength and, therefore, the deceleration strength, and it also cannot be adjusted as quickly during the experiment.

II.C.1. Alternating gradient deceleration of large molecules. The experimental setup of the alternating gradient decelerator is shown in Fig. 10. Using this setup we have AG focused and decelerated benzonitrile in its absolute ground state, and the obtained difference-time-of-flight profiles (see reference 5 for details) are shown in Fig. 12. Fig. 12a illustrates the focusing behavior of the 0_{00} ground state of benzonitrile for a constant velocity of the synchronous molecule of 320 m s^{-1}. The high voltages are switched on symmetrically around the centers of the AG lenses. Therefore, the molecular packet is focused in both transverse directions as well as in the longitudinal direction (bunching), but no change of the synchronous velocity occurs. In the experiments, three packets of focused molecules are observed. The central peaks of the TOF distributions occur 2.07 ms after the molecules exit the nozzle.

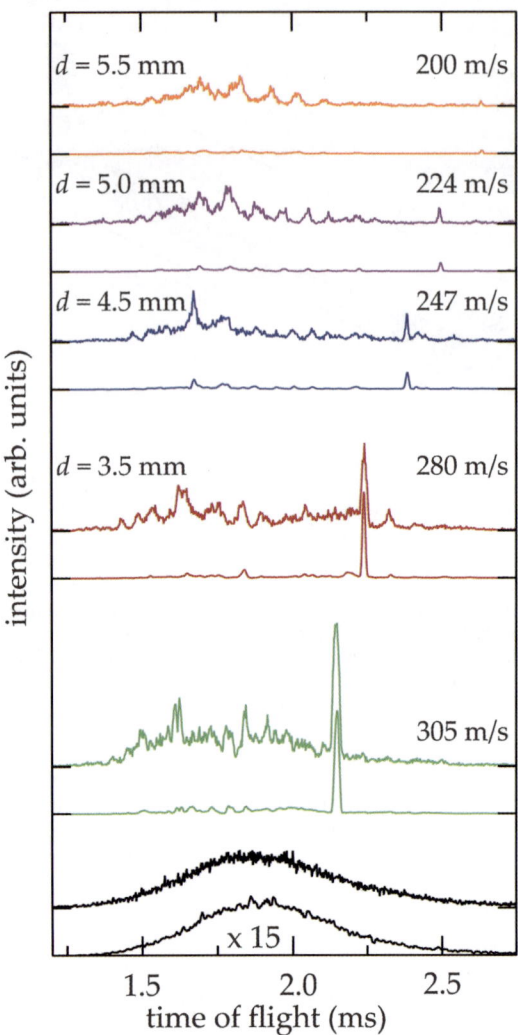

Fig. 11 AG focusing and deceleration of OH in its lfs state from 305 m s⁻¹ to varying velocities as specified next to the individual measurements. For a deceleration strength of $d = 5.5$ molecular packets with a centre velocity of 200 m s⁻¹ are obtained.

These packets contain the synchronous molecule. Hereafter, they are referred to as the "synchronous packets", and they are shaded in the simulated TOF distributions. The peaks at earlier and later arrival times correspond to molecular packets leading and trailing the synchronous packet by one AG lens (or 20 mm), respectively. These packets are also focused in all three dimensions. However, due to the lens pattern in our setup, they experience only 2/3 of the lenses at high voltage. This results in a reduced overall focusing for these packets.

For the focusing of the 0_{00} state it is seen that the synchronous packet is most intense for a focusing length of $f = 5$ mm. Under these conditions, approximately 10^5 molecules per quantum state per pulse are confined in the phase-stable central peak, corresponding to a density of 10^8 cm⁻³. For smaller focusing lengths a shallower time-averaged confinement potential is created, and less molecules are guided through the decelerator. For larger focusing lengths the molecular packet is over-focused, also resulting in a decreased transmission. For $f = 7$ mm the over-focusing

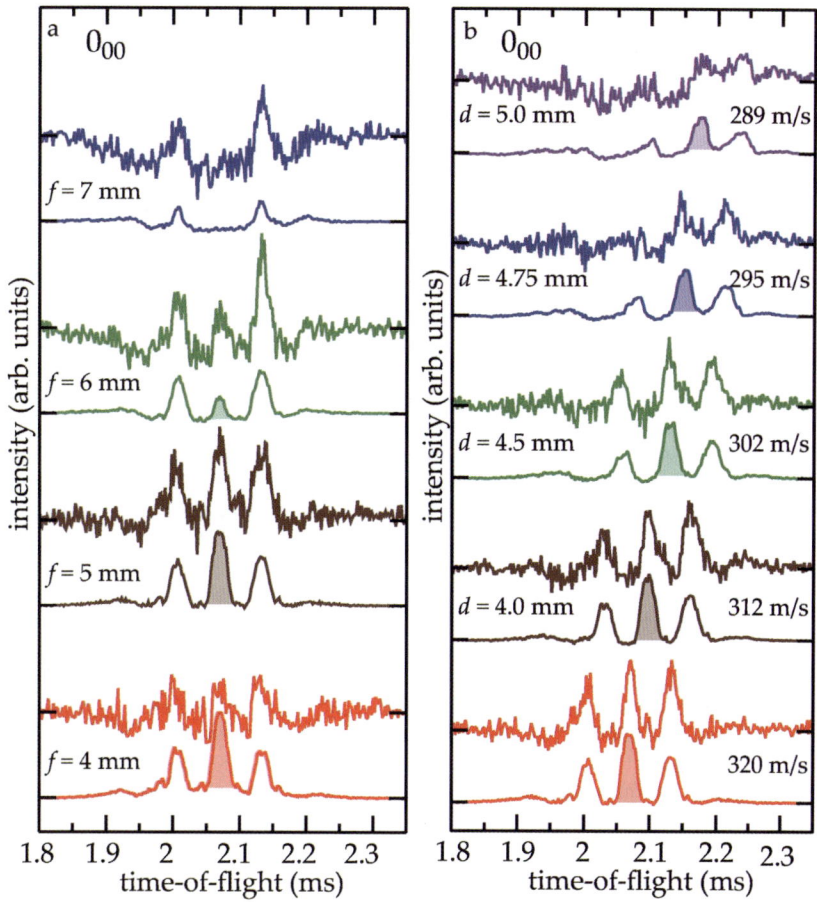

Fig. 12 Alternating-gradient focusing and deceleration of benzonitrile from a molecular beam of 320 m s^{-1}.[5] In the left column no overall deceleration is performed and the effects of a changing focusing strength are demonstrated. In the right column packets of benzonitrile molecules are decelerated from 320 m s^{-1} to successively lower final velocities as indicated in the figure.

is so severe that the synchronous packet completely disappears. As expected, the non-synchronous packets benefit from the increased focusing lengths.

Fig. 12b presents the results for the deceleration of benzonitrile in its 0_{00} state. The bottommost (red) trace shows focusing experiments for a constant velocity of the synchronous molecule of 320 m s^{-1} using the optimum focusing lengths. All other traces show experiments in which the synchronous packet is decelerated from 320 m s^{-1} to successively lower velocities, resulting in later arrival times at the detector. In these experiments the deceleration strength d is given as the position on the molecular beam axis behind the centre of the AG lens, where the electric field is switched off. Using $d = 5$ mm, the packet is decelerated to 289 m s^{-1}, corresponding to a reduction of the kinetic energy by 18%. The observed intensities of the non-synchronous packets decrease faster upon increasing d. Because the molecules in these packets miss every third deceleration stage, their trajectories are not stable and they are only observed due to the finite length of the decelerator. When deceleration to lower velocities is performed by increasing d the signal intensity decreases due to the reduction in phase-space acceptance. However, one could also decelerate to lower velocities by increasing the number of AG lenses for a given value of d. In this case, in principle, no decrease in the intensity of the synchronous packet is expected due to the phase stability of the deceleration process.[25] For all deceleration measurements, the relative intensities and the arrival times of the molecular packets at the detector are nicely reproduced by the trajectory simulations.

II.C.2. Towards the trapping of large molecules. Once large molecules in polar quantum states are decelerated down to slow velocities, *i. e.*, below approximately 25 m s^{-1} for molecules like benzonitrile or 3-aminophenol, they can be confined in ac electric traps. Such traps have already been demonstrated for slow ammonia molecules in hfs states,[6,7,76] which were produced from Stark decelerated ammonia in lfs states *via* a microwave transition.[77] Trapping lifetimes will be limited by collisions with background gas, by blackbody radiation,[78] and by non-adiabatic following of Stark curves. Even though the collisional cross sections for losses due to background gas will be different for large molecules compared to the ones for small molecules, the effects should be small. Nonadiabatic dynamics on the Stark potential energy curves do not occur for the lowest quantum state. For higher states it can be avoided, at least for low-lying rotational states, by providing finite minimum electric field strength of the order of 10 kV cm^{-1}. This also efficiently prohibits Majorana transitions. Since ac traps must provide a saddle point in the trap centre this requirement is practically automatically fulfilled.

The blackbody-radiation lifetimes of individual states can be calculated from the radiation density and the possible transitions. For the rovibronic ground state ($J_{K_aK_c} = 0_{00}$) two types of transitions have to be taken into account: rotational excitation into the rotational excited states that can be reached by an allowed transition and vibrational excitations. We have performed the calculation for the lifetime of benzonitrile in its rovibronic ground state as an example of a prototypical large polar molecule. At 300 K the rotational transition rate for the $1_{01} \leftarrow 0_{00}$ rotational a-type transition at 2.76 GHz^{12} is 10^{-5} Hz. In order to estimate the vibrational transition rate, we have performed a geometry optimisation and normal coordinate analysis at the B3LYP/aug-cc-pVTZ level of theory using Gaussian 2003,[79] which yields transition strengths for all 33 fundamental transitions. This results in an overall rate for loss of ground state molecules due to blackbody radiation at room temperature (300 K) of less than 1/4 Hz. The resulting trapping lifetime of 4 s is larger than currently achieved trapping times of small molecules in ac traps[6,7,76,80] and does not pose a major restriction on the trapping of such large molecules. However, the major loss rate is due to low-frequency vibrations with transition frequencies between 400 and 800 cm^{-1}. Therefore, the lifetime would increase to several hundred seconds when the trapping environment is cooled to liquid nitrogen temperatures (77 K). Moreover, one has to consider that many of the final states are also confined in the ac trap and, eventually, population will also be transferred back to the ground state. This shows that large molecules can be confined in traps once they are decelerated to slow enough velocities and that, while thermalization due to blackbody radiation has to be considered at room temperature, it will not prohibit trapping of large molecules and it is negligible at liquid nitrogen temperatures.

II.C.3. Conformer selection with the alternating-gradient decelerator. Generally, the quantum-state selectivity of the AG decelerator provides for the possibility to prepare clean samples of individual conformers of large molecules. Similar to the AG focuser, any polar conformer can be addressed. Additionally, even without deceleration to very slow velocities, the reduced velocity and the correspondingly longer flight-times to the detector allow for a practically complete temporal separation of the accepted molecules from the remaining beam. Therefore, the AG decelerator provides, in principle, the best selectivity of different conformers present in a molecular beam. Decelerating a single conformer to slow velocities and successively trapping it in ac traps provides additional selectivity and allows for the separation of quantum states and conformers with quite similar m/μ ratios. However, this strong selectivity comes at the price of a more complicated experimental setup.

III. Conclusions

We have compared and demonstrated different experimental approaches for the quantum-state selective manipulation of the motion of polar molecules. All methods are generally applicable for molecules in low- and high-field seeking states. However, the methods presented here are well adapted to the manipulation of large molecules, where all quantum states are high-field seeking.

Generally, all the different techniques described allow the spatial separation of different isomers of neutral species. While we have demonstrated this already for 3-aminophenol using the m/μ selector, it is clear that the quantum state selectivity of the deflector and the AG decelerator can be exploited in very much the same way. Moreover, both these alternative techniques can provide background-free samples of large molecules. Generally, each approach has its advantages and disadvantages, which shall be summarised here.

Using the deflector one disperses quantum states of particles of identical mass according to their effective dipole moments μ_{eff}, with the most polar states deflected the most. Therefore, using focused lasers or narrow slits for spatial discrimination of the sample, one can perform experiments using samples of these most polar quantum states. Contrary to the AG selector, these are pure samples without any background from unpolar states or conformers, or seed gas, as is shown by the simulations for *cis*- and *trans*-3-aminophenol in Fig. 4. In principle this background in experiments with the AG focuser can be removed by bending or tilting the device with respect to the molecular beam axis or by placing a beam-stop on the molecular beam axis in order to remove particles unaffected by the electric fields, but these changes will also considerably reduce the transmission of the selected particles. The AG focuser, however, has the advantage that it allows the selection of any conformer whose dipole moment differs (enough) from the other conformers.

Background reduction can, of course, also be performed in time. If one uses the alternating gradient decelerator to slow the accepted molecular packet – containing only a single conformer – down to reach the detector only after the original pulse has passed, one can also perform practically background free experiments with the accepted packets. Of course, the ultimate experiment in that respect would be the deceleration and subsequent trapping of a single conformer.

For future proof-of-principle experiments on single conformers, the deflector is surely the most promising setup. It is the simplest of the techniques presented, and it provides a background-free sample of the most polar quantum states of the most polar isomer, the states that can also be manipulated the most strongly in successive laser or dc electric field experiments, like mixed field orientation.[14] However, if one wants to employ the separation in routine experiments where specific or multiple isomers must be addressed individually, generally dynamic focusing is obligatory in order to obtain enriched samples of any but the most polar conformer. Whether the experiments require really pure samples or not determines whether the simpler focuser or the very complex decelerator shall be used.

Whereas the discussions of m/μ selection in this manuscript were restricted to the separation of isomers of molecules (fixed m), the selection can, of course, also be used to separate species from any other mixture based on the m/μ-ratio, *i.e.*, to separate polar clusters from the typically very broad size distributions.

Generally, the clean, well-defined samples provided by these experiments could aid, or even just allow, novel experiments with complex molecules, for instance, femtosecond pump–probe measurements, X-ray or electron diffraction in the gas-phase, or tomographic reconstructions of molecular orbitals. Such samples could also be very advantageous for metrology applications, such as, for example, matter-wave interferometry or the search for electroweak interactions in chiral molecules.

Appendix: Stark energies of asymmetric rotors

In this section we want to summarise the details of the calculations of adiabatic energy curves for asymmetric top molecules. The Hamiltonian matrix is set up in the basis of symmetric top wavefunctions. For the asymmetric rotor in a dc electric field only M is a good quantum number, as K is mixed by the molecular asymmetry and J is mixed by the field. Therefore, one can treat the different M levels individually, but one needs to set up the M matrices including all J and K levels. For the accurate description of higher rotational states it is also important to include centrifugal distortion constants. The Hamiltonian H of an asymmetric rotor molecule with dipole moment $\vec{\mu}$ in an electric field of strength E might be written as the sum of the Hamiltonian H_{rot} of an asymmetric rotor in free space and the contribution due to the Stark effect H_{Stark} as

$$H = H_{\text{rot}} + H_{\text{Stark}}$$

Following references 81 and 82, the corresponding matrix elements, using symmetric rotor basis functions, representation I^r, and Watson's A-reduction,[81] are:

$$\langle JKM|H_{\text{rot}}|JKM\rangle = \frac{B+C}{2}(J(J+1)-K^2) + AK^2$$
$$-\Delta_J J^2(J+1)^2 - \Delta_{JK} J(J+1)K^2 - \Delta_K K^4$$

$$\langle JK\pm 2M|H_{\text{rot}}|JKM\rangle = \left(\frac{B-C}{4} - \delta_J J(J+1) - \frac{\delta_K}{2}\left((K\pm 2)^2 + K^2\right)\right)$$
$$\cdot \sqrt{J(J+1)-K(K\pm 1)}\sqrt{J(J+1)-(K\pm 1)(K\pm 2)}$$

$$\langle JKM|\mu_a|JKM\rangle = -\frac{MK}{J(J+1)}\mu_a E$$

$$\langle J+1KM|\mu_a|JKM\rangle = \langle JKM|\mu_a|J+1KM\rangle$$
$$= -\frac{\sqrt{(J+1)^2-K^2}\sqrt{(J+1)^2-M^2}}{(J+1)\sqrt{(2J+1)(2J+3)}}\mu_a E$$

$$\langle JK\pm 1M|\mu_b|JKM\rangle = -\frac{M\sqrt{(J\mp K)(J\pm K+1)}}{2J(J+1)}\mu_b E$$

$$\langle J+1K\pm 1M|\mu_b|JKM\rangle = \langle JK\pm 1M|\mu_b|J+1KM\rangle$$
$$= \pm\frac{\sqrt{(J\pm K+1)(J\pm K+2)}\sqrt{(J+1)^2-M^2}}{2(J+1)\sqrt{(2J+1)(2J+3)}}\mu_b E$$

$$\langle JK\pm 1M|\mu_c|JKM\rangle = \pm i\frac{M\sqrt{(J\mp K)(J\pm K+1)}}{2J(J+1)}\mu_c E$$

$$\langle J+1K\pm 1M|\mu_c|JKM\rangle = \langle JK\pm 1M|\mu_c|J+1KM\rangle$$
$$= -i\frac{\sqrt{(J\pm K+1)(J\pm K+2)}\sqrt{(J+1)^2-M^2}}{2(J+1)\sqrt{(2J+1)(2J+3)}}\mu_c E$$

For the correct assignment of the states to the "adiabatic quantum number labels" $\tilde{J}_{\tilde{K}_a \tilde{K}_c} \tilde{M}$, i.e., to the adiabatically corresponding field-free rotor states, one has to classify the states according to their character in the *electric field symmetry group*.[83,84] This symmetry classification is performed by applying a Wang transformation[85] to the Hamiltonian matrix. If the molecule's dipole moment is along one of the principal axes of inertia, the matrix will be block diagonalised by this transformation according to the remaining symmetry in the field and the blocks are treated independently. For arbitrary orientation of the dipole moment in the inertial frame of the molecule the full matrix must be diagonalised. In any case, this process ensures that all states (eigenvalues and eigenvectors) obtained from a single matrix diagonalisation do have the same symmetry, and, therefore, no real crossings between these states can occur. Therefore, sorting the resulting levels by energy and assigning quantum number labels in the same order as for the field-free states of the same symmetry yields the correct adiabatic labels.

These calculations are performed for a number of electric field strengths – typically in steps of 1 kV cm^{-1} from 0 kV cm^{-1} to 250 kV cm^{-1} – and the resulting energies are stored for later use in simulations using the libcoldmol program package.[86]

References

1 K.-K. Ni, S. Ospelkaus, M. H. G. de Miranda, A. Pe'er, B. Neyenhuis, J. J. Zirbel, S. Kotochigova, P. S. Julienne, D. Jin and J. Ye, *Science*, 2008, **322**, 231.
2 J. Deiglmayr, A. Grochola, M. Repp, K. Mörtlbauer, C. Glück, J. Lange, O. Dulieu, R. Wester and M. Weidemüller, *Phys. Rev. Lett.*, 2008, **101**, 133004.
3 H. L. Bethlem, G. Berden, F. M. H. Crompvoets, R. T. Jongma, A. J. A. van Roij and G. Meijer, *Nature*, 2000, **406**, 491–494.
4 R. Weinkauf, J. Schermann, M. S. de Vries and K. Kleinermanns, *Eur. Phys. J. D*, 2002, **20**, 309–316.
5 K. Wohlfart, F. Grätz, F. Filsinger, H. Haak, G. Meijer and J. Küpper, *Phys. Rev. A*, 2008, **77**, 031404R.
6 J. van Veldhoven, H. L. Bethlem and G. Meijer, *Phys. Rev. Lett.*, 2005, **94**, 083001.
7 M. Schnell, P. Lützow, J. van Veldhoven, H. L. Bethlem, J. Küpper, B. Friedrich, M. Schleier-Smith, H. Haak and G. Meijer, *J. Phys. Chem. A*, 2007, **111**, 7411–7419.
8 N. F. Ramsey, Molecular Beams, *The International Series of Monographs on Physics*, Oxford University Press, London, UK, 1956.
9 J. Reuss in Atomic and molecular beam methods, vol. 1, ed. G. Scoles, Oxford University Press, New York, NY, USA, 1988, chapter 11, pp. 276–292.
10 H. L. Bethlem, G. Berden and G. Meijer, *Phys. Rev. Lett.*, 1999, **83**, 1558–1561.
11 S. Y. T. van de Meerakker, H. L. Bethlem and G. Meijer, *Nat. Phys.*, 2008, **4**, 595.
12 K. Wohlfart, M. Schnell, J.-U. Grabow and J. Küpper, *J. Mol. Spectrosc.*, 2008, **247**, 119–121.
13 M. Broyer, R. Antoine, I. Compagnon, D. Rayane and P. Dugourd, *Phys. Scr.*, 2007, **76**, C135–C139.
14 L. Holmegaard, J. H. Nielsen, I. Nevo, H. Stapelfeldt, F. Filsinger, J. Küpper and G. Meijer, *Phys. Rev. Lett.*, 2009, **102**, 023001.
15 M. Berninger, A. Stefanov, S. Deachapunya and M. Arndt, *Phys. Rev. A*, 2007, **76**, 013607.
16 H. J. Loesch and A. Remscheid, *J. Chem. Phys.*, 1990, **93**, 4779.
17 B. Friedrich and D. R. Herschbach, *Nature*, 1991, **353**, 412–414.
18 P. A. Block, E. J. Bohac and R. E. Miller, *Phys. Rev. Lett.*, 1992, **68**, 1303–1306.
19 B. Friedrich and D. Herschbach, *Phys. Rev. Lett.*, 1995, **74**, 4623–4626.
20 H. Stapelfeldt and T. Seideman, *Rev. Mod. Phys.*, 2003, **75**, 543–557.
21 V. Kumarappan, C. Z. Bisgaard, S. S. Viftrup, L. Holmegaard and H. Stapelfeldt, *J. Chem. Phys.*, 2006, **125**, 194309.
22 U. Buck and M. Fárník, *Int. Rev. Phys. Chem.*, 2006, **25**, 583–612.
23 S. Minemoto, H. Nanjo, H. Tanji, T. Suzuki and H. Sakai, *J. Chem. Phys.*, 2003, **118**, 4052–4059.
24 D. Auerbach, E. E. A. Bromberg and L. Wharton, *J. Chem. Phys.*, 1966, **45**, 2160.
25 H. L. Bethlem, M. R. Tarbutt, J. Küpper, D. Carty, K. Wohlfart, E. A. Hinds and G. Meijer, *J. Phys. B*, 2006, **39**, R263–R291.
26 F. Filsinger, U. Erlekam, G. von Helden, J. Küpper and G. Meijer, *Phys. Rev. Lett.*, 2008, **100**, 133003.

27 S. Deachapunya, P. J. Fagan, A. G. Major, E. Reiger, H. Ritsch, A. Stefanov, H. Ulbricht and M. Arndt, *Eur. Phys. J. D*, 2008, **46**, 307–313.
28 J. P. Simons, *Phys. Chem. Chem. Phys.*, 2004, **6**, E7.
29 M. S. de Vries and P. Hobza, *Annu. Rev. Phys. Chem.*, 2007, **58**, 585–612.
30 R. D. Suenram and F. J. Lovas, *J. Am. Chem. Soc.*, 1980, **102**, 7180–7184.
31 T. R. Rizzo, Y. D. Park, L. Peteanu and D. H. Levy, *J. Chem. Phys.*, 1985, **83**, 4819–4820.
32 R. Neutze, R. Wouts, D. van der Spoel, E. Weckert and J. Hajdu, *Nature*, 2000, **406**, 752–757.
33 J. Itatani, D. Zeidler, J. Levesque, M. Spanner, D. M. Villeneuve and P. B. Corkum, *Phys. Rev. Lett.*, 2005, **94**.
34 J. Itatani, J. Levesque, D. Zeidler, H. Niikura, H. Pepin, J. C. Kieffer, P. B. Corkum and D. M. Villeneuve, *Nature*, 2004, **432**, 867–871.
35 W. Gerlach and O. Stern, *Z. Phys.*, 1922, **9**, 349–352.
36 N. Vanhaecke, U. Meier, M. Andrist, B. H. Meier and F. Merkt, *Phys. Rev. A*, 2007, **75**, 031402R.
37 E. Narevicius, A. Libson, C. G. Parthey, I. Chavez, J. Narevicius, U. Even and M. G. Raizen, *Phys. Rev. Lett.*, 2008, **100**, 093003.
38 B. S. Zhao, H. S. Chung, K. Cho, S. H. Lee, S. Hwang, J. Yu, Y. H. Ahn, J. Y. Sohn, D. S. Kim, W. K. Kang and D. S. Chung, *Phys. Rev. Lett.*, 2000, **85**, 2705–2708.
39 R. Fulton, A. I. Bishop and P. F. Barker, *Phys. Rev. Lett.*, 2004, **93**, 243004.
40 *Atomic and molecular beam methods*, vol. 1, ed. G. Scoles, Oxford University Press, New York, NY, USA, 1988.
41 H. J. Loesch, *Annu. Rev. Phys. Chem.*, 1995, **46**, 555–594.
42 W. Kong, *Int. J. Mod. Phys. B*, 2001, **15**, 3471–3502.
43 K. J. Castle and W. Kong, *J. Chem. Phys.*, 2000, **112**, 10156–10161.
44 F. Dong and R. E. Miller, *Science*, 2002, **298**, 1227–1230.
45 B. Friedrich and D. Herschbach, *J. Chem. Phys.*, 1999, **111**, 6157.
46 B. Friedrich and D. Herschbach, *J. Phys. Chem. A*, 1999, **103**, 10280–10288.
47 T. Seideman and E. Hamilton, *Adv. At. Mol. Opt. Phys.*, 2005, **52**, 289–329.
48 H. Kallmann and F. Reiche, *Z. Phys.*, 1921, **6**, 352–375.
49 O. Stern, *Z. Phys.*, 1926, **39**, 751–763.
50 E. Wrede, *Z. Phys. A*, 1927, **44**, 261–268.
51 I. I. Rabi, S. Millman, P. Kusch and J. R. Zacharias, *Phys. Rev.*, 1939, **55**, 526–535.
52 R. Moro, X. Xu, S. Yin and W. A. de Heer, *Science*, 2003, **300**, 1265–1269.
53 M. Hillenkamp, S. Keinan and U. Even, *J. Chem. Phys.*, 2003, **118**, 8699–8705.
54 S. Stolte in Atomic and molecular beam methods, vol. 1, ed. G. Scoles, Oxford University Press, New York, NY, USA, 1988, chapter 25, pp. 631–652..
55 S. Y. T. van de Meerakker and G. Meijer, *Faraday Discuss.*, 2009, **142**, DOI: 10.1039/b819721k.
56 R. Moro, J. Bulthuis, J. Heinrich and V. V. Kresin, *Phys. Rev. A*, 2007, **75**, 013415.
57 F. Filsinger, K. Wohlfart, M. Schnell, J.-U. Grabow and J. Küpper, *Phys. Chem. Chem. Phys.*, 2008, **10**, 666–673.
58 Y. R. Wu and D. H. Levy, *J. Chem. Phys.*, 1989, **91**, 5278–5284.
59 H. G. Bennewitz and W. Paul, *Z. Phys.*, 1954, **139**, 489.
60 H. G. Bennewitz, W. Paul and C. Schlier, *Z. Phys.*, 1955, **141**, 6.
61 J. P. Gordon, H. J. Zeiger and C. H. Townes, *Phys. Rev.*, 1954, **95**, 282–284.
62 J. P. Gordon, H. J. Zeiger and C. H. Townes, *Phys. Rev.*, 1955, **99**, 1264–1274.
63 H. G. Bennewitz, K. H. Kramer, J. P. Toennies and W. Paul, *Z. Phys.*, 1964, **177**, 84.
64 P. R. Brooks and E. M. Jones, *J. Chem. Phys.*, 1966, **45**, 3449.
65 R. J. Beuhler, R. B. Bernstein and K. H. Kramer, *J. Am. Chem. Soc.*, 1966, **88**, 5331.
66 D. Kakati and D. C. Lainé, *Phys. Lett. A*, 1967, **24**, 676.
67 D. Kakati and D. C. Lainé, *Phys. Lett. A*, 1969, **28**, 786.
68 F. Günther and K. Schügerl, *Z. Phys. Chem.*, 1972, **80**, 155.
69 A. Lübbert, G. Rotzoll and F. Günther, *J. Chem. Phys.*, 1978, **69**, 5174–5179.
70 D. Kakati and D. C. Lainé, *J. Phys. E*, 1971, **4**, 269.
71 D. C. Lainé and R. C. Sweeting, *Entropie*, 1973, **42**, 165.
72 M. R. Tarbutt and E. A. Hinds, *New J. Phys.*, 2008, **10**, 073011.
73 K. Wohlfart, F. Filsinger, F. Grätz, J. Küpper and G. Meijer, *Phys. Rev. A*, 2008, **78**, 033421.
74 H. L. Bethlem, A. J. A. van Roij, R. T. Jongma and G. Meijer, *Phys. Rev. Lett.*, 2002, **88**, 133003.
75 M. R. Tarbutt, H. L. Bethlem, J. J. Hudson, V. L. Ryabov, V. A. Ryzhov, B. E. Sauer, G. Meijer and E. A. Hinds, *Phys. Rev. Lett.*, 2004, **92**, 173002.
76 H. L. Bethlem, J. van Veldhoven, M. Schnell and G. Meijer, *Phys. Rev. A*, 2006, **74**, 063403.

77 J. van Veldhoven, J. Küpper, H. L. Bethlem, B. Sartakov, A. J. A. van Roij and G. Meijer, *Eur. Phys. J. D*, 2004, **31**, 337–349.
78 S. Hoekstra, J. J. Gilijamse, B. Sartakov, N. Vanhaecke, L. Scharfenberg, S. Y. T. van de Meerakker and G. Meijer, *Phys. Rev. Lett.*, 2007, **98**, 133001.
79 M. J. Frisch, G. W. Trucks, H. B. Schlegel, G. E. Scuseria, M. A. Robb, J. R. Cheeseman, J. A. Montgomery, Jr., T. Vreven, K. N. Kudin, J. C. Burant, J. M. Millam, S. S. Iyengar, J. Tomasi, V. Barone, B. Mennucci, M. Cossi, G. Scalmani, N. Rega, G. A. Petersson, H. Nakatsuji, M. Hada, M. Ehara, K. Toyota, R. Fukuda, J. Hasegawa, M. Ishida, T. Nakajima, Y. Honda, O. Kitao, H. Nakai, M. Klene, X. Li, J. E. Knox, H. P. Hratchian, J. B. Cross, V. Bakken, C. Adamo, J. Jaramillo, R. Gomperts, R. E. Stratmann, O. Yazyev, A. J. Austin, R. Cammi, C. Pomelli, J. Ochterski, P. Y. Ayala, K. Morokuma, G. A. Voth, P. Salvador, J. J. Dannenberg, V. G. Zakrzewski, S. Dapprich, A. D. Daniels, M. C. Strain, O. Farkas, D. K. Malick, A. D. Rabuck, K. Raghavachari, J. B. Foresman, J. V. Ortiz, Q. Cui, A. G. Baboul, S. Clifford, J. Cioslowski, B. B. Stefanov, G. Liu, A. Liashenko, P. Piskorz, I. Komaromi, R. L. Martin, D. J. Fox, T. Keith, M. A. Al-Laham, C. Y. Peng, A. Nanayakkara, M. Challacombe, P. M. W. Gill, B. G. Johnson, W. Chen, M. W. Wong, C. Gonzalez and J. A. Pople, *GAUSSIAN 03 (Revision B.03)*, Gaussian, Inc., Wallingford, CT, 2004.
80 J. van Veldhoven, AC trapping and high-resolution spectroscopy of ammonia molecules, PhD thesis, Radboud Universiteit Nijmegen, The Netherlands, 2006.
81 J. K. G. Watson in *Vibrational Spectra and Structure*, ed. J. R. Durig, vol. 6, Marcel Dekker, 1977, chapter 1.
82 M. Abd El Rahim, R. Antoine, M. Broyer, D. Rayane and P. Dugourd, *J. Phys. Chem. A*, 2005, **109**, 8507–8514.
83 J. K. G. Watson, *Can. J. Phys*, 1975, **53**, 2210–2220.
84 P. R. Bunker and P. Jensen, *Molecular Symmetry and Spectroscopy*, NRC Research Press, Ottawa, Ontario, Canada, 2 ed., 1998.
85 S. C. Wang, *Phys. Rev.*, 1929, **34**, 243.
86 J. Küpper and F. Filsinger, *libcoldmol: A particle trajectory calculation framework*, 2003–2008.
87 D. C. Lainé and R. Sweeting, *Phys. Lett. A*, 1971, **34**, 144.
88 J. C. Helmer, F. B. Jacobus and P. A. Sturrock, *J. Appl. Phys.*, 1960, **31**, 458.
89 Nevertheless, the focusing of molecules in hfs states has been demonstrated using cylindrical capacitors with a central wire[87] and arrays of crossed wires[88].
90 Interestingly enough these studies were performed at the *Kaiser Wilhelm Institut für physikalische Chemie und Elektrochemie*, the predecessor of the Fritz Haber Institut in Berlin.
91 A similar field could, of course, be created by applying dc voltages in a quadrupole arrangement of ± 0.6 kV, although the trap depth and characteristics would be different due to the approximately quadratic Stark effect of NH_3 at the resulting low field strengths.

Sympathetic cooling by collisions with ultracold rare gas atoms, and recent progress in optical Stark deceleration

P. F. Barker,* S. M. Purcell, P. Douglas, P. Barletta, N. Coppendale, C. Maher-McWilliams and J. Tennyson

Received 28th October 2008, Accepted 5th February 2009
First published as an Advance Article on the web 28th May 2009
DOI: 10.1039/b819079h

We propose a general scheme for sympathetic cooling of molecules to μK temperatures on a timescale of seconds. Experimental parameters have been estimated from theory, which indicate the viability of the scheme. This method, which is particularly suited to optical Stark deceleration, utilises ultracold, laser cooled metastable rare gas atoms quenched to their ground state as collision partners to co-trapped molecular species within a deep optical trap (150 mK). We also describe the measurement of the role of laser-induced molecular alignment on the dipole force in optical Stark deceleration and outline progress towards the realisation of chirped optical Stark deceleration for producing slow molecular beams with mK energy spreads.

1 Introduction

Laser cooling, the cornerstone of cold atom physics, is capable of dissipatively cooling atomic gases to ultracold temperatures (μK).[1] This technique cannot, however, be applied to most molecules due to the lack of a single or even a few cycling transitions. While ultracold molecular species can be produced by association of laser cooled atomic species on Feshbach resonances[2] and by photoassociation,[3] the range of stable molecular species that can be produced in this way is limited by the small number of atomic species that can be laser cooled. New techniques have been developed to produce slow, cold molecules of much greater variety and complexity and these include buffer gas cooling,[4] and phase space filtering techniques such as electrostatic Stark deceleration[5] and more recently Zeeman deceleration.[6] These techniques are capable of producing cold trapped molecular ensembles and have now been used for a range of important low energy collision experiments[7] and precision spectroscopy.[8] However, they do not appear to be capable of reaching the ultracold regime (<1 mK), where much of the interesting physics and chemistry is to be found. To produce colder molecular gases without the losses of molecules inherent in filtering techniques, it is generally accepted that a dissipative cooling scheme will be required once the molecules are slowed and trapped using the techniques developed above. A number of important dissipative schemes are potentially available for cooling molecules below the 1 mK barrier, and these include stochastic, cavity,[9] and evaporative and sympathetic cooling.[10] Although no progress has been reported on stochastic cooling for atoms or molecules there is currently considerable theoretical effort and some experimental effort directed towards the study of cavity cooling of atoms and molecules. However, this technique is yet to be clearly demonstrated

Department of Physics and Astronomy, University College London, Gower Street, London, UK WC1E 6BT. E-mail: p.barker@ucl.ac.uk

experimentally. Evaporative and sympathetic cooling of one species *via* thermalising collisions with another colder species is conceptually simple and has been demonstrated using many cold atomic species and in a variety of traps. This method therefore appears to be a viable first choice to cool below the current 1 mK barrier. In this paper, we propose a promising route towards the sympathetic cooling of molecules which utilises ultracold, laser cooled rare gas atoms in their ground state co-trapped with molecules in an optical trap. We also describe recent developments in the optical Stark deceleration method which include the measurement of the role of laser-induced molecular alignment on the dipole force as well as progress towards the realisation of chirped optical Stark deceleration for producing slow molecular beams with mK energy spreads.

2 Sympathetic cooling with laser cooled rare gas atoms

The earliest sympathetic cooling experiments with stable cold molecules were carried out with helium gas.[4] This technique, called buffer gas cooling, uses collisions between helium gas at temperatures in the 100 mK range with paramagnetic molecular species, which are accumulated in a magnetic trap as they are cooled. This method has proven to be very successful for a number of molecular species but it is currently limited to temperatures in the 100 mK range. Sympathetic cooling of Stark decelerated species with laser cooled atoms is also a potential route towards μK temperatures. The collisional properties of a few molecular species, which can be subsequently electrostatically or magnetically trapped with laser cooled alkali metal atoms have been explored theoretically.[11,12] These studies show that maintaining a large elastic to inelastic collision ratio of rates is problematic over all of the temperature ranges in which the molecules must be sympathetically cooled. Molecules that can be electrostatically and magnetically trapped in a single quantum state can experience trap losses when inelastic collisions promote them to untrappable states. In addition, chemical reactions between the species can also lead to an effective trap loss.[12] We propose a scheme for sympathetic cooling of optical Stark decelerated molecules using ground state rare gas atoms as collision partners to minimise the possibility of reactions and avoid losses due to state changing collisions. An optical trap is an obvious choice of trap for sympathetic cooling experiments, since it is capable of trapping all ro-vibrational states in the electronic ground state, unlike magnetic and electrostatic traps. However, in order to make a trap deep enough to hold optical Stark decelerated molecules in the 10–100 mK range, it must have a small volume with length scales of the order of a hundred microns. This type of small volume trap does not, at first, appear attractive for electrostatic Stark and Zeeman deceleration techniques, which produce slowed single quantum state densities in the 10^8 cm^{-3} range over millimeter length scales, as very few molecules would be trapped. It does, however, appear to be a feasible method for trapping optical Stark decelerated molecules with densities in the 10^{12} cm^{-3} range, because the optical lattice used in these experiments is spatially well matched to optical trap dimensions. In addition, an optical trap is capable of trapping a molecular gas that is in a distribution of ro-vibrational states and these internal states may also be cooled using evaporative and sympathetic cooling within the trap.

2.1 State preparation for trapping and sympathetic cooling

Optical cooling of five rare gas atoms has been achieved by utilizing a closed transition from the lowest excited meta-stable state $ns[3/2]_2$ to the higher $np[5/2]_3$ state.[13–17] We have considered viable optical quenching routes for He, Ne, Ar, Kr and Xe, from the meta-stable 'cooling' states with special consideration given to the heating effect caused by the absorption of near infra-red photons and emission of VUV photons in the quenching process. Quenching of Ne*, Ar*, Kr* and Xe* can be achieved by optically pumping from the meta-stable state to a higher energy state

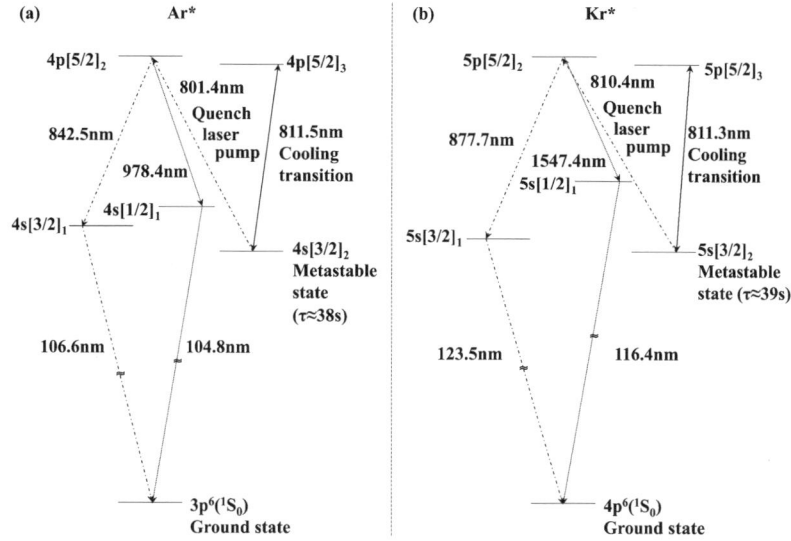

Fig. 1 Energy levels relevant for laser cooling and quenching of Ar* and Kr*.

which can undergo one or more dipole transitions to the ground state. Two such routes for Ar* and Kr* are illustrated in Fig. 1.

Argon can be optically quenched by pumping from the $4s[3/2]_2$ meta-stable level to the $4p[5/2]_2$ excited state using light at 801.4 nm. There are two possible two-photon decay routes from the $4p[5/2]_2$ level to the ground state; the intermediate levels are $4s[3/2]_1$ and $4s[1/2]_1$. This quenching process will give a maximum photon recoil temperature of 66 μK. A laser of wavelength 763.5 nm could also be used in an alternative quenching process exciting the $4p[3/2]_2$ state from which the atom could decay to the ground state *via* the same two intermediate levels as in the previous process. A maximum temperature increase of 68 μK will result from the photon recoil effect.

The quenching scheme for krypton is similar to that for argon; the $5s[3/2]_2$ meta-stable state is optically pumped, by 810.4 nm light, to the $5p[5/2]_2$ state from which dipole transitions to the ground state can proceed *via* either the $5s[3/2]_1$ or $5s[1/2]_1$ states. This process will produce a temperature increase of 25 μK. Laser light at 760.2 nm can also be used to quench Kr* by pumping to the $5p[3/2]_2$ state from which decays to the ground state result in a heating effect of 25 μK. Ne* and Xe* can be quenched using laser light at 633.4 nm and 979.9 nm respectively. The quenching process heats Ne by 261 μK and the heavier Xe by 11 μK.

We note that it is not possible to quench He* in the same way but this may be achieved by other methods.[18] However, the atomic cloud will be heated by more than 1.2 mK, due to the emission of a short wavelength (λ < 64.5 nm), high energy photon and thus He from this perspective appears to be an unsuitable species for sympathetic cooling.

2.2 A deep optical trap for molecules and rare gas atoms

Although the ultracold atomic species can easily be trapped in conventional quasi-electrostatic traps (QUEST),[19] one of the primary challenges is to create a continuous optical trap that is sufficiently deep to trap the initially 'hot' molecular species. To create the necessary trap depth we consider an optical potential that can be created within a high finesse optical buildup cavity.[20–22] For frequencies far-detuned (to the red) from any possible dipole-allowed transitions, a QUEST standing wave potential takes the general form

$$U(x,y,z) = -\frac{\alpha}{2}E(x,y,z)^2, \qquad (1)$$

where α is the ground state polarizability of the trapped particle, and $E(x,y,z)$ is the electric field distribution of an incident light field. In the specific case of a Gaussian beam in-coupled to a Fabry–Perot buildup cavity, a stationary periodic lattice potential is formed, which can be described by

$$U(r,z) = \frac{U_0}{1+(z/z_R)^2} \exp\left[-2(r/w(z))^2\right]\cos^2[kz], \qquad (2)$$

where $U_0 = \frac{2\alpha}{\varepsilon_0 c}I_c$, I_c is the one-way circulating peak intensity, z_R is the Rayleigh range of the intra-cavity field ($z_R = \pi w_0^2/\lambda$), w is the $1/e^2$ beam width in the radial direction, and $k = 2\pi/\lambda$. U_0 signifies the peak well depth. The beam waist evolves according to $w(z) = w_0[1 + (z/z_R)^2]^{1/2}$, where w_0 is the minimum beam waist at the cavity centre. A plot of the intra-cavity potential is given in Fig. 2. The variation in potential well depth for a QUEST is determined by the difference in polarizability of each species.

2.3 Cavity design

We describe a prototypical optical trap with a well depth of 150 mK for benzene molecules which requires a circulating intensity of $I_c = 2.5 \times 10^8$ W cm^{-2}. We consider a trap that could be constructed from a commercial, single frequency laser of 10 W output power (Verdi V10, Coherent), operating at a wavelength of 532 nm fed into an optical buildup cavity. In such a scheme it is possible that 65% of the initial laser power would be lost before it is coupled into the cavity. Damage to the mirror surfaces due to the high intra-cavity intensity necessitates a minimum beam spot size on the mirrors such that the intensity is below $\sim$ 100 MW cm^{-2}.[23] This leads to a minimum beam waist radius in the trapping region of the cavity of $w_0 = 55$ μm. This beam waist must also be sufficient to overlap well with our optical Stark deceleration volume so that the total number of trapped molecules is maximised. To produce the required intensity of $I_c = 2.5 \times 10^8$ W cm^{-2}, we consider

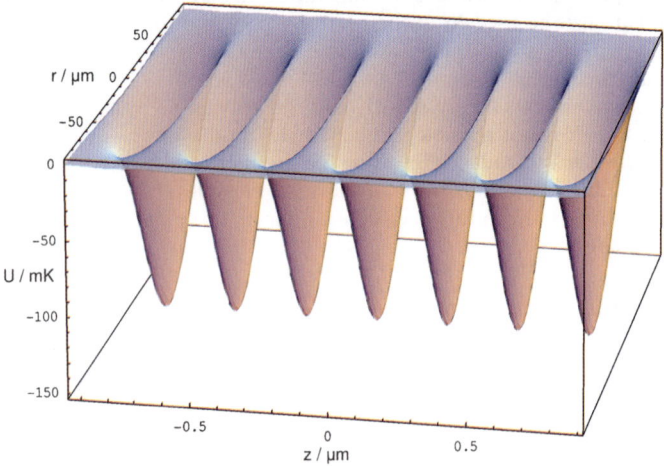

Fig. 2 A plot of the intra-cavity trapping potential for benzene molecules within the high finesse cavity.

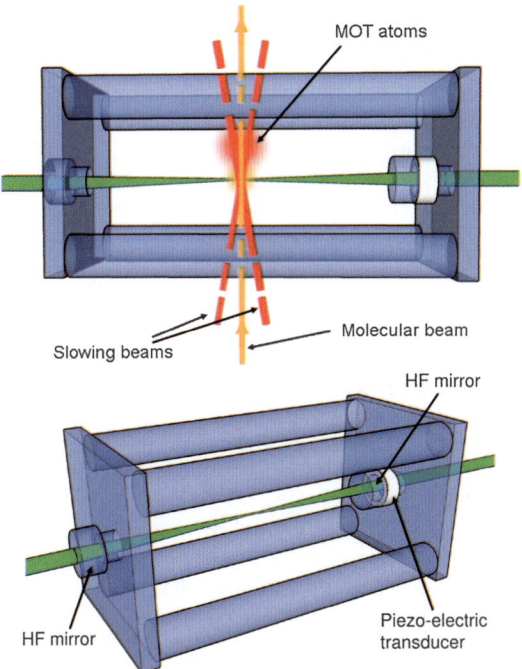

Fig. 3 Fused-silica cavity design. Molecular slowing beams are shown in red, intersecting at the beam waist of the 532 nm (green) dipole trap beam at the centre of the cavity (the trapping point).

a cavity consisting of two, high-reflectivity mirrors (radius of curvature $R = 5$ cm) which share the same optical axis, separated by a distance $L = 8.5$ cm. Fig. 3 is a diagram of the buildup cavity trap suitable for trapping optical Stark decelerated molecules. The mirrors are assumed to be supported on a structure made of a low thermal expansion material such as fused silica, but with optical access for the lattice deceleration beams and for the magneto-optical trap. The mirror transmission coefficients T must be chosen to enable a sufficiently large power build-up inside the cavity to produce the required intensity. For a perfectly impedance-matched cavity, the in-coupling mirror has transmissivity T_1 given by

$$T_1 = \frac{1}{G} = \frac{P_c}{P_1}, \qquad (3)$$

where G is the cavity power build-up factor, P_c is the intra-cavity intensity (one-way), and P_1 is the in-coupled laser power. The impedance matching condition fixes the in-coupling mirror transmission coefficient: $T_1 = T_2 + L$, where T_2 is the transmission coefficient of the second mirror, and L is the total loss coefficient for the cavity. This condition maximizes the power build-up. A one-way intra-cavity intensity of $I_c = 2.5 \times 10^{12}$ W m^{-2} gives a well depth of $U_0 = 150$ mK for benzene. With $w_0 = 55$ μm we obtained $P_c = 23.5 \times 10^3$ W, hence the required transmission coefficient is $T_1 = 149$ parts per million (ppm). Assuming realistic loss coefficients of 10 ppm for each mirror, we then find $T_2 = 129$ ppm. Using these values we calculate a suitable cavity's required spectral properties. These are a finesse of $F = 19800$, a free spectral range $FSR = 1.76$ GHz and a linewidth of $\Delta \nu = 89.3$ KHz, and the power buildup of $G = 6710$. All of these values have been achieved in many high finesse cavities[20–22] and therefore we conclude deep molecular traps constructed in this way are feasible.

We note that laser frequency instabilities, as well as thermal and vibrational instability in the cavity all lead to difficulties in maintaining a stable, single intra-cavity mode. For our purposes, a TEM$_{00}$ Gaussian mode is preferable for the formation of a deep QUEST trap. Therefore, to avoid higher order TEM-modes forming, the cavity length and/or laser frequency have to be actively stabilized using standard Pound–Drever–Hall techniques.[24,25]

2.4 Dipole trap potential for rare gas atoms and optical Stark decelerated molecules

The QUEST produces different well trap depths for each of the possible rare gas atoms and molecules due to their differing static polarizabilities. Table 1 shows the corresponding well depth U_0 for a one-way intra-cavity intensity of $I_c = 2.5 \times 10^8$ W cm^{-2}.

These well depths are compared with the temperature of the atoms after laser cooling and quenching to their ground state, since this has to be much lower (10×) than U_0 for sympathetic cooling to be effective.[20]

To estimate the number of molecules loaded into the lattice/dipole trap potential sites we assume that the molecular beam has a uniform density of $\rho = 10^{13}$ cm^{-3}, and that the volume of a lattice potential site is $V_{LS} = \pi w_0^2 \times a = 2.53 \times 10^{-9}$ cm^3, where a is the axial lattice spacing ($a = 0.266$ µm). Because of the small solid angle subtended by the lattice deceleration beam the transverse temperature ($T_x = T_y$) of benzene in a molecular beam is approximately 400 µK at 0.2 m from the orifice, while the longitudinal temperature T_z is approximately 1 K. The total number of molecules trapped in our lattice potential per lattice site thus depends on the density of the beam, the size of a potential site and also the proportion of the velocity distribution of molecules that can be captured by a potential depth of around 150 mK (for benzene). To simplify the calculation, we assume 100% trapping in the transverse direction (reasonable since $T_{x,y} \ll U_0$), and therefore model the beam as having a 1D velocity distribution given by $f(v_z) = \left(\dfrac{m}{2\pi k_B T_z}\right) \exp\left[-\dfrac{mv_z^2}{2k_B T_z}\right]$, where m is the molecular mass, and v_z is the axial velocity of a constituent molecule in the distribution. During the confinement process, evaporative cooling takes place, which rejects all molecules with temperatures approximately above 1/10th of the potential depth ($T > 15$ mK). The fraction, n, of molecules captured by the trap is given by

$$n = \int_{-v_z(T=15\ \mathrm{mK})}^{v_z(T=15\ \mathrm{mK})} \left(\dfrac{m}{2\pi k_B T_z}\right) \exp\left[-\dfrac{mv_z^2}{2k_B T_z}\right] dv_z, \quad (4)$$

where $v_z(T = 15$ mK$) = 1.26$ m s^{-1}, and $n = 0.097$. The total number of molecules trapped per lattice site is then $N_{LS} = \rho \times V_{LS} \times n = 2450$. With a molecular beam width of ≈ 100 µm, this gives a total number for the whole lattice of 0.92×10^6 molecules. Using the same arguments we would expect that 700 H$_2$ molecules could be trapped in each lattice site and used as a starting point for sympathetic cooling. This model neglects loss mechanisms that may be present during loading, and

Table 1 The calculated well depth for each rare gas species and the recoil temperature after quenching to its ground state

Rare gas	U_0/mK	Recoil temp./µK
He	3.08	1216.0
Ne	5.95	261.1
Ar	24.7	66.3
Kr	37.4	24.8
Xe	60.8	11.1

assumes a high value for the molecular beam density. This number is therefore an upper limit. Although we have considered benzene in these calculations the well depth for any species can be scaled by its polarizability. The well depths for H_2, NH_3 and CS_2 are 12 mK, 33 mK and 127 mK respectively.

2.5 A case study: sympathetic cooling of molecular hydrogen

Molecular hydrogen is an interesting first species to consider for sympathetic cooling since it is of astrophysical importance and is a candidate for ultracold chemistry experiments. Its collisional properties, in the absence of a trapping field, can be readily calculated, and it has a high polarizability to mass ratio which means that it is one of the most effective species to decelerate using optical Stark deceleration.[26,27] In order to estimate thermalisation times in the optical trap it is necessary to determine scattering lengths and collision cross sections for collisions between the rare gas atoms and the molecules of interest. Cold and ultracold calculations of cross sections for H_2 with He and Ar have been already considered in some detail[28,29] and we have extended these to include all the stable rare gas atoms that can be laser cooled and trapped.[30,31] H_2 is a simple first case to consider as the scattering energy is very small compared to the vibrational–rotational excitation energies of H_2, and thus the only allowed channel is the elastic one. Under these conditions, the scattering of Rg–H_2 was modelled as a two-body process.[32,31]

Table 2 shows the range of values obtained for the Rg–H_2 scattering length and the cross section. The spread in these values for all species except Xe–H_2 results from the variation in the published potential energy surfaces (PES).[31] For Xe–H_2, there is only one PES reported. The table illustrates that although there are large uncertainties in the collisional parameters, the physics of the scattering mechanisms is clear. The greatest cross-section is of He–H_2, due to the strong halo characteristics of this system.[33,32] It is possible to show that for those systems the scattering length, a_s depends on the square root of the inverse of the binding energy E, $a_s \approx 1/\sqrt{E}$. The smaller the binding energy the larger the scattering length and the five complexes can be interpreted in this way. The two He–H_2 isotopologues are the weakest bound and their scattering length is the greatest while the Ne–H_2 system is more strongly bound and its scattering length is smaller. The stabilization of a second vibrational band in Ar–H_2 makes this complex more reactive than its predecessor and the Xe–H_2 scattering length is the smallest of the whole group.

The large collisional cross-sections for ^{4}He–H_2 and Ar–H_2 are comparable to that utilized in sympathetic cooling experiments between cold alkalis[10] and these values are relatively constant over a large temperature range. We estimate a collision rate and thermalisation time for Ar–H_2 based on a cross-section of 1000 Å^2 and the axial and radial trap frequencies, assuming the harmonic approximation given as $\omega_{ax} = \sqrt{2\frac{U_0 k^2}{m}}$ and $\omega_{rad} = \sqrt{4\frac{U_0}{mw^2}}$ respectively. The radial and axial trap frequencies for Ar are $2\pi \times 15$ kHz and $2\pi \times 7$ MHz and for H_2 are $2\pi \times 41$ kHz and

Table 2 The scattering lengths a_s and the zero-energy elastic cross-sections σ calculated for all Rg–H_2 complexes

	$|a_s|$/Å	$\sigma(E=0)$/Å^2
^{3}He–H_2	67.6–90.6	57500–103000
^{4}He–H_2	22.7–24.7	6500–7800
Ne–H_2	3.30–3.85	140–190
Ar–H_2	8.71–10.1	950–1300
Kr–H_2	5.51–6.96	380–610
Xe–H_2	1.82	42

$2\pi \times 19$ MHz. We also assume that H_2 has evaporated to approximately one tenth of the well depth before sympathetic cooling occurs. We assume that each species has a density in the trap of the form $n = n_0 \exp[-U/kT]$ with peak densities of 5×10^9 cm^{-3} and 10^{11} cm^{-3} corresponding to 11 molecules and 1300 atoms per fringe of the trap at initial temperatures of 1.2 mK and 66 μK respectively. The Rg–H_2 collision rate is given by $\gamma = \sigma v_{rel} \int n_{Rg}(\mathbf{x}) n_{H_2}(\mathbf{x}) d\mathbf{x}$, v_{rel} is the mean relative velocity between collision partners and $n_{Rg}(\mathbf{x})$ and $n_{H_2}(\mathbf{x})$ are the densities of the rare gas and H_2. It is well known that approximately 3 collisions are required for thermalisation of collision partners of equal mass in a gas but for unequal masses this is approximated by $\frac{3}{\eta}$ where $\eta = 4\frac{m_{Rg} m_{H_2}}{(m_{Rg} + m_{H_2})^2}$. The initial time for thermalisation for two species within a trap is given by[10]

$$\tau = \frac{3\pi^2 k_b T_{H_2}}{(N_{H_2} + N_{Rg})\sigma \eta m_{H_2} \omega_{zH_2} \omega_{rH_2}^2}. \quad (5)$$

For the conditions given for Ar–H_2, a thermalisation time scale of 2 ms is determined. This fast thermalisation time can be accounted for by the large axial trap frequency and the large relative velocity between the hot molecular gas and the much colder atoms, despite the poor spatial overlap between the two species. For Xe–H_2 collisions, using the same initial densities and a collision cross-section of 42 Å^2, a thermalisation time of the order of seconds is estimated. Although the Xe–H_2 collision rate is much lower than for Ar–H_2, trapping of atomic species over this time scale has been demonstrated indicating that even Xe, which forms the deepest trap, could be used for sympathetic cooling.

3 Measuring the effect of molecular alignment on the dipole force

The optical dipole force on a ground state atomic species in a QUEST is proportional to isotropic polarizability. The force on molecules however, has an orientational dependence due to their non-spherical shape and anisotropic polarizability. In the relatively low intensity field of the QUEST described above, the rotational motion of a molecule averages to its static value. However, in the strong non-resonant laser fields (10^{11} W cm^{-2}) that are used to decelerate molecular species in optical Stark deceleration, molecules can be aligned with respect to the polarization of the field. For a molecule irradiated by a non-resonant pulsed optical field with an electric field of the form, $\mathbf{E}(r,t) = \frac{1}{2}\hat{\varepsilon} E(r,t)\exp^{i\omega_l t} + c.c.$, where $\hat{\varepsilon}$ is a unit vector of the field polarization, $E(r,t)$ is its amplitude and ω_l its frequency, the induced dipole potential or Stark shift is given by[34,35]

$$V(t) = -\frac{1}{4}\alpha_{\text{eff}} E(r,t)^2. \quad (6)$$

The effective polarizability and the optical Stark potential are dependent on the amplitude of the laser field and its polarization. For linearly polarized light incident on a linear or symmetric top molecule, the effective polarizability is given by[36] $\alpha_{\text{eff}} = [\Delta\alpha\cos^2\theta + \alpha_\perp]$, where $\Delta\alpha = \alpha_\parallel - \alpha_\perp$ and $\alpha_\parallel$ and $\alpha_\perp$ are the parallel and perpendicular polarizabilities with respect to the molecular symmetry axis. For a linear molecule the symmetry axis is along the bond axis as shown in Fig. 4. For linearly polarized light the polarization vector is perpendicular to the direction of the light field and θ is the angle between the symmetry axis and the polarization direction as shown in Fig. 4a. For circularly polarized light, whose polarization is in a plane orthogonal to the propagation direction, the angle θ is between the symmetry axis and the propagation direction as shown in Fig. 4b. The effective polarizability is $\alpha_{\text{eff}} = \frac{1}{2}[\alpha_\perp + \alpha_\parallel - \Delta\alpha\cos^2\theta]$. These alignment-dependent optical potentials produce

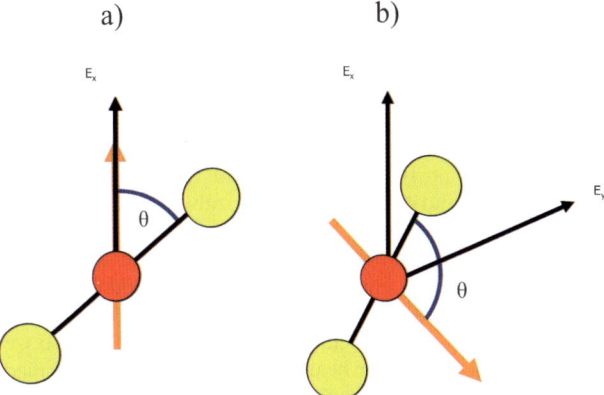

Fig. 4 The orientation of the linear molecule CS$_2$ with respect to the polarization vector for linearly and circularly polarized light defined by the angle θ.

a centre-of-mass force that is proportional to the gradient of the potential, $\mathbf{F} = -\nabla V(r,t)$, which acts to push molecules into the higher field regions of an optical field. For linearly polarized fields the maximum force is determined by the maximum effective polarizability which occurs when $\cos^2\theta = 1$ and the molecule is exactly aligned with the polarization vector leading to $\alpha_{\text{eff}} = \alpha_\parallel$. For circularly polarized light, the maximum polarizability occurs when $\cos^2\theta = 0$ and the molecule moves in the plane orthogonal to the polarization vector. These values will be less than these maximum values and this can be determined from the expectation value of $\cos^2\theta$ for a particular laser intensity and polarization by solution to the Schrödinger equation for a rigid rotor perturbed by the optical Stark potential. The aligned molecular wavefunction, $\Psi(t) = \sum_{JM} C_{J,M}(t)|JM\rangle$, is a superposition of the field free rotor wave-functions[37] $|JM\rangle$, where J are the quantum numbers for angular

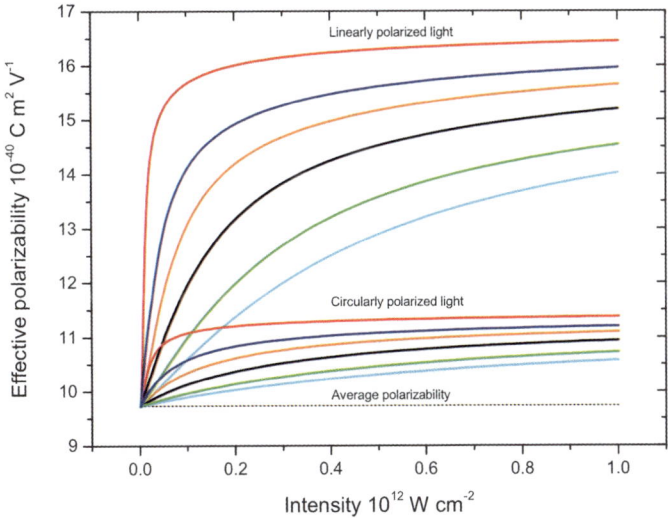

Fig. 5 The effective polarizability is plotted against the nonresonant laser intensity for rotational temperatures 2 K (red), 4 K (dark blue), 6 K (orange), 8 K (black), 10 K (green), 12 K (light blue), for linearly polarized light and circularly polarized light. The average polarizability is also shown for comparison (dashed line).

momentum and M is its projection onto a space-fixed axis. For a particular polarization and laser intensity an expectation value[38] of $\langle\cos^2\theta\rangle_{J,M}(t)$ can be determined and averaged according to an initial population distribution allowing a calculation of the effective polarizability as a function of laser intensity and rotational temperature. These values for both circular and linear polarization are plotted as a function of intensity for CS_2 in Fig. 5. The static polarizabilities of CS_2 are $\alpha_\parallel = 16.8 \times 10^{-40}$ C m² V^{-1} and $\alpha_\perp = 6.2 \times 10^{-40}$ C m² V^{-1} and the rotational constant is $B = 0.109$ cm^{-1}. Fig. 5 shows that the effective polarizability is higher for linear polarization and that all values saturate at higher intensities.

To measure the effect of molecular alignment on the dipole force we have measured the induced velocity imparted to carbon disulfide molecules in a cold molecular beam by an optical field of nanosecond duration and intensity in the 10^{12} W cm^{-2} range. This setup has been described elsewhere[35] and so we only briefly summarize it here. A cold molecular beam of carbon disulfide is formed by expanding CS_2 gas diluted in argon through a pulsed solenoid valve into a vacuum chamber. The centre of the molecular beam passes through an orifice (skimmer) and enters a second differentially pumped vacuum chamber at a pressure of 10^{-7} torr. The translational temperature of the beam is 2.1 K, travelling at an average velocity of 520 ms^{-1}. The molecules pass into a time-of-flight mass spectrometer, where they intersect a non-resonant infra-red (IR) optical field of wavelength 1064 nm, which induces a dipole force on the molecules. The focus of the optical field has a Gaussian spatial profile with an e^{-2} radius of 20 μm and Rayleigh range (constant spot diameter along the propagation direction) of 300 microns with a peak intensity of 6×10^{11} W cm^{-2}. The focusing lens is mounted on an XYZ translational stage outside the vacuum chamber which can position the beam with a resolution of ±1 μm. The single frequency IR laser beam is created by an injection seeded, Q-switched Nd:YAG laser which produces a temporally smooth intensity profile for the duration of the 15 ns pulse. The IR beam initially has a linear polarization of better than 1 part in 10^4 by passing it through two thin film polarizers. When required, the field polarization is converted to near circular polarization at the same intensity by passing the light through a $\frac{\lambda}{4}$ zero order wave-plate. To measure the velocity imparted to the molecules by the dipole force from the IR field we ionize the neutral molecules once the IR field is turned off. The velocity of the neutrals is measured in a Wiley-Maclaren time-of-flight mass spectrometer by converting the change in time-of-flight (TOF) to a change in velocity. Detection of the molecules is accomplished by ionization of the molecules after the IR field is applied using a (3 + 1) resonance enhanced multiple photon ionization (REMPI) scheme *via* the three photon transition $[\frac{1}{2}]np\sigma_u(^1\Pi_u) \leftarrow X^1\Sigma_g^+$.[39,40] Using a peak intensity of $7.6 \pm 2.3 \times 10^{11}$ W cm^{-2} for the linearly polarized light and $7.1 \pm 2.1 \times 10^{11}$ W cm^{-2} for near circular polarization we measure the variation in the dipole force measured at different locations across the center of the focused IR beam along the direction of the molecular beam. The measurements are made by alternating between both types of polarization by rotation of the waveplate so that the measurement is always taken at same position with respect to the IR beam. Velocity changes in this direction are either due to acceleration or deceleration of the molecules due to the dipole force. These measurements were made for both linear and circular polarization and each data point corresponds to a 1200 shot average of the TOF spectrum with the probe beam at different spatial positions across the IR beam. A typical TOF spectra for each polarization is shown in Fig. 6 where the intensity gradient of the IR beam is highest. For comparison the unperturbed TOF, where no IR beam is present, is also shown. By averaging a number of these measurements we find that the TOF and thus the velocity imparted to molecules changes by 25%. This occurs when changing from linearly to circularly polarized when the laser intensity for each case only changes by 6%. These results indicate that the change in the velocity shifts and thus the dipole force is due to field induced alignment. This difference is

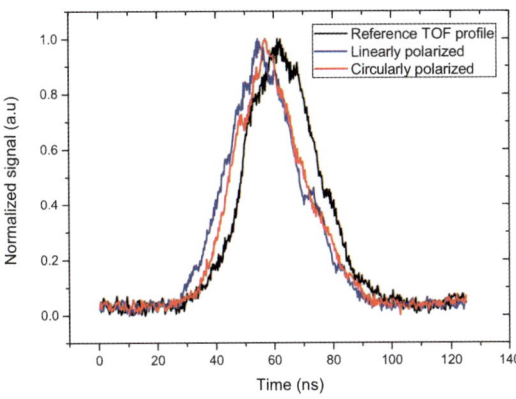

Fig. 6 Time-of-flight (TOF) spectra of CS_2 in the molecular beam when perturbed by a linearly and circularly polarized field as well as the unperturbed reference TOF when the field is turned off.

based on the peak velocity shifts of approximately 10 ms^{-1} for linearly polarized light and 7 ms^{-1} for circularly polarized light (Table 3). To compare our results with that predicted from field induced alignment we calculate the velocity changes due to both polarizations for rotational temperatures from 2 to 12 K. The velocity imparted to the molecules *via* the dipole force is determined by numerically solving the classical equation of motion of a molecule in the optical field from $F = m\dfrac{d^2x}{dt^2} = -\nabla V$ for the experimental conditions described above. The intensity of the laser field can be related to electric field by $I = \dfrac{\varepsilon_0 c}{2} E^2$. For the near impulsive kick to the center-of-mass motion, induced by the 15 ns duration of the laser, the induced velocity change is directly proportional to intensity and effective polarizability and we find that the differences in velocity do not strongly depend on temperatures above 8 K. As the uncertainty in the intensity is 30%, we find a best fit to the measured velocity shifts by allowing the intensity to vary within its uncertainty determined by experiment. Our best fit for both polarizations gives an intensity of 6.0×10^{11} W cm^{-2} for the linear case and 5.6×10^{11} W cm^{-2} for the circularly polarized case. The variation between these two values is approximately 6%, consistent with the variation in intensity between these two polarizations. Using this peak intensity and a rotational temperature of 12 K we obtain a maximum effective polarizability of 13.2×10^{-40} C m^2 V^{-1} for the linear case and 10.4×10^{-40} C m^2 V^{-1} for the circularly polarized case, with $<\cos^2\theta> = 0.66$ for linearly and $<\cos^2\theta> = 0.21$ for circularly polarized light. The average polarizability of a linear molecule that is not aligned by the field is 9.7×10^{-40} C m^2 V^{-1}. The field free alignment expectation value $<\cos^2\theta>$ is equal to $\frac{1}{3}$. The well depths for the linearly polarized beam are $U/k_b = 72$ K for linearly and $U/k_b = 57$ K for circularly polarized light.

Table 3 The measured peak velocity changes of CS_2 molecules accelerated and decelerated by the dipole force from a focused Gaussian beam, for different polarizations. The peak velocity changes occur where the focused beam has the highest intensity gradient

	Lin. Pol.	Circ. Pol.	Difference
Acceleration	9.9 ± 0.6	6.4 ± 0.5	3.5 ± 0.8
Deceleration	−9.9 ± 0.5	−7.2 ± 0.3	2.8 ± 0.6

The ability to control the effective polarizability *via* field polarization allows us to tailor the optical well depth and thus optical Stark deceleration. Of perhaps even more importance is that the effective polarizability is dependent on rotational state when strong fields are used. This may allow the use of these fields for spatial state selection of different rotational states when using a single focused beam before optical Stark deceleration. Such separation may be important in evaporative and sympathetic cooling where thermalisation of molecules in higher rotational states will act to heat molecules trapped in an optical trap.

4 Progress towards chirped deceleration experiments

Optical Stark deceleration is a phase space filtering technique which traps and decelerates cold molecules produced in a molecular beam and brings them to rest. In this typically two stage process the molecules are initially cooled by collisions with rare gas atoms in a supersonic expansion and reach temperatures in the 1 K range. The cold molecules in the molecular beam are, however, accelerated to supersonic speeds, and therefore a second stage is required to trap and transport some part of this initial phase space distribution, centered at the velocity of the molecular beam, back to the laboratory frame. In optical Stark deceleration the potential is created by an interaction between an induced dipole moment and the intense optical field that induced it. Since all molecules can be polarized in this way, in principle, any molecule or atom can be manipulated and slowed in the same manner. This capability has been demonstrated in previous experiments, where molecules were slowed using a moving lattice potential utilising a half phase space rotation or half oscillation in the lattice to rapidly decelerate or accelerate molecules.[26,27] This technique has been shown to produce a slowed distribution with a minimum energy spread that is at best equal to the initial energy distribution of the molecular beam. Although this is acceptable for loading a trap, because a narrower energy distribution can be achieved in a shallower trap, it is not suitable for a range of collision experiments where a narrow energy distribution is required. In addition, very fast switching of the optical fields is required. The creation of slowed energy distributions that are narrower than the molecular beam have been studied theoretically using chirped optical Stark deceleration.[41] This scheme uses a lattice that is initially

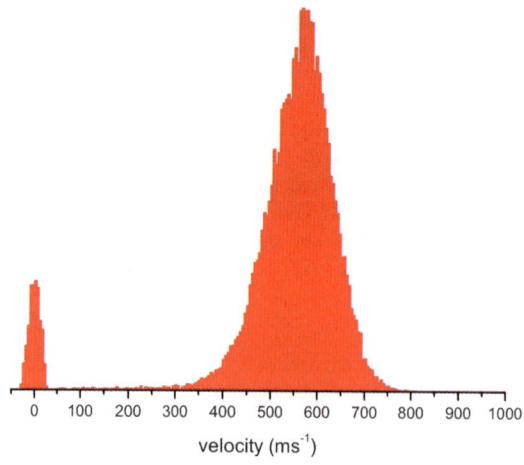

Fig. 7 Simulated velocity distribution function after chirped deceleration of H_2 seeded into a molecular beam with an initial temperature of Ar at 560 K with a FWHM energy spread of 70 mK. The laser is chirped over 1.1 GHz in 150 ns and requires an energy of 60 mJ per lattice beam.

travelling at the molecular beam velocity and traps a narrow energy distribution determined by the well depth of the potential. The lattice is constantly decelerated to zero velocity, transporting a fraction of the cold molecules to zero velocity in the laboratory frame. In practice the decelerating lattice is created by inducing a linear frequency chirp onto one of two near counter-propagating laser beams which interfere to create the moving lattice beams. A simulation of this process for molecular hydrogen is shown in Fig. 7. In this simulation we follow the trajectories of 20000 particles placed within a decelerating lattice with a constant chirp over 150 ns and a frequency excursion of 1.1 GHz. We assume a translational gas temperature of H_2 at 1 K seeded in a buffer gas of argon.

To undertake these experiments we have constructed an amplified laser system operating at a wavelength of 1064 nm. This system is largely constructed from conventional commercial nanosecond laser components but, is configured for chirped optical Stark deceleration. This system produces two output beams which can be used to form a constantly decelerated lattice over a timescale from 100 ns to a few microseconds. The laser and amplifier system that we have constructed to accomplish this task is shown schematically in Fig. 8. The amplifier is custom made by Continuum lasers and consists of two, separate amplifier arms each with three 100 mm long Nd:YAG amplifier rods that are flash lamp pumped. One arm is used to produce a constant frequency beam and the other arm for the chirped beam. The flash lamp pumped amplifiers are well established nanosecond Nd:YAG laser technology which are typically used in commercial Q-Switched Nd:YAG laser systems.[42] The unique feature of this system is that the two arms amplify two CW laser beams producing a flat-top temporal profile of constant laser intensity over most of the pulse duration. As the amplifier gain is non-linear as a function of time, the flat-top intensity profile is created by continuously varying the input laser intensity to produce a flat top profile after amplification. This is accomplished using an electro-optic attenuator consisting of a Pockels cell and a polarizer. The required input temporal profile to produce the flat-top is found for a particular pulse energy and length by trial and error. The CW laser input intensity can be modulated by up to 100% in increments of 7 ns for a total duration of 10 microseconds and a flat profile with variations of the order of 5% can be achieved. The CW laser system that is amplified in each arm is produced by a home built continuous wave (CW) Nd:YVO$_4$ laser which is similar in design to Li et al.[43] and produces a narrow line width (<1 MHz) which can also be rapidly tuned (chirped) in frequency. The CW laser system consists of an optical cavity consisting of a 5 mm thick Nd:YVO$_4$ crystal (doping of 3% atm) which forms the gain medium, as well as the highly reflecting back mirror. An intracavity electro-optical LiTaO$_3$ crystal is used to rapidly tune the laser and an output coupler with a reflectivity of 96% at 1064 nm. The physical

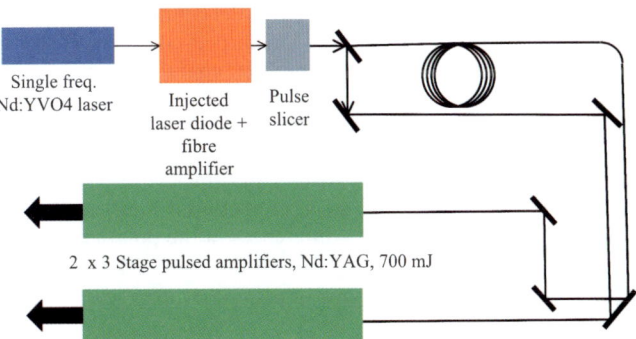

Fig. 8 Schematic diagram of the chirped optical Stark deceleration system which amplifies the output from a Nd:YVO$_4$ laser up to 700 mJ in each arm over a 1 µs interval.

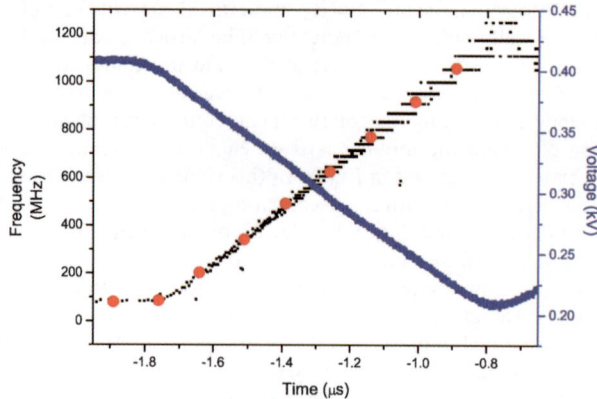

Fig. 9 Frequency of the microchip laser measured as a function of time by heterodyning a fixed frequency laser demonstrating the greater than 1 GHz μs^{-1} frequency chirp required for deceleration experiments.

length of this cavity is approximately 5 mm and is dominated by the 5 mm long by 1 mm thick y-cut electro-optic crystal. A voltage is applied across the 1 mm thick section to modulate the cavity length and thus the frequency of the laser. The Nd:YVO$_4$ laser crystal is anti-reflection (AR) coated for the 808 nm diode pump laser light and has a highly reflecting coating for the 1064 nm. The laser is pumped by an 808 nm laser diode. This Nd:YVO$_4$ laser can be linearly frequency chirped over timescales of less than 100 ns with a frequency change per volt of 7 MHz V^{-1}. The laser is maintained to within a few mK to ensure frequency stability. Deceleration experiments typically require frequency excursions of up 1 GHz and this requires a linear voltage ramp of 160 volts over the duration of the deceleration period, which is provided by amplification of a standard laboratory signal generator.

Fig. 9 is a graph of the instantaneous frequency of this laser system over approximately 1 μs. In order to create the chirped and unchirped laser beams for amplification to the energies (>400 mJ) required for deceleration, we firstly amplify the low power beam from the Nd:YVO$_4$ laser from approximately 10 mW up to 60 mW by injecting this light into a conventional InGaAs Fabry–Perot laser diode operating at 1064 nm.[44] Importantly, this step also serves to reduce intensity variations in the beam introduced by the chirping process. This beam is further amplified in a commercial 1 W Yb fibre amplifier (IPG Photonics). Finally, the preamplified CW beam is subsequently split into two beams of approximately equal intensity by a 50/50 beamsplitter. As the CW beam is unchirped most of the time the chirped temporal component can be superimposed in time with an unchirped component of the beam by sending one of the split beams into a delay line formed by a single mode optical fibre. When both the chirped and unchirped components are temporally co-incident they are subsequently directed into the separate amplifier arms described above to produce the final energies required for deceleration. We are now beginning experiments to utilise this laser system for chirped deceleration experiments and first plan to use this system to decelerate molecular hydrogen into the optical trap described above. We stress that this system is, in principle, capable of decelerating any molecular or indeed any atomic species introduced into the molecular beam, albeit with a different efficiency due to differences in the polarizability and mass.

5 Summary and conclusions

We have outlined a general scheme for cooling molecules to μK temperatures using sympathetic cooling within a deep optical trap and have applied this to the case of sympathetic cooling of H$_2$. The combination of optical trapping and the use of ultracold rare gas species is general in principle, and allows the trapping of essentially any species. It also has the potential to mitigate trap losses due to inelastic collisions and also reduces the potential for reactions between many ultracold atomic and molecular species. In addition, the deep optical trap produces high trap oscillation frequencies which help to reduce thermalisation times for species with very different initial temperatures. Such a method also appears to be particularly promising for optical Stark deceleration because the dimensions of the optical trap match well to those of the focused beams used in optical Stark deceleration. We have presented recent developments in the optical Stark deceleration method. This includes the measurement of the role of molecular alignment on the optical dipole force, where we have shown by variation of the polarization that continuous tuning of the well depth of an optical Stark decelerator is possible by changing from linear to circular polarization at a fixed intensity. Finally, we have also described a laser and amplifier system that has been constructed for chirped optical Stark deceleration experiments and have shown that it appears capable of decelerating a range of molecular species, including molecular hydrogen, with a narrow energy spread which can be used for future sympathetic cooling experiments and collision studies.

References

1 S. Chu, L. Hollberg, J. Bjorkholm, A. Cable and A. Ashkin, *Phys. Rev. Lett.*, 1985, **55**, 48.
2 T. Köhler, K. Góral and P. Julienne, *Rev. Mod. Phys.*, 2006, **78**, 1311–1361.
3 K. Jones, E. Tiesinga, P. Lett and P. Julienne, *Rev. Mod. Phys.*, 2006, **78**, 483–535.
4 J. D. Weinstein, R. deCarvalho, T. Guillet, B. Friedrich and J. M. Doyle, *Nature*, 1998, **395**, 148.
5 H. L. Bethlem, A. J. A. van Roij, R. T. Jongma and G. Meijer, *Phys. Rev. Lett.*, 2002, **88**, 133033.
6 N. Vanhaecke, U. Meier, M. Andrist, B. H. Meier and F. Merkt, *Phys. Rev. A*, 2007, **75**(3), 031402.
7 J. Gilijamse, S. Hoekstra, S. van de Meerakker, G. Groenenboom and G. Meijer, *Science*, 2006, **313**, 1617–1620.
8 E. R. Hudson, H. J. Lewandowski, B. C. Sawyer and J. Ye, *Phys. Rev. Lett.*, 2006, **96**, 143004.
9 M. Gangl and H. Ritsch, *Phys. Rev. A*, 2001, **64**, 063414.
10 A. Mosk, S. Kraft, M. Mudrich, K. Singer, W. Wohlleben, R. Grimm and M. Weidemuller, *Appl. Phys. B*, 2001, **73**, 791.
11 M. Lara, J. L. Bohn, D. Potter, P. Soldan and J. M. Hutson, *Phys. Rev. Lett.*, 2006, **97**, 183201.
12 P. Soldan and J. M. Hutson, *Phys. Rev. Lett.*, 2004, **92**, 163202.
13 A. Aspect, E. Arimondo, R. Kaiser, N. Vansteenkiste and C. C. Tannoudji, *Phys. Rev. Lett.*, 1988, **61**, 826.
14 F. Shimizu, K. Shimizu and H. Takuma, *Phys. Rev. A*, 1989, **39**, 2758.
15 H. Katori and F. Shimizu, *Jpn. J. Appl. Phys.*, 1990, **29**, L2124.
16 C. Y. Chen, Y. M. Li, K. Bailey, T. P. O'Connor, L. Young and Z. Lu, *Science*, 1999, **286**, 1139.
17 M. Walhout, H. J. L. Megens, A. Witte and S. L. Rolston, *Phys. Rev. A*, 1993, **48**, R879.
18 R. D. Knight and L. Wang, *Phys. Rev. A*, 1985, **32**(5), 2751.
19 T. Takekoshi, J. R. Yeh and R. J. Knize, *Opt. Commun.*, 1995, **114**, 421.
20 A. Mosk, S. Jochim, H. Moritz, T. Elsasser, M. Weidemller and R. Grimm, *Opt. Lett.*, 2001, **26**, 1837.
21 S. K. Lee, H. S. Lee, J. M. Kim and D. Cho, *J. Phys. B: At., Mol. Opt. Phys.*, 2005, **38**, 1381.
22 L. S. Cruz, M. Sereno and F. C. Cruz, *Opt. Express*, 2008, **16**, 2909.
23 L. S. Meng, J. K. Brasseur and D. K. Neumann, *Opt. Express*, 2005, **13**, 10085.
24 R. W. P. Drever, J. L. Hall, F. V. Kowalski, J. Hough, G. M. Ford, A. J. Munley and H. Ward, *Appl. Phys. B.*, 1983, **31**, 97.
25 E. D. Black, *Am. J. Phys.*, 2001, **69**, 79.
26 R. Fulton, A. I. Bishop and P. F. Barker, *Phys. Rev. Lett.*, 2004, **93**(24), 243004.
27 R. Fulton, A. I. Bishop, M. N. Shneider and P. F. Barker, *Nat. Phys.*, 2006, **2**, 465.
28 J. C. Flasher and R. C. Forrey, *Phys. Rev. A*, 2002, **65**, 032710.
29 A. Mack, T. K. Clark, R. C. Forrey, N. Balakrishnan, T.-G. Lee and P. C. Stancil, *Phys. Rev. A*, 2006, **74**, 052718.
30 P. Barletta, J. Tennyson and P. F. Barker, *Phys. Rev. A*, 2008, **78**, 052707.

31 P. Barletta, *Eur. Phys. J. D*, 2009, **53**, 33.
32 F. A. Gianturco, T. González-Lezana, G. Delgado-Barrio and P. Villareal, *J. Chem. Phys.*, 2005, **122**, 084308.
33 A. Kalinin, O. Kornilov, L. Yu Rusin and J. P. Toennies, *J. Chem. Phys.*, 2004, **121**, 625–627.
34 H. Sakai, A. Tarasevitch, J. Danilov, H. Stapelfeldt, R. W. Yip, C. Ellert, E. Constant and P. B. Corkum, *Phys. Rev. A*, 1998, **57**, 2794.
35 R. Fulton, A. Bishop and P. F. Barker, *Phys Rev. Lett.*, 2004, **93**, 243004.
36 T. Seideman, *J. Chem. Phys.*, 1997, **107**, 10420.
37 J. Ortigoso, M. Rodriguez, M. Gupta and B. Friedrich, *J. Chem. Phys.*, 1999, **110**, 3870.
38 E. Hamilton, T. Seideman, T. Ejdrup, M. D. Poulsen, C. Z. Bisgard, S. S. Viftrup and H. Stapelfeld, *Phys. Rev. A*, 2005, **72**, 043402.
39 J. Baker, M. Konstantaki and S. Couris, *J. Chem. Phys.*, 1995, **103**, 2436.
40 R. A. Morgan, M. A. Baldwin, A. J. Orr-Ewing, M. N. R. Ashfold, W. J. Burma, J. B. Milan and C. A. de Lange, *J. Chem. Phys.*, 1996, **104**, 6117.
41 P. Barker and M. N. Shneider, *Phys. Rev. A*, 2002, **66**, 065402.
42 A. Seigman, *Lasers*, University Science Books, New York, NY, 1986.
43 Y. Li, A. Viera, S. M. Goldwasser and P. Herczfeld, *IEEE Trans. Microwave Theory Tech.*, 2001, **49**, 2048.
44 M. J. Wright, P. L. Gould and S. D. Gensemer, *Rev. Sci. Instrum.*, 2004, **75**, 4718.

Prospects for sympathetic cooling of polar molecules: NH with alkali-metal and alkaline-earth atoms – a new hope

Pavel Soldán,[*a] Piotr S. Żuchowski[b] and Jeremy M. Hutson[b]

Received 17th December 2008, Accepted 23rd January 2009
First published as an Advance Article on the web 28th May 2009
DOI: 10.1039/b822769c

We explore the potential energy surfaces for NH molecules interacting with alkali-metal and alkaline-earth atoms using highly correlated *ab initio* electronic structure calculations. The surfaces for interaction with alkali-metal atoms have deep wells dominated by covalent forces. The resulting strong anisotropies will produce strongly inelastic collisions. The surfaces for interaction with alkaline-earth atoms have shallower wells that are dominated by induction and dispersion forces. For Be and Mg the anisotropy is small compared to the rotational constant of NH, so that collisions will be relatively weakly inelastic. Be and Mg are thus promising coolants for sympathetic cooling of NH to the ultracold regime.

I. Introduction

In recent years there has been growing interest in the production and properties of cold molecules. Possible applications, such as controlled ultracold chemistry,[1] quantum information and computing,[2] and high-precision measurements of the time-dependence of fundamental 'constants',[3–5] make cold molecules extremely interesting across many different fields.

Two main approaches to the production of cold molecules can be distinguished. One approach is based on the coherent formation of ultracold molecules such as Cs_2 or RbCs in trapped ultracold atomic gases.[6] The molecules may be formed either by photoassociation[7] or by Feshbach resonance tuning.[8] They inherit the μK–nK temperatures of the parent ultracold atomic cloud and usually need very little further cooling. Efforts in this area have led to the Bose–Einstein condensation of Feshbach molecules[9–11] and to the transfer of Feshbach molecules to low-lying states.[12–14] There have also been considerable successes in direct photoassociation to produce low-lying states.[15–18]

In the other approach, represented for example by Stark deceleration[19,20] or helium buffer-gas cooling,[21] preexisting molecules are decelerated either by external fields or by collisions with other particles and trapped in electrostatic or magnetic traps. The temperature of the resulting molecular cloud is usually in the K-mK region, and therefore new ways for cooling the molecules further are being sought. A promising route to cooling decelerated molecules down to the μK region is offered by sympathetic cooling.

---

[a]*Department of Chemical Physics and Optics, Faculty of Mathematics and Physics, Charles University in Prague, Ke Karlovu 3, CZ-12116 Prague 2, Czech Republic. E-mail: pavel.soldan@mff.cuni.cz*
[b]*Department of Chemistry, University of Durham, South Road, Durham, UK DH1 3LE. E-mail: piotr.zuchowski@durham.ac.uk; J.M.Hutson@durham.ac.uk*

Sympathetic cooling, in which one species is cooled by thermal contact with another much colder species, was originally developed as a cooling technique for trapped ions.[22] Diatomic[23] and polyatomic[24] molecular ions have been cooled to sub-Kelvin temperatures by thermal contact with cold ions, and sympathetic cooling is expected to be capable of cooling ions of very high mass, including those of biological relevance. Sympathetic cooling has also been successful in producing ultracold neutral atoms of species that are not themselves suitable for evaporative cooling; for example it was used to create the first Bose–Einstein condensates of potassium ^{41}K.[25]

Sympathetic cooling is effective only if the rate of elastic collisions is very large compared to the rate of inelastic collisions. Elastic collisions exchange kinetic energy between molecules and allow thermalization. However, inelastic collisions in which internal energy is converted into relative kinetic energy cause trap loss (if the energy released is greater than the trap depth) or heating (if the energy released is less than the trap depth). Magnetic and electrostatic traps always trap molecules in low-field-seeking states, which are not in their lowest state in the applied field. Since typical traps have depths in the μK or mK range, most inelastic collisions cause trap loss. A commonly stated rule of thumb is that the ratio of elastic to inelastic collision rates must be at least 100 for effective sympathetic cooling.

The obvious *coolants* for the sympathetic cooling of cold molecules are alkali-metal atoms, which can be cooled to ultra-low temperatures on demand. Sympathetic cooling of photoassociated alkali-metal dimers in triplet states by alkali-metal atoms was discussed by Cvitaš et al.,[26,27] who concluded that the dimers would need to be in their ground rovibrational state because of the unfavorable ratio of inelastic to elastic cross sections in the sub-mK temperature region.

Sympathetic cooling of decelerated molecules by alkali-metal atoms was first considered by Soldán and Hutson,[28] who studied the interactions of rubidium atoms with NH molecules. They showed that the $^2A''$ and $^4A''$ states of RbNH (bound by covalent and dispersion forces) are crossed by much deeper $^2A'$ and $^2A''$ ion-pair states in the energetically allowed region at linear geometries. They concluded that the ion-pair states are likely to have important consequences for the physics of sympathetic cooling of molecules such as CH, NH and OH, because they provide additional mechanisms for inelastic collisions and three-body recombination.

Lara et al.[29,30] subsequently focused on the interaction of OH molecules with ultracold Rb atoms. They developed full sets of coupled potential energy surfaces and carried out quantum collision calculations including spin–orbit and hyperfine coupling. Once again they found a deep ion-pair state ($^1A'$ for RbOH) that crossed the covalent states at energetically accessible geometries. However, even when the ion-pair state was excluded from the calculation, the anisotropy of the potential for the covalent states was enough to cause strong inelastic collisions that would prevent sympathetic cooling except for atoms and molecules in their absolute ground states. Lara et al. concluded in general that (i) light atomic partners are desirable as coolants, because the resulting high centrifugal barriers would suppress many inelastic channels; (ii) weak coupling of the electron to the axis, which occurs in Hund's case b molecules such as NH or CaH, would be beneficial; (iii) the anisotropy of the atom–molecule surface should be comparable to, or smaller than, the rotational constant of the molecule; and (iv) closed-shell coolants, such as alkaline-earth atoms, may produce more isotropic potential energy surfaces than open-shell coolants, such as alkali-metal atoms. Later, a detailed *ab initio* study of 2D adiabatic potential energy surfaces for NH interacting with Rb and Cs atoms was reported by Tacconi et al.,[31] followed by studies of the quantum dynamics of ultra-low-energy collision processes.[32,33]

Another set of potential coolants for sympathetic cooling are alkaline-earth atoms. Calcium[34] and strontium[35] atoms can be cooled and trapped at temperatures of the order of μK. Mehlstäubler et al.[36] have recently succeeded in cooling magnesium atoms to sub-Doppler temperatures of 500 μK. To our knowledge no attempt has yet been made to cool Be atoms.

Very recently, Żuchowski and Hutson[37] surveyed interactions of NH_3 molecules with alkali-metal and alkaline-earth atoms. All the systems exhibited deep potential wells (890 to 5100 cm^{-1}) when the atom was on the N side of the molecule, and shallow potential wells (100 to 130 cm^{-1}) when the atom was on the H side, resulting in very strong anisotropy of the surfaces. This will produce strong inelasticity in the molecular rotational and inversion degrees of freedom and sympathetic cooling is unlikely to be successful for molecules in low-field-seeking states.

In this paper we survey the possibilities for sympathetic cooling of NH molecules, which have very recently been cooled and magnetically trapped at 0.7 K in their ground $X^3\Sigma^-$ state[38] by buffer-gas cooling. NH molecules in their metastable $a^1\Delta$ state have also been Stark-decelerated[39] and electrostatically trapped at temperatures of 60–100 mK, and there is a proposal[40] to transfer $a^1\Delta$ molecules to the $X^3\Sigma^-$ state. In the present paper we investigate the interactions of NH($X^3\Sigma^-$) molecules not only with all the relevant alkali-metal (Alk) atoms, but also for the first time with the alkaline-earth (Ae) atoms. We characterize the potential energy surfaces of the covalent and dispersion-bound states of the AlkNH and AeNH systems and locate their conical intersections with the ion-pair states. We show that Be and Mg atoms are promising candidates for sympathetic cooling of NH($X^3\Sigma^-$) molecules.

II. Methods

To facilitate future quantum dynamics calculations, all results are reported in Jacobi coordinates. The intermolecular distance R is the distance between the alkali-metal or alkaline-earth atom and the center of mass of the NH molecule. The angle θ is measured at the center of mass and is zero for linear atom–HN geometries. In all our calculations the NH bond length r is fixed at the experimentally determined equilibrium value for the free monomer, 1.0367 Å.[41]

Supermolecular coupled-cluster calculations were carried out using a single-reference restricted open-shell variant[42] of the coupled cluster method[43] with single, double and non-iterative triple excitations [RCCSD(T)]. All electrons from the "outer-core" orbitals (1s, 2s2p, 3s3p, 4s4p, and 5s5p for Li and Be, Na and Mg, K and Ca, Rb and Sr, and Cs, respectively) were included in the RCSSD(T) calculations. All the *ab initio* calculations were performed using the MOLPRO package.[44]

To describe the interaction between the outer-core and valence electrons, and to reduce basis-set superposition errors, rather large basis sets are needed. We use the correlation-consistent polarized valence quintuple-ζ (cc-pV5Z) basis sets of Dunning[45] for hydrogen (without the g functions) and for nitrogen (without the h functions). Both these basis sets were augmented in an even-tempered manner and used in uncontracted form. For lithium, beryllium, sodium, magnesium, potassium, and calcium atoms, we use the correlation-consistent polarized core-valence quintuple-ζ cc-pCV5Z basis sets of Iron *et al.*,[46] again without the h functions. The Li, Be, Na, and Mg basis sets were used in fully uncontracted form and those for K and Ca were used partially contracted. These basis sets were also augmented by additional even-tempered diffuse functions. The resulting aug-cc-pCV5Z basis sets consist of (15s,10p,8d,6f,4g), (20s,13p,9d,7f,5g), and [13s,12p,9d,7f,5g] functions for Li, Na, and K, respectively, and of (15s,9p,8d,6f,4g), (21s,15p,9d,7f,5g), and [13s,12p,9d,7f,5g] functions for Be, Mg, and Ca, respectively. For rubidium, strontium, and cesium, we use the small-core scalar relativistic effective core potentials ECP28MDF and ECP46MDF,[47,48] together with the corresponding valence basis sets. These basis sets were augmented in the even-tempered manner and used in uncontracted form. The resulting basis sets consisted of (14s,11p,6d,4f,2g), (15s,12p,7d,5f,2g), and (13s,12p,6d,4f,3g) primitive Gaussian functions for Rb, Sr, and Cs, respectively.

All interaction energies are calculated with respect to the separated-atom–molecule limit, with both the atom and the molecule in their ground states. The full counterpoise correction of Boys and Bernardi[49] is used to compensate for basis set superposition errors. Optimizations of the counterpoise-corrected dimer interaction energies are performed making use of a general optimization algorithm implemented in MOLPRO.

In order to describe the dispersion-bound state of MgNH for all geometries, we use a version of symmetry-adapted perturbation theory (SAPT) based on a density-functional theory (DFT) description of the monomers. In the SAPT(DFT) method[50] (which we use here only for MgNH), the interaction energy is obtained as a sum of contributions,

$$E_{int}^{SAPT(DFT)} = E_{elst}^{(1)} + E_{exch}^{(1)} + E_{disp}^{(2)} + E_{ind}^{(2)} + E_{exch-disp}^{(2)} + E_{exch-ind}^{(2)} \quad (1)$$

where $E_{elst}^{(1)}$ is the electrostatic energy, $E_{exch}^{(1)}$ is the first-order exchange energy, $E_{disp}^{(2)}$ and $E_{ind}^{(2)}$ are the second-order dispersion and induction energies, and $E_{exch-disp}^{(2)}$ and $E_{exch-ind}^{(2)}$ are their exchange counterparts. The first-order terms are calculated using Kohn–Sham orbitals, while the dispersion, induction and exchange-induction terms are evaluated using coupled Kohn–Sham density susceptibilities. The exchange–dispersion term is estimated as described in ref. 50. However, in ref. 50, the second-order exchange corrections are given in the so-called S^2 approximation, which neglects terms of third and higher powers in the overlap matrix S. Since the overlap between Mg and NH is large, we scale the second-order exchange corrections by $E_{exch}^{(1)}(S^2)/E_{exch}^{(1)}$. This procedure was introduced by Patkowski and Szalewicz[51] to improve the performance of SAPT for systems with very diffuse monomer densities.

Our SAPT calculations use the PBE0 density functional,[52] with augmented correlation-consistent polarized valence quadruple-zeta (aug-cc-pVQZ) basis sets[45,46] supplemented with bond functions 0.9,0.3,0.1 s and p, 0.6,0.2 d and f placed at the midpoint between the Mg atom and the center of mass of NH. The Tozer–Handy asymptotic correction[53] of the exchange–correlation potential is used, with splicing parameters 3.5 and 4.7 Å.

III. Results and discussion

A. Alkali-metal atom + NH interactions

In discussing the electronic structure of NH interacting with alkali-metals (Alk), it is convenient to begin with linear arrangements. At linear geometries (point group $C_{\infty v}$), there are two covalent states $^4\Sigma^-$ and $^2\Sigma^-$, which correlate with the Alk(^2S) + NH($^3\Sigma^-$) dissociation limit. These are crossed by an ion-pair $^2\Pi$ state, which in a diabatic representation correlates with the Alk$^+$(^1S) + NH$^-$($^2\Pi$) dissociation limit. [In an adiabatic representation this state changes configuration at a long-range avoided crossing with a higher $^2\Pi$ state, and in the new configuration it correlates with Alk(^2P) + NH($^3\Sigma^-$); the ion-pair configuration is then carried up by a cascade of similar avoided crossings until it reaches the Alk$^+$(^1S) + NH$^-$($^2\Pi$) dissociation limit]. At non-linear geometries (point group C_s), the $^4\Sigma^-$ and $^2\Sigma^-$ states become $^4A''$ and $^2A''$ states, while the $^2\Pi$ state is subject to the Renner–Teller effect and splits into two states with the electron hole either in the triatomic plane ($^2A'$) or perpendicular to it ($^2A''$). In cuts through the potential at fixed N–H distance, the covalent $^2A''$ and ion-pair $^2A''$ states avoided-cross at nonlinear geometries but form a conical intersection at linear geometries. In the full three-dimensional picture they form a seam of conical intersections parameterized by the N–H distance.

We consider first the quartet states. The potential curves for the $^4\Sigma^-$ states at linear Alk–NH geometries are shown in Fig. 1. Results from geometry optimization are given in Table 1 for both linear configurations.

From the viewpoint of sympathetic cooling, the most important feature of the potential surface is its anisotropy. For sympathetic cooling to be successful for

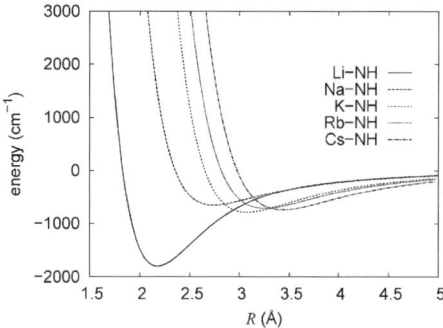

Fig. 1 One-dimensional cuts through the lowest $^4\Sigma^-$ potential energy surfaces of AlkNH systems at the linear Alk–NH arrangement.

Table 1 Lowest $^4\Sigma^-$ and $^2\Pi$ states of linear AlkNH: minima (R_{min}, V_{min}) and crossing points (R_x, V_x) at different arrangements (Alk–NH, Alk–HN). Energies are given in cm^{-1} and distances in Å

Alk	Alk-NH						Alk–HN	
	R^Σ_{min}	V^Σ_{min}	R_x	V_x	R^Π_{min}	V^Π_{min}	R^Σ_{min}	V^Σ_{min}
Li	2.176	−1799.1	3.09	−600	1.78	−21073	4.658	−115.3
Na	2.737	−651.3	3.21	−470	2.13	−13287	4.872	−98.9
K	3.073	−784.7	3.84	−432	2.42	−14235	5.386	−91.1
Rb	3.254	−709.3	3.99	−412	2.53	−13918	5.521	−87.3
Cs	3.435	−737.9	4.31	−381	2.65	−15116	5.761	−85.0

low-field-seeking molecular states, we need a potential surface where the anisotropy is small compared to the rotational constant of the monomer ($B_e = 16$ cm^{-1} for NH). The global minimum for quartet states is at linear Alk–NH geometries for all systems. For LiNH, the well depth at the global minimum is about 1800 cm^{-1}, while at the secondary minimum (Li–HN) it is only 115 cm^{-1}. The absolute well depths for the other AlkNH systems are a factor of 2 to 3 smaller than for Li–NH, while the depths of the secondary wells are comparable for all the alkali-metal atoms. Nevertheless, for all the AlkNH systems the anisotropy is very large compared to the rotational constant of NH.

The large anisotropy of the AlkNH quartet surfaces results from strong sp mixing of the alkali-metal orbitals in the Alk–NH arrangement. Such mixing is much weaker in the Alk–HN arrangement. To quantify this we have performed Mulliken population analysis.[54] The strongest effect is observed for LiNH, where the partial occupancy of the valence p_z orbital is 0.086 in the Li–NH arrangement compared to 0.004 in the Li–HN arrangement. The p_z occupancy for Alk–NH arrangements decreases down the periodic table: Na 0.068, K 0.037, Rb 0.027 and Cs 0.021. This hybridization tendency is similar to that found for the alkali-metal trimers in the quartet states.[55]

The present calculations give a quartet well depth for RbNH that is about 12% deeper than the value of 78 meV (630 cm^{-1}) obtained by Soldán and Hutson[28] using multireference configuration interaction (MRCI) calculations. The present work used RCCSD(T) calculations, which give a better treatment of dispersion effects, and also used much larger basis sets. Our well depths are also substantially deeper than those reported for Rb–NH and Cs–NH by Tacconi et al.[31] using methods and basis sets similar to those of ref. 28.

For all the Alk–NH systems there are also covalent states of $^2\Sigma^-$ symmetry ($^2A''$ at bent geometries) and an ion-pair state of $^2\Pi$ symmetry ($^2A'$ and $^2A''$ at bent geometries). As described above, there is an avoided crossing between the two $^2A''$ states. Table 1 gives the equilibrium positions and well depths of the $^2\Pi$ states for Alk–NH geometries, and it may be seen that for all systems the ion-pair well is more than 13000 cm^{-1} deep. The potential curve for the $^2\Sigma^-$ state cannot be obtained from RCCSD(T) calculations, but it is qualitatively similar to that for the $^4\Sigma^-$ state in the long-range region.[28] Table 1 includes the position and energy of the crossing point between the $^4\Sigma^-$ and $^2\Pi$ curves, and it may be seen that the crossing is always outside the minimum of the Σ state. Because of this, the lowest adiabatic surface of either $^2A'$ or $^2A''$ symmetry always has a very deep well of ion-pair character. This well is strongly anisotropic, so that any collision that samples the doublet surfaces is likely to be strongly inelastic.

In conclusion, it appears that both the doublet and quartet states of Alk–NH systems have sufficient anisotropy to prevent sympathetic cooling for low-field-seeking molecular states.

B. Alkaline-earth atom + NH interactions

The potential energy surfaces for alkaline-earth atoms (Ae) interacting with NH are substantially different from those for AlkNH systems. At linear geometries, there is one dispersion-bound state, $^3\Sigma^-$, which correlates with the Ae(^1S) + NH($^3\Sigma^-$) dissociation limit. This state is crossed by ion-pair $^3\Pi$ and $^1\Pi$ states, which in a diabatic representation correlate with the Ae$^+$(^2S) + NH$^-$($^2\Pi$) dissociation limit. At non-linear geometries (point group C_s), the $^3\Sigma^-$ state becomes a $^3A''$ state, and the $^3\Pi$ state is subject to the Renner–Teller effect and splits into two states with the electron hole either in the triatomic plane ($^3A'$) or perpendicular to it ($^3A''$). In cuts at fixed NH bond length, the dispersion-bound $^3A''$ and ion-pair $^3A''$ states form a conical intersection at linear geometries, while in the full three-dimensional picture they form a seam of conical intersections parameterized by the N–H distance.

The counterpoise-corrected equilibrium distances and well depths for AeNH systems are shown in Table 2 for both Ae–NH and Ae–HN linear geometries. The corresponding potential curves are shown in Fig. 2 and 3 for Ae–NH and Ae–HN geometries, respectively. It may be seen that the anisotropy of the dispersion-bound state is considerably smaller for the alkaline-earth atoms than for the alkali-metal atoms. However, for Ca and Sr the difference between the well depths at the two linear geometries (60.4 and 184.6 cm^{-1}, respectively) is still several times the NH rotational constant (B_e = 16 cm^{-1}). For Be and Mg, by contrast, the

Table 2 Lowest $^3\Sigma^-$ and $^3\Pi$ states of linear AeNH: minima (R_{min}, V_{min}) and crossing points (R_x, V_x) at different arrangements (Ae–NH, Ae–HN). Energies are given in cm^{-1} and distances in Å

	Ae-NH						Ae-HN	
Ae	R^Σ_{min}	V^Σ_{min}	R_x	V_x	R^Π_{min}	V^Π_{min}	R^Σ_{min}	V^Σ_{min}
Be	3.995	−84.5	2.30	2390	1.55	−20240	4.301	−95.4
Mg	4.157	−106.5^a	2.59	1510	1.95	−10120	4.636	−103.0^b
Ca	3.963	−165.7	3.19	−146	2.19	−17041	5.149	−104.5
Sr	3.175	−286.4	3.39	−267	2.32	−16734	5.340	−101.8

^{a} $E^{SAPT(DFT)}_{int}$ = −113.4 cm^{-1} and corresponding R^Σ_{min} = 4.24 Å. ^{b} $E^{SAPT(DFT)}_{int}$ = −91.7 cm^{-1} and corresponding R^Σ_{min} = 4.74 Å.

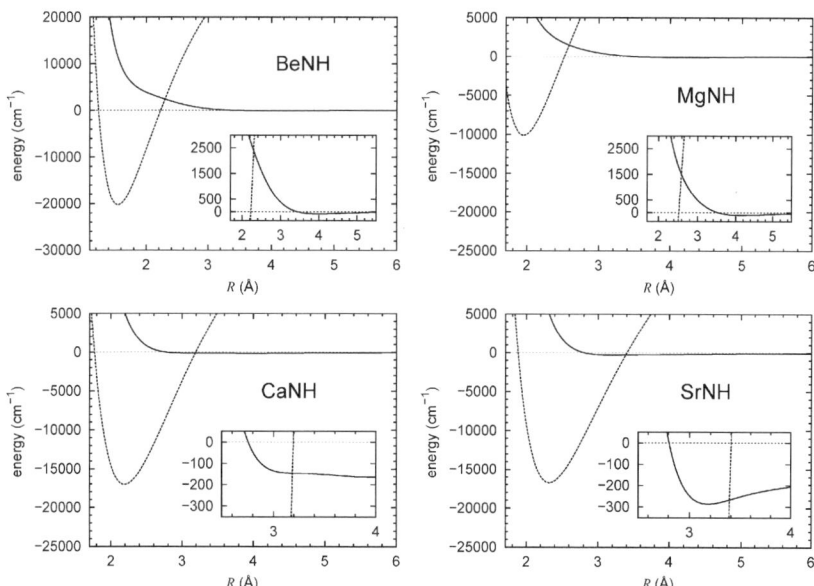

Fig. 2 One-dimensional cuts through the lowest $^3\Sigma^-$ (full) and $^3\Pi$ (dashed) potential energy surfaces of AeNH systems at the linear Ae–NH arrangement. For clarity, the region near the crossing of the dispersion-bound and ion-pair states is magnified.

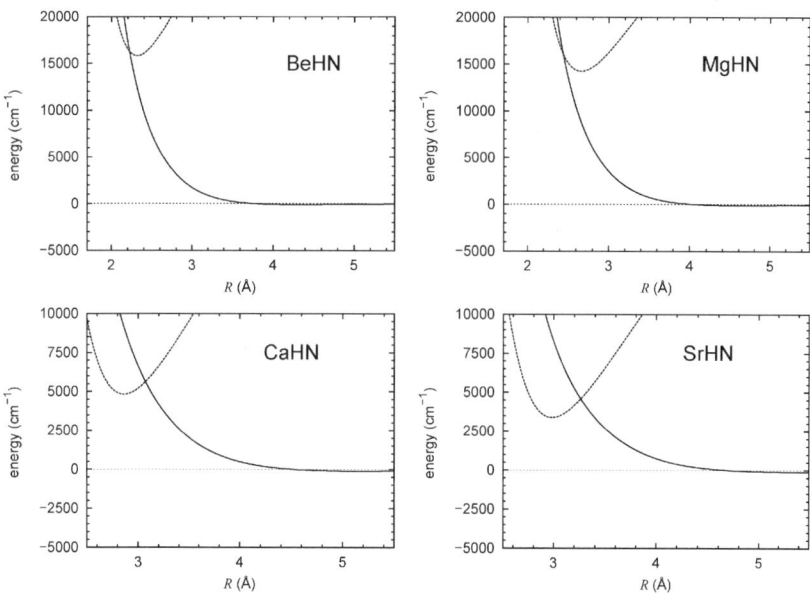

Fig. 3 One-dimensional cuts through the lowest $^3\Sigma^-$ (full) and $^3\Pi$ (dashed) potential energy surfaces of AeNH systems at the linear Ae–HN arrangement.

difference is only 10.5 cm^{-1} and 3.6 cm^{-1}, respectively, which is *smaller* than the NH rotational constant. The difference may be understood in terms of the smaller *s-p* excitation energies for Ca and Sr (1.9 and 1.8 eV, respectively) compared to those of Be and Mg (both 2.7 eV).

Another important feature is the position of the crossing between the dispersion-bound $^3\Sigma^-$ state and the ion-pair $^3\Pi$ states. These are also tabulated in Table 2 and shown in Fig. 2 and 3. For Be–NH and Mg–NH the crossing occurs fairly high on the repulsive wall of the $^3\Sigma^-$ state, while for Ca–NH and Sr–NH it occurs at negative energies (in the potential well). This may be crucial for collisional properties. If the crossing is located at negative energies, as for Ca–NH and Sr–NH, the deep, strongly anisotropic ion-pair well may be accessed in low-energy collisions and is likely to result in strong inelasticity. On the other hand, if the crossing occurs at a high energy in a classically inaccessible region, as for Be–NH and Mg–NH, the deep ion-pair well is accessible only by tunneling through a barrier and may not have a strong effect on collisions.

For Ae–HN geometries, shown in Table 2, the crossing occurs high on the repulsive wall for all the AeNH systems.

C. Mg–NH interaction potential

As shown above, the BeNH and MgNH systems have potential energy surfaces that appear promising for sympathetic cooling. However, Mg has been successfully laser-cooled[36] whereas Be has not. We therefore focus in this section on developing a complete potential energy surface for interaction of Mg with NH($^3\Sigma^-$).

For bent geometries near the conical intersection, the two lowest triplet states of A'' symmetry are near-degenerate. Under these circumstances single-reference coupled-cluster calculations are inappropriate. One alternative, which we have previously applied for RbOH,[29,30] is to carry out multireference configuration interaction calculations including single and double excitations (MR-CISD). However, for MgNH the contribution from triple excitations is extremely large: for example, the well depths of the linear $^3\Sigma^-$ state is underestimated by 40% in RCCSD calculations. Because of this, we use SAPT(DFT) calculations to study nonlinear configurations of MgNH. It has recently been demonstrated[50,51] that SAPT(DFT) gives reasonably good results for Mg_2, NH–He and MgHe, and the polarizabilities and Van der Waals coefficients for Mg_2 are reproduced with an accuracy of a few percent. A further advantage of the perturbation theory is that, by starting from zeroth-order wavefunctions corresponding to neutral monomers, we produce diabatic potential energy surfaces corresponding to neutral Mg–NH without contamination from ion-pair states.

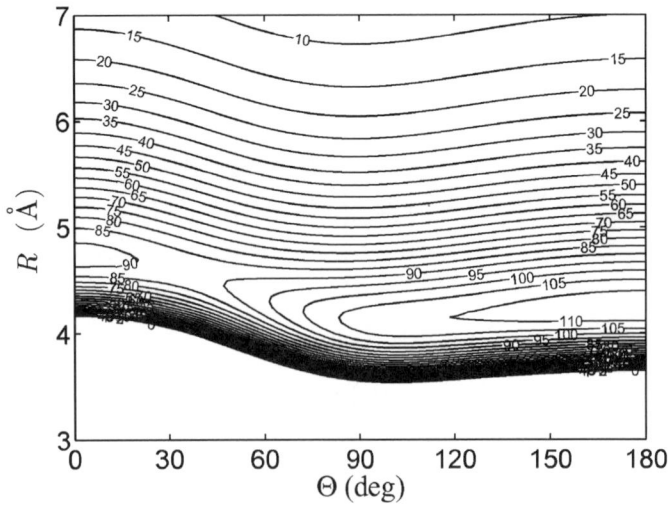

Fig. 4 Contour plot of the potential energy surface for the dispersion-bound state of Mg–NH, from SAPT(DFT) calculations (in cm^{-1}).

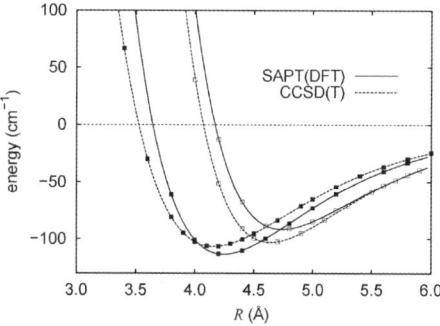

Fig. 5 Comparison of SAPT(DFT) and CCSD(T) interaction energies for linear MgNH configurations. Lines with black squares refer to Mg–NH geometries, while empty squares refer to Mg–HN geometries.

Fig. 4 shows the potential energy surface for Mg–NH obtained from SAPT(DFT) calculations. There are minima at both linear geometries and a saddle point between them, with a barrier of only 24 cm^{-1}. The agreement between CCSD(T) and SAPT(DFT) methods for linear configurations is very good for the Mg–NH geometry, where SAPT(DFT) overestimates the RCCSD(T) well depth by 6%. The agreement is slightly worse for the Mg–HN geometry, where SAPT(DFT) underestimates the well depth by 9% (see Fig. 5).

The ion-pair state does not reach negative energies until distances $R < 2.6$ Å. The dispersion-bound state is strongly repulsive at such distances. We have carried out multireference self-consistent field (MCSCF) calculations of the two states of $^3A''$ symmetry in the region of their avoided crossing for a range of angles using a cc-pVQZ basis set.[45,46] The lowest barrier for crossing onto the ion-pair state occurs for an angle $\sim 110°$ at a distance of $R \sim 2.4$ Å and an energy of +2200 cm^{-1} with respect to the atom–molecule threshold. The singlet ion-pair state is about 700 cm^{-1} shallower than the triplet near its equilibrium geometry and will therefore cross the dispersion-bound state at even higher energies.

The collision energies of importance to sympathetic cooling are in the range between 1 µK and 100 mK (0.07 cm^{-1}). At such energies we believe that the barrier separating the wells of the ion-pair and dispersion-bound states is wide and high enough to neglect the conversion from MgNH to Mg$^+$NH$^-$ and to perform collision calculations only on the dispersion-bound surface.

IV. Conclusions

In this paper we have presented an overview of the interaction potentials of alkali-metal and alkaline-earth atoms with NH molecules in their ground $^3\Sigma^-$ state. The interaction potentials of quartet states of AlkNH systems are strongly anisotropic, with deep wells at Alk–NH geometries. The bonding in the well region involves strong mixing of the s and p orbitals of the alkali-metal atom and is thus covalent in nature. For geometries close to Alk–NH configurations the quartet states are crossed by ion-pair doublet states in the energetically accessible region. Because of the presence of the ion-pair state, the lowest doublet adiabatic potential energy surface has a very deep well. The anisotropies for both doublet and quartet states are so large that it is unlikely that sympathetic cooling of NH by alkali-metal atoms will be successful for molecules in low-field-seeking states.

For alkaline-earth atoms the interaction potentials are much shallower and less anisotropic, especially for BeNH and MgNH. The sp mixing is much weaker and the bonding is dominated by induction and dispersion forces. For MgNH the anisotropy is comparable to or smaller than the rotational constant of NH, B_e. For CaNH and SrNH the anisotropy is rather larger, of the order of $4B_e$ for CaNH and $11B_e$ for SrNH. The relatively weak anisotropy raises the hope that alkaline-earth atoms could be used for sympathetic cooling of NH molecules.

The dispersion-bound $^3\Sigma^-$ states of AeNH systems are crossed by singlet and triplet ion-pair states $^1\Pi$ and $^3\Pi$. At nonlinear geometries the $^3\Sigma^-$ state becomes $^3A''$ and there is a component of the $^3\Pi$ state of the same symmetry. The absolute minimum thus has ion-pair character in all cases. For CaNH and SrNH the ion-pair state crosses the dispersion-bound state in the energetically accessible region, so that the adiabatic potential energy surfaces have a single-minimum structure with a deep potential well and strong anisotropy. For BeNH and MgNH, however, the crossing occurs on the repulsive wall of the potential of the dispersion-bound state and the deep ion-pair well is likely to be inaccessible in low-energy collisions. We have calculated a full 2-dimensional potential energy surface for MgNH and verified that the crossing occurs on the repulsive wall at all geometries.

The BeNH and MgNH systems are thus promising candidates for sympathetic cooling. In future work we will carry out collision calculations on MgNH to explore this further.

Acknowledgements

The authors are grateful to EPSRC and the Czech Science Foundation (grant No. QUA/07/E007) for funding under the collaborative projects QuDipMol and CoPoMol of the ESF EUROCORES Programme EuroQUAM. PS also acknowledges support from the Ministry of Education, Youth and Sports of the Czech Republic (Research project no. 0021620835).

References

1 R. V. Krems, *Phys. Chem. Chem. Phys.*, 2008, **10**, 4079.
2 D. DeMille, *Phys. Rev. Lett.*, 2002, **88**, 067901.
3 J. J. Hudson, B. E. Sauer, M. R. Tarbutt and E. A. Hinds, *Phys. Rev. Lett.*, 2002, **89**, 023003.
4 J. van Veldhoven, J. Küpper, H. L. Bethlem, B. Sartakov, A. J. A. van Roij and G. Meijer, *Eur. Phys. J. D*, 2004, **31**, 337.
5 T. Zelevinsky, S. Blatt, M. M. Boyd, G. K. Campbell, A. D. Ludlow and J. Yei, *ChemPhysChem*, 2008, **9**, 375.
6 J. M. Hutson and P. Soldán, *Int. Rev. Phys. Chem.*, 2006, **25**, 497.
7 K. M. Jones, E. Tiesinga, P. D. Lett and P. S. Julienne, *Rev. Mod. Phys.*, 2006, **78**, 483.
8 T. Köhler, K. Goral and P. S. Julienne, *Rev. Mod. Phys.*, 2006, **78**, 1311.
9 S. Jochim, M. Bartenstein, A. Altmeyer, G. Hendl, S. Riedl, C. Chin, J. Hecker Denschlag and R. Grimm, *Science*, 2003, **302**, 2101.
10 M. W. Zwierlein, C. A. Stan, C. H. Schunck, S. M. F. Raupach, S. Gupta, Z. Hadzibabic and W. Ketterle, *Phys. Rev. Lett.*, 2003, **91**, 250401.
11 M. Greiner, C. A. Regal and D. S. Jin, *Nature*, 2003, **426**, 537.
12 J. G. Danzl, E. Haller, M. Gustavsson, M. J. Mark, R. Hart, N. Bouloufa, O. Dulieu, H. Ritsch and H.-C. Nägerl, *Science*, 2008, **321**, 1062.
13 K.-K. Ni, S. Ospelkaus, M. H. G. de Miranda, A. Pe'er, B. Neyenhuis, J. J. Zirbel, S. Kotochigova, P. S. Julienne, D. S. Jin and J. Ye, *Science*, 2008, **322**, 231.
14 F. Lang, K. Winkler, C. Strauss, R. Grimm and J. Hecker Denschlag, *Phys. Rev. Lett.*, 2008, **101**, 133005.
15 J. M. Sage, S. Sainis, T. Bergeman and D. DeMille, *Phys. Rev. Lett.*, 2005, **94**, 203001.
16 E. R. Hudson, N. B. Gilfoy, S. Kotochigova, J. M. Sage and D. DeMille, *Phys. Rev. Lett.*, 2008, **100**, 203201.
17 M. Viteau, A. Chotia, M. Allegrini, N. Bouloufa, O. Dulieu, D. Comparat and P. Pillet, *Science*, 2008, **321**, 232.
18 J. Deiglmayr, A. Grochola, M. Repp, K. Mörtlbauer, C. Glück, J. Lange, O. Dulieu, R. Wester and M. Weidemüller, *Phys. Rev. Lett.*, 2008, **101**, 133004.
19 H. L. Bethlem and G. Meijer, *Int. Rev. Phys. Chem.*, 2003, **22**, 73.
20 H. L. Bethlem, M. R. Tarbutt, J. Küpper, D. Carty, K. Wohlfart, E. A. Hinds and G. Meijer, *J. Phys. B: At., Mol. Opt. Phys.*, 2006, **39**, R263.

21 J. D. Weinstein, R. deCarvalho, T. Guillet, B. Friedrich and J. M. Doyle, *Nature*, 1998, **395**, 148.
22 D. J. Larson, J. C. Bergquist, J. J. Bollinger, W. M. Itano and D. J. Wineland, *Phys. Rev. Lett.*, 1986, **57**, 70.
23 P. Blythe, B. Roth, U. Fröhlich, H. Wenz and S. Schiller, *Phys. Rev. Lett.*, 2005, **95**, 183002.
24 A. Ostendorf, C. B. Zhang, M. A. Wilson, D. Offenberg, B. Roth and S. Schiller, *Phys. Rev. Lett.*, 2006, **97**, 243005.
25 G. Modugno, G. Ferrari, G. Roati, R. J. Brecha, A. Simoni and M. Inguscio, *Science*, 2001, **294**, 1320.
26 M. T. Cvitaš, P. Soldán, J. M. Hutson, P. Honvault and J. M. Launay, *Phys. Rev. Lett.*, 2005, **94**, 200402.
27 M. T. Cvitaš, P. Soldán, J. M. Hutson, P. Honvault and J. M. Launay, *J. Chem. Phys.*, 2007, **127**, 074302.
28 P. Soldán and J. M. Hutson, *Phys. Rev. Lett.*, 2004, **92**, 163202.
29 M. Lara, J. L. Bohn, D. Potter, P. Soldán and J. M. Hutson, *Phys. Rev. Lett.*, 2006, **97**, 183201.
30 M. Lara, J. L. Bohn, D. E. Potter, P. Soldán and J. M. Hutson, *Phys. Rev. A*, 2007, **75**, 012704.
31 M. Tacconi, E. Bodo and F. A. Gianturco, *Theor. Chem. Acc.*, 2007, **117**, 649.
32 M. Tacconi, E. Bodo and F. A. Gianturco, *Phys. Rev. A*, 2007, **75**, 012708.
33 M. Tacconi, L. Gonzalez-Sanchez, E. Bodo and F. A. Gianturco, *Phys. Rev. A*, 2007, **76**, 032702.
34 T. Binnewies, G. Wilpers, U. Sterr, F. Riehle, J. Helmcke, T. E. Mehlstäubler, E. M. Rasel and W. Ertmer, *Phys. Rev. Lett.*, 2001, **8712**, 123002.
35 T. Mukaiyama, H. Katori, T. Ido, Y. Li and M. Kuwata-Gonokami, *Phys. Rev. Lett.*, 2003, **90**, 113002.
36 T. E. Mehlstäubler, K. Moldenhauer, M. Riedmann, N. Rehbein, J. Friebe, E. M. Rasel and W. Ertmer, *Phys. Rev. A*, 2008, **77**, 021402R.
37 P. S. Żuchowski and J. M. Hutson, *Phys. Rev. A*, 2008, **78**, 022701.
38 W. C. Campbell, E. Tsikata, H.-I. Lu, L. D. van Buuren and J. M. Doyle, *Phys. Rev. Lett.*, 2007, **98**, 213001.
39 S. Hoekstra, M. Metsala, P. C. Zieger, L. Scharfenberg, J. J. Gilijamse, G. Meijer and S. Y. T. van de Meerakker, *Phys. Rev. A*, 2007, **76**, 063408.
40 S. Y. T. van de Meerakker, R. T. Jongma, H. L. Bethlem and G. Meijer, *Phys. Rev. A*, 2001, **64**, 041401.
41 C. R. Brazier, R. S. Ram and P. F. Bernath, *J. Mol. Spectrosc.*, 1986, **120**, 381.
42 P. J. Knowles, C. Hampel and H. J. Werner, *J. Chem. Phys.*, 1993, **99**, 5219.
43 J. Čížek, *J. Chem. Phys.*, 1966, **45**, 4256.
44 R. D. Amos, A. Bernhardsson, A. Berning, P. Celani, D. L. Cooper, M. J. O. Deegan, A. J. Dobbyn, F. Eckert, C. Hampel, G. Hetzer, P. J. Knowles, T. Korona, R. Lindh, A. W. Lloyd, S. J. McNicholas, F. R. Manby, W. Meyer, M. E. Mura, A. Nicklass, P. Palmieri, R. Pitzer, G. Rauhut, M. Schütz, U. Schumann, H. Stoll, A. J. Stone, R. Tarroni, T. Thorsteinsson and H.-J. Werner, *MOLPRO, a package of* ab initio *programs designed by H.-J. Werner and P. J. Knowles, Version 2002.1*, 2002.
45 T. H. Dunning, Jr, *J. Chem. Phys.*, 1989, **90**, 1007.
46 M. A. Iron, M. Oren and J. M. L. Martin, *Mol. Phys.*, 2003, **101**, 1345.
47 I. S. Lim, P. Schwerdtfeger, B. Metz and H. Stoll, *J. Chem. Phys.*, 2005, **122**, 104103.
48 I. S. Lim, H. Stoll and P. Schwerdtfeger, *J. Chem. Phys.*, 2006, **124**, 034107.
49 S. F. Boys and F. Bernardi, *Mol. Phys.*, 1970, **19**, 553.
50 P. S. Żuchowski, R. Podeszwa, R. Moszynski, B. Jeziorski and K. Szalewicz, *J. Chem. Phys.*, 2008, **129**, 084101.
51 K. Patkowski and K. Szalewicz, *J. Phys. Chem. A*, 2007, **111**, 12822.
52 C. Adamo and V. Barone, *J. Chem. Phys.*, 1999, **110**, 6158.
53 D. J. Tozer and N. C. Handy, *J. Chem. Phys.*, 1998, **109**, 10180.
54 R. S. Mulliken, *J. Chem. Phys.*, 1955, **23**, 1833.
55 P. Soldán, M. T. Cvitaš and J. M. Hutson, *Phys. Rev. A*, 2003, **67**, 054702.

Continuous guided beams of slow and internally cold polar molecules

Christian Sommer, Laurens D. van Buuren, Michael Motsch, Sebastian Pohle, Josef Bayerl, Pepijn W. H. Pinkse* and Gerhard Rempe

Received 5th November 2008, Accepted 5th January 2009
First published as an Advance Article on the web 26th May 2009
DOI: 10.1039/b819726a

We describe the combination of buffer-gas cooling with electrostatic velocity filtering to produce a high-flux continuous guided beam of internally cold and slow polar molecules. In a previous paper (L.D. van Buuren *et al.*, *Phys. Rev. Lett.*, 2009, **102**, 033001) we presented results on density and state purity for guided beams of ammonia and formaldehyde using an optimized set-up. Here we describe in more detail the technical aspects of the cryogenic source, its operation, and the optimization experiments that we performed to obtain the best performance. The versatility of the source is demonstrated by the production of guided beams of different molecular species.

1 Introduction

The successful implementation of laser cooling of atoms in combination with evaporative cooling culminated in the spectacular realization of Bose–Einstein condensation more than a decade ago.[1–3] A natural next step seems to be the application of similar cooling techniques to molecules. Due to their rich internal structure, molecules can be employed to study phenomena which cannot be studied with atoms or are enhanced in molecular experiments.[4] Examples are long-range anisotropic electric dipole–dipole interactions in the case of polar molecules, cold collisions, where chemical reactivities are governed by quantum tunneling and resonances[5,6] or the measurement of the electron electric dipole moment (EDM)[7,8] in high-precision spectroscopy on heavy dipolar molecules. However, the complexity of the molecules makes it difficult to apply laser cooling techniques, due to the large amount of decay channels inherent in any electronic excitation.

For this reason different approaches for producing cold molecular gases are considered. These include so called indirect methods like photoassociation[9–11] or magnetic Feshbach resonances[12–14] to associate molecules from ultracold atomic samples. As these techniques have the disadvantage of being restricted to a few species and to smaller (mainly dimer) molecules, new methods to create cold samples from naturally occurring molecules are explored. Several direct methods have been developed so far: buffer-gas cooling,[15] electric,[16–18] magnetic[19,20] and optical[21,22] deceleration, rotating nozzles,[23] collisions of molecular beams[24] and collisions with moving surfaces,[25] molecules embedded in helium droplets[26] as well as velocity filtering by rotating mechanical filters,[27] and by electrostatic[28,29] or magnetic[30] guides.

Most experiments mentioned above will benefit from large and dense samples of cold molecules. The production of such samples is the main goal of these new direct methods. In our source, we have combined electrostatic velocity filtering and

Max-Planck-Institut für Quantenoptik, Hans-Kopfermann-Str. 1, 85748 Garching, Germany. E-mail: pepijn.pinkse@mpq.mpg.de

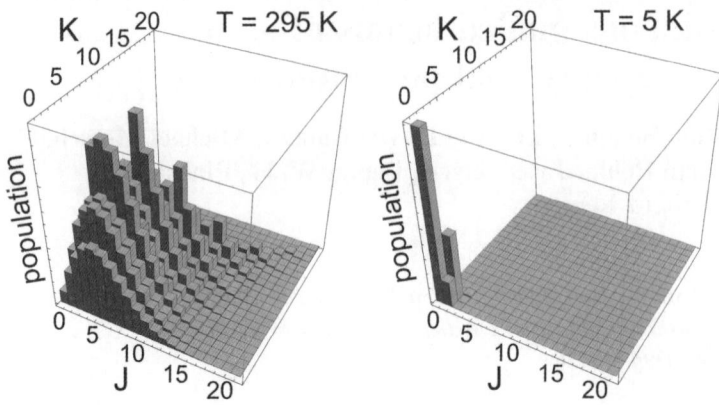

Fig. 1 Sketches of the state distributions of fully thermalized ND_3 ensembles at ~300 K (left) and at 5 K (right). J represents the rotational quantum number with K its projection on the molecular axis.

buffer-gas cooling to obtain a continuous beam of dense and internally cold molecules. Warm molecules are introduced into a cryogenic helium buffer-gas where all degrees of freedom are cooled by collisions with the helium atoms. As a result the number of populated states are strongly reduced. This is shown in Fig. 1, where state distributions of deuterated ammonia are presented for fully thermalized ensembles at room temperature and at a temperature of 5 K. Slow molecules are extracted out of the buffer-gas environment with an electrostatic quadrupole guide. In a recent experiment buffer-gas cooling was combined with a magnetic guide consisting of permanent magnets.[30] Buffer-gas cooling is a very general cooling method that can be applied to atoms and molecules alike to reach temperatures in the sub-Kelvin regime. In our set-up it is used as a cooling technique that delivers large quantities of slow and internally cold molecules into an electrostatic guide. In combination with laser depletion spectroscopy, applied in the near ultraviolet regime as decribed in ref. 31, the internal cooling was confirmed in our set-up for formaldehyde molecules.[32] In this paper, we present experiments performed with ND_3 molecules to optimize and characterize the performance of the cryogenic source. Several parameters of the set-up are varied to maximize the guided flux and internal cooling. To show the generality of the system, measurements with other species like trifluoromethane (CF_3H) and fluoromethane (CH_3F) are carried out.

The paper is organized in the following way. The set-up of the cryogenic source for cold polar molecules is presented in Section 2. In Section 3 the experimental control and data acquisition system is outlined. Section 4 presents the method to determine the extracted flux and density by the electrostatic guide. The steps taken to improve the performance of the system are discussed in Section 5. The guiding of different species is presented in Section 6 followed by the conclusions and outlook in Section 7. Additional information about the calibration of the gas densities in the cell is presented in the Appendix.

2 Experimental set-up

In our experiment, molecules from a room temperature source are injected into a cryogenic helium buffer-gas *via* a heated capillary. By collisions with the cold helium gas the internal (rotation) and external (motion) degrees of freedom of the molecules are cooled.† A fraction of the cold molecules is extracted by a quadrupole

---

† For ND_3 the lowest vibrational mode is at 749 cm^{-1}.[33] Therefore, only the ground-vibrational state is populated in these experiments.

guide[28,29] to an ultrahigh vacuum environment at room temperature, where they are available for further experiments (*e.g.* ref. 34 and 35)

The technical implementation of the cryogenic source is presented in detail in this section, accompanied by several drawings of the set-up. For more background information about buffer-gas cooling, references 15 and 36 can be consulted, whereas more information on velocity filtering can be found in refs 28, 29 and 37.

As shown in Fig. 2, the cryogenic source is located in a stainless steel vacuum vessel (height ∼60 cm and diameter ∼40 cm). It mainly consists of a cryogenic cooler to which two radiation shields and a gold-coated copper buffer-gas cell are mounted. The cell can be filled with helium gas *via* a gas supply line, which is thermally connected to the 5 K cooling stage of the refrigerator. Molecular gas can be introduced into the buffer-gas cell from a thermally isolated gas supply line. To extract the molecules from the buffer-gas cell, an aperture is made in the cell just in front of the electric quadrupole guide. Except for the turbo-molecular pump (∼550 l s^{-1}), all components are mounted to the top flange to allow for easy servicing of the system.

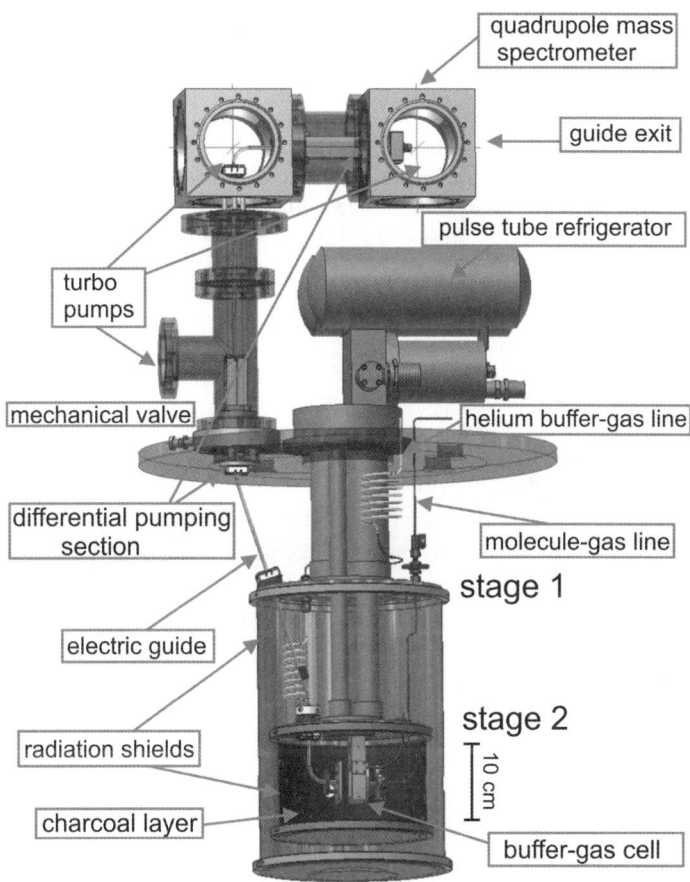

Fig. 2 Experimental set-up. The inner assembly is mounted directly to the refrigerator. With two capillaries, helium and molecular gases are injected into the buffer-gas cell. While the helium line is thermally connected to the first and the second stage of the pulse tube cooler, the molecular line is thermally isolated from these stages. With an electric quadrupole guide, molecules are extracted from the buffer-gas at the exit aperture of the buffer-gas cell and transported to a high-vacuum environment, positioned above the vessel. This part is separated from the vessel by differential pumping stages. During warm up, a mechanical valve can isolate these two regions completely.

2.1 Cryogenic system

To cool the buffer-gas cell to cryogenic temperature, we use a commercially available pulse-tube cooler (CRYOMECH PT410). For our purposes, a pulse-tube cooler is advantageous compared to cooling schemes based on cryogenic liquids. It is a closed system that delivers enough power to reach cryogenic temperatures in our set-up ($T_{min} \sim 3$ K) while vibrations are small (in the range of $\sim$20 µm). Other advantages are the fast cool down and warm up times ($\sim$2 hours cool down; $\sim$8 hours warm up). The two-stage design of the pulse-tube cooler allows for a configuration, where the cryogenic cell can be insulated from thermal radiation of the vacuum vessel as shown in Fig. 2. The first (upper) stage can maintain a temperature of $\sim$45 K with a heat load of 40 W and the second (lower) stage can reach temperatures less than $\sim$4 K if the heat load is reduced to less than 1 W. Therefore, round gold-plated copper radiation shields are installed on each stage of the cryogenic cooler to reduce the heat loads on both stages. To monitor the temperatures in our set-up we use silicon diode temperature sensors (LAKESHORE DT470) at various positions on the two stages. The sensors allow temperature measurements in the range of 1.4 K to 475 K with an accuracy of $\pm$0.5 K at low temperatures.

The inside of the radiation shield mounted to the second stage is covered with a layer of activated charcoal (Fig. 2 and Fig. 3). At cryogenic temperatures activated charcoal acts as a pump for helium. Large pumping speeds can be obtained due to its large effective area. Because of its simplicity it represents the optimal solution for our experiment. It is applied in our set-up to keep the pressure within the radiation shields as low as possible. From test experiments we concluded that for pressures lower than $\sim$10^{-6} mbar collisions of guided molecules with background gas are negligible. A reasonable vacuum is also required to prevent discharges between the guide electrodes. We used coconut-based granular activated carbon (Chemviron Carbon 207C). A surface of roughly 1000 cm^2 is covered with a layer of this material, resulting in a pumping speed of $\sim$5000 l s^{-1} at cryogenic temperatures. The gas (mainly helium) adsorbed by the charcoal is released during warm up of the cryogenic system after which it is pumped out of the vacuum vessel by the turbo-molecular pump.

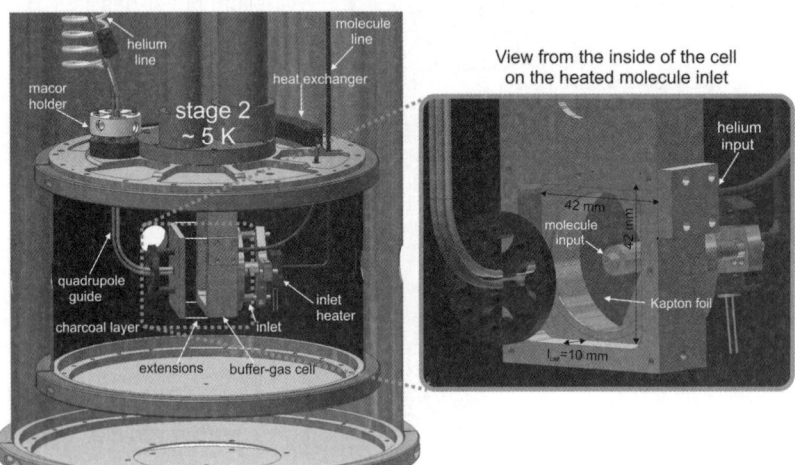

Fig. 3 The buffer-gas cell in the cryogenic environment. The helium line enters the copper cell from the side. The heated molecular line is connected to a special inlet assembly, which has a bad thermal connection to the buffer-gas cell. This allows for the heating of the inlet up to room temperature without affecting the temperature of the buffer-gas cell much. The length of the cell can be varied by placing 1 cm long copper extensions in between the main frame of the cell and the exit plate. Outside of the buffer-gas cell, a charcoal coated surface pumps away the helium and non-guided molecules. Note that the exit plate of the cell is omitted in the drawing on the right.

2.2 Buffer-gas cell and gas supply system

To bring the molecular gas as well as the helium gas into the cryogenic environment, a feed through for two 6 mm wide gas lines is installed on the top plate of the vessel (see Fig. 2). The buffer-gas line is attached to one of the feed through lines. The total buffer-gas line consists of four connected segments. For the first segment of the buffer-gas line, a stainless steel pipe with a 3 mm outer and 1 mm inner diameter is used and connected to the feed-through into the vacuum chamber. The stainless steel line contains several windings to increase the length, thereby reducing the heat conductance of the gas line. Since the buffer-gas needs to be cooled down on its way to the cell, the stainless steel segment of the line is connected to a subsequent copper line segment with equal dimensions. This copper part of the line is tightly connected to the first stage of the cryocooler for pre-cooling of the helium gas to 35 K. It then enters into the volume enclosed by the outer radiation shield, where it is connected to a stainless steel pipe with equal dimensions for thermal insulation. To cool the helium gas to 5 K, the stainless steel gas line is passed over into a copper segment, which is mounted to the second stage of the cryocooler with good thermal contact. From there, the helium gas line enters the inner radiation shield and is finally connected to the buffer-gas cell from the side (see Fig. 3). It is sealed with a thin indium foil and clamped *via* a small copper block to the cell wall.

The molecular gas line is heated to maintain a sufficiently high vapor pressure and to avoid freezing. For the molecules employed in this work, we used temperatures above 140 K (for ND_3 above 180 K). In most of the data presented in this paper, the molecular input capillary is kept at 295 K. Electric heaters and sensors (PT 100 and DT470) are placed at various positions to keep the line temperature fixed and prevent cold spots. To limit heat loads on the cell, the molecular gas line must be thermally disconnected from the cooling stages and the buffer-gas cell. The line is made of a 3 mm wide copper pipe that goes from the feed through to the second stage, where it is inserted into a gold-plated copper inlet with an inner diameter of 1 mm. A small Viton O-ring is used to seal the transition between the gas line and the inlet. To avoid heating by the molecular gas line, the inlet frame is clamped to the cell *via* small glass balls to minimize the contact surface. The conical inlet tip is glued on a 25 μm thick polyimide (Kapton) foil with a 1 mm opening. The Kapton foil forms the inlet front wall of the buffer-gas cell and reduces the heat load from the inlet to the copper cell due to its low thermal conductance. This guarantees that the heat conductance from the inlet to the buffer-gas cell is low enough to maintain cryogenic temperatures, even for a gas line temperature of $\sim$300 K.

The flow of the molecular and helium gas can be controlled *via* a gas handling system. To sensitively regulate the gas flows, electrically controlled needle valves adjust and maintain the inlet pressure with the assistance of a pressure gauge (capacitance gauges) mounted on the room temperature part of each line. The inner dimensions of the main body of the buffer-gas cell, which is mounted to the bottom plate of the cryostat, are $4.2 \times 4.2 \times (l_{cell} = 1)$ cm^3. To optimize performance the cell length l_{cell} can be varied. The buffer-gas cell can be shifted back and forth to allow changes of l_{cell} without altering the guide segment directly after the buffer-gas cell. To extend the cell length, additional 1 cm long extensions can be inserted. The exit aperture is formed by a 1 mm hole in a copper foil, which is glued on a Kapton foil to thermally disconnect it from the cell (see Section 5.3). This Kapton foil is clamped on the exit plate of the cell, which has a 10 mm opening.

2.3 Electric guide

To extract the molecules from the buffer-gas environment and obtain a pure sample of cold molecules, a bent electrostatic guide segment is placed in front of the exit aperture of the buffer-gas cell as shown in Fig. 3. The distance between the guide and the buffer-gas cell is around 1 mm to prevent discharges. A guide segment is

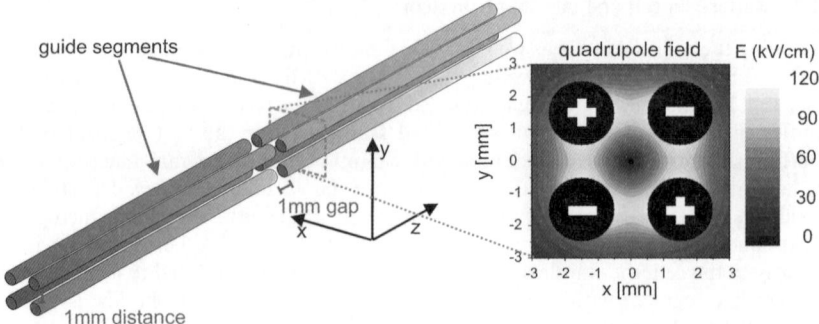

Fig. 4 Electrode arrangement. A guide segment consists of four stainless steel rods, each with a diameter of 2 mm, located at a distance of 1 mm to a neighboring electrode in a quadrupole configuration. The 1 mm gap between the guide segments keeps the guided molecules largely unaffected. The electric field distribution for ±5 kV on the electrodes is shown for a cross sectional area perpendicular to the propagation direction of the molecules.

made of four stainless steel rods, each with a diameter of 2 mm.[29] These electrodes are arranged in a way that a two-dimensional quadrupole field configuration is formed when positive and negative voltages are applied such that neighboring electrodes, separated by 1 mm, have opposite polarity (see Fig. 4). The electrodes are held by metallic holders which are mounted on an insulating ceramic material (Macor). High voltage is applied *via* Kapton insulated wires, connected to the holders. A bend of 90° with a 2.5 cm radius of curvature is made in the first quadrupole guide segment after less than 2 cm from the exit of the buffer-gas cell. The total guide consists of four quadrupole segments, separated by 1 mm gaps. Two segments are located in the vacuum vessel with a length of 12 cm for the first and 31.4 cm for the second segment. Two other segments with a length of 36.6 cm and 26.8 cm are in the high-vacuum region, placed above the vacuum vessel of the cryogenic source. These segments are used to deliver the guided beam to the detector. The last straight segment allows for depletion spectroscopy by overlapping the central axis of the guide with a laser beam.[32]

The high-vacuum region can be separated from the vessel by a valve. The valve consists of a stainless steel mechanical shutter which can be pushed through the 1 mm gap between the second and third guide segment. From below it is sealed by a Viton O-ring. The valve is used to separate the high-vacuum region from the vessel during warm up of the cryogenic part of the set-up. In this way a vacuum better than $\sim 10^{-8}$ mbar can be maintained, even when atmospheric pressure is applied inside the vessel. This enables us to adjust the source without polluting the high-vacuum region. To preserve a low background pressure ($< 10^{-9}$ mbar) in the detection chamber during measurements, when large amounts of helium gas are flown into the buffer-gas cell, we use two differential pumping stages. The first stage is placed above the mechanical valve and the other between the two cubes inside the high-vacuum region (see Fig. 2). The differential pumping sections consist of two Macor brackets which are clamped directly to the guide rods *via* metallic holders. It allows molecules to travel between the different vacuum regions only through the small opening in the middle of the quadrupole guide. To maintain a low pressure in the high-vacuum region, we use three turbo molecular pumps (two with pumping speeds of ~ 60 l s^{-1} and one with a pumping speed of ~ 210 l s^{-1} for the detection chamber) backed by another turbo pump and membrane pump. Pressures of 10^{-9} mbar in the region behind the valve and 10^{-10} mbar in the detection chamber can be obtained. Ion pressure gauges are installed in the vessel and near the detection unit.

For the detection of guided molecules we use a quadrupole mass spectrometer (QMS) with a crossbeam ion source, located in the detection chamber (PFEIFFER QMA 410). This analyzer provides an ion counter, which is used to determine the flux and density of the guided gas. The mass spectrometer is mounted on a translation stage to change the position of the ionization volume relative to the guide exit.

3 Data acquisition

To count the guided molecules with the QMS and to distinguish their signal from the background, we switch the high voltage on the guide on and off repeatedly. The difference in count rate reflects the contribution of guided molecules. Velocity distributions can be obtained from the time-of-flight (TOF) signal after switching on the high-voltage, as described in ref. 29. The first and second guide segments are connected to high-voltage transistor push-pull switches (BEHLKE HTS 151-03-GSM), each housed in a copper box to prevent the generated radio frequency from being radiated. TTL pulses from a pulse generator are applied to the switches to set the on (T_{ON} = 110 ms) and off (T_{OFF} = 100 ms) configurations. This timing scheme allows detection of molecular velocities down to about 10 m s^{-1}, well within the regime where noise and systematics become dominant. This is comparable to the effects found in our previous guiding experiments.[29] The third and fourth segment are permanently on high voltage to avoid pick-up currents on the QMS electrodes.

The QMS ion counter signal is sent to a TTL pulse shaper that generates ~100 ns TTL pulses if the amplitude of the incoming count signal exceeds a fixed threshold voltage. The threshold is used to eliminate noise. The TTL pulses are recorded by a MCS (Multi Channel Scalar) card, which creates histograms of the counts as a function of time. The trigger for the scalar card is generated by the pulse generator and starts 10 ms before the high voltage is switched on. The data from the temperature and pressure sensors are recorded with a LabVIEW interface.

4 Flux calibration

To determine the flux of guided molecules, the sensitivity of the QMS for the guided species has to be determined. To calibrate the QMS, we apply a constant flow of molecular gas (of the same species as the molecules that have been guided) to the detection chamber through a needle valve, mounted on a side flange. The density inside the chamber is determined from the pressure measured by an ion gauge (VARIAN Bayard/Alpert gauge). When the pressure has stabilized, we take a mass spectrum (1–100 amu) to determine the masses that contributed to the increase in pressure as well as the distribution of ionization fragments. The next step is to measure the count rate at the particular mass of the guided molecules. This is done for various densities in the detection chamber. For the mass spectrometer as well as the ion gauges the ionization probability of the investigated molecules has to be taken into account. Since the mass spectrometer is operated in a regime far from saturation for our guided molecules, the QMS signal is proportional to the density of the molecules in its ionization volume. We checked by variation of the ionization current that the ionization is not saturated, not even for slow molecules. Since the mass spectrometer detects only molecules that enter the ionization volume, we have to determine the fraction of the guided beam reaching the QMS ionization volume. By varying the position of the QMS ionization volume in the plane perpendicular to the guide exit, the beam spread is obtained (see Fig. 5). Since the width of the ionization volume is much smaller than the expanded beam, the beam shape can be resolved. The beam profile in the plane is well described by a Gaussian distribution. With this distribution the ratio of undetected to detected molecules is determined by a simple integration of the normalized Gaussian over its peak, with the integration boundaries given by the ionization region. Using this factor, the average velocity of the guided molecules (65 m s^{-1})‡ and the measured density, total fluxes of

‡ The average velocity is obtained from a velocity distribution measurement of the guided molecules.

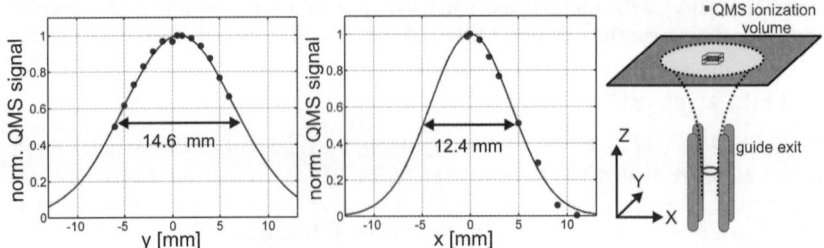

Fig. 5 Measurement of the beam shape in the *y*- and *x*-directions fitted by a Gaussian distribution. The distance between the guide exit and the ionization volume of the mass spectrometer is ∼25 mm along the *z*-direction. The profile in the *y*-direction is widened compared to the *x*-direction due to the centrifugal distortion of the trap potential in the bent path of the guide. For the beam in the guide a Gaussian profile is assumed with a standard deviation of $\sigma = 400$ µm, based on simulations presented in ref. 38.

guided molecules of $\sim 10^{10}$–10^{11} s^{-1} are obtained. Assuming a Gaussian profile of the beam in the guide, as supported by numerical simulations,[38] we obtain a peak density in the guide of $\sim 10^9$ cm^{-3}.[32]

5 Optimization of the source

To maximize the flux of slow molecules and to improve the cooling process, several parameters of the source are optimized. Different cell lengths, gas densities and exit apertures are tested in our set-up. In the end a compromise had to be found since optimal parameters for the flux, internal cooling and runtime of the source do not coincide perfectly with each other.

5.1 Flux optimization

In this section, the optimization of the flux with respect to the buffer-gas and molecular-gas densities is presented. Most of the discussed measurements are performed with a 2 cm long cell, although we also investigated other cell lengths. In Fig. 6

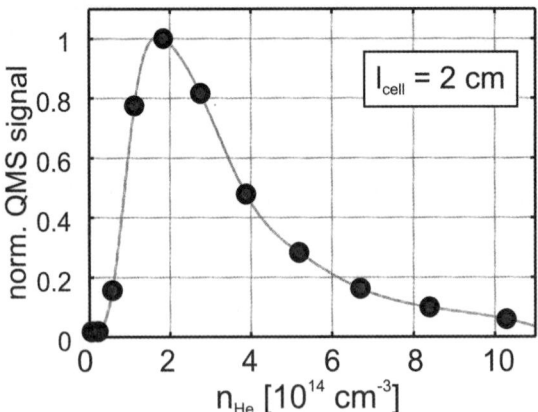

Fig. 6 Normalized QMS signal as a function of the helium density in a 2 cm cell with a fixed ND$_3$ density of $\sim 1.4 \times 10^{16}$ cm^{-3} at the input of the cell. The buffer-gas density is varied over about one order of magnitude. The rising edge indicates the region of insufficient cooling. The falling edge displays the effect of boosting as discussed in the text. At the peak position of n_{He} a flux of order of 10^{10}–10^{11} s^{-1} is obtained.[32] The curve connecting the data points is a guide for the eye.

the normalized QMS signal is shown for different buffer-gas densities. The characteristics of this curve can be explained in the following way. For low helium densities most of the molecules are still too fast and can escape from the guiding field of the bent quadrupole. With higher buffer-gas densities the cooling increases, and the signal rises to the maximum. If the density is raised to higher values the signal begins to drop. This is explained by an effect that is called "boosting".[36] Since the density of helium near the cell exit is higher than the molecular gas density at that position, molecules are most likely to collide with helium atoms. In this case, the probability of collisions between helium atoms and molecules, represented by the mean free path $\lambda_{mean} \sim 1/(n_{He} \times \sigma)$, predominantly depends on the density of the helium atoms n_{He}, where σ is the helium-molecular (elastic) cross section. If the exit aperture dimension (1 mm diameter) is larger than the mean free path, there is a high probability for molecules leaving the cell to be accelerated (boosted) in the forward direction. This happens due to collisions with helium atoms approaching the molecules from behind because of the pressure gradient and the large average velocity of helium ($\langle v_{He} \rangle \sim 230$ m s^{-1} at 5 K), much larger than the velocity of the molecules.

The effect of boosting is visible in velocity distributions obtained with high buffer-gas density. For very low buffer-gas densities (0.2×10^{14} cm^{-3}) the cooling is not optimal and can be improved (see Fig. 7). This can be seen from the peak position and the cut-off velocity, defined as the velocity where the measured velocity distribution is compatible with zero. Both shift to lower velocities when the helium density is increased to 0.6×10^{14} cm^{-3}. Information about the cooling for the internal degrees of freedom is obtained from the falling edge of the distribution at high velocities.

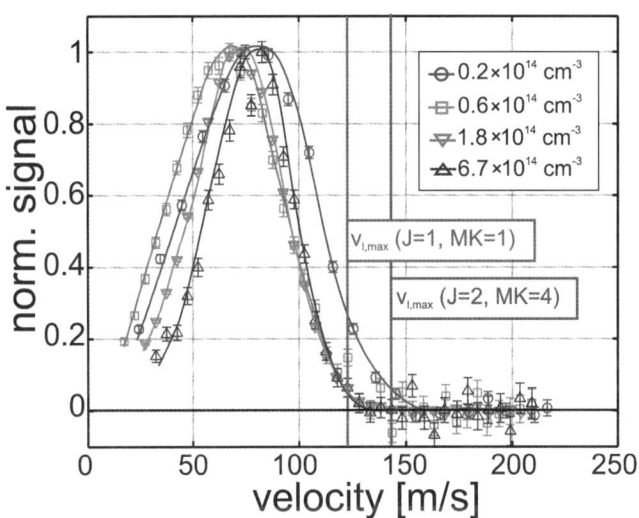

Fig. 7 Normalized velocity distribution for a 2 cm cell configuration with different helium densities. The (ND$_3$) molecular density at the cell entrance is fixed to $\sim 1.4 \times 10^{16}$ cm^{-3}. The cut-off velocities for the lowest-lying low-field-seeking rotation state ($J = 1$, $MK = 1$) and for the first higher-lying low-field-seeking state ($J = 2$, $MK = 4$) are marked by the straight lines. The data taken at the lowest density (0.2×10^{14} cm^{-3}) show contributions of velocities higher than the cut-off velocity for the ($J = 1$, $MK = 1$) state. This indicates the presence of higher-lying rotational states and, therefore, insufficient rotational cooling. Compared to this, the curve at 0.6×10^{14} cm^{-3} is shifted to lower velocities, indicating better cooling. The identical high-velocity sides of the highest two densities indicate that the internal state distribution has stabilized. However, the high helium densities above 0.6×10^{14} cm^{-3} do lead to 'boosting', the shift of the rising edge to higher velocities. The velocity distributions are obtained from TOF signals which are cut off if 99% of the steady-state value is reached, before noise and systematics become too dominant. The curves connecting the data points are guides for the eye.

This will be described in more detail in the upcoming Subsection 5.2. For helium densities higher than 0.6×10^{14} cm^{-3}, where collisions start to alter the shape of the velocity distribution, the rising edge is shifted to higher velocities, thereby supporting the argument of boosting.

From the ammonia density scans at fixed helium densities, in which various amounts of ND$_3$ are injected into the cell (Fig. 8), we can see that the molecular flux saturates. We believe that the saturation comes from the limited heat capacity of the helium gas. If more ammonia is injected, each molecule is simply cooled less. This explanation is supported by the fact that a larger helium density can support a larger ND$_3$ flux at a larger saturation density.

Similar examinations performed with a 1 cm and 3 cm cell length indicate that the maximal flux is obtained in a 2 cm long cell. This can be explained in the following way. In a shorter cell less collisions take place to cool down the molecules for equal buffer-gas densities. Therefore, higher buffer-gas densities have to be applied to a 1 cm long cell to obtain a maximal flux, in comparison to a 2 cm long cell. However, with a higher helium density the boosting is increased, which reduces the flux. For a 1 cm cell length the flux is by a factor of two lower than in the 2 cm configuration. If the cell is extended to 3 cm, we get a reduction of the flux due to the smaller ratio of exit aperture to inner surface of the cell. This means that less molecules find their way to the exit and a large fraction of them get stuck on the walls of the cell at cryogenic temperatures.

We come to the conclusion that the highest flux is obtained for a buffer-gas density of $n_{He} \sim 2 \times 10^{14}$ cm^{-3} and an ammonia density of $n_{ND_3} \sim 5 \times 10^{16}$ cm^{-3} in the inlet of a 2 cm long cell. These parameter settings are called the optimal flux settings (which can be different for other molecules) throughout the rest of

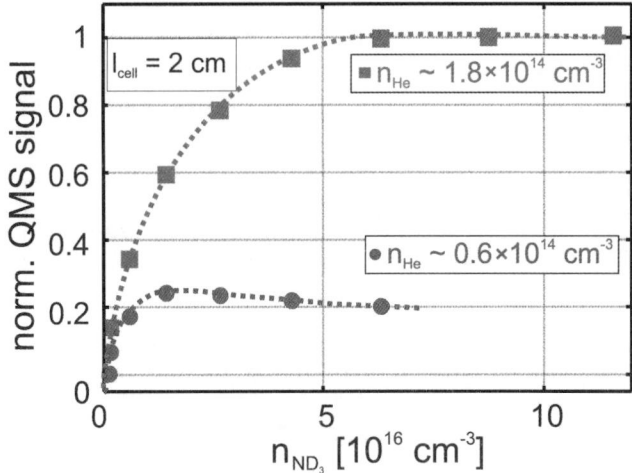

Fig. 8 Optimization of the ammonia density in the buffer-gas cell. The values for the ammonia density are given for the inlet tube exit. Since most of the ammonia is pumped away by the cell walls, the density close to the exit of the buffer-gas cell will be much less than its original value at the inlet. The helium density is fixed to 0.6×10^{14} cm^{-3} and 1.8×10^{14} cm^{-3}, the latter being the optimum value derived from the buffer-gas scan. In both cases the QMS signal reaches an optimum. For the low buffer-gas density case, we see that the signal starts to drop for higher ammonia densities, which might be explained by insufficient cooling power of the helium to cool down the molecules. For the higher buffer-gas density the signal saturates, whereby a signal drop is not visible even for much higher ND$_3$ densities. For all other measurements the ammonia density was set to a value of $\sim 1.4 \times 10^{16}$ cm^{-3}. At this setting a lower flux is obtained, but it allows for longer measuring times due to less ice formation. The dashed curves are guides for the eye.

this paper. For most of the measurements we used $n_{ND_3} \sim 1.4 \times 10^{16}$ cm^{-3} to have longer stable running periods as discussed in Subsection 5.3.

5.2 Optimization of internal cooling

From the velocity distributions for different buffer-gas densities we can extract information about the internal cooling of the molecules. This is possible because the cut-off velocities in the longitudinal direction (highest velocity that is populated in the velocity distribution) depend mainly on the radius of curvature of the bend in the guide ($R = 2.5$ cm) and the state-dependent Stark shift of the molecules. The Stark shift of ammonia is given by the approximation

$$\Delta W^s = \pm \sqrt{(W_{inv}/2)^2 + \left(\mu |\vec{E}| \frac{MK}{J(J+1)}\right)^2}, \quad (1)$$

which is valid for electric fields $|\vec{E}|$ in the range 0–100 kV cm^{-1}.[39] The dipole moment of ammonia $\mu = 1.5$ D, J represents the main rotation quantum number with its projections K on the molecular symmetry axis and M on the axis of the external electric field, and $W_{inv} = 0.053$ cm^{-1} is the inversion splitting of ND$_3$. We can determine the cut-off velocities[29]

$$v_{l,max} = \sqrt{\Delta W^s(E_{max}) \frac{R}{r \cdot m}}, \quad (2)$$

with the inner guide radius $r = 1.12$ mm specifying the location of the electric field maximum $E_{max} \sim 90$ kV cm^{-1} for ± 5 kV electrode voltage (see Fig. 4) and the mass of the molecule $m = 20$ amu. The cut-off velocity is a marker for the state with the highest Stark shift of the guided molecules. To obtain information on the cooling process for the internal degrees of freedom, we compare the calculated cut-off velocities (Table 1) with measured velocity distributions (see Fig. 7). For low temperatures, large populations in low-lying rotational states are expected. As can be seen from Fig. 7, the fraction of the normalized distributions above the cut-off velocity of the lowest-lying guidable rotational state ($J = 1$, $MK = 1$) is reduced if the helium density is increased above 0.2×10^{14} cm^{-3}. If the helium density is increased above 0.6×10^{14} cm^{-3} the cut-off velocity does not seem to change any more. Therefore we come to the conclusion that the population of higher rotational states and especially of those with a higher Stark shift than ($J = 1$, $MK = 1$) is negligibly small in these cases.

Table 1 Lowest lying low-field-seeking rotational states of ND$_3$ molecules up to $J = 2$. The Stark shifts of the rotation states are calculated with eqn (1) and (2) for an electric field strength of ~ 90 kV cm^{-1}, which corresponds to ± 5 kV on the electrodes. The maximal longitudinal velocities are evaluated for the same settings. The rotational constants of ammonia, which are needed to calculate the rotational energies, are taken from ref. 33

J	$|K|$	$|M|$	E_{rot}/cm^{-1}	Stark energy/cm^{-1}	$v_{l,max}$/m s^{-1}
1	1	1	8.29	1.13	123
2	1	1	28.85	0.38	71
2	1	2	28.85	0.76	100
2	2	1	22.88	0.76	100
2	2	2	22.88	1.51	142

In summary, the velocity distributions presented in Fig. 7 indicate that the molecules are rotationally cooled in the buffer-gas environment of a 2 cm set-up. Since this is just a qualitative measure of internal cooling, we established a depletion spectroscopy set-up, as described in ref. 31, for the cryogenic source. In this experiment, we obtained quantitative results for the population of internal states of formaldehyde. For this molecular species an ~80% pure beam was obtained at the optimal flux setting of n_{He}.[32] The disadvantage of the depletion method is, however, that different laser frequencies and possibly even different laser systems are required for different molecular species. The measurement of the cutoff velocity, in contrast, is simple and can readily be performed for any guidable molecule.

5.3 Runtime optimization

The measurement time in our set-up is limited by the sticking of molecular gas to the walls of the buffer-gas cell, which happens long before the saturation of the charcoal by helium gas. The charcoal can pump helium gas for several days in the density regimes we use in our system. However, we observe a steady decrease of the flux of guided molecules reaching half of the maximum in ~5 hours, which we attribute to frozen molecules clogging the entrance and the exit aperture of the cell. The ice formation has been observed directly after the installation of viewing ports in the vacuum vessel and radiation shields. Especially in the exit aperture, large blocks of molecular ice appear after several hours of operation; for a smaller opening of ~0.5 mm instead of 1 mm this is barely 30 min. The ice reduces the diameter of the exit aperture and with it the amount of molecules that can enter the guide. Since the warm up of the complete set-up takes ~8 hours, a different solution had to be found.

We use a Nichrome heater wire (10 Ω m^{-1}), attached with cyanoacrylate based fast-acting glue to a 4 mm diameter copper foil that has a 1 mm opening in the middle. The heater plate is glued to a Kapton foil with the wires pointing into the cell. The coil is additionally covered with Stycast 2850 epoxy for better adherence and thermal connection. A Pt100 temperature sensor is glued to the copper plate as well. It is needed to carefully monitor the temperature. Since the heater is glued to the Kapton foil and has a very low thermal contact with the copper frame we can easily alter the temperature of the exit aperture. From Fig. 9 we see that the signal indeed recovers each time a heater pulse is applied to the exit aperture. However, we see that the measurement periods after recovery get smaller. We attribute this to the growing ice layer within the cell. After 9 hours of measurement with

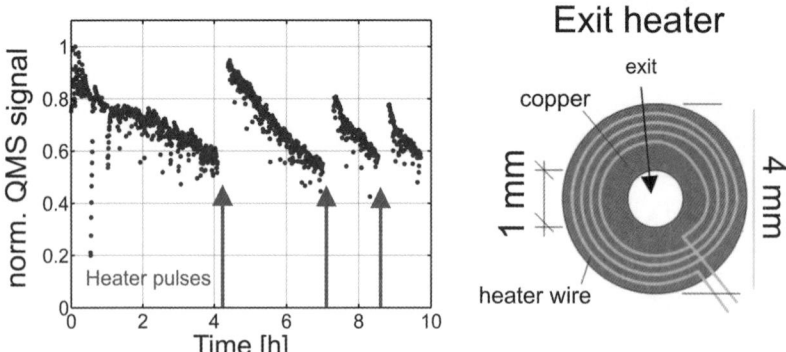

Fig. 9 Left: ND$_3$ guiding signal as a function of time showing the decrease in signal due to ice formation for a period of ~9 hours and three de-icing heater pulses. Right: a sketch of the heater plate, which is placed on the exit aperture of the buffer-gas cell. The plate consists of a 0.06 mm thin copper foil with a 0.22 mm thick heater wire attached to it. The signal can be restored several times, when heater pulses with ~0.3 W of ~1 min duration are applied.

optimal flux setting of the helium density and an ammonia density of ~1.4 × 10^{16} cm^{-3} at the inlet of the cell, we approach the time limit for continuous operation. At this ammonia density we then have ~50% of the maximum flux (see Fig. 8). Of course we could run at higher ammonia densities, but at the cost of shorter running times due to faster ice formation. After warming up the whole cell to release the frozen molecular gas, the initial flux can be fully recovered and the system is ready for operation again.

6 Cold beams of different molecular species

Most of our measurements are performed with deuterated ammonia (ND$_3$) and formaldehyde (H$_2$CO) because these molecules have been used in our previous room-temperature experiments[29,31] and because ND$_3$ has also been employed in buffer-gas cooling experiments.[41] Besides, as already mentioned in section 5.2, H$_2$CO has transitions in the near ultraviolet, with which we have quantified the purity of the produced beam using depletion spectroscopy.[32] These data gave direct evidence of internal cooling of the molecules by collisions with the buffer-gas. To demonstrate that our cryogenic source is a versatile tool for generating cold molecules, we also investigated the production of cold guided beams consisting of other symmetric top molecules such as trifluoromethane (CF$_3$H) and fluoromethane (CH$_3$F). The masses, dipole moments and cut-off velocities of the lowest guidable state ($J = 1$, $MK = 1$), calculated with eqn 2, for the molecules employed in this work are listed in Table 2.

Table 2 The masses, dipole moments and maximal longitudinal velocities of deuterated ammonia ND$_3$, fluoromethane CH$_3$F and trifluoromethane CF$_3$H

	Mass/amu	Dipole moment/D	$v_{l,max}^{(J=1,|MK|=1)}$/m s^{-1}
ND$_3$	20	1.5	123
CH$_3$F	34	1.86	105
CF$_3$H	70	1.65	69

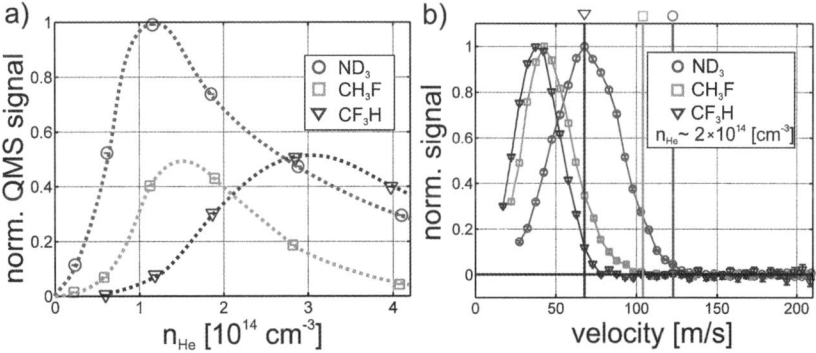

Fig. 10 (a) Buffer-gas scans for ND$_3$, CH$_3$F and CF$_3$H, normalized to the peak value of the ND$_3$ curve. For higher masses the optimal value of the buffer-gas scan is shifted to higher densities since more collisions are needed to slow down the heavier molecules. The dashed curves in (a) are guides to the eye. In (b) the velocity distributions for the three different species at a similar buffer-gas density ($n_{He} \sim 2 \times 10^{14}$ cm^{-3}) are compared. The cut-off velocities of the distributions are shifted to lower velocities for higher masses. The vertical lines indicate the calculated positions of the cut-off velocities for the $|J = 1, MK = 1\rangle$ low-field seeking state.

The measurements with CF$_3$H and CH$_3$F are performed without any change to the source except for the gas handling system. Due to the higher vapor pressure of these molecules, the molecular gas line was only heated to $T \sim 140$ K instead of $T \geq 180$ K for ND$_3$. In Fig. 10a buffer-gas scans are presented for the different species. To cool down heavy molecules, more collisions with helium are needed than for light molecules. Therefore, the position of the peak of the buffer-gas scan for the heavier CF$_3$H is shifted to higher n_{He} as compared to the other lighter molecules. The QMS signals obtained for CF$_3$H and CH$_3$F are only a factor of two lower than the ND$_3$ signal. Since we calibrated the QMS only for ND$_3$, we did not determine absolute numbers for the guided fluxes of the other species. However, we can conclude from these high signals that fluxes and densities comparable to those for ND$_3$ are feasible with this set-up, showing the universality of our source. In these measurements, we did not compensate for the change in the conductance of the molecular gas line due to the different molecular masses and different gas line temperatures. This change causes the molecular densities in the inlet to vary for the different data sets, but we expect no significant effects on the results. Note the shift in the peak position of the buffer-gas scan of ND$_3$ in Fig. 10a in comparison to Fig. 6. This probably results from a rebuild of the buffer-gas cell between these two measurements. As observed from several rebuilds, the peak position can shift by $\sim(\pm1) \times 10^{14}$ cm^{-3} after such an intervention. Therefore, buffer-gas scans are performed after each rebuild of the system to relocate the optimal setting for n_{He}.

The velocity distributions of the three different species are shown in Fig. 10b for a fixed helium density of $n_{He} = 2 \times 10^{14}$ cm^{-3}. The vertical lines indicate the cut-off velocities of the lowest guidable state of each molecule. The velocity distributions of both ND$_3$ and CH$_3$F are compatible with zero at the cut-off velocities of the lowest guidable state. This is not the case for CF$_3$H as discussed later. Since ΔW^s for states with $J^2 = MK$ increases with J for symmetric top molecules in the first order according to,[39]

$$\Delta W^s = \mu E \frac{MK}{J(J+1)}, \qquad (3)$$

the CH$_3$F and ND$_3$ data (for which eqn 1 is valid) indicate that the beams consist mainly of states with $J = 1$. Large populations in higher J states would show higher cut-off velocities due to contributions of states with $J^2 = MK > 1$. In other words, Fig. 10b shows that the ND$_3$ and CH$_3$F molecules are rotationally cooled by collisions with the buffer-gas. For ND$_3$, the onset of rotational cooling is demonstrated in Fig. 7b, by comparing the shapes and cut-off velocities of velocity distributions taken at different buffer-gas densities.

We also performed measurements at different buffer-gas densities for CF$_3$H and CH$_3$F. Fig. 11 shows two velocity distributions for both CH$_3$F and CF$_3$H. One distribution is taken at the optimal flux setting of n_{He} and the other for a lower value of n_{He}. The data taken at low n_{He} display a marginally smaller boosting than the data taken at the optimal flux settings for both gases. Although the low n_{He} data of CH$_3$F exhibit smaller boosting, the data show larger contributions at high velocities. This is explained by population in higher rotational states for which the cut-off velocities are larger than for the ($J = 1, MK = 1$) state. We have shown in ref. 32 that the internal cooling for H$_2$CO is already near its maximal value at the optimal flux setting of n_{He} and also the velocity distributions of ND$_3$ and CH$_3$F indicate qualitatively that most of the molecules are in the lowest guidable state ($J = 1, MK = 1$) at this setting. Therefore, we could expect that the CF$_3$H molecules are rotationally cold at their optimal flux setting of n_{He} as well. However, comparing data taken at different buffer-gas densities does not reveal any effect of internal cooling as shown in Fig. 11b. This can be explained by the small rotational constants of CF$_3$H ($A_0 = 0.19$ cm^{-1} and $B_0 = 0.35$ cm^{-1},[42]) causing population in higher states even at temperatures around 5 K. Since ΔW^s converges to μE for high J states

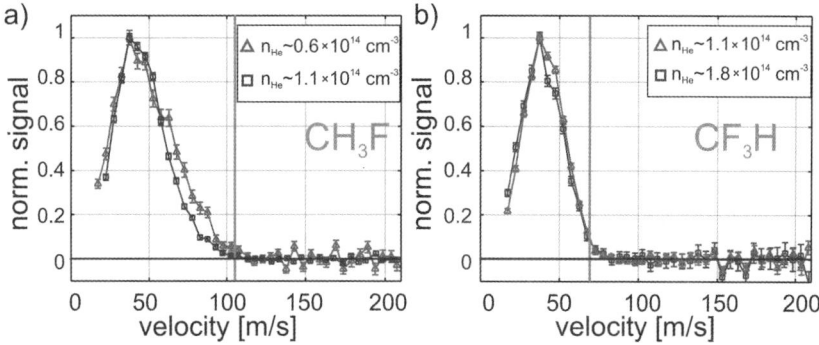

Fig. 11 Velocity distributions of CH_3F (a) and CF_3H (b), normalized to the maximum value. Helium densities resulting in maximum flux and below the optimal value are used to illustrate the effect of cooling. In the case of CH_3F the data taken at low $n_{He} \sim 0.6 \times 10^{14}$ cm^{-3} contain a marginally smaller boosting than the data taken at the optimal flux settings of $n_{He} \sim 1.1 \times 10^{14}$ cm^{-3}. Even with this smaller boosting, the low n_{He} data show higher contributions at high velocities. This indicates population in higher rotational states for which the cut-off velocities are higher than for the $J = 1$, $MK = 1$ state. The vertical lines indicate the calculated positions of the cut-off velocities for the ($J = 1$, $MK = 1$) low-field seeking state. A similar behavior is shown for ND_3 in Fig. 7. For CF_3H we do not observe such differences between velocity distributions taken at different n_{He}, except for a very small difference in boosting. This can be explained by the small rotational constants of CF_3H, causing populations in higher states even at temperatures around 5 K.

with $J^2 = MK$ (see eqn 3), a distinct shift towards smaller cut-off velocities is only visible between molecules populating mainly $J = 1$ states and molecules populating states with higher J. Therefore, we conclude that although the guided beam of CF_3H is not pure, it is likely to consist of a few states only with low rotational quantum numbers and low rotational energy. To increase the purity even more, one could deplete higher rotational states if light for the correct transitions is available. This would, however, cause an overall reduction of the flux. A better option would be to reduce the temperature of the buffer-gas even further. For CF_3H, temperatures around 1 K would suffice to increase the purity strongly. The cooling of CaH and CaF molecules in buffer gases with temperatures below 1 K has already been demonstrated.[15,43]

In summary, we were able to produce cold guided beams of several different molecular species. Our data suggest that these beams consist of molecules in low rotational state(s), where the number of populated states depends on the density of rotational states at low temperature.

7 Conclusion and outlook

With our cryogenic source we produce continuous guided beams of slow and internally cold polar molecules. The characteristics of the source are examined with respect to variations in the molecular density, helium buffer-gas density and cell length. From these studies, optimized settings for the system are found to produce maximal fluxes. Besides, signatures of internal cooling are obtained from the velocity distributions taken at different buffer-gas densities. The latter has been confirmed by depletion spectroscopy measurements on the guided beam emitted from the cryogenic source.[32] By comparing the cooling and guiding characteristics for four different molecular gases, we showed that our source is applicable to a wide variety of molecular species.

With the current set-up further experiments can be carried out. For example, molecules in the ro-vibrational ground state could be guided by applying alternating electric fields to the guide.[40] In this experiment one could use depletion spectroscopy for detection. The laser depletion technique itself could be used as a tool to further purify the guided beam. To increase the guided flux in an electrostatic quadrupole guide, Raman pulses or microwave fields could pump molecules from the high-field-seeking, and therefore not guidable ground state, into the first excited low-field-seeking state. This could be done in the small gap between the exit of the cell and the first guide segment. With an extension to a large-volume electric trap,[34] collision studies could be realized. Slow molecules from the same or different species could be brought into collision under controlled conditions.[35] With the ability to choose the rotational state of the molecules from our source, it would be possible to measure state-selective cross sections. Our cryogenic source could also be used as a new starting point for further cooling schemes.

Appendix: gas line calibration

In this appendix, we present the calibration of the helium and molecular gas lines. The calibrations are used to convert values from the room-temperature pressure gauges into densities in the buffer-gas cell. The measurement is done at room temperature with nitrogen (instead of helium, ammonia or other molecular gases) since the pumping efficiency of the turbo pump (on the bottom of the vessel) is specified by the manufacturer for nitrogen gas (S_{pump} = 550 l s^{-1} for N$_2$). We have removed the radiation shields and the buffer-gas cell in this experiment so that the conductance is mainly determined by the gas line itself. The nitrogen gas is fed through one of the gas lines into the vacuum vessel and is pumped by the turbo molecular pump. A pressure gauge in the gas handling system, another one in the vessel and a mass spectrometer mounted on the side of the vessel are used for detection. The mass spectrometer is needed to determine which masses contribute to the rise in pressure. With the pressure detected in the vessel we can determine the flux into the chamber via

$$\Phi = \frac{p_{vessel} S_{pump}}{k_B T_{room}},$$

where p_{vessel} is the pressure in the vessel, T_{room} = 293 K and k_B the Boltzmann constant. From this the conductance of the line can be obtained by taking into account the pressure in the gas handling system p_{gas}. This gives $C_{line} = \Phi k_B T_{room}/p_{gas}$. Next, correction factors for the different temperatures and masses must be applied. As a result the helium line conductance is

$$C_{line}^{He} = C_{line} \sqrt{\frac{5[K]}{295[K]} \times \frac{28[amu]}{4[amu]}}.$$

Similar correction factors are required for the molecular gas line. With the conductance of the 1 mm aperture C_{exit} estimated from[44] $C_{exit} = A\sqrt{R_0 T/(2\pi M_m)}$ with M_m the molar mass of molecular nitrogen used for calibration, the universal gas constant R_0 and the aperture area A at T = 5 K (valid for the molecular regime), we determine the values for the densities in the cell from $n_{cell} = \Phi/C_{exit}$.

References

1 M. H. Anderson, J. R. Ensher, M. R. Matthews, C. E. Wieman and E. A. Cornell, *Science*, 1995, **269**, 198.

2   C. C. Bradley, C. A. Sackett, J. J. Tollett and R. G. Hulet, *Phys. Rev. Lett.*, 1995, **75**, 1687; C. C. Bradley, C. A. Sackett, J. J. Tollett and R. G. Hulet, *Phys. Rev. Lett.*, 1997, **79**, 1170, Erratum.
3   K. B. Davis, M. O. Mewes, M. R. Andrews, N. J. van Druten, D. S. Durfee, D. M. Kurn and W. Ketterle, *Phys. Rev. Lett.*, 1995, **75**, 3969.
4   J. Doyle, B. Friedrich, R. V. Krems and F. Masnou-Seeuws, *Eur. Phys. J. D*, 2004, **31**, 149.
5   D. Herschbach, *Rev. Mod. Phys.*, 1999, **71**, 411.
6   N. Balakrishnan and A. Dalgarno, *Chem. Phys. Lett.*, 1999, **341**, 652.
7   E. A. Hinds, *Phys. Scr. T*, 1997, **70**, 34.
8   M. R. Tarbutt, H. L. Bethlem, J. J. Hudson, V. L. Ryabov, V. A. Ryzhov, B. E. Sauer, G. Meijer and E. A. Hinds, *Phys. Rev. Lett.*, 2004, **92**, 173002.
9   K. M. Jones, E. Tiesinga, P. D. Lett and P. S. Julienne, *Rev. Mod. Phys.*, 2006, **78**, 483.
10 U. Schlöder, C. Silber and C. Zimmermann, *Appl. Phys. B*, 2001, **73**, 801.
11 A. J. Kerman, J. M. Sage, S. Sainis, T. Bergeman and D. DeMille, *Phys. Rev. Lett.*, 2004, **92**, 33004.
12 T. Kohler, K. Goral and P. S. Julienne, *Rev. Mod. Phys.*, 2006, **78**, 1311.
13 S. Inouye, J. Goldwin, M. L. Olsen, C. Ticknor, J. L. Bohn and D. S. Jin, *Phys. Rev. Lett.*, 2004, **93**, 183201.
14 C. A. Stan, M. W. Zwierlein, C. H. Schunck, S. M. F. Raupach and W. Ketterle, *Phys. Rev. Lett.*, 2004, **93**, 143001.
15 J. D. Weinstein, R. de Carvalho, T. Guillet, B. Friedrich and J. M. Doyle, *Nature*, 1998, **395**, 148.
16 H. L. Bethlem, G. Berden and G. Meijer, *Phys. Rev. Lett.*, 1999, **83**, 1558.
17 H. L. Bethlem, F. M. H. Crompvoets, R. T. Jongma, S. Y. T. van de Meerakker and G. Meijer, *Phys. Rev. A*, 2002, **65**, 53416.
18 S. Y. T. van de Meerakker, H. L. Bethlem and G. Meijer, *Nat. Phys.*, 2008, **4**, 595.
19 E. Narevicius, A. Libson, C. G. Parthey, I. Chavez, J. Narevicius, U. Even and M. G. Raizen, *Phys. Rev. Lett.*, 2008, **100**, 93003.
20 N. Vanhaecke, U. Meier, M. Andrist, B. H. Meier and F. Merkt, *Phys. Rev. A*, 2007, **75**, 31402R.
21 R. Fulton, A. I. Bishop and P. F. Barker, *Phys. Rev. Lett.*, 2004, **93**, 243004.
22 R. Fulton, A. I. Bishop, M. N. Shneider and P. F. Barker, *Nat. Phys.*, 2006, **2**, 465.
23 M. Gupta and D. Herschbach, *J. Phys. Chem. A*, 1999, **103**, 10670.
24 M. S. Elioff, J. J. Valentini and D. W. Chandler, *Science*, 2003, **302**, 1940.
25 E. Narevicius, A. Libson, M. F. Riedel, C. G. Parthey, I. Chavez, U. Even and M. G. Raizen, *Phys. Rev. Lett.*, 2007, **98**, 103201.
26 F. Stienkemeier and A. F. Vilesov, *J. Chem. Phys.*, 2001, **115**, 10119.
27 S. Deachapunya, P. J. Fagan, A. G. Major, E. Reiger, H. Ritsch, A. Stefanov, H. Ulbricht and M. Arndt, *Eur. Phys. J. D*, 2008, **46**, 307.
28 S. A. Rangwala, T. Junglen, T. Rieger, P. W. H. Pinkse and G. Rempe, *Phys. Rev. A*, 2003, **67**, 043406.
29 T. Junglen, T. Rieger, S. A. Rangwala, P. W. H. Pinkse and G. Rempe, *Eur. Phys. J. D*, 2004, **31**, 365.
30 D. Patterson and J. M. Doyle, *J. Chem. Phys.*, 2007, **126**, 154307.
31 M. Motsch, M. Schenk, L. D. van Buuren, M. Zeppenfeld, P. W. H. Pinkse and G. Rempe, *Phys. Rev. A*, 2007, **76**, 61402R.
32 L. D. van Buuren, C. Sommer, M. Motsch, S. Pohle, M. Schenk, J. Bayerl, P. W. H. Pinkse and G. Rempe, *Phys. Rev. Lett.*, 2009, **102**, 033001.
33 G. Herzberg, *Infrared and Raman Spectra of Polyatomic Molecules*, Van Nostrand Reinhold, Inc., New York, 1945.
34 T. Rieger, T. Junglen, S. A. Rangwala, P. W. H. Pinkse and G. Rempe, *Phys. Rev. Lett.*, 2005, **95**, 173002.
35 S. Willitsch, M. T. Bell, A. D. Gingell, S. R. Procter and T. P. Softley, *Phys. Rev. Lett.*, 2008, **100**, 43203.
36 S. E. Maxwell, N. Brahms, R. deCarvalho, D. R. Glenn, J. S. Helton, S. V. Nguyen, D. Patterson, J. Petricka, D. DeMille and J. M. Doyle, *Phys. Rev. Lett.*, 2005, **95**, 173201.
37 M. Motsch, L. D. van Buuren, C. Sommer, M. Zeppenfeld, G. Rempe, and P. W. H. Pinkse, arXiv, 0809.1728v1.
38 T. Junglen, PhD Thesis, Max-Planck-Institut für Quantenoptik, 2006.
39 C. H. Townes and A. L. Schawlow, *Microwave Spectroscopy*, Dover Publications, Inc., New York, 1975.
40 T. Junglen, T. Rieger, S. A. Rangwala, P. W. H. Pinkse and G. Rempe, *Phys. Rev. Lett.*, 2004, **88**, 223001.

41 D. R. Willey, R. E. Timlin Jr., C. D. Ruggiero and I. A. Sulai, *J. Chem. Phys.*, 2004, **120**, 129.
42 W. A. Wensink, C. Noorman and H. A. Dijkerman, *J. Phys. B: At. Mol. Phys.*, 1979, **12**, 1687.
43 K. Maussang, D. Egorov, J. S. Helton, S. V. Nguyen and J. M. Doyle, *Phys. Rev. Lett.*, 2005, **94**, 123002.
44 J. M. Lafferty, *Foundations of Vacuum Science and Technology*, John Wiley & Sons, Inc., New York, 1998.

General discussion

Professor Julienne opened the discussion:
This question is addressed to all of the speakers in this session. It concerns the role of hyperfine structure associated with the presence of nuclear spins in many small molecules of interest. It is clear that when the collision energy is very low, as in the microkelvin domain of evaporatively cooled or quantum degenerate atomic gases, that such spin structure plays a very important role, since it has a much larger energy scale than typical thermal energy $k_B T$ values. The hyperfine and Zeeman spin substructure plays a crucial role in determining the properties of tunable near-threshold resonances that are so important in controlling low energy collisions of atoms. We thus might expect such structure to play a similar crucial role in molecular collisions.

The question I have is both for the experimental and theory groups working on collisions involving molecules in the cold domain, say on the order of 1 K and below, but not necessarily in the "deep ultracold" microkelvin domain. Hyperfine and Zeeman structure can clearly be important on this energy scale, since hyperfine splittings are typically on the order of GHz or larger, and $E/h = 1$ GHz corresponds to an approximate E/k_B of 50 mK. Of course, it is often simpler to think about doing theory without the complicating influence of nuclear spin structure. This is especially true if one is working with larger molecular species with several nuclear spins present. But such spin structure can lead to resonances that can be important in the energy ranges accessible to general cooling methods such as buffer gas cooling or Stark deceleration.

So I am interested in hearing from those doing either experiments or theory in the mK to K range as to the significance of nuclear spin hyperfine and Zeeman structure in this domain. What kind of opportunities or problems will such structure cause with larger molecules? Will it be possible to do realistic theory for such species, since it is already hard enough to do theory that accounts for electron spin and orbital angular momenta? It seems to me that it will be essential to do so.

Dr van de Meerakker replied: From the experimental point of view, in some cases the produced samples of cold molecules are prepared in only one single hyperfine quantum state. In our Stark deceleration experiments using OH radicals, for instance, the OH radicals are selected in the $F = 2$ hyperfine state of the $J = 3/2$, f rotational ground state. Measurements of hyperfine-structure-resolved state-to-state cross sections are certainly feasible in our experiments, although a dedicated narrow-band detection laser system would be required. It is noted that in other systems, like for instance CO molecules, there is no hyperfine structure, and the complicating influence of nuclear spin is absent.

Dr Groenenboom responded: Tscherbul *et al.*[1] reported a calculation of molecular hyperfine structure effects on collisions of molecules at collision energies up to 1 K. This paper clearly demonstrates that the hyperfine structure of molecules may lead to a manifold of Feshbach resonances that significantly modify the scattering dynamics at collision energies 0.1–1 K. These resonances can be tuned by both electric and magnetic fields.

The anisotropic (dipole–dipole) part of the hyperfine interaction couples the ground and the second excited rotational states of $^2\Sigma$ molecules, which leads to inelastic transitions between different Zeeman levels. As shown in the above reference, the hyperfine depolarization can be quite efficient for heavy polar molecules like YbF. The mechanism of this process is similar to that of the spin depolarization that occurs in collisions of $^3\Sigma$ molecules due to the spin–spin interaction.[2]

1 T. V. Tscherbul, J. Klos, L. Rajchel, and R. V. Krems, *Phys. Rev. A*, 2007, **75**, 033416.
2 R. V. Krems, H. R. Sadeghpour, A. Dalgarno, D. Zgid, J. Klos and G. Chalasinski, *Phys. Rev. A*, 2003, **68**, 051401.

Professor Hinds answered: In a Stark decelerator the density of molecules is low enough that one does not need to worry about unintended collisions between the molecules being decelerated. From this point of view the hyperfine interaction has no role to play during the Stark deceleration process. However, hyperfine interactions can cause a re-ordering of levels if the strength of the Stark interaction changes between low (compared with hyperfine interaction) and high. In such a region of electric field strength there are usually avoided crossings and these can lead to the loss of molecules travelling through the decelerator. In that case, one would design the decelerator so that the electric field is not allowed to drop to such a low level anywhere on the beam line. The same physics can also work in one's favour, with the hyperfine structure stabilising states that would otherwise be lost through Majorana transitions.

Dr Küpper remarked: Hyperfine structure is quite important in typical experiments. We have, for example, observed the crossing of two hyperfine states in the lower lambda level of the rovibronic ground state of OH radicals using the Alternating Gradient decelerator.[1] In these experiments, we typically switch between high-voltage and a small finite potential of the same polarity in order to avoid Majorana transitions at zero field. In our experiments on OH we have seen a sharp signal raise – a real threshold – when we increase this lower potential, such that the field on the beam axis "sweeps" over the field strength at which a real crossing of levels from different hyperfine manifolds ($F = 1$ and 2) occurs (see Fig. 9 of reference 1). We have also discussed this in terms of the parity mixing due to the Stark effect.[1]

1 Wohlfart *et al.*, *Phys. Rev. A*, 2008, **78**, 033421.

Dr Lara commented: The importance or not of introducing the hyperfine interaction in the calculations, when considering collisions at very low kinetic energies, has been repeatedly raised. For a hard worker in the field of reaction dynamics, the subject can sound non-realistic. However, even if the hyperfine interaction may be essential in the long range region, its magnitude is negligible (in comparison with kinetic energies and chemical potentials) once the system enters the short range.

As a consequence, two different spatial regions can be distinguished and frame transformation procedures may be applied. We used this strategy in our study of the Rb + OH inelastic collision, previously mentioned.[1] Other methods, like the hyperspherical quantum reactive scattering developed by Launay,[2] which explicitly splits the dynamics into short and long range regions, can allow an easy implementation of this idea in the reactive context.

1 M. Lara, J. L. Bohn, D. E. Potter, P. Soldan, and J. M. Hutson, *Phys. Rev. A*, 2007, **75**, 012704.
2 P. Honvault and J. M. Launay, *J. Chem. Phys.*, 2001, **114**, 1057.

Professor Hutson replied: Including hyperfine structure in full close-coupling calculations for atom–molecule and molecule–molecule collisions is a daunting task, especially when the potential is strongly anisotropic. The time taken for coupled channel calculations is proportional to the cube of the number of channels, and each nuclear spin multiplies the number of channels by a factor of $2I + 1$ (or a little less when there are additional symmetries). This is not too much of a problem for collisions with helium, even in magnetic fields, but for more complicated cases the number of channels needed can be enormous. For example, we have investigated ultracold Rb + OH collisions, including both electron and nuclear spins, in a basis set adapted to allow magnetic fields.[1,2] The system is strongly anisotropic and there are 5 coupled electronic states. Even a poorly converged basis set for this system required over 23 000 channels, which is computationally impractical. However, we were able to make the system just tractable by using a frame transformation procedure that neglected hyperfine couplings at short range but reintroduced them at long range (where the rotational basis set could be reduced). Even this gave over 1000 channels.

Rb + OH is in some ways quite simple: it has only two nuclei with spin and H is spin-half. In general, therefore, full close-coupling calculations in systems with several nuclear spins will be extremely expensive. Yet, as you say, hyperfine splittings can be quite large on the energy scales involved in ultracold collisions and including them is absolutely essential if we are to understand the resonance structures that can be used for control. This is not a problem that can be solved by waiting for computer power to increase, so it is essential to develop approximate methods that can handle these very large coupled-channel problems.

1 M. Lara, J. L. Bohn, D. E. Potter, P. Soldán and J. M. Hutson, *Phys. Rev. Lett.*, 2006, **97**, 183201.
2 M. Lara, J. L. Bohn, D. E. Potter, P. Soldán and J. M. Hutson, *Phys. Rev. A*, 2007, **75**, 012704.

Professor Meijer said: Hyperfine structure of molecules can often be used to advantage as it enables us to trap molecules in electrostatic traps with a true zero-field at the centre, yet without Majorana transitions. We have recently performed experiments on metastable CO for instance. In the most abundant isotopomer, $^{12}C^{16}O$, there is no hyperfine structure and there is a degeneracy between the $M = 0$ and $M\Omega = -1$ level in zero electric field, which can lead to trap loss. In $^{13}C^{16}O$, however, this degeneracy is lifted and the trappable state is exclusively low-field seeking at all fields. The states actually never come closer together than 53 MHz, and trap loss due to non-adiabatic transitions ('Majorana' transitions) can be completely avoided.

Professor Kotochigova opened the discussion of Dr van de Meerakker's paper: How many rotational states are formed in your chemical reaction?

Dr van de Meerakker replied: In our experiment, we have studied the state-to-state inelastic scattering of OH, prepared in the $J = 3/2$, $\Omega = 3/2$, f state, with Xe atoms as a function of the collision energy. The inelastic scattering channels that populate the $J = 3/2$, e state, both λ doublet components of the $J = 5/2$ state, and the $J = 1/2$, e state ($\Omega = 1/2$) of OH were studied. These are the scattering channels with the largest cross sections. Other states will, in principle, be populated as well, but these were not studied in our experiment. In the calculations that were performed by Dr Groenenboom, a larger number of inelastic channels were included.

Dr Groenenboom responded: We studied Xe–OH inelastic collisions. In the experiment four channels were probed, $F_1(3/2e)$, $F_1(5/2e)$, $F_1(5/2f)$, and $F_2(1/2e)$. At higher collision energies, up to 400 cm^{-1} in the experiment, other channels are open, but the cross sections for these channels were calculated to be very small.

Dr Tarbutt asked Dr van de Meerakker: In your paper you argue that molecular collision studies are better performed in a beam environment than in a trap, even though the interaction time in the trap is orders of magnitude larger. I think the argument is based on the difficulty of detection inside the trap environment, and the fact that the density in the trap will tend to be lower. Regarding the detection efficiency, I would have thought this could be as high in a trap as in a beam. For example, if cw laser-induced fluorescence is used as a detector, the trap could be filled with molecules in one state, and the probe laser tuned to excite molecules from another state. Then, the state-changing collision will be accompanied by a photon scattered from the probe laser, giving a sensitive and potentially background-free detection scheme. In cases like this, do you think the trap could be a better environment than the beam, or is the beam always better?

Dr van de Meerakker answered: For selected systems and appropriate detection schemes one could in principle exploit the long lifetimes of traps to experimentally observe collisions. Cw laser-induced fluorescence would indeed allow for the background-free detection of inelastically scattered molecules. However, there are a few problems with cw detection in traps from an experimental point of view. In elecrostatic traps, for instance, the UV photons from the laser could cause electrical breakdown, hampering the use of a cw laser to detect molecules inside the trapping potential. In addition, the (narrow band) cw detection laser will only be resonant with a subset of the trapped molecules due to broadening of spectral lines in the trapping potential (particularly important as most state-changing collisions will deposit recoil energy in the motion of the molecules, that will subsequently have larger excursions in the trapping potential). If experimental methods are developed that circumvent these obstacles (for instance by using magnetic traps and spectral lines that show a similar Zeeman effect in the ground and excited state), the cw quantum-state specific detection of molecules inside the trapping potential will offer interesting prospects for studying collisions between molecules in the trap.

Dr Groenenboom asked: For elastic cross sections is measurement in the trap in a storage ring preferable to the molecular beam experiment?

Dr van de Meerakker replied: In crossed beam experiments, one is usually not very sensitive to elastic scattering. The scattered products are deflected from their initial trajectory, but the deflection is too small to be detected in most experiments. Inelastic (state-to-state) processes can be sensitively detected in crossed beam experiments, provided that the state purity of the beam is high and a state-specific detection scheme is used. In this case, the scattered products are detected background-free. In storage rings, the argument is rather different. A storage ring does not bring any advantages for inelastic scattering. Inelastically scattered molecules in general do not experience the required potentials to stay confined in the ring after the collision. Storage rings, however, are ideal to measure the total (elastic + inelastic) scattering cross section, that is inferred from a measured loss of molecules stored in the ring. Even for elastic scattering, the deflection of the molecule's trajectory will in most cases be large enough to expel the molecule from a stable ring orbit.

Professor Hinds enquired of Dr van de Meerakker: Fig. 6(b) in your paper shows a stationary target of trapped molecules placed in a molecular synchrotron, the idea being to study the reaction between these target molecules and projectile molecules orbiting in the synchrotron. With the best available densities for target and projectile, how many product molecules would you hope to make per second?

Dr van de Meerakker answered: Also in this case, the benefits of a molecular synchrotron are probably best exploited when the loss of stored molecules is measured that results from elastic, inelastic or reactive scattering with the target atoms or molecules. When a density for the target species of 10^{11} cm^{-3} is assumed (as can, for instance, be obtained for atoms in a magneto optical trap), we expect that approximately 0.01%–0.1% of the molecules have left the synchrotron per round trip when one stationary target is placed in the ring. After having revolved the ring 100 times, which takes about one second in the present synchrotron, we thus expect a depletion of approximately 1 to 10%.

Professor Sims asked: The comparison of low temperature rate coefficients or low energy cross sections for reactive or inelastic collisions with the results of *ab initio* theoretical calculations ideally requires that the rate coefficient or cross section be measured on an absolute basis. Rate coefficients may be measured on an absolute basis under so-called pseudo-first-order conditions, as applied to cold collisions in the case of the CRESU (Cinétique de Réaction en Ecoulement Supersonique Uniforme or Reaction Kinetics in Uniform Supersonic Flow) technique.[1] However, it is notoriously difficult[2] to carry out absolute cross-section measurements when using two crossed molecular beams. What are the prospects for the measurement of absolute reactive and inelastic cross sections at very low collision energies using the ground-breaking techniques described in your article?

1 A. Canosa, F. Goulay, I. R. Sims and B. R. Rowe, Gas Phase Reactive Collisions at Very Low Temperature: Recent Experimental Advances and Perspectives, in *Low Temperatures and Cold Molecules*, ed. I. W. M. Smith, World Scientific, Singapore, 2008, p. 55.
2 G. Scoles, in *Atomic and Molecular Beam Methods*, ed. G. Scoles, Oxford University Press, New York, Oxford, 1988, vol. 1, p. 4.

Professor Meijer responded: The measurement of absolute reactive and inelastic cross-sections at very low collision energies will indeed face the same problems as the measurement of these quantities in conventional crossed molecular beam set-ups, as long as no independent method is available to accurately determine the absolute densities in either one of the beams. An advantage of using the Stark-decelerated beams is, however, that the phase-space distributions of both interacting beams are known (and can be manipulated) with extreme precision.

Dr van de Meerakker added: For the measurements of absolute cross sections, our experimental approach to study low temperature molecular collisions using crossed Stark decelerated molecular beams suffers from the same limitations as all other crossed beam experiments. Accurate absolute cross sections can only be obtained from the experimental data if the densities of the molecules in the beams are accurately known. Probably the most accurate experimental methods to measure the density of molecules in a molecular beam are laser absorption based techniques like cavity ring down spectroscopy. The implementation of these methods, however, require higher molecular densities than that are currently available in Stark decelerated beams.

Dr Żuchowski commented: Dr van de Meerakker has described the prospects for collision experiments using two Stark-decelerated molecular beams. Another interesting experiment is to use a velocity-controlled (and perhaps decelerated) beam to study collisions of molecules with ultracold atoms. There are experiments under way by Heather Lewandowski and coworkers at JILA, in which a decelerated beam of ND_3 molecules in the low-field-seeking component of the $J = 1$, $K = 1$ state collide with magnetically trapped Rb. The low-field seeking state correlates with the upper inversion component of the $J = 1$, $K = 1$ state at low field.

We have recently carried out quantum scattering calculations for low-energy collisions of Rb atoms with ND_3 molecules, with the ND_3 molecules initially in the upper inversion level.[1] The interaction potential for Rb–ammonia is shown in Fig. 1. It is strongly anisotropic, with anisotropy of thousands of cm^{-1}.[2] Because of this, we initially expected strongly inelastic collisions. However, the inelastic cross sections turned out to be lower than expected, much below the unitarity limit given by Langevin capture theory (see Fig. 2). As new channels open up at higher collision energies, the total inelastic cross section in the energy range studied (0–100 cm^{-1}) is dominated by far by the transition to the lower inversion component of the $J = 1$, $K = 1$ state, rather than transitions to excited J and/or K.

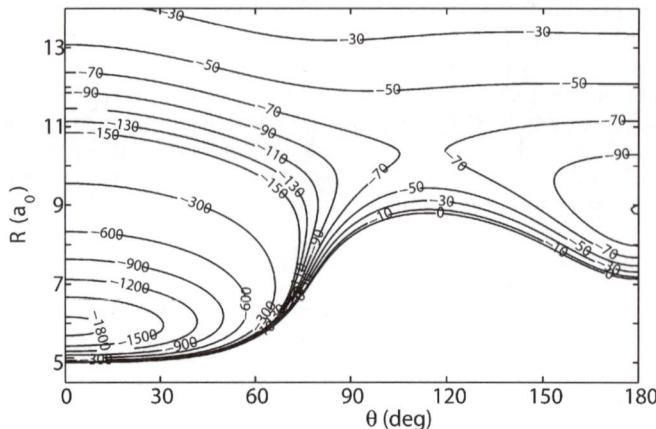

Fig. 1 The interaction potential for Rb–NH$_3$ from CCSD(T) calculations, averaged over the C_3 axis of ammonia. Contours are labelled in cm^{-1}. $\theta = 0$ corresponds to the approach toward the lone pair of NH$_3$.

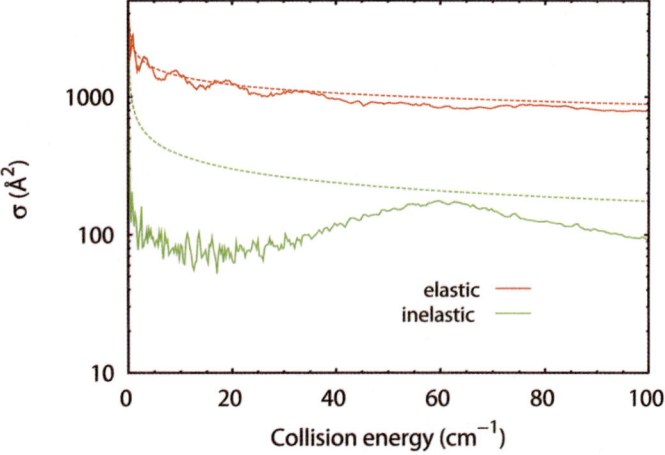

Fig. 2 Elastic and total inelastic cross sections for Rb–ND$_3$ scattering from the upper component of the inversion doublet for the $J = 1$, $K = 1$ state. The smooth dashed lines show the result of the semiclassical background formula for elastic cross sections and of Langevin capture theory for the total inelastic cross sections.

The low-energy inelastic cross sections have strong resonant structure, which is washed out for higher collision energies (above about 20 cm^{-1}). Moreover, the analysis of individual partial wave contributions to the inelastic cross sections shows that they are dominated by resonances and the contribution from the nonresonant background is small (see Fig. 3).

1 P. S. Żuchowski and J. M. Hutson, *Phys. Rev. A*, 2009, **79**, 062708.
2 P. S. Żuchowski and J. M. Hutson, *Phys. Rev. A*, 2008, **78**, 022701.

Dr Stoecklin contributed: An important issue when producing such ultracold complexes by tuning zero energy Feshbach resonances is to predict their lifetimes. We did such a study for He–NH($^3\Sigma$) collisions in the presence of parallel electric

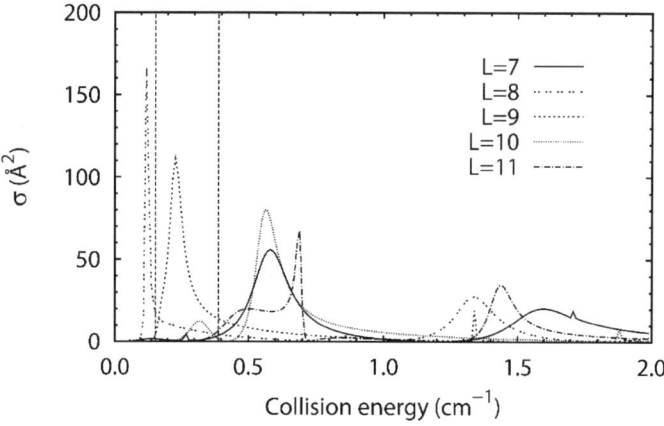

Fig. 3 Contributions from partial waves $L = 7\text{–}11$ to the inelastic cross sections for Rb–ND$_3$. The dashed vertical lines indicate the positions of centrifugal barriers for $L = 7$ and $L = 11$.

and magnetic fields using the Smith Q matrix, which is the tool of choice for the determination of resonance lifetimes. We focused on two magnetically tuned zero energy Feshbach resonances identified for this system by the Hutson team[1] around 7500 (ZEFR1) and 15000 Gauss (ZEFR2). We performed Close Coupling calculations using the extension of the formal theory of collisions between atom and molecule in the presence of a magnetic field of Krems and Dalgarno[2], developed by Tscherbul and Krems[3], to include the interaction with a superimposed parallel electric field. When the electric field is applied, each level of the field dressed $^3\Sigma$ diatomic molecule is subjected to a second order Stark effect, which results in quadratic shifts of the molecular energy levels $\Delta E_i = a_i \varepsilon^2$ as a function of the electric field strength ε. As a result, for a selected value ε of the applied electric field, the value of the magnetic field which needs to be applied to obtain a given Zero Energy Feshbach Resonance (ZEFR) is also quadratically shifted from $\Delta B_{\text{Res}}\varepsilon = B_{\text{Res}}\varepsilon - B^0_{\text{Res}} = a\varepsilon^2$ (where B^0_{Res} denotes the value of the magnetic field associated with the position of the ZEFR in the absence of any applied electric field, and $B_{\text{Res}}\varepsilon$ denotes its value for a given applied electric field. We report in Fig. 4a and 4b for ZEFR1 and ZEFR2 respectively, the results of our Close Coupling calculations for $\frac{\Delta B_{\text{Res}}(\varepsilon)}{\varepsilon}$ as a function of ε for the fundamental state of the field dressed diatomic molecule ($\alpha = 1$) and for the projection $M_T = -1$ of the total angular momentum. As can be seen in these figures, the magnetic field strengths associated with these two ZEFR are both shifted to higher values when a superimposed parallel electric field is applied (a is positive). However, the situation where the position of the ZEFR is shifted to lower values of the magnetic field when the electric field is applied (a is negative) can also be encountered as illustrated for the ZEFR2 in Fig. 4b. This behaviour is obtained when performing calculations this time for the first excited level of the field dressed diatomic molecule ($\alpha = 2$) and for $M_T = 0$ and has important consequences for the lifetimes as shown in Fig. 5 and 6. When the initial state of the field dressed diatomic molecule is the fundamental one ($\alpha = 1$), we found that the sign of a is positive for both ZEFR1 and ZEFR2. In both cases, the eigenlifetimes are maximum at zero field and decrease when the electric field strength increases as can be seen in Fig. 5 and in the lower panel of Fig. 6. Conversely for the first excited state ($\alpha = 2$) of field dressed NH, the sign of a is negative for ZEFR2. We see in the higher panel of Fig. 6 that, this time, the value of the eigenlifetime at zero field is a minimum and the eigenliftime increases when the applied electric field strength increases. This behaviour can be predicted using a model based on the scattering length approximation, to be published.[4] These results suggest that the lifetimes of magnetically tuned zero energy Feshbach resonances could be efficiently tuned using a superimposed parallel electric

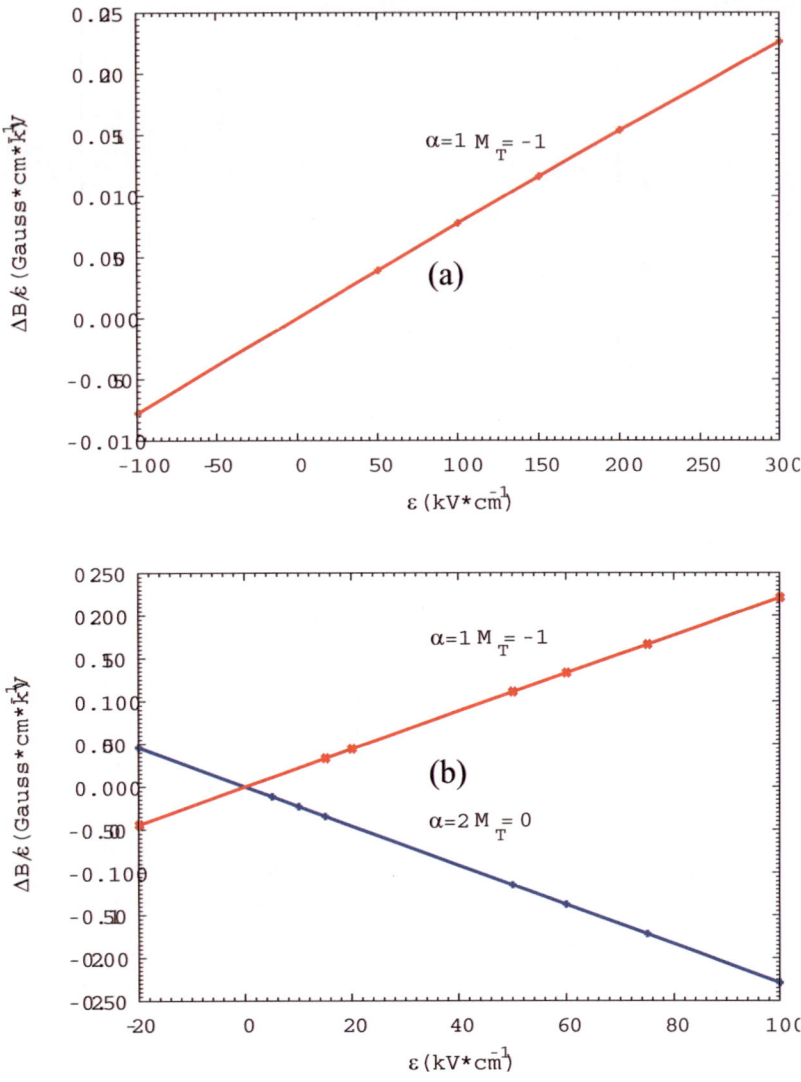

Fig. 4 Close Coupling Stark shift divided by the electric field strength of the positions of the two magnetically tuned zero energy Feshbach resonances ZEFR1 and ZEFR2 as a function of the electric field strength for the ^{3}He–NH($^3\Sigma$) collisions. The panels (a) and (b) are dedicated to ZEFR1 and ZEFR2, respectively. The value of the projection M_T of the total angular momentum on the space fixed z axis and the initial state of the field dressed diatomic molecule α are indicated on each curve.

field. They will have, in any case, to be checked in future experiments but may be generalized to any molecule submitted to a quadratic Stark effect.

1 M. L. Gonzáles-Martinez and J. M. Hutson, *Phys. Rev. A.*, 2007, **75**, 022702.
2 R. V. Krems and A. Dalgarno, *J. Chem. Phys.*, 2004, **120**, 2296.
3 T. V. Tscherbul and R. V. Krems, *J. Chem. Phys.*, 2006, **125**, 194311.
4 T. Stoecklin, arXiv, 2009, 0904.0108.

Dr Groenenboom asked Dr Stoecklin: What is the meaning of the vertical bars in your figures?

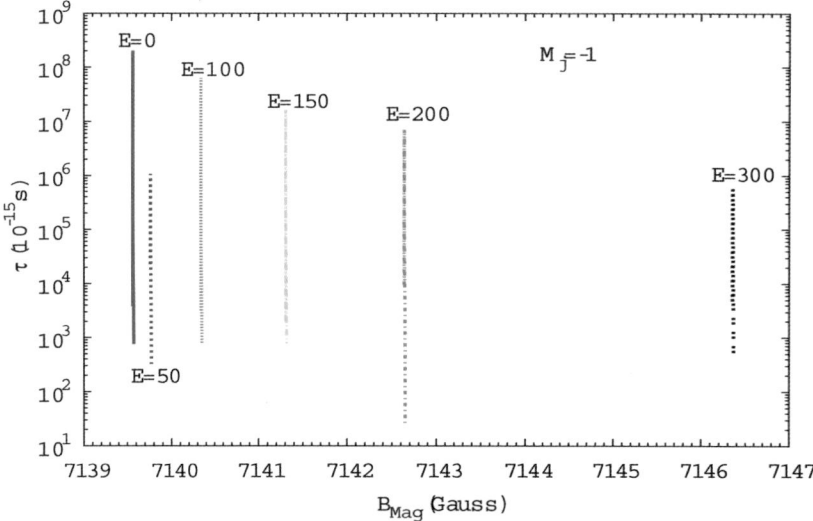

Fig. 5 Highest positive Close Coupling Q matrix eigenvalues as a function of the magnetic field for the collision of the fundamental field dressed state $\alpha = 1$ of NH($^3\Sigma$) with ^{3}He and for a projection $M_T = -1$ of the total angular momentum along the direction of the field. The value of the applied electric field is indicated on each curve and is given in kV cm^{-1}.

Dr Stoecklin responded: The vertical bars of Fig. 5 and 6 represent the highest positive Close Coupling Q matrix eigenvalues in the vicinity of a magnetically tuned zero energy Feshbach resonance. They are given in fs. These figures intend to show that the value of the maximum of the eigenlifetime can be efficiently tuned by applying a superimposed parallel electric field.

Professor Gianturco addressed Dr Stoecklin: As you discussed in your comment, the way to better analyse the presence of scattering resonances is to set up a delay matrix and then to search in its energy behaviour for signs of the presence of resonances. However, when one has to go down to vanishing values of the scattering energies of the processes at hand, the actual evaluation of the derivative of the scattering matrix with respect to energy might present problems: do you have any suggestion as to how one can avoid such problems?

Dr Stoecklin replied: That is a very pertinent question as in order to compute the **Q** matrix, one needs to know both the **S** matrix and its energy derivative and the accuracy of the evaluation of the **Q** matrix is then highly dependent on the accuracy of the evaluation of the derivative of the **S** matrix. Unfortunately most of the available scattering codes do not allow an analytical evaluation of the derivative of the **S** matrix as a function of the collision energy to be obtained but rely instead on a numerical procedure. The numerical evaluation of the derivative of the **S** matrix is not really appropriate, as its accuracy is questionable in the vicinity of the poles of the **S** matrix. We overcame this problem recently[1] by taking advantage of the simple expression of the sector adiabatic wave functions of the Magnus propagator[2] to obtain analytical values of the energy derivative of the **S** matrix, which in turn are used to get the Smith lifetime **Q** matrix. The procedure involves the simultaneous generation of both the **R** matrix and its energy derivative dR/dE which are propagated along the scattering coordinate. We then propagate both the wave function and its analytical energy derivative along the reaction coordinate.

1 G. Guillon and T. Stoecklin, *J. Chem. Phys.*, 2009, **130**, 144306.
2 G. A. Parker, T. G. Schmalz, and J. C. Light, *J. Chem. Phys.*, 1980, **73**, 1757.

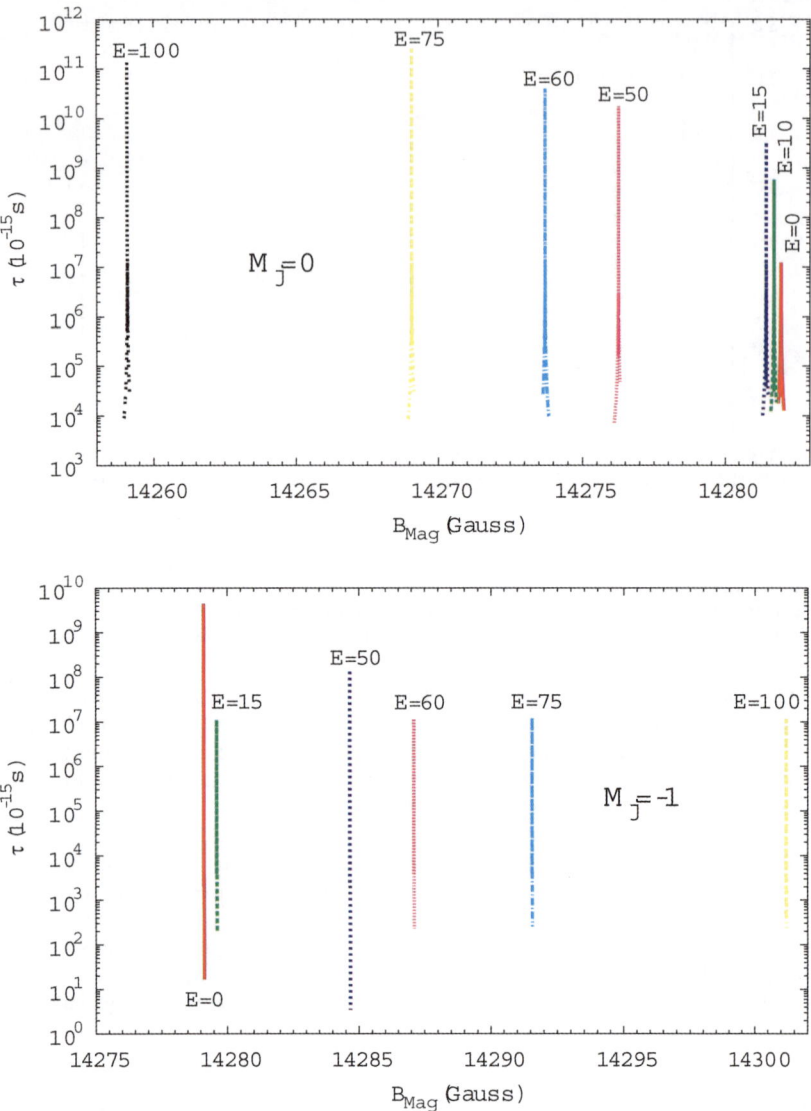

Fig. 6 Highest positive Close Coupling Q matrix eigenvalues as a function of the magnetic field for the collision of ^{3}He with NH($\alpha = 1$) and $M_T = -1$ on the lower panel and NH($\alpha = 2$) and $M_T = 0$ on the higher panel, respectively. The value of the applied electric field is indicated on each curve and is given in kV cm^{-1}.

Professor Gianturco responded: This is an interesting procedure and one which should be carefully analysed for ultralow energy collisions in different molecular systems. It would be useful to have it systematically applied to those systems where the stronger ionic interactions are more sensitive to small changes of the cross section features at ultralow energies.

Professor Hutson opened the discussion of Dr Groenenboom's paper: I'm very interested in your Fig. 4, which shows resonances in the field-dependence of spin relaxation cross sections. Your resonances look as if they are simple peaks in the inelastic cross sections. As you know, we have recently obtained analytic expressions

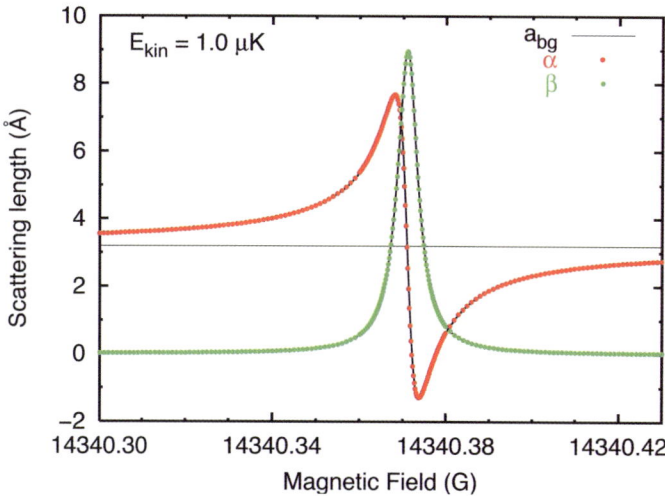

Fig. 7 Real and imaginary parts of the scattering length across a Feshbach resonance in He + NH ($n = 0$), showing a small symmetrical oscillation in the real part (red) and a peak in the imaginary part (green). Reproduced with permission from J. M. Hutson, *New J. Phys.*, 2007, **9**, 152.

for the way that scattering lengths and cross sections vary across such resonances.[1,2] In the presence of inelastic scattering, the real part of the scattering length shows an oscillation (instead of a pole). The amplitude of the oscillation is characterised by a resonant scattering length a_{res}, which can often be quite small (a few Å). In our initial work on He + NH,[1] there was very little background inelasticity and the imaginary part of the scattering length (and the inelastic cross section) showed a simple peak at resonance as shown in Fig. 7. However, we have recently investigated some more strongly coupled systems such as He + $^{16}O_2$, where there is relatively strong background inelasticity.[3] In some cases the inelastic cross sections show very deep troughs near resonance. This behaviour does in fact appear in our general theory[2] and is analogous to the dips seen in Fano lineshapes. Have you seen any cases where there are dips in inelastic cross sections near resonance?

1 M. L. González-Martínez and J. M. Hutson, *Phys. Rev. A*, 2007, **75**, 022702.
2 J. M. Hutson, *New J. Phys.*, 2007, **9**, 152.
3 J. M. Hutson, M. Beyene and M. L. González-Martínez, to be published.

Dr Groenenboom responded: We did not find such dips in the inelastic cross sections, but we only considered one particular case of Feshbach resonances at zero electric field and a collision energy of 0.001 cm^{-1}. The Feshbach resonances shown in Fig. 4 will be modified if either the electric field or collision energy (or both) are changed. This is indeed an interesting question, which deserves further study.

Mr Wallis commented: Following Gerrit Groenenboom's presentation of the excellent results for collisions of He–OH in combined electric and magnetic field, I shall present some unexpected results from collisions of He with SO($^3\Sigma$) and any comments Gerrit has would be very much welcome.

In Fig. 8, the calculated bound states of the He–SO complex are shown for each parity as a function of magnetic field; the dashed lines are the SO monomer thresholds. We can rationalize the bound-state pattern at zero field by coupling the total angular momentum of the SO monomer j, to the end-over-end angular momentum of the complex L to form a total angular momentum, J, with the resulting basis $|j\ L$

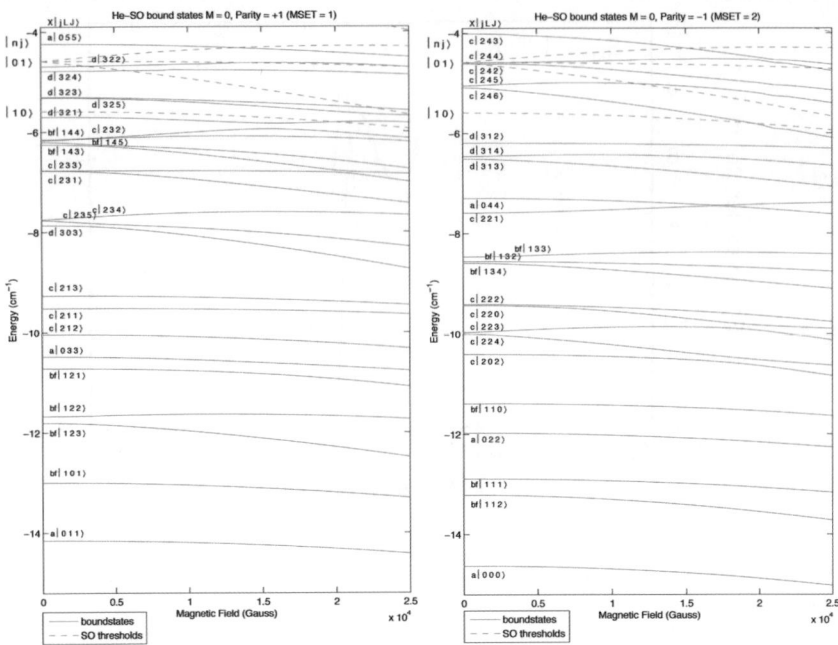

Fig. 8 Bound states of the HeSO($^3\Sigma$) complex shown for both parities (even parity in the left panel and odd parity in the right) as a function of magnetic field, for $M = 0$, where M is the projection of the total angular momentum onto the space fixed axis. The SO thresholds are shown as dashed lines.

$J \rangle$. In a magnetic field each bound state splits into $2J + 1$ states each labelled by a value of M, the projection of J onto the space-fixed magnetic field axis. The states shown on this figure are for $M = 0$.

Using the bound-state calculations we can predict the location of zero-energy Feshbach resonances. Depending on the parity of the bound-state crossing the threshold as a function of magnetic field the resonance will either be an s-wave ($l = 0$) or a p-wave ($l = 1$) resonance. In Fig. 9 we look more closely at bound states crossing the lowest SO threshold, which has zero monomer angular momentum, $j = 0$. We can see points at which bound states of both parities cross the lowest threshold. We expected to find an s-wave resonance at the location marked. The bound state $|j = 2\ L = 4\ J = 6\rangle$ and the threshold $|j = 0\ L = 0\ J = 0\rangle$ correlate with zero-field states with different values of J, but in a magnetic field this is not a conserved quantity. However, no resonance was found when only a magnetic field was applied. When an additional electric field was applied parallel to the magnetic field the resonance appeared. A log plot of the electric field strength against the resonance width is shown. The linearity of the width as a function of the electric field strength indicates that the resonance has zero width in the absence of an electric field. We do not at present understand why there is not coupling between this bound state and the threshold in the presence of a magnetic field but no electric field.

Dr Groenenboom answered: I do not have an explanation for this.

Dr Chandler asked: Can the effects of potential energy surfaces crossings associated with crossed magnetic and electric fields be observed by inducing dissociation of van der Waals molecules? More generally, are there experiments with van der Waals molecules that could be used to observe these induced crossings?

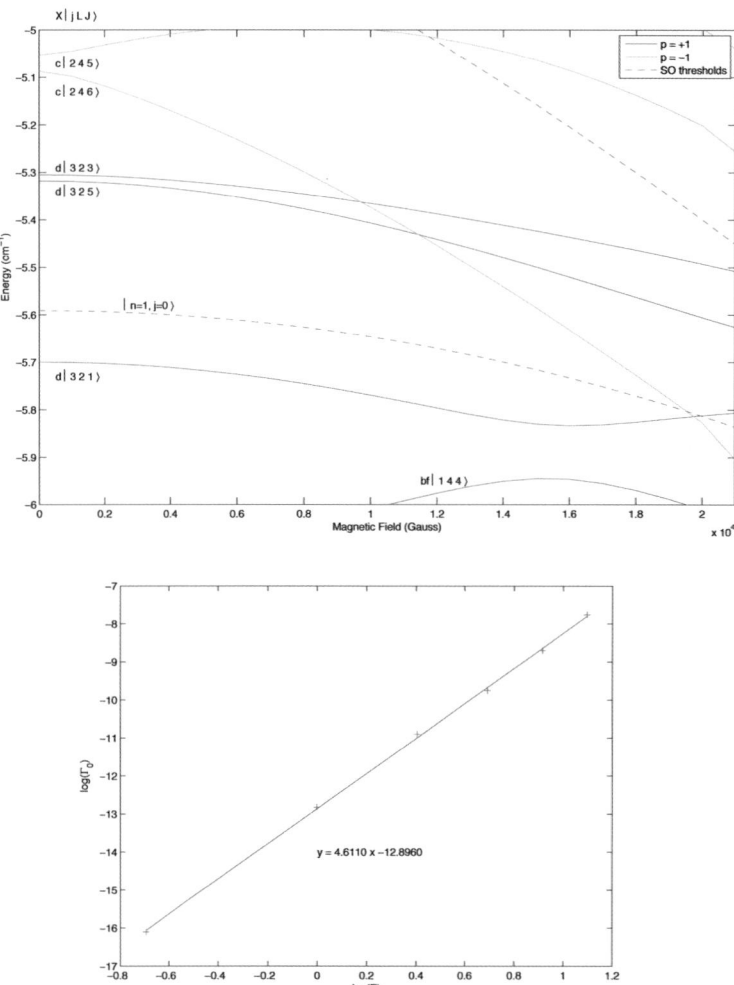

Fig. 9 Top panel: HeSO bound states near the lowest SO threshold, zero energy Feshbach resonances are predicted at points at which bound states cross threshold. The bound state with odd parity (green) was expected to result in an *s*-wave resonance as it crossed the $|n = 1, j = 0\rangle$ threshold. States of even and odd parity are shown in blue and green respectively, the SO thresholds are given as dashed lines. Bottom panel: Log-log plot of the *s*-wave Feshbach resonance width against applied electric field strength. The linearity of the plot indicates that the resonance has a vanishing width at a zero electric field.

Mr Wallis replied: In 2004, Krems looked at dissociating van der Waals molecules using magnetic fields.[1] Electromagnetic fields could be used to tune a near threshold atom–diatom van der Waals bound state above the threshold, inducing dissociation *i.e.* He–SO → He + SO. The timescale of the dissociation would depend on the coupling strength between the bound state and the open channel.

In combined magnetic and electric fields avoided crossings would be induced between bound states of opposite parity that cross as a function of just the magnetic field.[2] This may affect the bound state-threshold coupling and thus the dissociation time scale.

However, for Van der Waals states that are deeply bound the fields required to cause dissociation would be large and dissociation would not be an issue.

1 R. V. Krems, *Phys. Rev. Lett.*, 2004, **93**, 013201.
2 A. O. G. Wallis, S. A. Gardiner and J. M. Hutson, Conical intersections in laboratory coordinates with ultracold molecules, 2009, arXiv:0905.1052.

Dr Groenenboom responded: The dissociation of van der Waals molecules with magnetic fields was demonstrated in a theoretical paper.[1] This should also be possible with combined electric and magnetic fields, and that may reveal details about the energy levels near the dissociation threshold. The couplings between different Zeeman states are dramatically enhanced near the avoided crossings so the predissociation lifetimes of van der Waals molecules should significantly decrease in the vicinity of the avoided crossings. This enhancement of the predissociation may be used to observe the effect of these avoided crossings on the dynamics of van der Waals molecules.

1 R. V. Krems, *Phys. Rev. Lett.*, 2004, **93**, 013201.

Mr Lemeshko enquired: In Fig. 3 of your paper you present the energy dependence of the collision cross sections in combined fields. What happens with this resonant structure near avoided crossings points?

Dr Groenenboom answered: The energy levels of OH in the ground rotational state do not exhibit avoided crossings in combined electric and magnetic fields. This is because, in order to experience an avoided crossing, two Zeeman states must have the same total angular momentum projection M (otherwise it would be forbidden by symmetry). As shown in the upper panel of Fig. 1, the states with the same M occur in pairs (e/f) with opposite inversion parity. Such states can only be coupled by an external electric field (magnetic fields conserve parity).

However, once the electric field is applied, the e and f states begin to repel each other, so they never cross. As a result, all the crossings shown in the lower panel of Fig. 1 occur between the states of different M, and are therefore not avoided. This is the reason why there are no sharp peaks in Fig. 5, which would appear if the initial collision channel, $|J=3/2, M=3/2, e\rangle$ were involved in an avoided crossing.

The above does not preclude the possibility of the formation of avoided crossings between the Zeeman levels corresponding to different rotational states of OH. However, because of the large rotational splitting, the realization of such crossings between the two most closely spaced rotational levels, $J = 3/2$ and $J = 5/2$ would require magnetic fields of the order 75 T.

Professor Kotochigova asked Dr Groenenboom: Why is the loss to the $M = -3/2$ state so much smaller that to the other two states. Can you explain this in terms of selection rules?

Dr Groenenboom answered: The projection of the total angular momentum on the field axis is conserved. Hence, a change in the M quantum number of the OH radical requires a simultaneous change in the m_l quantum number of the overall rotation. This results in centrifugal barriers for the incoming and/or outgoing waves and suppression of the inelastic process. This was first observed by Volpi and Bohn.[1] The threshold laws for collisional reorientation of electronic angular momentum without external fields can be found in a paper by Krems and Dalgarno.[2]

Another possible explanation is that the coupling between the initial ($M = 3/2$) and final ($M' = -3/2$) states occurs due to the $\lambda = 3$ term in the expansion of the interaction potential in Legendre polynomials [Eqns. (25)–(27) in our paper]. The anisotropic terms $V_\lambda(R)$ decrease rapidly with increasing λ, so the $\lambda = 3$ term will be much smaller than the lower-order anisotropic terms with $\lambda = 1$ and 2. For this reason, the probability for the $M = 3/2 \rightarrow M' = -3/2$ transition should be smaller than for the transitions which change M by 1 or 2.

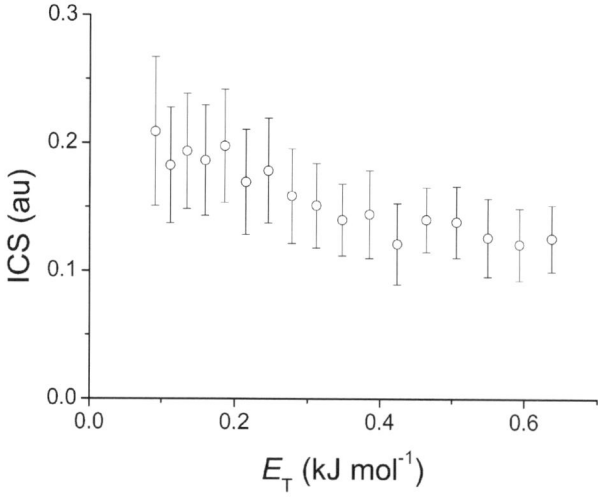

Fig. 10 Excitation function of the S(^{1}D$_2$) + H$_2$ (25% $J = 0$ and 75% $J = 1$) → SH + H reaction. The error bars correspond to one σ statistical uncertainty on 300 collision events for each point.

1 A. Volpi and J. L. Bohn, *Phys. Rev. A*, 2002, **65**, 052712.
2 R. V. Krems and A. Dalgarno, *Phys. Rev. A*, 2003, **67**, 050704.

Dr Stoecklin communicated: Dr van de Meerakker has emphasized that collision experiments will be feasible in the cold temperature regime (1 to 10 K) in the near feature by merging beams from a molecular synchrotron and a Stark decelerator.[1] I would like to communicate the very recent progress made in Bordeaux by Dr Costes and Prof. Naulin on the experimental investigation of elementary chemical reactions at very low collision energies. To date, the Bordeaux crossed molecular beam machine has already achieved the lowest collision energies in many integral cross section measurements ($E_T < 0.5$ kJ mol^{-1})[2] and also in one differential cross section study ($E_T = 0.7$ kJ mol^{-1})[3] of reactions between neutral (uncharged) species. The machine, which is characterized by a variable beam intersection angle down to 22.5°, has recently been upgraded by installation of cryo-cooled Even–Lavie pulsed valves.[4] Preliminary experiments have been conducted on the S(^{1}D$_2$) + H$_2$ ($J = 0$ and $J = 1$) → SH + H reaction. The H$_2$ beam was generated from a H$_2$–He mixture with a nozzle cooled down to 45 K. It had a velocity of 780 m s^{-1} and a speed ratio of 30. The S(^{1}D$_2$) beam was produced from UV photolysis at 196.6 nm of a CS$_2$–neon mixture. It had the same velocity as the H$_2$ beam and a speed ratio of 25. The H atoms produced in the reaction were probed by resonance-enhanced multiphoton ionisation with time-of-flight mass spectrometric detection. They were excited to the ^{2}P state by a photon tuned to the Lyman-alpha wavelength of 121.57 nm and subsequently ionised by a photon at 364.7 nm. The excitation function obtained when scanning the intersection angle between 62.5° and 22.5° by 2.5° steps is displayed in Fig. 10. It shows non-threshold behaviour down to $E_T = 0.085$ kJ mol^{-1} or 7 cm^{-1}, which corresponds to the mean energy of a Maxwell distribution (<E_T> = 3/2 **R**T) at $T = 7$ K.

1 S. Y. T. van de Meerakker and G. Meijer, *Faraday Discuss.*, 2009, 142, DOI: 10.1039/b819721k.
2 W. D. Geppert, D. Reignier, T. Stoecklin, C. Naulin, M. Costes, D. Chastaing, S. D. Le Picard, I. R. Sims and I. W. M. Smith, *Phys. Chem. Chem. Phys.*, 2000, **2**, 2873; D. Chastaing, S. D. Le Picard, I. R. Sims, I. W. M. Smith, W. D. Geppert, C. Naulin and M. Costes, *Chem. Phys. Lett.*, 2000, **331**, 170; W. D. Geppert, C. Naulin and M. Costes, *Chem. Phys. Lett.*, 2001, **333**, 51; W. D. Geppert, C. Naulin and M. Costes, *Chem. Phys. Lett.*, 2002, **364**, 121; W. D. Geppert, C. Naulin, M. Costes, G. Capozza, L. Cartechini, P. Casavecchia

and G. G. Volpi, *J. Chem. Phys.*, 2003, **119**, 10607; W. D. Geppert, F. Goulay, C. Naulin, M. Costes, A. Canosa, S. D. Le Picard and B. R. Rowe, *Phys. Chem. Chem. Phys.*, 2004, **6**, 566.
3 M. Costes, N. Daugey, C. Naulin, A. Bergeat, F. Leonori, E. Segoloni, R. Petrucci, N. Balucani and P. Casavecchia, *Faraday Discuss.*, 2006, **133**, 157.
4 U. Even, J. Jortner, D. Noy, N. Lavie, and C. Cossart-Magos, *J. Chem. Phys.*, 2000, **112**, 8068.

Professor Sims (with Dr Berteloite, Dr Lara, Dr Le Picard, Dr Dayou, Professor Launay and Dr Canosa) contributed: Remarkable progress has been achieved over the past few years in the *production* of cold and ultra-cold molecules at sub-1 K temperatures. However, the study of bimolecular *chemical reactions* (involving a real chemical transformation) between neutral species at such temperatures has not yet benefited from these advances, owing to a number of factors including notably the very low densities achieved in such experiments. The record of 13 K for an absolute measurement of the rate coefficient for a gas-phase chemical reaction between neutral species was established by Sims *et al.*[1] in the early 1990s using the CRESU (Cinétique de Réaction en Ecoulement Supersonique Uniforme, or Reaction Kinetics in Uniform Supersonic Flow) technique coupled with pulsed laser photochemical methods. We report here a further advance which has enabled the measurement of a reaction rate coefficient at temperatures as low as 5.8 K.

A joint project has been initiated involving low temperature reaction kinetics and quantum scattering calculations in Rennes, and low energy integral cross-section measurements in Bordeaux (in the group of M. Costes and C. Naulin), focusing on a number of atomic radical ($F(^2P_J)$, $O(^1D_2)$, $C(^1D_2)$, $S(^1D_2)$) – H_2 reactions. The reactions with ^1D atoms appear to possess no barrier to reaction on their ground state electronic potential energy surfaces, displaying rather deep wells. Owing to their simplicity they are very amenable to high level quantum mechanical calculations, and detailed measurements of both absolute rate constants in the CRESU and integral cross sections down to very low collision energies promise to provide a very interesting and insightful challenge to these calculations.

We report here on preliminary rate coefficient measurements and quantum scattering calculations on the reaction

$$S(^1D_2) + H_2 \rightarrow SH + H \quad (\Delta_r H^{\ominus}_{298\,K} = -21 \text{ kJ mol}^{-1})$$

at temperatures down to 5.8 K, a new low temperature record for a radical–molecule chemical transformation.[2] The CRESU technique combined with pulsed laser photochemical methods for the measurement of reactive and inelastic bimolecular rate coefficients at low temperatures has been described elsewhere.[1,3] Various Laval nozzles operating under specific conditions of pressure, flow and carrier gas gave temperatures down to 23 K. In order to achieve the lowest temperature in this study, a special nozzle was manufactured which possessed a double wall, enabling it to be pre-cooled to 77 K by the use of liquid nitrogen, along with the reservoir upstream of the nozzle. Impact pressure measurements indicated a temperature of 5.8 K in the uniform supersonic flow downstream of this nozzle. Full details will appear in a subsequent publication. $S(^1D)$ atoms were generated by 10 Hz 193 nm pulsed excimer laser photolysis of CS_2, and detected by pulsed vacuum ultraviolet laser-induced fluorescence (VUV LIF) at 166.67 nm, generated by two-photon resonant four-wave difference frequency mixing in Xe, as in previous studies on atomic carbon reaction kinetics.[4] Hydrogen reactant was taken directly from the gas bottle, and we estimate that its room temperature *ortho:para* ratio of 3 : 1 (so-called normal, *n*-H_2) will not be perturbed by the rapid cooling to 77 K, and subsequent adiabatic expansion to 5.8 K. The decay of VUV LIF, proportional to the concentration of $S(^1D)$ atoms was recorded at differing H_2 concentrations, all in great excess over the $S(^1D)$ concentration (pseudo-first-order conditions)[5] enabling the determination of rate coefficients for the removal of $S(^1D)$ by *n*-H_2 at temperatures down to 5.8 K.

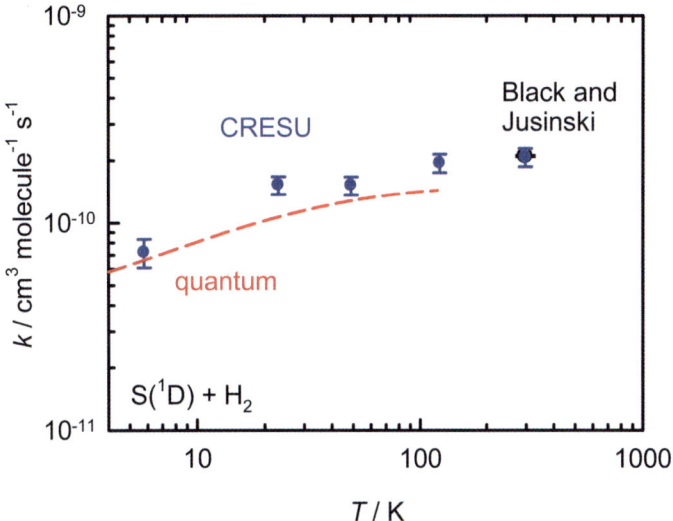

Fig. 11 Comparison of experimental total removal (filled circles) and theoretical reactive (dashed line) rate coefficients for S(^{1}D) + n-H$_2$ collisions. The 300 K experimental result of Black and Jusinski[6] (open circle) is also shown.

Results are shown in Fig. 11, along with the theoretical calculations and the room temperature measurement of Black and Jusinski.[6]

Quantum reactive cross-sections for collisions of S(^{1}D) with *ortho-* and *para-*hydrogen, in the energy range 0–120 K, have been calculated using the hyperspherical quantum reactive scattering method developed by Launay and Honvault.[7] The short range dynamics were described using the *ab initio* potential energy surface of Ho *et al.*[8] complemented by new and accurate *ab initio* calculations of the long range interactions between the open-shell S(^{1}D) atoms and H$_2$ molecules. The dispersion ($\sim R^{-6}$) and electrostatic quadrupole–quadrupole, ($\sim R^{-5}$) parts of the interaction were taken into account. The quantum cross-sections were extrapolated to higher energies, enabling the calculation of thermally-averaged rate constants using the room-temperature ratio of *ortho-* to *para-*hydrogen. Low-temperature calculated values are shown along with the experimental results in Fig. 11.

Excellent agreement was obtained at room temperature between our experimental rate coefficient and the previous measurement of Black and Jusinski.[6] The experimental results probe the total removal of S(^{1}D) by n-H$_2$ (reaction plus relaxation), and thus represent an upper limit to the reactive rate coefficient. In view of this, the theoretical results are in rather good agreement with experiment. The quadrupole–quadrupole part of the interaction appears to be important in describing cold collisions of S(^{1}D) with *ortho-*hydrogen. Further kinetics experiments are planned to confirm these results, and also to determine the branching ratio (reaction to relaxation) by determination of the S(^{3}P) yield from the relaxation of S(^{1}D) by n-H$_2$. In parallel, a better description of the cold collisional process, including non-adiabatic couplings, is under development.

1 I. R. Sims, J. L. Queffelec, A. Defrance, C. Rebrion-Rowe, D. Travers, P. Bocherel, B. R. Rowe and I. W. M. Smith, *J. Chem. Phys.*, 1994, **100**, 4229.
2 C. Berteloite, M. Lara, S. D. LePicard, F. Dayou, J.-M. Launay, A. Canosa and I. R. Sims, Chemical reactivity at extremely low temperatures: rate coefficients for S(^{1}D) and H$_2$ down to 5.8 K, *Faraday Discuss.*, 2009, **142**, poster presentation.
3 P. L. James, I. R. Sims, I. W. M. Smith, M. H. Alexander and M. B. Yang, *J. Chem. Phys.*, 1998, **109**, 3882.

4 D. Chastaing, S. D. Le Picard and I. R. Sims, *J. Chem. Phys.*, 2000, **112**, 8466.
5 A. Canosa, F. Goulay, I. R. Sims and B. R. Rowe, Gas Phase Reactive Collisions at Very Low Temperature: Recent Experimental Advances and Perspectives, in *Low Temperatures and Cold Molecules*, ed. I. W. M. Smith, World Scientific, Singapore, 2008, p 55.
6 G. Black and L. E. Jusinski, *J. Chem. Phys.*, 1985, **82**, 789.
7 P. Honvault and J. M. Launay, *J. Chem. Phys.*, 2001, **114**, 1057.
8 T. S. Ho, T. Hollebeek, H. Rabitz, S. Der Chao, R. T. Skodje, A. S. Zyubin and A. M. Mebel, *J. Chem. Phys.*, 2002, **116**, 4124.

Dr Lara commented: Regarding the integral total cross sections obtained by the group of M. Costes (Bordeaux) for the cold collision $S(^1D_2) + H_2$, their theoretical counterparts,[1] obtained using the hyperspherical quantum reactive scattering method of Launay and Honvault[2] (Rennes), show oscillating features. Interestingly enough, these structures can be attributed to particular partial waves. Although the current error bars do not allow conclusive confirmation of their presence, new measurements are going to be performed to assess the question.

1 C. Berteloite, M. Lara, S. D. Le Picard, F. Dayou, J.-M. Launay, A. Canosa and I. R. Sims, *Faraday Discuss.*, 2009, **142**, poster presentation P51.
2 P. Honvault and J. M. Launay, *J. Chem, Phys.*, 2001, **114**, 1057.

Dr Raston asked Professor Sims: Is there much of a dependence of the rate coefficient on the nuclear spin isomer of hydrogen involved in the low temperature $(S(^1D) + H_2)$ reaction?

Professor Sims replied: Our experimental measurements of the rate coefficient for the total removal of $S(^1D)$ by H_2 do not probe this effect as so-called *normal* hydrogen (with a fixed *ortho*/*para* ratio corresponding to that of room temperature hydrogen) was used. The quantum calculations show virtually no dependence of the rate coefficient on the hydrogen *ortho*/*para* ratio. The calculations for *ortho* hydrogen included an accurate description of the quadrupole–quadrupole long-range term in the potential, which appears to be important in describing the dynamics.

Dr Dulieu continued the discussion of Dr Groenenboom's paper: To what extent can Fig. 1 help to predict results for other systems in combined electric and magnetic fields? In other words, could the existence (but not necessarily the location) of the abrupt drop in the spin relaxation cross section of Fig. 5, attributed to the bending of some of the Zeeman levels leading to avoided crossing, be generally predicted by a simple analysis of the Stark/Zeeman energy diagram?

Dr Groenenboom answered: The pattern in Fig. 1 is quite general for degenerate states. However, the field-free splitting between opposite parity states is very different for diatomics in Σ states. In that case it involves different rotational states, and the crossings occur at much larger fields.[1,2]

1 T. V. Tscherbul and R. V. Krems, *Phys. Rev. Lett.*, 2006, **97**, 083201.
2 T. V. Tscherbul and R. V. Krems, *J. Chem. Phys.*, 2006, **125**, 194311.

Professor Meijer enquired: The proposal to have parallel electric and magnetic fields of a certain strength-ratio to suppress inelastic channels is highly interesting, and should experimentally be picked up. This would imply 'nesting' a quadrupole magnetic trap with a quadruple electric trap, and in the experiment you will never get the electric and magnetic fields parallel throughout the whole trap. So the question is, how far an angle of a few degrees between the electric and magnetic field is detrimental to the expected (hoped for) suppression, and whether calculations for different angles would be feasible?

Dr Groenenboom responded: The effect will most likely be present for non-zero angles between the electric and magnetic field. Recent calculations of atomic[1] and molecular[2] collisions indicate that rotating the electric field with respect to the magnetic field modifies the structure of the collision complex but does not change the qualitative features of the collision dynamics. This means that the same effects should be observable at arbitrary orientations of the electric and magnetic fields, although the magnitudes of the electric and magnetic fields leading to a particular effect should be different depending on the angle between the fields. For He + OH the calculation would be more difficult than the previous calculations mentioned above, but should be feasible.

1 Z. Li and K. W. Madison, *Phys. Rev. A*, 2009, **79**, 042711.
2 E. Abrahamsson, T. V. Tscherbul, and R. V. Krems, *J. Chem. Phys.*, 2007, **127**, 044302.

Professor Julienne opened the discussion of the paper by Dr Chandler: What is the number of slowed molecules that you are able to work with and what volume do they occupy?

Dr Chandler replied: We have performed density measurements for two different kinematically-cooled systems, and have estimated a density of cold NO molecules (produced by NO/Ar scattering) of about 108 molecules per cm^3, a density of cold Kr atoms (produced by Kr/Kr scattering) of approximately 2×10^9 atoms per cm^3. Note that for the NO/Ar system it is an inelastic collision process that generates the cold molecules, whereas for the Kr/Kr system it is an elastic process (with an unfavorable differential cross section) that produces cold Kr. Our collision volume is approximately 1 mm^3, so the actual number of cold molecules or atoms we make is approximately 106–107 per cm^3.

Professor Barker commented: With Stark deceleration of NO molecules, we estimate that number densities in the range of 10^{11} cm^{-3} are produced, distributed over a number of rotational levels.

Dr Wrede commented on Dr Chandler's paper: Chandler and co-workers have demonstrated how collisions in a crossed molecular beam experiment can be exploited to stop and kinematically cool molecules in the laboratory frame. Dr Carty and I would like to present a new, accessible and economical technique, dubbed PhotoStop, for producing molecules with zero velocity in the laboratory frame utilising a half-collision. The method is based on the controlled breaking of one chemical bond, *via* photodissociation, in a precursor molecule initially cooled in a supersonic molecular beam. The result is an atomic or molecular fragment that recoils with a velocity that cancels out the initial velocity of the precursor molecule, as illustrated in Fig. 12.

A pulsed molecular beam of NO$_2$ seeded in Xe is formed with a velocity of 415 m s^{-1}. The molecular beam is intersected by a linearly polarised pulsed photodissociation laser beam. As can be seen in Fig. 12(b), the excess energy after the dissociation of an NO–O bond is partitioned into the internal energy of the photofragments NO and O (E_{int}) and the total kinetic energy release (TKER). The TKER is partitioned among the fragments according to their mass ratio by conservation of momentum. By varying the photon energy, the recoil velocity of one of the fragments can be tuned to cancel out the velocity of the molecular beam resulting in fragments that are standing still in the laboratory frame. In this study we use a pulsed laser to ionise the NO fragments state selectively using 1 + 1 resonance enhanced multi-photon ionisation (REMPI). The laser beams are counter-propagating and cross the molecular beam at the center of an ion imaging lens system operated under velocity mapping conditions.

The recorded ion images are shown in Fig. 13 at various time delays between the photodissociation and probe laser pulses. The probe laser wavelength was set to

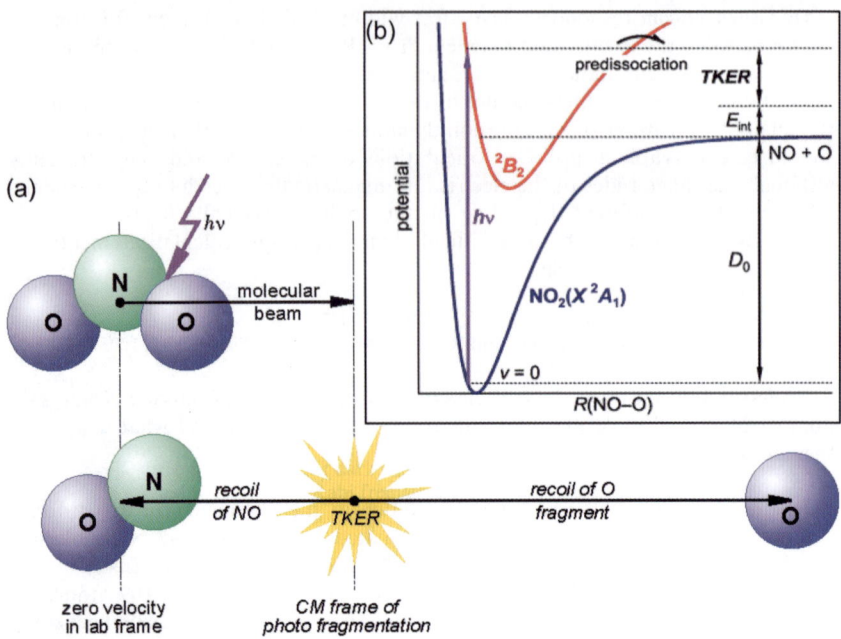

Fig. 12 (a) Schematic outline of the PhotoStop technique. The recoil velocity of NO fragments due to the total kinetic energy release (TKER) from the photodissociation of NO_2 molecules cancels the velocity of the molecular beam. As such, NO molecules are produced at a standstill in the laboratory frame. (b) Schematic diagram of the NO_2 potential energy surfaces. Ground state NO_2 molecules are photo excited into the $^2B_2$ state, which predissociates onto the ground state surface to form $NO(X^2\Pi_\Omega)$ molecules and $O(^3P_J)$ atoms. By tuning the photon energy, $h\nu$, the TKER of the dissociating fragments can be adjusted for given NO and O internal states.

ionise NO molecules in the $X^2\Pi_{3/2}$, $v = 0$, $J = 1.5$ state. The images represent the velocity distributions of the probed NO molecules in the plane of the molecular beam (v_z) and laser beams (v_x). The image recorded at a time delay of 0 μs is centred at the molecular beam velocity. The ion signal appears on a ring with a radius that corresponds to the velocity of the recoiling NO fragments in the moving frame of the molecular beam. The wavelength of the dissociation laser was tuned to 386.4 nm, the photon energy at which the $NO(X^2\Pi_{3/2}, J = 1.5)$ fragments recoil from ground state $O(^3P_2)$ at 415 m s^{-1}, the velocity of the molecular beam. The angular anisotropy (β-parameter of 1.2) is determined by the complex NO_2 photodissociation dynamics.[1] The polarisation plane of the dissociation laser beam was chosen to lie in the plane of the ion images in order to maximise the number of NO molecules stopped at lab velocity origin, $(v_z,v_x) = (0,0)$, marked by the cross-hairs in Fig. 13.

As the time delay between the laser pulses is increased, the recorded ion signal decreases because NO fragments with high enough velocities in the laboratory frame fly out of the small cylindrical probe volume defined by the probe laser beam (diameter ∼0.15 mm). Analysis of the signal projected onto the velocity axis along the molecular beam, v_z, shows the narrowing velocity distributions as shown below each ion image in Fig. 13. The velocity distributions along v_x are broader because the probe volume expands along the probe laser axis and fragments with velocities along the x-axis can stay in the probe volume for longer. The final ion image was recorded at a time delay of 10 μs. Slow NO molecules can clearly be identified around the laboratory velocity origin at the centre of the image, which demonstrates the validity of the PhotoStop technique. The measured full-width-half-maximum (FWHM) along v_z of 50 m s^{-1} corresponds to a translational temperature of 1.6 K for the probed NO fragments.

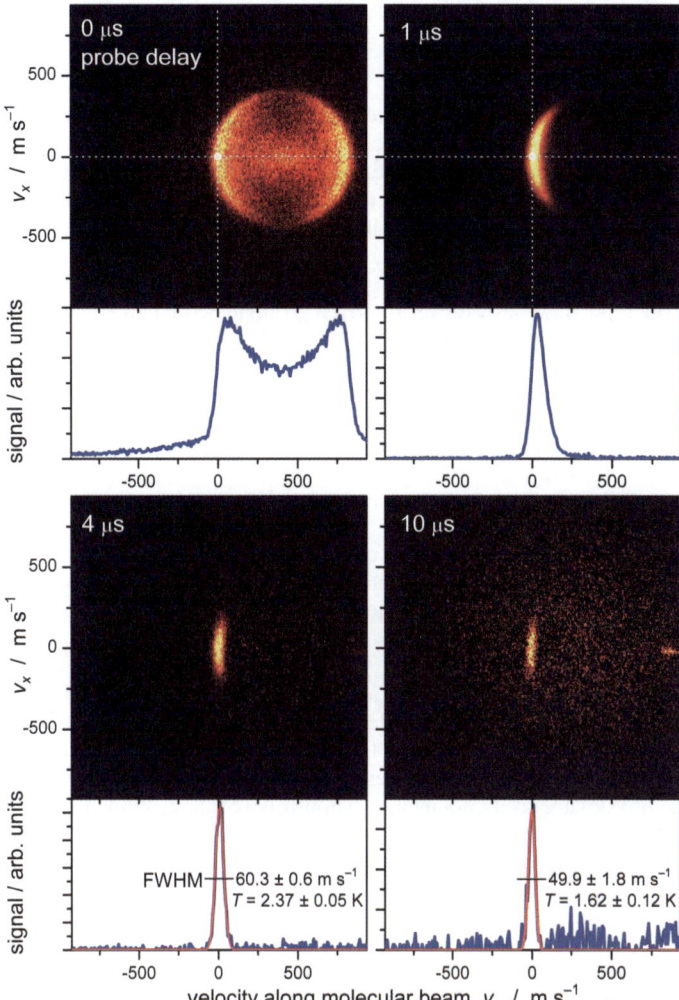

Fig. 13 Velocity mapped ion images of NO($X^2\Pi_{3/2}$, $v = 0$, $J = 1.5$) fragments from the photodissociation of NO_2 at 386.4 nm at different time delays between the dissociation and probe lasers. The centre of the images corresponds to the velocity origin in the laboratory frame, v_z and v_x correspond to the velocity components along the directions of the molecular beam and the laser beams, respectively. The velocity distribution projected onto the v_z axis is shown underneath each ion image.

We regard the PhotoStop technique as a useful addition to the armoury of complementary techniques that can produce cold molecules at standstill in the laboratory frame.[2] Like kinematic cooling in crossed molecular beams[3] and buffer gas cooling,[4] PhotoStop allows molecules and atoms to be stopped inside an electrostatic, magnetostatic, or optical trap and density to be accumulated over many molecular beam and laser pulses. However, unlike kinematic and buffer gas cooling, PhotoStop is not a cooling technique as it does not collapse the velocity distribution of the stopped molecules, the initial cooling takes place in the supersonic expansion. As such, the temperature of the trapped sample will be determined by the trap depth.

1 S. J. Matthews, S. Willitsch and T. P. Softley, *Phys. Chem. Chem. Phys.*, 2007, **9**, 5656.
2 M. T. Bell and T. P. Softley, *Mol. Phys.*, 2009, **107**, 99.

3 M. S. Elioff, J. J. Valentini and D. W. Chandler, *Science*, 2003, **302**, 1940.
4 W. C. Campbell, E. Tsikata, H. I. Lu, L. D. van Buuren and J. M. Doyle, *Phys. Rev. Lett.*, 2007, **98**, 213001.

Professor Meijer addressed Dr Wrede: Some time ago, I heard a presentation from Dr Bum Suk Zhao, originally from the University of Seoul, South Korea, who presented measurements on the production of slow photofragments after dissociation. He did not show velocity map images, but rather time-of-flight measurements, and I do not know whether or not this work has been published, but I just wanted to bring this to your attention.

Dr Wrede replied: Yes, we are aware that a group in Seoul had attempted a similar experiment, however, prior to this meeting we had no knowledge whether they have been successful or not. To the best of our knowledge, no results have been published yet, however, since the meeting, the group have made their results available on the arXiv server.[1] The idea to slow photofragments by the recoil from a dissociation is not new and has been contemplated by several research groups within the community.

1 B. S. Zhao, S. E. Shin, S. T. Park, X. Sun and D. S. Chung, 2009, arXiv:0905.0786.

Dr Pinkse asked: 'Cooling' is being used very carefully in the field of cooling and trapping and is usually restricted to phase-space density increasing processes. This must involve dissipation. In my opinion this is not the case in this work. Do you agree?

Dr Chandler answered: No, I do not agree. As you note, cooling refers to processes that increase phase-space density while slowing is simply a rotation of phase space density. With that, it can be shown that any process that employs a velocity dependent force can increase phase-space density. The two most common examples from cold atom physics are laser cooling and evaporative cooling. While the laser cooling example is clear,[1] the evaporative cooling process is more subtle.

Evaporative cooling experiments typically begin with a trapped system of atoms with a defined temperature/energy distribution. The hot tail of the distribution is then actively removed from the trap. The remaining atoms thermalize and reach a lower temperature with a higher phase-space density. Removing the hot tail of the Boltzmann distribution from the trap acts as the velocity-dependent force, since it only affects the atoms with the highest velocities. If the energy of the atom is high enough that it can escape the trap, the atom is lost, removing entropy even though the collision itself is not inherently non-conservative. The exclusion of the hot atom from the system (removal by collision with a wall or by a pump) is the non-reversible step and hot atoms are therefore selectively excluded, decreasing the entropy of the system in the trap. In the same manner we can consider the atom–molecule collisions in the kinematic cooling technique taking place inside of a confining volume (like a trap). Collisions with more energy transfer more energy than lower energy collisions. Therefore, when considering just the molecules that are side-scattered in the center-of-mass reference frame, the force to bring a molecule to rest is velocity dependent. This gives rise to the observed compression of the velocity spread in the side-scattered molecules. Any molecule that is brought to rest from a collision with an atom, imparts translational energy to that atom. As before, if the post-collision energy of the atom is enough to allow it to escape from our confining region, the atom is lost, and entropy is removed, even though there is nothing inherently non-conservative about the collision. The exclusion of the hot atom from the system is the non-reversible step.

We should point out that, unlike the evaporative cooling technique, we make no claim to have a nascent thermal distribution of cold molecules from this technique. We do note that if the cold molecules are trapped and held long enough they will thermalize. Also, for a rigorous mathematical derivation of this proof please see our chapter in Ian Smith's book, Low Temperatures and Cold Molecules.[2]

1 H. J. Metcalf and P. van der Straten, *Laser Cooling and Trapping*, Springer, 1999.
2 I. W. M. Smith (ed.), *Low Temperature and Cold Molecules*, World Scientific Publishing, ISBN 978-1-84816-209-9, 2008.

Professor Hutson said: I understand that in cryophysics the term "cooling" is traditionally restricted to dissipative processes. However, it's quite common for technical terms to be used differently in different fields and I want to question whether it is helpful to insist on this definition in the field of cold molecules. In our field, it is a common objective to produce a sample of cold molecules from a sample of hot ones: in natural language the production of "cold" from "hot" may sensibly be called "cooling", and the English language does not really have another word for it.

Since the supply of hot molecules is often almost unlimited, it does not matter much if molecules are lost in the process. In techniques such as Stark deceleration and velocity filtering, a subset of molecules with low relative or absolute velocities is selected from a larger ensemble and isolated from it, in such a way that the "cold fraction" no longer exchanges energy with the rest. Such a process has produced a sample of cold molecules from a sampler of hotter molecules, which is what we wanted. It seems unhelpful and cumbersome to me to define "cooling" so restrictively as to exclude it.

Dr Pinkse remarked: Following up on Professor Hinds' comment during the discussion, understanding the physics behind a cooling mechanism is important and will reveal much about its applicability and its limitations. For instance, deceleration or velocity filtering are not dissipative and will therefore not be able to make quantum-degenerate samples by themselves, whereas evaporative cooling in principle can. It is good to be aware of the different meanings. I would like to encourage everybody to specify the use of the word "cooling" with phrases like "reducing temperature", "increasing phase-space density" and "dissipative", where appropriate.

Professor Hinds contributed by saying: I agree that this discussion stems from a difficulty of language, but it really is more than a pedantic distinction. If the "cooling" does not involve any dissipative process, then it cannot increase the phase space density of the molecules. For many applications — e.g. making a BEC or loading a trap or guide - it is the phase space density that is the figure of merit. From this perspective the non-dissipative methods are fundamentally limited in comparison with the dissipative ones and it is helpful to identify them as such. I don't think we need to ban the word "cooling" but we should try to distinguish "cooling by selection" from "cooling by dissipation".

Dr Hudson added: There is debate over the meaning of "cooling". In particular, much is made of whether "cooling" should mean simply a narrowing of a velocity distribution in some region of phase-space, or whether it should be, more strictly, defined as phase-space compression. Jeremy Hutson has argued that we should spend less of our time with such pernickety debate. I argue that the debate is important.

One must remember that the goal of the field is not simply to make molecules cold. Rather there is almost always an application in mind, for instance: the study of degenerate quantum gases, few- and many-body physics, controlled, low-energy scattering and reaction studies, spectroscopy, or precision measurement. The requirements of these diverse applications are not the same and can not always be stated in terms of temperature alone. Rather, it is often necessary to consider the full phase-space distribution. When one carefully considers experimental details the situation can be complex. Such crushingly prosaic concerns as the height of the laboratory ceiling, the breakdown-voltage rating of a capacitor, or the tolerance of a machine tool can ultimately determine the true optimum phase-space distribution.

It is important, therefore, to properly describe a "cooling" technique's effect on the full phase-space distribution. The current ambiguous use of the word "cooling" hinders clear description and it might be better not to use it at all. It is probably too late, but I suggest we make a start by renaming the field to the altogether less memorable "phase-space-manipulated molecules".

Professor Barker continued the discussion of Dr Chandler's paper by asking: What molecular density do you expect to attain when the molecules are trapped?

Dr Chandler replied: In our reply to Professor Julienne's earlier question we state that we can reach a density of 108 to 109 molecules per cm^3 in free space (without trapping). Because we can (in principle) load the trap over several molecular beam pulses, we hope to achieve a final density approximately one order of magnitude higher. Since our trap is larger than the actual collision volume, the cold molecules will correspondingly occupy a larger volume as well.

Dr Wester asked Dr Chandler: The velocity spread of 20 m s^{-1}, obtained from the analysis of the velocity map images as an upper bound, is much larger than the estimated velocity spread of 1 m s^{-1}, based on a calculation of the scattering kinematics (see Table 1 of your paper). How large do you estimate the effect of imaging resolution on the upper bound will be and what do you expect is the actual velocity spread of the ND_3 molecules? Since the opportunities for trapping the scattered molecules in electrostatic or optical traps depends crucially on their velocity spread it would be useful to derive this spread not only from the hold time of the signal of slow molecules on the imaging detector, but also directly from high resolution imaging. What do you think are the ultimate limits in resolution for velocity map imaging, more than 1 m s^{-1} or even less?

Dr Chandler answered: We expect that the actual velocity spread of the cold molecules lies somewhere between the measured and calculated values. As we discuss in the paper, we know from the outset that several experimental effects (electron recoil, finite velocity resolution of the detector, and projection asymmetry) distort the measured velocity spread. The experimentally measured velocity spread represents a strict upper bound, while the calculated value is a strict lower bound. We recently looked at the question of how well we can measure the velocity with the Velocity Mapped Ion Imaging technique using the NO–Ar system[1], by both measuring the velocity distribution *via* velocity mapped ion imaging and directly observing the fly-out of the cooled molecules from our interaction region. Due to a combination of Coulomb repulsion, image distortions, and electron recoil affecting the imaging we observed an estimated 6 m s^{-1} image blurring. We predict that around 0.5 m s^{-1} should be obtainable, for NO, under ideal circumstances. This limits the usefulness of the ion-imaging techniques for sub-mK temperatures and heavier molecules. Spectroscopic measurements utilizing high resolution microwave sources to measure the Doppler profile of the cold molecules could yield more accurate information about the velocity distribution of the cold molecules, and we are currently pursuing such measurements in our laboratory. We will also be able to learn about our velocity distribution by varying the trap depth and monitoring the number of observed molecules when we are able to trap them using external electric or magnetic fields.

1 D. W. Chandler and K. E. Strecker, *Phys. Rev. A*, 2008, **78**, 063406.

Professor Softley asked Dr Chandler: How generally applicable do you expect the technique of kinematic cooling to be? Do you envisage a wide range of molecules and quantum states can be decelerated?

We have seen that this technique can decelerate ammonia molecules by inelastic collisions and this therefore allows a direct comparison with the technique of Stark deceleration. How do the properties of the cooled molecular samples compare in the two experiments?

Could you envisage a form of crossed beam collisional experiment using two kinematic cooling setups – is it possible to steer the cold molecules in a chosen direction with a selected velocity as proposed in the two Stark decelerator experiments in Dr van de Meerakker's paper?

Dr Chandler responded: 1. In principle any molecule can be stopped with this technique. The only requirement is that the molecule collide with a species of similar mass. Elastic collisions with a species of equal mass will produce cold molecules with no change in quantum state, and inelastic collisions with a species of greater mass will produce cold molecules with a selectable change in quantum state. However, the greater the number of available product channels, the lower the density of cold molecules will be at the end. The best choice is therefore a diatomic or small polyatomic molecule and an atomic collision partner. The final state can be chosen by adjusting the velocity of the molecular beam.

2. Both techniques can achieve a density of about 10^8 molecules of ammonia per cm^3. The ammonia molecules leaving the Stark decelerator are in a selected m_j low-field seeking state, but the molecules stopped by kinematic cooling are not m_j selected. Because kinematic cooling can be performed within the trap volume, cold molecules can be added until an equilibrium is reached, in which the number of cold molecules generated during each cycle is equal to the number that are expelled. We are hopeful that this will allow us to reach densities approximately one order of magnitude higher than we can reach in free space.

3. In our experiment only molecules of a particular quantum state, scattering in a particular direction in the center-of-mass reference frame, come to rest in the laboratory. Molecules or atoms scattered at "almost the right direction" are moving slowly in the laboratory, and we observe them above the molecular/atom beams. These can be steered with hexapole fields or simply allowed to expand into another sample of cold molecules. Molecules that are scattered in the proper direction but into a nearly degenerate internal energy state to the one that is stopped in the lab are also slowly moving and could be used in subsequent collision experiments. Since the molecule to be cooled is generally seeded in the atomic collision partner, cold atoms are typically produced in the same volume as the cold molecules, and it may be possible to trap both of them with (for example) an off-resonant laser trap and observe low-energy collisions between them. If we were to cross two pulses approximately 1 cm in length, and with a density of 10^8 atoms or molecules per cm^3, we would expect only about 10^2 to 10^4 collisions to occur per molecular beam pulse. The chances of observing the impact of these collisions would be low. Dr van de Meerakker is building two crossed Stark-decelerated beams for the purpose of studying such collisions. If that experiment succeeds, it should be possible to cross the Stark decelerated beam with a kinematically-cooled sample. Probably the best opportunity to observe very low collision energy events is to co-trap the molecules for a long period of time. Trap loss would then provide evidence of collisions. At the present time no group has been able to either trap a high enough density of molecules, or hold them long enough, to see such effects.

Professor Whitaker enquired: What are the prospects for being able to accumulate a sufficient number of kinematically cooled Kr atoms in a trap such that three-body collisions leading to dimer formation might be observable? Put another way, what is the approximate density required for the onset of three-body collisions in comparison to the number densities of cold atoms currently achievable by kinematic cooling?

Dr Chandler answered: We have estimated a density of cold molecules in the NO/Ar system of about 10^8 molecules per cm^3, and a similar number for the K/Kr system. This is the number density we are able to make with a single molecular beam pulse colliding with a single atomic beam pulse. The greatest advantage of this technique is that the trap can be "continuously" loaded. We hope that losses due to secondary collisions will be limited, so that samples with a high enough density to observe two-body collisions may be produced. It may then be possible to further cool the sample to the point where many-body affects arise. We would have to achieve densities similar to those produced in degenerate alkali systems ($>10^{13}/cm^3$) using Feshbach resonances[1] to observe these many-body effects.

1 T. Kraemer *et. al.*, *Nature*, 2006, **440**, 315.

Professor Grimm opened the discussion of Professor Barker's paper: How well can you suppress state-changing collisions by putting the rare-gas atoms into the ground state? Is it a complete suppression? How does the suppression depend on the particular molecule you want to cool?

Professor Barker replied: The suppression of inelastic collisions is important for the experiment. Our technique has been to focus on systems which are in their ground states and where inelastic collisions are almost totally suppressed by high energy thresholds. Thus in the centre-of-mass frame, the first inelastic channel for H_2 is only open at above 100 K, whereas for benzene it is open at above 2 K. As collisions will occur at energies lower than approximately 100 mK, these inelastic thresholds should be inaccessible. We note, however, that so far we have not considered effects due to nuclear spins; we would expect nuclear spin couplings to be extremely weak (see ref. 1 for example) and therefore not lead to significant numbers of inelastic collisions.

1 A. Miani and J. Tennyson, *J. Chem. Phys.*, 2004, **120**, 2732.

Professor Gianturco asked: During the presentation of your work, it was mentioned that one of the systems preliminarily analysed computationally as one of the possible candidates was that of benzene molecules interacting with rare gases. Since such systems have a long history of being studied in VdW spectroscopy, I was wondering how well you have managed to obtain reliable potential energy surfaces and how much you expect that the preliminary calculations reflect the best available surfaces, especially for their very long-range regions which are usually not accessed by spectrocopic studies.

Professor Barker responded: We are well aware of the earlier work on rare gas (Rg)–benzene systems and indeed have used it as the starting point for our own studies.

Checking the stability of the molecular properties of interest with respect to the uncertainties of the potential energy surface (PES) used in the calculation is very important as it is strictly connected to the reliability of the results presented. In some cases even small changes in the PES can have dramatic consequences in the zero-energy cross section. The PES employed for our Rg–benzene scattering calculations are the empirical ones due to Pirani *et al.*[1]

We expect these PESs to be accurate for ultra-low energy collisions as they were extracted from scattering data albeit at higher energies than the range we are interested in. There are two different ways to check the stability of the calculations with respect to the PES. Following our work on Rg–H_2(ref. 31 in our paper), we can compare results obtained with different PESs from the literature, or it is possible to arbitrarily modify the PES. Such work is currently underway.

1 F. Pirani, M. Porrini, S. Cavalli, M. Bartolomei, and D. Cappelletti, *Chem. Phys. Lett.*, 2002, **367**, 405.

Professor Julienne asked: In Table 2 of your paper where the scattering lengths and cross sections of H_2 with various rare gases are given, there is an interesting trend of decreasing cross section with increasing mass of the rare gas atom. However, Ne is a notable outlier. Is there a simple explanation for this? How many bound states does the potential bind in going down the list? For example, the scattering length is related to the binding energy of the last s-wave bound state in the potential. Thus, if the number of bound states supported by the shallow potential between the rare gas atom and H_2 changes, that could interrupt the trend.

Professor Barker said: The trend in the scattering lengths can be explained in terms of the bound states of each complex. Both HeH_2 and NeH_2 have one vibrational bound state, whereas the heavier complexes, ArH_2, KrH_2 and XeH_2 have two. Furthermore, for ArH_2 the excited vibrational state lies very close to dissociation, resulting in an enhanced zero-energy cross-section. We have investigated several PESs available in the literature and found excellent qualitative agreement between the results obtained; the only exception is XeH_2 for which only one surface is available. Details of these calculations are given in ref. 31 of our paper.

Dr Portier enquired: Will the size of the metastable atoms MOT be a problem to load the dipole trap?

Professor Barker answered: The size of the metastable argon MOT is expected to be on the order of a few hundreds of microns. The trapping region of interest of the dipole trap is set by the size of the molecular beam width, which will be around 100 microns in the axial direction of the dipole trap. Therefore, the overlap between the MOT and the trapped molecules will exist for a fraction of the extent of the MOT. The MOT density is expected to be around 10^{10}–10^{11} cm^{-3} (measured in reference 15 of our paper), which will lead to around 1000 atoms trapped per lattice site. If need be, fine position control could be implemented using magnetic field bias coils in the x,y,z directions to shift the MOT centre relative to the dipole trap.

Dr Küpper asked Professor Barker: You surely know our recent work[1,2] on laser alignment of quantum-state selected molecules. This selection could help you in various ways: (a) you will immediately reduce the number of quantum states to a small number, even for polyatomic molecules; (b) these states are the most polar and also the most efficiently aligned, so they would be the most decelerated ones in your experiment; (c) it removes the atomic seed gas, avoiding potential detrimental effects of the seed atoms. How far, do you think, would this help your optical deceleration experiments?

1 F. Filsinger, J. Küpper, G. Meijer, L. Holmegaard, J. H. Nielsen, I. Nevo and H. Stapelfeldt, *Phys. Rev. Lett.*, 2009, **102**, 023001.
2 F. Filsinger, J. Küpper, G. Meijer, L. Holmegaard, J. H. Nielsen, I. Nevo and H. Stapelfeldt, 2009, arXiv:0903.5413.

Professor Barker replied: We are aware of the very nice work on laser alignment and state selection in these references. Such a scheme does appear to be useful for state selection in optical Stark deceleration of polar molecules and would be a good direction to explore for removing seed gas and for loading traps. As demonstrated in our paper, we have shown that the dipole force on molecules can be strongly modified using laser induced alignment with the same field. We plan to extend our current work to explore laser induced alignment for state selection by utilizing the difference in effective polarizability for different rotational states. This is attractive because it would extend state selection to non-polar species and it uses the same field for alignment and deceleration or deflection.

Dr Tarbutt enquired of Professor Barker: As you point out in your paper, the intracavity optical dipole trap looks like a promising trap for sympathetic cooling because it can trap both atoms and molecules in their ground states. One concern with this type of trap is heating due to intensity noise. The cavity will transform frequency noise and the pointing and mode-shape instabilities of the laser into intensity noise and this will shake the trap and heat the molecules. In a high finesse cavity this problem could be severe. Are their any estimates of how high the heating rate is likely to be?

Professor Barker responded: Parametric heating in the dipole trap occurs at twice the harmonic trap frequencies in the axial and radial directions. This corresponds to frequency/intensity fluctuations taking place at frequencies in the 10 kHz range for the radial motion and in the 10 MHz for the axial motion. We anticipate intensity and frequency noise can be corrected on the kHz scale using an acousto optical modulator (AOM), which has been included in our optical locking scheme. Noise at much higher frequencies in the 10s of MHz range cannot be corrected using this scheme. However, as this is about the relaxation oscillation frequencies of these lasers (0.5–1 MHz) noise at these frequencies is not expected to be significant. We intend to study the stability of the Verdi V10 laser system using a test cavity with a linewidth and finesse close to that of the designed dipole trap cavity, so that we can investigate instability and locking strategies before implementing the final design. We cannot make any quantitative statements on heating rates in the optical trap until such a stability study has been carried out.

Professor Weidemüller remarked: A possible way to create an optical dipole trap inside an optical resonator without suffering from atom loss caused by intensity fluctuations due to the transformation of frequency noise into intensity noise by the resonator, is to place the dipole trap within the active cavity of a laser resonator. We have successfully demonstrated trapping of atoms in such a scheme some while ago[1] by integrating a quasi-electrostatic optical dipole trap in the active cavity of a CO_2 laser.

1 M. Eichhorn, M. Mudrich and M. Weidemüller, *Opt. Lett.*, 2004, **29**, 1147.

Professor Barker replied: This may be a promising avenue for the creation of a deep trap for cold molecules without using a high finesse buildup cavity. However, scaling up to ~20 kW of circulating power appears to require a large laser (many metres long) if a DC discharge is to be used. Although compact high power RF CO_2 lasers are available, it is not clear if the required intra-cavity intensity could be reached.

Dr Marian asked Dr Barker: How many quenched ground-state Ar/Ne atoms do you expect to have as a starting point for your sympathetic cooling experiment? What fraction thereof do you expect to load into your optical trap?

Professor Barker responded: We expect quenching to take the MOT atoms to their ground state during a time period of a few 100s of nanoseconds. For this reason, we therefore expect the MOT cloud to retain its density and size immediately after quenching. The MOT volume will be approximately four times larger than the dipole trapping region of interest, therefore overlap between these would lead to ~ 25% of quenched atoms to be loaded. More detailed calculations in our paper point towards a trapped quenched atom number of around 1000 atoms per trap fringe. This number entirely depends on the MOT density after quenching, and does not take into account unidentified loss mechanisms associated with the dipole trap loading scheme. We will be able to make a more quantitative estimate of the quenched density once the metastable MOT is built and characterised.

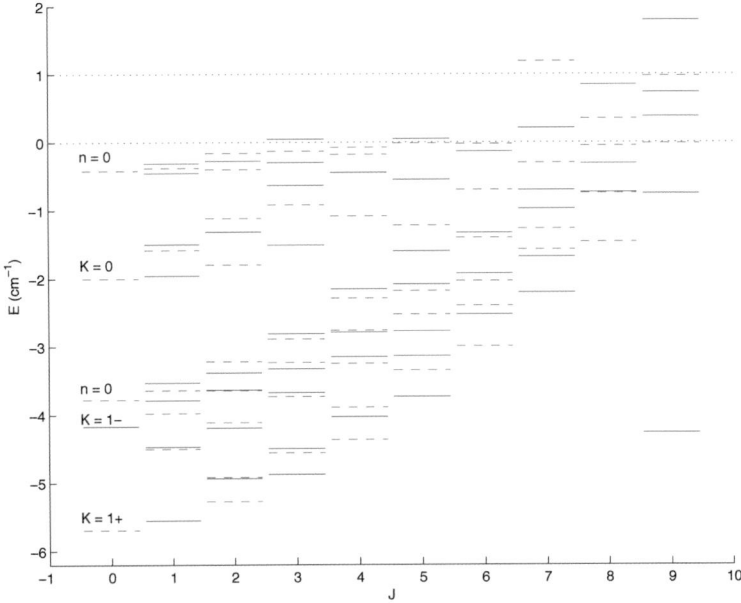

Fig. 14 The near-threshold bound states of Mg–NH shown as a function of total angular momentum J. The ground state energy of Mg–NH is -87.2 cm^{-1}. States of odd parity are shown as a solid line, whereas states of even parity are shown as a dashed line.

Mr Wallis opened the discussion of Professor Hutson's paper: Professor Hutson's paper presented potential energy surfaces for the interaction of NH with alkali-metal and alkaline-earth atoms. We have now carried out bound-state and collision calculations on Mg + NH using the SAPT(DFT) potential shown in Fig. 4 of the paper.

Mg–NH three-body bound-state calculations have been performed in the absence of electromagnetic fields. Shown in Fig. 14 are the near-threshold bound states as a function of the total angular momentum J. The zero of energy is the lowest zero field threshold of NH.

We can rationalize the bound states in terms of the projection of n onto the body-fixed molecular axis K, where n is the angular momentum of the NH monomer without electronic spin. The $J = 0$ states labelled $n = 0$ are from the $n = 0$ NH ground state. The bound states labelled by a value of K originate from the $n = 1$ first rotationally excited state of NH.

In a magnetic field the NH $n = 0$ ground state is split into three thresholds $m_s = -1, 0, 1$. Calculations of the scattering cross sections from the low-field-seeking $m_s = +1$ threshold as a function of magnetic field shows very little resonant structure. This can be explained by noticing that the low-field-seeking threshold is only crossed by high J bound states as a function of magnetic field. These high J bound states are so weakly coupled to the threshold that the Feshbach resonances are vanishingly small.

Shown in Fig. 15 are the total elastic and inelastic Mg–NH cross sections as a function of collision energy at a magnetic field of 0.5 Tesla. In the micro to millikelvin range the elastic cross section is two orders of magnitude greater than the inelastic cross section, indicating that sympathetic cooling may be possible.

Professor Grimm asked Professor Hutson: How general is the 'new hope' for sympathetic cooling? You analysed the particular situation of NH. What does one learn from this for cooling other molecules?

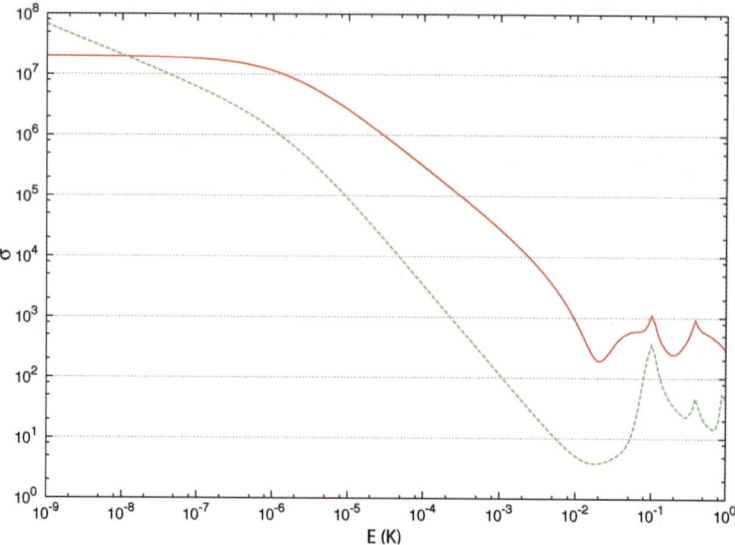

Fig. 15 Elastic (solid) and inelastic (dashed) MgNH low-field-seeking state cross sections as a function of collision energy at a magnetic field of 0.5 T. The cross sections incorporate s, p, and d partial wave contributions and are in Å^2.

Professor Hutson responded: Inelastic collisions are the enemy of sympathetic cooling because they cause trap loss. Closed-shell atoms have two advantages over alkali-metal atoms in this respect. First, there is no possibility of spin exchange collisions involving the nonexistent atomic electron spin. Second, the anisotropy of the interaction is often lower, and the anisotropy is an important factor in driving collisions that change the state of the molecule (including its Zeeman state) without changing the atomic state. In this sense the preference for alkaline-earth atoms is fairly general. However, alkaline earth atoms retain one disadvantage of alkali metals, which is that they have low ionisation potentials. For molecules with vacant valence orbitals, such as OH, NH and CH, this creates low-lying charge-transfer states that can intefere with sympathetic cooling. In our study of NH, these states are accessible for Ca and Sr but are probably inaccessible for Be and Mg. This is quite a delicate balance, and for other molecules the boundary might come somewhere else in the series. However, the alkaline earths do have some general advantages over alkali-metal atoms for cooling molecules that are not in their absolute ground state.

The situation is different for molecules in closed-shell singlet states that are trapped in their absolute ground state in the applied field. For these molecules, we now think that the prospects for sympathetic cooling with magnetically trapped alkali-metal atoms are quite good.[1] That is because, even if the anisotropy of the potential is large, the molecular degrees of freedom are only weakly coupled to the atomic degrees of freedom. I think Dr Żuchowski is planning to say some more about this later in the Discussion.

1 P. S. Żuchowski and J. M. Hutson, *Phys. Rev. A*, 2009, **79**, 062708.

Professor Grimm continued: An idea would be to use laser-coolable atoms with an inner-shell cooling transition, such as erbium. The outer closed shells may strongly suppress inelastic decay. Do you think this might work?

Professor Hutson answered: Yes, I think that is quite a good prospect. It may still be possible for an electron to hop from the molecule to the atom to create a charge-transfer state, but if the outer electrons create enough repulsion the charge transfer state may cross the ground state some way up on its repulsive wall, as in Mg–NH. In that case there is a good prospect that the charge-transfer state may not interfere with sympathetic cooling.

A caveat is that, even for atoms with "buried" open shells, there can be coupling between the atomic and molecular angular momenta arising from a variety of sources. Some of these are long-range and quite strong, such as the interaction between an atomic quadrupole moment and a molecular dipole moment. Each case needs careful analysis to decide which terms might cause strong inelasticity and whether it is possible to choose atomic and molecular states that avoid them (such as spin-stretched states).

Dr Żuchowski commented: We have used our potential energy surface for Rb–NH$_3$[1] to investigate the possibility of sympathetic cooling of ND$_3$ molecules with Rb atoms.[2] Our calculations were carried out in the absence of external fields and neglected hyperfine coupling. For molecules that are initially in the upper inversion component u of the $J = 1$, $K = 1$ state, which correlates with the low-field-seeking state that is usually used in deceleration experiments, the only inelastic channel corresponds to the lower inversion component l of the same state. Inelastic collisions will release internal energy into kinetic energy and cause trap loss. Fig. 16 shows the ultracold limit of the cross sections for elastic scattering and for relaxation from the upper to the lower inversion component. Since the interaction potential is highly anisotropic and strongly couples rotation-inversion energy levels, the inelastic collisions are surprisingly slow. However, they are probably still too fast for sympathetic cooling to succeed for molecules in low-field-seeking states, since the ratio of elastic-to-inelastic cross sections is never bigger than 100.

Another interesting question is whether ND$_3$ molecules in high-field-seeking states, for example stored in an AC or optical trap, can be sympathetically cooled by collisions with an ultracold gas of magnetically trapped Rb atoms, with the atoms in their low-field-seeking state. For Rb + OH collisions, Lara *et al.*[3,4] found that

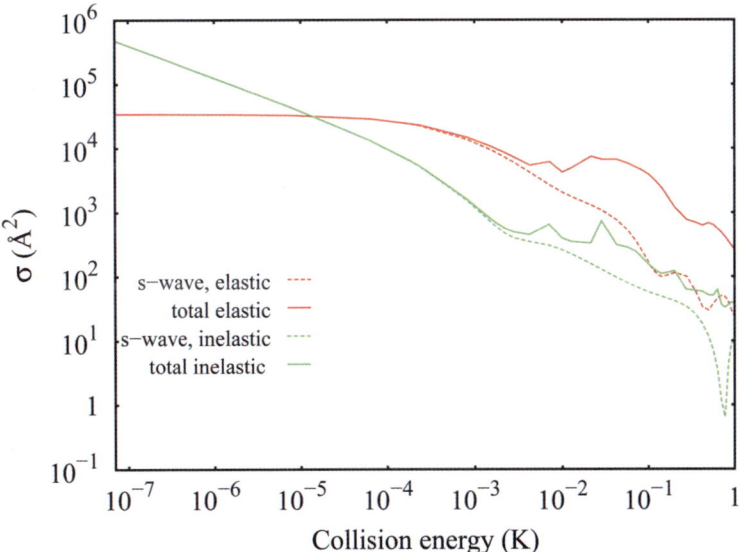

Fig. 16 Ultracold limit of the elastic $J = 1$, $K = 1$, u to $J = 1$, $K = 1$, u and inelastic $J = 1$, $K = 1$, u to $J = 1$, $K = 1$, l cross sections and the contribution from s-wave scattering.

inelastic collisions that change the hyperfine state of the Rb atom are very fast. However, careful inspection of the collision Hamiltonian for Rb + NH_3[2] suggests that in this case the only terms which couple the atomic spins to molecular degrees of freedom and can drive inelastic collisions that change the atomic hyperfine state are very weak. We therefore expect that relaxation between atomic energy levels due to the collisions with ammonia will be slow and sympathetic cooling of NH_3 or ND_3 in high-field seeking states with magnetically trapped atoms may well succeed. However, quantitative calculations are needed to verify this and present a substantial theoretical and computational challenge.

The hyperfine splittings of most isotopologues of NH_3 (except $^{15}NH_3$) are dominated by nuclear quadrupole splittings. For $^{14}ND_3$, for example, the splittings are around 100 µK. If collisions that change the hyperfine levels of ND_3 are fast, the transfer of internal hyperfine energy into translation may limit sympathetic cooling. This provides a further motivation for developing computational methods capable of calculating the rates of hyperfine-changing collisions.

1 P. S. Żuchowski and J. M. Hutson, *Phys. Rev. A*, 2008, **78**, 022701.
2 P. S. Żuchowski and J. M. Hutson, *Phys. Rev. A*, 2009, **79**, 062708.
3 M. Lara, J. L. Bohn, D. Potter, P. Soldán, and J. M. Hutson, *Phys. Rev. Lett.*, 2006, **97**, 83201.
4 M. Lara, J. L. Bohn, D. E. Potter, P. Soldán, and J. M. Hutson, *Phys. Rev. A*, 2007, **75**, 012704.

Professor Gianturco asked Professor Hutson: In your presentation you mentioned calculations for the potential energy surfaces for the interactions of NH molecules with alkaline-earth atoms as new hopes for finding possible systems which may exhibit rather weak coupling for directing flux into the inelastic channels. I was therefore wondering if you have also explored the size of inelastic channels for the quenching of the more highly excited rotational states of NH. Although one hopes and expects to have it prepared in its lowest rotational state , it may be useful to have an idea of the quenching probabilities as we have already found in our published calculations[1] that those cross sections are rather large for the case of NH with Rb and Cs.

1 M. Tacconi, L. Gonzaléz-Sanchez, E. Bodo and F. A. Gianturco, *Phys. Rev. A*, 2007, **76**, 032702.

Professor Hutson replied: No, we haven't calculated the rotationally inelastic cross sections. Our focus was on the M-changing transitions that will determine whether sympathetic cooling is possible for molecules in low-field-seeking states. However, it would be quite easy to do it if it becomes experimentally interesting.

Professor Meijer opened the discussion of Professor Rempe's paper: The density in the optical Stark-decelerated sample, as well as in the (electrostatic) Stark-deceleration and in the Zeeman deceleration is, in the end, determined by the density in the pulsed supersonic source, so it should be rather similar for these different approaches. In the continuous source there is a high flux, but this is only useful when one can accumulate the slow molecules somehow in the trap; otherwise it might be better to quote (state-selected) densities instead. So I question the statement that the continuous source is 'the most intense source of slow and internally cold molecules' and I would argue that a pulsed supersonic source is superior; one starts with a high phase-space density and just needs to get rid of the overall velocity.

Professor Rempe replied: Our cold molecules source transfers the fraction of slow molecules out of a reservoir into an UHV chamber with background gas pressure in the 10^{-11} mbar range. We find that the molecule density at the exit of the guide equals the corresponding density of slow molecules in the reservoir within a factor of order unity. We argue that this remarkably high filtering efficiency together with the simplicity of the setup makes our source ideal for a wide range of experiments, in particular for those where a continuous beam is advantageous. This is the case for collision studies with immobile targets such as trapped molecules of another species, trapped atoms or surfaces where only the average number of particles per second per surface area counts. Our source produces continuous beams with fluxes of 10^{11} s^{-1} and densities of 10^9 cm^{-3} in a useful area of about 1 mm^2. The internal state purity of the beam can be close to unity by implementing cryogenic buffer gas cooling.[1] To increase the phase-space density even further, we have recently proposed a dissipative scheme for accumulating molecules in an electrostatic trap.[2] This trap fits ideally to the end of the quadrupole guide and can be loaded continuously.

1 L. D. van Buuren, C. Sommer, M. Motsch, S. Pohle, M. Schenk, P. W. H. Pinkse and G. Rempe, *Phys. Rev. Lett.*, 2009, **102**, 033001.
2 M. Zeppenfeld, M. Motsch, P. W. H. Pinkse and G. Rempe, 2009, arXiv:0904.4144.

Professor Barker commented: I agree that the state selected density in optical Stark deceleration or electrostatic Stark deceleration should be essentially the same and only dependent on how close we can work to the exit of the pulsed valve that produces the molecular beam. However, in optical Stark deceleration we can also decelerate all the occupied rotational states for a particular molecule. When trapped in an optical trap (as described in our paper), we may also be able to cool rotational degrees of freedom and increase the state density by sympathetic cooling with a laser cooled, rare gas atomic species.

Dr Vanhaecke asked Professor Rempe: In your proposed scheme for accumulation and cooling of polar molecules in an electrostatic trap, you show only spontaneous emission which leads to low-field seeking Stark states. How is it possible to avoid spontaneous emission which leads to high-field seeking states? What is the trick to do so?

Professor Rempe answered: The cooling scheme we propose in is expected to work best for symmetric top molecules.[1]
This is mainly due to the angular momentum selection rules which exist for such molecules, ($\Delta K = 0$). Thereby the molecules can be kept in low-field seeking states in both the vibrational ground state and the vibrationally excited state. We assume, of course, that the trap is initially filled with molecules in low-field seeking states, as is automatically the case when loaded with our electrostatic guide.

1 M. Zeppenfeld, M. Motsch, P. W. H. Pinkse and G. Rempe, 2009, arXiv:0904.4144.

Dr Tarbutt asked Professor Rempe and Dr Pinkse: My question concerns the usefulness of the bend in the guide. For some experiments, *e.g.* collision experiments, the bend is useful because it allows for the control of the velocity range in the experiment. But for trapping experiments, it is not clear that the bend is so useful – the trap already serves as a velocity filter. A straight guide of sufficient length would seem to fulfill the same function of separating the molecules of interest from the background gas and delivering them to a region of high vacuum. For the purpose of filling a trap, would a straight guide be equally good?

Professor Rempe and Dr Pinkse responded: A straight guide would support a large flux of molecules which are transversally slow but longitudinally too fast to be captured in a trap. The bend prevents these molecules from reaching the trap, thereby suppressing the bombardment of trapped molecules by fast molecules from the source. Note that for the same reason, the bend reduces the background vapour pressure in the science region at the end of the guide. In general, once a slow molecule has made it around the bend, it can no longer collide with fast molecules emerging from the source.

Professor Softley asked: The use of the helium buffer gas cooled source has successfully led to cooling of the internal degrees of freedom of the molecules, but this appears to be at the expense of the lowest translational temperatures. Your guided beam velocities are typically of order 60 ms^{-1}. Is this limited by the helium velocity, or can you see possibilities to produce lower velocities while maintaining the low internal temperature? Could you employ some form of deceleration stage after the quadrupole guide?

Professor Rempe answered: A certain number of collisions between the molecules and the cryogenic helium gas are required to provide translational and rotational cooling. Note that the light molecules used in our experiment are already in the vibrational ground state. A consequence of the required helium density is that collisions occur between relatively fast cryogenic helium atoms and slow molecules in the vicinity of the exit hole of the buffer gas cell. In our past experiments, this 'boosting' effect imposes probably the most severe limitations on the achievable flux of slow and guided molecules. The 'boosting' is much less pronounced in our experiments without a helium buffer gas[1]. With the cryogenic buffer gas present, the highest flux of slow molecules and lowest internal temperatures are achieved after optimization of the helium density. It is always possible to compromise between high flux at slow velocities and low rotational temperature. Alternatively, one can attempt to get rid of the boost with a special two-cell arrangement[2] or by further deceleration by means of static or time-dependent forces.

1 M. Motsch, C. Sommer, M. Zeppenfeld, L. D. van Buuren, P. W. H. Pinkse and G. Rempe, 2008, arXiv:0812.2850.
2 D. Patterson and J. M. Doyle, *J. Chem. Phys.*, 2007, **126**, 154307.

Dr Küpper enquired: On your last slide you state that the presented source is applicable to 'any' molecule. How far do you assume you can take this at reasonable fluxes? Would it work with substituted benzenes, *i.e.* benzonitrile? With considerably larger molecules? In how far would you expect to produce cold and slow bare molecules and not van der Waals clusters of molecules with the buffer gas atoms?

What are, after all, the relevant molecular properties to assure efficient cooling without condensation to clusters? How important is the molecular polarizability for the multipole moment?

Professor Rempe replied: In our experiments employing the cryogenic source[1] we have so far used many different molecules ranging from light species such as ammonia (symmetric rotor), or formaldehyde (asymmetric rotor), to heavier ones like trifluoropropyne and fluoroform (symmetric rotor). Our method relies only on a favourable ratio of (positive) Stark shift over mass. Changing the molecular gas requires hardly any changes to the apparatus as long as the molecules can be brought into the buffer-gas cell in the gas phase. In that sense, the cold molecules source is readily available for "any molecule". No experiments exist for larger molecules with – most likely – smaller Stark shifts. Testing such molecules would definitely be interesting. Concerning your question about the possibility of helium van der Waals clusters forming around the molecule, I can answer that in our experiments with ammonia, such clusters have not been detected with our mass spectrometer at the end of the guide. In fact, it seems unlikely that helium clusters are produced under our experimental conditions: the pressure in the buffer-gas cell we employ is a fraction of a mbar, which is several orders of magnitude lower than the pressure normally used in various techniques for the formation of clusters between a molecule and noble gas atoms, typically ranging from a few bar to a few tens of bar.[2]

1 L. D. van Buuren, C. Sommer, M. Motsch, S. Pohle, M. Schenk, P. W. H. Pinkse and G. Rempe, *Phys. Rev. Lett.*, 2009, **102**, 033001.
2 M. Hartmann, R. E. Miller, J. P. Toennies and A. Vilesov, *Phys. Rev. Lett.*, 1995, **75**, 1566; R. M. Helm, H.-P. Vogel and H. J. Neusser, *Chem. Phys. Lett.*, 1997, **270**, 285.

Broadband lasers to detect and cool the vibration of cold molecules

Matthieu Viteau,[a] Amodsen Chotia,[a] Dimitris Sofikitis,[a] Maria Allegrini,[ab] Nadia Bouloufa,[a] Olivier Dulieu,[a] Daniel Comparat[a] and Pierre Pillet[*a]

Received 4th November 2008, Accepted 9th February 2009
First published as an Advance Article on the web 27th May 2009
DOI: 10.1039/b819697d

By using broadband lasers, we demonstrate the possibilities for control of cold molecules formed *via* photoassociation. Firstly, we present a detection REMPI scheme (M. Viteau *et al.*, *Phys. Rev. A*, 2009, **79**, 021402) to systematically investigate the mechanisms of formation of ultracold Cs_2 molecules in deeply bound levels of their electronic ground state $X^1\Sigma_g^+$. This broadband detection scheme could be generalized to other molecular species. Then we report a vibrational cooling technique (M. Viteau *et al.*, *Science*, 2008, **321**, 232) through optical pumping obtained by using a shaped mode locked femtosecond laser. The broadband femtosecond laser excites the molecules electronically, leading to a redistribution of the vibrational population in the ground state *via* a few absorption–spontaneous emission cycles. By removing the laser frequencies corresponding to the excitation of the $v = 0$ level, we realize a dark state for the so-shaped femtosecond laser, leading, with the successive laser pulses, to an accumulation of the molecules in the $v = 0$ level, *i.e.* a laser cooling of the vibration. The simulation of the vibrational laser cooling allows us to characterize the criteria to extend the mechanism to other molecular species (R. V. Krems, *Int. Rev. Phys. Chem.*, 2005, **24**, 99). We finally discuss the generalization of the technique to laser cooling of the rotation of the molecule.

1 Introduction

The control of the dynamics of a quantum system is a crucial goal in both physics and chemistry.[1] The precise control of both the internal and external degrees of freedom of a molecule should open up fascinating new fields of research, which motivate the important activities developed in the cold-molecule domain. To prepare robust samples of trapped ultracold ground-state molecules with neither vibration, nor rotation is expected to lead to significant advances in molecular spectroscopy, molecular clocks, fundamental tests in physics such as electron-dipole moment, chirality or variation of some fundamental constants, collision, super or controlled photochemistry studies, and also in quantum computation with the use of polar molecules.[2,3] The difficulties of extending laser cooling techniques from atoms to molecules[4] is linked to the lack of isolated and closed two-level schemes in the molecular systems. The large number of ro-vibrational levels of a molecule makes the addition of repumper lasers, as in the case of atoms, to close such a transition unreasonable. An interesting alternative could be the use of a laser with a large

---

[a]*Laboratoire Aimé Cotton, CNRS, Univ Paris-Sud, Bât. 505, 91405 Orsay, France. E-mail: pierre.pillet@lac.u-psud.fr*
[b]*CNISM, Physics Department, Pisa University, Largo Pontecorvo, 3, 56127 Pisa, Italy*

spectral band able to excite several transitions between rovibrational levels of two different electronic potentials of the molecules simultaneously.

The interest of the use of such broadband lasers has already been demonstrated for two applications with translationally cold molecules formed *via* photoassociation of cold cesium atoms. Both applications are linked to the distribution of the populations of the molecules in a large number of different ro-vibrational levels. The first one concerns the REMPI (Resonantly-Enhanced Multi-Photon Ionization) detection of the translationaly cold molecules. In this case the laser is a dye laser pumped by the second harmonic of a pulsed Nd:YAG laser with a repetition rate of 10 Hz, a temporal duration of the pulse of 10 ns, and a spectral bandwidth of 25 cm^{-1}. The second application is the laser cooling of the internal degree of freedom of the vibration of the molecules, to prepare a molecular sample in the ground state with no vibration.[5] The laser used is a shaped mode locked femtosecond laser, with a repetition rate of 80 MHz, a pulse duration of 100 fs, and a σ-Gaussian bandwidth of 54 cm^{-1}.

The article is organized as follows. We first recall the dynamics of photoassociation of cold atoms leading to the formation of translationally cold molecules in several ro-vibrational levels of the ground state or of the lowest triplet state. In the next section, we expose an efficient broadband REMPI process well-adapted to the detection of such cold molecules. Then we report the use of broadband shaped femtosecond sources for the laser cooling of the vibration of these translationally cold molecules. We analyze the criteria for extending the method to any molecular species and the possibilities for realizing the cooling of the rotation of the molecule. In the conclusion, we discuss further applications for the manipulation of molecules by broadband lasers.

2 Cold molecules formed *via* photoassociation

Photoassociation has been demonstrated for a very large number of homonuclear or heteronuclear systems. Cold ground-state molecules have been observed in many cases. The experiments reported here concern the case of the cesium atom. Our experiment uses Cs, but similar experiments, or theoretical investigations, have been performed on several other systems, including heteronuclear ones. However, in this article, we use cesium as an example to describe photoassociation and refer to more general reviews, such as refs 2, 6 and 7 for detailed investigations concerning the other systems. Cesium offers many configurations in the formation of cold molecules, corresponding for a pair of atoms to a reaction

$$Cs(6s,F) + Cs(6s,F) + h\nu_L \rightarrow Cs_2^*(\Omega(6s + 6p_j); v, J) \quad (1)$$

Two colliding Cs atoms in a hyperfine level F of their ground state $6s$ absorb one laser photon at frequency $h\nu_L$ red-detuned compared to the atomic resonance frequency ($6s + 6p_j, j = 1/2$ or $3/2$) to form a molecule in a well defined rovibrational level (v, J) of an excited molecular state Ω correlated to one of the asymptotes ($6s + 6p_{1/2}$) or ($6s + 6p_{3/2}$). In cold thermal samples, the resolution of the photoassociation process is limited by the width of the statistical distribution of the relative kinetic energy of the colliding atoms, which is of the order of k_BT (k_B is the Boltzman constant and T the temperature of the atomic sample). Due to the extremely narrow width of the thermal distribution of ultracold atoms ($k_BT \sim 2$ MHz at $T \sim 100$ μK), smaller than any other relevant energy of the system like the molecular level spacing, photoassociation with cold atoms has proven to be a powerful tool for high-resolution molecular spectroscopy, in particular for the whole class of alkali atoms from Li to Cs.[2,6,7] It has given access to the previously unexplored domain of molecular dynamics at distances well beyond those of well-known chemical bonds. Indeed, a pair of identical ground state atoms interact at large interatomic distances R through their Van der Waals interaction behaving as R^{-6}; when the atom pair is excited, the dipole–dipole interaction is dominant, and varies as R^{-3}. Vibrational levels with a very large elongation (from a few tens up to a few hundred atomic units)

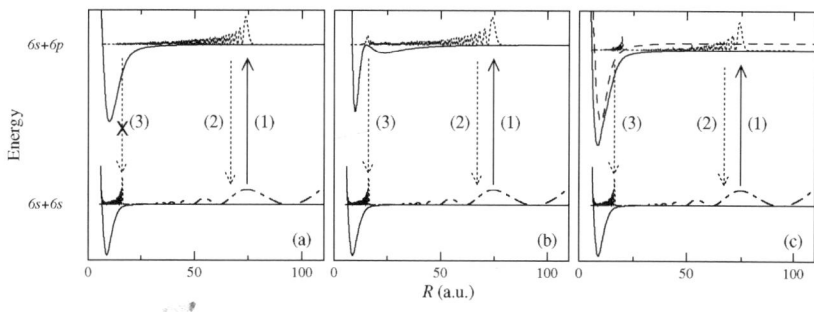

Fig. 1 Photoassociation from the cesium 6s + 6s continuum [reaction (1)] to (a) a single excited state; (b) a double-well state [e.g. 0_g^-, $1_u(6s + 6p_{3/2})$]; (c) two coupled states [e.g. $0_u^+(6s + 6p_{1/2, 3/2})$]. The system decays by spontaneous emission either back to the continuum [reaction (2)] or to a bound level of the ground state [reaction (3)]. For case (a), reaction (3) is unlikely.

are then efficiently populated by photoassociation, as can be understood from Fig. 1(a) where typical radial wavefunctions for the initial collision state and for the final vibrational level are represented. In a perturbative approach and under the assumption of an R-independent dipole transition moment, the photoassociation rate is proportional to the squared overlap integral between radial wavefunctions. Due to the very different oscillatory pattern of the wavefunctions at large distances, we can show that the overlap integral is proportional to $|\psi_\alpha(R_T)|^2$ where ψ_α is the initial collision state radial wavefunction, and R_T is the outer classical turning point of the photoassociated level. In other words, the photoassociation process occurs at a large distance, roughly at the classical outer turning point of the final vibrational level. The R^{-3} behavior of the electronically excited potential curve makes the photoassociation process quite efficient.

The electronically excited molecules created by photoassociation have a short lifetime of a few tens of nanoseconds due to spontaneous emission. Most often they decay back to a pair of "hot" atoms (i.e. with a large relative kinetic energy). Most experimental photoassociation setups use a magneto-optical trap illuminated by a photoassociation laser. The spontaneous decay provides the detection method of the photoassociation signal: it results in decreasing intensity of the trap fluorescence due to the escape of hot atoms from the trap. As suggested by Fig. 1(a), deeply bound levels of the electronic states correlated to the ground state asymptote cannot be easily populated by spontaneous decay due to the poor overlap of the corresponding radial wave functions, meaning poor Franck–Condon factors. The issue is now to find situations where the probability density of the radial motion can be transferred towards small interatomic distances, to ensure a good spatial overlap with the vibrational wave functions of the deeply bound levels of the lowest potential curves. From a classical point of view, the ideal vibrational motion of the PA molecule should slow down at least in the intermediate distance range (say, around 15–20 atomic units) to let spontaneous decay occur before going back towards the long distance range. This solution was demonstrated for the first time in 1998, starting from an ensemble of cold cesium atoms.[8] Fig. 1(b) shows a scheme for a double-well configuration corresponding to the 0_g^- and 1_u double-well potential curves correlated to the $Cs(6s)+Cs(6p_{3/2})$ limit. These peculiar states, known as pure long-range molecules, result from the competition of spin–orbit interaction and long-range dipole–dipole interaction. The slow R-variation of the left edge of the outer potential well induces a "speed bump" in the distance range appropriate for radiative decay towards stable vibrational levels. Fig. 1(c) illustrates another efficient formation mechanism involving resonances between vibrational levels of different electronic states coupled through a non-Born–Oppenheimer interaction. Such a situation involving internal couplings has been demonstrated for the 0_u^+ molecular symmetry in Cs_2 but also in Rb_2 and KRb.[9] Such a pattern is very common in molecular systems and opens up many routes for the formation of cold molecules in photoassociation experiments.

3 Broadband detection of cold molecules

The details of the experimental setup for photoassociation of cold atoms in a Cs vapor-cell magneto-optical trap can be found in reference 8. The photoassociation of cold cesium atoms is achieved with a cw Titanium:Sapphire laser (intensity 300 W cm^{-2}), pumped by an Argon-ion laser, exciting molecules which can decay by spontaneous emission into vibrational levels of the molecular ground state $X^1\Sigma_g^+$, (hereafter referred to as X), or of the lowest triplet state $a^3\Sigma_u^+$. A key point of these experiments is the detection mechanism. Formation of ultracold molecules *via* photoassociation is demonstrated by using very sensitive REMPI (Resonant Enhanced MultiPhoton Ionization) detection. Fig. 2(a) shows the potential[5] involved in the REMPI process. The cold molecules are ionized to Cs_2^+ ions by using a pulsed dye laser (7 ns duration, 1 mJ energy, focused to a 1 mm^2, 15 GHz resolution) pumped by the second harmonic of a Nd-YAG laser, at a 10 Hz repetition rate. Fig. 3 shows the REMPI spectrum obtained by scanning the frequency of the pulsed REMPI laser in the range 13500–14500 cm^{-1}, when the molecules are formed *via* the photoassociation of the vibrational level, $v = 79$, of the state $0_g^-(6s + 6p_{3/2})$. For frequencies above 13850 cm^{-1}, the scan consists of many densely located molecular lines. The REMPI process consists of a first resonant step corresponding to transitions between ro-vibrational levels of the lowest triplet state $a^3\Sigma_u^+$ and the $2^3\Pi_g$ state converging to the dissociation limit $6s + 5d$. The second step corresponds to a one-photon ionization of the intermediate ro-vibrational level of the $2^3\Pi_g$ state. After the photoionization laser pulse, a pulsed high electric field (3 kV cm^{-1}, 0.5 μs) is applied at the trap position by means of a pair of grids spaced 15 mm apart to accelerate the ions. The ions are expelled from the photoassociation region in a 6 cm free field zone constituting a time-of-flight mass spectrometer, useful to separate the Cs_2^+ ions from spurious Cs^+ ions. A pair of microchannel plates detects the ions. This detection scheme is quite sensitive, allowing one to observe up to one unique ion. However, the global efficiency of the process is relatively poor, limited by the ion recollection rate (80%), the microchannel plate efficiency (35%) and the REMPI efficiency itself, which is unknown. An estimation of 10% for the REMPI efficiency is obtained by

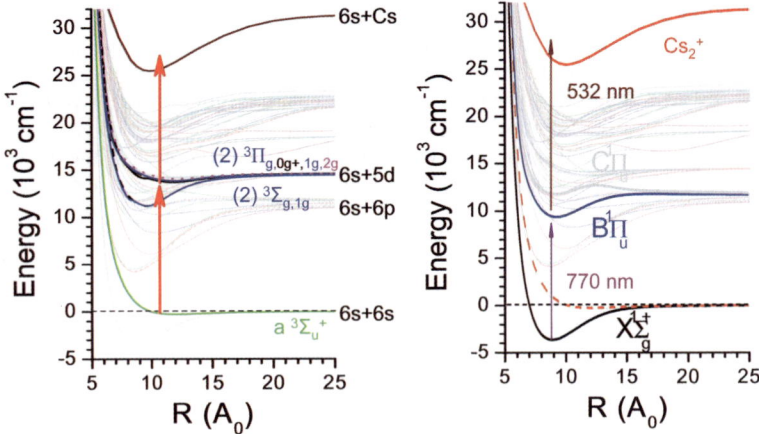

Fig. 2 (a) REMPI scheme of the first triplet state with a pulsed dye laser (~720 nm); (b) REMPI detection scheme of deeply-bound ground state Cs_2 molecules with a broad-band laser (770 nm and 532 nm) *via* the $B^1\Pi_u$ state.

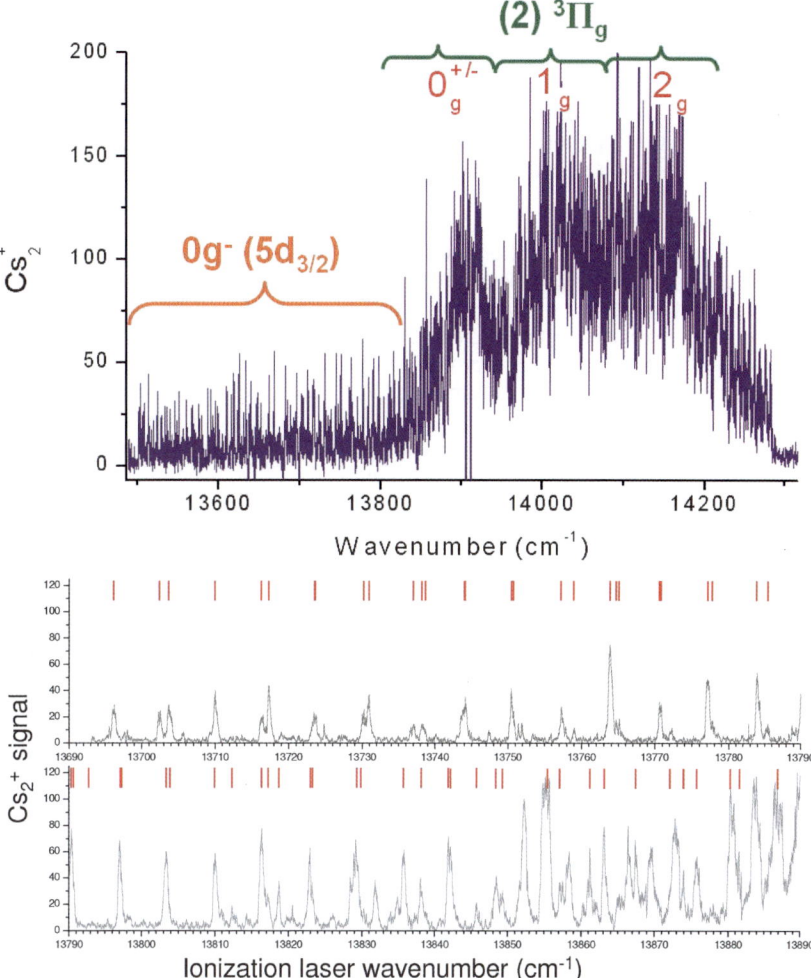

Fig. 3 (a) Photoionisation spectrum of Cs_2 cold molecules formed after spontaneous emission from the vibrational level, $v = 79$, in the $0_g^-(6s + 6p_{3/2})$ state. (b) Part of the spectrum where each transition identified is indicated by a line. The ionisation, in this part of the spectrum is obtained via the state $0_g^-(6s + 5d_{3/2})$.

comparing the Cs_2^+ ion signal with the trap loss fluorescence signal, by using calculated branching ratios between bound–bound and bound–free transitions for the photoassociated molecules.[10] This relatively low rate for the REMPI process is linked to its resonant character, leading to the fact that only a few initially populated ro-vibrational levels, for which the two-photon process is resonant, are efficiently ionized. Nevertheless, the envelope of the photoionization spectrum stays roughly the same for any vibrational levels, which makes the considered REMPI process here well-adapted to the analysis of the formed molecules, without knowing *a priori* the details of the molecular structure.

For frequencies below 13850 cm^{-1}, the scan consists of a resolved spectrum. The first resonant step of the REMPI process corresponds closely to transitions between ro-vibrational levels of the lowest triplet state $a^3\Sigma_u^+$ and the $2^3\Sigma_g^+$ state converging to the dissociation limit $6s + 5d$. The analysis of the lines allows us to conclude that the vibrational levels between $v = 11$ and 20 of the lowest triplet state, $a^3\Sigma_u^+$, are populated. This part of the spectrum is complementary to the latter one, but if these REMPI resonances give information on the population of the different vibrational levels, they cannot be used for research into mechanisms of the formation of cold molecules. To conclude at this point of the article, the detected cold molecules are vibrationally excited in the lowest triplet state. Molecules in the singlet ground state, such as those obtained *via* photoassociation in the $1_u(6s + 6p_{3/2})$ and $0_u^+(6s + 6p_{1/2})$ states, are also detected. In both cases, they are in highly excited vibrational levels.

Research into molecule formation schemes based on photoassociation necessitates a general and systematic detection method, without knowing *a priori* the details of the molecular structure. The proposed method is based on a detection procedure which does not select the population of a particular bound level. In order to observe deeply-bound molecules in the X ground state, which could result from an *a priori* unknown mechanism, we set up a broadband REMPI detection through vibrational levels v_B of the spectroscopically known $B^1\Pi_u$ excited state[11] (refereed to as the B state). The two-photon transition is induced by a pulsed dye laser (LDS751 dye, wavelength ~770 nm, pulse energy ~1 mJ, focused waist ~500 μm) and by the pump laser (532 nm wavelength) as illustrated in Fig. 2(b). The major advance of the method compared to previous one is the broadband detection of the formed ultracold molecules. We replaced the grating in the pulsed dye laser cavity by a less dispersive prism, which broadens its line width from ~0.05 cm^{-1} to ~25 cm^{-1}. We display in Fig. 4 the results of a model of the ionization process, for both narrowband (panel a) and broadband (panel b) schemes. We assume that the ionization probability due to the second (532 nm) photon is independent of the specific v_B level, so that it is proportional to the population of this level induced by the first photon at 770 nm. The excitation probabilities of the v_X levels toward the

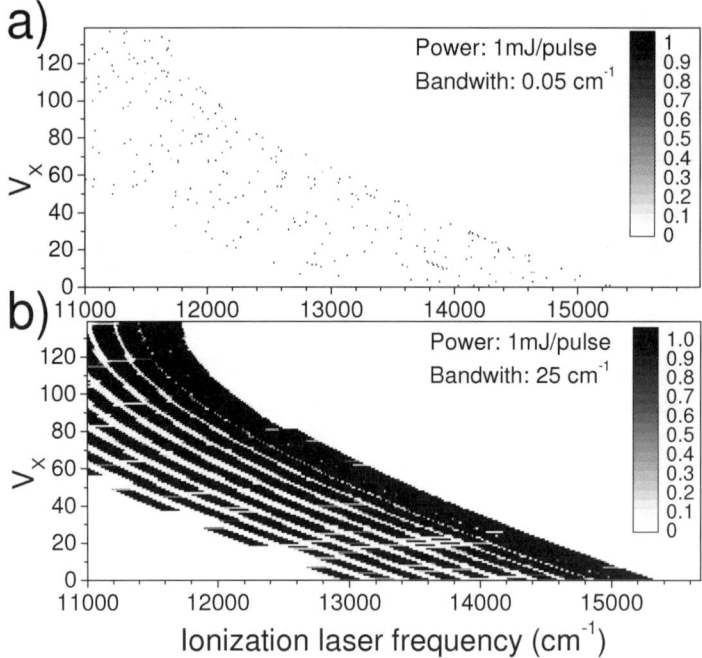

Fig. 4 (a) and (b) Comparison of transition probabilities (in shades of grey) for the transitions at energy E_{X-B} between vibrational levels of the $X(v_X)$ and B states, for a laser line width of 0.05 cm^{-1} (a) and of 25 cm^{-1} (b), with identical power (1 mJ/pulse). The probability is set at unity for a saturated transition.

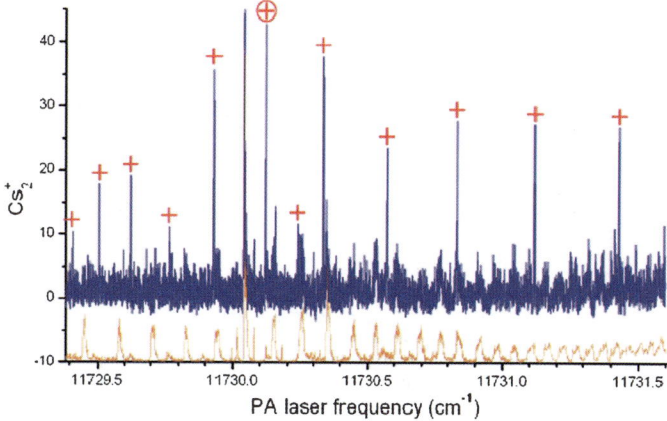

Fig. 5 Upper trace: Cs_2^+ ion spectrum recorded after scanning the frequency of the PA laser below the $6s + 6p_{3/2}$ dissociation limit, and using the broadband REMPI detection laser with energy around 13000 cm^{-1}. The crosses label the previously unobserved PA lines. Lower trace: Cs_2^+ ion spectrum obtained using the conventional narrowband REMPI detection,[8] displayed with an offset of 10 ions for clarity.

v_B levels are obtained from Franck–Condon factors computed for the experimentally known X and B potential curves,[12,11] assuming a constant dipole transition moment. As expected, the narrowband ionization scheme allows the ionization of a single v_X level at a given frequency (Fig. 2(a)). In contrast, the broadband scheme involves a laser pulse width of the order of the vibrational spacing of both the X and B states (up to 40 cm^{-1}), so that many vibrational v_X levels can be ionized in a single shot (Fig. 4(b)). For instance, a laser pulse at ~11730 cm^{-1} or at ~13000 cm^{-1} would excite almost all molecules lying in vibrational levels $v_X > 37$ or $v_X < 70$ respectively.

Choosing the latter energy range for the broadband detection laser (~13000 cm^{-1}) and scanning the photoassociation laser frequency over a few wavenumbers below the $6s + 6p_{3/2}$ dissociation limit, we discovered several intense photoassociation lines, labeled with crosses in Fig. 5, revealing a large number of ultracold molecules formed in low ($v_X < 70$) vibrational levels of the X state. These detected singlet molecules were present in our previous experiments performed in the same photoassociation energy range,[8] but our previous narrowband REMPI detection scheme (wavelength ~720 nm), optimized to detect triplet $a^3\Sigma_u^+$ molecules, was blind to these singlet molecules (see lower spectrum of Fig. 5). A new mechanism for the formation of cold molecules has therefore been identified. We depict the process as follows: the PA laser excites the atom pair into a bound level of the lowest $1_g(6s + 6p_{3/2})$ long-range potential curve (Fig. 6(a)), which is coupled at short distances to the lowest $1_g(6s + 6d_{5/2})$ potential curve, through several avoided crossings induced by spin–orbit interaction (situation similar to that of Fig. 1(c)).

4 Laser cooling of the vibration of molecules

To determine precisely the internal state of these formed molecules we performed a more conventional REMPI spectrum using a narrowband (DCM) dye laser (wavelength ~627 nm). We fixed the PA laser energy on the most intense line of Fig. 5 (at 11730.1245 cm^{-1}, about 2 cm^{-1} below the $6s + 6p_{3/2}$ asymptote. Then we scanned the ionization laser frequency to record the ionization spectrum of the ground state molecules excited through the intermediate $C^1\Pi_u$ state (see Fig. 6(b)). Taking advantage of the spectroscopic knowledge of the X–C transitions,[12,13] the lines, shown in Fig. 7, were easily assigned to transitions from ground state vibrational levels

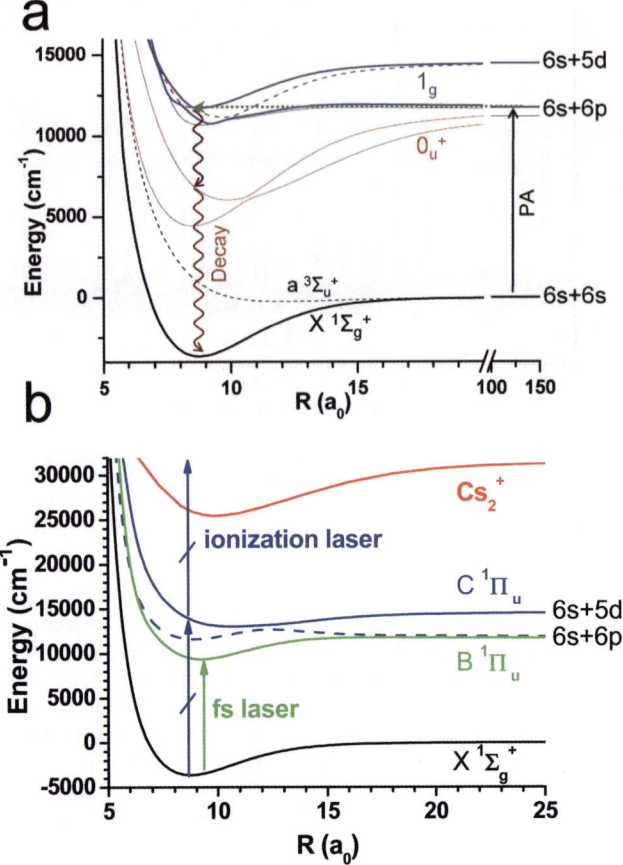

Fig. 6 Relevant molecular states of the cesium dimer, converging towards the dissociation limits $6s + 6s$, $6s + 6p$, and $6s + 5d$. (a) Photoassociation of cold atoms and formation of cold molecules. For the potentials 1_g, the long-range radial wavefunction is coupled to the short range radial wavefunction by internal coupling of the potentials. The ground-state molecules, $X^1\Sigma_g^+$, are formed in a cascading spontaneous emission via the 0_u^+ potentials. (b) REMPI detection process via the $C^1\Pi_u$ state, and transition ($X^1\Sigma_g^+$ towards $B^1\Pi_u$) of applying the femtosecond laser.

restricted to the range $v_X = 1$ to $v_X = 9$. Taking into account the efficiency of the detection,[8] the ion signal corresponds to a cumulative formation rate for the $v_X < 10$ molecules close to 10^6 per second. The next challenge was to prepare these molecules in a well-defined vibrational level.

Different schemes have been proposed to favor the formation of cold molecules in their lowest vibrational level. By using a two photon process for photoassociation, a few $v = 0$ (no vibration) ultracold ground-state potassium dimers have been observed,[14] but several other vibrational levels are populated as well. By transferring a given vibrational level to the lowest vibrational level, cold, ground state molecules have been prepared.[15,23–25] Raman photoassociation for preparing ultracold molecules in a well-defined level has been studied by different groups. Its efficiency is unfortunately limited, because the so-prepared molecules can be excited again, and spontaneously decay towards other vibrational levels. To go further, an exciting direction is vibrational laser cooling, which would consist of transferring all the populations of the different vibrational levels to the lowest one.[18] Such a goal is not possible without carrying away the entropy from the molecule via a dissipative

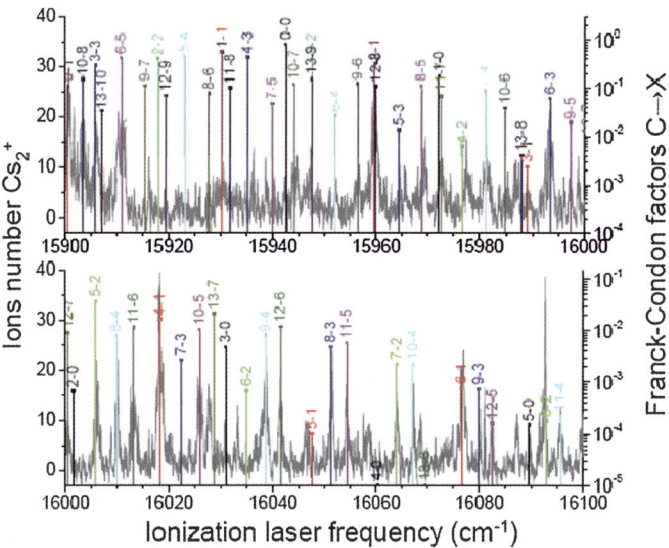

Fig. 7 Cs_2^+ ion count (left vertical axis) resulting from a narrowband REMPI detection (frequency ~627 nm). The PA laser energy is fixed at 11730.1245 cm^{-1} corresponding to the position of the most intense line marked with a circled cross in Fig. 5. Transition labels v_C–v_X are extracted from computed Franck–Condon factors (right vertical axis) between vibrational levels of the spectroscopically known C and X states.

spontaneous emission process. Before this work, laser cooling of vibrational degrees of freedom of cold molecules has never been experimentally realized. Several theoretical approaches have been proposed to favor spontaneous emission towards the lowest ro-vibrational level, such as, for instance, the use of an external cavity.[16]

The first step in vibrational laser cooling is the realization of the optical pumping applied to the molecules by using a femtosecond laser. We have demonstrated the transfer of population from the vibrational levels of cold singlet-ground-state Cs_2 molecules prepared *via* photoassociation, towards the level, $v = 0$, with no vibration. The main idea is to use a broadband laser tuned to the transitions $X^1\Sigma_g^+(v_X)$ to $B^1\Pi_u(v_B)$ between the different vibrational levels of the ground state and of the electronically excited state. With the successive laser pulses, the absorption–spontaneous emission cycles lead, through optical pumping, to a redistribution of the vibrational population in the ground state. By removing the laser frequencies corresponding to the excitation of the $v_X = 0$ level, we realize a dark state for the shaped femtosecond laser, leading to an accumulation of the molecules in the $v_X = 0$ level, meaning vibrational laser cooling. A similar concept of a dark state is introduced in theoretical articles,[17,18] where the vibration of the molecule is manipulated through quantum interferences between the different transitions. Interplay of controlled laser fields and spontaneous emission has been investigated for rotational or vibrational cooling. The mechanism proposed is called, molecular vibration selective coherent population trapping (MVSCPT), in analogy to the corresponding mechanism of velocity selective coherent population trapping (VSCPT) in atoms for sub-Doppler cooling of translations.[18]

The broadband laser used is a Tsunami mode locked femtosecond laser from Spectra Physics. The repetition rate is 80 MHz, the average power 2 W, the pulse duration 100 fs, and the σ-Gaussian bandwidth 54 cm^{-1}. The femtosecond laser is tuned (wavelength 767 nm) to the electronic transitions $X^1\Sigma_g^+(v_X)$ towards $B^1\Pi_u(v_B)$ of the cesium dimer (see Fig. 6(b)). The low vibrational levels ($v_X < 10$) can be excited to vibrational levels, v_B, of the state inducing through optical pumping a redistribution of the populations in the different vibrational levels of the ground state. Without shaping the femtosecond laser, we can observe, depending on the wavelength, a modification of the molecular resonance lines interpreted as a transfer of population between vibrational levels through optical pumping. To control the optical pumping of the molecules, we have shaped the femtosecond laser by suppressing the frequencies corresponding to the excitation of the vibrational level, $v_X = 0$. Fig. 8(b) shows the Condon parabola of the Franck–Condon factors for the transitions between the vibrational levels of the states X and B. The importance of the Franck–Condon factors is indicated by the level of grey. Fig. 8(c) shows the overlap between the radial wavefunctions, showing how the Franck–Condon factors depend, as the square of the overlap, on the relative positions of the wells X and B. The transitions from the $v_X = 0$ level towards v_B levels are at frequencies greater than 13030 cm^{-1}. The spectral shape of the laser does not contain any transitions from $v_X = 0$ (see Fig. 8(a)). We have used a home made shaper using a grating (1800 lines per mm) for diffracting the laser beam and for screening a part of the frequencies. In this way, the vibrational level, $v_X = 0$, becomes a dark state for the so-shaped laser. If we consider, for instance, $v_X = 4$, it is essentially excited

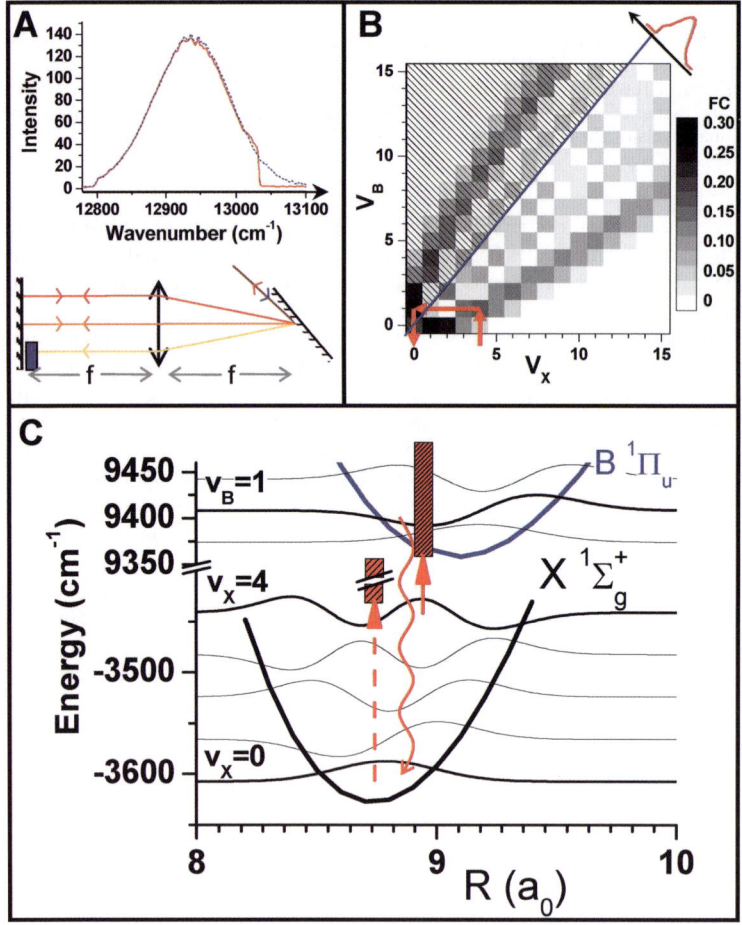

Fig. 8 (a) Shaping of the femtosecond laser (see text). (b) Franck–Condon parabola indicating the importance of the Franck–Condon factors (level of grey). (c) Scheme of the optical pumping.

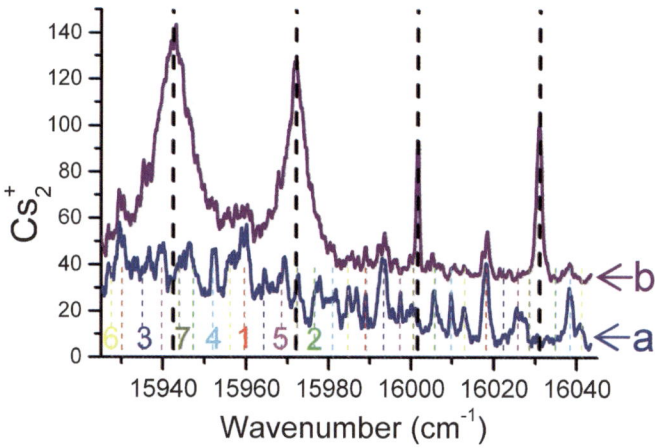

Fig. 9 Cs_2^+ ion spectra: (a) without the shaped laser, (b) with the shaped laser. The vibrational levels, v_X, of the ground state are labeled by different colors. The dashed lines indicate the resonance lines for vibrational transitions (v_X towards v_C) between the ground state, $X^1\Sigma_g^+$, and the electronically excited one, $C^1\Pi_u$. The color of the dashed line correspond to that of the involved v_X. The observed transitions correspond to $\Delta v = |v_C - v_X| = 0, 1, 2$ and 3.

into $v_B = 1$, which decays with a rate of about 30% in the dark level, $v_X = 0$, and with a rate of 70% essentially into the levels $v_X = 3$, 4 or 5. More generally we understand easily that after a few cycles of absorption of laser light then spontaneous emission that a large fraction of the molecules can be accumulated in the lowest vibrational level, $v_X = 0$.

Fig. 9 shows the experimental results. Adding the shaped laser completely modifies the resolved REMPI spectrum. The resonance lines corresponding to the transition $v_X = 0 \rightarrow v_C = 0$–3, mostly absent in the original spectrum, are now very strong. Their broadening corresponds to the saturation of the resonance in the REMPI process. The intensity of the lines indicates a very efficient transfer of the molecules into the lowest vibrational level, meaning a vibrational laser-cooling of the molecules. Taking into account the efficiency of the detection (<10%), the detected ion signal corresponds to about one thousand molecules in the $v_X = 0$ level, corresponding to a flux of $v_X = 0$ molecules of a few 10^5 per second. Fig. 10(a) shows the experimental time evolution of the population in the different vibrational levels. We show that the transfer of population into the $v_X = 0$ level is almost saturated after the application of 10000 pulses, which requires one hundred microseconds.

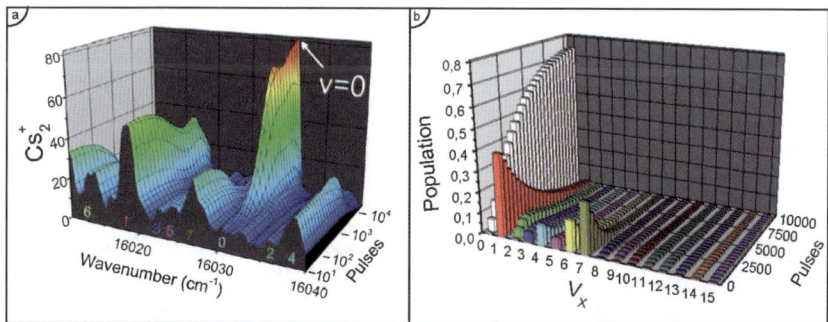

Fig. 10 Temporal evolution of the population transfer. (a) Compilation of experimental spectrum for different durations of the femtosecond laser (number of pulses). (b) Simulation of the vibrational laser cooling.

5 Simulation of vibrational laser cooling

We have modeled the optical pumping in a very simple way. Using the known $X^1\Sigma_g^+$ and $B^1\Pi_u$ potential curves and their rotational constants,[12,11] we have calculated the ro-vibrational energy levels. In the perturbative regime, we assume that the excitation probability is simply proportional to the laser spectral density at the transition frequencies, to the Franck–Condon factor,[11] and to the Hönl–London factor.[19] We assume a laser spectrum shape very close to the experimental one: average intensity of 150 mW cm^{-2} Gaussian shape center at 12940 cm^{-1} with a Gaussian linewidth $\sigma = 54$ cm^{-1} and we removed all spectral components, due to the shaping, above 13030 cm^{-1}. After being excited by a pulse and before the shot of the next pulse, we assume a total decay of the excited state population with branching ratio given by the Franck–Condon factors, and the Hönl–London factors. The perturbative regime and the lifetime of the electronically state $\sim$15 ns, close to the period of the pulses 12.5 ns, make reasonable the hypothesis of neglecting any accumulation of coherence due to excitation by a train of ultrashort pulses.[20] Applying the simple theoretical model shows that the molecules make a random walk, mostly in low vibrational levels, until reaching the $v_X = 0$ vibrational level. The accumulation of many molecules in the lowest vibrational level occurs with near unit transfer efficiency. Fig. 10(b) shows a simulation of the transfer of the 70% population into the $v_X = 0$ level, after 10000 pulses when the molecules are initially in a distribution of vibrational levels simulating the experimental one. The theoretical model gives a good agreement with the data. It indicates that only about 5 cycles of absorption–spontaneous emission (the number of necessary laser pulses depends on its intensity) are necessary for molecules to be transferred to the $v_X = 0$ level, if the molecules are initially in $v_X < 10$ vibrational levels. The mechanism could be called, incoherent molecular vibration selective population trapping (incoherent MVSPT), in analogy to the proposed mechanism of reference 18. The limitation of the mechanism is in the optical pumping towards high vibrational levels. A broader bandwidth laser could probably increase the population in $v_X = 0$. Nevertheless the Franck–Condon factors favor the accumulation of population in low vibrational levels.

6 Simulation for rotational laser cooling

Can the laser cooling of the vibration be extended to the rotation of the molecules? For that we need to consider the transitions between the different rotational levels, restricted to $\Delta J = J_B - J_X = 0, \pm 1$. A possibility for laser cooling of the rotation of the molecule, is to shape the laser by removing the frequency corresponding to the transitions $\Delta J = J_B - J_X = 0, +1$, leading with the absorption–spontaneous emission cycles to a decrease of the average principal quantum number J_X: i.e. a laser cooling

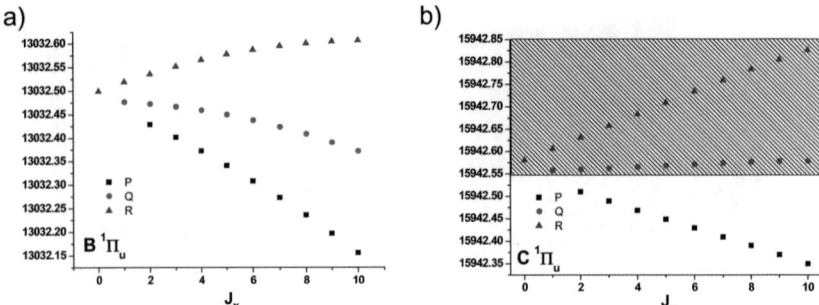

Fig. 11 Energy of the different ro-vibrational transitions ($\Delta J = 0 \pm 1$ for the P, Q and R branches) (a) with the B state, (b) with the C state. The hatched areas represent a shaping of the laser, to realize rotational cooling.

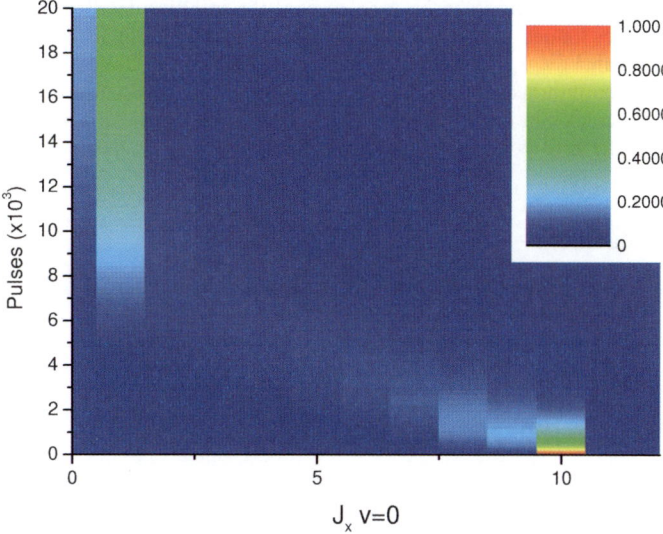

Fig. 12 Simulation of the temporal evolution of ro-vibrational cooling. Rotational cooling, transfer of the populations obtained after the vibrational cooling to the two last rotational levels $v_X = 0 - J = 0, 1$, via the state C.

of the rotation. Fig. 11 shows the energies corresponding to the transitions between the vibrational levels $v_X = 0$ and $v_B = 0$. By a blue-shaping of the laser, it is not possible to suppress the transitions $\Delta J = J_B - J_X = 0$. The situation is different, if we no longer consider the state, $B^1\Pi_u$, but the state, $C^1\Pi_u$. Fig. 11 shows that the transitions $\Delta J = J_C - J_X = 0$ can be easily suppressed. Fig. 12 shows the results of a simulation where the molecules are first vibrationally cooled by considering the excitation of the state, $B^1\Pi_u$, then rotationally cooled by considering the excitation of the state $C^1\Pi_u$.

7 Conclusion

We pointed out the interest of the use of broadband sources for cold molecules formed *via* photoassociation. We demonstrated the ability of a new broadband ionization procedure to detect most of the bound ground state molecules formed in a cold gas. This approach, combined with PA in excited electronic states, provides a general method for the search for novel paths for formation of cold molecules. Applied to a Cs MOT, this novel approach allowed us to detect deeply bound Cs_2 molecules in the $X^1\Sigma_g^+$ state. We show the possibility of the optical pumping of molecules, to modify the distribution of the populations in the vibrational levels. By making the $v = 0$ level a dark state, we enable the accumulation of the population in this level, demonstrating laser cooling of molecular vibration. The broadband character of the femtosecond laser is however the essential property of our experiment. The possibility to easily shape such a laser is also an important argument for using it, but a continuous broadband laser such as a fiber laser or diode laser could maybe offer the same potentiality and should be tested in the near future.

In our experiment, we do not presently have the capability of analyzing the rotational populations of the molecules in the vibrational level, $v_X = 0$, although a specific high resolution detection can be developed. Rotational cooling can be performed in a similar way provided that the laser bandwidth and the experimental ability to shape the laser match the rotational energy spread. The optical pumping of molecules has important other implications.[4] The possibility of controlling the vibration and the rotation can be used to cool the internal degrees of freedom of polar molecules loaded in an electrostatic trap, after velocity filtering of an effusive molecular beam.[21,22]

The trapping of molecules can provide the way to prepare a sample of one million or more molecules. Similar results could also be reached for hetero-nuclear systems and the formation of polar molecules, opening up exciting perspectives in quantum information.

Acknowledgements

This work is supported by the "Institut Francilien de Recherche sur les Atomes Froids" (IFRAF). M. A. thanks the EC-Network EMALI.

References

1 H. Rabitz, R. de Vivie-Riedle, M. Motzkus and K. Kompa, *Science*, 2000, **288**, 824–828.
2 J. Doyle, B. Friedrich, R. V. Krems and F. Masnou-Seeuws, *Eur. Phys. J. D*, 2004, **31**, 149–164.
3 R. V. Krems, *Int. Rev. Phys. Chem.*, 2005, **24**, 99–118.
4 J. T. Bahns, W. C. Stwalley and P. L. Gould, *J. Chem. Phys.*, 1996, **104**, 9689–9697.
5 M. Viteau, A. Chotia, M. Allegrini, N. Bouloufa, O. Dulieu, D. Comparat and P. Pillet, *Science*, 2008, **321**, 232–234.
6 O. Dulieu, M. Raoult and E. Tiemann, *J. Phys. B*, 2006, **39**(19).
7 K. M. Jones, E. Tiesinga, P. D. Lett and P. S. Julienne, *Rev. Mod. Phys.*, 2006, **78**, 483–535.
8 A. Fioretti, D. Comparat, A. Crubellier, O. Dulieu, F. Masnou-Seeuws and P. Pillet, *Phys. Rev. Lett.*, 1998, **80**, 4402–4405.
9 C. M. Dion, C. Drag, O. Dulieu, B. Laburthe Tolra, F. Masnou-Seeuws and P. Pillet, *Phys. Rev. Lett.*, 2001, **86**, 2253–2256.
10 C. Drag, B. L. Tolra, O. Dulieu, D. Comparat, M. Vatasescu, S. Boussen, S. Guibal, A. Crubellier and P. Pillet, *IEEE J. Quantum Electron.*, 2000, **36**, 1378–1388.
11 U. Diemer, R. Duchowicz, M. Ertel, E. Mehdizadeh and W. Demtröder, *Chem. Phys. Lett.*, 1989, **164**, 419–426.
12 W. Weickenmeier, U. Diemer, M. Wahl, M. Raab, W. Demtröder and W. Müller, *J. Chem. Phys.*, 1985, **82**, 5354–5363.
13 M. Raab, G. Höning, W. Demtröder and C. R. Vidal, *J. Chem. Phys.*, 1982, **76**, 4370–4386.
14 A. N. Nikolov, E. E. Eyler, X. T. Wang, J. Li, H. Wang, W. C. Stwalley and P. L. Gould, *Phys. Rev. Lett.*, 1999, **82**, 703–706.
15 J. M. Sage, S. Sainis, T. Bergeman and D. Demille, *Phys. Rev. Lett.*, 2005, **94**(20), 203001.
16 G. Morigi, P. W. H. Pinkse, M. Kowalewski and R. de Vivie-Riedle, *Phys. Rev. Lett.*, 2007, **99**(7), 073001.
17 S. G. Schirmer, *Phys. Rev. A*, 2000, **63**(1), 013407.
18 A. Bartana, R. Kosloff and D. J. Tannor, *Chem. Phys.*, 2001, **267**, 195–207.
19 G. Herzberg, *Atomic Spectra and Atomic Structure*, Dover Publications, New York, 1944, p. 208.
20 D. Felinto, C. A. C. Bosco, L. H. Acioli and S. S. Vianna, *Opt. Commun.*, 2003, **215**, 69–73.
21 S. A. Rangwala, T. Junglen, T. Rieger, P. W. Pinkse and G. Rempe, *Phys. Rev. A*, 2003, **67**(4), 043406.
22 T. Rieger, T. Junglen, S. A. Rangwala, P. W. Pinkse and G. Rempe, *Phys. Rev. Lett.*, 2005, **95**(17), 173002.
23 F. Lang, K. Winkler, C. Strauss, R. Grimm and J. Hecker Denschlag, *Phys. Rev. Lett.*, 2008, **101**, 133005.
24 K.-K. Ni, S. Ospelkaus, M. H. G. de Miranda, A. Pe'er, B. Neyenhuis, J. J. Zirbel, S. Kotochigova, P. S. Julienne, D. S. Jin and J. Ye, *Science*, 2008, **322**, 231.
25 J. Deiglmayer, A. Grochola, M. Repp, K. Mörtlbauer, C. Glück, J. Lange, O. Dulieu, R. Wester and M. Weidemuller, *Phys. Rev. Lett.*, 2008, **101**, 133004.

Dark state experiments with ultracold, deeply-bound triplet molecules

Florian Lang,[a] Christoph Strauss,[a] Klaus Winkler,[a] Tetsu Takekoshi,[a] Rudolf Grimm[ab] and Johannes Hecker Denschlag[*a]

Received 27th October 2008, Accepted 15th January 2009
First published as an Advance Article on the web 8th May 2009
DOI: 10.1039/b818964a

We examine dark quantum superposition states of weakly bound Rb_2 Feshbach molecules and tightly bound triplet Rb_2 molecules in the rovibrational ground state, created by subjecting a pure sample of Feshbach molecules in an optical lattice to a bichromatic Raman laser field. We analyze, both experimentally and theoretically, the creation and dynamics of these dark states. Coherent wavepacket oscillations of deeply bound molecules in lattice sites, as previously observed by Lang et al. (Phys. Rev. Lett., 2008, **101**, 133005), are suppressed due to laser-induced phase locking of molecular levels. This can be understood as the appearance of a novel multilevel dark state. In addition, the experimental methods developed help to determine important properties of our coupled atom/laser system.

1 Introduction

Very recently, several groups have produced dense, ultracold ensembles of molecules that are deeply bound[1-5] and in a ro-vibrational ground state.[1,2,4,5] This was achieved by binary association of alkali atoms in ultracold ensembles via two different pathways: (1) photoassociation[6,7] and (2) magneto-association at Feshbach resonances[7,8] combined with stimulated Raman adiabatic passage (STIRAP),[9] a special coherent optical transfer method. In contrast to photoassociation, magneto-association only produces weakly-bound Feshbach molecules.[7,8] STIRAP can then be used to transfer these weakly-bound molecules to the rovibrational ground state. This method is coherent, efficient, fast, reversible, and highly selective. STIRAP is based on a counter-intuitive light pulse sequence giving rise to a dynamically changing dark superposition state (Fig. 1a)

$$|DS\rangle = (\Omega_2|f\rangle - \Omega_1|g\rangle) \Big/ \sqrt{\Omega_1^2 + \Omega_2^2}. \quad (1)$$

In this paper, we deliberately replace the efficient but complex STIRAP transfer of ref. 1 with a simple square laser pulse scheme. This reveals interesting fundamental processes and dynamics in the coupled atom/laser system, that would otherwise be hidden. In addition, this procedure allows us to determine important properties and parameters of our system and to check for consistency with our theoretical model. We study the creation and lifetime of dark superposition states that contain

[a] Institut für Experimentalphysik und Zentrum für Quantenphysik, Universität Innsbruck, A-6020 Innsbruck, Austria. E-mail: johannes.denschlag@uibk.ac.at
[b] Institut für Quantenoptik und Quanteninformation der Österreichischen Akademie der Wissenschaften, A-6020 Innsbruck, Austria

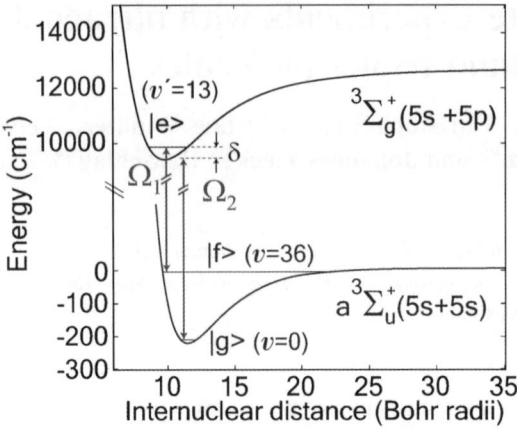

Fig. 1 Λ-type three-level scheme for dark states. The lasers 1 and 2 couple the molecular levels $|f\rangle$, $|g\rangle$ to the excited level $|e\rangle$ with Rabi frequencies $\Omega_{1,2}$, respectively.

a sizeable fraction of deeply bound molecules. These molecules are held in a 3D optical lattice. Because the lattice potential is much shallower for the deeply bound molecules than for the Feshbach molecules, and because the transfer is fast, the deeply bound molecules coherently populate several Bloch bands. In contrast to ref. 1, where similar circumstances lead to coherent oscillations in the lattice, oscillations are suppressed in the experiment described here due to phase locking of all quantum levels involved. A novel dark state appears which is a superposition of up to 8 quantum levels. We investigate the limiting conditions under which oscillations set in.

2 Experimental setup and initial preparation of molecules

We carry out our dark state experiments with a 50 μm-size pure ensemble of 3×10^4 weakly bound Rb_2 Feshbach molecules. The molecules are trapped in the lowest Bloch band of a cubic 3D optical lattice with no more than a single molecule per lattice site[10] and an effective lattice filling factor of about 0.3. The lattice depth for the Feshbach molecules is 60 E_r, where $E_r = \pi^2\hbar^2/2ma^2$ is the recoil energy, with m the mass of the molecules and $a = 415.22$ nm the lattice period. Such deep lattices suppress tunneling between different sites. A pure ensemble of Feshbach molecules has been produced as follows. We prepare a cold cloud of 6×10^5 ^{87}Rb atoms that are either Bose condensed or *nearly*† Bose condensed in a Ioffe-type magnetic trap with trap frequencies $\omega_{x,y,z} = 2\pi \times (7, 19, 20)$ Hz. Within 100 ms we adiabatically load the atoms into the 3D optical lattice. After turning off the magnetic trap, we flip the spins of our atoms from their initial state $|F = 1, m_F = -1\rangle$ to $|F = 1, m_F = +1\rangle$ by suddenly reversing the bias magnetic field of a few G. This spin state features a 210 mG-wide Feshbach resonance at 1007.40 G.[11] By adiabatically ramping over this resonance, we efficiently convert atoms at multiply occupied lattice sites into Rb_2 Feshbach molecules. After conversion, inelastic collisions occur at lattice sites that contain more particles than a single Feshbach molecule, leading to vibrational relaxation of these molecules, release of binding energy into kinetic energy and removal of all particles from these sites. A subsequent combined microwave and optical purification pulse removes all remaining chemically unbound atoms, creating a pure sample of 3×10^4 Feshbach molecules. Afterwards, the magnetic field is set to 1005.8 G, where the Feshbach molecules are in a quantum

---

† It turns out that this increases the number of Feshbach molecules.

state $|f\rangle$, which correlates with $|F = 2, m_F = 2, f_1 = 2, f_2 = 2, v = 36, l = 0\rangle$ at 0 G. Here, F and $f_{1,2}$ are the total angular momentum quantum numbers for the molecule and its atomic constituents, respectively, and m_F is the total magnetic quantum number; v is the vibrational quantum number for the triplet ground state potential ($a^3\Sigma_u^+$) and l is the quantum number for rotation.

The bichromatic Raman laser field for the creation of the molecular dark states is based on two lasers (1 and 2), which connect the Feshbach molecule level $|f\rangle$, *via* an excited level $|e\rangle$, to the absolute lowest level in the triplet potential $|g\rangle$ (Fig. 1a). Laser 1 is a Ti:Sapphire laser and laser 2 is a grating-stabilized diode laser. Both lasers are Pound–Drever–Hall locked to a single cavity, which itself is locked to an atomic ^{87}Rb-line. From the lock error signals, we estimate frequency stabilities on a ms-timescale of 40 kHz and 80 kHz for lasers 1 and 2, respectively. Both laser beams have a waist of 130 μm at the location of the molecular sample, propagate collinearly, and are polarized parallel to the direction of the magnetic bias field. Thus, the lasers can only induce π transitions.

The ground state $|g\rangle$ has a binding energy of 7.03806(3) THz $\times$ h and can be described by the quantum numbers $|F = 2, m_F = 2, S = 1, I = 3, v = 0, l = 0\rangle$ where S and I are the total electronic and nuclear spins of the molecule, respectively. At 1005.8 G $|g\rangle$ is separated by hundreds of MHz from any other bound level, so that there is no ambiguity as to which level is addressed. The level $|e\rangle$ is located in the vibrational $v = 13$ manifold of the electronically excited $^3\Sigma_g^+$ (5s + 5p) potential and has 1_g character. It has an excitation energy of 294.62610(6) THz $\times$ h with respect to $|f\rangle$, and a width $\Gamma = 2\pi \times 8$ MHz. The Rabi frequencies $\Omega_{1,2}$ of the two lasers depend on their respective intensities $I_{1,2}$, *i.e.*, $\Omega_1 = 2\pi \times 0.4$ MHz $\sqrt{I_1/(\text{Wcm}^{-2})}$ and $\Omega_2 = 2\pi \times 30$ MHz $\sqrt{I_2/(\text{Wcm}^{-2})}$, and are typically chosen to be in the MHz regime.

3 Dark state evolution within a square pulse

Our square pulse projection experiments are carried out as follows. We expose the Feshbach molecules $|f\rangle$ in the lattice to square pulses of Raman lasers 1 and 2 of variable pulse duration. Laser 2 is switched on about 1 μs before laser 1 to avoid excitation from $|f\rangle$ to $|e\rangle$ due to jitter in the laser pulse timing. The Raman lasers are resonant ($\delta = 0$) and the Rabi frequency $\Omega_2 \approx 2\pi \times 7$ MHz while Ω_1 is varied (Fig. 2). After the pulse, we measure the fraction of molecules remaining in state $|f\rangle$ by dissociating them into pairs of atoms at the Feshbach resonance, releasing them from the lattice and applying standard absorption imaging. It is important to note that we actually only count atoms in the lowest Bloch band of the lattice. The release from the optical lattice is done as described in ref. 12, where after 13 ms of ballistic expansion we map out the Bloch bands in momentum space (see Appendix for details).

Fig. 2 shows the remaining fraction of molecules in state $|f\rangle$ *versus* pulse duration. Within 1 μs we observe a rapid loss of molecules that depends on the ratio Ω_2/Ω_1. The remaining molecules are stable on a much longer timescale. This can be understood in terms of formation of a dark state $|DS\rangle$. We can write

$$|f\rangle = (\Omega_2 | DS\rangle + \Omega_1 | BS\rangle) / \sqrt{\Omega_1^2 + \Omega_2^2} \qquad (2)$$

where

$$|BS\rangle = (\Omega_1 |f\rangle + \Omega_2 |g\rangle) / \sqrt{\Omega_1^2 + \Omega_2^2} \qquad (3)$$

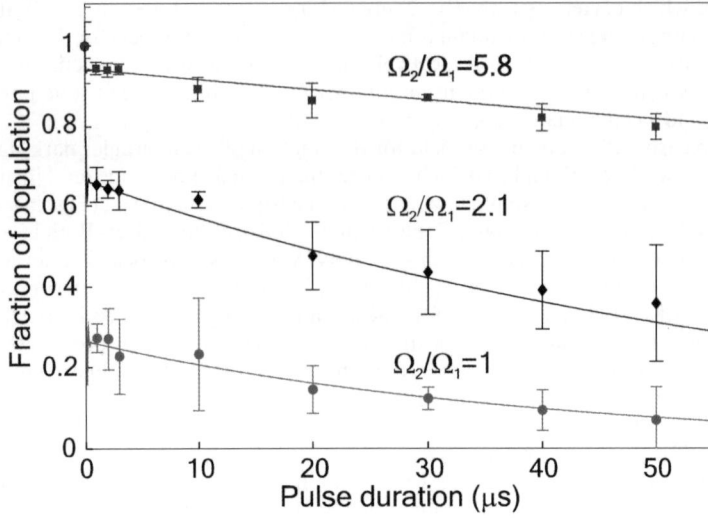

Fig. 2 Dark state formation and lifetime. Shown is the fraction of Feshbach molecules remaining after subjecting them to a square pulse of Raman laser light of varying length for various Rabi frequency ratios Ω_2/Ω_1 ($\Omega_2 \approx 2\pi \times 7$ MHz). After switching on the lasers, a certain fraction of molecules is lost within 1 μs and a dark state has formed which has a much longer lifetime. The solid lines represent model calculations (Section 4), which can be used to determine the Rabi frequencies and short-term laser linewidths.

is a bright state which quickly decays *via* resonant excitation to level $|e\rangle$. The dark state remains after the lasers are switched on and can be detected as a fraction $\Omega_2^4/(\Omega_1^2 + \Omega_2^2)^2$ of molecules projected back to $|f\rangle$ after switching off the lasers‡.

Also, after the pulse, a fraction $\Omega_1^2\Omega_2^2/(\Omega_1^2 + \Omega_2^2)^2$ of the initial molecules are in state $|g\rangle$ with a maximum of 25% for $\Omega_1 = \Omega_2$. Thus, a sizeable fraction of the molecules can be coherently transferred to the ground state. Remarkably, this transfer takes place in less than 1 μs! Such short transfer times cause Fourier broadening, resulting in considerably reduced laser stability requirements. In addition, due to the formation of a dark state, there is still a well-defined phase relation between the $|f\rangle$ and $|g\rangle$ molecules.

As can be seen from Fig. 2, the dark state slowly decays. Its lifetime is shortest for $\Omega_1 = \Omega_2$, where we measure it to be ≈ 50 μs. The decay of the dark state is likely due to phase fluctuations of the Raman lasers. Phase fluctuations lead to an admixture of a bright state component to the otherwise dark state, which causes losses. In Section 4 we will show that these fluctuations can be expressed in terms of the short-term relative linewidth of the lasers, γ, which we find to be about $2\pi \times 20$ kHz. In principle, the decay of the dark state could be due to other effects, such as coupling to levels other than $|f\rangle$, $|e\rangle$, and $|g\rangle$. However, we have verified that this is not the case, because losses due to optical excitation are completely negligible on the 100 μs-timescale when we expose a pure ensemble of $|f\rangle$ ($|g\rangle$) molecules to only laser 2 (1).

We also searched for laser power dependent shifts of the two-photon resonance. Using the Raman square pulse measurements, we scanned the relative detuning of the lasers for a fixed pulse duration and various laser powers. Within the accuracy of our measurements of $2\pi \times 200$ kHz, we could not detect any shifts of the resonance.

---

‡ This fact can be used to conveniently calibrate the Rabi frequency ratio Ω_1/Ω_2. We found good agreement with other calibration methods for the Rabi frequencies.

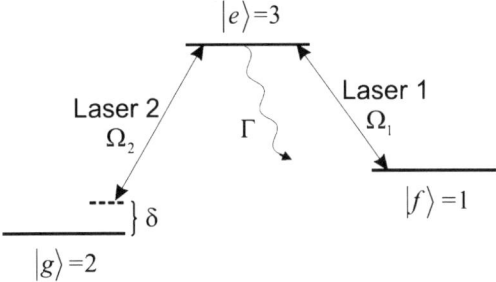

Fig. 3 Level scheme for the master equation.

The behavior in Fig. 2 is described well by a closed three-level model (a Λ system) and its dynamics can be simulated with a master equation which we describe in the following.

4 Three-level model and master equation

Neglecting lattice effects, we can describe the internal dynamics of the molecules as they are subjected to the Raman laser fields with a three-level model. We use a master equation,[13,14] which takes into account decoherence due to phase fluctuations of the Raman lasers. We consider the case where laser 1 is kept on resonance and laser 2 has a detuning δ (Fig. 3). Identifying the levels $|f\rangle$, $|g\rangle$, $|e\rangle$ with numbers 1, 2, 3, respectively, we can write the master equation as,

$$\frac{d\rho}{dt} = -i\delta[\sigma^{22}, \rho] - \frac{i}{2}\sum_{k=1}^{2}\Omega_k\left[\sigma_{-}^{3k} + \sigma_{+}^{3k}, \rho\right]$$

$$- \frac{1}{2}\Gamma\left(\sigma^{33}\cdot\sigma^{33}\cdot\rho + \rho\cdot\sigma^{33}\cdot\sigma^{33}\right) \quad (4)$$

$$+ \frac{1}{2}\gamma\left(2\sigma^{22}\cdot\rho\cdot\sigma^{22} - \sigma^{22}\cdot\rho - \rho\cdot\sigma^{22}\right),$$

where ρ is the density matrix, $\Omega_{1,2}$ are the Rabi frequencies, Γ is the spontaneous decay rate of the excited level $|e\rangle$, and γ is the relative linewidth of the two Raman lasers. The matrices σ_{-}^{rs} and σ_{+}^{rs} are ladder operators and each is the transpose of the other. For example

$$\sigma_{-}^{32} = \begin{pmatrix} 0 & 0 & 0 \\ 0 & 0 & 0 \\ 0 & 1 & 0 \end{pmatrix} = \left(\sigma_{+}^{32}\right)^{T}. \quad (5)$$

Setting the linewidth of the excited level $\Gamma = 8$ MHz, the detuning $\delta = 0$ and Rabi frequencies $\Omega_2 = 2\pi \times 7$ MHz and Ω_1 to give the ratios in Fig. 2, we fit all the data with a single fit parameter γ. As a best fit, we obtain a relative linewidth of the two Raman lasers $\gamma = 2\pi \times 20$ kHz, which is a reasonable value for our laser system.

5 Coherent oscillations and their suppression

In ref. 1 coherent oscillations of molecular wavepackets of $|g\rangle$ molecules in the optical lattice were observed. We now investigate how these observations fit together with the experimental results of the square pulse projection experiments presented here. For clarity, the oscillation data from ref. 1 are presented again in Fig. 4 and briefly discussed.

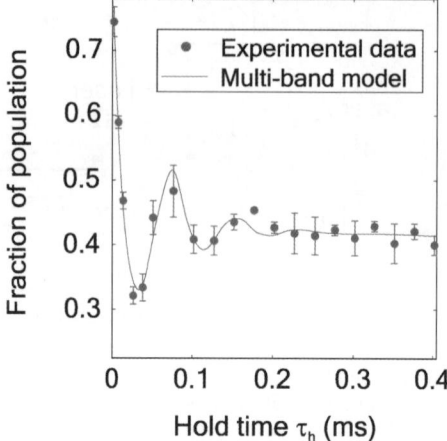

Fig. 4 We plot the transfer efficiency for the round-trip STIRAP process as a function of the hold time τ_h between the two STIRAP pulses. With our procedure we only count molecules whose constituent atoms end up in the lowest Bloch band after transfer. The oscillations in the transfer efficiency are due to breathing oscillations of localized spatial wavepackets of molecules in the lattice sites. The solid line is from a multi-band model calculation (Section 6). This plot is taken from ref. 1.

Using a STIRAP pulse, Feshbach molecules are efficiently transferred to level $|g\rangle$. The Raman lasers are extinguished and the molecules are held for a time τ_h, after which they are transferred back to $|f\rangle$ with a reverse STIRAP pulse. The number of recovered Feshbach molecules is counted. However, we only detect atoms that end up in the lowest Bloch band after dissociation of the Feshbach molecules (see Appendix). The oscillation can be understood as follows. We consider the localized spatial center-of-mass (c.o.m.) wavepacket of a Feshbach molecule at a particular lattice site in the lowest Bloch band. The first STIRAP transfer projects this wavepacket onto the much shallower§ lattice potential felt by the $|g\rangle$ molecules (Fig. 5) without changing its shape. As a consequence, $|g\rangle$ molecules are coherently spread over various Bloch bands, and the wavepacket undergoes "breathing" oscillations with the lattice site trap frequency ω_t. These coherent oscillations (period $\approx$ 80 μs) are damped by tunneling of $|g\rangle$ molecules in higher Bloch bands to neighboring lattice sites. The reverse STIRAP transfer maps this periodic oscillation back to the Feshbach molecule signal in Fig. 4. Higher Bloch bands are populated here as well, but are at most partially counted in our scheme (see Appendix), which leads to an apparent decrease in transfer efficiency.

The question arises why similar oscillations are not observed in our square pulse projection measurements shown in Fig. 2, especially for the case $\Omega_1 = \Omega_2$ where 50% of the population is in state $|g\rangle$. One might assume that the spatial wavepackets of the $|g\rangle$ molecules undergo similar breathing oscillations. These oscillations would then periodically break up the dark superposition state and lead to corresponding losses. They would also periodically produce population in higher Bloch bands of the Feshbach molecule lattice. As we will see, the oscillations are suppressed because the Raman lasers phase lock the involved quantum levels which stops, in a sense, the free evolution of the wavepackets. We can understand this behavior in detail with the help of a multi-band model, which we describe in the following.

---

§ Due to a smaller dynamic polarizability, the lattice depth for the tightly bound $|g\rangle$ molecules is shallower than for the Feshbach molecules by a factor of ≈ 10.

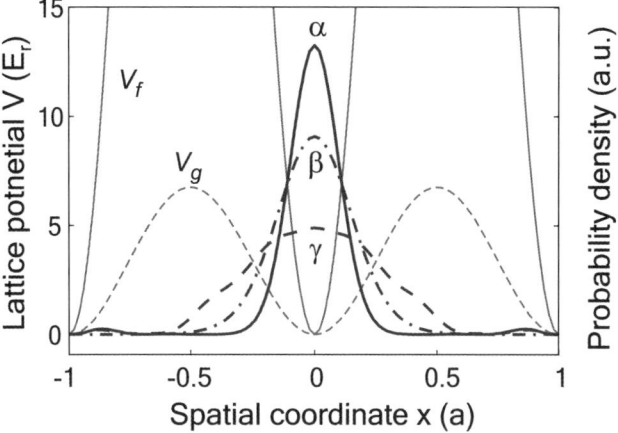

Fig. 5 Wavepacket dynamics. Directly after the STIRAP transfer of a molecule from $|f\rangle$ to $|g\rangle$ the shape of its wavepacket (thick solid line α) essentially corresponds to the vibrational ground state of the sinusoidal lattice potential for Feshbach molecules V_f (thin solid line). In the much weaker potential felt by the ground state molecules V_g (thin dashed line) the wavepacket starts to oscillate. After 1/8 of the oscillation period $\tau_{osc} = 2\pi/\omega_t$ its shape roughly corresponds to the vibrational ground state for $|g\rangle$ (thick dash-dotted line β) and reaches its maximum extension after $\tau_{osc}/4$ (thick dashed line γ).

6 Multi-band model

In an optical lattice the molecular levels $|f\rangle$, $|g\rangle$ and $|e\rangle$ from the previous model have a substructure given by the lattice Bloch bands. Because the lattice depths for the levels $|f\rangle$, $|g\rangle$ and $|e\rangle$ are, in general, different, the respective band structures will also vary. This combination of external (c.o.m. motion in the lattice) and internal degrees of freedom gives rise to a number of new quantum levels which are coupled by the laser fields (Fig. 6). We assume each Feshbach molecule to be initially localized in a singly-occupied lattice site. The corresponding localized molecular wavepacket can be described by Wannier functions,[15] which form a complete set of orthonormal functions. In the following we will denote the Wannier function for

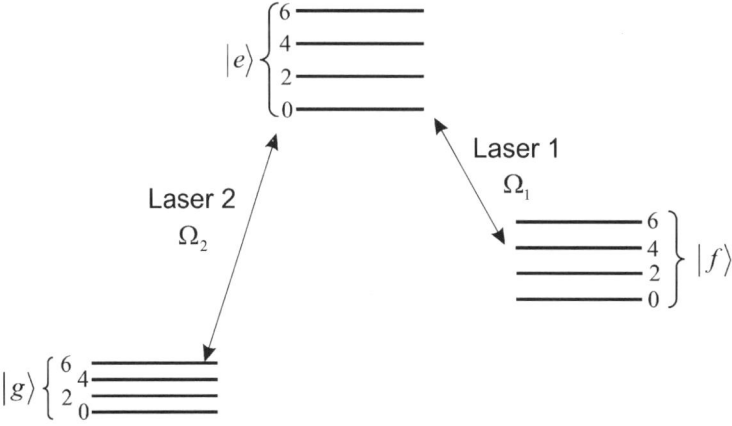

Fig. 6 Multi-band model. The three molecular levels $|f\rangle$, $|e\rangle$ and $|g\rangle$ have a Bloch band substructure due to the optical lattice. We restrict the model to the 4 lowest Bloch bands with even symmetry (band index $n = 0, 2, 4, 6$).

level $|\alpha\rangle$ and band n as $|\Psi_{\alpha n}\rangle$. We note that for deep lattices, these Wannier functions closely resemble harmonic oscillator wavefunctions.

The Raman lasers couple different $|\Psi_{\alpha n}\rangle$ according to the respective wavefunction overlaps (Fig. 6). Since the initial wavepackets of the Feshbach molecules are symmetric, only even bands will be populated. We restrict our calculations to the four lowest Bloch bands with even symmetry,¶ corresponding to the band indices $n = 0, 2, 4, 6$. The dynamics in each of the three lattice directions is then described by a 12-level model, which can in principle be solved in terms of a master equation (Section 4). However, we have used a Schrödinger equation-based model since the numerical code is less involved. In this approach, laser phase fluctuations are neglected, and we introduce a lattice site tunnel rate for each band. These tunnel rates are chosen to match the expected tunnel rates for the different bands and are slightly adjusted for a better fit of the data in Fig. 4. We note that the results of the model calculations are essentially independent of the excited state lattice depth, which is not well known.

The Hamiltonian $\hat{H}$ of our time dependent Schrödinger equation

$$i\hbar \frac{\partial}{\partial t}|\Phi\rangle = \hat{H}|\Phi\rangle \qquad (6)$$

has the form of a 12×12 matrix,

$$\hat{H} = \hbar \begin{pmatrix} E_{f0} - \frac{i}{2}J_{f0} & 0 & \frac{1}{2}\Omega_1(t) \cdot M_{f0,e0} & 0 & \cdots \\ 0 & E_{g0} + \delta - \frac{i}{2}J_{g0} & \frac{1}{2}\Omega_2(t) \cdot M_{g0,e0} & 0 & \cdots \\ \frac{1}{2}\Omega_1(t) \cdot M_{e0,f0} & \frac{1}{2}\Omega_2(t) \cdot M_{e0,g0} & E_{e0} - \frac{i}{2}\Gamma - \frac{i}{2}J_{e0} & \frac{1}{2}\Omega_1(t) \cdot M_{e0,f2} & \cdots \\ 0 & 0 & \frac{1}{2}\Omega_1(t) \cdot M_{f2,e0} & E_{f2} - \frac{i}{2}J_{f2} & \cdots \\ \vdots & \vdots & \vdots & \vdots & \end{pmatrix}$$

(7)

Here $E_{\alpha n}$ and $J_{\alpha n}$ are the energy and tunnel matrix elements respectively for the Wannier function $|\Psi_{\alpha n}\rangle$ in band n of level $|\alpha\rangle$. $M_{\alpha n,\beta k} = \langle \Psi_{\alpha n}|\Psi_{\beta k}\rangle$ is the overlap integral of the respective Wannier functions.

Diagonalizing this Hamiltonian, we find twelve "eigenstates" of the coupled system, which in general have complex eigenvalues. In the following, we study the case of strong coupling ($\Omega_{1,2} \gg \omega_t$),∥ which is the regime for phase locking. In this regime, four of these eigenstates have negligible contribution from the excited level $|e\rangle$ and thus a long lifetime. These 4 quasi-dark states essentially correspond to the 4 lattice bands in our model and will be denoted as $|DS_n\rangle$ with $n = 0, 2, 4, 6$. We now study the spatial waveforms of these dark states (Fig. 7) and compare the components with $|f\rangle$ and $|g\rangle$ character. Neglecting a small $|e\rangle$ component the dark superposition state, $|DS_n\rangle$ can be written as

$$|DS_n\rangle = |g\rangle\langle g|DS_n\rangle + |f\rangle\langle f|DS_n\rangle \qquad (8)$$

As an example ($\Omega_1 = \Omega_2$) Fig. 7 shows that the wavepackets $\langle g|DS_n\rangle$ and $\langle f|DS_n\rangle$ have the same shape. This is not surprising since this ensures that the ratio of the $|f\rangle$ and $|g\rangle$ amplitudes equals Ω_2/Ω_1 everywhere, as in eqn 1.

¶ The effect of including higher bands into the model was found to be negligible.
∥ For our experiments where $\Omega_{1,2} \gtrsim 2\pi \times 1$ MHz and $\omega_t \sim 2\pi \times 10$ kHz this condition is satisfied.

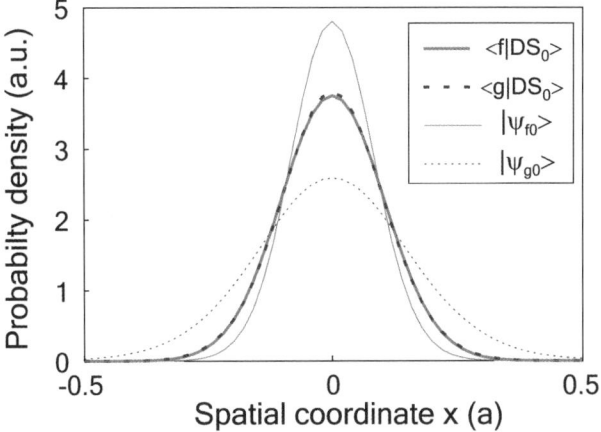

Fig. 7 Spatial wavepackets of the dark state $|DS_0\rangle$ and the Wannier functions $|\Psi_{f0}\rangle, |\Psi_{g0}\rangle$. The dark state $|DS_0\rangle$ has two components, one having $|f\rangle$ character ($\langle f|DS_n\rangle$) and the other one having $|g\rangle$ character ($\langle g|DS_n\rangle$). The wavepackets of $\langle f|DS_n\rangle$ and $\langle g|DS_n\rangle$ essentially have the same shape. They are mainly composed of the lattice ground states $|\Psi_{f0}\rangle$ (thin solid line) and $|\Psi_{g0}\rangle$ (thin dotted line). All depicted states are normalized. The parameters used are $\Omega_1 = \Omega_2 = 2\pi \times 7$ MHz, $V_f = 60\ E_r$ and $V_g = 6\ E_r$, as in our experiments.

Let us now discuss the formation and evolution of the dark state that we have observed in the square pulse experiments of Section 3. A dark state $|DS\rangle$ is formed in less than 1 μs by subjecting Feshbach molecules to a square Raman laser pulse. As in the STIRAP transfer (discussed in Section 5) the initial projection onto $|DS\rangle$ will not change the shape of the Feshbach molecule wavepacket, given by the Wannier function $|\Psi_{f0}\rangle$. The dark state can be expressed as a coherent superposition of the four dark eigenstates $|DS_n\rangle$ of the 12-level Hamiltonian

$$|DS\rangle = \sum_{n=0,2,4,6} c_n |DS_n\rangle. \qquad (9)$$

The subsequent coherent evolution of these dark states will again in principle lead to breathing oscillations. The amplitude of these oscillations depends on the extent to which higher bands (i.e., $|DS\rangle_n$, $n > 0$) are excited. The excitation increases with increasing deviation of $|DS\rangle$ from the initial state $|f\rangle$, i.e., with rising Ω_1/Ω_2.

This can also be understood from another point of view. The effective lattice potential felt by the molecules in such a superposition state is the weighted average of the potentials for the two contributing states $|f\rangle$ and $|g\rangle$. For the case $\Omega_1 = \Omega_2$ this effective potential is about half as deep as that for the Feshbach molecules. Compared to the case of pure ground state molecules (Fig. 4) where the lattice potential is reduced by a factor of 10, the oscillations of the wavepacket are strongly suppressed and cannot be observed with our current experimental precision. For $\Omega_1 \gg \Omega_2$, the dark state $|DS\rangle$ has a dominant contribution from state $|g\rangle$, and the effective lattice potential essentially corresponds to the one for ground state molecules. In this case oscillations appear despite the strong coupling, a fact which we also have experimentally verified**.

** For these experiments we have ramped into the dark state and back in a fashion similar to STIRAP transfer pulses to avoid strong losses caused by direct projection into $|DS\rangle \approx |g\rangle\langle g|DS\rangle$.

Conclusion

We have analyzed coherent wavepacket dynamics and their suppression in a 3D optical lattice. We observed optically induced phase locking of a number of quantum levels, which can also be viewed as the appearance of a novel multi-level dark state. The experiments were carried out with tightly bound molecules as a component of a dark quantum superposition state. Thus, the experiments demonstrate control of molecular motion in an optical lattice for the first time. In addition, different models have been introduced and discussed in detail, with which the lattice dynamics can be understood and quantitatively described.

Appendix: Theoretical band population analysis

As stated before, our signals only include molecules for which the constituent atoms end up in the lowest Bloch band of the lattice. A controlled lattice rampdown in a few milliseconds maps the bands and quasi-momentum distribution of the atoms into momentum space.[16,12] We image these distributions after 13 ms of time-of-flight *via* absorption imaging. Fig. 8 shows a typical distribution. The dotted square region corresponds to the lowest Bloch band and is dominantly populated.

An important question is how the Bloch bands for the Feshbach molecules map onto the Bloch bands for the atoms. In other words, if we measure the atomic population of the Bloch bands, do we know what the band population for the molecules was? As the lattice is very deep for the Feshbach molecules and atoms, we can approximate the potential at an individual lattice site as harmonic with trap frequency ω_t. In one dimension, the eigenfunctions of the harmonic oscillator are

$$|\Phi_n\rangle = \frac{1}{\sqrt{2^n n! \sqrt{\pi} x_0}} \exp\left(-\frac{1}{2}\left(\frac{x}{x_0}\right)^2\right) H_n\left(\frac{x}{x_0}\right), \tag{10}$$

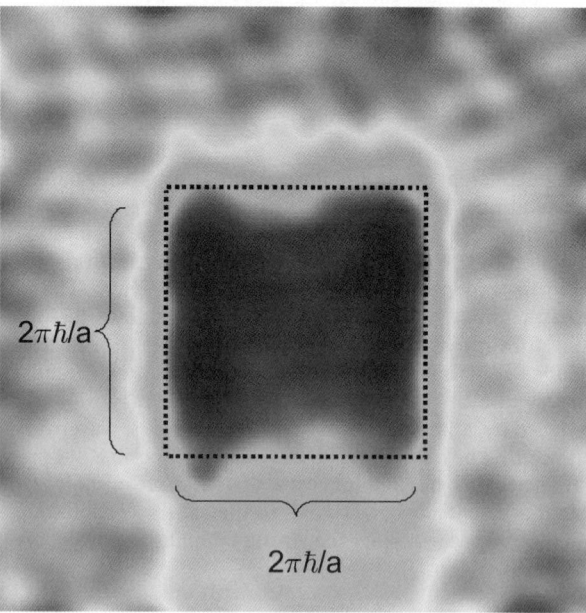

Fig. 8 Shown is a typical absorption image which displays the atomic quasi-momentum distribution in the optical lattice after exposure to the Raman laser beams and subsequent adiabatic molecule dissociation. Atoms inside the square region come from the lowest Bloch band. $2\pi\hbar/a$ is the modulus of the reciprocal lattice vector.

where $x_0 = \sqrt{\hbar/\omega_t m}$ is the oscillator length and H_n is the nth Hermite polynomial. We assume that we have two atoms in a lattice site with coordinates $x_{1,2}$. The relative and c.o.m. coordinates of the atom pair are

$$x_r = 1/\sqrt{2}(x_1 - x_2) \tag{11}$$

$$x_c = 1/\sqrt{2}(x_1 + x_2) \tag{12}$$

The c.o.m. potential V_c for the pair will be harmonic with trap frequency ω_t and the potential V_r for the relative coordinate will be a sum of the harmonic potential and the interaction potential (Fig. 9).

When we form or dissociate a molecule by adiabatically ramping across a - Feshbach resonance, only the quantum level in the V_r potential will change – from a molecular bound state to an unbound atomic pair state in the lowest Bloch band. The wavefunction in the c.o.m. coordinate remains unchanged. We can now calculate how band populations of Feshbach molecules convert to atomic band populations by using the coordinate transformations. As an example: a Feshbach molecule in the lowest Bloch band (*i.e.*, center-of-mass coordinate) will produce an atom pair with the following wavefunction: $|\Psi\rangle \propto \exp(-½x_c^2) \exp(-½x_r^2) = \exp(-½x_1^2) \exp(-½x_2^2)$. This means that both atoms will also end up in the lowest Bloch band of the lattice. This analysis can be extended to any band. Table 1 gives the conversion amplitudes from molecular to atomic bands for the four lowest symmetric molecular bands. Correlations between the two constituent atoms of a molecule are not discussed here.

We finally note that when we apply absorption imaging, the optical density of the atomic sample is integrated in the direction of observation. Thus in this direction no band population analysis is possible. We accounted for this in our multi-band model described in Section 6.

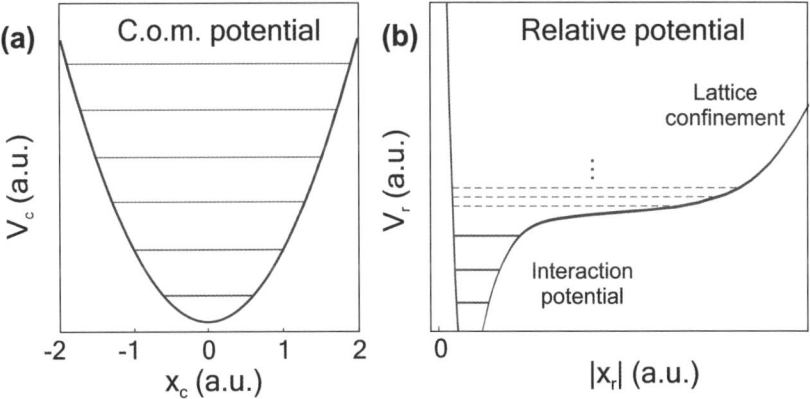

Fig. 9 Potentials for the center-of-mass and relative coordinate of two atoms trapped at a site of the optical lattice. (a) The bound states (solid lines) of the harmonic center-of-mass potential V_c correspond to the molecular Bloch bands. (b) At short interatomic distances the relative potential V_r is dominated by the interaction potential, which allows the formation of bound molecular states (solid lines). Unbound atoms are trapped by the lattice potential at larger separation (dashed lines). With the help of a Feshbach resonance the lowest trap state can be converted into a high molecular state. Note that in this schematic view both energy and distance for the two contributions to the relative potential are not to scale.

Table 1 Band conversion amplitudes in the harmonic oscillator approximation. Each line gives the amplitudes for a constituent atom of a molecule in a certain band to populate various atomic bands after dissociation. Note that the squares of the amplitudes correspond to the binomial coefficients

Molecular band	Atomic Bloch band							
	0	1	2	3	4	5	6	7
0	1	0	0	0	0	0	0	0
2	½	$-\sqrt{\frac{1}{2}}$	½	0	0	0	0	0
4	¼	$-\sqrt{\frac{4}{16}}$	$\sqrt{\frac{6}{16}}$	$-\sqrt{\frac{4}{16}}$	¼	0	0	0
6	⅛	$-\sqrt{\frac{6}{64}}$	$\sqrt{\frac{15}{64}}$	$-\sqrt{\frac{20}{64}}$	$\sqrt{\frac{15}{64}}$	$-\sqrt{\frac{6}{64}}$	⅛	0

Acknowledgements

The authors thank Helmut Ritsch for very helpful discussions and support in model calculations. We also thank Gregor Thalhammer for early assistance in the lab, and Florian Schreck for loaning us a Verdi V18 pump laser. This work was supported by the Austrian Science Fund (FWF) within SFB 15 (project part 17).

References

1 F. Lang, K. Winkler, C. Strauss, R. Grimm and J. Hecker Denschlag, *Phys. Rev. Lett.*, 2008, **101**, 133005.
2 M. Viteau, A. Chotia, M. Allegrini, N. Bouloufa, O. Dulieu, D. Comparat and P. Pillet, *Science*, 2008, **321**, 232.
3 J. G. Danzl, E. Haller, M. Gustavsson, M. J. Mark, R. Hart, N. Bouloufa, O. Dulieu, H. Ritsch and H.-C. Nägerl, *Science*, 2008, **321**, 1062.
4 K.-K. Ni, S. Ospelkaus, M. H. G. de Miranda, A. Pe'er, B. Neyenhuis, J. J. Zirbel, S. Kotochigova, P. S. Julienne, D. S. Jin and J. Ye, *Science*, 2008, **322**, 5899.
5 J. Deiglmayr, M. Repp, A. Grochola, M. Mörtlbauer, C. Glück, O. Dulieu, J. Lange, R. Wester and M. Weidemüller, *Phys. Rev. Lett.*, 2008, **101**, 133004.
6 K. M. Jones, E. Tiesinga, P. D. Lett and P. S. Julienne, *Rev. Mod. Phys.*, 2006, **78**, 483.
7 J. M. Hutson and P. Soldán, *Int. Rev. Phys. Chem.*, 2006, **25**, 497.
8 T. Köhler, K. Goral and P. S. Julienne, *Rev. Mod. Phys.*, 2006, **78**, 1311.
9 K. Bergmann, H. Theuer and B. W. Shore, *Rev. Mod. Phys.*, 1998, **70**, 1003.
10 G. Thalhammer, K. Winkler, F. Lang, S. Schmid, R. Grimm and J. Hecker Denschlag, *Phys. Rev. Lett.*, 2006, **96**, 050402.
11 T. Volz, S. Dürr, S. Ernst, A. Marte and G. Rempe, *Phys. Rev. A*, 2003, **68**, 010702R.
12 K. Winkler, G. Thalhammer, F. Lang, R. Grimm and J. Hecker Denschlag, *Nature*, 2006, **441**, 853.
13 D. F. Walls and G. J. Milburn, in *Quantum Optics*, Springer-Verlag, Berlin, 1994.
14 T. Haslwanter, H. Ritsch, J. Cooper and P. Zoller, *Phys. Rev. A*, 1988, **38**, 5652.
15 W. Kohn, *Phys. Rev. B*, 1973, **7**, 4388.
16 J. Hecker Denschlag, J. E. Simsarian, H. Häffer, C. McKenzie, A. Browaeys, D. Cho, K. Helmerson, S. L. Rolston and W. D. Phillips, *J. Phys. B: At., Mol. Opt. Phys.*, 2002, **35**, 3095.

PAPER

Precision molecular spectroscopy for ground state transfer of molecular quantum gases

Johann G. Danzl,[*a] Manfred J. Mark,[a] Elmar Haller,[a] Mattias Gustavsson,[a] Nadia Bouloufa,[b] Olivier Dulieu,[b] Helmut Ritsch,[c] Russell Hart[a] and Hanns-Christoph Nägerl[a]

Received 8th September 2008, Accepted 17th December 2008
First published as an Advance Article on the web 12th May 2009
DOI: 10.1039/b820542f

One possibility for the creation of ultracold, high phase space density quantum gases of molecules in the rovibronic ground state relies on first associating weakly-bound molecules from quantum-degenerate atomic gases on a Feshbach resonance and then transferring the molecules *via* several steps of coherent two-photon stimulated Raman adiabatic passage (STIRAP) into the rovibronic ground state. Here, in ultracold samples of Cs_2 Feshbach molecules produced out of ultracold samples of Cs atoms, we observe several optical transitions to deeply-bound rovibrational levels of the excited 0_u^+ molecular potentials with high resolution. At least one of these transitions, although rather weak, allows efficient STIRAP transfer into the deeply-bound vibrational level $|v = 73\rangle$ of the singlet $X^1\Sigma_g^+$ ground state potential, as recently demonstrated (J. G. Danzl, E. Haller, M. Gustavsson, M. J. Mark, R. Hart, N. Bouloufa, O. Dulieu, H. Ritsch, and H.-C. Nägerl, *Science*, 2008, **321**, 1062). From this level, the rovibrational ground state $|v = 0, J = 0\rangle$ can be reached with one more transfer step. In total, our results show that coherent ground state transfer for Cs_2 is possible using a maximum of two successive two-photon STIRAP processes or one single four-photon STIRAP process.

1 Introduction

Ultracold and dense molecular samples in specific deeply-bound rovibrational levels are of high interest for fundamental studies in physics and chemistry. They are expected to find applications in high-resolution spectroscopy and fundamental tests,[2,3] few-body collisional physics,[4,5] ultracold chemistry,[6] quantum processing,[7] and in the field of dipolar quantum gases and dipolar Bose–Einstein condensation.[8,9] Ideally, full control over the molecular wave function is desired, *i.e.* full (quantum) control over the internal and external degrees of freedom of the molecules. High phase space densities are needed for molecular quantum gas studies. For many of the envisaged studies and applications, initial preparation of the molecular sample in the rovibronic ground state, *i.e.* the lowest energy level of the electronic ground state, is desired. Only in this state can one expect sufficient collisional stability.

But how is it possible to produce dense samples of ultracold molecules in the rovibrational ground state? Laser cooling of atoms, which has lead to the production

[a]*Institut für Experimentalphysik und Zentrum für Quantenphysik, Universität Innsbruck, Technikerstraße 25, A-6020 Innsbruck, Austria. E-mail: johann.danzl@uibk.ac.at*
[b]*Laboratoire Aimé Cotton, CNRS, Université Paris-Sud XI, Bât. 505, 91405 Orsay Cedex, France*
[c]*Institut für Theoretische Physik und Zentrum für Quantenphysik, Universität Innsbruck, Technikerstraße 25, A-6020 Innsbruck, Austria*

of quantum degenerate atomic Bose and Fermi gases,[10] cannot, so far, be adapted to the case of molecular systems, as suitable cycling transitions are not available. Versatile non-optical cooling and slowing techniques such as buffer gas cooling and Stark deceleration in combination with molecule trapping[11–13] have been developed, but high molecular densities and in particular high phase space densities are yet to be reached. An alternative route to producing ultracold molecular samples is given by first producing ultracold atomic samples and then associating molecules from the atomic sample. While this technique is so far limited to the production of selected species of dimer molecules, it has the advantage that ultra-low temperatures and high particle densities are easily inherited from the atomic precursor sample. There are essentially two association techniques, photoassociation[14] and magnetically induced Feshbach association.[15,16] In photoassociation experiments,[17–20] ultracold samples of deeply-bound molecules have been created. Additional techniques such as vibrational cooling[19] should allow selective pumping into the rovibrational ground state and open up the prospect of high molecular phase space densities. In Feshbach association experiments,[21,22] high-density samples of weakly-bound molecules are produced. For dimer molecules composed of fermions, collisional stability of the highly-excited molecules is assured as a result of a Pauli blocking effect, and molecular Bose–Einstein condensation could be achieved in the limit of extremely weak binding.[23]

Here, we are interested in combining the techniques of Feshbach association and coherent molecular state transfer to produce quantum gases of molecules in the rovibrational ground state $|v = 0, J = 0\rangle$ of the lowest electronic state. As usual, v and J are the vibrational and rotational quantum numbers, respectively. The molecules, produced on a Feshbach resonance and hence initially very loosely bound, are to be transferred in a few successive steps of coherent two-photon laser transfer to the rovibrational ground state, acquiring more and more binding energy in each step. The general idea is sketched in Fig. 1A for the case of Cs_2. Each two-photon step involves an excited state level. Population transfer into this level needs to be avoided to prevent loss due to spontaneous emission. One possibility is to use the technique of stimulated Raman adiabatic passage (STIRAP),[24] which is very robust and largely insensitive to laser intensity fluctuations. The scheme has several advantages. First, production of Feshbach molecules out of a quantum degenerate atomic sample can be very efficient.[25] Second, the optical transition rate on the first transition starting from the Feshbach molecules is greatly enhanced in comparison to the free atom case. Further, the scheme is fully coherent, not relying on spontaneous processes, allowing high state selectivity, and involving only a comparatively small number of intermediate levels. A ground state binding energy of typically 0.5 eV for an alkali dimer can be removed essentially without heating the molecular sample, as the differential photon recoil using pairwise co-propagating laser beams driving the two-photon transitions is very small. If losses and off-resonant excitations can be avoided, the scheme essentially preserves phase space density and coherence of the initial particle wave function, allowing the molecular sample to inherit the high initial phase space density from the atomic precursor sample.

Certainly, several challenges have to be met: going from weakly-bound Feshbach to tightly bound ground state molecules corresponds to a large reduction in internuclear distance. Consequently, the radial wave function overlap between successive levels is small, and a compromise has to be found between the number of transitions and the minimum tolerable wave function overlap. To keep the complexity of the scheme low, one or, at most, two two-photon transitions are desirable. Accordingly, suitable intermediate levels have to be identified that allow a balanced division of wave function overlap, as given by the Franck–Condon factors, between the different transitions. For example, for a four-photon transition scheme with Cs_2 as shown in Fig. 1A, the Franck–Condon factors are all of the order of 10^{-6}. We emphasize that the identification of the first excited level and hence of the first transition starting from the Feshbach molecules is of crucial importance. Detailed

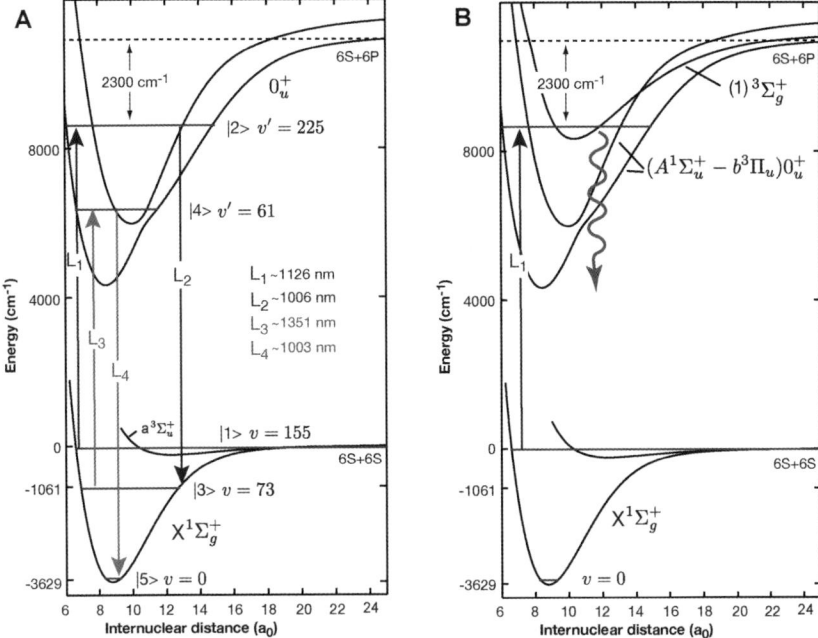

Fig. 1 (A) Simplified molecular level scheme for Cs_2 showing the relevant ground state and excited state potentials involved in rovibrational ground state transfer. Molecules in a weakly-bound Feshbach level $|1\rangle = |v \approx 155\rangle$ (not resolved near the $6S_{1/2} + 6S_{1/2}$ two-atom asymptote, but shown in Fig. 2) are to be transferred to the rovibrational ground state $|5\rangle = |v = 0, J = 0\rangle$ of the singlet $X^1\Sigma_g^+$ potential with a binding energy of 3629 cm^{-1} by two sequential two-photon STIRAP processes involving lasers L_1 and L_2 near 1126 nm and 1006 nm and lasers L_3 and L_4 near 1351 nm and 1003 nm. The intermediate ground state level $|3\rangle = |v = 73, J = 2\rangle$ has a binding energy of 1061 cm^{-1}. (B) Probing candidate levels for $|2\rangle$ belonging to the electronically excited coupled $(A^1\Sigma_u^+ - b^3\Pi_u) 0_u^+$ potentials. Here, we search for $|2\rangle$ in loss spectroscopy with laser L_1 in a region near 8890 cm^{-1} above the $6S_{1/2} + 6S_{1/2}$ asymptote, corresponding to an excitation wavelength range of 1118 to 1134 nm. The wiggly arrow indicates loss from the excited levels due to spontaneous emission. Also shown is the excited $(1)^3\Sigma_g^+$ potential, for which we find several levels.

calculations determining the wave function overlap are generally missing, and estimates on the Franck–Condon factors using hypothetical last bound states of either the singlet or triplet potentials of an alkali dimer molecule do not necessarily reflect the transition dipole moments adequately. In addition, for electronic molecular states or energy regions where spectroscopic data is missing, the precise energy of the excited state levels above the atomic threshold is known only with a large uncertainty, which can approach the vibrational spacing of up to a few nanometers. Hence, considerable time has to be spent on searching for weak transitions starting from the initial Feshbach molecules.

In a pioneering experiment, Winkler et al.[26] demonstrated that the STIRAP technique can efficiently be implemented with quantum gases of weakly-bound Feshbach molecules. In this work, the transferred molecules, in this case Rb_2, were still weakly-bound with a binding energy of less than 10^{-4} of the binding energy of the rovibronic ground state, and the intermediate excited state level was close to the excited-atom asymptote. Here, we observed several optical transitions starting from a weakly-bound Feshbach level to deeply-bound rovibrational levels of the mixed excited $(A^1\Sigma_u^+ - b^3\Pi_u) 0_u^+$ molecular potentials of the Cs_2 molecule in a wavelength range from 1118 to 1134 nm, far to the red of the atomic D_1 and D_2 transitions. The Cs_2 molecular potentials are shown in Fig. 1A. We observed the levels as loss from an ultracold sample of Cs_2 Feshbach molecules as shown in Fig. 1B. We observed two progressions, one that we attribute to the $(A^1\Sigma_u^+ - b^3\Pi_u)\ 0_u^+$ potentials and one that we associate with the triplet $(1)^3\Sigma_g^+$ potential. From the loss measurements, we determined the transition strengths and find that the stronger transitions should be suitable for STIRAP to an intermediate, deeply-bound rovibrational level of the singlet $X^1\Sigma_g^+$ potential with $v = 73$. Recently, we implemented STIRAP into $|v = 73, J = 2>$.[1] For the case of the dimer molecule KRb, Ni et al.[27] could demonstrate quantum gas transfer all the way into the rovibrational ground state $|v = 0, J = 0>$ of the singlet $X^1\Sigma^+$ molecular potential. Here, the transfer could be achieved in only a single step as a result of the favorable shape of the excited state potentials, which is generally the case for heteronuclear molecules composed of alkali atoms.[28] Also recently, transfer to the rovibrational ground state level of the lowest triplet $a^3\Sigma_u^+$ state of Rb_2 could be achieved.[29]

2 Preparation of a sample of weakly-bound Feshbach molecules

We produced ultracold samples of molecules on two different Feshbach resonances, one near 1.98 mT and one near 4.79 mT.[30] In both cases, essentially following the procedure detailed in ref. 31, we first produced an ultracold sample of typically 2×10^5 Cs atoms in the lowest hyperfine sublevel $F = 3$, $m_F = 3$ in a crossed optical dipole trap. As usual, F is the atomic angular momentum quantum number, and m_F its projection on the magnetic field axis. The trapping light at 1064.5 nm is derived from a single-frequency, highly-stable Nd:YAG laser. The offset magnetic field value for evaporative cooling is 2.1 mT. We support optical trapping by magnetic levitation with a magnetic field gradient of 3.1 mT cm^{-1}. We then produced weakly-bound Feshbach molecules out of the atomic sample.[22] We produced a sample every 8 s, i.e. our spectroscopic measurements were performed at a rate of one data point every 8 s. In order to be able to search for optical transitions over large frequency ranges it is advantageous to work with the shortest possible sample preparation times. For this reason we stopped evaporative cooling slightly before the onset of Bose–Einstein condensation (BEC), which also made sample preparation somewhat less critical. The temperature of the initial atomic sample is then typically about 100 nK. At higher temperatures and hence lower phase space densities the molecule production efficiency is reduced, so that there is a trade-off between ease of operation and molecule number. We note that for our ground state transfer experiments reported in ref. 1 we produced a pure atomic BEC at the expense of longer sample preparation times.

The spectrum of weakly-bound Feshbach levels near the two-free-atom asymptote is shown in Fig. 2.[30] For molecule production at the Feshbach resonance at 4.79 mT, we first ramped the magnetic field from the BEC production value to 4.9 mT, about 0.1 mT above the Feshbach resonance. We produced the molecular sample on a downward sweep at a typical sweep rate of 0.025 T s^{-1}. The resulting ultracold sample contained up to 11000 molecules, immersed in the bath of the remaining ultracold atoms. The resonance at 4.79 mT is a d-wave resonance,[30] and hence the molecules are initially of d-wave character, i.e. $\ell = 2$, where ℓ is the quantum number associated with the mechanical rotation of the nuclei. However, there is a weakly-bound s-wave Feshbach state ($|s> = |\ell = 0>$) belonging to the open scattering channel right below threshold. This state couples quite strongly to the initial d-wave state, resulting in an avoided state crossing (as shown in the inset to Fig. 2), on which the molecules are transferred to the s-wave state $|s>$ upon lowering the magnetic field.[30,1] Upon further lowering of the magnetic field to less than 2.0 mT, the molecules acquire more and more character of a closed-channel s-wave state on a second, very broad avoided crossing. Here, we performed spectroscopy in this transition range from open channel to closed channel s-wave character. At a magnetic field value of 2.0 mT, the binding energy of the molecules is near

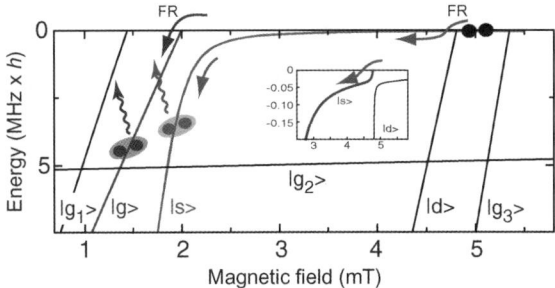

Fig. 2 Initial Feshbach molecule production: Zeeman diagram showing the energy of weakly-bound Feshbach levels[30] and the Feshbach resonances (FR) used in the present work. The binding energy is given with respect to the $F = 3$, $m_F = 3$ two-atom asymptote. The molecules are produced either on a d-wave Feshbach resonance at 4.79 mT (see inset) and then transferred to the weakly-bound s-wave state $|s\rangle$ on an avoided state crossing, or on a g-wave Feshbach resonance at 1.98 mT, resulting in molecules in level $|g\rangle$. In the first case, further lowering of the magnetic offset field to below 2.0 mT changes the character of the $|s\rangle$ level from open-channel to closed-channel dominated.[30] The levels $|s\rangle$ and $|g\rangle$ are both candidate levels for the initial level $|1\rangle$ shown in Fig. 1. For completeness, further g-wave Feshbach levels, $|g_1\rangle$, $|g_2\rangle$, and $|g_3\rangle$ are shown. Level $|g_2\rangle$ connects $|g\rangle$ to $|s\rangle$ and can be used for Feshbach state transfer.[30] Level $|g_3\rangle$ is a further interesting candidate level for $|1\rangle$ with low nuclear spin contribution.[30]

5 MHz × h with respect to the $F = 3$, $m_F = 3$ two-atom asymptote, where h is Planck's constant.

For molecule production at the Feshbach resonance at 1.98 mT, we simply ramped the magnetic field down from the initial BEC production value. Again, we produced an ultracold molecular sample with about 11 000 molecules. The molecules in $|g\rangle$ have g-wave character, i.e. $\ell = 4$. When we lowered the magnetic field to 1.6 mT, the binding energy of the molecules was also near 5 MHz × h with respect to the $F = 3$, $m_F = 3$ two-atom asymptote.

For spectroscopy, we released the molecules from the trap after magnetic field ramping was completed and performed all subsequent experiments in free flight without any other light fields on except for the spectroscopy laser.

For molecule detection in both cases, we reversed the magnetic field ramps.[22] The g-wave molecules are dissociated on the g-wave Feshbach resonance at 1.98 mT, and the s-wave molecules are dissociated on the d-wave Feshbach resonance at 4.79 mT. Prior to the reverse magnetic field ramp, we applied a magnetic field gradient of 3.1 mT cm^{-1} for about 5 ms to separate the molecular sample from the atomic sample in a Stern–Gerlach-type experiment. Finally, we detected atoms by standard absorption imaging. The minimum number of molecules that we can detect is of the order of 200 molecules.

3 Spectroscopy

We performed optical spectroscopy on Feshbach molecules in the wavelength region around 1125 nm. Based on selection rules, there are two sets of electronically excited states that we address in the spectroscopic measurements presented here, namely the $(A^1\Sigma_u^+-b^3\Pi_u)\ 0_u^+$ coupled state system and the purely triplet $(1)^3\Sigma_g^+$ state. We first discuss transitions to the 0_u^+ coupled state system. Transitions to the latter state are discussed in Section 3.2.

3.1 Transitions to the $(A^1\Sigma_u^+-b^3\Pi_u)\ 0_u^+$ coupled electronically excited states

For ground state transfer, we are primarily interested in transitions from Feshbach levels to rovibrational levels of the $(A^1\Sigma_u^+-b^3\Pi_u)\ 0_u^+$ electronically excited states. In the heavy alkali dimers, most notably in Cs_2, the $A^1\Sigma_u^+$ state and the $b^3\Pi_u$ state are strongly coupled by resonant spin–orbit interaction,[32,33] yielding the 0_u^+ coupled states in Hund's case (c) notation. The singlet component of the 0_u^+ states allows us to efficiently couple to deeply-bound $X^1\Sigma_g^+$ state levels, specifically to the $|v = 73, J = 2\rangle$ level of the ground state potential, as has recently been shown in a coherent transfer experiment.[1] We chose to do spectroscopy in the wavelength range of 1118 nm to 1134 nm above the $6S_{1/2} + 6S_{1/2}$ dissociation threshold of the Cs_2 dimer. This corresponds to a detuning of roughly 2300 cm^{-1} from the cesium D_1 line and to an energy range of approximately 12572 cm^{-1} to 12450 cm^{-1} above the rovibronic ground state $X^1\Sigma_g^+ |v = 0, J = 0\rangle$. This region was chosen in order to give a balanced distribution of transition strengths in a 4-photon transfer scheme to the rovibronic ground state. In addition, the wavelengths of the four lasers used in the transfer experiments were chosen such that they lie within the wavelength range covered by the infrared fiber-based frequency comb that we use as a frequency reference in the state transfer experiments.

The transitions of interest here lie outside the energy regions for which Fourier transform spectroscopic data was obtained at Laboratoire Aimé Cotton from transitions to the $X^1\Sigma_g^+$ state.[34] The vibrational progression of the 0_u^+ states is highly perturbed by the resonant spin–orbit coupling and exhibits an irregular vibrational spacing. Molecular structure calculations are complicated by the spin–orbit coupling and calculated term values are highly sensitive to the coupling. Prior to the experiments discussed here the absolute energies of the vibrational levels of the $(A^1\Sigma_u^+ - b^3\Pi_u) 0_u^+$ excited state were poorly known in the region of interest from 1118 nm to 1134 nm. We therefore performed a broad-range search by irradiating the weakly-bound Feshbach molecules at a fixed wavelength for a certain irradiation time τ of up to $\tau = 6$ ms and by recording the number of remaining molecules as a function of laser frequency. In one run of the experiment one particular laser frequency is queried. We thus took data points at the repetition rate of our experiment, which is given by the sample preparation time of 8 seconds. Based on the available laser intensity from L_1 and an estimate of the dipole transition moments for the strongest expected lines, we chose a frequency step size of about 100 MHz to 150 MHz for initial line searching. We obtained the laser light at 1118 nm to 1134 nm from a grating-stabilized external cavity diode laser. For coarse frequency scanning, the laser is free running and tuned *via* a piezoelectric element on the grating of the laser. For more precise measurements, we locked the laser to a narrow-band optical resonator that can be tuned *via* a piezoelectric element. Fig. 3A shows a typical loss spectrum starting from Feshbach state $|s\rangle$ for excitation near 1126 nm, measured at a magnetic field of 1.98 mT. In this particular case we find three resonances, which we associate with the rotational splitting of the excited state level, $J = 5, 3, 1$, where J is the rotational quantum number. Based on molecular structure calculations we identified this level as the 225th level of the 0_u^+ progression with an uncertainty of about two in the absolute numbering. We zoom in on these three transitions in Fig. 3B, C, and D and recorded loss resonances at reduced laser intensity in order to avoid saturation of the lines. For these measurements, the laser is locked to the narrow-band optical resonator and the resonator in turn is stabilized to the optical frequency comb to assure reproducibility and long term frequency stability. As one can expect, the loss is strongest on the transition to the $|J = 1\rangle$ level, and it is weakest on the transition to $|J = 5\rangle$. The width of all lines gives an excited state spontaneous decay rate of around $2\pi \times 2$ MHz, in agreement with the typical expected lifetimes of excited molecular levels. The transition to $|J = 1\rangle$ shown in Fig. 3D is of special interest to the current work. It has been used as the intermediate excited state level for coherent transfer to $X^1\Sigma_g^+ |v = 73, J = 2\rangle$ in our recent experiments.[1]

By fitting to a series of such measurements, obtained with different laser intensities, a two-level model that takes into account decay from the upper level, we determine the transition strength as given by the normalized Rabi frequency. As a result

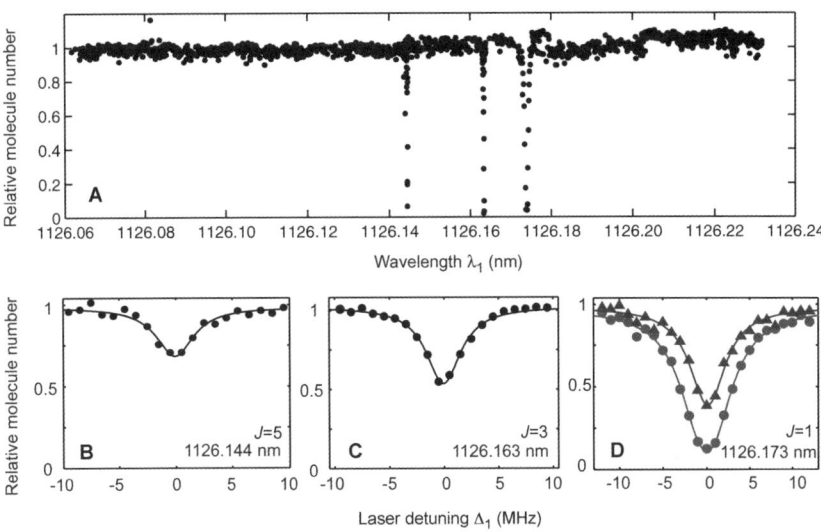

Fig. 3 Loss resonances for excitation from the initial Feshbach level $|s\rangle$ to the 0_u^+ system. (**A**) Typical scan showing the relative number of molecules in $|s\rangle$ as a function of laser wavelength λ_1 near 1126 nm. Three resonances can be identified, corresponding to $|J=5\rangle$, $|J=3\rangle$, and $|J=1\rangle$, from left to right. The sample is irradiated with laser light at an intensity of 1×10^6 mW cm^{-2} for $\tau = 200$ μs. The laser is locked to a narrow-band optical resonator that is tuned *via* a piezoelectric element with a step size of approximately 40 MHz. Wavelength is measured on a home-built wavemeter. The molecule number is normalized to the atom number measured in the same individual realization of the experiment to cancel out fluctuations that stem from shot-to-shot atom number fluctuations and then normalized to unity. (**B**), (**C**), and (**D**) show measurements of the three individual lines with $|J=5\rangle$, $|J=3\rangle$, and $|J=1\rangle$ at reduced intensity in order to avoid saturation. The solid lines represent fits as described in the text. The spectroscopy laser is stabilized to an optical resonator and the resonator is in turn referenced to an optical frequency comb, which allows precise and reproducible tuning of the frequency. The transition to $|J=1\rangle$ in panel (**D**) is recorded at an intensity of 1.5×10^4 mW cm^{-2} (circles) and 6×10^3 mW cm^{-2} (triangles), (**B**) and (**C**) are recorded at 1×10^6 mW cm^{-2} and 2×10^5 mW cm^{-2}, respectively. The pulse duration is $\tau = 10$ μs.

of optical excitation, for small saturation the number N of Feshbach molecules decays as a function of laser detuning Δ_1 according to $N(\Delta_1) = N_0 \exp(-\tau\Omega_1^2/(\Gamma(1 + 4\pi^2\Delta_1^2/\Gamma^2)))$, where N_0 is the molecule number without laser irradiation and τ is the irradiation time. From the fit we obtain the Rabi frequency on resonance Ω_1 and the excited state spontaneous decay rate Γ. We determine the normalized Rabi frequency to $\Omega_1 = 2\pi \times 2$ kHz $\sqrt{I/(\text{mW cm}^{-2})}$ for $|J=1\rangle$, where I is the laser intensity. This value is sufficient to perform STIRAP given the available laser power.[1] The corresponding transition strengths for $|J=3\rangle$ and $|J=5\rangle$ are $\Omega_1 = 2\pi \times 0.3$ kHz $\sqrt{I/(\text{mW cm}^{-2})}$ and $\Omega_1 = 2\pi \times 0.1$ kHz $\sqrt{I/(\text{mW cm}^{-2})}$, respectively. The absolute values of these transition strengths bear an estimated uncertainty of 20% because the laser beam parameters for the spectroscopy laser are not well determined.

We also recorded the time dependence of the molecular loss on some of the stronger lines. For this, we stepped the laser irradiation time τ from 0 to 150 μs, while laser L_1 was kept on resonance. The result is shown in Fig. 4A for the transition at 1126.173 nm for two different values of the excitation laser intensity.

We note that the transition strength for a particular line starting from Feshbach level $|s\rangle$ strongly depends on the value of the magnetic field, as evidenced in Fig. 4B. Loss resonances for the transition at 1126.173 nm at 1.9 mT and 2.2 mT are shown.

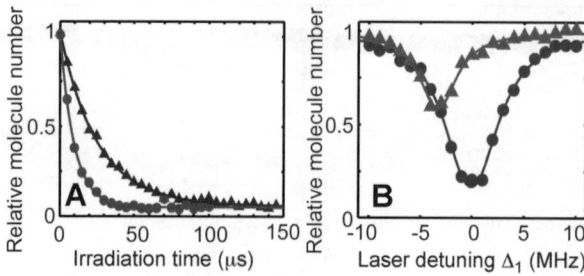

Fig. 4 Loss of molecules for excitation near 1126.173 nm from Feshbach level |s>. (**A**) Time dependence of molecular loss on resonance at 1126.173 nm for two different laser intensities, 5.7×10^5 mW cm^{-2} (circles) and 2.1×10^5 mW cm^{-2} (triangles). The magnetic offset field is 1.9 mT. The fitted exponential decay gives the decay constants $\tau_d = 9.7 \pm 0.6$ μs (circles) and $\tau_d = 25.5 \pm 1$ μs (triangles). (**B**) Loss of molecules in |s> as a function of laser detuning Δ_1 near 1126 nm with an irradiation time of $\tau = 10$ μs for two values of the magnetic field, 1.9 mT (circles) and 2.2 mT (triangles). In both cases, the excited state spontaneous decay rate was determined to $\approx 2\pi \times 2$ MHz. At higher magnetic fields, Feshbach level |s> acquires more open-channel character, reducing radial wave function overlap with the excited rovibrational levels. The shift in transition frequency is essentially the result of the change in binding energy as seen in Fig. 2.

For ground state transfer,[1] we chose a magnetic field of around 1.9 mT, which is somewhat below the magnetic field region where state |s> is strongly curved, but above the avoided state crossing with state $|g_2>$, as seen in Fig. 2. The pronounced bending of |s> is the result of a strong avoided crossing between two s-wave Feshbach levels.[30] For magnetic field values beyond 3.0 mT the level |s> can be associated with the $F_1 = 3$, $F_2 = 3$ asymptote, where F_i, $i = 1, 2$, is the atomic angular momentum quantum number of the i-th atom, respectively. Below 2.0 mT the level |s> can be associated with the $F_1 = 4$, $F_2 = 4$ asymptote. It is hence of closed-channel character and much more deeply-bound with respect to its potential asymptote, effectively by twice the atomic hyperfine splitting. This improves the radial wave function overlap with the excited state levels and hence increases the transition strength. Trivially, the resonance frequency is shifted as the binding energy is reduced for larger magnetic field values. Coupling to the excited state level is reduced from $\Omega_1 = 2\pi \times 2$ kHz $\sqrt{I/(\text{mW cm}^{-2})}$ to $\Omega_1 = 2\pi \times 1$ kHz $\sqrt{I/(\text{mW cm}^{-2})}$ when the magnetic field is changed from 1.9 mT to 2.2 mT.

As will be discussed in Section 4 it is advantageous to be able to choose different Feshbach states as a starting state for ground state transfer experiments. Therefore, we probed transitions from Feshbach level |g> to $(A^1\Sigma_u^+ - b^3\Pi_u) 0_u^+$ levels. Fig. 5 shows loss resonances to the same excited state levels as shown in Fig. 3, only that now the initial Feshbach level is |g> instead of |s>. In this case, the transition to $|J = 3>$ is the strongest, while the transition to $|J = 1>$ is very weak, but can be detected. A comparison of the transition strengths from |g> to the excited state level $|J = 3>$, giving $\Omega_1 = 2\pi \times 1$ kHz $\sqrt{I/(\text{mW cm}^{-2})}$ versus |s> to $|J = 1>$ giving $\Omega_1 = 2\pi \times 2$ kHz $\sqrt{I/(\text{mW cm}^{-2})}$ shows that level |g> could also be potentially used as a starting level for coherent population transfer to deeply-bound levels of the ground state but requires longer STIRAP times in order to assure sufficient adiabaticity.[24] The $|J = 3>$ excited state level in turn couples to $|J = 2>$ in the ground state, the level used in our previous work.[1]

In addition to the transition near 1126 nm we find a series of other excited state levels that we assign to the $(A^1\Sigma_u^+ - b^3\Pi_u) 0_u^+$ coupled state system. These are listed in Table 1. The assignment to either the $(A^1\Sigma_u^+ - b^3\Pi_u) 0_u^+$ system or to the $(1)^3\Sigma_g^+$

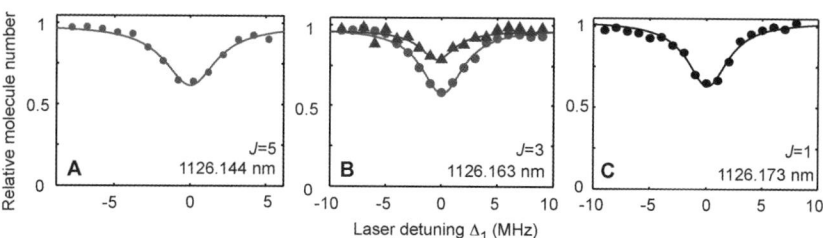

Fig. 5 Loss resonances for excitation from the initial Feshbach level |g>. (A), (B), and (C) show the loss for excitation to |J = 5>, |J = 3>, and |J = 1>, corresponding to the resonances shown in Fig. 3. The laser intensities are 1.5×10^4 mW cm^{-2} for panel (A) and for the circles in panel (B). The second resonance in (B) (triangles) is measured with 5.6×10^3 mW cm^{-2}. (C) The line at 1126.173 nm is measured at 1×10^6 mW cm^{-2}. All measurements are done with an irradiation time of $\tau = 10$ μs. From a series of such measurements at different intensities we determine the line strengths for |J = 5>, |J = 3>, and |J = 1> as $\Omega_1 = 2\pi \times 1$ kHz $\sqrt{I/(\text{mW cm}^{-2})}$, $\Omega_1 = 2\pi \times 1$ kHz $\sqrt{I/(\text{mW cm}^{-2})}$, and $\Omega_1 = 2\pi \times 0.1$ kHz $\sqrt{I/(\text{mW cm}^{-2})}$, respectively.

Table 1 Observed excited state levels in the wavelength range from 1118 nm to 1134 nm. Transitions were measured from Feshbach state |s> to the first electronically excited state, addressing both (A$^1\Sigma_u^+$–b$^3\Pi_u$)0$_u^+$ levels and (1)$^3\Sigma_g^+$ levels. Levels are given according to the excitation wavelength (WL) from |s>, which essentially corresponds to the $F = 3$, $m_F = 3$ two-atom asymptote. The data is taken at a magnetic field of 1.98 mT. Wavemeter accuracy is about 0.001 nm. The energy of these levels above the rovibronic ground state X$^1\Sigma_g^+$ |v = 0, J = 0> is given in the second column, where the binding energy of the rovibronic ground state is taken from ref. 1. The assignment to either the coupled (A$^1\Sigma_u^+$–b$^3\Pi_u$) 0$_u^+$ system or to the (1)$^3\Sigma_g^+$ is based on the vibrational spacing and similarities in the substructure of the levels. The levels marked with * have been used for dark resonance spectroscopy coupling to deeply-bound levels of the X$^1\Sigma_g^+$ state.[1] The ability to couple to such levels unambiguously reflects an important singlet component stemming from the A$^1\Sigma_u^+$ state and therefore clearly assigns these levels to the 0$_u^+$ system. The quantum numbers given for the 0$_u^+$ levels are coupled channel quantum numbers derived from molecular structure calculations and bear an uncertainty of two in the absolute numbering. The calculations show that these levels have about 70% A$^1\Sigma_u^+$ state contribution. Two further levels observed near 1120.17 nm and 1117.16 nm that belong to the 0$_u^+$ progression are not given in the table since no further measurements have been done on these levels. The level near 1129.5 nm exhibits a somewhat richer structure than the other levels assigned to 0$_u^+$ and than exemplified in Fig. 3. Levels assigned to the (1)$^3\Sigma_g^+$ state form a regular vibrational progression and show a more complex substructure than the levels attributed to the 0$_u^+$ system, as exemplified in Fig. 6. For these levels, the transition wavelength to one of the most prominent features is given, since an in-depth analysis of the rotational and hyperfine structure remains to be done. The vibrational numbering for the (1)$^3\Sigma_g^+$ levels is the same as in ref. 35

WL/nm	Energy above X$^1\Sigma_g^+$ |v = 0>/cm^{-1}	Assignment
1132.481	12458.875	0$_u^+$ |v' = 221, J = 1>
1129.492	12482.245	0$_u^+$
1126.173*	12508.332	0$_u^+$ |v' = 225, J = 1>
1123.104*	12532.598	0$_u^+$ |v' = 226, J = 1>
1133.680	12449.536	(1)$^3\Sigma_g^+$ |v' = 32>
1130.510	12474.274	(1)$^3\Sigma_g^+$ |v' = 33>
1127.379	12498.838	(1)$^3\Sigma_g^+$ |v' = 34>
1124.274	12523.334	(1)$^3\Sigma_g^+$ |v' = 35>
1121.196	12547.756	(1)$^3\Sigma_g^+$ |v' = 36>
1118.155	12572.013	(1)$^3\Sigma_g^+$ |v' = 37>

electronically excited state discussed below is primarily based on the spacing between neighboring vibrational levels and, in addition, on the pattern of loss resonances associated with each particular vibrational level. Resonant spin–orbit coupling in the case of the 0_u^+ states leads to an irregular vibrational spacing. In contrast, the $(1)^3\Sigma_g^+$ state is not perturbed by spin–orbit interaction and therefore has a regular vibrational progression. The levels near 1126 nm and near 1123 nm have been used to detect dark resonances with deeply-bound levels of the $X^1\Sigma_g^+$ state.[1] The ability to couple to these essentially purely singlet ground state levels unambiguously assigns the corresponding excited state levels to the 0_u^+ system. The data given in Table 1 does not represent a fully exhaustive study of the $(A^1\Sigma_u^+ - b^3\Pi_u)\, 0_u^+$ coupled states in the wavelength range of interest. In fact, for the most part we observe those levels of the 0_u^+ system that have a dominant $A^1\Sigma_u^+$ state contribution, as determined from molecular structure calculations.

3.2 Transitions to the $(1)^3\Sigma_g^+$ electronically excited state

The Feshbach levels that serve as starting levels for the spectroscopy are of mixed $X^1\Sigma_g^+$ and $a^3\Sigma_u^+$ character. In the wavelength range explored here, excitation to the $(1)^3\Sigma_g^+$ electronically excited triplet state is possible from the $a^3\Sigma_u^+$ component of the Feshbach molecules. In fact, for a heavy molecule such as Cs$_2$, the $(1)^3\Sigma_g^+$ state is better described by the two separate electronic states 0_g^- and 1_g, denoted by Hund's case (c) notation. The $(1)^3\Sigma_g^+$ state has been previously studied by Fourier transform spectroscopy.[35] This state is not of prime interest for the present work

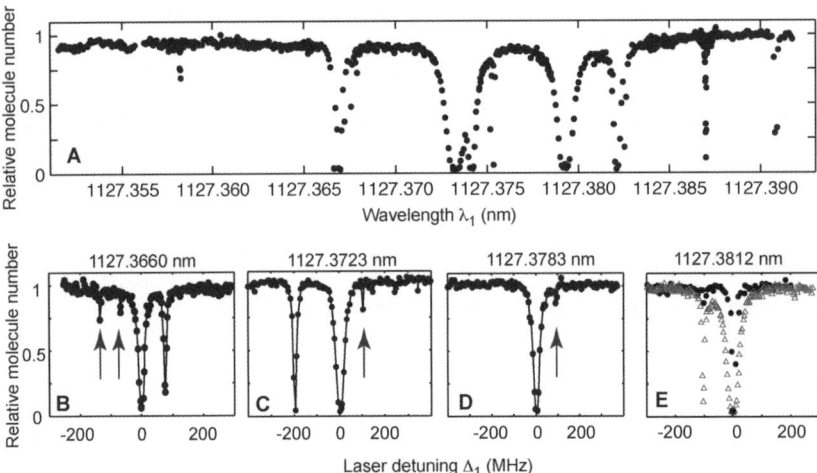

Fig. 6 Loss of molecules for excitation near 1127.17 nm from Feshbach level |s> to the triplet $(1)^3\Sigma_g^+$ state. (**A**) represents a broad scan with laser irradiation at an intensity of 5×10^5 mW cm^{-2} for $\tau = 100$ μs at a step size of 20 MHz. A rich structure due to rotation and excited state hyperfine splitting can be seen, which is qualitatively different from the spectrum shown in Fig. 3. The lines are greatly broadened by the high intensity and long irradiation time. The spectroscopy laser is locked to a narrow-band optical resonator that is stepped *via* a piezoelectric element. Scans of about 750 MHz were recorded as a function of piezo voltage on the resonator. Voltage was converted to wavelength for each scan by a linear interpolation. (**B**)–(**E**) represent scans over some of the observed features at a reduced intensity of 8×10^4 mW cm^{-2} and an irradiation time of $\tau = 10$ μs in order to reduce broadening of the lines. The step size is about 7 MHz. Resonator piezo voltage is converted to frequency with an estimated error of 10%. The absolute wavelength accuracy is limited by wavemeter calibration to 0.001 nm, the relative accuracy is about a factor of 10 better. The vertical arrows indicate weak lines that have been verified in additional scans with higher power. In panel (**E**) the power was somewhat increased for an additional measurement (triangles) that emphasizes such a weak line.

as transitions from this state down to the $X^1\Sigma_g^+$ ground state are expected to be strongly suppressed, but would be important for STIRAP transfer into the rovibrational ground state level of the shallow triplet $a^3\Sigma_u^+$ potential.[29] Certainly, it is important to be able to distinguish rovibrational levels belonging to the $(1)^3\Sigma_g^+$ state from the ones belonging to the 0_u^+ system, because otherwise time would be wasted in searching for ground state dark resonances that are very weak or do not even exist. Fig. 6A shows a typical loss spectrum for one of the lines that we detected near 1127.37 nm. Due to hyperfine splitting, levels of triplet character exhibit a much richer substructure than the 0_u^+ levels used for ground state transfer. Several components can be identified as a result of rotational and excited state hyperfine splitting. Zoomed-in regions are shown in Fig. 6B, C, D, and E. We have observed a regularly spaced series of optical transitions which we attribute to the $(1)^3\Sigma_g^+$ excited state as listed in Table 1. The level energies are well reproduced by the Dunham coefficients determined in ref. 35. The vibrational numbering used here is the same as in that work. However, it relies on the absolute energy position of the potential, T_e, which was not determined precisely in ref. 35. By fixing T_e to the value given in ref. 35 we get good agreement with our data.

4 Conclusion

We have performed optical spectroscopy starting from weakly-bound Cs_2 Feshbach molecules into deeply-bound rovibrational levels of the mixed excited state 0_u^+ system and the excited triplet $(1)^3\Sigma_g^+$ state. At least one of the observed transitions, namely the one at 1126.173 nm starting from the Feshbach level $|s\rangle$ to the excited level $|v' = 225, J = 1\rangle$ of the 0_u^+ system, at an offset magnetic field value of 1.9 mT, is strong enough to allow efficient STIRAP transfer into deeply-bound rovibrational levels of the singlet $X^1\Sigma_g^+$ ground state potential. The use of this transition for STIRAP has recently been demonstrated in ref. 1. In that work, the deeply-bound rovibrational level $|v = 73, J = 2\rangle$ of the $X^1\Sigma_g^+$ ground state potential was populated in the molecular quantum gas regime with 80% efficiency. The rovibrational ground state $|v = 0, J = 0\rangle$ of the $X^1\Sigma_g^+$ ground state potential can thus be reached from the atomic threshold with a maximum of two two-photon STIRAP transfers. Dark resonances connecting $|v = 73, J = 2\rangle$ to $|v = 0, J = 0\rangle$ have recently been observed,[36] and two-step STIRAP into $|v = 0, J = 0\rangle$ has recently been implemented.[37] For future experiments, the use of Feshbach level $|g\rangle$ as the initial state might be advantageous. Level $|g\rangle$ can be more easily populated, as the Feshbach resonance connected to this level is at a low magnetic field value of 1.98 mT,[30] where the atomic background scattering length has a moderate value of 155 a_0, where a_0 is Bohr's radius. The use of this resonance avoids excitation of collective motion of the atomic BEC as a result of a large mean field interaction near the Feshbach resonance at 4.79 mT,[1] where the atomic background scattering length is about 935 a_0. The transition starting from level $|g\rangle$ appears to be strong enough to allow STIRAP, this time via the excited state level $|v' = 225, J = 3\rangle$ of the 0_u^+ system. An attractive strategy for the production of a BEC of ground state molecules relies on the addition of a three-dimensional optical lattice. Starting from the atomic BEC, pairs of atoms at individual lattice sites can be produced in a superfluid-to-Mott-insulator transition[38] with high efficiencies of almost 50%.[39] These pairs can then be very efficiently associated on a Feshbach resonance[40] and subsequently transferred to the rovibronic ground state with STIRAP. The lattice has the advantage of shielding the molecules against inelastic collisions during the association process and subsequent state transfer. In particular, it should allow long STIRAP pulse durations, allowing us to resolve the weak hyperfine structure of ground state molecules.[41] As proposed by Jaksch et al.,[42] dynamical melting of the Mott-insulator state should ideally result in the formation of a BEC of molecules in the rovibronic ground state in a Mott-insulator-to-superfluid-type transition.

5 Acknowledgements

We are indebted to R. Grimm for generous support and we thank T. Bergeman, H. Salami, J. Hutson, J. Aldegunde, and E. Tiemann for valuable discussions. We gratefully acknowledge funding by the Austrian Ministry of Science and Research (BMWF) and the Austrian Science Fund (FWF) in the form of a START prize grant and by the European Science Foundation (ESF) in the framework of the Euro-QUAM collective research project QuDipMol. R. H. acknowledges support by the European Union in form of a Marie Curie International Incoming Fellowship (IIF).

References

1 J. G. Danzl, E. Haller, M. Gustavsson, M. J. Mark, R. Hart, N. Bouloufa, O. Dulieu, H. Ritsch and H.-C. Nägerl, *Science*, 2008, **321**, 1062.
2 T. Zelevinsky, S. Kotochigova and J. Ye, *Phys. Rev. Lett.*, 2008, **100**, 043201.
3 D. DeMille, S. Sainis, J. Sage, T. Bergeman, S. Kotochigova and E. Tiesinga, *Phys. Rev. Lett.*, 2008, **100**, 043202.
4 P. Staanum, S. D. Kraft, J. Lange, R. Wester and M. Weidemüller, *Phys. Rev. Lett.*, 2006, **96**, 023201.
5 N. Zahzam, T. Vogt, M. Mudrich, D. Comparat and P. Pillet, *Phys. Rev. Lett.*, 2006, **96**, 023202.
6 R. V. Krems, *Int. Rev. Phys. Chem.*, 2005, **24**, 99.
7 D. DeMille, *Phys. Rev. Lett.*, 2002, **88**, 067901.
8 K. Góral, L. Santos and M. Lewenstein, *Phys. Rev. Lett.*, 2002, **88**, 170406.
9 M. Baranov, L. Dobrek, K. Góral, L. Santos and M. Lewenstein, *Phys. Scr., T*, 2002, **102**, 74.
10 K. Southwell (ed.), Ultracold matter, *Nature (Insight)*, 2002, **416**, 205.
11 J. Doyle, B. Friedrich, R. V. Krems and F. Masnou-Seeuws, *Eur. Phys. J. D*, 2004, **31**, 149.
12 R. V. Krems, *Phys. Chem. Chem. Phys.*, 2008, **10**, 4079.
13 S. Y. T. van de Meerakker, H. L. Bethlem and G. Meijer, *Nat. Phys.*, 2008, **4**, 595.
14 For a review on photoassociation, see: K. M. Jones, E. Tiesinga, P. D. Lett and P. S. Julienne, *Rev. Mod. Phys.*, 2006, **78**, 483.
15 T. Köhler, K. Góral and P. S. Julienne, *Rev. Mod. Phys.*, 2006, **78**, 1311.
16 F. Ferlaino, S. Knoop, and R. Grimm, in: Cold Molecules: Theory, Experiment, Applications, ed. R. V. Krems, B. Friedrich, and W. C. Stwalley (publication expected in July 2009), preprint at arXiv:0809.3920.
17 A. N. Nikolov, J. R. Ensher, E. E. Eyler, H. Wang, W. C. Stwalley and P. L. Gould, *Phys. Rev. Lett.*, 2000, **84**, 246.
18 J. M. Sage, S. Sainis, T. Bergeman and D. DeMille, *Phys. Rev. Lett.*, 2005, **94**, 203001.
19 M. Viteau, A. Chotia, M. Allegrini, N. Bouloufa, O. Dulieu, D. Comparat and P. Pillet, *Science*, 2008, **321**, 232.
20 J. Deiglmayr, A. Grochola, M. Repp, K. Mörtlbauer, C. Glück, J. Lange, O. Dulieu, R. Wester and M. Weidemüller, *Phys. Rev. Lett.*, 2008, **101**, 133004.
21 C. A. Regal, C. Ticknor, J. L. Bohn and D. S. Jin, *Nature*, 2003, **424**, 47.
22 J. Herbig, T. Kraemer, M. Mark, T. Weber, C. Chin, H.-C. Nägerl and R. Grimm, *Science*, 2003, **301**, 1510.
23 For an overview, see: Ultracold Fermi Gases, *Proceedings of the International School of Physics Enrico Fermi, Course CLXIV*, ed. M. Inguscio, W. Ketterle, and C. Salomon, IOS Press, Amsterdam, 2008.
24 K. Bergmann, H. Theuer and B. W. Shore, *Rev. Mod. Phys.*, 1998, **70**, 1003.
25 M. Mark, T. Kraemer, J. Herbig, C. Chin, H.-C. Nägerl and R. Grimm, *Europhys. Lett.*, 2005, **69**, 706.
26 K. Winkler, F. Lang, G. Thalhammer, P. v. d. Straten, R. Grimm and J. Hecker Denschlag, *Phys. Rev. Lett.*, 2007, **98**, 043201.
27 K.-K. Ni, S. Ospelkaus, M. H. G. de Miranda, A. Pe'er, B. Neyenhuis, J. J. Zirbel, S. Kotochigova, P. S. Julienne, D. S. Jin and J. Ye, *Science*, 2008, **322**, 231.
28 W. C. Stwalley, *Eur. Phys. J. D*, 2004, **31**, 221.
29 F. Lang, K. Winkler, C. Strauss, R. Grimm and J. Hecker Denschlag, *Phys. Rev. Lett.*, 2008, **101**, 133005.
30 M. Mark, F. Ferlaino, S. Knoop, J. G. Danzl, T. Kraemer, C. Chin, H.-C. Nägerl and R. Grimm, *Phys. Rev. A*, 2007, **76**, 042514.
31 T. Weber, J. Herbig, M. Mark, H.-C. Nägerl and R. Grimm, *Science*, 2003, **299**, 232.

32 O. Dulieu and P. Julienne, *J. Chem. Phys.*, 1995, **103**, 60.
33 C. Amiot, O. Dulieu and J. Vergès, *Phys. Rev. Lett.*, 1999, **83**, 2316.
34 H. Salami, T. Bergeman, O. Dulieu, D. Li, F. Xie, and L. Li, manuscript in preparation (2008).
35 C. Amiot and J. Vergès, *Chem. Phys. Lett.*, 1985, **116**, 273.
36 M. J. Mark, J. G. Danzl, E. Haller, M. Gustavsson, N. Bouloufa, O. Dulieu, H. Salami, T. Bergeman, H. Ritsch, R. Hart and H.-C. Nägerl, *Appl. Phys. B*, 2008, **95**, 219.
37 J. G. Danzl, M. J. Mark, E. Haller, M. Gustavsson, N. Bouloufa, O. Dulieu, H. Salami, T. Bergeman, H. Ritsch, R. Hart, H.-C. Nägerl, manuscript in preparation.
38 M. Greiner, O. Mandel, T. Esslinger, T. W. Hänsch and I. Bloch, *Nature*, 2002, **415**, 39.
39 S. Dürr, 2008, private communication.
40 G. Thalhammer, K. Winkler, F. Lang, S. Schmid, R. Grimm and J. Hecker Denschlag, *Phys. Rev. Lett.*, 2006, **96**, 050402.
41 J. Aldegunde and J. M. Hutson, *Phys. Rev. A*, 2009, **79**, 013401.
42 D. Jaksch, V. Venturi, J. I. Cirac, C. J. Williams and P. Zoller, *Phys. Rev. Lett.*, 2002, **89**, 040402.

Rotational spectroscopy of single carbonyl sulfide molecules embedded in superfluid helium nanodroplets

Rudolf Lehnig,[ab] Paul L. Raston[a] and Wolfgang Jäger[*a]

Received 6th November 2008, Accepted 4th February 2009
First published as an Advance Article on the web 27th May 2009
DOI: 10.1039/b819844f

The pure rotation spectrum of carbonyl sulfide embedded in superfluid helium nanodroplets was measured in the frequency range from 4 to 15.5 GHz. Four lines, corresponding to the $J = 1\text{–}0$, $J = 2\text{–}1$, $J = 3\text{–}2$, and $J = 4\text{–}3$ transitions, were found. The line widths of the transitions increase with increasing rotational quantum number J, which is indicative of a distribution of the effective B rotational constant. A comparison of the pure rotational spectrum with the microwave-infrared double resonance spectrum [S. Grebenev, M. Havenith, F. Madeja, J. P. Toennies, and A. F. Vilesov, *J. Chem. Phys.*, 2000, **113**(20), 9060] reveals that the double resonance measurement scheme probes predominantly rotational transitions within the vibrationally excited state.

Introduction

Spectroscopy of molecules embedded in superfluid helium droplets provides information about the interaction between the dopant molecule and the helium environment.[1–5] The rotation of the dopant can be described by a free rotor Hamiltonian. However, the rotational constants are smaller than those of the isolated molecules. It was found that for most of the light molecules (B rotational constant larger than 1 cm^{-1}) the rotational constants are at least at 80% of the respective gas phase values. For the majority of heavy molecules (B rotational constant smaller than 1 cm^{-1}), the rotational constant in helium droplets is about 30% of that in the gas phase. For those heavy rotors, significant helium density follows the rotating dopant adiabatically and thereby increases the moment of inertia.[6] In addition to the rotational constants, the line shapes of the observed transitions also provide information about the dopant–helium interaction.[7–10] In general, one distinguishes between homogeneously and inhomogeneously broadened lines.[11] In the homogeneous case, all probed molecules experience the same influence by the environment. Homogeneously broadened lines often have a Lorentzian shape whose width is determined by the lifetimes of the initial and final states of the transition. From homogeneously broadened transitions it was found that rotational relaxation in helium droplets is usually much faster than vibrational relaxation.[1] Vibrational relaxation also depends on the symmetry of the vibration. For pentacene, for example, it was found that the vibrational relaxation time can be quite different for vibrations that possess a similar vibrational frequency but a different vibrational symmetry.[12] For inhomogeneously broadened lines, the probed sample consists of dopants that all experience

---

[a]*Department of Chemistry, University of Alberta, Edmonton, Alberta T6G 2G2, Canada. E-mail: wolfgang.jaeger@ualberta.ca; Fax: +1 780 492 8231; Tel: +1 780 492 5020*
[b]*BASF, D-67056 Ludwigshafen, Germany*

slightly different surroundings and that possess, therefore, slightly different transition frequencies. Inhomogeneously broadened lines can have arbitrary line shapes.

A large number of molecules embedded into helium droplets have been studied with infrared spectroscopy.[1,3,4] Usually, the rotational fine structure can be resolved and the band origin v_0 of the particular vibration as well as the rotational constant of the molecule within the helium droplet can therefore be determined. For the understanding of the interaction between the guest molecule and the helium droplet it is important to distinguish between the effects arising from the vibration and the rotation of the dopant. Therefore, it is interesting to measure pure rotational spectra of molecules embedded in helium droplets. So far, there are only two examples of pure rotational spectra in helium droplets in the literature. The first one is cyanoacetylene (HCCCN).[7,13] This molecule was studied with single and double resonance microwave spectroscopy as well as with microwave-infrared double resonance spectroscopy. It was found that the observed transitions of the pure rotation are inhomogeneously broadened.[7] By studying the saturation behavior of the $J = 3$–2 and the $J = 5$–4 transitions it could be shown that the inhomogeneous broadening is dynamic, which means that the influence of the helium environment on the transition frequency changes on a time scale that is estimated to be similar to that of the rotational relaxation.[13] Here, J is the quantum number corresponding to the rotational angular momentum $\mathbf{J}$. The second example of pure rotational spectroscopy in helium droplets is hydrogen cyanide (HCN) and deuterated hydrogen cyanide (DCN).[9] For both isotopologues one pure rotational transition was observed, i.e. the $J = 1$–0 transition. The lines are inhomogeneously broadened and, under the assumption of a static homogeneous broadening, the rotational relaxation time could be estimated from the saturation behavior of the observed transition to be 12 ns for HCN and 17 ns for DCN.[9]

In the present paper, the pure rotational spectrum of carbonyl sulfide (OCS) embedded in helium droplets is reported. The goal of the current study was to investigate the line shapes of the pure rotational transitions of OCS in helium droplets, and thereby to gain insight in the interaction of the OCS molecule with the helium surroundings. The rotationally resolved infrared spectrum around 2061 cm^{-1} has been measured in helium droplets before, and the B rotational constant as well as the centrifugal distortion constant D of OCS in the vibrational ground state $v = 0$ are already known.[14] Also, microwave-infrared double resonance measurements on OCS in helium droplets are reported in the literature.[15] Two different types of microwave-infrared double resonance measurements have been performed. For the first type, the microwave frequency was kept fixed at the peak of a pure rotational transition and the infrared frequency was scanned. For the second type, the infrared frequency was tuned to a ro-vibrational transition and the microwave frequency was varied. For this second type of double resonance measurements, a spectrum is presented in Fig. 6 of reference 15 that was obtained by keeping an infrared laser at the frequency of the R(1) transition ($v = 1$, $J = 2 \leftarrow v = 0$, $J = 1$) and scanning the microwave frequency over the range from 8 to 18 GHz. Signals were found for the $J = 1$–2, $J = 2$–3, $J = 3$–4 and $J = 4$–5 rotational transitions. It was concluded that the ro-vibrational transition starting form the $v = 0$, $J = 1$ state affects the population of all rotational states within the vibrational ground state $v = 0$.[15] Furthermore, splittings of the order of 200 MHz have been observed for the $J = 2$–3 transition. The measurement of a pure rotational spectrum will clarify if this substructure has to do with the double resonance measurement scheme used in reference 15 or if it is inherent to the rotation of the OCS molecule within the helium droplet.

Experimental

The helium nanodroplet spectra reported in this paper were measured using the depletion technique.[16,17] This method relies on the dissipation of the excitation energy absorbed by the dopant into the droplet. As a consequence of this energy transfer, the temperature of the droplet rises and leads to an increase of the evaporation rate of helium atoms from the droplet.[18] The resulting reduction of the droplet size can be detected either with a mass spectrometer or a bolometer. Consequently, the strength of a depletion signal depends on the energy of the excitation of the dopant. In order to evaporate one helium atom from a helium droplet an energy of $\sim$5 cm^{-1} (150 GHz) is necessary.[19] Since the transitions studied in this paper fall in the frequency range from 4 to 15.5 GHz, multiple excitations are required to evaporate at least one helium atom. It is therefore crucial to provide high power microwave radiation. To achieve that, we have installed a microwave resonator into our helium droplet apparatus as described in the following.

Fig. 1 shows a schematic diagram of the instrument built for measuring microwave depletion spectra of molecules embedded in helium nanodroplets. Except for the microwave resonator, the design of this set-up follows the typical design of a helium droplet depletion spectrometer.[16,17] Helium droplets are formed in a continuous supersonic expansion of helium gas (research grade, 99.99990% purity) through a nozzle with a diameter of 5 μm into a vacuum. The nozzle assembly is directly mounted to the cold head of a closed cycle helium cryostat (Sumitomo RDK 408S). By choosing a certain backing pressure P_0 and a certain nozzle temperature T_0 the mean size of the droplets and the droplet size distribution can be varied.[20] Typical values for T_0 are 8 to 25 K and for P_0 20 to 80 bar. At a distance of 20 mm behind the nozzle the helium droplet beam passes through a skimmer with a diameter of 0.5 mm (Beam Dynamics, Inc.) and enters a second differentially pumped chamber. Therein, a 70 mm long doping cell is mounted which is traversed by the helium droplet beam. The entrance aperture of this cell is at a distance of 330 mm from the nozzle. The helium droplets pick up foreign molecules while passing through the doping cell. Inside the doping cell a pressure of the order of 10^{-5} mbar is sufficient for doping the droplets with, on average, one molecule.

At a distance of 570 mm from the nozzle, the doped droplets enter a third differentially pumped chamber which houses a microwave resonator. The resonator is formed by two spherical aluminium mirrors with radii of curvature of 13.2 cm. The helium droplets enter and leave the resonator through holes with a diameter of 5 mm in each of the two mirrors. These holes are 15 mm off the center of the two mirrors. The distance between the two mirrors can be adjusted to tune the cavity into resonance with the microwave radiation. A computer program was developed which enables automatic adjustment of the cavity length to tune to the desired frequency. Typical mirror distances are 10 to 20 cm; this distance determines the interaction time between the droplet beam and microwave radiation. Microwave

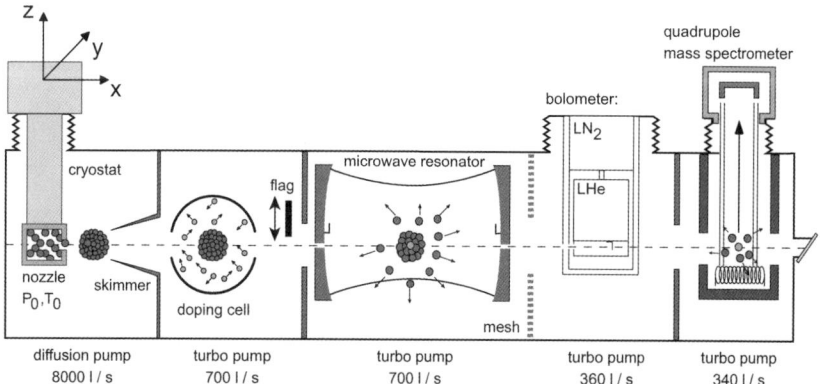

Fig. 1 Helium nanodroplet spectrometer for measuring microwave depletion spectra.

radiation is coupled into the resonator *via* a wire hook antenna located in the center of the first mirror. A second wire hook antenna in the center of the second mirror is connected to a zero bias Schottky detector (Krytar) to monitor the cavity modes. The resonator modes typically have a full width at half maximum of 6 MHz. From this value, together with the free spectral range of the cavity of about 1 GHz, the power stored between the two mirrors can be estimated to be about 50 times the power P_{in} that is coupled into the resonator. A synthesizer (Agilent E8257C) is used as source for the microwave radiation. The output of the synthesizer is amplified with a traveling wave tube amplifier (CPI, 6900K4 series) to about 20 W to 57 W depending on the microwave frequency. Consequently, power levels of 1 kW to 2.8 kW can be achieved inside the cavity. In order to maintain a constant microwave power throughout a frequency scan, the amplitude of the resonator mode is constantly monitored with the Schottky detector and the power P_{in} is adjusted to keep the amplitude of the mode at a constant level.

Two detection devices – a liquid ^{4}He-cooled silicon bolometer (Infrared Labs, Inc.) and a quadrupole mass spectrometer (ABB Extrel) – are available for measuring the microwave depletion spectra. The bolometer can be moved into and out of the droplet beam axis, thereby allowing the use of either of the two detection devices. The bolometer is located at a distance of 1.3 m from the nozzle and the mass spectrometer at a distance of 1.75 m. Since the bolometer is also a very sensitive detector of the microwave radiation leaking out of the resonator a metal mesh was mounted between the resonator and the bolometer to block this stray microwave radiation. Alternatively, an aluminium plate was used to replace this mesh, but no improvement in preventing the microwave radiation from reaching the bolometer was achieved. For the case that not all of the stray radiation can be blocked with the mesh or the aluminium plate, the measured spectra were background corrected. This was done by blocking and unblocking the helium droplet beam with a beam flag in front of the entrance aperture to the chamber housing the microwave resonator. At each frequency step two measurements were taken, one without blocking the beam and another one with blocking of the droplets. The difference between the two detector signals obtained was taken as the actual depletion signal. The microwave radiation was square wave modulated with a frequency of 400 Hz and the signal from the bolometer or the mass spectrometer was processed with a phase sensitive detector (SR830 DSP) referenced to the chopping frequency of the microwave radiation. The output of the phase sensitive detector was recorded with a computer.

Typical pressures in the different chambers are 2×10^{-5} mbar for the nozzle chamber, 4×10^{-7} mbar for the chamber housing the doping cell, 2×10^{-8} mbar for the microwave chamber, 1×10^{-8} mbar for the bolometer chamber, and 6×10^{-10} mbar for the mass spectrometer chamber. These values are for a nozzle temperature T_0 of 20.0 K and a backing pressure P_0 of 44 bar.

Results

The infrared spectrum of the antisymmetric stretch vibrational band ν_3 of OCS in helium droplets shows a rotational P- and R-branch, while no Q-branch was observed.[14] The conventional energy level expressions of a linear rotor have been used to analyze this ro-vibrational spectrum:

$$E(J, v = 0) = B_0 J(J + 1) - D_0 [J(J + 1)]^2 \quad (1a)$$

$$E(J, v = 1) = \nu_0 + B_1 J(J + 1) - D_1 [J(J + 1)]^2 \quad (1b)$$

J is the rotational quantum number, and B and D are the rotational and centrifugal distortion constants, respectively. The subscripts 0 and 1 denote the vibrational ground state ($v = 0$) and the vibrationally excited state ($v = 1$), respectively. ν_0 is

Table 1 Transition frequencies of the first four pure rotational transitions of OCS in helium droplets within the vibrational ground state $v = 0$ and the first vibrationally excited state $v = 1$ of the antisymmetric stretch. The frequencies were calculated from the rotational constant B and the centrifugal distortion constant D of OCS in helium droplets (mean size $<N>$ of 3×10^3 helium atoms) as determined from the infrared spectrum.[14]

	$v = 0$	$v = 1$
$J = 1\text{–}0$	4.34(2) GHz[a]	4.41(2) GHz
$J = 2\text{–}1$	8.41(4) GHz	8.56(5) GHz
$J = 3\text{–}2$	11.93(6) GHz	12.15(7) GHz
$J = 4\text{–}3$	14.6(1) GHz	14.9(1) GHz
B	2.194(9) GHz	2.23(1) GHz
D^b	0.0114(3) GHz	

[a] Values in parentheses are the standard variations of the last digit. [b] The same value of the distortion constant D was assumed for both vibrational states, $v = 0$ and $v = 1$.[14]

the vibrational band origin. The B constant was determined for both the vibrational ground state, $v = 0$, and the vibrationally excited state, $v = 1$. The D constant could not be obtained separately for both states and the same value was assumed for both $v = 0$ and $v = 1$. The frequencies of pure rotational transitions within $v = 0$ and $v = 1$ can be calculated using the values determined for B and D from the infrared study.[14] In Table 1 the predicted frequencies of the first four pure rotational transitions $J = 1\text{–}0, J = 2\text{–}1, J = 3\text{–}2$, and $J = 4\text{–}3$ are listed together with the B and D constants of the $v = 0$ and $v = 1$ states.

Because of the helium droplet temperature of 0.4 K[21] only the vibrational ground state of OCS is populated and therefore only rotational transitions within $v = 0$ can be observed. The same source conditions as in the infrared study[14] were chosen for the search for the pure rotational transitions ($P_0 = 44$ bar, $T_0 = 20$ K). Consequently, the cluster size distribution is identical for both experiments and the mean cluster size amounts to $<N> = 3 \times 10^3$. Indeed, four transitions were found at the predicted transition frequencies and are therefore assigned as the $J = 1\text{–}0$, $J = 2\text{–}1$, $J = 3\text{–}2$, and $J = 4\text{–}3$ transitions within the vibrational ground state ($v = 0$) of single OCS molecules in helium droplets. Fig. 2 shows all four transitions.

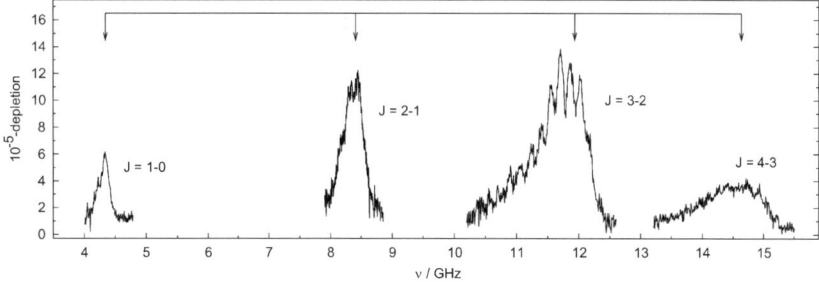

Fig. 2 Overview spectrum of the four lowest pure rotational transitions of OCS in helium droplets ($<N> = 3 \times 10^3$ helium atoms). The arrows indicate the expected line positions calculated from the constants as determined from the infrared measurements (see Table 1).[14] The experimental conditions were as follows: $J = 1\text{–}0$: frequency increment: $\Delta \nu = 8$ MHz, accumulation time per point $t_{\text{acc.}} = 350$ s, $P_{\text{in}} = 2.9$ W; $J = 2\text{–}1$: $\Delta \nu = 4$ MHz, $t_{\text{acc.}} = 80$ s, $P_{\text{in}} = 5.0$ W; $J = 3\text{-}2$: $\Delta \nu = 10$ MHz, $t_{\text{acc.}} = 140$ s, $P_{\text{in}} = 6.8$ W; $J = 4\text{-}3$: $\Delta \nu = 8$ MHz, $t_{\text{acc.}} = 90$ s, $P_{\text{in}} = 4.8$ W.

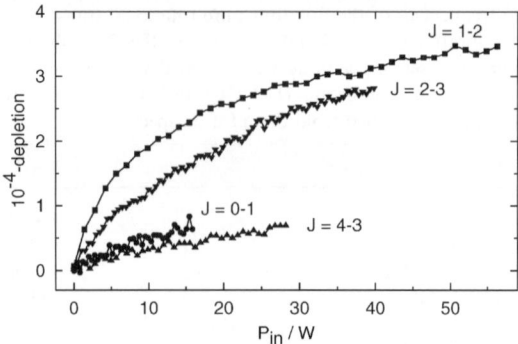

Fig. 3 Depletion signal as a function of the microwave input power P_{in} for all four transitions studied. The curves were measured at 4.17 GHz ($J = 1$–0), 8.4 GHz ($J = 2$–1), 11.88 GHz ($J = 3$–2), and 14.66 GHz ($J = 4$–3). All curves were recorded up to the maximum available power of the traveling wave tube amplifier at the particular frequency.

The saturation behavior was measured for all four transitions (see Fig. 3). The gain of the traveling wave tube amplifier varies over the range from 4 to 15.5 GHz and the saturation behavior of the four transitions was always measured up to the maximum available power of the amplifier. All transitions saturate between $P_{in} = 5$ to 10 W. The spectra shown in Fig. 2 were all measured under non-saturating conditions.

The signal intensity within the $J = 2$–1 transition was measured as function of the pressure of OCS within the doping cell. The obtained values could be fitted well to a Poisson distribution typical for doping of single molecules into helium droplets.[22,23] This confirms that the observed transition is due to single OCS molecules in helium droplets. The maximum signal was observed for a pressure of about 1×10^{-5} mbar and all four lines shown in Fig. 2 were measured with this OCS pressure in the doping cell.

The line widths of the first four pure rotational transitions increase with J: 180($\pm$20) MHz ($J = 1$–0); 350($\pm$30) MHz ($J = 2$–1); 730($\pm$40) MHz ($J = 3$–2); 1100($\pm$120) MHz ($J = 4$–3). All values are full widths at half maximum and the uncertainties are due to the noise of the spectrum. The uncertainties in the determination of the input power P_{in} prevented us from making a quantitative comparison of the intensities of the observed lines.

The observed depletion of about 1×10^{-4} corresponds to a loss of about 0.3 helium atoms per droplet (for a mean droplet size of $<N> = 3 \times 10^3$). At most

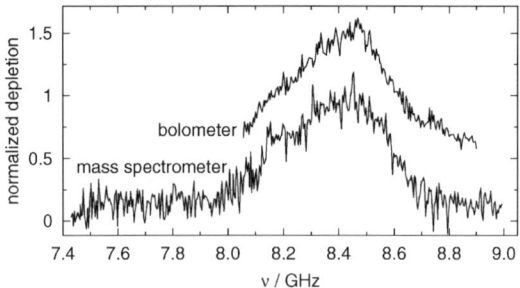

Fig. 4 Comparison of the detection devices: The $J = 2$–1 transition of OCS in helium droplets measured with the mass spectrometer and the bolometer. The droplets had a mean size $<N>$ of 3000 atoms ($P_0 = 44$ bar, $T_0 = 20.0$ K).[20] The total accumulation time for the spectrum measured with the mass spectrometer was 196.5 hours and for the spectrum detected with the bolometer 12.3 hours. Both spectra were measured with the highest available microwave power and are therefore power saturated.

37% of all droplets are doped with one OCS molecule; therefore, the average loss for singly doped droplets amounts to about 1 helium atom. The weakness of the signal explains why the pure rotational spectrum of OCS could not be observed previously[15] since in that study the detection limit was estimated to correspond to a depletion signal of 1×10^{-3}. Fig. 4 shows the $J = 2-1$ transition measured with the mass spectrometer and the bolometer as detection devices for beam depletion. Both spectra were measured with a microwave input power P_{in} of 43.8 W. The signal to noise ratio is better by about a factor of two for the spectrum detected with the bolometer compared to the one measured with the mass spectrometer. Additionally, the averaging time for each point of the spectrum was eight times longer for the mass spectrometer experiment than when the bolometer was used. Therefore, it can be estimated that the bolometer is about ten times more sensitive than the mass spectrometer for detecting microwave depletion spectra. This consideration does not take into account the slightly longer distance from the nozzle to the mass spectrometer, which results in some additional attenuation of the helium droplet beam. Previously, it was estimated for the case of infrared depletion spectra that the bolometric detection is about 50 times more sensitive than the mass spectrometric detection.[17]

Discussion

Comparison with microwave-infrared double resonance spectrum

We begin the discussion of our results with a comparison of the microwave spectrum with the microwave-infrared double resonance spectrum.[15] In Fig. 6 of reference 15 a microwave-infrared double resonance spectrum is shown that was measured by parking the frequency of an infrared laser at the frequency of the R(1) transition of the antisymmetric stretch of OCS at 2061.93 cm^{-1}. The microwave frequency was scanned from 8 to 18 GHz and the change of the depletion signal of the R(1) transition caused by the microwave radiation was recorded.[15]

In Fig. 5 of this work the microwave-infrared double resonance spectrum is shown together with the microwave spectrum. The frequencies of pure rotational transitions within the vibrational ground state ($v = 0$) and the first vibrationally excited state of the antisymmetric stretch of OCS ($v = 1$) are indicated by dotted vertical lines. These frequencies were calculated using the constants determined from the infrared spectrum[14] and are listed in Table 1. From Fig. 5 it can be seen that the

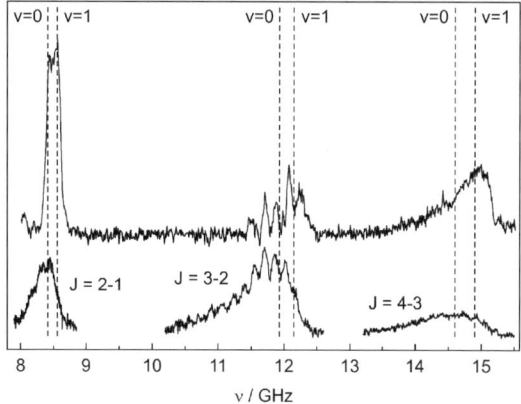

Fig. 5 Comparison of the microwave-infrared double resonance spectrum (upper trace, adapted from Fig. 6 in reference 15) with the microwave transitions (lower trace, same spectra as in Fig. 2). The dotted lines indicate the frequencies of pure rotational transitions within the vibrational ground state ($v = 0$) and the first vibrationally excited state ($v = 1$) of the antisymmetric stretch vibration (see Table 1).

transitions of the microwave spectrum are shifted to lower frequencies compared to the transitions of the microwave-infrared double resonance spectrum. Also, the peak maxima of the microwave transitions coincide with the calculated frequencies for the pure rotational transitions within $v = 0$. From Fig. 5, it appears that the line peaks from the double resonance experiment coincide better with the transition frequencies predicted for the excited, $v = 1$, vibrational state. It is difficult to discern from the double resonance spectra if there are also components at the positions of the ground state rotational transitions. We conclude that the double resonance experiment probes predominantly pure rotational transitions within the excited $v = 1$ vibrational state and to a lesser degree perhaps also those within the vibrational ground state. This, and the fact that rotational transitions were observed which are not connected to the pump transition, requires that the vibrationally excited state lives long enough to allow for rotational relaxation to take place.

Since the double resonance spectrum[15] probes mainly transitions within the excited vibrational state, the applied measurement scheme can also be described as a pump–probe experiment in which the infrared laser serves as the pump and populates a manifold of rotational states that was not populated before and that can therefore be probed with the microwave radiation. In that sense the experiment is similar to other pump–probe experiments performed in helium droplets, for example with tetracene,[24] phthalocyanine[25] or the HCN-HF complex.[26]

A pronounced substructure like the one observed for the $J = 3$–2 transition of the double resonance measurements[14] was initially also observed for the pure rotational transitions, in particular the $J = 3$–2 transition (see Fig. 6). It transpired, after much trial and error, that this substructure is an instrumental artefact, which resulted from a two foot long microwave cable between the receiving antenna mounted into one of the cavity mirrors and the Schottky diode detector. The interference effects disappeared when the detector was mounted into the vacuum system and connected directly to the antenna (see Fig. 6). It is unclear if the substructure observed in the double resonance experiments[15] is the result of a similar interference effect.

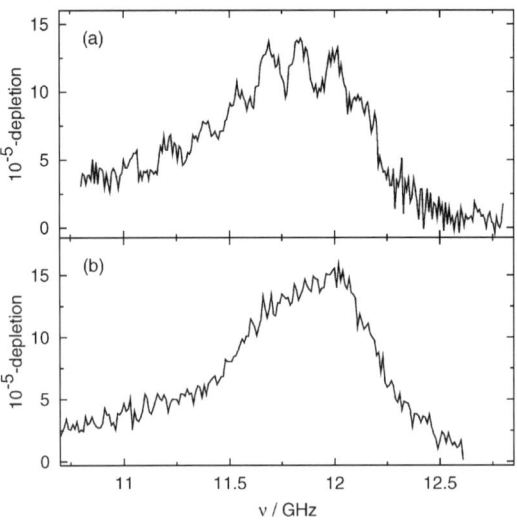

Fig. 6 The $J = 3$–2 transition of OCS in helium droplets measured with (a) and without (b) a two-foot long microwave cable between the receiving antenna and Schottky diode detector. The substructure in trace (a) is a result of interference effects in the cable.

Line widths and line shapes of the rotational transitions

The increased linewidths and particular lineshapes observed in helium nanodroplet spectra have been the topic of a number of experimental and theoretical investigations. It is now generally accepted that the line widths of heavy rotors are not determined by limited lifetimes and that instead the lines are inhomogeneously broadened. Several mechanisms have been proposed to account for the observed line broadening. These include: (i) Distribution in the magnitudes of the effective B-rotational constants of the embedded molecules. Since in a helium droplet experiment a droplet size distribution is probed, rather than droplets of uniform size, a droplet size dependence of the B-constant would give rise to inhomogeneously broadened lines. (ii) Coupling of the molecular rotation with other degrees of freedom of the molecule in the droplet, which leads to a splitting of molecular eigenstates into sublevel manifolds. It was suggested, for example, that the confining potential of the helium droplet leads to particle-in-a-box like states of the dopant molecule.[28,8] The assumption of such sublevel manifold resulted in a successful qualitative explanation of the line shape of the ammonia inversion transition in helium droplets.[10] (iii) Another possibility is coupling of the droplets to surface excitations, which are known as ripplons.[18] (iv) A Coriolis type interaction between the molecular rotation and the overall rotation of the doped droplet.[14] (v) Hydrodynamic effects associated with the molecular rotation.[8]

The shapes of the observed pure rotational transitions (Fig. 2) show a distinct asymmetry, with a tail to the low frequency side and a steeper slope towards higher frequency. In addition, the line width increases with increasing J value. Line widths that increase with J and asymmetric line shapes were also observed in the infrared spectrum of OCS in helium nanodroplets.[14] In Fig. 7 the line widths of the ro-vibrational transitions[14] are compared with the line widths of the pure rotational transitions. The line widths and shapes of the rotational $J = 1–0$ transition and the $R(0)$ infrared transitions are virtually the same. If we assume that the dip on top of the $J = 2–1$ transition is an instrumental artefact (see above) the true intensity is higher, and also this line would overlap well with the corresponding $R(1)$ infrared transition. Similar linewidths of pure rotational and ro-vibrational transitions were also found for the case of cyanoacetylene.[13] This indicates, as was pointed out in ref. 13, that the same broadening mechanisms are dominant for rotational and ro-vibrational transitions, and that vibrational decoherence processes do not contribute significantly. The rotational $J = 3–2$ and 4–3 transitions are broader than the corresponding ro-vibrational $R(2)$ and $R(3)$ transitions; this may be attributable to a different B-value distribution for the excited vibrational state (see discussion below).

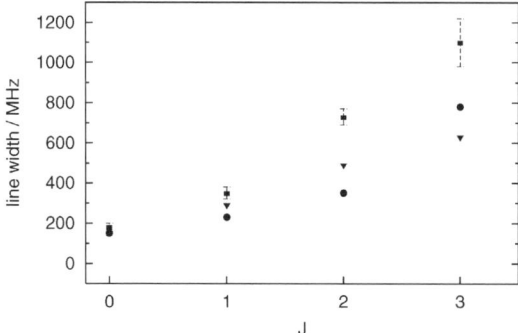

Fig. 7 Line widths of the ro-vibrational transitions[14] (circles for the R-branch, triangles for the P-branch) and pure rotational transitions (squares) plotted against the quantum number J indicating the state from which the transition starts.

Lehmann[8] has developed a model involving a coupling of the molecular rotation with translational modes of the molecule within the droplet. Also hydrodynamic effects were included, and with this model it was possible to explain the line shape of the $R(0)$-transition of OCS in helium droplets. However, the increase of the line widths with J could not be explained.[8]

The line shapes are reminiscent of the log-normal distribution of the helium droplet sizes in the helium expansion. The shapes could be explained by a droplet size dependence of the B rotational constant, where dopants embedded in larger droplets have smaller B values. In this picture, the line shape of the $J = 3-2$ transition, for example, is a result of overlap of transitions of OCS molecules in droplets of different sizes. It is thus possible to assign a B-constant to every frequency step ν_{32} of the transition:

$$B = \tfrac{1}{6}(\nu_{32} + 108D) \qquad (2)$$

This distribution of B-constants can be used to predict the other three measured transitions:

$$\begin{aligned}\nu_{10} &= \tfrac{1}{3}(\nu_{32} + 108D) - 4D \\ \nu_{21} &= \tfrac{2}{3}(\nu_{32} + 108D) - 32D \\ \nu_{43} &= \tfrac{4}{3}(\nu_{32} + 108D) - 256D\end{aligned} \qquad (3)$$

For the centrifugal distortion constant D, the value determined from the infrared spectrum will be used (0.0114(3) GHz,[14] see Table 1). In Fig. 8 the predicted lines (upper trace) are compared with the measured lines (lower trace). In Fig. 8(c) the original $J = 3-2$ transition is shown for reference. It can be seen from Fig. 8 that the predicted line widths and the overall line shapes (sharp rise at the higher frequency side and tail extending to lower frequencies) of the $J = 1-0$ and $J = 4-3$ transitions are in reasonable agreement with experiment. For the $J = 2-1$ transition, the sharp rise is also predicted well, however, the tail is predicted to be about 80 MHz broader than in the measured transition. The comparison in Fig. 8 shows that the assumption of an inhomogeneously broadened line resulting from the overlap of transitions with different B-constants allows us to quantitatively explain the increase of the line widths with increasing J. This result supports the hypothesis that an ensemble of embedded OCS molecules with different B-constants is probed in the experiment.

To test if the apparent B-value distribution can be attributed to a dependence of the rotational constant on droplet size (model (i)), we measured the $J = 3-2$ transition for different droplet sizes. The result (see Fig. 9) shows that the general line shape does not change significantly, but that the maximum of the line shifts to lower frequencies for larger droplets sizes. However, the shift is quite small, in the order of

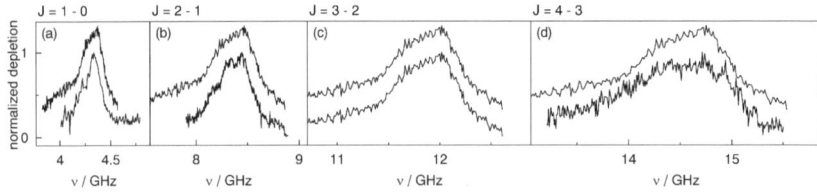

Fig. 8 Comparison of the predicted transitions (upper trace) with the actually measured transitions (lower trace). The predictions were done by calculating a B_0-constant for every frequency step of the $J = 3-2$ transition (see eqn (2)) and using this B_0-constant for calculating the $J = 1-0$, $J = 2-1$, and $J = 4-3$ transitions (see parts (a), (b), and (d), respectively). In part (c) the original $J = 3-2$ transition is shown for reference.

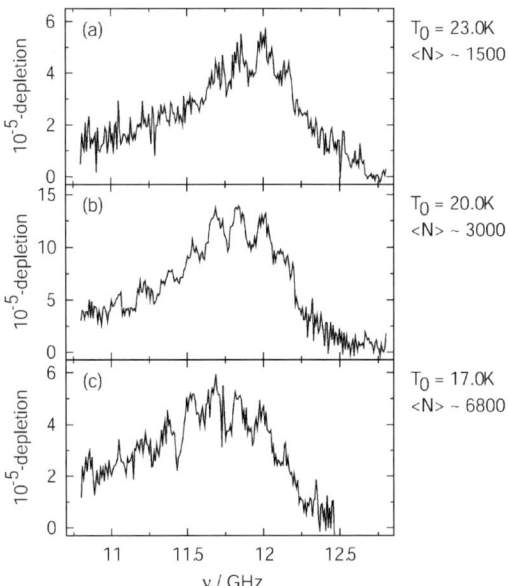

Fig. 9 $J = 3$–2 transition of OCS in helium droplets measured with different droplet size distributions. The backing pressure P_0 was set to 44 bar for all scans. The nozzle temperature T_0 and the corresponding mean droplet size $<N>^{20}$ are indicated for each spectrum. The pressure of OCS in the doping cell was 1.6×10^{-5} mbar for (a), 1.1×10^{-5} mbar for (b), and 6.0×10^{-6} mbar for (c).

only 250 MHz, when going from $<N> \sim 1500$ to $<N> \sim 6800$, and insufficient to account for the overall widths of the lines. It is unclear which mechanism could lead to the apparent distribution of B values considered. Further evidence against a direct droplet size dependence comes from the double resonance experiments.[15] In ref. 15, the laser line width was estimated to be about 40 MHz. The laser radiation should therefore only address a 40 MHz wide slice in the droplet size distribution, and this would also result in a 40 MHz wide double resonance signal. Instead, much broader signals were observed (see Fig. 5) albeit somewhat narrower than the transitions within the ground vibrational state measured in this work.

Summary and concluding remarks

Pure rotational transitions of OCS embedded in helium nanodroplets were studied in the microwave frequency range. Four transitions were measured and were assigned to the $J = 1$–0, $J = 2$–1, $J = 3$–2, and $J = 4$–3 transitions. The comparison of the rotational spectrum with the microwave-infrared double resonance spectrum[15] reveals that the measurement scheme used for the double resonance allows one to probe rotational transitions mainly within the excited vibrational state. This, and the fact that rotational transitions not connected to the pump transition were observed, requires that the lifetime of the vibrationally excited state is long enough for rotational relaxation to occur within the vibrationally excited state.

The line widths of the four transitions increase with increasing J. It is possible to explain this increase in the line width quantitatively by assigning individual B-constants to every frequency step, which means that the observed line shape can be attributed to an inhomogeneous broadening caused by the presence of a B-value distribution.

It is unclear, however, which mechanism is responsible for this possible distribution of B-values. In smaller clusters, with N up to 72, an oscillation of B-value with cluster size was observed.[29] It is tempting to assume a continuation of these oscillations to larger droplet sizes, but the oscillation amplitude appears to be too small to explain the relatively large line widths obtained in the helium droplet work. In addition, the observed lineshapes are reminiscent of the log-normal distribution of droplet sizes. However, the rather small line shifts observed in our experiments with different mean droplet sizes also indicate that a simple droplet size dependence of the rotational constant cannot account for the line widths. In their infrared studies of carbonyl sulfide embedded in helium droplets, Grebenev et al.[14] found a reversal of the asymmetric line shapes of the $R(0)$ and $R(1)$ transitions with increasing droplet size. There may be indication of a similar effect in the pure rotational transitions (see Fig. 9), where the line shape appears to become more symmetric for smaller droplet sizes. This observation is also difficult to justify with a droplet size dependent distribution of B-values. In earlier work, we had measured a microwave inversion transition of ammonia ($J''-J' = 1-1$) in helium droplets, which resulted in a peculiar line shape, namely a broad background with a sharp peak (full-width half-height only 15 MHz) on top.[10] This was interpreted in terms of an energy level substructure which could result from a coupling of ammonia rotation and particle-in-box type motion. How can this observation and its interpretation can be reconciled with the apparent B-value distribution in the current measurements? The observed increase in line width with higher J quantum numbers could perhaps also be related to a lifting of the m_J degeneracy, as proposed by Lehmann[8] and observed by Nauta and Miller.[31] Lehmann predicts a J-dependence of the coupling between molecule rotation and helium motion, which initially increases with J.[32] The resulting larger effective moment of inertia could lead to a larger m_J splitting. An increased population of higher excited states, beyond what is expected from the droplet temperature, would also have significant effects on the line widths and shapes. Lehmann and Dokter[33] have found such increased excitation by modeling the distribution of internal excitations in doped helium droplets, taking into account angular momentum conservation. At the present time it appears that there is no single mechanism which can explain the various line shapes observed in the spectra of doped helium droplets.

Acknowledgements

This work was funded by the Natural Sciences and Engineering Research Council of Canada (NSERC), the Canada Foundation for Innovation (CFI), the Alberta Science and Research Investments Program, the University of Alberta (UofA), and the Department of Chemistry at the UofA.

References

1 M. Y. Choi, D. E. Douberly, T. M. Falconer, W. K. Lewis, C. M. Lindsay, J. M. Merritt, P. Stiles and R. E. Miller, *Int. Rev. Phys. Chem.*, 2006, **25**, 15.
2 F. Stienkemeier and K. K. Lehmann, *J. Phys. B*, 2006, **39**, R127.
3 J. P. Toennies and A. F. Vilesov, *Angew. Chem., Int. Ed.*, 2004, **43**, 2622.
4 C. Callegari, K. K. Lehmann, R. Schmied and G. Scoles, *J. Chem. Phys.*, 2001, **115**(22), 10090.
5 F. Stienkemeier and A. F. Vilesov, *J. Chem. Phys.*, 2001, **115**(22), 10119.
6 Y. Kwon, P. Huang, M. V. Patel, D. Blume and K. B. Whaley, *J. Chem. Phys.*, 2000, **113**(16), 6469.
7 I. Reinhard, C. Callegari, A. Conjusteau, K. K. Lehmann and G. Scoles, *Phys. Rev. Lett.*, 1999, **82**(25), 5036.
8 K. K. Lehmann, *Mol. Phys.*, 1999, **97**(5), 645.
9 A. Conjusteau, C. Callegari, I. Reinhard, K. K. Lehmann and G. Scoles, *J. Chem. Phys.*, 2000, **113**(12), 4840.
10 R. Lehnig, N. V. Blinov and W. Jäger, *J. Chem. Phys.*, 2007, **127**, 241101.
11 W. Demtröder, *Laser Spectroscopy*, Springer-Verlag, Berlin, 3rd edn, 2003.
12 R. Lehnig and A. Slenczka, *J. Chem. Phys.*, 2005, **122**, 244317.

13 C. Callegari, I. Reinhard, K. K. Lehmann, G. Scoles, K. Nauta and R. E. Miller, *J. Chem. Phys.*, 2000, **113**(11), 4636.
14 S. Grebenev, M. Hartmann, M. Havenith, B. Sartakov, J. P. Toennies and A. F. Vilesov, *J. Chem. Phys.*, 2000, **112**(10), 4485.
15 S. Grebenev, M. Havenith, F. Madeja, J. P. Toennies and A. F. Vilesov, *J. Chem. Phys.*, 2000, **113**(20), 9060.
16 J. P. Toennies and A. F. Vilesov, *Annu. Rev. Phys. Chem.*, 1998, **49**, 1.
17 C. Callegari, A. Conjusteau, I. Reinhard, K. K. Lehmann and G. Scoles, *J. Chem. Phys.*, 2000, **113**(23), 10535.
18 D. M. Brink and S. Stringari, *Z. Phys. D*, 1990, **15**, 257.
19 S. Stringari and J. Treiner, *J. Chem. Phys.*, 1987, **87**(8), 5021.
20 M. Lewerenz, B. Schilling and J. P. Toennies, *Chem. Phys. Lett.*, 1993, **206**(1–4), 381.
21 M. Hartmann, R. E. Miller, J. P. Toennies and A. Vilesov, *Phys. Rev. Lett.*, 1995, **75**(8), 1566.
22 M. Lewerenz, B. Schilling and J. P. Toennies, *J. Chem. Phys.*, 1995, **102**(20), 8191.
23 M. Hartmann, R. E. Miller, J. P. Toennies and A. F. Vilesov, *Science*, 1996, **272**, 1631.
24 A. Lindinger, J. P. Toennies and A. Vilesov, *Phys. Chem. Chem. Phys.*, 2001, **3**, 2581.
25 R. Lehnig and A. Slenczka, *J. Chem. Phys.*, 2004, **120**(11), 5064.
26 G. E. Douberly, J. M. Merritt and R. E. Miller, *Phys. Chem. Chem. Phys.*, 2005, **7**, 463.
27 F. Madeja, P. Markwick, M. Havenith, K. Nauta and R. E. Miller, *J. Chem. Phys.*, 2002, **116**(7), 2870.
28 J. P. Toennies and A. F. Vilesov, *Chem. Phys. Lett.*, 1995, **235**, 596.
29 A. R. W. McKellar, Y. Xu and W. Jäger, *Phys. Rev. Lett.*, 2006, **97**, 183401.
30 A. R. W. McKellar, Y. Xu and W. Jäger, *J. Phys. Chem. A*, 2007, **111**, 7329.
31 K. Nauta and R. E. Miller, *Phys. Rev. Lett.*, 1999, **82**, 4480.
32 K. K. Lehmann, *J. Chem. Phys.*, 2001, **114**, 4643.
33 K. K. Lehmann and A. M. Dokter, *Phys. Rev. Lett.*, 2004, **92**, 173401.

Self-organisation and cooling of a large ensemble of particles in optical cavities

Yongkai Zhao,[a] Weiping Lu,[*a] P. F. Barker[b] and Guangjiong Dong[c]

Received 21st October 2008, Accepted 5th January 2009
First published as an Advance Article on the web 29th April 2009
DOI: 10.1039/b818653g

We present an investigation of the dynamics of centre-of-mass of a neutral particle cloud in a cavity pumped by an optical field. We derive an expression for the pump threshold for spatial self-organization of the particles and analyze its scaling laws in terms of the system parameters. Using a newly developed statistical model, we simulate the dynamics of the particles and numerically obtain the scaling laws. We show good agreement between the analytic formulae and simulations. We further use the scaling relation to discuss the operating conditions for cavity cooling a large ensemble of particles. Finally, we study cavity cooling of an ensemble of molecules with an initial temperature of around 10 mK. We show that 35% of the molecules are trapped by the optical field intensity in the cavity and a final temperature below 1 mK is reached.

1. Introduction

It has been observed that the coupled atom-field dynamics in an optical cavity can lead to a friction force that damps atomic motion.[1,2] This scheme is commonly called cavity cooling where dissipation takes place *via* cavity loss rather than by spontaneous emission. Cavity cooling avoids or reduces several problems of the laser-cooling scheme, such as photon re-absorption and recoil heating. Moreover, as there is no requirement in cavity cooling for a closed multilevel system, it is an attractive approach for creating ultracold molecules. A range of techniques is now capable of producing and trapping stable cold molecules at temperatures in the 10–100 mK range.[3-6] Cavity cooling appears to be a promising route towards further cooling a large range of species[1] to the ultracold regime below 1 mK.[7-11]

In cavity cooling, the commonly used mathematical model comprises a set of equations of motion for each particle that is coupled to the same intracavity field.[9,10] This many-particle model can be represented quantum mechanically or semi-classically. The latter has shown to be in good agreement with the former in recent extensive studies for temperatures higher than the cavity cooling limit. While the model is effective, it becomes impractical when a large ensemble is involved. Our understanding so far of cooling an ensemble of particles is based on the scaling laws which are obtained using various approximations which have not yet been fully tested.[9] In this paper we explore the scaling laws with a view to the self-organisation and cooling of a large ensemble of particles. We show that the scaling laws we obtain differ from the current ones in the case when a large ensemble of particles are involved. To test these laws we have further developed a new statistical model based on the

[a] *Physics, School of Engineering and Physical Sciences, Heriot-Watt University, Edinburgh, UK EH14 4AS*
[b] *Department of Physics and Astronomy, University College London, London, UK WC1E 6BT*
[c] *State Key Laboratory of Precision Spectroscopy, Physics Department, East China Normal University, Zhongshan Road, Shanghai, China*

Boltzmann equation. We show that there is good agreement between the theory and numerical simulation. Finally, we study the cooling of a CN molecular cloud of density 10^{13} per cm^3, with an initial temperature at 10 mK in an optical cavity. We find that more than a third of the molecules are stably trapped by the intracavity field and the final temperature reached is below 1 mK.

2. Scaling laws

We study the centre-of-mass motion of N two-level particles in an optical cavity pumped by a laser field that is transverse to the cavity axis, as shown in Fig. 1. For simplicity, we use a one-dimensional model in the cavity axis direction. The equations of motion in the semiclassical limit are given as[9,10]

$$\dot{\alpha} = [i\Delta_c - \kappa]\alpha - [\Gamma_0 + iU_0]\sum_j \cos^2(kx_j)\alpha - \eta_{\text{eff}}\sum_j \cos(kx_j) + \xi_\alpha$$

$$\dot{p}_j = \hbar k U_0 \left(|\alpha|^2 - \frac{1}{2}\right)\sin(2kx_j) + i\hbar k \left(\eta_{\text{eff}}^*\alpha - \eta_{\text{eff}}\alpha^*\right)\sin(kx_j) + \xi_{pj} \quad (1)$$

$$\dot{x}_j = p_j/m \quad (j = 1 \, \& \, N)$$

where $\alpha(t)$ is the amplitude of the intracavity field, $x(t)$, $p(t)$ and t are particle position, momentum and time. The parameters

$$U_0 = \frac{\Delta_A g^2}{\gamma^2 + \Delta_A^2}, \quad \Gamma_0 = \frac{\gamma g^2}{\gamma^2 + \Delta_A^2}, \quad \eta_{\text{eff}} = \frac{\eta g}{-i\Delta_A + \gamma} \quad (2)$$

describe the dispersion and absorption of the particles and the effective external pumping respectively, $\Delta_c = \omega - \omega_c$ and $\Delta_A = \omega - \omega_A$ are the pump–cavity and pump–particle detuning. We restrict our investigation to the case of large detuning, $|\Delta_A| \gg \gamma,\kappa,g$, for which Γ_0 can be neglected. Eqn (1) include a set of noise terms that are defined as

$$\langle \xi_\alpha^* \xi_\alpha \rangle = \kappa + \Gamma_0 \sum_{j=1}^N \cos^2(kx_j) \quad (3a)$$

$$\langle \xi_{pj}^* \xi_\alpha \rangle = -i\hbar k \Gamma_0 \alpha \sin(kx_j)\cos(kx_j) \quad (3b)$$

$$\langle \xi_{pj}^* \xi_{pj} \rangle = 2\hbar^2 \Gamma_0 k^2 u^2 |\cos(kx_j)\alpha + \eta/g|^2 + 2\hbar^2 k^2 \Gamma_0 |\alpha|^2 \sin^2(kx_j). \quad (3c)$$

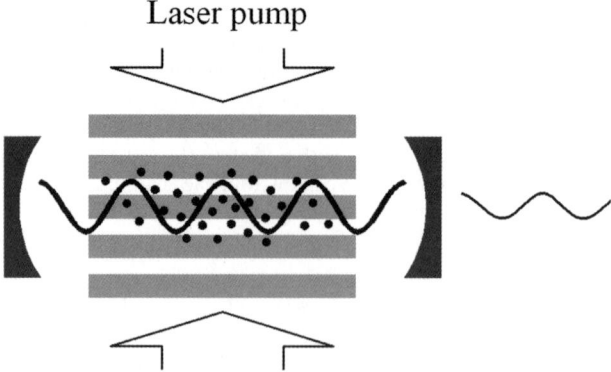

Fig. 1 Schematic of a cavity cooling system.

We first investigate the pump threshold for the build-up of the intracavity field, which also marks the onset of self-organisation (localisation) of particles in space. Particles are initially uniformly distributed in space with statistical fluctuations. When the external pump is applied to them, the intracavity field emerges through a scattering process benefiting from the statistical fluctuations. The intracavity field then produces a periodic potential well along the cavity direction, which in turn localises the particles around $kx = 2n\pi$ and $(2n + 1)\pi$ positions. Such a run-away process occurs only when the pump intensity reaches a certain level. The threshold for the build-up of a sustaining intracavity field can be defined at the point where the potential depth produced by the pump and intracavity fields equals the average kinetic energy of the particles,[9] *i.e.*, $\Delta E = k_\text{B} T/2$. In the following we use this relation to derive an expression for the threshold.

From the second equation of eqn (1), the potential is given by

$$U \approx \hbar \left[U_0 \left(|\alpha|^2 - \frac{1}{2} \right) \cos^2(kx) + \frac{2\eta g}{\Delta_a} \alpha_\text{r} \cos(kx) \right] \quad (4)$$

where α_r is the real part of the intracavity field amplitude. The intracavity field amplitude (and its real part) can be estimated under the steady state solution of the first equation of eqn (1),

$$|\alpha|^2 = \frac{|\eta_\text{eff}|^2 \left| \sum_j \cos(kx_j) \right|^2}{\left| [i\Delta_c - \kappa] - U_0 \sum_j \cos^2(kx_j) \right|^2} \quad (5)$$

The relations $\sum \cos(kx_j) \approx 2\delta N/\pi$ and $\sum \cos^2 kx_j \approx N/2$ hold well in the vicinity of the threshold, where δN is the deviation from the uniform distribution, *i.e.*, $(N - \delta N)/2$ particles around $kx = 2n\pi$ and $(N + \delta N)/2$ around $(2n + 1)\pi$. Since $\eta \gg |\alpha|$ in the vicinity of the threshold, we can neglect the first term in eqn (4) and the amplitude of the potential wells is therefore $U_\text{amp} \approx 2\hbar\eta g \alpha_\text{r}/\Delta_A$. Particles around the $(2n + 1)\pi$ positions and with energy

$$\frac{p^2}{2m} \leq \bar{U} \equiv \frac{2}{\pi} U_\text{amp} \quad (6)$$

can be trapped by the potential. Here $\bar{U}$ is the mean value of the potential U around the $(2n + 1)\pi$ positions. The inequality (6) gives the maximum momentum that can be trapped

$$p_\text{trap} = \sqrt{2m\bar{U}} \approx \sqrt{m\hbar\kappa} \frac{4}{\pi} \left(\frac{\eta g}{\Delta_a \kappa} \right) \sqrt{\delta N} \sqrt{\left(1 - \frac{N}{2} \frac{U_0}{\kappa} \right)} \bigg/ \sqrt{1 + \left[-1 + \frac{N}{2} \frac{U_0}{\kappa} \right]^2} \quad (7)$$

If the initial momentum distribution of the particles is Gaussian, the number of the trapped particles is given by

$$\delta N = \frac{1}{2} N \int_{-p_\text{trap}}^{p_\text{trap}} \frac{1}{\sigma\sqrt{2\pi}} \exp\left(-\frac{p^2}{2\sigma^2} \right) \text{d}p \approx \frac{p_\text{trap} N}{\sigma\sqrt{2\pi}} \quad (8)$$

The factor of a ½ in the above results from the fact that only those molecules around the $(2n + 1)\pi$ positions can be possibly trapped. By substituting eqn (7) into eqn (8), we have

$$\delta N \approx N^2 \frac{8}{\pi^3} \frac{m\hbar\kappa}{\sigma^2} \left(\frac{\eta g}{\Delta_a \kappa} \right)^2 \left(1 - \frac{N}{2} \frac{U_0}{\kappa} \right) \bigg/ \left\{ 1 + \left[-1 + \frac{N}{2} \frac{U_0}{\kappa} \right]^2 \right\} \quad (9)$$

The deviation, δN, thus depends on the external pump as well as other parameters of the system. We note that the value given by eqn (9) in the vicinity of the threshold is significantly larger than that of statistical fluctuations, $\delta N = \sqrt{N}$, when a large number of particles are placed in the cavity. We have verified this through numerical simulations. Now we return to the threshold condition, $\Delta E = k_B T/2$. The trap depth is twice the amplitude of the potential produced by the intracavity field,[9] and we therefore have

$$\hbar\kappa \frac{8}{\pi}\left(\frac{\eta g}{\Delta_a \kappa}\right)^2 \delta N \left(1 - \frac{N}{2}\frac{U_0}{\kappa}\right) \Big/ \left\{1 + \left[-1 + \frac{N}{2}\frac{U_0}{\kappa}\right]^2\right\} = \frac{1}{2}k_B T \quad (10)$$

By substituting eqn (9) into eqn (10), we obtain the threshold

$$\eta_{\text{thr}} = \frac{\pi\sqrt[4]{2}}{4}\sqrt{\frac{k_B T}{\hbar\kappa}}\frac{\Delta_a}{\sqrt{N}g}\sqrt{\left\{1 + \left[-1 + \frac{N}{2}\frac{U_0}{\kappa}\right]^2\right\} \Big/ \left(1 - \frac{N}{2}\frac{U_0}{\kappa}\right)} \quad (11)$$

We briefly discuss this expression. When the collective dispersion width is much smaller than cavity line width, $U_0 N \ll \kappa$, eqn (11) is simplified to

$$\eta_{\text{thr}} = \frac{\pi\sqrt[4]{8}}{4}\sqrt{\frac{k_B T}{\hbar\kappa}}\frac{\Delta_a}{\sqrt{N}g} \quad (12)$$

eqn (12) having the same scaling as that obtained under the mean-field approximation in terms of the system parameters $\Delta_A, N,$ and g.[9] We note that there was a different scaling law derived in the same paper based on the statistical fluctuation consideration using the same threshold criterion as ours, $\Delta E = k_B T/2$. The reason for the difference is that they applied the simple finite-size fluctuation relation, $\delta N \approx \sqrt{N}$ to eqn (10), whereas we derive a new expression (eqn (9)). The physics behind the build-up of the intracavity field is therefore different between the two approaches. In a certain sense, our approach unifies the scaling laws obtained by the statistical and mean-field models in the $U_0 N \ll \kappa$ limit. When $U_0 N/\kappa$ is not negligible, the scaling given by eqn (11) deviates from that of eqn (12). Generally speaking, the pump threshold η_{thr} increases faster than $\Delta_A/(\sqrt{N}g)$. This will be discussed further when compared to numerical results.

3. Statistical model and numerical results

When the particle number in the cavity is large, their positions and velocities can be described by a continuous distribution function $f(x,v,t)$. The dynamics of the intracavity field in this statistical model is given by

$$\dot{\alpha} = [i\Delta_c - \kappa]\alpha - [\Gamma_0 + iU_0]N\int\int f(x,v,t)dv\,\cos^2(x)dx\,\alpha - \eta_{\text{eff}}N\int\int f(x,v,t)dv\,\cos(x)dx + \xi_\alpha \quad (13)$$

eqn (13) is a straightforward extension from the first equation of eqn (1). The distribution function obeys the collisionless Boltzmann Equation

$$\frac{\partial f(x,v,t)}{\partial t} + v\frac{\partial f(x,v,t)}{\partial x} + \frac{F(x,t)}{m}\frac{\partial f(x,v,t)}{\partial v} = 0 \quad (14)$$

where $F(x,t)$ is the force exerted on the particles,

$$F(x,t) = \hbar k\{U_0(|\alpha|^2 - 1/2)\}\sin(2x) - i(\eta_{\text{eff}}^*\alpha - \eta_{\text{eff}}\alpha^*)\sin(x)\} + \xi_p \quad (15)$$

which comprises three sources: the dipole force, the force resulting from pump–intracavity field interference and noise ξ_p due to the recoil. In our study we are interested in the case of large detuning, i.e., $\Delta_A \gg \gamma,\kappa,g$, in which only the noise term in the field equation needs to be considered.

To validate the statistical model, we have compared its simulation results with those from the many-particle model eqn (1). The initial particle distribution is Gaussian, placed in the central region of the cavity. The results show a good agreement between the two models. Fig. 2 shows the evolution of momentum and position distributions. Our simulations establish that as few as hundreds of particles can be described well by the statistical model.

Now we compare the numerically computed scaling laws to the analytic expression eqn (11). By substituting eqn (11) into eqn (9), we obtain the value of the population deviation at the threshold, $\delta N_{thr} = 22.5\%$, for the parameter set used in Fig. 3. The defects at the threshold, given as $(N - \delta N_{thr})/2N$, are 38.8%. We have observed numerically that this value of defects indeed corresponds to the beginning of a sustaining intracavity field. We thus use this value to determine the pump threshold in our simulations. Fig. 3 shows η_{thr} vs. N while

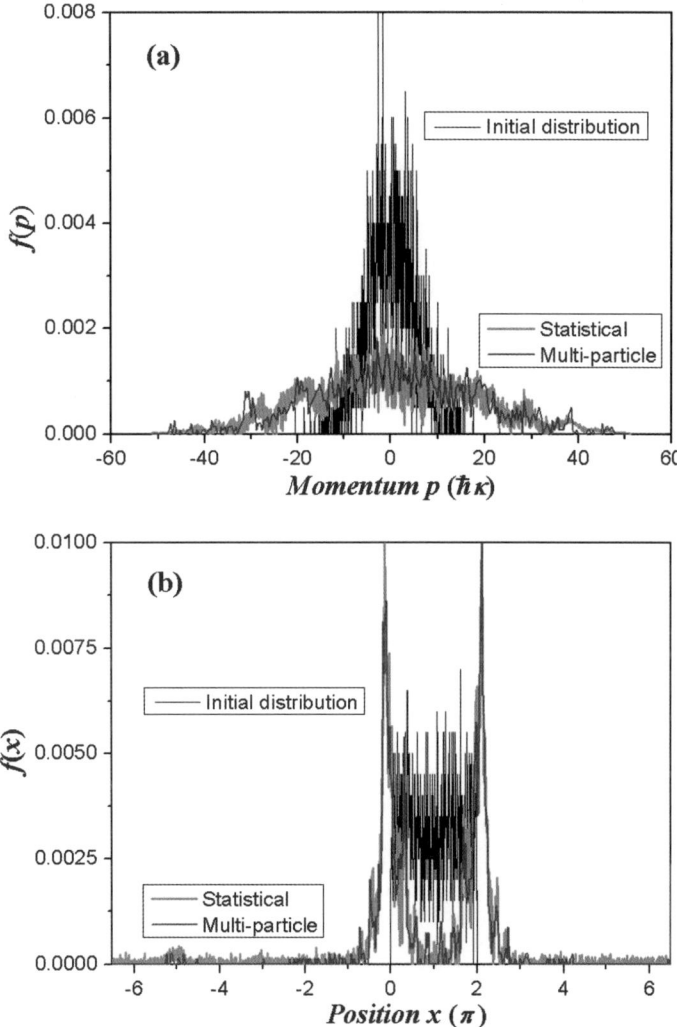

Fig. 2 Comparison of the results between the statistical and many-particle models. Fig. 2(a) shows the initial and final momentum distributions whereas Fig. 2(b) gives the position distribution. The parameters used are $N = 2000$, $\kappa = 20$ MHz, $g = 0.866\,\kappa$, $\eta = 300\,\kappa$, $\Delta_A = -3160\,\kappa$.

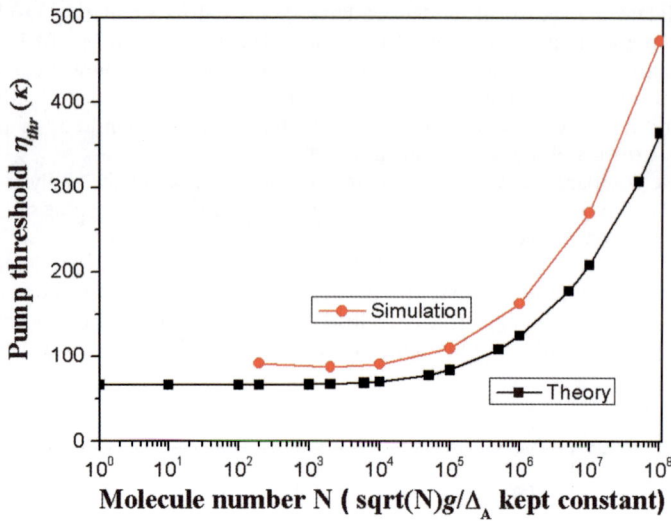

Fig. 3 The scaling law for the pump in terms of the system parameters. A good agreement is obtained between the analysis and simulation. The parameters used in both theory and simulation are $g = 0.866\ \kappa$, $\Delta_A = -1000\ \kappa$, $\kappa = 20$ MHz. The ratios $\Delta_A/(\sqrt{N}g) = -81.7$, and $\sqrt{k_B T/\hbar\kappa} = 1$.

$\Delta_A/(\sqrt{N}g)$ is kept a constant. This ratio is constant up to $N = 1000$, confirming the scaling given by eqn (12) for $U_0 N < \kappa$. When the molecular number increases above this value, this relation breaks and the curve follows the trend as described by eqn (11). The numerical results fit the well with the analytical predictions quite well in that the latter is derived using the steady state solutions and involving various approximations. We note that stable defects have been shown to appear when the molecular number $N > N_{thr} = \kappa/|U_0|$ and they can destroy spatial self-organisation.[9,10] Since $U_0 \approx g^2/\Delta_A$, must increase the detuning Δ_A with the molecular number proportionally to avoid this effect. This indicates that if a large ensemble of molecules is to be cooled in an optical cavity, a strong, far off resonant pump should be utilised.

4. Trapping and cooling of a molecular cloud

The above sections deal with a general case of spatial self-organisation of an ultra cold ensemble in an optical cavity. Here we study the cooling and trapping effects of a large ensemble of particles with a temperature much higher than the cavity cooling temperature limit. As an example we consider CN molecules, pumped by a far detuned optical beam. As studied in our earlier work, the molecules can be approximated as a three-level system under far off detuning pumping.[10] The statistical model gives us an effective tool to simulate a large CN ensemble. Fig. 4 (a) shows the time evolution of the intracavity field intensity and the temperature of the CN cloud. The results can be divided into three time zones. In the first zone, a small window from the initial field turn-on, the intracavity field fluctuates but has no sustained build-up, whereas the temperature stays at the initial value at around 11 mK. This region corresponds to molecules undergoing a free expansion from their initial position along the axis direction. No molecular localisation has been observed. The duration of this process depends on the cavity length and the initial position of the molecules. The second region begins when a molecular group with the fastest velocity escapes as they reach the cavity mirrors. The intracavity field

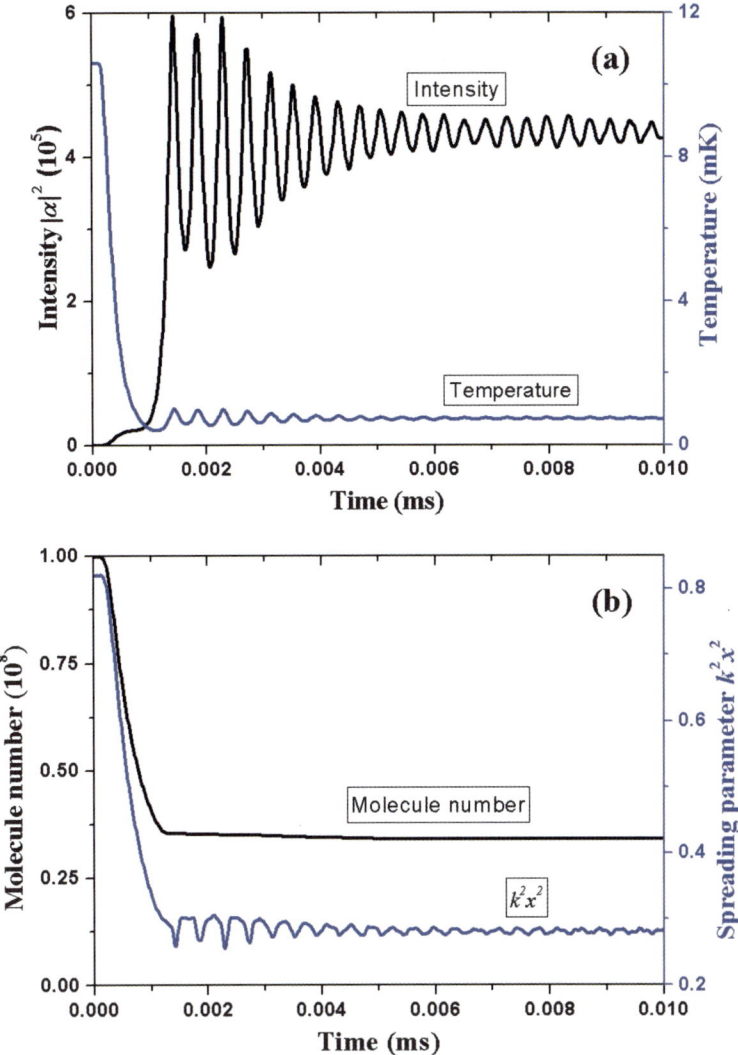

Fig. 4 Cavity cooling of a large ensemble of CN molecules. The parameters are set as $g = 0.055\ \kappa$, $\Delta_A = -1.1 \times 10^4\ \kappa$, $N = 10^8$, $\eta = 380\kappa$ and $\kappa = 62.8$ MHz. They correspond to a CN cloud of 3×10^{13} per cm^3 in an optical cavity of length $L = 1.5$ mm, waist $w_o = 25$ μm, and finesse $F = 10^4$. The pump intensity is 12 mW cm^{-2}.

intensity then rapidly builds up while the temperature drops sharply. A rapid spatial self-organisation of the remaining molecules is confirmed by the measurement of the spreading parameter (Fig. 4(b)), from initial uniform distribution (($k^2x^2 \approx 0.83$) to localisation ($k^2x^2 \approx 0.28$). The molecules continue to escape until the full build-up of the intracavity field occurs. This marks the beginning of the third region where around 35% of the molecules are stably trapped in the cavity by the potential produced by the intracavity field intensity (Fig. 4(b)). The temperature is now reduced to just below 1 mK. Molecules are further cooled in this region as the cavity non-adiabatic cooling process continues, but on a much slower scale. In general, the final temperature of the molecular cloud and the percentage number of trapped molecules depends on the pump intensity.

5. Conclusions and remarks

The cavity cooling concept was first proposed a decade ago[1] and was confirmed through a single experiment.[2] However, little experimental progress has been made in extending cavity cooling from a single particle to a dense cloud and from atoms to molecules, while optical manipulations of molecules using strong far off resonant lattices have been demonstrated in experiments.[12] The main activity in cavity cooling is so far still theoretical.[7,11] Lack of further progress may come from several factors. The realisation of cavity cooling for a large molecular cloud is more complicated than conventional laser cooling, which makes it difficult to provide an accurate prediction on the operation conditions for experiments. However, as there are few methods available to cool molecules below 1 mK it is becoming an important technique that will see further development. The purpose of our present work was to address the key issue of the pump threshold scaling law for the onset of cavity cooling for a large ensemble of particles. We have shown a new scaling law applicable to a large ensemble. In addition, we have developed a statistical model that offers an effective tool for extending cavity cooling to a molecular ensemble of any number. We have applied it to study the cooling of a dense CN molecular cloud. The agreement between the analysis and simulation is very good, providing a platform for further theoretical and experimental studies.

References

1 P. Horak, G. Hechenblaikner, K. M. Gheri, H. Stecher and H. Ritsch, *Phys. Rev. Lett.*, 1997, **79**, 4974.
2 P. Maunz, T. Puppe, I. Schuster, N. Syassen, P. W. H. Pinkse and G. Rempe, *Nature (London)*, 2004, **428**, 50.
3 J. D. Weinstein, R. deCarvalho, T. Guillet, B. Friedrich and J. M. Doyle, *Nature (London)*, 1998, **395**, 148.
4 J. M. Doyle, B. Friedrich, Jinha Kim and David Patterson, *Phys. Rev. A*, 1995, **52**, R2515.
5 H. L. Bethlem, G. Berden, F. M. H. Crompvoets, R. T. Jongma, A. J. A. van Roij and G. Meijer, *Nature (London)*, 2000, **406**, 491.
6 M. R. Tarbutt, H. L. Bethlem, J. J. Hudson, V. L. Ryabov, V. A. Ryzhov, B. E. Sauer, G. Meijer and E. A. Hinds, *Phys. Rev. Lett.*, 2004, **92**, 173002.
7 P. Domokos and H. Ritsch, *Phys. Rev. Lett.*, 2002, **89**, 253003.
8 H. W. Chan, A. T. Black and V. Vuletic, *Phys. Rev. Lett.*, 2003, **90**, 063003.
9 J. K. Asboth, P. Domokos, H. Ritsch and A. Vukics, *Phys. Rev. A*, 2005, **72**, 053417.
10 W. Lu, Y. Zhao and P. F. Barker, *Phys. Rev. A*, 2007, **76**, 013417.
11 B. L. Lev, A. Vukics, E. R. Hudson, B. C. Sawyer, P. Domokos, H. Ritsch and J. Ye, *Phys. Rev. A*, 2008, **77**, 023402.
12 R. Fulton, A. I. Bishop, M. N. Shneider and P. F. Barker, *Nat. Phys.*, 2006, **2**, 465, and references within.

General discussion

Professor Ye opened the discussion of the paper by Dr Lu: Spatial density is very important for the threshold of the collective behaviour. In the simulation, a temperature of 10 mK is assumed, but what is the spatial density?

Dr Lu replied: 10^{10} cm^{-3}

Mr Lemeshko asked: In your model, you do not take into account intermolecular interactions, is that correct? Do you have any estimate of when such interactions might be significant?

Dr Lu responded: We do not consider intermolecular interactions in this model. Molecules interact indirectly through the intracavity field. Intermolecular interaction will become important when they form an ultracold dense gas.

Professor Stwalley commented: There are three approaches for the formation of molecules in the lowest rovibrational level of the X ground electronic state of a molecule [$X(0,0)$ for short]: direct photoassociation (DPA), stimulated Raman, and broadband optical pumping, which have been discussed in several of the papers at this meeting. Two new variants of DPA have recently been developed, based on the mixing of near resonant electronic states. One variant involves the mixing of the initial collisional state of photoassociation with a Feshbach resonance, so-called Feshbach-optimized photoassociation.[1] The second involves mixing in the final state of photoassociation.[2] Both involve coupling of a short range state with a long range state, the long range component providing the high photoassociation probability and the short range state providing the strong overlap with the X(0,0) level.

1 P. Pellegrini, M. Gacesa, and R. Cote, *Phys. Rev. Lett.*, 2008, **101**, 053201.
2 J. R. Majumder, M. Bellos, R. Carollo, M. Recore, M. Mastrianni, V. Tagliamonti and W. C. Stwalley, Resonant coupling in the heteronuclear alkali dimers for direct photoassociative formation of X(0,0) ultracold molecules, *Faraday Discuss.*, 2009, **142**, poster presentation.

Professor Kosloff opened the discussion of Dr Pillet's paper by commenting: Traditional laser cooling relies on a closed cycle of excitation and spontaneous emission. For molecules, such a transition is difficult to find, therefore an alternative cooling scheme has been suggested theoretically based on the use of a broad band laser. Initially together with Allon Bartana and David Tannor we studied schemes of laser cooling of vibrational motion based on coherent control.[1,2] We set a molecular model with ground and excited electronic potentials. Transitions between these surfaces are induced by a shaped radiation field:

$$\hat{H} = \begin{pmatrix} \hat{H}_e & -\hat{\mu} \cdot \varepsilon \\ -\hat{\mu}^* \varepsilon^* & \hat{H}_g \end{pmatrix} \quad (1)$$

Where $\hat{H}_{e/g} = \hat{P}^2/2m + \hat{V}_{e/g}$ is the ground and excited state Hamiltonian, $\hat{\mu}$ is the transition dipole and $\varepsilon(t)$ is the time dependent electromagnetic field. The dissipation was taken as spontaneous emission. We modeled this process within the formalism of quantum open systems:

$$\frac{d\hat{\rho}}{dt} = -\frac{i}{\hbar}[\hat{H}, \hat{\rho}] + \mathcal{L}_D(\hat{\rho}) \quad (2)$$

where $\hat{\rho}$ is the density operator and $\mathcal{L}_D$ is the dissipative superoperator describing spontaneous emission.[2]

$$\mathcal{L}_D(\hat{\rho}) = \hat{F}\hat{\rho}\hat{F}^\dagger - \frac{1}{2}\{\hat{F}^\dagger\hat{F}, \hat{\rho}\} \qquad (3)$$

where $\hat{F}$ is an operator defined on the Hilbert space of the system. Specifically, for spontaneous emission this operator, $\hat{F} = \Gamma(\hat{r}) \otimes \hat{S}_-$, describes a transition from the excited to the ground electronic surface. where: $\hat{S}_- = |g\rangle\langle e|$ and $\hat{S}_+ = |e\rangle\langle g|$. The density of states of the photon bath in an isotropic space is proportional to ν^3, the cube of the emission frequency reflecting the density of states of the photon bath. Assuming a constant dipole function, the operator $\Gamma(\hat{r})$ obeys the relation: $\Gamma(\hat{r}) = \gamma\Delta(\hat{r})^{3/2} = \gamma(\hat{V}_e - \hat{V}_g)^{3/2}$. Other dissipative processes which contain transitions to the dark state are also sufficient for closing the cooling loop.

The dark state: The target of cooling is any single vibrational quantum state $v = k$ on the ground electronic state: $|k\rangle\langle k| \otimes \hat{P}_g$. The change in time in the occupation of this state using the Heisenberg equation of motion becomes:

$$\frac{d\langle |k\rangle\langle k| \otimes \hat{P}_g\rangle}{dt} = \frac{i}{\hbar}\langle [\hat{H}, |k\rangle\langle k| \otimes \hat{P}_g]\rangle + \langle \mathcal{L}_D^\dagger(|k\rangle\langle k| \otimes \hat{P}_g)\rangle. \qquad (4)$$

For this state to be dark, the commutation relation with the Hamiltonian should vanish. The dark state projection operator commutes with the stationary part of the Hamiltonian, while the commutation relation with the transient part of the Hamiltonian leads to the condition:

$$2Imag\{\langle \hat{\mu}|k\rangle\langle k| \otimes \hat{S}_+\rangle \cdot \varepsilon(t)\} = 0. \qquad (5)$$

Eqn (5) imposes a condition on the locking of the phase of the field to the phase of the projection of the transition dipole on the ground vibrational state:

$$2|\langle \hat{\mu}|k\rangle\langle k| \otimes \hat{S}_+\rangle||\varepsilon(t)|\sin(\phi_{\mu|k\rangle\langle k|} + \phi_\varepsilon) = 0. \qquad (6)$$

Eqn 6 vanishes if the sum of the phase angles, $\phi_{\mu|k\rangle\langle k|} + \phi_\varepsilon$, is equal to $n\pi$ where n is an integer. This condition is fulfilled by the explicit relation:

$$\varepsilon(t) = \frac{\langle \hat{\mu}|k\rangle\langle k| \otimes \hat{S}_-\rangle}{|\langle \hat{\mu}|k\rangle\langle k| \otimes \hat{S}_-\rangle|} c(t) \qquad (7)$$

where $c(t)$ is a real function of time which can be either positive or negative. Eqn (5–7) are equivalent statements of the dark state condition. We then used optimal control theory to find the optimal field $\varepsilon(t)$ which will lead from an initial thermal mixed density operator $\hat{\rho}_T$ to a final pure ground state $\hat{\rho} = |0\rangle\langle 0|$ in the ground electronic state.

We employed two versions of the optimal control theory, a local approach which determines the field at each instant to increase the projection of the state on the target. Then a global approach using the full power of optimal control theory, which seeks the optimal solution in a finite time interval. In both cases we found significant cooling leading to a final state very close to the target. When analyzing the spectrum of the optimal fields, we found that both possessed a common feature, all the transition amplitudes from the final target state were removed from the pulse spectrum. The method of cooling then became apparent: the field excites all molecules which are not in the target state which is dark. Eventually, after a few cycles of excitation and spontaneous emission, all molecules will accumulate in the target state. This idea has been realized in a remarkable experiment involving cooling Cs_2 using a broad band 100 femotosecond laser.[3,4] The authors state correctly that no coherence was involved in the weak field conditions in the experiment. Therefore a sufficient condition for cooling is to remove the transition of the target state from the broad excitation spectrum. In ref. 4, target states beyond the ground vibrational state were achieved. When the weak field scheme was applied to cooling the rotation of the molecule, success was limited. In addition, when the scheme was applied to other molecules or other combinations of ground and excited electronic states no cooling was observed. One obvious limitation is the bandwidth of the laser that should be able to excite all transitions that result from the spontaneous emission process. To overcome these difficulties the full power of optimal control theory should be employed. The necessary condition for coherent control is interfering pathways. To achieve this a strong field which goes beyond the one-photon excitation is required.

Rotational cooling: To demonstrate the power of coherent control a rotational model is used as an example. The following Hamiltonian is used:

$$\hat{H} = \hat{H}_0 + \hat{V}_t; \hat{H}_0 = \begin{pmatrix} \hat{H}_e & 0 \\ 0 & \hat{H}_g \end{pmatrix} \tag{8}$$

where: $\hat{H}_{e/g} = 2\pi \hat{B}_{e/g} \hbar l(l+1) + V_{e/g}$, and

$$\hat{V}_t = \begin{pmatrix} 0 & \mu_x \cdot \varepsilon_x(t) + \mu_y \cdot \varepsilon_y(t) + \mu_z \cdot \varepsilon_z(t) \\ \mu_x \cdot \varepsilon_x^*(t) + \mu_y \cdot \varepsilon_y^*(t) + \mu_z \cdot \varepsilon_z^*(t) & 0 \end{pmatrix} \tag{9}$$

A schematic view of the model is presented in Fig. 1.

The rotational Hamiltonian is influenced by the fact that the electromagnetic field can change the angular momentum of the system.

When optimal control theory was applied to this model significant cooling was observed. Analysis of the mechanism of cooling shows a very different picture (Fig. 2). Unlike vibrational cooling, the projection on the target state does not increase monotonically. For most of the period it stays constant and only in the final time does it increase to the final value close to one. On the other hand, the purity $tr\{\hat{\rho}_g^2\}$ increases monotonically meaning that a coherent superposition is being created. A different interpretation could be that the field breaks the degeneracy of the original Hamiltonian, aligning the molecule with the polarization of the field. The cooling is performed with respect to the dressed set of rotational levels. At the final step the pure dressed state is transformed by a unitary transformation to the final target state.

Conclusions: The application of optimal control theory to cooling molecules has revealed on the one hand the instantaneous phase locking of the laser pulse to the phase

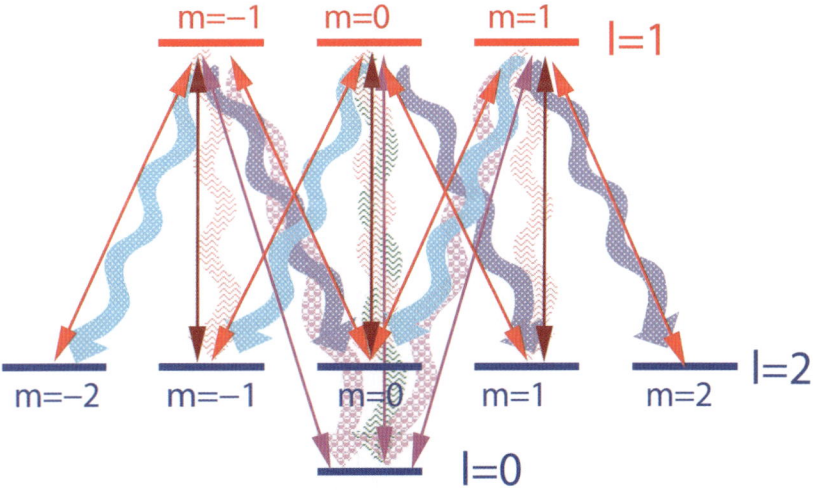

Fig. 1 The rotational cooling model. The solid arrows indicate stimulated excitations and the wavy arrows are the spontaneous emission.

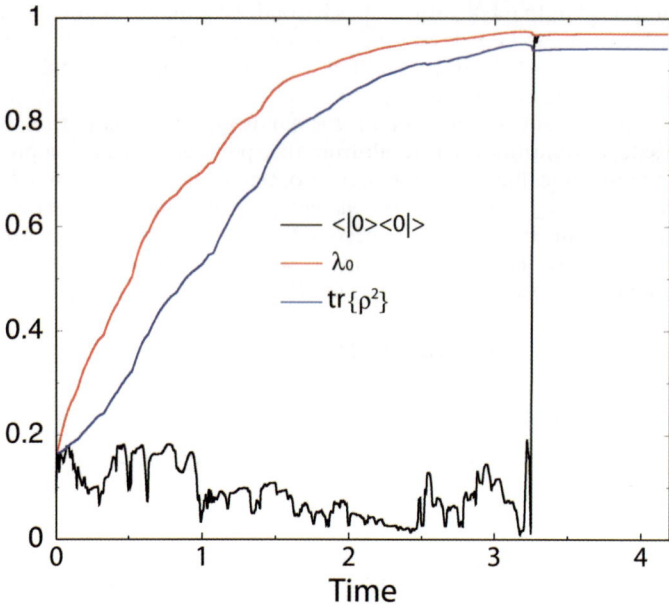

Fig. 2 The projection on the ground state $tr\{\hat{\rho}_g|0\rangle\langle 0|\}$ the purity $tr\{\hat{\rho}_g^2\}$, and the largest eigenvalue of the density operator $\hat{\rho}_g$, λ_0 as a function of time, for an optimal control solution.

of the transition dipole moment of $v = 0$ with the excited population. These conditions lead to a dark vibrational target state which eventually accumulates all the molecular population. On the other hand, control of the rotational cooling process is carried out through the transient modification of the Hamiltonian. This means that the control part is unitary. Since cooling requires a nonunitary step,[5,6] the control of the process is not direct. The analysis shows three mechanisms of control.

• Passive control through the creation of a dynamical dark state and optimizing the spontaneous emission to this state.

• Quasi-adiabatic modification of states of the system so that the rate of purity increase is optimized.

• Modifying the non unitary process by controlling the system–bath coupling, which in the present case is the dipole operator.

1 D. J. Tannor, R. Kosloff and A. Bartana, *Faraday Discuss.*, 1999, **113**, 365.
2 A. Bartana, R. Kosloff and D. J. Tannor, *Chem. Phys.*, 2001, **267**, 195.
3 M. Viteau, A. Chotia, M. Allegrini, N. Bouloufa, O. Dulieu, D. Comparat, and P. Pillet, *Science*, 2008, **321**, 232.
4 M. Viteau, A. Chotia, D. Sofikitis, M. Allegrini, N. Bouloufa, O. Dulieu, D. Comparat and P. Pillet, *Faraday Discuss.*, 2009, **142**, DOI: 10.1039/b819697d.
5 W. Ketterle and D. Pritchard, *Phys. Rev. A*, 1992, **46**, 4051.
6 A. Bartana, R. Kosloff and D. J. Tannor, *J. Chem. Phys.*, 1997, **106**, 1435.

Dr Pillet responded: The comment by Ronnie Kosloff investigates the role of coherence in broad band molecular cooling. The remarks are very interesting. Taking into account the coherence in such problems can offer new possibilities. The problem of vibrational cooling, which we treat, is nevertheless slightly different because it is mostly non coherent based on optical pumping with cycles of absorption + spontaneous emission in a weak laser field regime. The recent use (results not yet published) of a broad band laser diode working below the laser threshold for vibrational cooling confirms the non coherent character of the process and its possibilities. The aspects discussed in the comment concern the limitations of the weak regime in the generalization of the laser cooling of molecules and the role of the coherence for a possible improvement in the mechanism. We are convinced by the second point, for which the experimental conditions need to be precise for the chosen molecule. We answer essentially the first point, which concerns the limitations of weak field broadband optical pumping for the laser cooling of molecules.

Several problems are pointed out in the comment in particular for extending vibrational cooling to rotational cooling.

"When the weak field scheme was applied to cooling the rotation of the molecule success was limited." In our paper we are showing a simulation where rotational cooling is demonstrated (we transfer all population into the rotational level $j = 1$). At the time we were thinking of state C as the intermediate state, but finally, we were able to simulate rotational cooling *via* the B state as well. We treat the rotation as we do the vibration, only that the shaping resolution required is much higher. An experiment is now in progress to test the validity of our simulation.

"In addition, when the scheme was applied to other molecules or other combinations of ground and excited electronic states no cooling was observed." The application of our method to another molecule, is not immediate, but we believe that several schemes of shaped broadband laser can be established, at least theoretically, for different molecules. Further theoretical and experimental investigations allow us to precisely determine the universality and the strength of the method.

"One obvious limitation is the bandwidth of the laser that should be able to excite all transitions that result from spontaneous emission" This is true, but this will be true for any scheme, coherent or non coherent. We need reasonable FC factors and reasonable laser bandwidth to transfer the molecules towards lower v levels, down to $v = 0$ after a few cycles.

"To overcome these difficulties the full power of optimal control theory should be employed. The necessary condition for coherent control is interfering pathways. To achieve this a strong field which goes beyond the one photon excitation is required." The proposed technique is different compared to our approach. It can be powerful, but the use of the strong field regime can nevertheless be a problem. The molecules are quite sensitive to the intensity of the laser field. In our case, increasing the laser field "destroys" the molecules (probably due to two-photon excitation).

Dr Koch commented: In their paper on the vibrational cooling of Cs_2 molecules[1], Viteau *et al.* rely on the effect of resonant coupling between two excited state Born–Oppenheimer potentials to form molecules in low vibrational levels of the electronic ground state. I would like to comment on the generality of the resonant coupling mechanism. Resonant coupling in the photoassociation of ultracold molecules has been studied experimentally for cesium[2] and rubidium.[3] It leads to vibrational wave functions that show peaks at the four classical turning points of the two potentials, significantly modifying the Franck–Condon factors of these wave functions. In principle, such a coupling may also be induced by an external field.[4] A note of caution is, however, in order. The coupling might actually be too strong leading to adiabatic behaviour. The wave functions show then just two peaks at the classical turning points of the adiabatic potential.[5] The R-dependence of the coupling function might play a role. For photoassociation, it seems that ideally the coupling should be large at long range and show a minimum near the crossing point of the potentials.

1 M. Viteau, A. Chotia, D. Sofikitis, M. Allegrini, N. Bouloufa, O. Dulieu, D. Comparat and P. Pillet, *Faraday Discuss.*, 2009, **142**, DOI: 10.1039/b819697d.
2 C. Dion, C. Drag, O. Dulieu, B. Laburthe Tolra, F. Masnou-Seeuws and P. Pillet, *Phys. Rev. Lett.*, 2001, **86**, 2253.
3 H. K. Pechkis, D. Wang, Y. Huang, E. E. Eyler, P. L. Gould, W. C. Stwalley and C. P. Koch, *Phys. Rev. A*, 2007, **76**, 022504.
4 C. P. Koch and R. Moszynski, *Phys. Rev. A*, 2008, **78**, 043417.
5 S. Ghosal, R. J. Doyle, C. P. Koch and J. M. Hutson, *New J. Phys.*, 2009, **11**, 055011.

Professor Ye said: 1) Traditionally STIRAP has been applied to population transfer between, for example, two hyperfine states in atoms. The frequency gap is small and one can use EOM's or AOM's to bridge the gap. For transfer of population between two molecular levels spaced by thousands of wave numbers, then a frequency comb is very useful, and sometimes necessary, to establish phase coherence between two optical fields.

2) >92% efficiency for STIRAP transfer has been achieved for KRb ground state molecules. There are some intrinsic limitations to STIRAP due to the existence of neighbouring levels.

3) Coherent control concepts can be applied to the STIRAP process.

Professor Barker asked Dr Pillet: The method appears to be promising for translation laser cooling of molecules. For example, a two step scheme could be envisioned where the molecules could be laser cooled using a suitable ground state to excited electronic transition then, in the second step, repumped utilising your vibrational cooling scheme. What are the prospects for this type of, or a similar, scheme?

Dr Pillet replied: The use of a shaped broadband laser as a repumper for translational laser cooling of molecules is exciting. However, this prospect is still a long-term goal. Before realizing this ambitious goal we need to demonstrate the possibility of performing a rotational cooling at the same time as the vibrational one. The efficiency of the repumping technique is also limited, for instance by the spectral width of the laser. If we assume we will be able to reach 99.9% efficiency for the vibrational cooling, we limit the energy transfer to about 1000 photon recoils. The method cannot be applied to molecules at room temperature, but pre-cooled molecules at temperatures below one kelvin could be considered.

Mr Deiglmayr remarked: We have simulated the method of vibrational cooling described in your paper for the case of LiCs molecules in the $X^1\Sigma^+$ state using the excited states $A^1\Sigma^+$ and $B^1\Pi$. In this very simple simulation we apply a 50 cm^{-1} broad Gaussian spectrum incoherently to ground state molecules, spectrally cut at the blue wing to avoid excitation of $v'' = 0$ molecules.

We start with an ensemble of molecules in the ground state with populated levels v'' less than 10, corresponding to the population after spontaneous decay from the excited state $B^1\Pi$, $v' = 0$. Fig. 3 shows the initial population of ground state levels (black dots) and the population after several hundred absorption–emission cycles (red and blue symbols). While some molecules are pumped into the $v'' = 0$ level after many cycles, the majority of molecules are pumped into higher lying levels. Different spectral shapes of the pump light did not have a significant influence on the resulting population. The reason for this becomes immediately clear from the shape of the Franck–Condon parabola shown for the case of $A^1\Sigma^+$ in Fig. 4: $v'' = 0$ in the ground state is populated efficently from decay at the inner turning point of the excited levels $v' = 3$–11. However these levels decay at the outer turning point into levels $v'' = 10$ and above with high probability. For a realistic light source with finite bandwith, most of these levels are out of resonance for the broad band pump. Although we did not perform a systematic study for more systems, we got the impression that the scheme relies critically on a favourable shape of the Frank–Condon parabola and the difference in the vibrational spacing of the ground and excited states. We were therefore wondering, if you have identified other promising candidates for vibrational cooling to support the generality of vibrational cooling using broad band laser light?

Dr Pillet responded: In order to perform a simple "blue cut" optical pumping, the Franck–Condon parabola should basically be similar to the one we show in the cesium dimer case: *i.e.* showing a good overlap between the vibrational $v = 0$ of the two molecular states (ground and electronically excited).

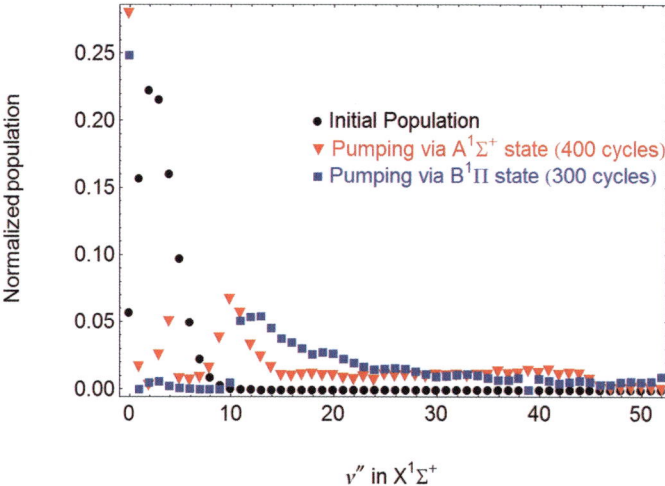

Fig. 3 Simulated redistribution of population in the case of LiCs, using the scheme of "vibrational cooling" described in paper 13. A Gaussian light spectrum with a width of 50 cm^{-1}, cut at higher frequencies to avoid excitation of $v'' = 0$, is applied to an initial population of vibrational levels corresponding to the distribution after spontaneous decay from $B^1\Pi$, $v' = 0$ (black dots). Resulting populations for pumping *via* different excited states are shown in blue and red.

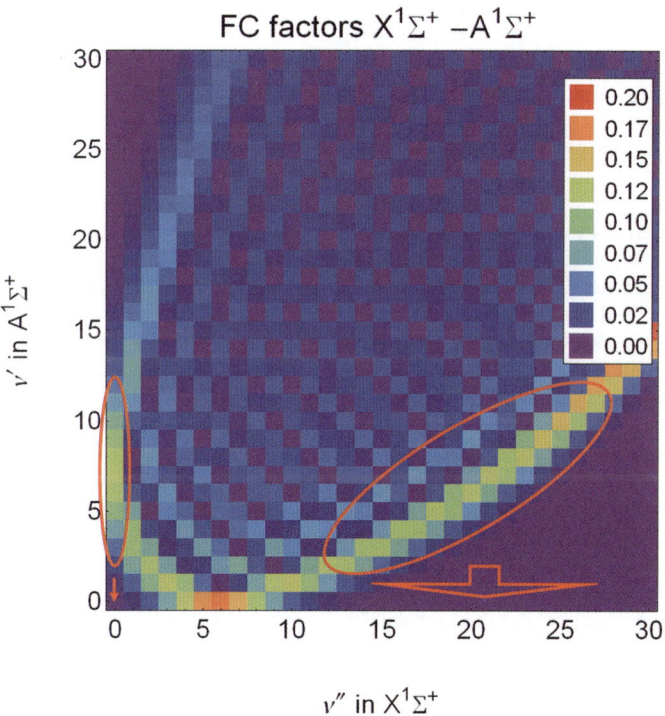

Fig. 4 Franck–Condon factors for transitions between vibrational levels in $X^1\Sigma^+$ and $A^1\Sigma^+$. The red arrows indicate the two different channels which have to be considered in the broadband optical pumping of ground state molecules; left arrow: pumping into $v'' = 0$ ("cooling"); right arrow: pumping into higher lying levels ("heating").

Concerning the difference between the vibrational spacing of the ground and excited states, this is just a matter of how broad the light source should be to address all the possible transitions. As an extreme example, a supercontinuum generated from the femtosecond laser will solve this problem.

Professor Pichler contributed: Although the vibrational cooling scheme works very well in the Cs_2 $X \to B$ transition, it was not successful for the triplet molecules using $a^3\Sigma_u^+ \to (2)^3\Pi_g$ or the $(2)^3\Sigma_g$ transitions. This was important to try due to the large number of ground state triplet molecules formed by photoassociation on "gigant" resonances (in the pure long-range 0_g^- state). So indeed the vibrational cooling depends strongly on the particular states involved *i.e.* their relative position and FC factors. However, other mechanisms leading to loss of molecules from the vibrational cooling transition can play a significant role and need to be carefully considered in each case. In addition, for the case of heteronuclear molecules (such as your LiCs system) since the u–g symmetry is removed, there can be even more loss channels disabling the vibrational cooling process.

Dr Pillet responded: We do not try to identify systematically other systems (for sure the method is applicable to Rb_2 for instance). We strongly believe that the method is quite general. The basic requirements are 1) a broad light source for optical pumping able to excite all the levels of interest (a supercontinuum, a laser diode or even a LED can replace a mode locked femtosecond laser).

2) an efficient frequency shaping optimizing the path. In the LiCs case the optimized shaping should probably correspond to the excitation of the levels of only the "outer" branch of the parabola, by keeping the $v = 0$ always as a dark state. In this case the excited molecules can only decrease their vibrational quantum number, v. Such a process can be realized by using Spatial Light Modulator as demonstrated in the paper: arXiv:0903.1222.

For a complete answer one needs to make the full calculation with the Franck–Condon factors and energy levels of the LiCs dimer.

Professor Hutson commented: It is wonderful to see that all the work that has gone into producing molecules in their rovibrational ground states is succeeding at last. But we need to learn to control not only rovibrational quantum numbers but also hyperfine quantum numbers. The lowest hyperfine state is the only one that cannot undergo collisions that release translational energy (and then only if there are no exoergic chemical reactions possible). One way to control hyperfine levels has already been demonstrated in the Innsbruck Cs_2 work, where magnetic fields were used to "drive around" on the network of Cs_2 bound states just below dissociation,[1] shown in Fig. 5. There are many avoided crossings between bound states, and at each one it is experimentally possible either to tune the magnetic field slowly over the crossing, so that the molecule follows the energy level adiabatically, or to tune fast so that the molecule "jumps the track" and makes a nonadiabatic transition. In the Cs_2 experiments it was important to find a state that could be tuned deeper into the well to enhance its short-range character before performing STIRAP. It will usually be possible and often necessary to select a molecular state with the properties needed to allow the desired transitions, and to do it we need to be able to characterise the near-dissociation bound states and the avoided crossings between them. However, the theoretical method that has been used to do this until recently, based on a discrete variable representation (DVR) for the radial coordinate,[2] is computationally expensive and can handle only a limited number of coupled channels – perhaps 25 or so – before running out of memory. However, there is an alternative approach,[3] based on propagating coupled differential equations without using a radial basis set, that I have been using for some time for the bound states of Van der Waals complexes. I have recently generalised my BOUND program[4] to implement the coupled equations needed for Cs–Cs interactions in a magnetic field,

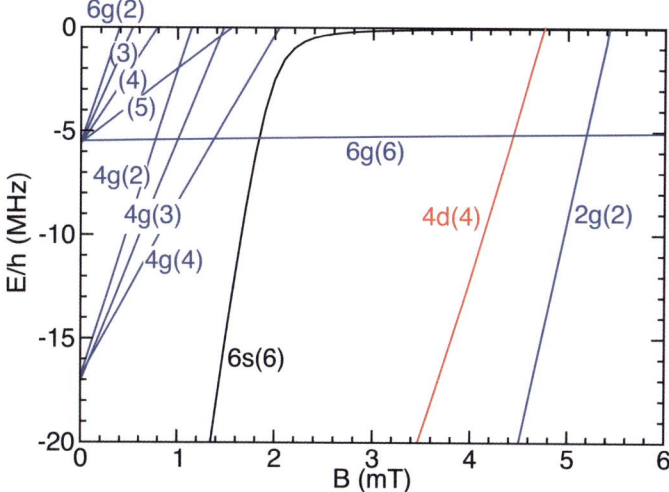

Fig. 5 Bound-state energy E/h as a function of B for levels of the Cs_2 molecules with even $L \leq 4$ and $M_{tot} = +6$. Energies are given relative to the energy of two Cs atoms in their ground Zeeman sublevel ($f = 3$, $m_f = +3$). The $FL(M_F)$ labeling scheme is shown for each level. Off-diagonal coupling between levels with different $FL(M_L)$ quantum numbers is neglected in this calculation.

and made a number of changes to allow propagation to very large distances. We have then compared the resulting method with the DVR approach.[5] For cases that the DVR approach can handle, we obtain exact agreement. However, the propagation approach can handle much larger sets of coupled equations, such as Cs–Cs for $M_{tot} = +6$ including channels up to $L = 8$ (225 channels). It can also handle levels extremely close to dissociation, such as the levels bound by less than 50 kHz shown in Fig. 6. The avoided crossing near 4.8 mT (48 G) in this Figure is the one that is actually used to enter the 6s(6) state in the current Cs_2 experiments.

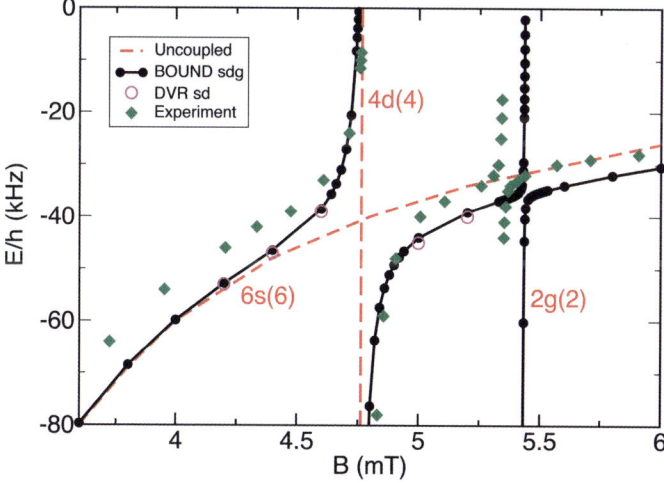

Fig. 6 Expanded view of the crossing in Fig. 5 of the 4d(4) and 6s(6) levels near 4.8 mT. The long dashed line shows the uncoupled calculation with the s(6) and d(4) basis sets. The solid lines show the propagator calculations with an sdg basis. The upper crossing near 5.4 mT is due to a 2g(2) level. The open circles show DVR calculations with a full sd basis in a finite box of 5000 a_0. The diamonds show experimental results.

There are some significant discrepancies between experiment and theory for the positions of $L = 8$ levels in Cs_2. The new computational method is actually cheap enough and stable enough that it will be possible to include it inside a least-squares fitting program to refine the Cs_2 potentials.

1 M. Mark, F. Ferlaino, S. Knoop, J. G. Danzl, T. Kraemer, C. Chin, H.-C. Nägerl, and R. Grimm, *Phys. Rev. A*, 2007, **76**, 042514.
2 E. Tiesinga, C. J. Williams, and P. S. Julienne, *Phys. Rev. A*, 1998, **57**, 4257.
3 J. M. Hutson, Comput. Phys. Commun., 1994, **84**, 1.
4 J. M. Hutson, *BOUND: A program for calculating bound-state energies for weakly bound molecular complexes*, version 5 (1993), distributed *via* Collaborative Computational Project No. 6 of the Engineering and Physical Sciences Research Council, on Molecular Quantum Dynamics.
5 J. M. Hutson, E. Tiesinga and P. S. Julienne, *Phys. Rev. A*, 2008, **78**, 052703.

Dr Aldegunde contributed: The experiments described by Danzl *et al.*[1] produce molecules in the ground rovibrational state but not necessarily in the lowest hyperfine state. We have done calculations to try to understand the hyperfine levels[2,3] and see which ones are likely to be produced in the experiment.

For the Cs_2 molecule in the absence of external fields and in the ground rotational state ($N = 0$), the hyperfine splitting is shown in Fig. 7. Instead of having a single energy level, we find 4 levels corresponding to the values of the total nuclear spin I allowed by nuclear exchange symmetry. The splitting is mainly due to the scalar spin-spin term $c_4 I_1 \cdot I_2$ that represents an electron-mediated interaction between the magnetic moments of the nuclei. We have calculated the value of the hyperfine coupling constants (including c_4) using DFT calculations with relativistic corrections.[1] Our estimate for c_4 is 13 kHz and the corresponding hyperfine splitting for $Cs_2(N = 0)$ amounts to 300 kHz.

In the presence of a magnetic field, each of the zero-field levels splits into $2I + 1$ components. The hyperfine Zeeman splitting for $Cs_2(N = 0)$ is presented in Fig. 8. The magnetic field does not destroy the total nuclear spin I, which remains a nearly good quantum number at all fields. Energy levels corresponding to the same value of M_I (the projection of I on the laboratory axis) remain parallel as a function of the field. An example is given by the $M_I = 0$ levels drawn in red.

The first step for the Cs_2 experiments performed in Innsbruck is to produce Feshbach molecules[4] in a state with $M_{tot} = 6$. Once these molecules have been formed, two STIRAP (stimulated Raman adiabatic passage) cycles are used to transfer the dimers into the ground rovibrational state. As the polarization of the four lasers

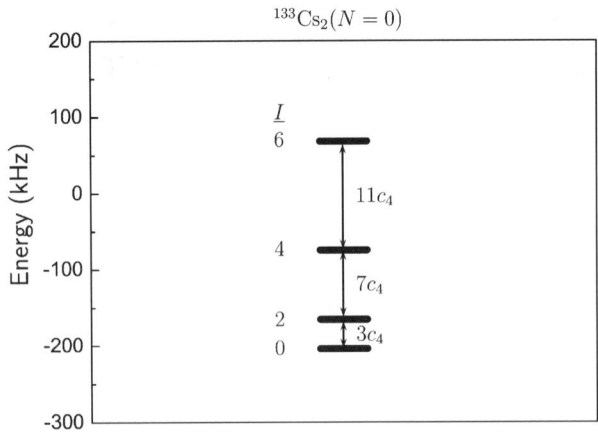

Fig. 7 Zero-field splitting for the $N = 0$ state of Cs_2.

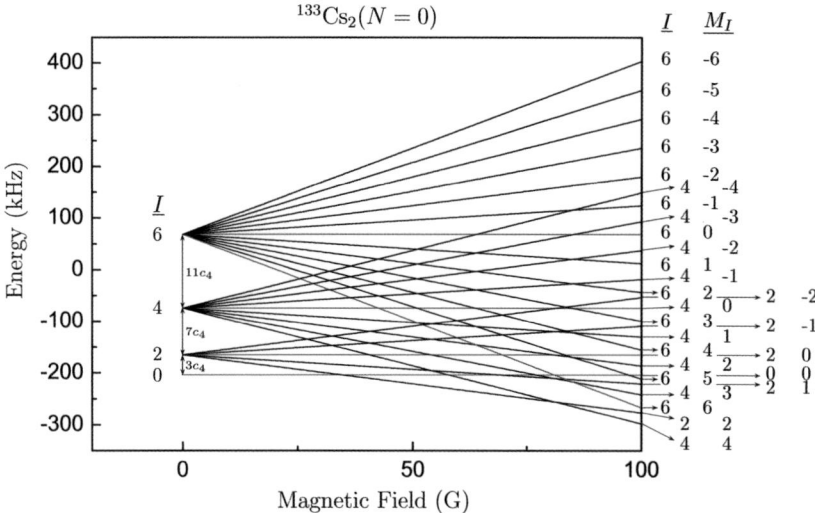

Fig. 8 Zeeman splitting of hyperfine levels for Cs_2 ($N = 0$). Red (blue) lines represent the energy levels corresponding to $M_I = 0(6)$.

involved in the process is parallel to the magnetic field, the selection rule that applies to the transfer is $\Delta M_{tot} = 0$ and the Cs_2 molecules formed will have a M_{tot} quantum number equal to 6. Only the $|I = 6, M_I = 6\rangle$ hyperfine state (highlighted in blue) of Cs_2 ($N = 0$) fulfils this condition, and therefore this will be the state in which the dimers are prepared.

It is important to understand whether the hyperfine state of the molecules can be changed in collisions. These changes could be caused by molecule–molecule collisions even at very low temperatures where only *s*-wave processes can occur because $|I = 6, M_I = 6\rangle$ is not the ground hyperfine state. In particular, collisions where the exit channel is *d*-wave could lead to the formation of dimers in $M_I = 4$ states. In order to ascertain whether this is possible we have estimated the height of the barrier for Cs_2–Cs_2 *d*-wave collisions as approximately 1 MHz.

As this value is at least three times larger than the energy differences between the $|I = 6, M_I = 6\rangle$ and the $M_I = 4$ states, we think that the Cs_2 dimers will be stable to collisions with other dimers and that they will not change their hyperfine state during the event.

1 J. G. Danzl, E. Haller, M. Gustavsson, M. J. Mark, R. Hart, N. Bouloufa, O. Dulieu, H. Ritsch and H.-C. Nägerl, *Science*, 20008, **321**, 1062.
2 J. Aldegunde, B. A. Rivington, P. S. Żuchowski and J. M. Hutson, *Phys. Rev. A*, 2008, **78**, 033434.
3 J. Aldegunde and J. M. Hutson, *Phys. Rev. A*, 2009, **79**, 013401.
4 M. Mark, F. Ferlaino, S. Knoop, J. G. Danzl, T. Kraemer, C. Chin, H.-C. Nägerl and R. Grimm, *Phys. Rev. A*, 2007, **76**, 042514.

Professor Julienne opened the discussion of Dr Hecker Denschlag's and Dr Danzl's papers: Is there evidence that the Cs_2 ground state molecules occupy multiple Bloch bands of the lattice, as is the case for Rb_2 molecules studied in Dr Hecker Denschlag's paper? Would one in general expect to get multiple Bloch band occupancy unless you form the lattice at a "magic wavelength" where the initial Feshbach molecule and final target molecule have the same polarizability? Did you, or can you, do any active control by locating such a magic wavelength for the Cs system? Should experimentalists try to take advantage of such wavelengths, or will it often be satisfactory to use any wavelength available?

Dr Hecker Denschlag responded: If the polarizability is different for the weakly and the deeply bound molecules then in general multiple Bloch bands will be populated. Therefore it is useful to use a laser wavelength for the optical lattice that is at least close to the magical one. When the STIRAP process is much faster than the trap frequency, dynamically changing the lattice depth right after the transfer can also circumvent the problem of populating various lattice bands. In principle it would also be possible to carry out very slow STIRAP transfers that would then resolve the lattice band structure.

Dr Danzl replied: In Cs_2, it turns out that the polarizability in the rovibronic ground state is very similar to the polarizability of the Feshbach molecules at the wavelength of our lattice light of 1064 nm. Therefore, we operate close to such a "magic wavelength" where we populate only the lowest Bloch band and do not expect to observe dynamics due to multiple Bloch band occupation similar to the observations for the Rb_2 lowest triplet state in Dr Hecker Denschlag's paper. We measure the polarizabililty by either phase- or amplitude-modulating the optical lattice and comparing the excitation spectrum to a calculation of the band structure. Nadia Bouloufa and Olivier Dulieu have calculated the polarizability of Cs_2 for $|v = 0\rangle$ of the $X^1\Sigma_g^+$ electronic ground state as a function of trapping laser wavelength by summing up the contributions from all the levels of the different $^1\Sigma_u$ and $^1\Pi_u$ states (including spin-orbit coupling for the 0_u^+ states). At 1064 nm, the lattice light is about 40 nm red detuned from the bottom of the $A^1\Sigma_u^+$ state, the lowest singlet state. Therefore, in this region, the polarizability has a smooth dependence on wavelength that can indeed be used to tune to a wavelength where the polarizability of the rovibronic ground state molecules exactly matches the polarizability of the Feshbach molecules. Below this region, there are only a few resonances arising from rovibrational levels of the $b^3\Pi_u$ state that, due to spin-orbit coupling, has some admixture of the $A^1\Sigma_u^+$ state. In regions of higher energy, *i.e.* above the minimum of the $A^1\Sigma_u^+$ state, there is a rich structure in the polarizability reflecting the presence of excited state rovibrational levels. In principle, these resonances could also be exploited to set the polarizability to a desired value but near resonance the issue of photon scattering from the lattice becomes more serious.

We feel that it is highly advantageous to choose a wavelength for the optical lattice where the polarizabilities are similar for the initial and the final state in a state transfer experiment. Theoretical knowledge of the polarizabilities should be taken into account when planning an experiment.

Professor Stwalley asked Dr Danzl: Could you please comment on the understanding of the polarizability change in going from a Feshbach molecule to $v = 73$ and then to $v = 0$? In particular, are there theoretical calculations of the polarizability to compare with your experimental results?

Dr Danzl answered: The polarizability of a Feshbach molecule essentially corresponds to the sum of the polarizabilities of the two constituent atoms. When considering more deeply bound levels, the polarizability needs to be calculated by summing up all the contributions from the relevant molecular excited state rovibrational levels. In our experiment on Cs_2, the lattice light is generated by a highly stable laser at 1064 nm. For the rovibronic ground state $X^1\Sigma_g^+ |v = 0, J = 0\rangle$, the lattice is red detuned relative to the bottom of the lowest electronically excited singlet state, namely the $A^1\Sigma_u^+$ state. Therefore, essentially all the rovibrational levels in the different excited state potentials add up constructively and contribute to the polarizability. In contrast, for the intermediate level $|v = 73\rangle$, the lattice light at 1064 nm reaches the 0_u^+ states much higher up in the potential curve. Accordingly, for many of the rovibrational levels in the electronically excited state the lattice light is blue detuned. Consequently the polarizability is effectively reduced. Due to the presence of resonances associated with individual rovibrational levels, the precise value of the polarizability for $|v = 73\rangle$ very sensitively depends on the wavelength. We find that the polarizability for $|v = 73\rangle$ is about 30% of the polarizability of Feshbach molecules at the particular wavelength we are using.

The corresponding calculations for $|v = 0\rangle$ and for $|v = 73\rangle$ by Nadia Bouloufa and Olivier Dulieu mirror this behaviour and agree well with the experiment.

Professor Kotochigova addressed Dr Danzl and Professor Stwalley: In the case of the KRb molecule, polarizability of the Feshbach molecule and the molecule in the absolute ground state are very similar. They differ by less than an order of magnitude.

Professor Julienne asked Dr Danzl: This question is relevant to the Cs_2 molecule production experiment, and more generally, to any experimental group trying transfer molecular populations using STIRAP. Of course, STIRAP is appealing because it is a very robust process and can allow nearly 100 per cent efficiency in transferring population from one state to another. However, this robustness is based on the specific time-dependent dynamics of a three-state system, and may be partially lost for a system with a more complex structure of levels. For example, STIRAP will not in general work for scattering states because of their continuum instead of discrete structure. In the Cs_2 case, a four-color STIRAP scheme is used, and there is unknown hyperfine structure in the intermediate state (possibly unresolved because of the relatively small hyperfine splittings associated with an $\Omega = 0$ intermediate molecular state). There is also hyperfine structure in the ground state. Could you comment on the expected robustness of STIRAP in the presence of such structure (I also invite comments by other experimentalists working with these schemes)? Could a multiplicity of intermediate levels with similar detunings destroy the "dark state" properties of the simple three-state model? The reason this question is significant is that one wishes to know in general if the STIRAP technique can be made to approach near 100 per cent efficiency or not. For example, if one wished to make a lattice of molecules by first preparing a region of lattice cells which contain only the desired two atoms to be associated, the STIRAP efficiency will determine the number of defects (*i.e.*, cells with no molecules) in a resulting molecular lattice. How close to 100 per cent efficiency in molecular population transfer is it realistic to expect? Do you expect to be able to greatly increase the Cs_2 production efficiency?

Dr Danzl replied: STIRAP is indeed a very robust technique. We feel that currently the transfer efficiency is mainly limited by technical issues like finite available laser power to drive the extremely weak transitions in our four-photon scheme and finite laser coherence. We are thus confident that we will be able to improve the STIRAP efficiency to close to unity.

Nevertheless, the question of additional levels possibly interfering with the ground state transfer is a very important and interesting one. There is a series of papers by Klaas Bergmann and coworkers that analyze transfer in the presence of multiple sublevels and also transfer involving a continuum state. With favorable parameters, highly efficient transfer can be achieved, even in the case of several Lambda-systems existing in parallel.

In our experiment on Cs_2 ground state molecules, the initial Feshbach level and the particular hyperfine quantum state in the rovibronic ground state $X^1\Sigma_g^+ |v = 0, J = 0\rangle$ are well defined, in the latter case due to angular momentum selection rules, since the total angular momentum projection $M_{tot} = 6$ is conserved by driving $\Delta M = 0$ transitions with the STIRAP lasers. This is also true for $|v = 73, J = 0\rangle$ as an intermediate electronic ground state level in the four-photon scheme. In contrast, for $|v = 73, J = 2\rangle$ and for the intermediate levels belonging to electronically excited states, there could be more than one sublevel involved. The hyperfine substructure in the electronically excited state is not resolved, since the Zeeman splittings between the sublevels are small at the low magnetic fields of around 20 G applied in the STIRAP experiments. This field is given by the particular Feshbach level used as starting state. It is conceivable that the STIRAP lasers weakly couple to other hyperfine sublevels in the excited intermediate levels, causing photon scattering or unwanted Stark shifts, and therefore lead to loss out of the three- or five-level system. If this turns out to be a severe limitation, the transfer could be done at higher magnetic fields by choosing a different Feshbach level as starting state, effectively separating the different sublevels. Clearly, an in-depth theoretical analysis of the excited state hyperfine structure and its influence on the transfer process in our experiment would be highly desirable. The experiments on ground state transfer in KRb (Professor Ye's paper) and transfer to the lowest triplet state in Rb_2 (Dr Hecker Denschlag's paper) are carried out at much higher magnetic fields of around 550 G and 1010 G, respectively, where the individual sublevels are clearly separated from each other.

In our experiment, Feshbach molecules are produced at the individual lattice sites out of an atomic Mott-insulator state with a filling factor close to one, meaning that practically every lattice site is occupied by one molecule. We currently transfer about 60% of molecules to the rovibronic ground state, *i.e.* on average more than half of the lattice sites are filled with a molecule in the absolute ground state. The remaining lattice sites are mostly filled with molecules in arbitrary rovibrational levels after excitation to an electronically excited level and spontaneous decay to the electronic ground state. We expect to achieve filling factors close to one for absolute ground state molecules with optimized STIRAP efficiencies.

Professor Stwalley asked: Could you please comment on the relative merits of a single four-photon transfer compared to a sequence of two two-photon transfers in preparing the lowest rovibrational level?

Dr Danzl replied: The main difference between a four-photon transfer and two sequential two-photon transfers is that in the former, the population in the intermediate level can be minimized during the transfer by choosing the Rabi frequencies for the second and third laser transitions much higher than for the first and fourth. In principle, this means that loss in the intermediate ground state level can be avoided. However, loss in the intermediate level is not a major issue in the current experiments because the timescale of the STIRAP transfer is short, on the order of microseconds. The molecules are not likely to undergo inelastic collisions on this timescale when held in a dipole trap. In the presence of an optical lattice, as employed in the Cs_2 ground state experiments, collisional loss during the transfer is suppressed. The observed lifetimes for the intermediate level in the optical lattice are on the order of 20 ms, limited by photon scattering from the lattice, much longer than the timescale of the transfer. There are three reasons why we still prefer to use four-photon transfer. First, avoiding population in the intermediate level means that we can largely avoid population of higher Bloch bands in the lattice that could arise due to a difference in polarizability of the intermediate level relative to the Feshbach level or to rovibrational ground state molecules. Second, when the four-photon dark state is expressed as a linear combination of the bare states, the prefactors for the individual bare states are pairwise products of the Rabi frequencies of the different transitions. Therefore, there is some freedom to compensate for a rather weak transition with the other laser transition that enters into the same prefactor. Third, a 4-photon dark state allows the observation of coherence and an exotic superposition state between the initial Feshbach state and the rovibronic ground state in a Ramsey-type experiment, spanning a binding energy difference of roughly 110 THz in frequency units. In addition, from a very technical side, it is convenient to optimize just two laser ramps instead of four.

Mr Lemeshko asked Dr Hecker Denschlag: Are the Bloch bands in the optical lattice continuous or they are sets of discrete levels, as you presented in Fig. 6 of your paper? Another, technical, question: what is the intensity of the laser beams, forming the lattice?

Dr Hecker Denschlag responded: The Bloch bands have a certain width which is not indicated in Fig. 6. This width is related to the tunneling between lattice sites. In our model we ignore the finite width of the bands. Tunneling is included artificially *via* a decay width $J_{\alpha,n}$.

The intensity of the lattice laser beams is on the order of 1 kW cm^{-2}.

Professor Weidemüller asked Dr Danzl and Dr Hecker Denschlag: The efficiencies achieved so far for STIRAP from magnetoassociated molecules to the ground state are of the order of 90%. I wonder whether there are fundamental limitations to the transformation efficiency of ultracold atoms into molecules, *e.g.* due to molecular states spuriously coupled to the laser radiation, or whether technical reasons limit the efficiency of molecule formation *via* STIRAP.

Dr Hecker Denschlag answered: We believe, that the transfer efficiencies are limited mostly by technical reasons. We estimate that half of our losses are due to relative phase fluctuations of our two Raman lasers. These phase fluctutations lead to an imperfect dark state, which leads to losses during transfer. The other half of our losses is due to non-adiabaticity of our transfer which takes place in a few micro-seconds.

Dr Danzl replied: We believe that currently our efficiencies for ground state transfer in Cs$_2$ are largely limited by technical reasons, such as laser coherence and finite available laser power to drive the extremely weak transitions in the Cs$_2$ four-photon scheme. Franck–Condon factors are considerably smaller for the Cs$_2$ singlet ground state experiment compared to transfer to the rovibronic ground state in heteronuclear molecules or to $|v = 0\rangle$ of the lowest triplet state, making the issues of laser coherence and available laser power especially important for ensuring adiabatic transfer. Coupling to spurious levels can in principle occur but can be largely avoided by a careful choice of the intermediate levels, laser polarizations, and the magnetic field at which the transfer is performed. Therefore, efficiencies close to unity should be possible.

Dr Dulieu asked Dr Hecker Denschlag: Are the Rb$_2$ molecules created in a single quantum state of the lowest triplet state?

Dr Hecker Denschlag responded: Yes, the molecules have a well defined inner quantum state with quantum numbers $F = 2$, $m_F = 2$, $S = 1$, $I = 3$, $v = 0$, $l = 0$.

Professor Hinds opened the discussion of Dr Lu's paper: In your paper you present results for cavity-cooling a cloud, starting from an initial temperature of 10 mK, which is evidently low enough for the cooling to take effect. I presume this would not work for a 1 K cloud. In general, is the temperature threshold for the onset of cooling the same as the threshold for the build-up of the intracavity field?

Dr Lu answered: In general, once the intracavity field is formed, molecules are cooled through the cavity dynamics. However, the threshold pump for the build-up of the intracavity field depends on the initial temperature of the molecular sample placed in the cavity, the threshold pump for cooling a 1 K cloud is much higher compared to cooling a 10 mK cloud. In fact, if you place a molecular sample with initial temperature much higher than 10 mK in the cavity, those molecules at higher velocities will escape from the cavity (thus the temperature of the sample in the cavity is reduced), before the intracavity field can build up. As a result, only a percentage of molecules (say, 30–35%) in the sample can be trapped and subsequently cooled by the cavity dynamics. In cavity cooling, trapping and cooling happen at the same time and cooled molecules are confined in a small volume, which are the advantage points of the scheme. We have investigated the higher temperature cases in detail as they are important for practical experimental work.

Professor Grimm commented: The "holy grail" is the production of a degenerate quantum gas of molecules. I want to make a clear distinction between the two association approaches that are followed. Both start with a standard sample of cold atoms (prepared in a magneto-optical trap) at phase-space densities several orders of magnitude below degeneracy. There is the photoassociation approach, where the first step is to form (ground state) molecules by optical association. Far from degeneracy, such a sample is already interesting for spectroscopy and some other applications. To the degenerate regime, however, it is a very long way to go. There is the quantum-gas approach, where the first step is to prepare a degenerate atomic sample. The important intermediate step is then the magneto-association of "Feshbach molecules" (which themselves offer a lot of interesting physics). The next step is coherent transfer to the ground state, efficiently done with the STIRAP method. This leads to a near-degenerate sample, and the way into the degenerate regime is quite short.

PAPER

Formation of ultracold dipolar molecules in the lowest vibrational levels by photoassociation

J. Deiglmayr,[ab] M. Repp,[a] A. Grochola,†[a] K. Mörtlbauer,[a] C. Glück,[a] O. Dulieu,[b] J. Lange,[a] R. Wester[a] and M. Weidemüller*[a]

Received 20th October 2008, Accepted 12th December 2008
First published as an Advance Article on the web 29th April 2009
DOI: 10.1039/b818391k

We recently reported the formation of ultracold LiCs molecules in the rovibrational ground state $X^1\Sigma^+, v'' = 0, J'' = 0$ (J. Deiglmayr et al., Phys. Rev. Lett., 2008, **101**, 133004). Here we discuss details of the experimental setup and present a thorough analysis of the photoassociation step including the photoassociation line shape. We predict the distribution of produced ground state molecules using accurate potential energy curves combined with an *ab initio* dipole transition moment and compare this prediction with experimental ionization spectra. Additionally we improve the value of the dissociation energy for the $X^1\Sigma^+$ state by high resolution spectroscopy of the vibrational ground state.

1 Introduction

Ultracold molecular gases are promising candidates for studying diverse systems reaching from tests of the standard model to ultracold chemistry and quantum computing. Of special interest is the preparation of these molecules in the absolute rovibrational ground state, as this state is stable against inelastic collisions and allows, therefore, the creation of a stable, dense gas of molecules. A number of experimental approaches are currently being studied to prepare and manipulate ultracold molecules.[1,2] Magneto-association using interspecies Feshbach resonances followed by adiabatic transfer *via* two or more optical transitions has been recently used to produce dense gases of deeply bound molecules of Cs_2,[3] KRb^4 and Rb_2,[5] even reaching the rovibrational ground state. Photoassociation (PA), a long established technique, has been successfully employed in the production of K_2,[6] $RbCs$,[7] Cs_2,[8] and $LiCs^9$ in the lowest vibrational level of the electronic ground state. In contrast to the homonuclear molecules, heteronuclear alkali dimers exhibit strong dipole moments ranging from 0.6 Debye for LiNa and KRb to 5.5 Debye for LiCs.[10] The formation of these dipolar molecules in their lowest vibrational level opens the way to the exploration of quantum phases in dipolar gases,[11,12] the development of quantum computation techniques,[13] precision measurements of fundamental constants,[14] and the investigation and control of ultracold chemical reactions.[15]

We recently reported the formation of ultracold LiCs molecules in the rovibrational ground state $X^1\Sigma^+, v'' = 0, J'' = 0$.[9] In Section 2 of this article, we discuss details of the experimental setup with a special focus on the reduction of interspecies losses

[a]*Physikalisches Institut Albert-Ludwigs-Universität Freiburg, Germany. E-mail: weidemueller@physi.uni-heidelberg.de*
[b]*Laboratoire Aimé Cotton, CNRS, University Paris-Sud XI, Orsay, France*
† Also at the Institute of Experimental Physics, Warsaw University, Poland.

in overlapped magneto-optical traps. After analyzing the PA step and modeling the PA line shape in Section 3, we focus on the distribution of populated vibrational ground state levels in Section 4. In the discussion of the detection step, these predictions are compared with experimental ionization spectra and qualitative agreement is found (Section 5). Finally we present high-resolution spectroscopy of the vibrational ground state and improve the value of the dissociation energy of the lowest $X^1\Sigma^+$ level in Section 6.

2 Experimental setup

In this section, the experimental setup for the formation and detection of ultracold polar molecules is described. We outline the general principles and focus on recent important additions and improvements of the setup. Further details can be found in references 9, 16 and 17.

2.1 Trapping a large number of atoms in overlapped magneto-optical traps

For the formation of dipolar ground state molecules, we cool and trap ^{7}Li and ^{133}Cs atoms simultaneously in two overlapped magneto-optical traps (MOTs). The atoms are evaporated in a double species oven and are decelerated in a single Zeeman slower‡. In overlapped magneto-optical traps for different atomic species, high loss rates due to inelastic interspecies collisions are a well known experimental difficulty.[18] Fig. 1(a) demonstrates the importance of these losses for the setup used in previous experiments.[16] The dominant loss channel has been identified as collisions between excited $Cs(6P_{3/2})$ and ground state $Li(2S_{1/2})$ atoms.[19] It was therefore straightforward to reduce these losses by lowering the number of cesium atoms in the excited state. This was achieved by implementing a dark magneto-optical trap (also called "dark spontaneous force optical trap" or "dark SPOT"[20]) for cesium. In the center of such a trap, the repumping light is blocked, so that already cold atoms are pumped by the cooling light off-resonantly into the lower hyperfine ground state which is now a dark state. Only if they leave the central part of the trap they are pumped back by repumping light and take part in the cooling cycle again. This has been widely used to achieve increased numbers and densities of trapped atoms which we also observe. However we mainly profit from the large fraction of atoms in the dark hyperfine ground state (typically 97%), which reduces interspecies lossrates by the same fraction. The central part of the repumping light is blocked by imaging the shadow of a 3.5 mm large piece of plastic§ with a 1 : 1 telescope onto the trap center. As the laser beam used for Zeeman-slowing of the atoms also contains repumping light and passes through the center of the trap, we additionally image a dark spot (3.5 mm, 1 : 1 imaging ratio) in the Zeeman beam onto the trap center. The Rayleigh range of the imaging focus is on the order of only a few centimeters, so that the slowing of the atomic beam is not significantly reduced by the shadow in the center of the Zeeman beam. We note that the reduced absolute power in the Zeeman beam leads to a reduced loading rate, which is however compensated by the reduced single species losses in a dark SPOT. In order to reach a high darkstate fraction, we detune the repumping laser by $\sim 5\Gamma$ and shine an additional depumping laser on the center of the trap (100 µW, $\omega_0 = 1.5$ mm). This setup is called a detuned forced dark SPOT.[21] Fig. 1(b) demonstrates the increase in the number of simultaneously trapped lithium and cesium atoms after implementing the detuned forced dark SPOT for cesium. With this setup we typically trap 4×10^7 Cs atoms

---

‡ When both traps are loaded simultaneously, the configuration of the Zeeman fields is optimized for slowing lithium atoms. With this field configuration, the loading rate for cesium atoms is reduced to 1×10^7 atoms s^{-1}, which is still sufficient for the experiments described here.
§ It turned out that the best extinction ratio of 1 : 10^4 at the trap position was obtained by using the magnetic plastic from an old floppy disk.

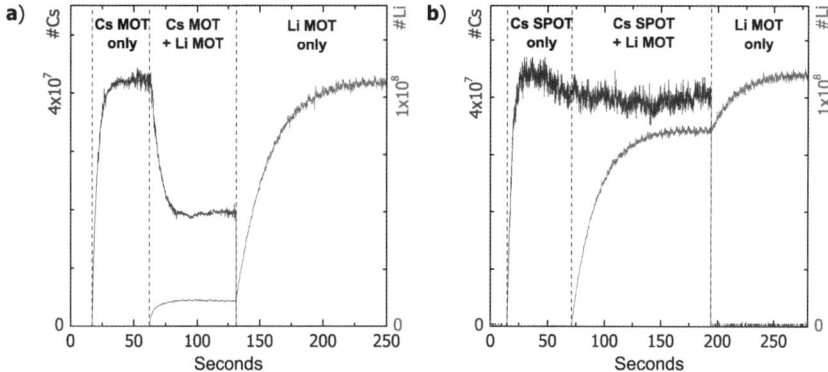

Fig. 1 Loading curves for overlapped lithium and cesium traps: (a) First only the cesium MOT is loaded, then additionally the lithium MOT. The number of trapped cesium atoms is reduced by roughly a factor of two. After blocking the cesium MOT, roughly a 9-fold increase in the number of trapped lithium atoms is observed, indicating very high losses due to Li–Cs collisions. (b) The same sequence with overlapped forced dark SPOT for cesium and identical lithium MOT. Only a weak suppression of trapped atom numbers is observed when loading both traps simultaneously.

and 10^8 Li atoms at densities of 3×10^9 cm^{-3} and 10^{10} cm^{-3} respectively. Using absorption images and time-of-flight expansion we measure a Cs temperature of 250(50) μK. The temperature of the Li atoms lies significantly higher (hundreds of μK) due to the unresolved hyperfine structure of the excited state and the large photon recoil.

2.2 Molecule production and detection

In order to form ultracold molecules, colliding pairs of laser cooled atoms are transferred into bound molecules by photoassociation (PA).[22] Light from a tunable laser is continuously shone on the overlapped atomic clouds. Two colliding atoms can absorb a photon from this laser and form an excited bound molecule. The spontaneous decay of these excited molecules can either lead back to pairs of free atoms (usually with additional kinetic energy) or into bound molecules in the lowest singlet or triplet states. In the system described here, the latter is the dominant process as it will be shown later (Section 4). The light for the PA is provided by a commercial Ti:Sa laser (Coherent MBR-110, pumped by 9 to 12 Watts from a Coherent Verdi V18) and is delivered to the UHV chamber by a high-power optical fibre. After the fibre, the light is collimated to a waist of 1.0 mm, matched to the size of the lithium MOT. It is aligned to pass through the center of the overlapped cesium and lithium MOTs by optimizing the depletion of trapped cesium atoms with resonant light from the Ti:Sa laser. Typical laser powers after the chamber are 400–500 mW for wavelengths between 946 and 852 nm. The wavelength is measured using a home-built wavemeter, calibrated to an atomic cesium resonance with a relative accuracy of 10^{-7}. Additionally we monitor Ti:Sa frequency scans with a reference cavity, which is stabilized *via* an offset-locked diode laser to an atomic cesium resonance. This reference cavity is also used for monitoring long-term drifts of the Ti:Sa laser and to lock it to an arbitrary frequency with a remaining fluctuation of ≤ 2 MHz day^{-1}.

In order to detect the produced ground state molecules, they are first ionized state-selectively by a pulsed laser. The resulting LiCs$^+$ ions are then separated by time-of-flight mass spectrometry from other atomic or molecular ions and are finally detected on a microchannel plate in a single ion counting setup. For details of the time-of-flight mass spectrometer we refer to a previous work.[16] We only note that it allows us to clearly separate Cs⁺ ions from LiCs⁺ ions which have a mass difference of 5%. For ionization, two photons of one color from a pulsed dye laser (Radiant Dyes NarrowScan, Rhodamin B/6G pumped by 532 nm, typically 4 mJ in a beam with a waist of 5 mm and a pulse length of 7 ns, bandwith 0.1 cm^{-1}) are used in a resonant-enhanced multi-photon ionization (REMPI) scheme. The pulsed laser beam passes roughly one beam diameter below the trapped atom clouds in order to reduce excessive ionization of cesium and lithium atoms, which would saturate the detector. Additionally the number of cesium ions is reduced by turning off the repumping light 0.6 ms before the ionization pulse. The depumper used in the forced dark-spot setup quickly pumps all trapped cesium atoms into the lower hyperfine state $F = 3$, so that the trapping laser is now off-resonant and the atoms remain in the electronic ground state from where only three-photon ionization is possible. In contrast to two-photon ionization from the excited $6P_{3/2}$ state, this process is strongly suppressed and therefore the possible effect on the detected LiCs⁺ signal is minimized. After the ionization pulse the repumper is turned on again within 300 μs and nearly all cesium atoms are recaptured in the MOT.

3 Photoassociation spectroscopy

In Fig. 2 we show an individual resonance in a scan of the PA laser. As the PA laser becomes resonant with a transition from a colliding atom pair to an excited molecular level, the number of detected LiCs⁺ ions increases, indicating the formation of ground state molecules. We can indeed be sure that the ion signal arises from ground state molecules and not from excited molecules, as the PA and ionization laser are spatially separated and the molecules do not move significantly over the lifetime of the excited level. We note, however, that in these first experiments the wavelength of the ionization laser was chosen based on a rough estimate only, relying on *ab initio* potentials for the intermediate levels and for the ion potential. The precise ionization mechanism at wavelengths around 14700 cm^{-1} is still under investigation.

The resonance shown in Fig. 2 does not show further substructure, *i.e.* no molecular hyperfine structure, indicating zero electronic angular momentum and therefore an $\Omega = 0^{\pm}$ character. As only a single rotational component was observed for this

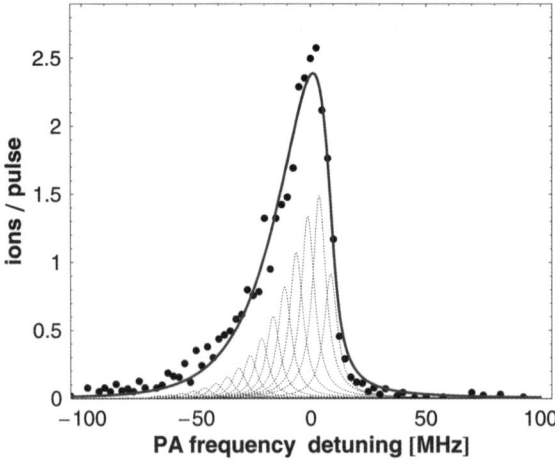

Fig. 2 Temperature-broadened PA resonance (dots) together with a fit of eqn (1) (solid line). Best fit parameters are $T_{rel} = 580(80)$ μK and $\Gamma = 10(3)$ MHz. The thin dashed lines illustrate the contribution to the PA line shape from atom pairs with discrete collisions energies. The frequency axis shows detuning from the setpoint of 11632.11 cm^{-1}. The molecules are ionized by two photons at 14693.7 cm^{-1}.

line, no further assignment is possible. Nevertheless an accurate analysis of the PA line shape can yield important information about the atomic scattering state. More specifically, the asymmetric broadening of the line towards red detunings indicates the influence of the collision energy on the PA line shape and can therefore be used as a measure for the relative temperature of the two species. As discussed by Jones et al.,[23] an accurate model of PA line shapes can be derived from Wigner's law. It states that for low collision energies ε, the amplitude of the scattering wavefunction with angular momentum ℓ scales as $\varepsilon^{(\ell+1/2)/2}$. With the probability of having a collision with energy ε given by the Boltzmann factor $e^{-\varepsilon/T}$, the shape of a PA resonance is then described by

$$W(f, f_0) = B \int_0^\infty e^{-\varepsilon/T} \varepsilon^{(\ell+1/2)} L_\Gamma(f, f_0 - \varepsilon) \, d\varepsilon \qquad (1)$$

where $L_\Gamma(f, f_0)$ is the discrete energy Lorentzian with natural linewidth γ and central frequency f_0. Fig. 2 shows the temperature broadened PA resonance together with a fit of the model from Eq. (1) taking only s-wave scattering ($\ell = 0$) into account. The fit yields realistic values for $T = 580(80)$ µK and $\Gamma = 10(3)$ MHz. Adding higher partial waves did not improve the fit.

Jones et al.[23] also derive a limit on the collision energy ε, for which Wigner's law holds:

$$\frac{\hbar \varepsilon}{k_B} \ll \frac{E_{col}}{k_B} \frac{2}{(0.6 - A_0/R_0)^2}$$

where

$$R_0 = \left(\frac{2\mu C_6}{\hbar^2}\right)^{1/4}$$

and

$$E_0 = \frac{\hbar}{2\mu R_0^2}.$$

For ^{7}Li^{133}Cs the reduced mass μ and the C_6 dispersion coefficient[24] are well known. However there are only two experimental values for the scattering length A_0, namely $A_{0,\text{singlet}} = 50(20)a_0$ from high-resolution spectroscopy of the singlet ground state potential[25] and $A_{0,\text{eff}} = 180(40)a_0$ from the thermalization of ^{7}Li and ^{133}Cs atoms in an optical dipole trap.[26] While the first value yields an upper limit of roughly 0.8 K, well above our experimental conditions, the latter yields 1.7 mK, closer to the fitted temperature. As a second condition the long-range wavefunction of the ground state has to be linear near the outer turning point of the PA level. The PA for the fitted resonance was performed at a detuning of more than 100 cm^{-1} from the asymptote where the excited potential energy curve is still quite steep. Therefore the vibrational wavefunction is very localized at the outer turning point and this second condition should be well fulfilled. We conclude that if the relevant scattering length for our experiment lies between $A_{0,\text{singlet}}$ and $A_{0,\text{eff}}$, the derived collision energy of $T = 580(80)$ µK should be a realistic estimate for our experimental conditions.

This relatively high temperature has to be seen in relation to the centrifugal barrier E_ℓ for collisions with higher angular momentum ℓ:

$$E_\ell = \left(\frac{\hbar^2 \ell(\ell+1)}{3\mu(2C_6)^{1/3}}\right)^{3/2}. \qquad (2)$$

For p-wave scattering this yields $E_{\ell=1} = 1.6$ mK (barrier peaked at $103a_0$), for d-wave scattering $E_{\ell=2} = 8.5$ mK (barrier peaked at $78a_0$), and for f-wave scattering $E_{\ell=3} = 24.7$ mK (barrier peaked at $66a_0$). Our measured collision temperature is well below the p-wave limit, so that one can expect to observe dominantly s-wave scattering. Assuming a Boltzmann distribution of kinetic energies, in roughly 6% of the collisions the energy lies above the p-wave barrier, so that we can expect a small contribution from $\ell = 1$ but none from higher partial waves.

Using the above measured value of the collision temperature, we can draw further conclusions about the temperature of the lithium atoms and give an estimate on the temperature of the formed molecules. As a simple approximation we assume an average collision angle between lithium and cesium atoms of 90°. The relative kinetic energy is then given by $E_{\text{rel}} = \frac{1}{2}\frac{m_{\text{Li}} m_{\text{Cs}}}{m_{\text{Li}} + m_{\text{Cs}}}\left(v_{\text{Li}}^2 + v_{\text{Cs}}^2\right)$ and therefore the relative temperature $T_{\text{rel}} = \frac{m_{\text{Cs}} T_{\text{Li}} + m_{\text{Li}} T_{\text{Cs}}}{m_{\text{Li}} + m_{\text{Cs}}}$ is dominated by the lighter lithium. Solving this equation for T_{Li} with $T_{\text{Cs}} = 250$ µK and $T_{\text{rel}} = 580$ µK yields $T_{\text{Li}} = 600$ µK, which seems reasonable. For the center-of-mass motion of the formed molecules one finds $E_{\text{c.m.}} = \frac{1}{2}(m_{\text{Li}} + m_{\text{Cs}})\left(\frac{m_{\text{Li}} \vec{v}_{\text{Li}} + m_{\text{Cs}} \vec{v}_{\text{Cs}}}{m_{\text{Li}} + m_{\text{Cs}}}\right)^2$. Therefore the molecular temperature $T_{\text{c.m.}} = \frac{m_{\text{Li}} T_{\text{Li}} + m_{\text{Cs}} T_{\text{Cs}}}{m_{\text{Li}} + m_{\text{Cs}}}$ is dominated by the heavier cesium. From the above determined temperatures, T_{Li} and T_{Cs}, we calculate a molecular temperature of 270(60) µK.

3.1 Photoassociation into the $B^1\Pi$ state

The effort of searching for PA resonances, especially at large detunings, can be substantially reduced when spectroscopic information on the excited states of the system under investigation is available. Most experimentally determined potential energy curves are based on the observation of absorption and fluorescence between bound levels, leading to a relative uncertainty of the order of 0.01 cm^{-1} in addition to an uncertainty of the dissociation energy of the molecular ground state of up to 1 cm^{-1} and therefore a corresponding uncertainty in the PA wavelength which refers to the atomic asymptote. Once, however, a few levels have been observed and assigned, this dissociation energy is known with high precision and all other levels can be found easily. If no spectroscopic data is available, *ab initio* calculated molecular potentials can be helpful. With up-to-date methods, the typical uncertainty of the calculated potentials is of the order of 100 cm^{-1}, while the vibrational spacings, related to the shape of the potentials, are determined much more precisely. Therefore, *ab initio* potential curves provide a good estimate of the typical number of expected PA resonances in a given energy range, with an uncertainty for hitting one of them of the order of the vibrational spacing. Once a vibrational progression is observed, they are of importance for the final assignment and the prediction of further vibrational levels.

In Fig. 3(a) *ab initio* curves for selected electronic states of LiCs correlating to the lowest three asymptotes are shown (spin–orbit coupling is neglected in this calculation). For two excited states, the $B^1\Pi$ and the $D^1\Pi$ state, Stein *et al.* have published potential curves derived from high-resolution laser-induced fluorescence spectroscopy.[27] They report strong perturbations of the rovibrational levels in the $D^1\Pi$ state due to spin–orbit coupling to nearby triplet states. These perturbations reduce the probability of decay into the lowest singlet state and also reduce the accuracy of an extrapolation from the highly-excited rotational levels observed in ref. 27 to the lowest rotational levels addressable by PA. In contrast to the $D^1\Pi$ state, the $B^1\Pi$ state showed no significant perturbations, allowing for an extrapolation of level energies to low rotational states with high precision and making spontaneous decay into the $X^1\Sigma^+$ ground state more likely. For the formation of ground state molecules we therefore focused on the $B^1\Pi$ state. Extensive PA scans were performed and PA resonances corresponding to rovibrational levels in $B^1\Pi$ from the atomic asymptote

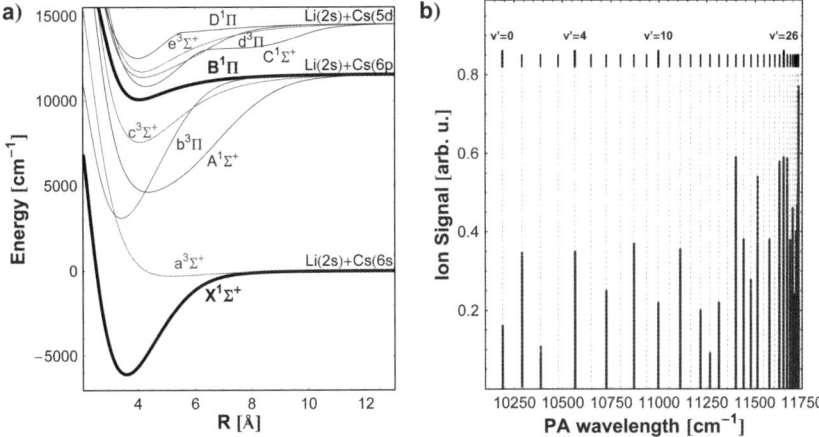

Fig. 3 (a) Potential energy curves for Σ^+ and Π states of LiCs correlating to the three lowest asymptotes. The relevant states $B^1\Pi$ and $X^1\Sigma^+$ are marked bold. (b) Overview over all observed vibrational levels in the $B^1\Pi$ state.

down to $v' = 0$ at ~ 1540 cm^{-1} detuning from this asymptote were observed. For levels below $v' = 25$ we found very good agreement with the energies reported in ref. 27. As the authors of the cited work could only extrapolate from the last observed level around $v' = 25$ towards the asymptote, the predictions became less accurate as we approached the asymptote. Here, additional vibrational levels up to $v' = 35$ were found, yielding a data set for the $B^1\Pi$ state with level energies ranging from the bottom of the potential at $v' = 0$ up to the last level below the asymptote, bound only by ~ 1 GHz. A publication describing the complete $B^1\Pi$ potential energy curve is in preparation.[28] We note that we did not observe line broadening due to predissociation in the region between the fine structure asymptotes Li($2^2S_{1/2}$) + Cs($6^2P_{3/2}$) and Li($2^2S_{1/2}$) + Cs($6^2P_{1/2}$) as suggested by ref. 27. We are currently studying theoretically and experimentally how far predissociation could be relevant for other electronic states correlated to the upper asymptote Li($2^2S_{1/2}$) + Cs($6^2P_{3/2}$).[29]

In Fig. 3(b) an overview over all observed PA resonances in the $B^1\Pi$ state is given. It is not straightforward to compare the strength of the different PA lines, as not all PA resonances could be detected at the same ionization wavelength. Additionally, the ion yield strongly depends on the populated ground state levels, which vary between different PA resonances. However, the observed rate of molecule formation for PA into deeply-bound vibrational levels of $B^1\Pi$ is surprisingly high, as these transitions occur at very short internuclear separations around $R = 4$ Å, where the amplitude of the free scattering wavefunction is already greatly reduced. It is noteworthy, that the Condon points of these deeply-bound levels coincide with the inner turning point of the $a^3\Sigma^+$ potential, which could be an important clue towards a full explanation of the observed high PA rates.¶

The observed PA resonances in the $B^1\Pi$ state show a rich substructure due to molecular hyperfine interactions as expected for a state with electronic angular momentum. Exemplary resonances are shown in Fig. 4. The observed hyperfine splitting generally decreases with increasing J'. We do not expect to observe broadening of the lines due to stray fields, as the electric extraction fields for the detection are switched on only 500 µs before the dye laser pulse, which ionizes all molecules that have been produced within the last 20 ms. Therefore the PA is performed mainly under electric-field-free conditions. Also magnetic fields should not influence the

---

¶ Note added in proof: The observed high PA rates at large detunings are due to the proximity of a Feshbach resonance in the entrance channel, as discussed in ref. 45.

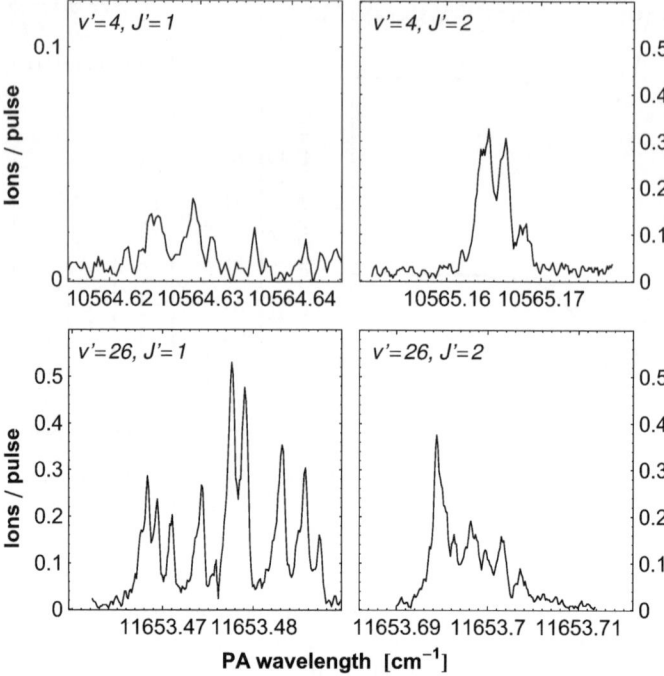

Fig. 4 Photoassociation resonances for different rovibrational levels $B^1\Pi, v', J'$. Abscissas and ordinates are on the same scale for all panels except for $v' = 4$, $J' = 1$, where the ordinate is scaled by a factor of 5 for visibility.

measured PA spectra. The magnetic field gradient for trapping the atoms is applied constantly, but residual fields in the trap center (where the gradient field reaches zero) have been compensated to <1 Gauss.

For low lying levels $v' < 25$, rotational components $J' = 1$ and 2 were observed, while for levels $v' \geq 25$ a weaker $J' = 3$ component and for the last bound levels rotational components up to $J' = 5$ were visible. As the Condon points for the last bound levels lie well within the centrifugal barrier for higher angular momenta in the scattering continuum, the higher rotational components are unlikely to be due to PA from atom pairs with higher angular momentum. The observation of these higher J' components is probably due to the change of coupling between angular momenta as the molecules become less tightly bound. From Fig. 4 one can further learn, that while for the higher lying state the $J' = 1$ component is the stronger one, for $v' = 4$ the $J' = 2$ component is clearly dominant. This also indicates a change in the free–bound coupling from PA at short distances around 5 Å to PA at larger internuclear separations around 9 Å, which is currently under investigation.

4 Calculated probabilities for decay into the $X^1\Sigma^+$ state

After PA, the formed excited molecules decay within a few nanoseconds either into bound molecules or into free atom pairs. In order to estimate the branching ratio between these two competing channels and to predict the distribution of populated vibrational ground state levels, we calculated Franck–Condon (FC) factors and Einstein A coefficients for transitions between the excited and the ground state levels.

As mentioned before, accurate experimental potential energy curves have been published for the $X^1\Sigma^+$ state[25] and for vibrational levels $v' \leq 25$ of the $B^1\Pi$ state[27] of $^7Li^{133}Cs$. The authors of the latter paper included our observed PA resonances in the fit describing the full potential energy curve of the $B^1\Pi$ state,[30] yielding very accurate potential energy curves covering all binding energies. We calculate rovibrational levels in these potentials using the Mapped Fourier Grid Hamiltonian (MFGH) method, which is used to solve the time-independent radial Schrödinger equation numerically. A detailed description of the method may be found elsewhere.[31] As expected, the resulting term energies are in very good agreement with experimental observations (residuals ≤ 0.02 cm^{-1}). These transition energies are used in the identification of ionization resonances (Section 5) and depletion resonances (Section 6). The resulting wavefunctions are used to calculate FC factors for the bound–bound transitions. It is worth mentioning, that for excited states $B^1\Pi, v' \leq 26$ the sum of FC factors with all vibrational levels of the ground state is close to unity. Molecules in these $B^1\Pi$ levels will therefore predominantly decay into bound molecular ground state levels. We also note, that the authors of ref. 27 report only for levels $v' \geq 17$ the observation of weak decay into the $a^3\Sigma^+$ lowest triplet state. One can therefore expect, that levels $B^1\Pi, v' < 17$ decay almost completely towards $X^1\Sigma^+$ levels and that higher levels exhibit only a weak additional decay channel towards $a^3\Sigma^+$ levels.

Using the approach described in ref. 10, the R-dependent dipole transition moment $\mu(R)$ was calculated for the electronic transition $X^1\Sigma^+$–$B^1\Pi$. The transition dipole moment does not vary by more than 10% around a value of 9.9 Debye over the relevant range of distances. By calculating the Einstein A coefficients for spontaneous decay $A_{ij} = \frac{2}{3}\frac{\omega_{ij}^3}{\varepsilon_0 hc^3}|\langle\Psi_i|\mu(R)|\Psi_j\rangle|^2$ with ω_{ij} being the transition frequency and Ψ_i the nuclear wavefunction of the vibrational level i, one can derive the relative population of ground state vibrational levels after PA *via* different excited states. In Fig. 5(b) the normalized population of $X^1\Sigma^+$ levels after PA *via* $B^1\Pi, v' = 4$ and $B^1\Pi, v' = 26$ is compared. While in both cases low-lying vibrational levels are populated, we note that after PA *via* $B^1\Pi, v' = 4$, more than 20% of the excited molecules should decay into the lowest vibrational level $v'' = 0$. Fig. 5(a) shows the small influence of a calculation with the R-dependent *versus* a constant dipole moment function for this calculation.

5 Ionization spectroscopy of $X^1\Sigma^+$

The formation of ground state molecules is probed by ionizing them *via* resonantly-enhanced multiphoton ionization (REMPI) and subsequent detection of the molecular ions (see Section 2.2). Once the formation of molecules by the PA laser has been detected, a scan of the ionization laser yields information about the distribution of populated vibrational levels. Extracting this information requires accurate knowledge of the involved ground and intermediate electronic states. In the case of LiCs, deeply-bound $X^1\Sigma^+$ levels $v'' \leq 4$ can be ionized *via* intermediate levels in the $B^1\Pi$ state, which allows us to assign observed ionization resonances to ground and intermediate state vibrational levels.

As discussed in the previous section, different PA resonances are expected for different relative populations of ground state levels. This can be seen very clearly in the three REMPI scans of Fig. 6 covering the same range of ionization energies after PA *via* different levels in the $B^1\Pi$ state. Although the assignment of resonances to ground state levels is often ambiguous and a number of unpredicted resonances indicate the importance of intermediate states other than $B^1\Pi$, quite a few clear progressions are visible. In Fig. 6(a), resonances from $X^1\Sigma^+, v'' = 0$ to $B^1\Pi, v' = 13$–17 and $X^1\Sigma^+, v'' = 2$ to $B^1\Pi, v' = 17$–21 can be identified; in Fig. 6(b), in addition to the resonances originating from $v'' = 2$, resonances from $X^1\Sigma^+, v'' = 1$ to $B^1\Pi, v' = 14$–21 are visible; and in Fig. 6(c), in addition to molecules in $v'' = 1$ and $v'' = 2$, also molecules in $v'' = 3$ are ionized through $B^1\Pi, v' = 22$–29. The appearance of these "ionization windows" in the intermediate $B^1\Pi$ state demonstrates the importance of the bound–bound transition dipole moment and of the level-dependent ionization probability of the $B^1\Pi$ state, which lead to a selection of ionization pathways.

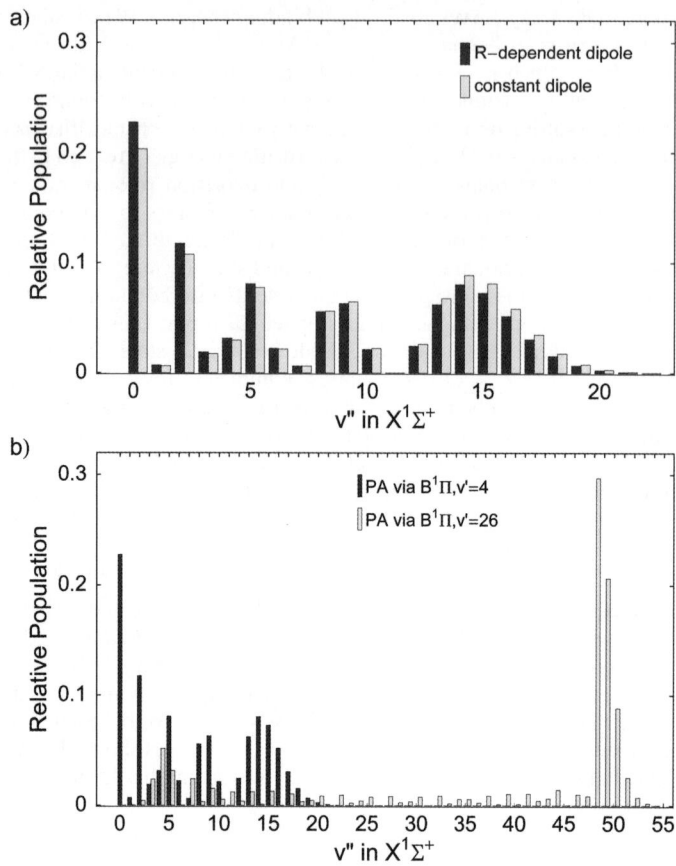

Fig. 5 Calculated $X^1\Sigma^+$ level populations after PA, normalized to unity: (a) influence of the R-dependence of the transition dipole moment on the expected distribution after PA via $B^1\Pi, v' = 4$; (b) comparison of the expected level population after PA via $B^1\Pi, v' = 4$ and $v' = 26$.

The different and unknown ionization probabilities for different intermediate $B^1\Pi$ levels as well as near-degeneracy of many transitions complicate the analysis of the relative populations of vibrational ground state levels from the observed ionization spectra. However, as shown in Fig. 7, we can establish at least a qualitative agreement with the predictions of Section 4. In Fig. 7, the average peak value for each of the above identified transition bands originating from ground state levels $v'' = 0$–3 is shown. While the unknown ionization probabilities of intermediate levels impede the direct extraction of the level population from the measured spectra, we can still deduce the population strength of a single level after PA via different excited levels and compare it with our prediction. As can be seen in Fig. 7, the relative ordering of the line strengths for each level v'' is described correctly by the theory, while the relative ratios show strong deviations. This is very likely due to saturation and a non-optimal signal-to-noise ratio.

The identification of the observed ionization resonances also yields information about the lowest state of the ion LiCs$^+$, the $X^2\Sigma^+$ state. The energy of two photons from $X^1\Sigma^+, v'' = 0$ (via intermediate level $B^1\Pi, v' = 13$) leads to ionization at a detuning of -3385 cm^{-1} from the Li(2 $^2S_{1/2}$) + Cs$^+$($^1S_0$) threshold at 31406.71 cm^{-1}.[32] Therefore, the $X^2\Sigma^+$ state of the ion has a depth of at least 3385 cm^{-1}, favoring the predictions of Von Szentpaly *et al.* ($D_e = 3468$ cm^{-1})[33], Korek *et al.*

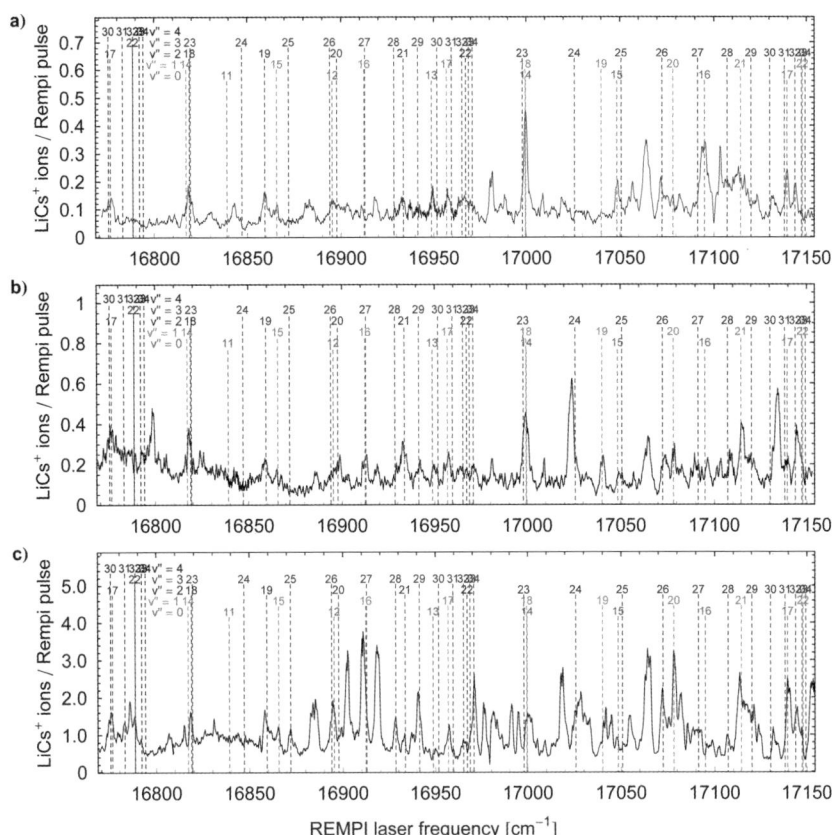

Fig. 6 Resonantly-enhanced two-photon ionization spectra of ground state molecules after photoassociation *via* (a) $B^1\Pi, v' = 4, J' = 2$, (b) $v' = 10, J' = 2$, and (c) $v' = 26, J' = 2$ over the same range of ionization energies. In the upper part of the graph, expected transition series originating from all levels v'' in $X^1\Sigma^+$ to all levels v' in $B^1\Pi$ are marked (note that this cannot yield a full assignment of the observed lines, as different ionization probabilities and other intermediate states are neglected). Wavelengths calibrated with a High Finesse WS7 wavemeter with an accuracy $\ll 0.1$ cm^{-1}.

($D_e = 3578$ cm^{-1})[34] and Azizi *et al.* ($D_e = 3564$ cm^{-1})[35] over calculations of Patil and Tang ($D_e = 2662$ cm^{-1}),[36] and Bellomonte *et al.* ($D_e = 2984$ cm^{-1}).[37]

6 High resolution spectroscopy of the vibrational ground state

After having identified vibrational ground state levels by selective ionization of these levels, we can further probe the ground state molecules formed by applying depletion spectroscopy.[38] While the Ti:Sa laser is stabilized to a PA resonance (see Section 3) and the pulsed dye laser is kept on resonance for the ionization of a certain ground state level, a narrow-band laser is used to optically pump the molecules out of this specific level and therefore reduce the number of detected ions. As this laser (Radiant Dyes cw dye laser, Rhodamin 6G pumped at 532 nm) has a bandwidth of the order of a few MHz, it allows one to resolve rotational transitions between ground and excited molecular states. The light from the laser is coupled into a fibre and, after being collimated to a waist of 0.7 mm, is aligned collinear with the PA-laser by maximizing the depletion of the ion signal on resonance. The frequency of this laser is measured with a wavemeter (Burleigh WA-1000, relative stability 500 MHz) and

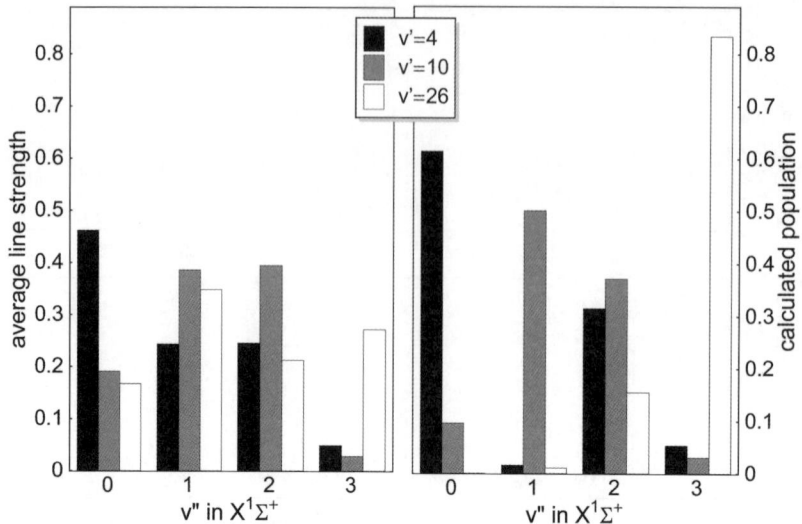

Fig. 7 Comparison of detected ground state levels for different photoassociation resonances with predictions. Left panel: averaged peak value for the transition bands originating in $X^1\Sigma^+, v' = 0$–3 as discussed in the text (normalized to unity for every PA resonance); Right panel: calculated ground state population, normalized to the summed population in levels $v' = 0$–3 (see Section 4).

scans are monitored additionally with a reference cavity (FSR = 750 MHz), which is actively stabilized to an atomic cesium resonance *via* an offset-locked diode laser. This high-resolution spectroscopy of ground state molecules allowed us to confirm assignments made in the interpretation of REMPI spectra and to confirm the production of $X^1\Sigma^+, v'' = 0$ molecules after a single PA step.[9]

The depletion spectroscopy makes it possible to gain further information about the $X^1\Sigma^+$ potential energy curve. As shown in Fig. 8, the $B^1\Pi, v' = 12, J' = 1$ level was observed in PA spectroscopy starting from the $Li(2\ ^2S_{1/2}, F = 2) + Cs(6\ ^2S_{1/2}, F = 3)$ asymptote as well as in depletion spectroscopy starting from $X^1\Sigma^+, v'' = 0, J'' = 2$. Using the calculated rotational constant $B_{v''=0} = 0.1874$ cm^{-1} for the ground state, this allows us to give a hyperfine averaged value for the binding energy of the lowest vibrational level $D_0^X = 5783.53(3)$ cm^{-1}. This agrees within the error bars with the value of $D_0^X = 5783.4(1)$ cm^{-1} given by Staanum *et al.*[25]

7 An efficient route towards a dense gas of absolute ground state molecules

In this article we have discussed in detail the steps in the formation and detection of ultracold polar LiCs molecules in the vibrational ground state. We have shown that a single absorption–emission cycle is sufficient to produce the molecules in the lowest vibrational state. Previously we have demonstrated the formation of ultracold LiCs molecules in the rotational and vibrational ground state $X^1\Sigma^+, v'' = 0, J'' = 0^9$ in a single step of PA. In the future we will repeat these experiments with lithium and cesium atoms trapped in a quasi-electrostatic trap (QUEST) formed by a focused CO_2 laser at 10.6 μm. This very large detuning from any atomic or molecular resonance makes the absorption of a trap laser photon by LiCs molecules very unlikely: a transition from deeply-bound molecules into the first excited state requires roughly ten photons from the trap laser. Due to the permanent dipole moment of the polar molecules, transitions within an electronic state are, in principle, also possible; however, they are very unlikely due to extremely small Franck–Condon factors for such transitions.

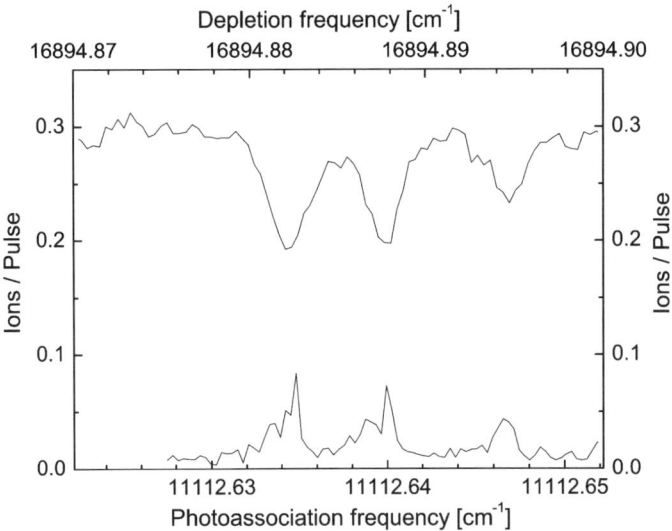

Fig. 8 Depletion scan of $X^1\Sigma^+, v''=0, J''=2$ molecules via $B^1\Pi, v'=12, J'=1$ (upper trace) aligned to a photoassociation scan of the same excited level (lower trace). Depletion laser intensity 26 mW cm^{-2}, photoassociation laser intensity 25 W cm^{-2}. Molecules detected at different ionization wavelengths.

From a calculation of the static polarizability of LiCs molecules in $X^1\Sigma^+, v''=0$[39] and previous results obtained with this setup,[40] we expect a trap depth of the order of 600 μK for the ground state molecules in $v''=0$. The calculation shows that the trap depth increases for higher vibrational levels, so that all the ground state molecules produced are expected to be trapped. While the molecules in $v''=0, J''=0$ are expected to be stable against collisions with other atoms and molecules in the same state, collisions with vibrationally excited molecules can lead to a loss of $v''=0$ molecules. It will therefore be useful to apply the PA laser only as long as the number of atoms in the trap significantly exceeds the number of produced molecules, in which case any molecule would most likely collide with a trapped atom and not with a molecule. As demonstrated for Cs$_2$ + Cs collisions[41,42] and RbCs + Cs/Rb collisions,[43] such collisions lead to a fast removal of vibrationally excited molecules from the trap. After an accumulation time of the order of seconds, one would end up with a relatively pure sample of molecules in the rovibrational ground state. Optimistically, 20% of the trapped atoms could be transferred into molecules in the rovibrational ground state. Starting from atomic densities of the order of 10^{11} cm^{-3} and atomic temperatures around 20 μK,[26] one can therefore hope to reach molecular samples with densities of the order of 10^{10} cm^{-3} and temperatures also around 20 μK. While this is far away from a degenerate gas of polar molecules, the long-range dipolar interactions will already lead to a strongly interacting system at these densities. After stabilizing the polar gas against inelastic collisions due to these interactions, *e.g.* via a "blue shield" for molecules[44] or by preparing aligned, 2-dimensional samples in combined static electric fields and strong laser fields,[39] the realization of a new quantum phase like a dipolar crystal[12] seems feasible.

Acknowledgements

We thank S. D. Kraft and P. Staanum for contributions at the early stage of the experiment. We also thank E. Tiemann and A. Pashov for providing experimental LiCs potentials before publication and, together with J. Hutson, for fruitful discussions. This work is supported by the DFG under WE2661/6-1 in the framework of the Collaborative Research Project QuDipMol within the ESF EUROCORES EuroQUAM program. JD acknowledges partial support of the French-German University. AG is a postdoctoral fellow of the Alexander von Humboldt-Foundation.

References

1 J. Doyle, B. Friedrich, R. V. Krems and F. Masnou-Seeuws, *Eur. Phys. J. D*, 2004, **31**, 149.
2 O. Dulieu, M. Raoult and E. Tiemann, *J. Phys. B*, 2006, **39**(19), Introductory review.
3 J. G. Danzl, E. Haller, M. Gustavsson, M. J. Mark, R. Hart, N. Bouloufa, O. Dulieu, H. Ritsch and H.-C. Nägerl, *Science*, 2008, **321**, 1062; H.-C. Nägerl, private communication.
4 K.-K. Ni, S. Ospelkaus, M. H. G. de Miranda, A. Pe'er, B. Neyenhuis, J. J. Zirbel, S. Kotochigova, P. S. Julienne, D. S. Jin and J. Ye, *Science*, 2008, **322**, 231.
5 F. Lang, K. Winkler, C. Strauss, R. Grimm and J. H. Denschlag, *Phys. Rev. Lett.*, 2008, **101**, 133005.
6 A. N. Nikolov, J. R. Ensher, E. E. Eyler, H. Wang, W. C. Stwalley and P. L. Gould, *Phys. Rev. Lett.*, 2000, **84**, 246.
7 J. M. Sage, S. Sainis, T. Bergeman and D. DeMille, *Phys. Rev. Lett.*, 2005, **94**, 203001.
8 M. Viteau, A. Chotia, M. Allegrini, N. Bouloufa, O. Dulieu, D. Comparat and P. Pillet, *Science*, 2008, **321**, 232.
9 J. Deiglmayr, A. Grochola, M. Repp, K. Mörtlbauer, C. Glück, J. Lange, O. Dulieu, R. Wester and M. Weidemüller, *Phys. Rev. Lett.*, 2008, **101**, 133004.
10 M. Aymar and O. Dulieu, *J. Chem. Phys.*, 2005, **122**, 204302.
11 A. Micheli, G. K. Brennen and P. Zoller, *Nat. Phys.*, 2006, **2**, 341.
12 G. Pupillo, A. Griessner, A. Micheli, M. Ortner, D.-W. Wang and P. Zoller, *Phys. Rev. Lett.*, 2008, **100**, 050402.
13 P. Rabl, D. DeMille, J. M. Doyle, M. D. Lukin, R. J. Schoellkopf and P. Zoller, *Phys. Rev. Lett.*, 2006, **97**, 033003.
14 T. Zelevinsky, S. Kotochigova and J. Ye, *Phys. Rev. Lett.*, 2008, **100**, 043201.
15 T. V. Tscherbul and R. V. Krems, *Phys. Rev. Lett.*, 2006, **97**, 083201.
16 S. D. Kraft, P. Staanum, J. Lange, L. Vogel, R. Wester and M. Weidemüller, *J. Phys. B*, 2006, **39**, S993.
17 S. D. Kraft, J. Mikosch, P. Staanum, J. Deiglmayr, J. Lange, A. Fioretti, R. Wester and M. Weidemüller, *Appl. Phys. B*, 2007, **89**, 453.
18 M. W. Mancini, A. R. L. Caires, G. D. Telles, V. S. Bagnato and L. G. Marcassa, *Eur. Phys. J. D*, 2004, **30**, 105.
19 U. Schlöder, H. Engler, U. Schünemann, R. Grimm and M. Weidemüller, *Eur. Phys. J. D*, 1999, **7**, 331.
20 W. Ketterle, K. B. Davis, M. A. Joffe, A. Martin and D. E. Pritchard, *Phys. Rev. Lett.*, 1993, **70**, 2253.
21 C. G. Townsend, N. H. Edwards, K. P. Zetie, C. J. Cooper, J. Rink and C. J. Foot, *Phys. Rev. A*, 1996, **53**, 1702.
22 K. M. Jones, E. Tiesinga, P. D. Lett and P. S. Julienne, *Rev. Mod. Phys.*, 2006, **78**, 483.
23 K. M. Jones, P. D. Lett, E. Tiesinga and P. S. Julienne, *Phys. Rev. A*, 1999, **61**, 012501.
24 M. Marinescu, H. R. Sadeghpour and A. Dalgarno, *Phys. Rev. A*, 1994, **49**, 982.
25 P. Staanum, A. Pashov, H. Knöckel and E. Tiemann, *Phys. Rev. A*, 2007, **75**, 042513.
26 M. Mudrich, S. Kraft, K. Singer, R. Grimm, A. Mosk and M. Weidemüller, *Phys. Rev. Lett.*, 2002, **88**, 253001.
27 A. Stein, A. Pashov, P. Staanum, H. Knöckel and E. Tiemann, *Eur. Phys. J. D*, 2008, **48**, 177.
28 A. Grochola, A. Pashov, J. Deiglmayr, M. Repp, R. Wester, and M. Weidemüller, submitted.
29 A. Grochola, J. Deiglmayr, R. Wester, M. Weidemüller, and O. Dulieu, in preparation.
30 A. Pashov, private communication.
31 V. Kokoouline, O. Dulieu, R. Kosloff and F. Masnou-Seeuws, *J. Chem. Phys.*, 1999, **110**, 9865.
32 C. E. Moore, *Atomic Energy Levels, vol. III*, National Bureau of Standards, Washington, DC, 1958.
33 L. Von Szentpaly, P. Fuentealba, H. Preuss and H. Stoll, *Chem. Phys. Lett.*, 1982, **93**, 555.
34 M. Korek, A. Moghrabi, A. Allouche and M. Aubert Frécon, *Can. J. Phys.*, 2006, **84**, 959.
35 S. Azizi, M. Aymar and O. Dulieu, *AIP Conf. Proc.*, 2007, **935**, 164.
36 S. H. Patil and K. T. Tang, *J. Chem. Phys.*, 2000, **113**, 676.
37 L. Bellomonte, P. Cavaliere and G. Ferrante, *J. Chem. Phys.*, 1974, **61**, 3225.

38 D. Wang, J. T. Kim, C. Ashbaugh, E. E. Eyler, P. L. Gould and W. C. Stwalley, *Phys. Rev. A*, 2007, **75**, 032511.
39 J. Deiglmayr, M. Aymar, R. Wester, M. Weidemüller and O. Dulieu, *J. Chem. Phys.*, 2008, **129**, 064309.
40 A. Mosk, S. Kraft, M. Mudrich, K. Singer, W. Wohlleben, R. Grimm and M. Weidemüller, *Appl. Phys. B*, 2001, **73**, 791.
41 P. Staanum, S. D. Kraft, J. Lange, R. Wester and M. Weidemüller, *Phys. Rev. Lett.*, 2006, **96**, 023201.
42 N. Zahzam, T. Vogt, M. Mudrich, D. Comparat and P. Pillet, *Phys. Rev. Lett.*, 2006, **96**, 023202.
43 E. R. Hudson, N. B. Gilfoy, S. Kotochigova, J. M. Sage and D. DeMille, *Phys. Rev. Lett.*, 2008, **100**, 203201.
44 A. V. Gorshkov, P. Rabl, G. Pupillo, A. Micheli, P. Zoller, M. D. Lukin and H. P. Büchler, *Phys. Rev. Lett.*, 2008, **101**, 073201.
45 J. Deiglmayr, P. Pellegrini, A. Grochola, M. Repp, R. Côté, O. Dulieu, R. Wester and M. Weidemüller, *New J. Phys.*, 2009, **11**, in press.

PAPER | www.rsc.org/faraday_d | Faraday Discussions

Ultracold polar molecules near quantum degeneracy

S. Ospelkaus,[a] K.-K. Ni,[a] M. H. G. de Miranda,[a] B. Neyenhuis,[a] D. Wang,[a] S. Kotochigova,[b] P. S. Julienne,[c] D. S. Jin[*a] and J. Ye[*a]

Received 27th November 2008, Accepted 16th December 2008
First published as an Advance Article on the web 27th May 2009
DOI: 10.1039/b821298h

We report the creation and characterization of a near quantum-degenerate gas of polar ^{40}K–^{87}Rb molecules in their absolute rovibrational ground state. Starting from weakly bound heteronuclear KRb Feshbach molecules, we implement precise control of the molecular electronic, vibrational, and rotational degrees of freedom with phase-coherent laser fields. In particular, we coherently transfer these weakly bound molecules across a 125 THz frequency gap in a single step into the absolute rovibrational ground state of the electronic ground potential. Phase coherence between lasers involved in the transfer process is ensured by referencing the lasers to two single components of a phase-stabilized optical frequency comb. Using these methods, we prepare a dense gas of 4×10^4 polar molecules at a temperature below 400 nK. This fermionic molecular ensemble is close to quantum degeneracy and can be characterized by a degeneracy parameter of $T/T_F = 3$. We have measured the molecular polarizability in an optical dipole trap where the trap lifetime gives clues to interesting decay mechanisms. Given the large measured dipole moment of the KRb molecules of 0.5 Debye, the study of quantum degenerate molecular gases interacting *via* strong dipolar interactions is now within experimental reach. PACS numbers: 37.10.Mn, 37.10.Pq.

I. Introduction

Ultracold molecular quantum systems promise to open exciting new scientific directions ranging from ultra-cold chemistry[1,2] and precision measurements[3–5] to the realization of novel quantum many-body systems.[6] In particular, polar molecules, which possess permanent electric dipole moments, have very unique perspectives as systems with long-range and anisotropic interactions.[7] This is in sharp contrast to the short-range contact interaction dominating atomic Bose–Einstein condensates and degenerate Fermi gases. The strong dipole–dipole interaction between polar molecules adds intriguing and entirely new aspects to the physics of ultracold quantum matter. By means of external static electric fields, dipoles can be oriented along the field axis. This allows for precise control of interactions between the constituents of the quantum gas, resulting in novel dynamics such as anisotropic collision resonances.[8,9] When these dipoles are lined up in a lattice geometry provided by optical potentials, the long-range interaction introduces complex quantum dynamics that can be

[a]*JILA, National Institute of Standards and Technology, Department of Physics, University of Colorado, Boulder, CO 80309-0440, USA. E-mail: junye@jilau1.colorado.edu; jin@jilau1.colorado.edu*
[b]*Physics Department, Temple University, Philadelphia, PA 19122-6082, USA*
[c]*Joint Quantum Institute, NIST and University of Maryland, Gaithersburg, MD 20899-8423, USA*

precisely controlled *via* the field strength and orientation. These types of control of interactions have been the basis for numerous exciting theoretical proposals ranging from quantum phase transitions[6] to novel systems for quantum information processing[10–12] and quantum control with external magnetic and electric fields.[13]

However, the observation of dipolar interactions in a molecular gas is challenging and requires the dipolar interaction energy of the molecular gas to be at least comparable to its kinetic energy. This requirement can be fulfilled only with an ultracold, high-density gas of polar molecules with a reasonably large electric dipole moment, namely a polar molecular gas near quantum degeneracy.

Experimental approaches towards the realization of a quantum degenerate gas of ultracold polar molecules have generally followed two different strategies. The first approach is to develop experimental techniques for direct cooling of ground-state polar molecules. In recent years, research in this direction has resulted in powerful techniques such as buffer gas cooling,[14] Stark deceleration,[15,16] or velocity filtering.[17] However, direct cooling strategies have typically been restricted to the millikelvin temperature range and very low phase space densities that are a trillion times away from quantum degeneracy. For further advances in this direction, laser cooling or collisional cooling of molecules will have to be introduced.[18] The second approach is to make use of the powerful cooling techniques for atoms and subsequently convert dual-species atom pairs to deeply bound polar molecules. The challenges in this context are to make this conversion process efficient and to effectively remove the binding energy of the deeply bound molecules without heating the resulting molecular ensemble. In brief, the desirable high phase space density of the initial atomic ensemble has to be preserved during the conversion process. Photoassociation of ultracold atoms provides an efficient way of removing the binding energy of molecules by means of spontaneously emitted photons.[19–21] However, the spontaneous processes result in a large spread of population among many continuum and bound molecular states, leading to significant dilution of the initial atomic phase space density. As a result, typical molecular phase space densities at the end of this process are comparable to the current best results from direct cooling of ground state molecules.

Recently, we reported a novel approach to efficiently convert a near quantum-degenerate atomic gas into a near quantum-degenerate polar molecular gas in the absolute rovibrational ground state.[22] A heteronuclear quantum gas mixture of fermionic ^{40}K and bosonic ^{87}Rb is first efficiently converted into weakly bound Feshbach molecules in the vicinity of a Fano–Feshbach resonance.[23–25] This initial conversion process is efficient. The system is now prepared in a single molecular quantum state, even though it is a highly excited vibrational level. Following this initial association process, we then use highly phase-coherent laser fields to precisely and efficiently transfer populations between internal states of specific electronic, vibrational and rotational degrees of freedom.[26,27] For our specific goal, we convert the weakly bound Feshbach molecules into the absolute ground state, namely the zero vibration and zero rotation quantum level in the singlet electronic ground-state molecular potential. Given the large energy gap between the near-threshold Feshbach molecules and the rovibrational ground state, phase coherence between the lasers used in the transfer process can only be ensured by referencing to individual components of a phase-stabilized optical frequency comb.[28] In this article we report the further refinement of our technique after the initial demonstration[22] and the preliminary characterization of some of the properties for absolute ground-state polar molecules near quantum degeneracy.

II. Experimental techniques and challenges

The starting point for our approach is a near quantum degenerate gas mixture of fermionic ^{40}K atoms and bosonic ^{87}Rb atoms confined in a single beam far-off-resonance optical dipole trap. A DC magnetic-field is ramped across a Fano–Feshbach resonance at 546.7 G to associate pairs of K and Rb atoms into extremely weakly bound heteronuclear KRb molecules.[24,25] This association process converts up to 25% of the atoms into weakly bound heteronuclear molecules. We create up to 5 × 10^4 heteronuclear Feshbach molecules with a binding energy of $h \times 230$ kHz, where h is Planck's constant. The resulting trapped gas of fermionic near-threshold molecules has a density of $n = 5 \times 10^{11}$ cm^{-3} and an expansion energy of $T = 400(15)$ nK. The gas of Feshbach molecules is thus close to quantum degeneracy with $T/T_F \approx 3$. Here T_F is the Fermi temperature of the molecules.

The next challenge is to transfer the weakly bound molecules into their absolute ground state. We emphasize that our goal is transfer to a single internal quantum state without any heat dissipation to external motional states. Hence, a fully coherent conversion process is key. The conversion efficiency is also of paramount importance in order to preserve the initial high phase-space density. As shown in Fig. 1, a coherent Raman transfer process involving an electronically excited intermediate state can be envisioned for the transfer process between the initial and final vibration levels. Due to the vastly mismatched vibrational wave-functions of the weakly bound and the absolute ground-state molecules, it was initially regarded as nearly impossible to find a suitable intermediate state that can provide reasonable resonance strengths for both the upward and downward transitions. In fact, a theory proposal was made to employ a train of two-color, phase-coherent pulses that would allow coherent accumulations of a pump-dump process to implement a fully coherent molecular conversion process.[29] The properly time-delayed pump-dump sequence takes advantage of the excited vibration dynamics to enhance the

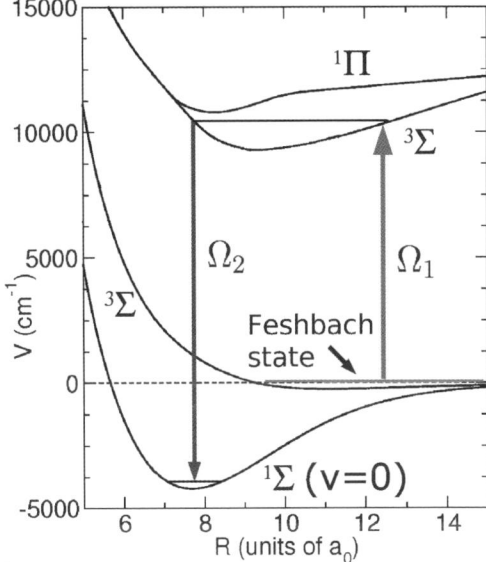

Fig. 1 Sketch of the KRb electronic ground and excited molecular potentials involved in the two-photon coherent state transfer from the Feshbach molecules to the rovibrational ground state $X^1\Sigma_0(v = 0)$. We choose the vibrational level $v = 23$ of the $2^3\Sigma_1$ excited state molecular potential as the intermediate coupling state. This state has favorable Franck–Condon factors to both the initial Feshbach state and the $X^1\Sigma_0$ rovibrational ground state. In addition, its proximity to the bottom of the $1^1\Pi_1$ excited state molecular potential ensures the necessary singlet triplet mixing to convert predominantly triplet character Feshbach molecules to singlet rovibrational ground state molecules. Laser 1, connecting the Feshbach state to the intermediate excited state, is operating at 970 nm. Laser 2, connecting the intermediate state to the rovibrational ground state, operates at 690 nm. The Raman laser system therefore bridges the binding energy difference between Feshbach and rovibrational ground state molecules of $\Delta E_B \approx h \times 125$ THz.

transition probability while the coherent accumulation gradually selects a single final state. Systematic and detailed single photon spectroscopy ensued, connecting the initial Feshbach state to specific electronically excited states with a CW laser referenced to an optical frequency comb. An intense theory-experiment collaboration soon led us to the realization that we can vastly reduce the complexity of the problem by using a single stationary intermediate state instead of dynamic wave-packets. As indicated in Fig. 1, the key is to find an intermediate state where the inner turning point closely matches the Condon point for the downward transition (labeled by the field Ω_2) and the outer turning point lies close to the Condon point for the upgoing transition (labeled by Ω_1). In doing so, the chosen intermediate state provides favorable Franck–Condon factors to both the initial weakly bound Feshbach state and the rovibrational ground state. In addition, the intermediate state lies in proximity to the crossing of $2^3\Sigma$ and $1^1\Pi$ excited potentials, which ensures the necessary singlet–triplet mixing to couple Feshbach molecules with a predominantly triplet character to the singlet rovibrational ground state. We believe that such a suitable intermediate state can always be found for any bi-alkali heteronuclear molecular system, thus making our approach a universally applicable technique for the production of different kinds of bi-alkali polar molecules.

A Raman laser system consists of two different-colored CW lasers of reliable mutual phase coherence. The difference frequency between the two Raman lasers needs to bridge an energy gap of 125 THz, corresponding to the binding energy of the absolute rovibrational ground state of the KRb molecule. This challenging requirement can only be met by referencing and individually phase-locking the two lasers to a femtosecond optical frequency comb,[28] which itself is phase locked to an ultra-stable Nd:YAG laser at 1064 nm. We use the dark resonance technique[30] to search and precisely determine the energy position of the rovibrational ground state. In particular, we determine the binding energy of the rovibrational ground state $X^1\Sigma_0$ ($v = J = 0$) to be 125.319703(1) THz.[22] The relevant single-photon transition strength is also determined in the dark resonance two-photon spectroscopy.

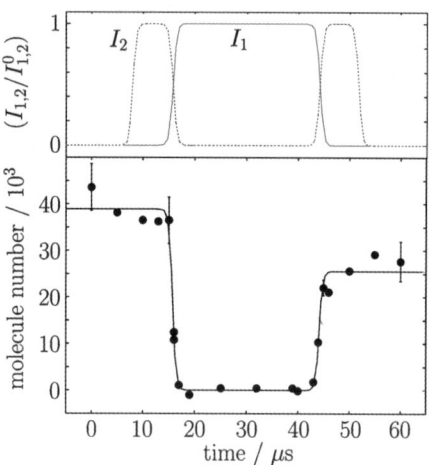

Fig. 2 Time evolution of the coherent two-photon transfer (STIRAP) from Feshbach molecules to the absolute rovibrational ground state $X^1\Sigma_0$ ($v = 0$). (a) Counterintuitive STIRAP pulse sequence, here I_1 and I_2 are the intensities of laser 1 (red solid line) and laser 2 (blue dashed line), respectively (see Fig. 1). (b) Measured population in the initial Feshbach state during the STIRAP pulse sequence. Starting with about 4×10^4 Feshbach molecules, the molecules are coherently transferred to the rovibrational ground state $X^1\Sigma_0$ ($v = 0$) by the first pulse sequence ($t = 15$ to 20 μs). The rovibrational ground state molecules are invisible to the detection light. Reversing the pulse sequence, $X^1\Sigma_0$ ($v = 0$) molecules are converted back to weakly bound Feshbach molecules ($t = 45$ μs to $t = 50$ μs).

For population transfer, we use a single step of STIRAP (Stimulated Raman Adiabatic Passage)[31] to convert Feshbach molecules into rovibrational ground state $X^1\Sigma_0$ ($v = J = 0$) molecules, as shown schematically in Fig. 1. Fig. 2 displays the coherent two-photon transfer (STIRAP) process from Feshbach molecules to the absolute rovibrational ground state. The top panel depicts the counterintuitive STIRAP pulse sequence, where the field of Ω_2 is turned on first to establish coherence between the intermediate and the final target states. As Ω_2 ramps down, field Ω_1 ramps up, transferring the population adiabatically from the initial Feshbach state to the final ground-state target. To monitor the population transfer process, we measure the population in the initial Feshbach state during the STIRAP pulse sequence by direct resonant absorption imaging.[24,32] The rovibrational ground state molecules are invisible to the detection light. The population evolution is shown in the bottom panel of Fig. 2. About 4×10^4 Feshbach molecules are coherently transferred to the rovibrational ground state $X^1\Sigma_0$ ($v = 0$) by the first pulse sequence ($t = $ 15 to 20 μs), resulting in a near-zero signal for detection of Feshbach molecules after the first pulse. To demonstrate that these molecules are indeed hidden in the absolute ground state, we reverse the pulse sequence and convert the $X^1\Sigma_0$ ($v = 0, J = 0$) molecules back to weakly bound Feshbach molecules ($t = 40$ μs to $t = 45$ μs), allowing for direct detection again.

III. Characterization of the transfer process

The fraction of the returned molecules gives the efficiency of the round-trip STIRAP process. To optimize the STIRAP process, we precisely scan the frequency difference between the two lasers so as to search for the two-photon Raman resonance. Fig. 3 shows a typical STIRAP lineshape, where the number of Feshbach molecules returning after a full round-trip STIRAP sequence is measured as a function of the two-photon Raman laser detuning δ. Starting with 4.5×10^4 Feshbach molecules (indicated by the blue dashed line in the figure), we recover about 3.6×10^4 Feshbach molecules after a round-trip STIRAP. This corresponds to a transfer efficiency of 80% for the round-trip transfer, suggesting a transfer efficiency to ground-state polar molecules of 90%. To prove that this transfer process is fully coherent and does not result in any motional heating of the molecular sample, we compare the kinetic energy of the absolute ground-state molecules *versus* the kinetic energy of the initial Feshbach molecules. This comparison is accomplished by time-of-flight expansion measurement on the Feshbach molecules before the STIRAP transfer and after a round-trip STIRAP process. Fig. 4 documents the time-of-flight expansion

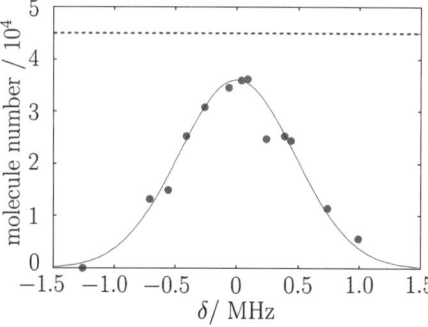

Fig. 3 STIRAP lineshape. Plotted is the number of Feshbach molecules returning after a full round-trip STIRAP sequence as a function of the two-photon Raman laser detuning δ. Starting with 4.5×10^4 Feshbach molecules (blue dashed line), we recover about 3.6×10^4 Feshbach molecules after a round-trip STIRAP. This corresponds to a transfer efficiency of 80% for the round-trip transfer suggesting a creation efficiency of ground state polar molecules of 90%.

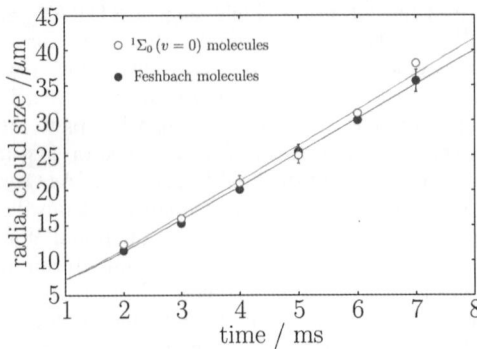

Fig. 4 Comparison of the kinetic energy of the Feshbach molecules before STIRAP transfer (filled circles) and after a round-trip STIRAP process (open circles). The latter can be interpreted as an upper limit on the kinetic energy of the rovibrational ground state molecules. The temperature of both clouds is extracted by time-of-flight expansion analysis. We extract $T = 400(15)$ nK for the Feshbach molecules and $T = 430(20)$ nK for rovibrational ground state molecules. The analysis suggests that the transfer process does not cause any noticeable heating of the molecules.

measurement results, with blue filled circles corresponding to the Feshbach molecules before STIRAP transfer and the red open circles for those after a round-trip STIRAP process. Clearly, the latter molecules are expanding at nearly the same kinetic energy as the former ones, indicating that the rovibrational ground-state molecules are translationally as cold as the initial population. In fact, by extracting the temperature from the expansion analysis, we find $T = 400(15)$ nK for the Feshbach molecules and $T = 430(20)$ nK for rovibrational ground state molecules. We have thus experimentally established that our transfer process is highly efficient, allowing 90% of the initial phase-space density to be transferred into the absolute ground state.

IV. Properties of KRb ground state polar molecules

The rovibrational level structure of the entire $X^1\Sigma_0$ electronic ground potential can be mapped out using the frequency comb-assisted two-photon dark resonance spectroscopy. In ref. 22, we measured the rotational constant for the $v = 0$ manifold. The dipole moment of the KRb rovibrational ground state can be measured *via* Stark spectroscopy performed on the two-photon dark resonance. Since the initial Feshbach state has no appreciable electric dipole moment, the frequency shift of the two-photon Raman resonance under an external DC electric field arises solely from the level shift of the absolute ground state. Since we already know the rotational structure, the dipole moment can thus be precisely determined to be $d = 0.566(17)$ Debye (see ref. 22). The hyperfine structure[33] of the KRb absolute ground-state can be mapped out using high-resolution dark resonance spectroscopy, and the molecules can be prepared in a single hyperfine level.

Another interesting property to measure for these ground-state molecules is their AC polarizability. In fact, the existence of the permanent dipole moment allows us to align these molecules under an external DC electric field of a few kilovolts per centimeter. As these molecules are confined inside an optical dipole trap, we can vary the polarization of the trapping light with respect to the DC electric field and measure the anisotropic property of the polarizability.[34] We have now taken the first step in measuring the average value of the AC polarizability $\alpha(X^1\Sigma_0)$. Theoretically, the AC polarizability of the $X^1\Sigma_0$ $v = 0$ ground state of the KRb molecule was first investigated in ref. 35. Experimentally, the AC polarizability can be determined from

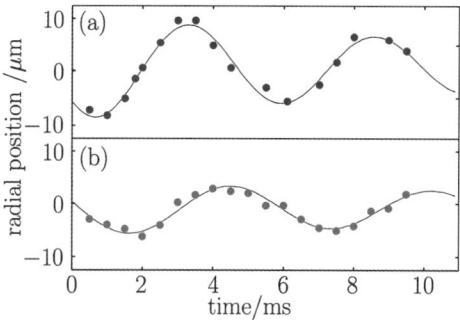

Fig. 5 Comparison of the radial slosh of Feshbach molecules (a) *vs.* ground state molecules (b) in the optical dipole trap at $\lambda = 1090$ nm. The slosh in the trap is excited by perturbing the optical potential for 0.5 ms and measured after 2 ms of expansion. We extract a trap frequency of $2\pi \times 190(3)$ Hz for Feshbach molecules (a) as compared to a trap frequency of $2\pi \times 175(5)$ Hz for ground state molecules. From the frequency ratio, we derive a preliminary polarizability ratio of $\alpha(X^1\Sigma_0)/\alpha_F = \omega^2(X^1\Sigma_0)/\omega_F^2 = 0.85(5)$ at $\lambda = 1090$ nm.

the oscillation frequencies of trapped molecules provided we know the trapping light intensity. Fig. 5 shows a preliminary comparison of the radial slosh of Feshbach molecules and ground-state molecules in the optical dipole trap at $\lambda = 1090$ nm. The slosh in the trap is excited by perturbing the optical potential for 0.5 ms. We observe a pronounced difference between the trap frequencies for Feshbach molecules and ground state molecules. We extract a trap frequency of $\omega_F = 2\pi \times 190(3)$ Hz for Feshbach molecules as compared to a trap frequency of $\omega(X^1\Sigma_0) = 2\pi \times 175(5)$ Hz for ground-state molecules. From the frequency ratio, we derive a polarizability ratio of $\alpha(X^1\Sigma_0)/\alpha_F = \omega^2(X^1\Sigma_0)/\omega_F^2 = 0.85(5)$ at $\lambda = 1090$ nm. The polarizability of the Feshbach molecules is simply the sum of the Rb and K atomic polarizabilities[36] $\alpha_F = 5.7 \times 10^{-5}$ MHz W^{-1} cm^{-2}.

One important property of $v = 0$ molecules is their expected long lifetime, which will allow further cooling to quantum degeneracy. The background pressure of our vacuum chamber and the black body excitation rate of the rovibrational transitions both support a lifetime longer than a few tens of seconds. The light scattering by the trapping beam should also be fairly weak, and we do not expect it to limit the lifetime below 10 s. The lifetime could be limited by inelastic molecule–molecule and atom–molecule collisions. If the rovibrational ground state ^{40}K^{87}Rb molecules are created in a single hyperfine state, then the Fermi nature of these molecules will preclude their collisions at this ultracold temperature, unless a sufficiently large electric field is turned on to induce dipolar interactions. A possible loss mechanism is the collision of a K or a Rb atom with a KRb molecule, resulting in either ultracold

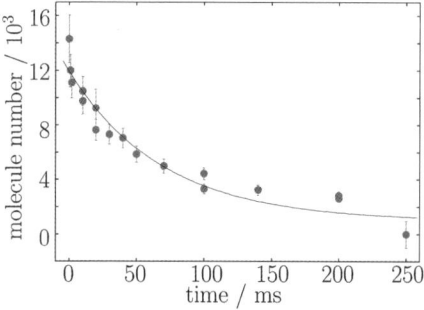

Fig. 6 Lifetime of ground-state polar molecules in the optical dipole trap. We extract $\tau = 70(8)$ ms.

chemical reactions or loss *via* harpooning mechanisms or three-body bound states.[37] By removing most of the K and Rb atoms, we have measured a typical lifetime of $\tau =$ 70(8) ms for the ground-state polar molecules sitting in the optical dipole trap. Fig. 6 shows one such measurement. When we purposely allow K and Rb atoms to remain in the same optical trap and when their number is similar to that of the KRb molecules, we observe a decrease of the trap lifetime to below 10 ms. At this time, we are still searching for the reason behind the 100 ms limit for the lifetime. Planned measurements in the future include trap lifetime measurements with molecules prepared in a purified single hyperfine level to prevent possible KRb–KRb collisions and the measurement of incoherent scattering of ground-state molecules due to the trapping light.

V. Conclusions

The successful production of high phase-space density ultracold polar molecules has now prepared us to explore an exciting range of scientific topics including simulation of quantum phase transitions, quantum information processing, precision measurement, and novel collisions and chemical reactions at ultralow energies. The field of ultracold molecules is poised to explode with many exciting new results in the near future and we are eager to be part of this endeavor.

We thank A. Pe'er and J. Zirbel for their contributions to the work reported here. Funding support is provided by NSF, NIST, AFOSR, and ARO. S. O. acknowledges support from the Alexander von Humboldt Foundation, K.-K. N. and B. N. from NSF, and M. H. G. de.M. from the CAPES/Fulbright.

References

1 R. V. Krems, *Phys. Chem. Chem. Phys.*, 2008, **10**, 4079.
2 E. R. Hudson, C. Ticknor, B. C. Sawyer, C. A. Taatjes, H. J. Lewandowski, J. R. Bochinski, J. L. Bohn and Jun Ye, *Phys. Rev. A*, 2006, **73**, 063404.
3 P. G. H. Sandars, *Phys. Rev. Lett.*, 1967, **19**, 1396–1398.
4 M. G. Kozlov and L. N. Labzowsky, *J. Phys. B*, 1995, **28**, 1933–1961.
5 E. R. Hudson, H. J. Lewandowski, B. C. Sawyer and J. Ye, *Phys. Rev. Lett.*, 2006, **96**, 143004.
6 G. Pupillo, A. Micheli, H. P. Büchler, and P. Zoller, arXiv:0805.1896 (2008).
7 L. Santos, G. V. Shlyapnikov, P. Zoller and M. Lewenstein, *Phys. Rev. Lett.*, 2000, **85**, 1791–1794.
8 A. V. Avdeenkov, M. Kajita and J. L. Bohn, *Phys. Rev. A*, 2006, **73**, 022707.
9 C. Ticknor, *Phys. Rev. Lett.*, 2008, **100**, 133202.
10 D. DeMille, *Phys. Rev. Lett.*, 2002, **88**, 067901.
11 A. Andre, D. DeMille, J. M. Doyle, M. D. Lukin, S. E. Maxwell, P. Rabl, R. J. Schoelkopf and P. Zoller, *Nat. Phys.*, 2006, **2**, 636–642.
12 S. F. Yelin, K. Kirby and R. Cote, *Phys. Rev. A*, 2006, **74**, 050301(R).
13 R. V. Krems, *Internat. Rev. Phys. Chem.*, 2005, **24**, 99–118.
14 J. D. Weinstein, R. deCarvalho, T. Guillet, B. Friedrich and J. M. Doyle, *Nature*, 1998, **395**, 148–150.
15 H. L. Bethlem, G. Berden and G. Meijer, *Phys. Rev. Lett.*, 1999, **83**, 1558–1561.
16 B. C. Sawyer, B. L. Lev, E. R. Hudson, B. K. Stuhl, M. Lara, J. L. Bohn and J. Ye, *Phys. Rev. Lett.*, 2007, **98**, 253002.
17 S. A. Rangwala, T. Junglen, T. Rieger, P. W. H. Pinkse and G. Rempe, *Phys. Rev. A*, 2003, **67**, 043406.
18 B. K. Stuhl, B. C. Sawyer, D. Wang and J. Ye, *Phys. Rev. Lett.*, 2008, arXiv:0808.2171, in press.
19 K. M. Jones, E. Tiesinga, P. D. Lett and P. S. Julienne, *Rev. Mod. Phys.*, 2006, **78**, 483.
20 D. Wang, J. Qi, M. F. Stone, O. Nikolayeva, H. Wang, B. Hattaway, S. D. Gensemer, P. L. Gould, E. E. Eyler and W. C. Stwalley, *Phys. Rev. Lett.*, 2004, **93**, 243005.
21 J. M. Sage, S. Sainis, T. Bergeman and D. DeMille, *Phys. Rev. Lett.*, 2005, **94**, 203001.
22 K.-K. Ni, S. Ospelkaus, M. H. G. de Miranda, A. Peer, B. Neyenhuis, J. J. Zirbel, S. Kotochigova, P. S. Julienne, D. S. Jin and J. Ye, *Science*, 2008, **322**, 231.
23 T. Kohler, K. Goral and P. S. Julienne, *Rev. Mod. Phys.*, 2006, **78**, 1311.

24 J. J. Zirbel, K.-K. Ni, S. Ospelkaus, J. P. D'Incao, C. E. Wieman, J. Ye and D. S. Jin, *Phys. Rev. Lett.*, 2008, **100**, 143201.
25 J. J. Zirbel, K.-K. Ni, S. Ospelkaus, T. L. Nicholson, M. L. Olsen, P. S. Julienne, C. E. Wieman, J. Ye and D. S. Jin, *Phys. Rev. A*, 2008, **78**, 013416.
26 S. Ospelkaus, A. Pe'er, K.-K. Ni, J. J. Zirbel, B. Neyenhuis, S. Kotochigova, P. S. Julienne, J. Ye and D. S. Jin, *Nat. Phys.*, 2008, **4**, 622.
27 K. Winkler, F. Lang, G. Thalhammer, P. v. Straten, R. Grimm and J. Hecker Denschlag, *Phys. Rev. Lett.*, 2007, **98**, 043201.
28 S. T. Cundiff and J. Ye, *Rev. Mod. Phys.*, 2003, **75**, 325.
29 A. Pe'er, E. A. Shapiro, M. C. Stowe, M. Shapiro and J. Ye, *Phys. Rev. Lett.*, 2007, **98**, 113004.
30 E. Arimondo and G. Orriols, *Lett. Nuovo Cimento*, 1976, **17**, 333.
31 K. Bergmann, H. Theuer and B. W. Shore, *Rev. Mod. Phys.*, 1998, **70**, 1003–1025.
32 C. Ospelkaus, S. Ospelkaus, L. Humbert, P. Ernst, K. Sengstock and K. Bongs, *Phys. Rev. Lett.*, 2006, **97**, 120402.
33 J. Aldegunde, B. A. Rivington, P. S. Zuchowski and J. M. Hutson, *Phys. Rev. A*, 2008, **78**, 033434.
34 J. Deiglmayr, M. Aymar, R. Wester, M. Weidemüller and O. Dulieu, *J. Chem. Phys.*, 2008, **129**, 064309.
35 S. Kotochigova and E. Tiesinga, *Phys. Rev. A*, 2006, **73**, 041405(R).
36 B. Arora, M. S. Safronova and C. W. Clark, *Phys. Rev. A*, 2007, **76**, 052516.
37 J. L. Bohn, 2009, private communications.

PAPER | www.rsc.org/faraday_d | Faraday Discussions

Ultracold molecules from ultracold atoms: a case study with the KRb molecule

Paul S. Julienne*

Received 24th November 2008, Accepted 15th January 2009
First published as an Advance Article on the web 28th May 2009
DOI: 10.1039/b820917k

Ultracold collisions of cold atoms or molecules make the bound states of the collision complex formed from the two colliding species accessible for control and manipulation of the cold species or the complex. Such resonances are best treated by a resonant scattering theory, which in the ultracold domain can take advantage of the properties of the long-range potential and the methods of multichannel quantum defect theory. Coupled channels calculations on the threshold scattering states and bound states of the $^{40}K^{87}Rb$ molecule illustrate the ideas and methodology of quantum defect theory using the long-range potential and also demonstrate the spin properties of the bound states throughout the spectrum.

1. Introduction

The success of cooling and trapping of neutral atoms in the ultracold regime of the order of μK or less has led to a broad range of advances in multidisciplinary research with quantum degenerate gases of bosons[1–3] or fermions.[4–6] Recent work has emphasized strongly correlated many-body physics and reduced dimensional physics.[7,8] The collisional interaction between cold atoms is a crucial factor in determining the properties and stability of an ultracold gas. These properties can, in many cases, be precisely controlled using tunable scattering resonances known as Feshbach resonances.[9,10]

Ultracold atom research is especially promising with atoms confined in optical lattices, which can provide confinement in one, two, or three dimensions.[11–13] Three dimensional lattices are comprised of a periodic array of trapping cells and offer the possibility of realizing quantum phase transitions whereby each cell holds an individual atom, or a pair of atoms. Cell trapping depths can be of the order of μK and intercell spacings of the order of hundreds of nm. Given the ability to control lattice depth and spacing and the strength of interatomic interactions, lattices of cold atoms or molecules can realize a rich variety of physical systems for a variety of applications.

Molecules are more difficult to cool than atoms, because their much more complex internal structure does not allow simple laser cooling methods[14–16] to be applied to them.[17] While other methods to cool molecules are being developed, such as molecular beam deceleration,[18] they have not yet succeeded in reaching the ultracold regime; see the review by Krems.[19] So far, the only successful route to getting ultracold molecules in the μK domain is to associate two already cold atoms to make a dimer molecule. The first work in this area, reviewed in references 20 and 21, was carried out by a number of groups that used photoassociation to make molecules in magneto-optical traps at relatively low phase space density and

Joint Quantum Institute, National Institute of Standards and Technology and The University of Maryland, 100 Bureau Drive Stop 8423, Gaithersburg, Maryland 20899-8423, USA. E-mail: paul.julienne@nist.gov

translational temperatures of the order of 100 μK. This early work relied on spontaneous emission from excited electronic states, which resulted in a distribution of ground state vibrational and rotational levels and negligible population in the vibrational ground state of the molecule.

The most effective way to make an ultracold molecular gas with high phase space density is to convert atom pairs in an already cold and dense atomic gas to bound molecular states. This can be done, for example, by using a time-dependent magnetic field to sweep a Feshbach resonance across threshold from higher to lower energy.[22] Kokkelmans et al.[23] suggested such a sweep could be combined with optical Raman population transfer to enhance the production of deeply bound molecules in the vibrational ground state. The vibrationally excited, weakly bound threshold molecules are in most cases not collisionally stable but undergo rapid vibrational relaxation upon collision with an atom or another molecule. On the other hand, $v = 0$ levels would not suffer from collisional vibrational relaxation.

Jaksch et al.[24] suggested that cold ground state $^{87}Rb_2$ molecules could be made by photoassociation of pairs of ^{87}Rb atoms in optical lattice cells to make a weakly bound state, followed by a series of Raman population transfer steps to transfer population stepwise to the $v = 0$ ground state. Upon turning off the lattice these vibrational ground state molecules could be expected to form a molecular Bose–Einstein condensate. Damski et al.[25] suggested a dipolar superfluid could be made in a similar way by associating K and Rb atoms in optical lattice cells and optically converting them to ground state KRb polar molecules. The essential step in any of these schemes is the first step of associating two atoms to make a weakly bound molecular state. Once associated, well-established optical techniques should be capable of transferring the population of the highly vibrationally excited molecular state to the ground state.

Sage et al.[26] showed that a single stimulated Raman step could be used to transfer population from a weakly bound excited vibrational level to the $v = 0$ ground state

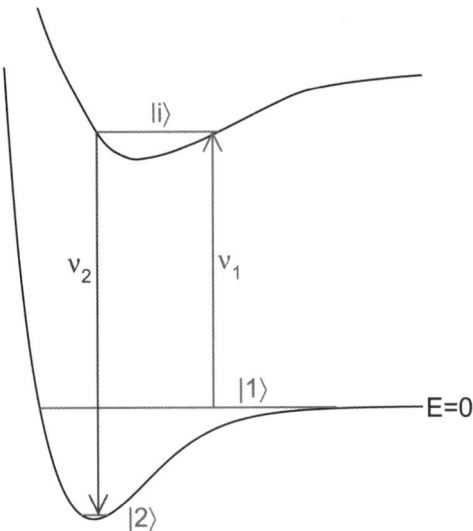

Fig. 1 Schematic figure of the potential energy curves $V(R)$ of two interacting atoms A and B versus interatomic distance R, with the zero of energy E set as the energy of two separated atoms in the states in which they are prepared. The horizontal line just below threshold indicates the energy of a weakly bound molecular level $|1\rangle$, which is prepared by associating two cold atoms and then converted to the target deeply bound vibrational level $|2\rangle$ by a Raman transition through an excited intermediate molecular state $|i\rangle$. The Raman process could also be initiated starting from a scattering state with $E > 0$, although this has proved difficult in practice. Schemes of this general type were proposed by references 23–25 and realized in references 26, 40 and 41 by starting from associated atoms in a weakly bound state.

of the RbCs molecule. Fig. 1 schematically illustrates this kind of process. The initial molecular state in ref. 26 was formed by spontaneous emission in a low phase space density gas. However, it is much better to use magnetoassociation to convert atom pairs in a high phase space density gas to very weakly bound molecules known as Feshbach molecules. Köhler et al.[9] have extensively reviewed research that uses magnetically tunable Feshbach resonances to achieve such association. Molecules formed in this way are just as translationally cold as the atoms that are initially present, and their density is similar to that of the atoms when the transfer efficiency is high. It is now apparent that the key to making deeply bound ultracold molecules is to take advantage of a tunable Feshbach resonance to pair two unbound atoms into a weakly bound molecular state, which in turn can be converted by optical methods to the more deeply bound state.

A number of homonuclear alkali dimer Feshbach molecules have been made by magnetoassociation: ^{6}Li$_2$,[27–29] ^{23}Na$_2$,[30] ^{42}K$_2$,[31] ^{87}Rb$_2$,[32] and ^{133}Cs$_2$,[33,34] including molecular Bose–Einstein condensates made of pairs of fermionic atoms.[35–37,31] Thalhammer et al.[38] demonstrated high efficiency in converting pairs of ^{87}Rb atoms in optical lattice cells to make weakly bound ^{87}Rb$_2$ Feshbach molecules, which are also trapped in the lattice cells. Winkler et al.[39] were also able to demonstrate efficient coherent optical transfer of population to a lower energy weakly bound ^{87}Rb$_2$ vibrational level using the STImulated Raman Adiabatic Passage (STIRAP) technique. Danzl et al.[40] succeeded in using the STIRAP method to convert a weakly bound ^{133}Cs$_2$ Feshbach molecule to a much more deeply bound level with a binding energy of around 1000 cm^{-1}.

Ni et al.[41] have succeeded in using magnetoassociation plus STIRAP to convert cold atom pairs in a dense gas of ^{40}K fermions and ^{87}Rb bosons at 350 nK to $v = 0$ ^{40}K^{87}Rb ground vibrational state molecules in both the a$^3\Sigma_u^+$ state and the X$^1\Sigma_g^+$ electronic ground state. Previously Zirbel et al.[42] had characterized the magnetoassociation to a Feshbach molecule near 54.6 mT, and Ospelkaus et al.[43] demonstrated STIRAP to form a dense, ultracold molecular gas of levels bound by about $E/h = 10$ GHz. Since formation of ^{40}K^{87}Rb in lattice cells has previously been demonstrated by association of two atoms in a lattice cell,[44] it should be straightforward to make a lattice of $v = 0$ ^{40}K^{87}Rb molecules. Polar molecules in lattices have a number of possible applications, including the realization of exotic condensed matter phases[45–47] and quantum computation.[48]

Collisions are an essential aspect of understanding and applying ultracold atomic or molecular gases or lattices. Collisions can be beneficial coherent ones that allow control of the system or destructive ones that cause loss or decoherence of trapped atoms or molecules. The goal of this paper is to illustrate a number of features of ultracold collisions of atoms and molecules using calculations on the bound and scattering states of the ^{40}K^{87}Rb molecule as an example system. We use coupled channels calculations with accurate potential energy curves in calculations that extend from the $E = 0$ collision threshold to the energy of the deeply bound $v = 0$ ground state vibrational level. It is clear that near-threshold scattering resonances and bound states play a crucial role in ultracold molecule formation. In particular, we would like to demonstrate the usefulness of a resonant scattering viewpoint, amplified by the insights of generalized multichannel quantum defect theory based on the long-range potential between the colliding species.

Section 2 develops the resonant scattering view of an ultracold collision, whereby a collision allows access to a wide range of bound states of a collision complex. Section 3 describes the scattering channels of the ^{40}K^{87}Rb system, the coupled channels method, and the near-threshold scattering resonances in this system. Section 4 shows how the properties of the van der Waals long-range potential determine the qualitative and semi-quantitative features of the near-threshold bound and scattering states. Section 5 describes the character of the molecular levels as energy decreases below threshold to the domain of "normal" molecular levels, including the vibrational ground state level. The final section provides a summary of the results of the paper.

2. The resonant scattering viewpoint

2.1 Ultracold collisions for precise spectroscopy and state control

We will begin with some general considerations that are generic to ultracold atomic or molecular collisions. First, let us assume that the species A and B, each of which could be an atom or molecule, are prepared in specific internal states $|p\rangle_A$ and $|q\rangle_B$ in an ultracold gas at temperature T, where relative collision energies tend to be of the order of $E/k_B \approx 1$ μK, where k_B is the Boltzmann constant; in other units, 1 μK corresponds to $E/h = 21$ kHz, $E/hc = 7.0 \times 10^{-7}$ cm^{-1}, or $E = 86$ peV. Consequently, the colliding species are prepared in a very precise energy state relative to the energy scale associated with the "collision complex" AB, where the fragments A and B interact strongly when they are close together. The energy scale associated with AB is of the order of typical chemical bond strengths, say $E = 1$ eV, equivalent to $E/h = 242$ THz or $E/hc = 8066$ cm^{-1}. Thus, the complex AB can be prepared with a precise energy spread that may be only 1 part of 10^{10} of its ground state binding energy.

Fig. 2 shows a schematic view of an ultracold collision, indicating that the species A and B have internal structure and could be prepared in one of several energy states $E_\alpha = E_p + E_q$ of the pair. For simplicity, we take the energy of the state that is prepared as the zero of the energy scale, illustrated in the Figure as $E_\alpha = E_1 = 0$. Atoms typically have hyperfine and Zeeman spin structure in their ground states,

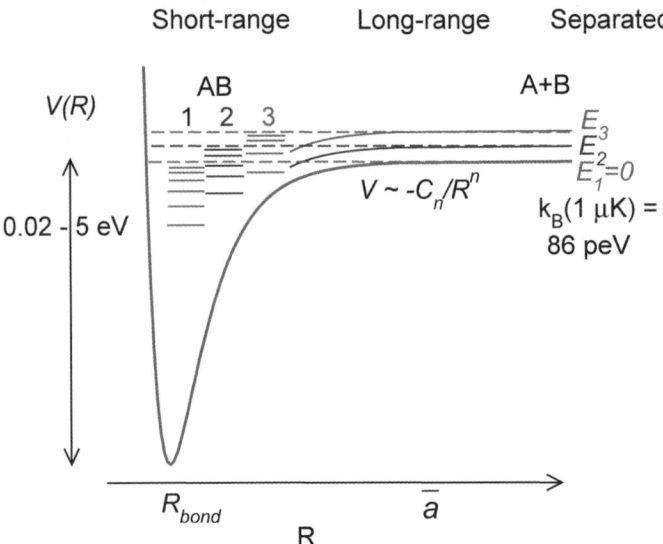

Fig. 2 Schematic view of an ultracold collision to form a molecular "complex". The species A and B are each prepared in internal states labeled collectively by $\alpha = 1$ and with a very small relative collision energy near the separated species energy $E_1 = 0$. Two closed channels 2 and 3 with different internal energies E_2 and E_3 of the separated species are also schematically indicated, along with the spectrum of associated bound state levels (short horizontal lines) in channels 1, 2 and 3. The energy scale for an ultracold collision is of the order of $k_B T = 86$ peV or $k_B T/h = 21$ kHz for $T = 1$ μK. The energy scale for the short-range part of the potential, where $R \approx R_{bond}$ in the order of the chemical bond length R_{bond}, is given by the binding energy of the $v = 0$ level of the ground state potential, of the order of $E/h = 5$ THz for a weakly bound van der Waals molecule (0.02 eV) to 1 PHz for a strong chemical bond (5 eV). The near-threshold bound and scattering states of the complex are sensitive to the long-range part of the potential that varies asymptotically as $-C_n/R^n$ and has a characteristic length $\bar{a}$. Near-threshold bound states associated with closed channels 2 and 3 of the collision form threshold scattering resonances in channel 1 that can be tuned across $E = E_1$ by varying magnetic, electric, or electromagnetic fields.

whereas molecules will have rotational and vibrational degrees of freedom as well. Collision channels β with $E_\beta > E > E_\alpha$ are "closed channels" at the collision at energy E, whereas channels β with $E > E_\alpha \geq E_\beta$ are "open channels." Collision products can exit the collision in open channels but not closed ones, due to energy conservation. If more than two atoms are present in the AB species, reactive channels may also be open.

The long-range potential of the AB complex will generally support a series of bound levels $E_{n\beta}$ leading up to the dissociation limits at E_β corresponding to the various internal states β of the separated species. Such levels are indicated schematically for three channels in Fig. 2. Levels in closed channels with $E_{n\beta} > E_\alpha$ are quasi-bound levels that can decay to channel α, depending on the presence of a coupling term in the Hamiltonian of the AB complex. Such levels give rise to scattering resonances when collision energy E is near $E_{n\beta}$. Such resonances can be tuned or coupled to threshold scattering states by external fields. Magnetically tunable resonances, such as the 54.6 mT ^{40}K^{87}Rb resonance studied in this paper, can be used to control elastic and inelastic collision rates, and to form weakly bound molecular states by time-dependent manipulation of the magnetic field.[9,10] Optically tunable resonances, or photoassociation resonances,[20] can be tuned or turned on and off by varying the respective frequency or intensity of the driving electromagnetic radiation. Magnetically or optically tunable resonances are treated by formally equivalent theory.

If the atoms are assumed to start in a scattering state with $E > 0$, Fig. 1 gives examples of optically tunable resonances, namely the one-color photoassociation process at frequency ν_1 or the two-color photoassociation process involving ν_1 and ν_2. Photoassociation is most naturally treated as a resonant scattering process with a decaying resonance level.[49–51,52] The theory of threshold resonant scattering of a decaying resonance can be readily extended to tightly confining optical lattices of reduced dimension geometries[53] and can be readily adapted to magnetically tunable or other kinds of resonances.

Extraordinary success has been achieved with ultracold atoms with high resolution spectroscopic probing of the states of the AB molecule starting from the precisely prepared states of the atoms. The collision complex AB can be prepared in a sharp energy state defined by the spread in energy $k_B T$ of the colliding atoms. Tunable magnetic or radiofrequency probes can tune states within a few GHz of threshold.[9,10] One-color photoassociation is especially successful in probing excited states of the AB complex, while two-color Raman photoassociation is especially useful for probing the ground state.[20] Optical methods are capable of tuning to bound states that are removed even by hundreds of THz from threshold with a precision determined by the linewidths of the lasers involved.

Perhaps, even more importantly than spectroscopic probing, a time-dependent field can be used to associate an A + B pair to make a stable, weakly bound AB complex that does not dissociate back to A + B.[9] This stable complex can then, in turn, be coupled to other bound states in the near-threshold domain. Population can then be transferred coherently to other weakly bound near-threshold states using time dependent magnetic,[54,55] radiofrequency,[56] or optical fields.[39,42] Time-dependent optical fields can achieve coherent population transfer to deeply bound states.[40,41] We have every reason to expect that these techniques will be extended to new species and other domains of frequency, including microwave, THz, and infrared.

Thus, we see that a sample of ultracold atoms – and presumably now samples of ultracold molecules – can be prepared in specific internal states at a precisely defined energy near $E = 0$. The collision of the prepared species then makes available a large part of the entire bound state spectrum of the collision complex for high resolution probing and coherent population transfer. Bound states of the complex that can be brought into resonance with the near $E = 0$ separated species then serve as "gateway" states into the rich spectrum of the complex. This permits very state-specific control over all degrees of freedom of the complex: electronic, vibrational, rotational, spin, and translation. By confining the stable species AB in a single trapping cell of an optical lattice, even the energy of relative motion is quantized and even more sharply defined. Furthermore, a molecule in a lattice cell is protected from collisions with A or B atoms or other AB molecules, thus prolonging its lifetime.[38]

2.2 Importance of the long-range potential

Given the sensitivity of the near-threshold bound state spectrum to the properties of the long-range form of the potential $V(R)$ between A and B, much can be gained by trying to understand the states of the molecule AB associated with the long-range $V(R)$, which varies with some lead power of R as $1/R^n$. This contrasts with "normal chemistry", where one usually seeks to understand the bound states from the ground state level up to the dissociation limit. In the ultracold domain, collisions are normally much more understandable and even treated quantitatively by starting with the states of the separated species A and B and the states of their complex AB due to the long-range interaction between them. In this way, one does not need to know the full spectrum of the AB molecule in order to characterize the near-threshold domain quite precisely. This approach has been extraordinarily successful with ultracold atoms;[9,10] see refs. 57–59 for some examples.

For neutral S-state atoms, the long-range potential has the van der Waals form with $n = 6$. Molecules can have dipole or quadrupole moments, corresponding to $n = 3$ and 5 for two dipoles or two quadrupoles respectively. These potentials are anisotropic, but have vanishing diagonal matrix elements for s-wave interactions in free space. In the absence of an external field which breaks the symmetry of free space, an isolated polar molecule in a definite state of total angular momentum has a vanishing dipole moment. However, a strong electric field can induce a dipole moment.

It is convenient to introduce a characteristic length and energy scale associated with the long-range potential. For this purpose, we use the scale length defined by Gribakin and Flambaum for a potential varying as $-C_n/R^n$,[60]

$$\bar{a}(n) = \cos\left(\frac{\pi}{n-2}\right)\left(\frac{2\mu C_n}{\hbar^2(n-2)^2}\right)^{\frac{1}{n-2}}\Gamma\left(\frac{n-3}{n-2}\right)/\Gamma\left(\frac{n-1}{n-2}\right), \qquad (1)$$

where μ is the reduced mass of the AB pair, Γ is the Gamma function, and $\hbar$ is Planck's constant divided by 2π. This length defines a corresponding energy scale

$$\bar{E}(n) = \frac{\hbar^2}{2\mu\bar{a}(n)^2}. \qquad (2)$$

For a van der Waals potential with $n = 6$, this simplifies to $\bar{a} = 0.477989(2\mu C_6/\hbar^2)^{1/4}$. Jones et al.[20] and Chin et al.[10] review the properties of the van der Waals potential relevant to ultracold physics, and Friedrich and Trost[61] adapt semiclassical theory to obtain the threshold properties for $n = 6$ and other cases of n.

2.3 Ultracold resonant scattering theory

Since molecules have more complex internal structure than atoms, subject to electric as well as magnetic and electromagnetic field control, the collisions of cold molecules should also have numerous scattering resonances; for example, see the work of refs 62–64 and 65. Consequently, the theory of resonant scattering is both necessary and useful for understanding cold molecular collisions. As noted earlier, resonant scattering theory has been extensively developed for photoassociative collisions, which represent optical Feshbach resonances.[49–53,66,67]

A particularly insightful set of articles on a resonant scattering viewpoint of molecule association and chemical reactions has been given by Mies.[68,69] He considered the role of molecular resonances in the association of two atomic or molecular fragments A and B to form an AB molecule. He considered both the case of radiative association, which is formally equivalent to the resonant scattering theory of molecule formation by photoassociation, and the case of collisonal association, where a third body deactivates a resonant state of the complex to stabilize it. In the latter case, the decay of the resonance is simulated by a complex energy with an imaginary term, just as spontaneous emission is represented in radiative association. The formalism is useful, since it gives a general S-matrix resonant scattering theory of association and resonant-enhanced reactions, bounded by the unitarity limit of the S-matrix, even when there is a complex set of overlapping resonances. While the Mies theory was developed for a high temperature gas, where k_BT is large compared to the spacing between the dense set of molecular resonances, there is no reason why the formalism cannot carry over directly to the ultracold case, if the threshold properties of the S-matrix are incorporated into the theory. This will be especially useful if the insights of generalized multichannel quantum defect theory[70,71] are brought to bear in the ultracold regime.[22,72–74]

The thermally averaged expression from Mies[68,69] for the inelastic collision rate coefficient $K_{in}(T)$ has a very instructive general form. Assume species A and B are prepared in channel α in a gas described by a Maxwellian thermal distribution of collision energies at temperature T. Then

$$K_{in}(T) = \frac{1}{Q_T}\frac{k_BT}{h}f_D(T), \quad (3)$$

where the dimensionless dynamical factor $f_D(T)$ is

$$f_D(T) = \sum_{\ell m_\ell}\left(1 - |S_{\alpha\alpha}(\ell m_\ell)|^2\right)e^{-E/(k_BT)}dE/(k_BT). \quad (4)$$

Here Q_T is the translational partition function, $1/Q_T = (2\pi\mu k_BT/h^2)^{-3/2} = \Lambda_T^3$ where Λ_T is the thermal de Broglie wavelength for the relative motion of A and B. The sum defining $f_D(T)$ runs over the contributing partial waves ℓ and the $2\ell + 1$ projections m_ℓ of ℓ. Inelastic collisions are those that remove A and B from channel $\alpha\ell m_\ell$, such as loss to a different channel $\beta\ell_\beta m_{\ell_\beta}$, which could represent removal of the resonant state through decay, relaxation, or reaction. If removal of A and B results in formation of an AB molecule, the removal rate is the same as the association rate. Note that in time-independent scattering theory, production of a resonant quasibound state that only decays back to the entrance channel is not a molecule formation process that can be represented by an S-matrix element, since the quasibound state does not persist into the asymptotic domain. However, if the quasibound resonant state is irreversibly removed to a loss "channel" so that it cannot decay back to the entrance channel, then this process can be represented by S-matrix scattering theory. Thus, S-matrix resonant scattering theory can be used to represent two-color photoassociation to bound states of the ground state potential, as long as such states have some "decay width".[52,75]

The removal rate of the density n_A of species A or density n_B of species B is $\dot{n}_A = \dot{n}_B = -K_{in}n_An_B$ (we assume nonidentical species; otherwise identical particle properties would need to be taken into account). We thus see that the respective removal rates $K_{in}n_B$ and $K_{in}n_A$ of species A and B are proportional to $n_B\Lambda_T^3$ and $n_A\Lambda_T^3$. One thus finds that the removal rate of a species is proportional to the dimensionless phase space density of its collision partner. The removal rate is also proportional to a thermal factor k_BT/h, which sets a basic time scale for the dynamics (e.g., 21 kHz at 1 μK). The only part that depends on the specific collision dynamics of the particular system is encapsulated in the dimensionless dynamical factor $(1 - |S_{\alpha\alpha}(\ell m_\ell)|^2)$ that describes the loss of flux from the $\alpha\ell m_\ell$ entrance channel. In the case of elastic scattering this factor is replaced by $|1 - S_{\alpha\alpha}(\ell m)|^2$ in a corresponding definition of a dynamical $f_D(T)$ expression for elastic scattering.

The structure of the S-matrix elements in the expression for the f_D term in the Mies theory is determined by a set of scattering resonances. However, the dimensionless $f_D(T)$ factor for inelastic collisions is bounded by the unitarity property of the S-matrix so that the contribution from each ℓm_ℓ term in the sum has a maximum value of unity. Thus, if ℓ_{max} partial waves contribute to the sum, then the bounds on $f_D(T)$ are $0 \leq f_D(T) \leq (\ell_{max} + 1)^2$, so that $0 \leq f_D \leq 1$ for s-waves. The analogous f_D factor for elastic collisions has an upper bound due to unitarity that is 4 times larger, $4(\ell_{max} + 1)^2$.

For s-waves in the $E \to 0$ threshold limit, $S = [1 - ik(a - ib)]/[1 + ik(a - ib)] \approx \exp[-2ik(a - ib)]$ is represented by a complex scattering length $a - ib$, where $\hbar k$ is the relative collision momentum, b is nonnegative, and the condition $k|a - ib| \ll 1$ applies. In this limit, $1 - |S|^2 = 4kb$, and K_{in} reduces to the usual threshold law expression,

$$K_{in} = 2(h/\mu)b = 0.42 \times 10^{-10} \frac{b[\text{au}]}{\mu[\text{amu}]} \text{ cm}^3 \text{ s}^{-1}. \tag{5}$$

Here $b[\text{au}]$ is in atomic units and $\mu[\text{amu}]$ is in atomic mass units. Similarly, the $E \to 0$ s-wave elastic scattering cross section reduces to the standard form $4\pi(a^2 + b^2)$. Given that $k|a - ib| \ll 1$, the f_D factor for elastic or inelastic collisions remains much less than its upper bound with a value determined by a, b and T. Threshold resonance structure can also modify the value of $f_D(T)$, requiring the use of full energy-dependent resonant scattering expressions.[52,53]

The expression for K_{in} in eqn (3) is based on very general thermodynamic and dynamical considerations at equilibrium and applies to atomic or molecular collisions. We may expect eqn (3) to give a guide to the rate of molecular association even in more general nonequilibrium cases. It is gratifying that the rate of association of a given species A is proportional to the phase space density of its collision partner, as in semi-empirical treatments of atom association to a Feshbach molecule with time-dependent fields.[42,76] This is to be expected for a fast process at constant entropy. Also, time-dependent processes would not be expected to beat the fundamental upper bound to f_D set by the unitarity limit of the time-independent S-matrix. If the dimensionless phase space and dynamical factors in the association rate are both less than unity, as they would be for a Maxwellian gas with an s-wave association process in the threshold law domain, then the $k_B T/h$ factor in eqn (3) sets a limiting upper bound to the rate (and a lower bound to the time scale) for the association of A and B pairs to form AB molecules.

Using the actual threshold resonance form for the S-matrix as a function of collision energy[52,53] would allow one to work out the general case of resonant elastic and inelastic cross sections and rate constants, including the effect of multiple or overlapping resonances if they are present. See Machholm et al.,[66] who develop a coupled channels S-matrix resonant scattering theory for decaying resonances, including the effect of overlapping resonances, using a form similar to the Mies theory. They also develop approximations for describing isolated resonances, using the factorization of matrix elements associated with quantum defect theory. Coupled channels resonant scattering theory, in its various numerical or approximate analytical representations, should prove to be a very powerful tool for characterizing ultracold molecular collisions, especially if it can take advantage of the long-range properties of the interaction potentials to characterize near-threshold states.

3. Multichannel scattering of ^{40}K^{87}Rb

Since ultracold ground state ^{40}K^{87}Rb polar molecules have now been made using resonant association followed by optical population transfer, we will concentrate on using this system to illustrate the features of this process. We calculate the bound and scattering states of ^{40}K^{87}Rb all the way from threshold to the ground state, including the effect of the long-range potentials in determining the states near threshold. Fortunately, an excellent set of adiabatic Born–Oppenheimer electronic potentials is available for the $X^1\Sigma^+$ and $a^3\Sigma^+$ states of this species, based on the high resolution spectroscopic analysis of Pashov et al.[77] These are the potentials needed to describe the interactions of two ground state atoms. Using these potentials, and standard representations of the full Hamiltonian of the interactions, including the angular momenta from electron and nuclear spins and internuclear axis rotation, we have carried out full coupled channels calculations[9,10,22,59,78] of the bound and scattering states of this species.

Fig. 3 and 4 show the Zeeman substructure of the ^{40}K and ^{87}Rb ^{2}S ground state atoms as a function of magnetic field. Each has an electron spin quantum number of $S = 1/2$ and respective nuclear spin quantum numbers of $I = 4$ and 3/2. Thus, ^{40}K is a composite fermion and ^{87}Rb is a composite boson, and the ^{40}K^{87}Rb molecule is a fermion. The figures show how the two zero-field hyperfine levels with total angular momentum $F = I - 1/2$ and $F = I + 1/2$ split with increasing magnetic field strength B. We use an alphabetical notation to designate each level of the Zeeman manifold by an italic Roman letter a, b, c..., starting with the lowest energy state at each B and increasing in order of energy. In the Figures, the zero of energy is that of the spinless (nonrelativistic) atom. Thus, the lowest energy hyperfine level at $B = 0$ has an energy of $-4E_{hf}(^{40}K)/9$ for ^{40}K and $-5E_{hf}(^{87}Rb)/8$ for ^{87}Rb, where $E_{hf}(^{40}K)/h = 1.285790$ GHz and $E_{hf}(^{87}Rb)/h = 6.8346826$ GHz are the hyperfine splittings of the two atoms.[79]

Fig. 5 shows the adiabatic Born–Oppenheimer potentials for the $X^1\Sigma^+$ and $a^3\Sigma^+$ electronic states that correlate with the two ground state separated atoms.[77] The inset to the figure shows the adiabatic potentials (i.e., those that diagonalize the full electronic plus spin Hamiltonian) on an expanded scale showing the long-range region. Since the atoms are S-state atoms, all of the long-range potentials have the same van der Waals C_6 coefficient, which has a value of $4299.51 E_h a_0^6$,[77] where $E_h = 4.359744 \times 10^{-18}$ J and $a_0 = 0.05291772$ nm. The characteristic van der Waals length

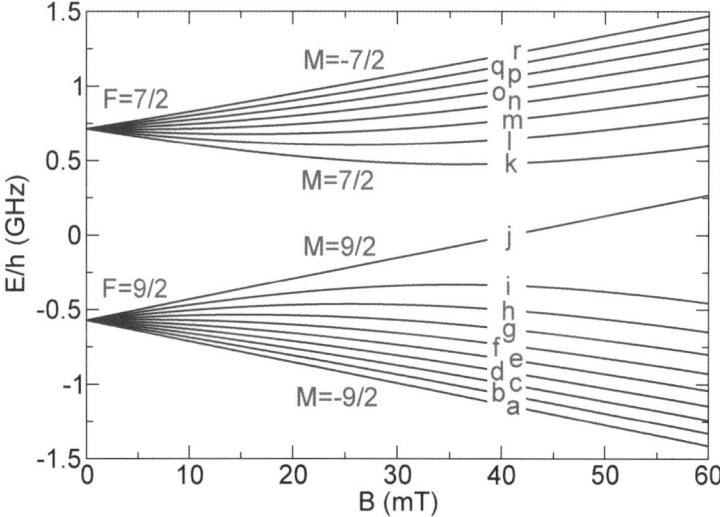

Fig. 3 The energy of the magnetic Zeeman sublevels of the fermionic ^{40}K atom versus magnetic field B. The zero of energy is the nonrelativistic energy center of gravity of the multiplet. The levels, labeled a, b, ... in order of increasing energy correlate at $B = 0$ with the two hyperfine levels with total electron spin plus nuclear spin angular momentum $F = 9/2$ and $F = 7/2$. The M quantum number specifies the angular momentum projection on the magnetic field axis and, unlike F, remains a good quantum number as B increases.

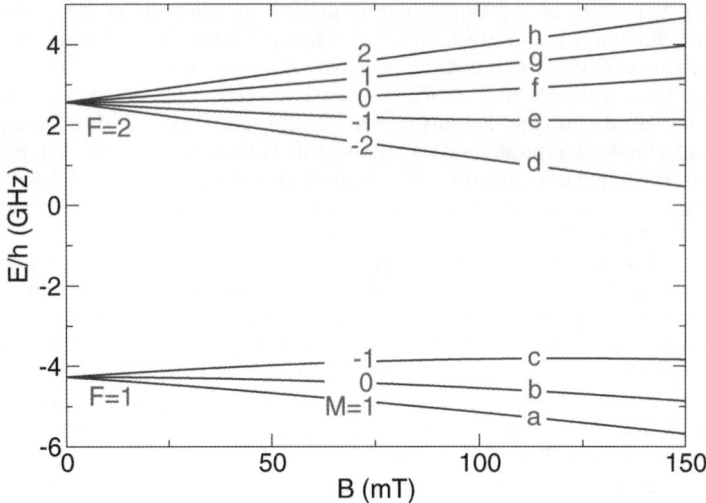

Fig. 4 The energy of the magnetic Zeeman sublevels of the bosonic [87]Rb atom *versus* magnetic field B. The zero of energy is the nonrelativistic energy center of gravity of the multiplet. The levels, labeled a, b, ... in order of increasing energy correlate at $B = 0$ with the two hyperfine levels with total electron spin plus nuclear spin angular momentum $F = 1$ and $F = 2$. The M quantum number specifies the angular momentum projection on the magnetic field axis and, unlike F, remains a good quantum number as B increases.

for the long-range potential is $\bar{a} = 68.8\ a_0$ and the characteristic energy is $\bar{E}/h = 13.9$ MHz.

Since we are concerned with *s*-wave resonances in the *aa* channel, where both atoms are in their lowest energy state, it is necessary to consider all of the *s*-wave

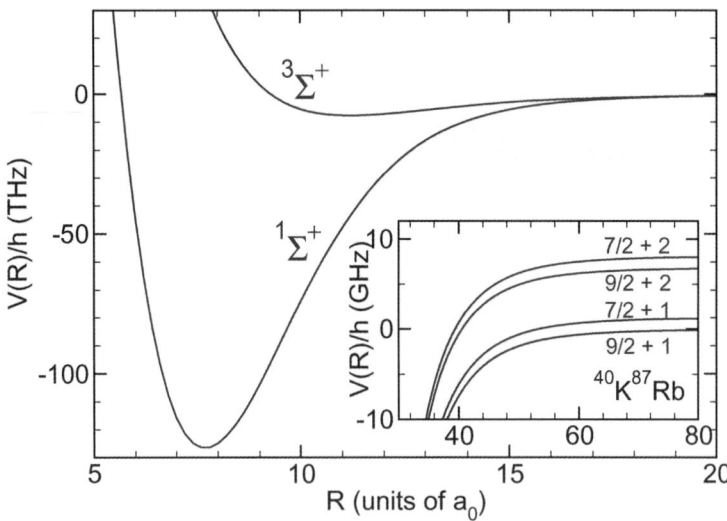

Fig. 5 Adiabatic Born–Oppenheimer $X^1\Sigma^+$ and $a^3\Sigma^+$ potential energy curves $V(R)$ of the KRb molecule correlating with ^2S ground state K and Rb atoms. These curves dissociate to $E = 0$ at the nonrelativistic energy center of gravity of the $B = 0$ atomic hyperfine levels. The inset shows the long-range adiabatic curves for the $M = -7/2$ projection states of the [40]K[87]Rb molecule at $B = 0$. These curves separate asymptotically to the atoms in one of their ground hyperfine levels. In the inset, the energy zero is set as the energy of the lowest 9/2 + 1 set of levels at $B = 0$. The characteristic van der Waals length is $\bar{a} = 68.8 a_0$.

states with projection quantum number $M_{tot} = -9/2 + 1 = -7/2$, where the M_{tot} quantum number also includes the projection m_ℓ of partial wave (m_ℓ is trivially zero for an s-wave). The projection quantum number is a conserved quantum number at finite field, so that only states with the same value of M_{tot} are coupled through terms in the Hamiltonian. While it is possible to include other partial waves in the expansion basis that are coupled to s-waves through anisotropic spin-dependent terms in the Hamiltonian,[59] such coupling terms are small and have a small effect here and need not be included (but see below for the effect of coupling d-waves to s-waves).

Fig. 6 illustrates the 12 s-wave channels that have $M_{tot} = -7/2$. The dotted zero-field curves are the same as in Fig. 5, but these split into 12 different curves at finite B. These channels separate themselves into four different groups A, B, C, and D associated with the four different zero field hyperfine separated atom limits, as described in the caption. Letting each channel be labeled by the quantum numbers $\beta = ij$, where ij represents the alphabetic label of the two atoms, with the ^{40}K label first, the coupled channels expansion of the wave function for atoms initially prepared in channel $\alpha = aa$ is

$$\Psi_\alpha(R, E) = \sum_\beta f_{\beta\alpha}(R, E)|\beta\rangle/R. \qquad (6)$$

The coupled channels Schrödinger equation then determines the solutions with scattering boundary conditions for $E > E_\alpha$ or the discrete set of bound state solutions with bound state boundary conditions for $E < E_\alpha$. The bound state solutions are found with the discrete variable method described in ref. 59, whereas a standard propagator method is used for the scattering solutions.

Fig. 7 shows the results of coupled channels scattering and bound state calculations with the 12 s-wave basis functions with $M_{tot} = -7/2$ in the expansion. The positions of the four scattering length singularities are in excellent agreement with the

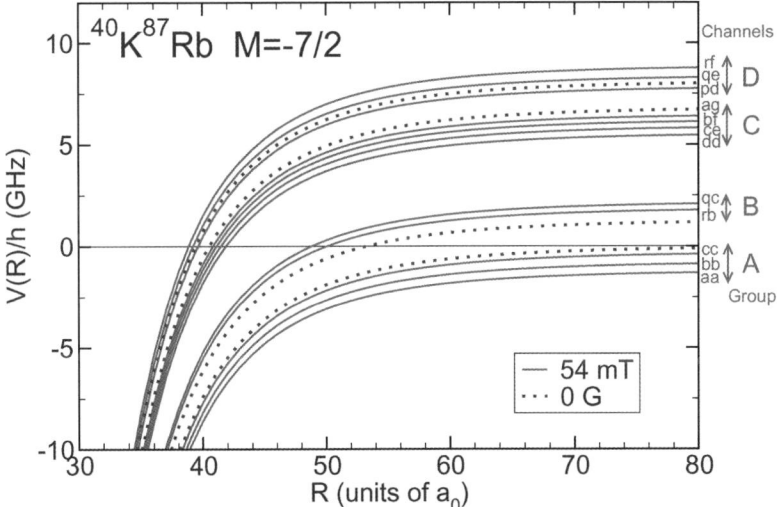

Fig. 6 Long-range adiabatic curves for the 12 $M = -7/2$ projection states of the ^{40}K^{87}Rb molecule. The dashed lines show the same curves as the inset of Fig. 3 for $B = 0$ and the solid lines show the 12 curves for $B = 54$ mT, labeled according to the Zeeman levels of each of the separated atoms. The zero of energy is taken to be the energy of the lowest hyperfine levels $9/2 + 1$ at zero field. The groupings into 4 sets of states labeled by A, B, C, and D correspond to the sets of states (aa, bb, cc), (rb, qc), (dd, ce, bf, ag), and (pd, qe, rf). These respective groupings are associated with the zero field separated atom hyperfine levels $9/2 + 1$, $7/2 + 1$, $9/2 + 2$ and $7/2 + 2$.

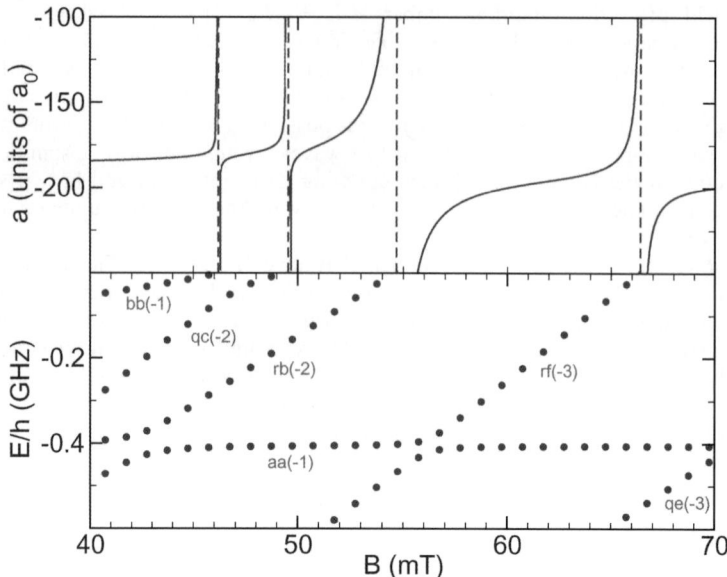

Fig. 7 Scattering length (upper panel) and bound state energy (lower panel) for the $M_{tot} = -7/2$ s-wave channels of the $^{40}K^{87}Rb$ molecule. The zero of energy is the energy E_α of the aa channel at each B field. Thus, the bound state energies give the binding energies of the levels relative to the aa separated atom energy. The dashed vertical lines show the points of singularity B_0 of the scattering length, calculated to be at 46.239 mT, 49.563 mT, 54.694 mT and 65.969 mT, where the binding energy of a molecular bound state becomes zero as the state reaches threshold. The solid circles show the bound state energies calculated for a discrete set of B values. The labels $\beta(n)$ show the dominant spin character of the bound eigenstate, where β indicates a separated atom closed channel and n gives the vibrational quantum number of the vibrational level in that channel, counting down from the dissociation limit of the channel at E_β.

measured positions.[80–82] The resonance near 54.6 mT is well represented near its singularity by the standard resonance formula[9,10]

$$a(B) = a_{bg}\left(1 - \frac{\Delta}{B - B_0}\right), \quad (7)$$

where the background scattering length is $a_{bg} = -190.6 a_0$, the resonance width $\Delta = -0.3103$ mT, and the resonance position is $B_0 = 54.6937$ mT, compared to measured positions of 54.69 mT[80] and 54.67 mT.[82]

The bound levels in Fig. 7 are labeled by the index β of the dominant spin component in the coupled channels expansion in eqn (6) and by the vibrational quantum number n counting down from $n = -1$, which designates the last bound state below the dissociation limit at $E = E_\beta$. Thus, the line near -0.4 GHz parallel to the $E = 0$ axis is the last $n = -1$ bound state in the aa entrance channel, whereas the bound state that crosses threshold to make the 54.69 mT resonance is the next to last $n = -2$ bound state of the rb channel, which is associated with the group B of Fig. 6. The two $n = -3$ levels in the Figure are associated with the highest energy group D.

Fig. 8 shows an expanded view of Fig. 7 near the 54.69 mT resonance. When d-waves with $\ell = 2$, $m_\ell = 0$ are added to the basis set, the weak coupling between the s- and d-waves due to the spin-dipolar interaction shifts B_0 from 54.694 mT to 54.687 mT (not shown). In addition, a new narrow resonance appears on the shoulder of the 54.69 mT resonance at 54.8305 mT with a width of only 4.6 µT. The narrow resonance has not been observed. It is due to the entrance channel s-wave being coupled to a bound state of d-symmetry. The bound state is a $\beta(n\ell m_\ell)$

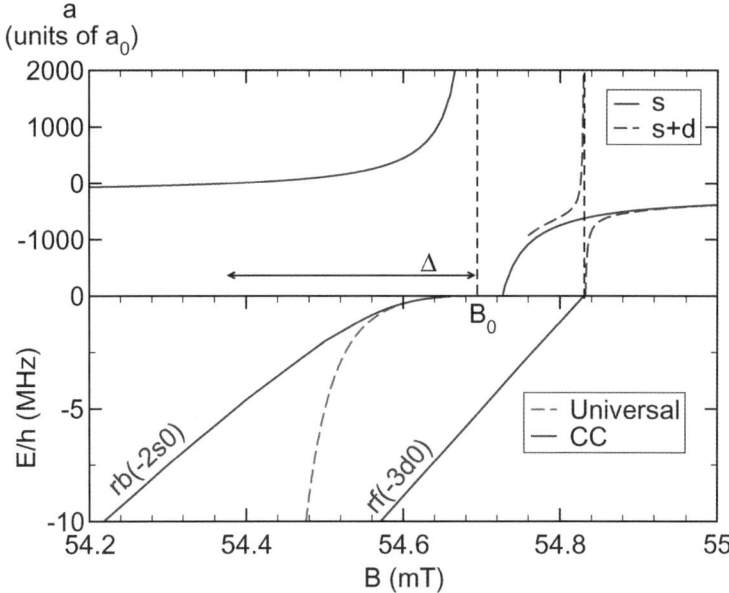

Fig. 8 Expanded view of the 54.69 mT resonance in Fig. 7. The upper panel shows the scattering length and the lower panel the bound state energies *versus* B. The solid line in the upper panel shows the scattering length calculated using an *s*-wave basis set only, the vertical dashed line marks the resonance position B_0, and the double headed arrow shows the magnitude of the resonance width Δ. The dashed curve in the upper panel shows an extra resonance that appears when *s* and *d* basis functions are both used in the calculation. The bound state labeled rb(−2s0) is the same as in Fig. 7. The bound state label includes the $\ell m_\ell = s0$ designation. The new bound state labeled rf(−3d0) is a *d*-wave bound state that appears when a *d*-wave basis set is added to the calculation with $M_{\text{tot}} = -7/2$. The dashed line in the lower panel indicates the universal energy derived from the scattering length using eqn (7).

= rf(−3d0) resonance with the same spin and *n* character as the rf(−3) *s*-wave resonance in Fig. 7, but with $\ell = 2$ units of axis rotation angular momentum. To a good approximation its energy can be obtained by adding the *d*-wave rotational energy for the $n = -3$ level of the $^3\Sigma^+$ potential to the energy of the *s*-wave rf(−3) level. Finally, Fig. 8 shows the universal energy that is derived from the scattering length,[9]

$$E = -\frac{\hbar^2}{2\mu a(B)^2}. \tag{8}$$

This universal formula only applies very close to resonance where $a(B) \gg \bar{a}$.

4. The quantum defect approach with a long-range potential

While coupled channels calculations have proven to be an excellent and reliable tool for understanding the near threshold domain of bound and scattering states of ultracold atoms, they are very computer intensive and give a "black box" representation of the physics. Therefore, it is also very desirable to have alternative methods for analysis, understanding, and developing approximations. In this regard, the general form of multichannel quantum defect theory (MQDT) provides a very powerful set of tools and concepts for the ultracold domain, which takes advantage of the separation between the short-range and long-range aspects of the collision, and exploits the analytic properties of the wavefunction as a function of interatomic separation *R* and energy *E*. The concepts of MQDT are implicit in the accumulated phase method pioneered by the Eindhoven group for characterizing ground state interactions of cold atoms[83–86] and are explicitly developed for neutral atoms in refs 72–74, 87, 88 and for ion–atom collisions in ref. 89. Gao has developed analytic MQDT for potentials with the long-range form $-C_n/R^n$.[90–93] We will show here several specific ways in which MQDT theory has been applied or still needs to be developed for application to the ultracold domain, especially when implemented using the form of the long-range potential,

4.1 The near-threshold spectrum

There is a close relationship between the near-threshold bound state spectrum and the near-threshold scattering properties of ultracold collisions. Not only is there a general relationship between the scattering length and the last bound state binding energy, as illustrated by the limiting expression in eqn (8), but the number and properties of Feshbach resonance states are related to the density and tuning properties of near-threshold bound states. In view of the importance of the near threshold spectrum for precision spectroscopic probing in Section 2.1 and collision dynamics in Section 2.3, simple ways of understanding the spectrum are desirable.

Gao[92,93] has worked out analytic bound state properties for the van der Waals potential for $n = 6$. We will characterize the long-range $n = 6$ potential by the length parameter $\bar{a}$ instead of the $\beta = 2.092099\bar{a}$ parameter used by Gao, since the formulas of the theory are simpler when $\bar{a}$ is used. The spectrum is completely determined for all partial waves ℓ by specifying only three parameters: the reduced mass μ, the C_6 constant, and the short-range QDT K_c-matrix element, or equivalently, the s-wave scattering length in units of $\bar{a}$ that uniquely determines the value of K_c. The value of K_c is related to the "quantum defect" for the vibrational levels of the potential.

Fig. 9 shows the spectrum of the last two bound states below threshold for the ^{40}K^{87}Rb molecule for different partial waves ℓ. The Figure shows several specific cases. The blue lines show the spectrum of bound states that one gets for the special

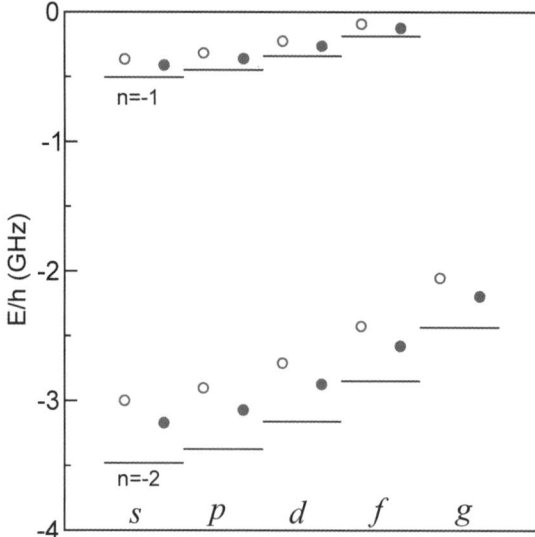

Fig. 9 Bound state spectrum of the last two vibrational levels of ^{40}K^{87}Rb with quantum numbers $n = -1$ and -2 for partial waves $\ell = 0, 1, 2, 3, 4$ (s,p,d,f,g). The horizontal lines show the levels for a pure van der Waals potential with the C_6 constant of the ^{40}K^{87}Rb molecule and with an infinite s-wave scattering length. The open and solid circles respectively show the calculated energy levels for the X$^1\Sigma^+$ and a$^3\Sigma^+$ potentials, for which the scattering lengths are -111.8 a_0 and -216.2 a_0 respectively.[77] The next lowest $n = -3$ levels of these potentials are at -10.24 GHz and -10.56 GHz.

case that the s-wave scattering length $a = \infty$. This corresponds to having an s-wave bound state at $E = 0$. The Gao theory shows in this case for a pure $-C_6/R^6$ potential that all partial waves with ℓ divisible by 4 also have a bound state at $E = 0$, as for example, the g-wave in Fig. 9. In the $a = \infty$ case the first s-wave level with $E < 0$, labeled by $n = -1$ in the Figure, lies at $-36.1\bar{E} = -0.503$ GHz, and the next $n = -2$ level lies at $-249\bar{E} = -3.47$ GHz. Varying the scattering length over its full range between $-\infty$ and $+\infty$ will produce one $n = -1$ and one $n = -2$ level in the "bins" demarked by these values. Fig. 9 also shows the actual energy levels for the $X^1\Sigma^+$ and $a^3\Sigma^+$ potentials, relative to their nonrelativistic separated atom limit at $V(\infty) = 0$. These are near the bottom of their "bins", due to their respective negative scattering lengths of $-111.8\ a_0$ and $-216.2\ a_0$.[77] The actual binding energies for the $n = -1$ and $n = -2$ levels are respectively about 1 percent and 2 percent larger than the same level calculated for a pure van der Waals potential with the same scattering length. This small difference indicates the small effect of other terms in the potential that contribute to the binding energy of near threshold levels. It also shows that the van der Waals theory alone is a very good approximate theory, although real potentials need to be used in fitting binding energy data that is more accurate than 1 percent.[94]

Fig. 9 also shows the bound states for other partial waves. The energies of these levels can be approximated by adding the rotational energy $\langle n, s|\hbar^2\ell(2\ell + 1)/(2\mu R^2)|n, s\rangle$ to the energy of the s-wave $|n, s\rangle$ level. In the angular momentum insensitive version of the long-range quantum defect theory,[92,93] the energies of the levels with $\ell \geq 1$ are also determined from a knowledge of the s-wave scattering length alone (plus μ and C_6, of course). However, there will always be a small error since the short-range form of the potential never corresponds to $1/R^6$ all the way to $R = 0$ but has a finite depth and inner turning point. For any given entrance channel in the multichannel problem, the near-threshold bound states can be calculated to a good approximation by using the scattering length for that channel to get the K_c matrix for that channel.

Fig. 10 shows the vibrational wave function $f_n(R, E_n)$ for the last three s-wave bound states of the $^3\Sigma^+$ potential with energy E_n for $n = -1, -2$, and -3. The wave functions show distinct differences for distances of the order of $\bar{a}$. However, the three wave functions are remarkably similar in shape at small R, where the potential is very deep and the local de Broglie wavelength, which governs the spacing between oscillations, is small compared to $\bar{a}$. These three unit-normalized wave functions have the same phase at small R and differ only in amplitude. In fact, these wave functions would have almost identical energy-insensitive amplitudes at short-range if they were given a "quantum defect" normalization per unit energy by multiplying $f_n(R, E_n)$ by $|\partial n/\partial E|^{1/2}$ for level n, where n is viewed as a continuous function of E that takes on integer values at an eigenvalue $E = E_n$, and $|\partial E/\partial n|$ is the mean vibrational spacing between levels at level n[70–72] (alternatively $|\partial n/\partial E|$ is the density of states per unit energy). The wave function can be understood using its energy-normalized semiclassical JWKB form in the short-range classical region, namely $C\sin\beta(R)/\sqrt{k(R, E)}$, where $\beta(R)$ is the phase, $\hbar^2 k(R, E)^2 = 2\mu(E - V(R))$ determines the "local" momentum wave number $k(R, E)$, and $C = [2\mu/(\pi\hbar^2)]^{1/2}$ is a constant. In the deep part of the potential, where $E - V(R) \gg \bar{E}$ is very large and $k(R, E) \approx k(R, 0)$, the semiclassical phase and amplitude are not at all sensitive to the value of $E \approx 0$. This insensitivity leads to the basic concept of MQDT or the accumulated phase method that the short-range physics, such as wave function phases or coupling matrix elements due to short-range interactions, can be quite adequately represented by energy-insensitive quantities.[70–72]

4.2 Feshbach resonance properties

The simple van der Waals form of MQDT also gives an excellent way to represent near-threshold scattering and bound state properties of Feshbach resonances In

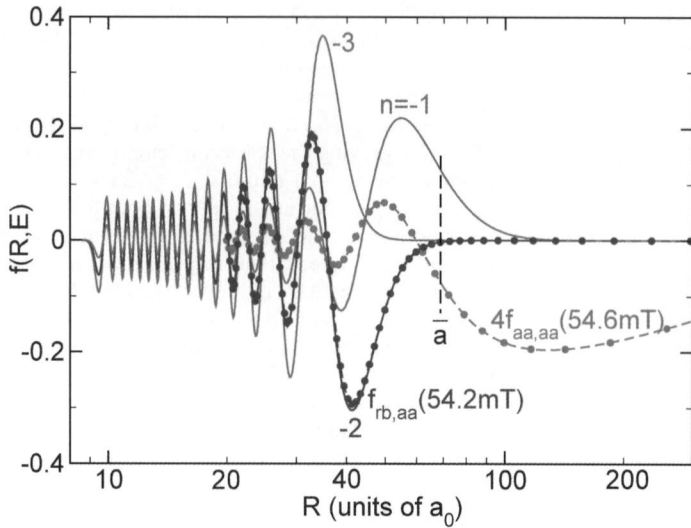

Fig. 10 Calculated wave functions for near-threshold bound states of the ^{40}K^{87}Rb molecule. The solid lines labeled $n = -1, -2, -3$ show the unit-normalized wave functions $f_n(R)$ for the last three levels of the a$^3\Sigma^+$ potential. The vertical line indicates the value of $\bar{a}$. The dots and dashed lines show results from the coupled channels calculation of the aa channel bound state of eqn 6 near the 54.69 mT resonance, namely, the rb component $f_{rb,\,aa}(R)$ at $B = 54.2$ mT and the aa channel component $f_{aa,\,aa}(R)$ at $B = 54.6$ mT. The latter represents a "halo molecule" with open channel aa spin character and an extension large compared to $\bar{a}$. The former represents a closed channel molecule with rb spin character and a vibrational function very close to the $n = -2$ a$^3\Sigma^+$ vibrational wave function.

addition to $\bar{a}$ and $\bar{E}$, the only other quantities that are needed in the near-threshold region are the background s-wave scattering length a_{bg} for the open entrance channel, the resonance width $\varDelta$, and the difference $\delta\mu$ in magnetic moments between the separated atoms in the entrance channel and the magnetic moment of the "bare" resonance state in the closed channel (the "bare" state is the approximate eigenstate in the closed channel before coupling to the entrance channel is turned on). Julienne and Gao[74] have shown how the Mies version of MQDT[70,71] can be can be adapted for the Feshbach resonance scattering states. MQDT theory first develops a set of uncoupled "reference" states with which to analyze the problem. Each 'reference' channel β for a given partial wave is characterized by a single reference potential that dissociates to energy E_β as $R \to \infty$ and which has a reference scattering phase shift $\eta_\beta(E)$ for $E \geq E_\beta$ and a reference bound state phase shift $\nu_\beta(E)$ for $E < E_\beta$. Bound states exist in the reference channel for a discrete set of energies for which $\tan \nu_\beta(E) = 0$. Near threshold, two auxiliary MQDT functions $C_\beta(E)$ and $\tan \lambda_\beta(E)$ are needed for each reference channel to characterize the quantum threshold behavior. These auxiliary functions can be physically interpreted using semiclassical concepts[70-72] and have the property that $C_\beta(E) \to 1$ and $\tan \lambda_\beta(E) \to 0$ when collision energy is sufficiently large. For s-waves in a van der Waals reference potential, this means $E - E_\beta \gg \bar{E}$; semiclassical connections between the short and long-range parts of the wave function break down when $E - E_\beta < \bar{E}$, and the quantum connections expressed by the $C_\beta(E)$ and $\tan \lambda_\beta(E)$ functions need to be used.

Even in a complex multichannel problem, it is usually sufficient to represent the closed channel bound state as a single bound state whose properties do not change with magnetic field detuning over a modest range of detuning.[9,22,95] Thus we can reduce the tunable Feshbach resonance problem to an effective 2-channel problem with a single "bare", or uncoupled, closed channel bound state $|c\rangle$ and a single entrance channel we call the "background" channel, $\alpha = bg$. With this framework and using the background channel van der Waals potential as the reference channel, the scattering phase shift for the fully coupled problem, where the "bare" closed channel with bound state energy $E_c = \delta\mu(B - B_c)$ interacts with the entrance channel with background phase shift $\eta_{bg}(E)$ to make a scattering resonance, the phase shift $\eta(E, B)$ for $E > 0$ is[74]

$$\eta(E, B) = \eta_{bg}(E) - \tan^{-1}\left(\frac{\frac{1}{2}\bar{\Gamma}_c C_{bg}(E)^{-2}}{E - \delta\mu(B - B_c) - \frac{1}{2}\bar{\Gamma}_c \tan\lambda_{bg}(E)}\right), \quad (9)$$

where the coupling between the closed and background reference channels is completely contained in the coupling parameter $\bar{\Gamma}_c$, which is independent of energy E and magnetic field B as these vary near the resonance over ranges of the order of $\bar{E}$ and Δ respectively. The resonant phase shift is characterized by an energy-dependent width $\Gamma_c(E) = \bar{\Gamma}_c C_{bg}(E)^{-2}$ in the numerator and an energy-dependent shift $\delta E_c(E) = \delta\mu\delta B_c(E) = (\bar{\Gamma}_c/2)\tan\lambda_{bg}(E)$ in the denominator of the resonance term in eqn (9). The general form of eqn (9) has been shown to be in excellent agreement with near-threshold coupled scattering calculations for a number of examples of Feshbach resonances in the literature.[74] Close to threshold the MQDT functions have the following s-wave limiting forms, which permit analytic limiting expressions to be given for Feshbach scattering as $E \to 0$[87]: $C_{bg}(E)^{-2} \to k\bar{a}(1 + (r - 1)^2)$, $\tan\lambda_{bg}(E) \to 1 - r$, where $r = a_{bg}/\bar{a}$ is the scattering length in dimensionless $\bar{a}$ units.

In a similar way, the MQDT coupled channels bound state equation from refs 70 and 71 can be put in the form

$$(E - \delta\mu(B - B_c))\tan\nu_{bg}(E) = \frac{\bar{\Gamma}}{2}. \quad (10)$$

where $\nu_{bg}(E)$ is the background reference channel MQDT phase function for $E < 0$, which has the property near threshold[87] that $\nu_{bg}(E) \to \tan^{-1}[1/(r - 1)] - \bar{a}\kappa$, where $E = -\hbar^2\kappa^2/(2\mu)$ for $E < 0$. Using this threshold property, it is straightforward to show that the coupled channels bound state crosses threshold at $B_0 = B_c + \delta B_c$, where the shift $\delta B_c = -\Delta r(r - 1)/(1 + (r - 1)^2)$ is the same shift that eqn (9) gives for the singularity at $B = B_0$ in the scattering length $a(B)$; see eqn (7).

One advantage of the MQDT approach is the factorization of the resonance coupling into a part associated with the long-range physics, $C(E)^{-2}$, and a reduced coupling parameter, $\bar{\Gamma}$, associated with the short-range physics. The $\bar{\Gamma}$ factor is proportional to the dimensionless MQDT energy-insensitive $Y_{c,bg}$ matrix element which gives the strength of the coupling:

$$|Y_{c,bg}|^2 = \frac{\bar{\Gamma}}{2}\frac{\partial\nu_c(E)}{\partial E}, \quad (11)$$

where $\partial\nu_c(E)/\partial E \approx \pi/\Delta E_c$ and ΔE_c is the mean vibrational spacing between adjacent vibrational levels near $E = E_c$. For all ultracold atom Feshbach resonances in the literature, $Y_{c,bg} \ll 1$ represents weak coupling, that is, the width of the resonance is small compared to the spacing between adjacent bound vibrational levels in the closed channel (mixing sometimes occurs with other closed channels; see Fig. 8).

In order to more explicitly show the connection between the long-range potential and Feshbach resonance properties, Chin et al.[10] introduced a dimensionless resonant strength parameter s_{res} by expressing a_{bg} and $\delta\mu\Delta$ in dimensionless units of $\bar{a}$ and $\bar{E}$ respectively:

$$s_{res} = \frac{a_{bg}}{\bar{a}} \frac{\delta\mu\Delta}{\bar{E}}. \qquad (12)$$

This is the inverse of the η parameter used by ref. 9 to characterize Feshbach resonances. The resonance width $\Gamma_c(E) = (k\bar{a})(\bar{E}s_{res})$ exhibits the threshold energy-dependence near $E = 0$, and the energy-insensitive MQDT resonance strength $\bar{\Gamma}$ is

$$\frac{\bar{\Gamma}}{2} = \bar{E}\frac{s_{res}}{1 + (r-1)^2}. \qquad (13)$$

This expression can be used in eqn (10) to get a general dimensionless MQDT equation for bound states. The bound state equation can be used to show that as $\kappa \to 0$ the bound state energy is given by $\kappa = 1/(a(B) - \bar{a})$, which is the same relationship for large positive scattering length as found by Gribakin and Flambaum[60] for a single van der Waals potential, namely, $\kappa_{bg} = 1/(a_{bg} - \bar{a})$ when $a_{bg} \gg \bar{a}$. In the Feshbach resonance case, the relationship applies when $a(B) \gg \bar{a}$ when $s_{res} \gg 1$, but only applies over a much more restricted range where $a(B) \gg 4\bar{a}/s_{res}$ when $s_{res} \ll 1$. Solving eqn (10) in general gives the right coupled channels bound states; see ref. 10. When $B - B_0$ is sufficiently far from resonance, the equation also recovers the "bare" bound state energy varying as $E = \delta\mu(B - B_c)$. The variation with B of the bound state energy of the coupled Feshbach bound state can be used to determine the norm Z of the closed channel piece in the unit normalized coupled channels bound state wave function: $Z = \delta\mu^{-1}\partial E/\partial B$.[9,10]

The dimensionless parameter s_{res} permits the classification of resonances into two basic types,[9,10] depending on how their character changes as B is tuned over a range of the order of Δ near $B = B_0$. One type with $s_{res} \gg 1$ is "open channel dominated" and the other type with $s_{res} \ll 1$ is "closed channel dominated". In either case this pertains to the character of the bound state over a tuning range that is a significant fraction of Δ. In the open channel dominated case, the wave function takes on the spin character of the entrance channel, has a universal energy given by eqn (7), and closed channel norm $Z \ll 1$. In contrast, the closed channel dominated case has only a very small domain of universality very close to B_0 and takes on the spin character of the closed channel bound state with $Z \approx 1$. Some of the more interesting experimental resonances are open channel dominated, namely the ^{6}Li 83.4 mT, ^{40}K 20.2 mT, and ^{85}Rb 15.5 mT resonances, but many others are closed channel dominated.[9,10]

The ^{40}K^{87}Rb 54.69 mT resonance, for which $\delta\mu = 2.40\mu_B$ and $s_{res} = 2.07$, is an example of a resonance tending to open channel dominance. Thus, Fig. 8 shows that it has a universal domain over about a third of its width, and in fact, a calculation of Z shows that $Z < 0.4$ for detuning up to 0.1 mT. At 54.6 mT near the field of the experiment of Ni et al.[41] the energy is $E/h = -329$ kHz, $Z = 0.28$, and the scattering length is 443 a_0, predicting a "universal" energy of -336 kHz. Fig. 10 shows that the open channel component $f_{aa,aa}(R)$ of the coupled channels wave function at 54.6 mT extends to very large $R \gg \bar{a}$. This entrance channel component has a norm of 0.718 and is the dominant part of the wave function. The wave function becomes even more "open channel dominated" as B increases towards resonance at B_0. It will show more closed channel character as B tunes farther away from resonance.

Fig. 10 shows the $f_{rb,aa}(R)$ closed channel component of the bound state wave function at 54.2 mT, where the energy of the bound state is $E/h = -10.6$ MHz, the detuning from resonance is about 1.6 resonance widths $|\Delta|$, and the scattering length is -71 a_0. This level is far from resonance, the bound state is far from universality, and the norm of the $f_{rb,aa}(R)$ closed channel component is 0.90. It is evident that the $f_{rb,aa}(R)$ component has a very strong overlap with the $n = -2$ bound vibrational level of the long-range potential, so that it is legitimate to characterize the level to a good approximation as being an $n = -2$ bound state of the rb channel. The other rf(−3d0) resonance in Fig. 8 is a closed channel dominated resonance with s_{res} = 0.12. It is a very narrow resonance with a very small (≪μT) domain of universality, and the below-threshold bound state has d-wave $n = -3$ rf channel character.

Using the full MQDT in its angular momentum insensitive form[73] should permit the prediction of the positions and widths of the various Feshbach resonances in all the channels of the problem. The only additional information needed to develop this theory is the analytic basis set transformation ("frame transformation") between the short-range and long-range basis sets for representing the approximate spin eigenstates of the system. This method is similar to the asymptotic bound state (ABM) method used by Wille et al.[58] to characterize ^{6}Li^{40}K resonances. The MQDT method is an "on-the-energy-shell" coupled channels method, whereas the ABM method relies on an expansion in a basis set of bound states of the $X^1\Sigma^+$ and $a^3\Sigma^+$ potentials. Either are capable of giving an approximate coupled channels representation of the bound states of the system, and finding the threshold resonances within the framework of those approximations. Either method could provide a useful alternative to full numerical coupled channels calculations.

4.3 Inelastic collision rates

Finally, there is an additional insight from MQDT that is very helpful in estimating, and in some cases quantitatively calculating, the rate coefficients for ultracold atomic or molecular collisions. The basic idea of the method is given in ref. 72 and was implemented for atomic collisions with Penning ionization by refs 96 and 97, and for molecular vibrational quenching collisions by ref. 98. It takes advantage of eqns (3)–(4) together with the MQDT concept of factoring the S-matrix into separate parts due to short- and long-range interactions, as we saw with the resonant width in the last section.

Let us assume that the probability for loss of the colliding species A and B is unity, given that the species are close together and strongly interacting at distances that are small compared to the scale length $\bar{a}$ of the long-range potential. If the collisional wave function were semiclassical at all R and not affected by quantum threshold effects, then one could calculate a semiclassical rate coefficient in a Langevin model, summing over all contributing partial waves that reach short distance, taking $1 - |S_{\alpha\alpha}|^2 = 1$ in eqn (4) to represent maximal loss and no reflection for the contributing partial waves. However, in the threshold regime, the threshold law associated with the long De Broglie wavelength needs to be taken into account. Let us assume that collision energy E is low enough and no electric field is present to induce an actual molecular dipole moment so that only s-waves contribute. The total s-wave loss is calculated by taking the loss probability $1 - |S_{\alpha\alpha}|^2$ to be the transmission probability for reaching short-range, namely, $1 - P_r$, where P_r is the quantum reflection probability from the long-range potential. This gives the needed quantum correction to semiclassical theory as $E \to 0$. The transmission factor can be calculated numerically[96,98] or, in some cases, analytically.[61,97,99] Given the analytic result $1 - P_r = 4k\bar{a}$ for s-wave transmission through the long-range van der Waals potential,[61,99] the rate coefficient for collisional loss from eqn (5) with $b = \bar{a}$ is

$$K_{in} = 2(h/\mu)\bar{a} \qquad (14)$$

for a "strong" molecular loss collision with unit short-range probability of an inelastic event. This formula is an example of "inelastic universality", where there is a universal rate constant depending only on the long-range potential and not on the scattering length, since there is no "back reflection" from the short-range region. Typical values of μ and $\bar{a}$ give an order of magnitude of 10^{-10} cm^3 s^{-1} for K_{in} for such collisions.

Consequently, if the probability for a short-range loss process is very high, then knowing the long-range van der Waals coefficient is sufficient to make quite accurate estimates of the threshold collision rate coefficient, as demonstrated by applying the above model to vibrational quenching of excited vibrational levels of the RbCs molecule.[98] This factorization procedure of MQDT theory[72] can be generalized to include other partial waves or resonance structure[66] and should be capable of giving much insight into molecular collisions. If the probability $\bar{P}$ of short-range reaction or loss is near unity and one has a way to calculate or estimate it, then that factor can be included in the expression for the s-wave rate constant by replacing $\bar{a}$ in eqn (14) by $\bar{a}\bar{P}$. On the other hand, if $\bar{P} \ll 1$ so there is significant reflection from the short-range region, K_{in} will depend on the scattering length a_α of the entrance channel. In any case, we see that if a collision leads to a strong, highly probable short-range loss event, the influence of the long-range potential is critical in determining how threshold modifications to Langevin theory occur in the ultracold domain.

5. Deeply bound states of ^{40}K^{87}Rb

So far we have concentrated on bound states that are very close to threshold with binding energies of less than 1 GHz. Fig. 11–13 show an expanded view of the $M = -7/2$ s-wave bound states with binding energies up to 30 GHz, or about 1 cm^{-1}, between $B = 0$ and 120 mT. The character of the vibrational levels is relatively easy to explain in this region of the spectrum, where the splitting between adjacent vibrational levels of $^1\Sigma^+$ and $^3\Sigma^+$ symmetry is much smaller than the splitting of the atomic hyperfine manifold. The singlet and triplet vibrational bound states have similar binding energies because they (accidentally) have similar scattering lengths. Each vibrational level, away from avoided crossings where different levels mix, is characterized by the dominant spin character of one of the 12 separated atom channels β shown in Fig. 6. They are also characterized by a vibrational wave function characteristic of $n = -1, -2, \ldots$ long-range levels, so that levels are located at energies that are below the corresponding separated atom limits by the binding energies of

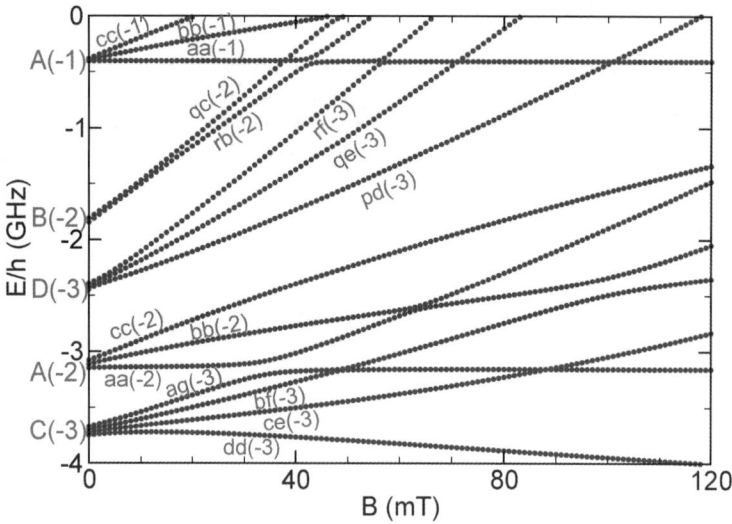

Fig. 11 Bound state energy for the $M_{tot} = -7/2$ s-wave bound states of the ^{40}K^{87}Rb molecule down to 4 GHz binding energy. The zero of energy is the energy E_α of the aa channel at each B field. The levels are labeled according to the channel index of their dominant spin component and by the vibrational quantum number n counting down from the dissociation limit of the channel. The Roman letters indicate the group in Fig. 6 with which the $B = 0$ channels are associated.

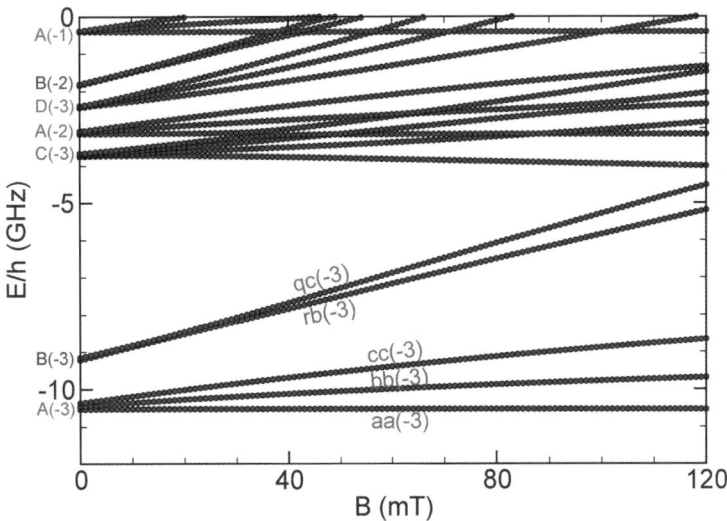

Fig. 12 Bound state energy for the $M_{tot} = -7/2$ s-wave bound states of the ^{40}K^{87}Rb molecule down to 12 GHz binding energy. The zero of energy is the energy E_α of the aa channel at each B field. Labels are the same as in Fig. 11.

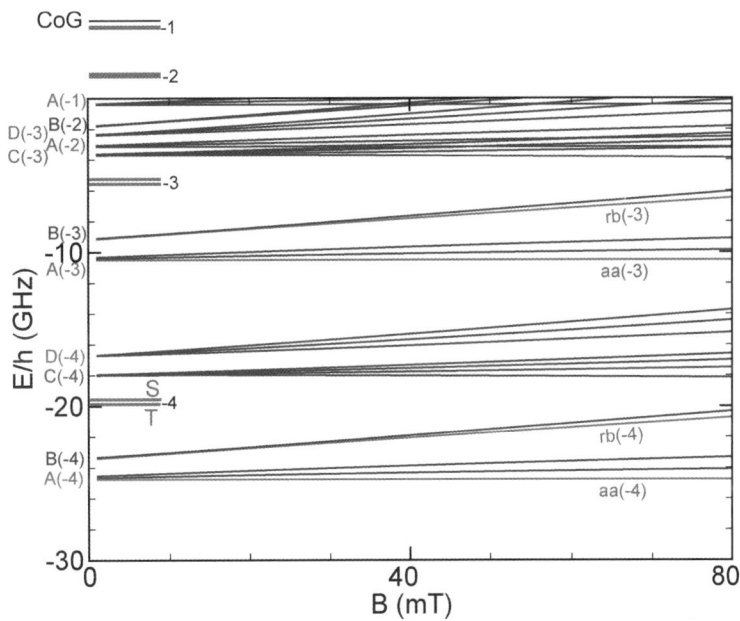

Fig. 13 Bound state energy for the $M_{tot} = -7/2$ s-wave bound states of the ^{40}K^{87}Rb molecule down to 30 GHz binding energy. The zero of energy is the energy E_α of the aa channel at each B field. Labels are the same as in Fig. 11. The short horizontal lines next to the $B = 0$ axis labeled by vibrational quantum number n show the energies of the levels of the $X^1\Sigma^+$ and $a^3\Sigma^+$ adiabatic Born–Oppenheimer potentials relative to the energy center of gravity (CoG) of the separated atom multiplet at $E/h = 4.843439$ GHz. The $X^1\Sigma^+$ and $a^3\Sigma^+$ potentials support $N = 100$ and 32 vibrational levels respectively. The normal vibrational quantum number v counting up from $v = 0$ as the lowest level is $v = N + n$.

the n levels in these channels. Fig. 11 indicates these groupings, and shows how the levels in the close-up views in Figs. 6 and 7 relate to neighboring levels. Levels of different channels and n are intermixed because of the similarity in vibrational spacings and hyperfine spacings in some cases.

Fig. 12 extends the broader view to a scale of 10 GHz. The A and B groups cluster in the 10 GHz region, since the binding energy of the $n = -3$ levels are of the order of 10 GHz. The aa (−3) level parallel to the $E = 0$ axis has the spin character of the aa entrance channel and is the level made by Ospelkaus et al.[43] by 2-color STIRAP transfer of population from the Feshbach molecule state at 54.6 mT.

Fig. 13 gives an even more expanded view down to 30 GHz binding energy, which includes the full manifold of the 12 $n = -4$ levels associated with each of the separated atom channels β. The figure also shows the energy levels of the $X^1\Sigma^+$ and $a^3\Sigma^+$ potentials, referenced to the nonrelativistic energy of the separated atoms at $B = 0$. This energy is 4.843439 GHz above the energy of two a atoms in the aa channel at $B = 0$. In this domain of binding energy, the figure shows that the splitting between the two adjacent singlet and triplet levels is much less than the splittings of the atomic hyperfine manifold. Each molecular level, except near avoided crossings of levels of different channels, thus has the βn character of two weakly bound separated atoms in channel β with the long-range vibrational wave function of level n. If the singlet and triplet potentials had very different scattering lengths, so the corresponding singlet and triplet vibrational levels were not so close in energy as in the figure, then one would begin to see a breakdown in this atomic coupling scheme in this range of energy.

Fig. 14 and 15 examine the transition from the threshold domain, where the singlet and triplet levels are mixed by spin-dependent interactions, and the more deeply bound region, where the vibrational levels are to a good approximation

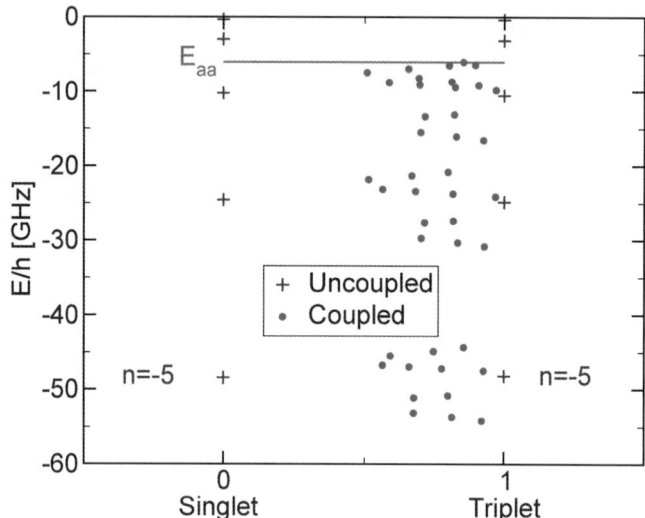

Fig. 14 Bound state energy for the $M_{tot} = -7/2$ s-wave bound states of the ^{40}K^{87}Rb molecule at $B = 54.6$ mT. All energies are relative to the energy center of gravity of the atomic hyperfine multiplets. The horizontal scale gives a measure of the singlet or triplet character of the eigenstate, with 0 representing a pure singlet state and 1 representing a pure triplet state. The horizontal line at $E/h = -4.834$ GHz labeled E_{aa} marks the energy of two a state atoms on this scale. The crosses labeled "Uncoupled" mark the energies of the $X^1\Sigma^+$ and $a^3\Sigma^+$ energy levels; $n = -5$ indicates the fifth level down in each potential. The red dots labeled "Coupled" locate the actual eigenstates of the coupled channels calculation. No levels have pure singlet character in this domain.

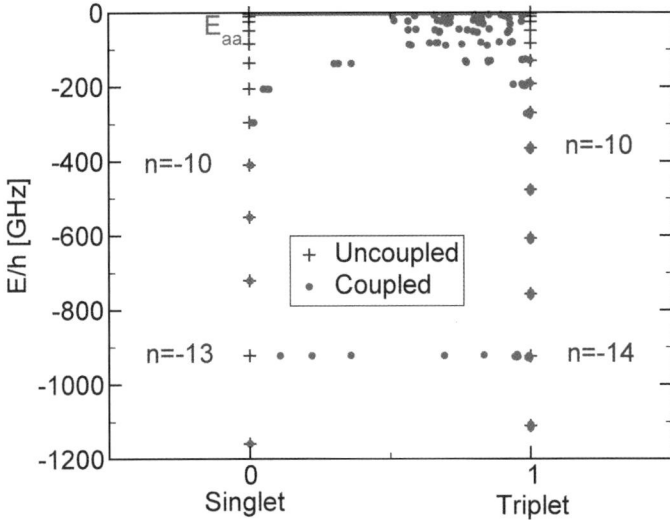

Fig. 15 Bound state energy for the $M_{tot} = -7/2$ s-wave bound states of the ^{40}K^{87}Rb molecule at $B = 54.6$ mT. The labeling is the same as in Fig. 14, but the energy scale extends to $E/h = 1200$ GHz binding energy. Uncoupled singlet levels with $|n| \leq 7$ are strongly mixed with triplet levels of the same n. The eigenstates separate into two sets of eigenstates of nearly pure singlet or triplet character when the binding energy becomes larger than around 200 GHz. There is an accidental near-degeneracy of the $n = -13$ ($v = 87$) singlet level and $n = -14$ ($v = 18$) triplet level near 900 GHz binding that results in strong mixing between singlet and triplet levels in that energy range. There are no more accidental degeneracies with strong perturbations for any levels with lower energy.

dominantly singlet or triplet in character. This transition occurs for binding energies of around 200 GHz, where the splitting between adjacent $n = -8$ singlet and triplet vibrational levels at -205.987 GHz and -192.174 GHz becomes of the order of the atomic hyperfine splittings. Levels with less binding tend to be of mixed singlet and triplet character. More deeply bound levels are clearly identified as being dominantly singlet or triplet, with the exception that the $n = -13$ singlet level at -922.325 GHz has an accidental near-degeneracy with the $n = -14$ triplet level at -924.070 GHz. This gives rise to a strong mixing of the two levels that is very local in energy. Singlet–triplet mixing is very small for all levels with larger binding energy.

Finally, Fig. 16 shows the spin structure of the $v = 0$ $J = 0$ $X^1\Sigma^+$ ground state molecule. Since the two active electrons are paired into a singlet state with no net spin, and since the projection on the molecular axis $\Lambda = 0$ for a Σ molecular state, there is no coupling of the nuclear spins to the electrons or the molecular axis. If there were no coupling to distant triplet electronic states, all nuclear spin components of the $v = 0$ $J = 0$ $X^1\Sigma^+$ level would be degenerate. Fig. 16 shows a slight zero-field splitting at $B = 0$ of 41 kHz between the lowest energy $I_{tot} = 11/2$ and highest energy $I_{tot} = 5/2$ nuclear spin components, where I_{tot} is the resultant of the ^{40}K and ^{87}Rb nuclear spins. This very small splitting in our model is due to second-order coupling through the distant $a^3\Sigma^+$ state. This model calculation based on atomic spin coupling constants may not be accurate in the molecular environment of the spins in the $v = 0$ level. However, the nuclear spin structure at the 54.6 mT field of the experiment of ref. 41 represents essentially uncoupled nuclear spins in the large B field, where the energies are very close to being the energies of the isolated separate atoms in the same field. Thus, the ^{40}K atom splits into nine Zeeman components and the ^{87}Rb atoms splits into four components with a total spread of 3.4 MHz across the manifold of levels at 54.6 mT.

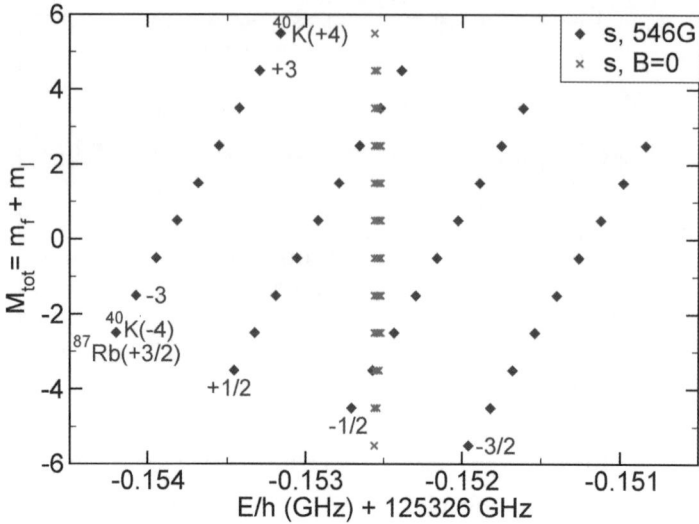

Fig. 16 Bound state energy for all M_{tot} $v = 0$ $J = 0$ bound states of the ^{40}K^{87}Rb molecule at $B = 0$ (x symbols) and 54.6 mT (diamond symbols). Energy is relative to the energy center of gravity of the atomic hyperfine multiplets. At $B = 0$ the levels divide into four degenerate groups with total nuclear spin quantum number $I_{tot} = 11/2, 9/2, 7/2$, and $5/2$ in order of increasing energy, with a spread of 41 kHz between the highest and lowest energies. At $B = 54.6$ mT the nuclear spins become uncoupled from one another and represent two independent nuclear spins in the strong B field. The labels indicate the projection quantum numbers for the four projection levels of the ^{87}Rb nuclear spin and the nine levels of the ^{40}K nuclear spin.

It is a very interesting question to determine to what extent individual nuclear spin states can be prepared and manipulated using the very precise optical control that may be possible with the ultra high precision frequency comb technology that went into the STIRAP experiment of Ni et al.[41] This may be possible using a combination of polarization and frequency control. There also is the question of the collisional stability of the $v = 0$ ground state molecules. If reactive collisions are inhibited by a reaction barrier at µK temperatures, then the only destructive collisions that change the state are ones that depolarize or relax the nuclear spins. Even at zero field, the spread of energy of 41 kHz corresponds to E/k_B of 2 µK. Due to the absence of coupling of the nuclear spins to the electrons or molecular axis in the problem, spin depolarization due to molecule–molecule collisions probably has a very small inelastic loss rate constant (perhaps too small to be measurable). However, a collision with a ^{40}K or ^{87}Rb atom could have a non-negligible inelastic collision rate coefficient, due to the coupling of the nuclear spins in the molecule to the unpaired atomic electron during the collision. It will be important to determine if the nuclear spins can be controlled in a polar molecule gas or lattice. For example, the spins might make good quantum memory, subject to weak decoherence, but capable of rapid optical manipulation and measurement.

It is worth noting the excellent quality of the potentials in ref. 77 that were used in this calculation. We calculate the energy of the $v = 0$ $J = 0$ $X^1\Sigma^+$ level relative to the energy of two a state atoms at 54.588 mT to be $E/h = -125320.10$ GHz. This differs by only 0.4 GHz from the measured value of $-125319.703(1)$ GHz, which corresponds to an error of only 3 parts in 10^6 in the absolute binding energy. The calculations for the $a^3\Sigma^+$ state are not as accurate. The lowest spin component of the $v = 0$ $N = 0$ triplet state was measured[77] to be at $-7180.4180(5)$ GHz, compared to a calculated value of -7195.6 GHz. The much larger error of 16 GHz is likely due to the lack of accurate spectroscopic data for the lowest vibrational levels to constrain the minimum of the $a^3\Sigma^+$ potential. The new, very precise data on the absolute molecular binding energies should aid the construction of much more accurate potentials for future use.

6. Concluding remarks

This paper has presented some very general considerations for ultracold atomic and molecular collisions and interactions as well as some specific results for the ^{40}K^{87}Rb fermionic molecule. A general consideration is that ultracold collisions very precisely prepare the collision complex of two colliding species in a very sharp energy state, which then serves as a gateway for various kinds of spectroscopic probing and manipulation of the complex. Thus, one can manipulate the properties of an ultracold gas or lattice using tunable scattering resonances, one can use such resonances to make a near-threshold bound state, and one can probe or populate levels far from threshold by optical manipulation. Thus, resonant scattering theory and an account of near-threshold bound states are of crucial importance to ultracold atomic and molecular physics.

Much progress can be made by understanding the spectrum and resonances associated with the long-range potential. Good theoretical progress in understanding the more complex world of molecular collisions can be made by developing theories that explicate the near-threshold domain of resonant scattering. Much of the physics associated with strong short-range interactions can be parameterized by energy-insensitive parameters of multichannel quantum defect theory. A simple example of applying these ideas is in understanding the magnitude of rate constants for inelastic molecular vibrational relaxation, where in the ultracold domain, one only needs to know the quantum reflection off the long-range potential between the two colliding species, given that relaxation has unit probability if the colliding species reach the short-range region. Such ideas can be generalized to include resonances and weak short-range processes. It remains to be seen how successful such an approach can be in the more complex environment of molecular collisions.

The bound and threshold scattering states of the ^{40}K^{87}Rb molecule illustrate some important aspects of molecular physics related to the formation and use of a vibrational ground state molecule. The domain from near-threshold to the most deeply bound levels can be understood quite precisely from coupled channels calculations. These require the full Hamiltonian for the problem, including accurate molecular adiabatic Born–Oppenheimer potentials and spin coupling constants. In general these can only be obtained from a combination of accurate *ab initio* calculations and precise spectroscopy. While excellent quality potentials are available for KRb, these will need to be developed for other systems of future interest. Much insight can be gained and practical calculations can be done in the near-threshold domain from a combination of methods involving the properties of the long-range potential and various implementations and approximations based on the long-range potential. These methods require a semi-empirical approach where parameters of the theoretical models, such as scattering lengths for the potentials, need to be extracted from fitting experimental data. Such methods are already highly developed for atomic collisions, but still need to be developed for molecular collisions, if possible. It is clear, however, that accurate theoretical models of the spin structure of ground state molecules will be needed, including the effect of external magnetic, electric, or electromagnetic fields. Ultracold resonant scattering theory of colliding molecular species with structure must first understand the various scattering channels associated with the internal structure of the colliding species, and then try to understand the resonances or other scattering properties associated with the long-range potential. Finally, the threshold resonances and dynamics associated with strong short-range interactions of the molecular collision complex will be difficult to predict accurately from first principles, but may be accessible to a combination of theoretical modeling and experimental probing of the complex.

References

1. W. Ketterle, *Rev. Mod. Phys.*, 2002, **74**, 1131–1151.
2. E. A. Cornell and C. E. Wieman, *Rev. Mod. Phys.*, 2002, **74**, 875–893.
3. F. Dalfovo, S. Giorgini, L. P. Pitaevskii and S. Stringari, *Rev. Mod. Phys.*, 1999, **71**, 463–512.
4. *Ultracold Fermi Gases*, ed. M. Inguscio, W. Ketterle, and C. Salomon, IOS Press, Amsterdam, 2008.
5. S. Giorgini, L. P. Pitaevskii and S. Stringari, *Rev. Mod. Phys.*, 2008, **80**, 1215–1274.
6. I. Bloch, *Science*, 2008, **319**, 1202–1203.
7. I. Bloch, J. Dalibard and W. Zwerger, *Rev. Mod. Phys.*, 2008, **80**, 885.
8. V. A. Yurovsky, M. Olshanii and D. S. Weiss, *Adv. At. Mol. Opt. Phys.*, 2008, **55**, 61–138.
9. T. Köhler, K. Góral and P. S. Julienne, *Rev. Mod. Phys.*, 2006, **78**, 1311–1361.
10. C. Chin, R. Grimm, P. S. Julienne, and E. Tiesinga, arXiv:0812.1496, 2008.
11. P. S. Jessen and I. H. Deutsch, *Adv. At. Mol. Opt. Phys.*, 1996, **37**, 95–136.
12. I. Bloch, *Nat. Phys.*, 2005, **1**, 23–30.
13. M. Greiner and S. Fölling, *Nature*, 2008, **435**, 736–738.
14. W. D. Phillips, *Rev. Mod. Phys.*, 1998, **70**, 721–741.
15. S. Chu, *Rev. Mod. Phys.*, 1998, **70**, 685–706.
16. C. N. Cohen-Tannoudji, *Rev. Mod. Phys.*, 1998, **70**, 707–719.
17. J. T. Bahns, W. C. Stwalley and P. L. Gould, *J. Chem. Phys.*, 1996, **104**, 9689–9697.
18. H. L. Bethlem, G. Berden and G. Meijer, *Phys. Rev. Lett.*, 1999, **83**, 1558–1561.
19. R. V. Krems, *Phys. Chem. Chem. Phys.*, 2008, **10**, 4079–4092.
20. K. M. Jones, E. Tiesinga, P. D. Lett and P. S. Julienne, *Rev. Mod. Phys.*, 2006, **78**, 483–535.
21. J. M. Hutson and P. Soldán, *Int. Rev. Phys. Chem.*, 2006, **25**, 497–526.
22. F. H. Mies, E. Tiesinga and P. S. Julienne, *Phys. Rev. A*, 2000, **61**, 022721.
23. S. J. J. M. F. Kokkelmans, H. M. J. Vissers and B. J. Verhaar, *Phys. Rev. A*, 2001, **63**, 031601R.
24. D. Jaksch, V. Venturi, J. I. Cirac, C. J. Williams and P. Zoller, *Phys. Rev. Lett.*, 2002, **89**(4), 040402.
25. B. Damski, L. Santos, E. Tiemann, M. Lewenstein, S. Kotochigova, P. S. Julienne and P. Zoller, *Phys. Rev. Lett.*, 2003, **90**(11), 110401.
26. J. M. Sage, S. Sainis, T. Bergeman and D. DeMille, *Phys. Rev. Lett.*, 2005, **94**, 203001.
27. J. Cubizolles, T. Bourdel, S. J. J. M. F. Kokkelmans, G. V. Shlyapnikov and C. Salomon, *Phys. Rev. Lett.*, 2003, **91**, 240401.
28. S. Jochim, M. Bartenstein, A. Altmeyer, G. Hendl, C. Chin, J. Hecker Denschlag and R. Grimm, *Phys. Rev. Lett*, 2003, **91**, 240402.
29. K. E. Strecker, G. B. Partridge and R. G. Hulet, *Phys. Rev. Lett.*, 2003, **91**, 080406.
30. K. Xu, T. Mukaiyama, J. Abo-Shaeer, J. Chin, D. Miller and W. Ketterle, *Phys. Rev. Lett.*, 2003, **91**, 210402.
31. M. Greiner, C. A. Regal and D. S. Jin, *Nature*, 2003, **412**, 537–540.
32. S. Dürr, T. Volz, A. Marte and G. Rempe, *Phys. Rev. Lett.*, 2004, **92**, 020406.
33. J. Herbig, T. Kraemer, M. Mark, T. Weber, C. Chin, H.-C. Nägerl and R. Grimm, *Science*, 2003, **301**, 1510–1513.
34. M. Mark, T. Kraemer, J. Herbig, C. Chin, H.-C. Nägerl and R. Grimm, *Europhys. Lett.*, 2005, **69**, 706–712.
35. T. Bourdel, L. Khaykovich, J. Cubizolles, J. Zhang, F. Chevy, M. Teichmann, L. Tarruell, S. J. J. M. F. Kokkelmans and C. Salomon, *Phys. Rev. Lett.*, 2004, **93**, 050401.
36. S. Jochim, M. Bartenstein, A. Altmeyer, G. Hendl, S. Riedl, C. Chin, J. Hecker Denschlag and R. Grimm, *Science*, 2003, **302**, 2101–2103.
37. M. W. Zwierlein, C. A. Stan, C. H. Schunck, S. M. F. Raupach, S. Gupta, Z. Hadzibabic and W. Ketterle, *Phys. Rev. Lett.*, 2003, **91**, 250401.
38. G. Thalhammer, K. Winkler, F. Lang, S. Schmid, R. Grimm and J. Hecker Denschlag, *Phys. Rev. Lett.*, 2006, **96**, 050402.
39. K. Winkler, F. Lang, G. Thalhammer, P. van der Straten, R. Grimm and J. Hecker Denschlag, *Phys. Rev. Lett.*, 2007, **98**, 043201.
40. J. G. Danzl, E. Haller, M. Gustavsson, M. J. Mark, R. Hart, N. Bouloufa, O. Dulieu, H. Ritsch and H.-C. Nägerl, *Science*, 2008, **321**, 1062–1066.
41. K.-K. Ni, S. Ospelkaus, M. H. G. de Miranda, A. Pe'er, B. Neyenhuis, J. J. Zirbel, S. Kotochigova, P. S. Julienne, D. S. Jin and J. Ye, *Science*, 2008, **322**, 231–235.
42. J. J. Zirbel, K.-K. Ni, S. Ospelkaus, T. L. Nicholson, M. L. Olsen, C. E. Wieman, J. Ye, D. S. Jin and P. S. Julienne, *Phys. Rev. A*, 2008, **78**, 013416.
43. S. Ospelkaus, A. Pe'er, K.-K. Ni, J. J. Zirbel, B. Neyenhuis, S. Kotochigova, P. S. Julienne, J. Ye and D. S. Jin, *Nat. Phys.*, 2008, **4**, 622–626.
44. C. Ospelkaus, S. Ospelkaus, L. Humbert, P. Ernst, K. Sengstock and K. Bongs, *Phys. Rev. Lett.*, 2006, **97**, 120402.

45 A. Micheli, G. K. Brennen and P. Zoller, *Nat. Phys.*, 2006, **2**, 341.
46 M. Lewenstein, *Nat. Phys.*, 2006, **2**, 309.
47 H. P. Büchler, E. Demler, M. Lukin, A. Micheli, N. Prokof'ev, G. Pupillo and P. Zoller, *Phys. Rev. Lett.*, 2007, **98**, 060404.
48 D. DeMille, *Phys. Rev. Lett.*, 2002, **88**, 067901.
49 H. R. Thorsheim, J. Weiner and P. S. Julienne, *Phys. Rev. Lett.*, 1987, **58**, 2420–2423.
50 R. J. Napolitano, J. Weiner, C. J. Williams and P. S. Julienne, *Phys. Rev. Lett.*, 1994, **73**(10), 1352–1355.
51 P. O. Fedichev, Y. Kagan, G. V. Shlyapnikov and J. T. M. Walraven, *Phys. Rev. Lett.*, 1996, **77**(14), 2913–2916.
52 J. L. Bohn and P. S. Julienne, *Phys. Rev. A*, 1999, **60**(1), 414–425.
53 P. Naidon and P. S. Julienne, *Phys. Rev. A*, 2006, **74**, 062713.
54 M. Mark, F. Ferlaino, S. Knoop, J. G. Danzl, T. Kraemer, C. Chin, H.-C. Nägerl and R. Grimm, *Phys. Rev. A*, 2007, **76**, 113201.
55 M. Mark, T. Kraemer, P. Waldburger, J. Herbig, C. Chin, H.-C. Nägerl and R. Grimm, *Phys. Rev. Lett.*, 2007, **99**.
56 F. Lang, P. van der Straten, B. Brandstätter, G. Thalhammer, K. Winkler, P. S. Julienne, R. Grimm and J. Hecker Denschlag, *Nat. Phys.*, 2008, **4**, 223.
57 A. Marte, T. Volz, J. Schuster, S. Dürr, G. Rempe, E. G. M. van Kempen and B. J. Verhaar, *Phys. Rev. Lett.*, 2002, **89**, 283202.
58 E. Wille, F. M. Spiegelhalder, G. Kerner, D. Naik, A. Trenkwalder, G. Hendl, F. Schreck, R. Grimm, T. G. Tiecke, J. T. M. Walraven, S. J. J. M. F. Kokkelmans, E. Tiesinga and P. S. Julienne, *Phys. Rev. Lett.*, 2008, **100**, 053201.
59 J. M. Hutson, E. Tiesinga and P. S. Julienne, *Phys. Rev. A*, 2008, **78**, 052703.
60 G. F. Gribakin and V. V. Flambaum, *Phys. Rev. A*, 1993, **48**, 546–553.
61 H. Friedrich and J. Trost, *Phys. Rep.*, 2004, **397**, 359449.
62 J. L. Bohn, A. Avdeenkov and M. P. Deskevich, *Phys. Rev. Lett.*, 2003, **89**, 043006.
63 A. V. Avdeenkov and J. L. Bohn, *Phys. Rev. Lett.*, 2003, **90**, 203202.
64 A. V. Avdeenkov, D. C. E. Bortolotti and J. L. Bohn, *Phys. Rev. A*, 2004, **69**, 012710.
65 J. Aldegunde, B. A. Rivington, P. S. Zuchowski and J. M. Hutson, *Phys. Rev. A*, 2008, **78**, 033434.
66 M. Machholm, P. S. Julienne and K.-A. Suominen, *Phys. Rev. A*, 2001, **64**, 033425.
67 R. Ciuryło, E. Tiesinga and P. S. Julienne, *Phys. Rev. A*, 2005, **71**, 030701R.
68 F. H. Mies, *J. Chem. Phys.*, 1969, **51**, 787–797.
69 F. H. Mies, *J. Chem. Phys.*, 1969, **51**, 798–807.
70 F. H. Mies, *J. Chem. Phys.*, 1984, **80**, 2514–2525.
71 F. H. Mies and P. S. Julienne, *J. Chem. Phys.*, 1984, **80**, 2526–2536.
72 P. S. Julienne and F. H. Mies, *J. Opt. Soc. Am. B*, 1989, **6**, 2257.
73 B. Gao, E. Tiesinga, C. J. Williams and P. S. Julienne, *Phys. Rev. A*, 2005, **72**, 042719.
74 P. S. Julienne and B. Gao in *Atomic Physics 20*, ed. C. Roos, H. Häffner and R. Blatt, AIP, Melville, New York, 2006, pp. 261–268.
75 P. S. Julienne, K. Burnett, Y. B. Band and W. C. Stwalley, *Phys. Rev. A*, 1998, **58**, 797R–800, R.
76 E. Hodby, S. T. Thompson, C. A. Regal, M. Greiner, A. C. Wilson, D. S. Jin and E. A. Cornell, *Phys. Rev. Lett.*, 2005, **94**, 120402.
77 A. Pashov, O. Docenko, M. Tamanis, R. Ferber, H. Knöckel and E. Tiemann, *Phys. Rev. A*, 2007, **76**, 022511.
78 H. T. C. Stoof, J. M. V. A. Koelman and B. J. Verhaar, *Phys. Rev. B*, 1988, **38**, 4688–4697.
79 E. Arimondo, M. Inguscio and P. Violino, *Rev. Mod. Phys.*, 1977, **49**, 31–76.
80 C. Klempt, T. Henninger, O. Topic, J. Will, W. Ertmer, E. Tiemann and J. Arlt, *Phys. Rev. A*, 2007, **76**, 020701R.
81 F. Ferlaino, C. D'Errico, G. Roati, M. Zaccanti, M. Inguscio, G. Modugno and A. Simoni, *Phys. Rev. A*, 2006, **73**, 040702R.
82 F. Ferlaino, C. D'Errico, G. Roati, M. Zaccanti, M. Inguscio, G. Modugno and A. Simoni, *Phys. Rev. A*, 2006, **74**, 039903E.
83 C. C. Tsai, R. S. Freeland, J. M. Vogels, H. Boesten, B. J. Verhaar and D. J. Heinzen, *Phys. Rev. Lett.*, 1997, **79**(7), 1245–1248.
84 F. A. van Abeelen, B. J. Verhaar and A. J. Moerdijk, *Phys. Rev. A*, 1997, **55**, 4377–4381.
85 F. A. van Abeelen and B. J. Verhaar, *Phys. Rev. A*, 1999, **59**, 578–584.
86 J. M. Vogels, B. J. Verhaar and R. H. Blok, *Phys. Rev. A*, 1998, **57**, 4049–4052.
87 F. H. Mies and M. Raoult, *Phys. Rev. A*, 2000, **62**, 012708.
88 M. Raoult and F. H. Mies, *Phys. Rev. A*, 2004, **70**, 012710.
89 Z. Idziaszek, T. Calarco, P. S. Julienne and A. Simoni, *Phys. Rev. A*, 2009, **79**, 010702R.
90 B. Gao, *Phys. Rev. A*, 1998, **58**, 1728–1734.
91 B. Gao, *Phys. Rev. A*, 1998, **58**, 4222–4225.

92 B. Gao, *Phys. Rev. A*, 2000, **62**, 050702.
93 B. Gao, *Phys. Rev. A*, 2001, **64**, 010701.
94 M. Kitagawa, K. Enomoto, K. Kasa, Y. Takahashi, R. Ciurylo, P. Naidon and P. S. Julienne, *Phys. Rev. A*, 2008, **77**, 012719.
95 N. Nygaard, B. I. Schneider and P. S. Julienne, *Phys. Rev. A*, 2006, **73**, 042705.
96 C. Orzel, M. Walhout, U. Sterr, P. S. Julienne and S. L. Rolston, *Phys. Rev. A*, 1999, **59**, 1926–1935.
97 A. S. Dickinson, *J. Phys. B*, 2007, **40**, F237–F240.
98 E. R. Hudson, N. B. Gilfoy, S. Kotochigova, J. M. Sage and D. DeMille, *Phys. Rev. Lett.*, 2008, **100**, 203201.
99 R. Côté, H. Friedrich and J. Trost, *Phys. Rev. A*, 1997, **56**, 1781–1787.

Two-photon coherent control of femtosecond photoassociation

Christiane P. Koch,[*a] Mamadou Ndong[a] and Ronnie Kosloff[b]

Received 20th October 2008, Accepted 15th December 2008
First published as an Advance Article on the web 6th May 2009
DOI: 10.1039/b818458e

Photoassociation with short laser pulses has been proposed as a technique to create ultracold ground state molecules. A broad-band excitation seems the natural choice to drive the series of excitation and deexcitation steps required to form a molecule in its vibronic ground state from two scattering atoms. First attempts at femtosecond photoassociation were, however, hampered by the requirement to eliminate the atomic excitation leading to trap depletion. On the other hand, molecular levels very close to the atomic transition are to be excited. The broad bandwidth of a femtosecond laser then appears to be rather an obstacle. To overcome the ostensible conflict of driving a narrow transition by a broad-band laser, we suggest a two-photon photoassociation scheme. In the weak-field regime, a spectral phase pattern can be employed to eliminate the atomic line. When the excitation is carried out by more than one photon, different pathways in the field can be interfered constructively or destructively. In the strong-field regime, a temporal phase can be applied to control dynamic Stark shifts. The atomic transition is suppressed by choosing a phase which keeps the levels out of resonance. We derive analytical solutions for atomic two-photon dark states in both the weak-field and strong-field regime. Two-photon excitation may thus pave the way toward coherent control of photoassociation. Ultimately, the success of such a scheme will depend on the details of the excited electronic states and transition dipole moments. We explore the possibility of two-photon femtosecond photoassociation for alkali and alkaline-earth metal dimers and present a detailed study for the example of calcium.

I. Introduction

Coherent control was conceived to solve the problem of optimizing the outcome of a chemical reaction.[1,2] Initial work was devoted to photodissociation.[3,4] In this study, we address the inverse problem of controlling the free-to-bound transition in photoassociation,[5] where atoms are assembled into a molecule by laser light. The underlying principle of coherent control is interference of different pathways that lead from the initial state to the final outcome. Typically, in a free-to-bound transition, the relative phase between the reactants is not defined:[2,6] If the wavefunction of the initial state can be written as a product of the wavefunctions of the atoms, the outcome of the binary reaction cannot be controlled. At very low temperatures, these considerations do not hold. Ultracold collisions are characterized by threshold effects.[7,8] The kinetic energy is very small, and the inner part of the scattering

---

[a]*Institut für Theoretische Physik, Freie Universität Berlin, Arnimallee 14, 14195 Berlin, Germany. E-mail: ckoch@physik.fu-berlin.de*
[b]*Institute of Chemistry and The Fritz Haber Research Center, The Hebrew University, Jerusalem 91904, Israel*

wavefunction is dominated by the potential energy. As a result, the colliding pair becomes entangled. Ultracold atoms are therefore the best candidates for coherent control of a binary reaction. The formation of molecules in ultracold atomic gases is presently attracting significant interest.[9]

Previous proposals for short-pulse photoassociation were all based on one-photon transitions. Initially the goal was to optimize the excitation process in photoassociation by introducing a chirp.[10,11] In order to produce molecules in their ground electronic state, a pump-dump scheme was studied:[12–14] A first pulse excites two atoms and creates a molecular wave packet. This is followed by free dynamics on the electronically excited state bringing the wave packet to shorter internuclear distances. There a second pulse stabilizes the product by transferring amplitude to the ground electronic state. The pump-dump scheme can be supplemented by a third field to engineer the excited state wave packet dynamics to create favorable conditions for the dump pulse.[15]

Experiments aiming at femtosecond photoassociation with a one-photon near-resonant excitation were faced with the obstacle that weakly bound molecular levels are very close to an atomic resonance. The challenge is to populate these levels without exciting the atomic transition. The weakly bound molecular levels possess the biggest free-bound Franck–Condon factors. However, the atomic transition matrix elements are several orders of magnitude larger. Excitation of the atomic transition is followed by spontaneous emission and trap loss, *i.e.* it leads to a depletion of the trap.[16,17] Pulses with bandwidths of a few wavenumbers corresponding to transform-limited pulse durations of a few picoseconds were proposed as a remedy.[12,13] However, for such narrow bandwidths, pulse shaping capabilities have yet to be developed. Experimenters employing femtosecond pulses with bandwidths of the order of 100 cm^{-1} have resorted to suppressing the spectral amplitude at the atomic resonance frequency by placing a razor knife into the Fourier plane of the pulse shaper.[16–18] Due to the finite spectral resolution, this leads also to the suppression of spectral amplitude which could excite the molecular levels with the largest free-bound Franck–Condon factors as illustrated in Fig. 1 (right-hand side). On the other hand, the spectral amplitude from the peak of the spectrum as well as from just below the cut finds almost no ground state population to excite since the probability of finding two colliding atoms at short internuclear distance is extremely small (left-hand side of Fig. 1). As a result, only a tiny part of the spectral amplitude contributes to the photoassociation signal.[18]

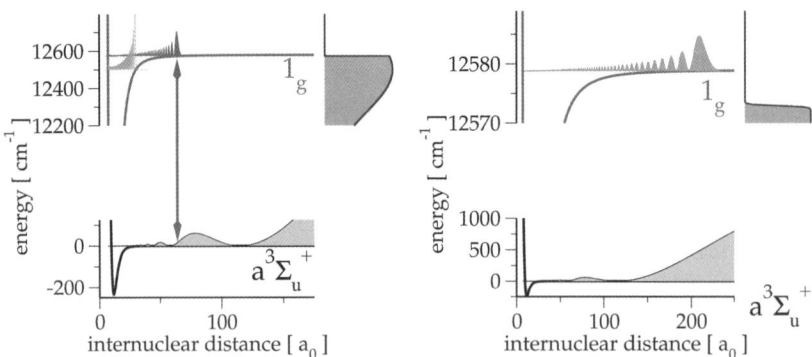

Fig. 1 In one-photon photoassociation, blocking the spectral amplitudes at, and close to, the atomic resonance prevents weakly bound molecular levels with large free-bound Franck–Condon factors from being excited (right-hand side). The majority of spectral components do not excite any ground state population since the probability of finding two colliding atoms at short internuclear distance is tiny (left-hand side, data shown here for Rb_2). In femtosecond photoassociation employing a one-photon transition, the large majority of spectral components pass through the sample without any effect.

Here we suggest the employment of two-photon transitions for femtosecond photoassociation. The main idea consists of rendering the atomic transition dark by applying spectral or temporal phase control while exciting population into weakly bound molecular levels by two-photon transitions. The spectral resolution is determined by that of the pulse shaper, typically of the order of one wavenumber. This should allow for the population of weakly bound molecular levels with the best free-bound Franck–Condon factors.

Two-photon control is based on constructive or destructive interference of all two-photon pathways adding up to the two-photon transition frequency.[19,20] It allows for the excitation of a very narrow atomic transition by a broad band femtosecond pulse.[19] In particular, an anti-symmetric spectral phase optimizes non-resonant two-photon population transfer, while a symmetric spectral phase yields a dark pulse.[19] For weak fields, the two-photon absorption yield can be calculated within perturbation theory, allowing for an analytical derivation of suitable phase functions. At intermediate fields, four-photon pathways additionally contribute to the two-photon absorption, requiring a higher order in the perturbation treatment.[21,22] In both the weak-field and intermediate-field regime, rational pulse shaping is based on frequency domain arguments.[19–22] For strong fields, a time-domain picture becomes more adequate since the dynamics are dominated by time-dependent Stark shifts.[23–25] The atom accumulates a phase due to the dynamic Stark shift. A suitable control strategy consists of eliminating this phase by applying a time-dependent phase function, *i.e.* by chirping the pulse.[24] This allows for maintenance of a π-pulse condition despite the levels being strongly Stark shifted.

These results for coherent control of two-photon absorption in atomic systems serve as our starting point to derive pulses which are dark at the atomic transition but optimize two-photon photoassociation, *i.e.* population transfer into molecular levels. The shape of the pulses yielding a dark resonance of the atoms are derived analytically: in the weak-field and intermediate-field regimes, the two-photon spectrum is required to be zero at the atomic transition frequency, while for strong fields a time-dependent phase allows the maintenance of an effective 2π-pulse condition for the atoms. The shaped pulses are tested numerically for molecule formation. The paper is organized as follows: an effective two-state Hamiltonian is derived in Section II by adiabatically eliminating all off-resonant levels and invoking the two-photon rotating-wave approximation (RWA). In Section III, conditions for pulses that are dark at the two-photon atomic transition are derived analytically. Photoassociation is studied in Sections IV and V, with Section IV reviewing possible two-photon excitation schemes for alkali and alkaline-earth dimers and Section V presenting the numerical study of two-photon photoassociation for calcium. Section VI concludes.

II. Theoretical framework for multi-photon excitations

We extend the treatment of multi-photon transitions in atoms[19–23,25] to include the vibrational degree of freedom, $\hat{R}$, of a diatomic molecule. To this end, we write the nuclear Hamiltonian for each electronic state, $\hat{H}_i = \hat{T} + V_i(\hat{R}) + \omega_i$ (setting $\hbar = 1$), *i.e.* all potentials, $V_i(\hat{R})$, go to zero asymptotically and the atomic excitation energies are contained in ω_i. $\hat{T}$ denotes the vibrational kinetic energy. Assuming the dipole approximation for the matter-field interaction, the total Hamiltonian is given by

$$\hat{H} = \hat{H}_0 + \hat{\mu} E(t), \qquad (1)$$

where $\hat{\mu}$ denotes the dipole operator and $E(t)$ the laser field with envelope $S(t)$, $E(t) = \frac{1}{2} E_0 S(t)(e^{\frac{i}{2}\varphi(t)} e^{-i\omega_L t} + c.c.)$. For weak and intermediate fields, the dynamics under the Hamiltonian, $\hat{H}$, can be solved by perturbation theory. To the lowest order, the two-photon absorption is given in terms of the two-photon spectral amplitude on resonance,[19–21] $S_2 \sim \left| \int_{-\infty}^{+\infty} \tilde{E}(\omega_{ge}/2 + \omega)\tilde{E}(\omega_{ge}/2 - \omega)d\omega \right|^2$, where ω_{ge} is the two-photon resonance frequency and $\tilde{E}(\omega)$ the Fourier transform of $E(t)$. For an atomic system, conditions for a dark two-photon resonance are derived analytically by setting the integral to zero, see below in Section III A.

For strong fields, we follow the derivation of Trallero-Herrero et al.[23,25] The matter Hamiltonian is expanded in the electronic basis, $|i\rangle\langle i|$,

$$\hat{H}_0 = \sum_i \left(\hat{T} + V_i(\hat{R}) + \omega_i\right)|i\rangle\langle i|, \tag{3}$$

and the electronic states are separated into initial ground state, final excited state and intermediate states. If the one-photon detunings of the intermediate states, $\omega_{m,g/e} - \omega_L$, are large compared to the spectral bandwidth of the laser pulse, $\Delta\omega_L$, the intermediate states can be adiabatically eliminated.[23] Since the electronic excitation energies are much larger than the vibrational energies, the adiabatic elimination in the molecular model proceeds equivalently to the atomic case. Invoking a two-photon RWA, an effective two-state Hamiltonian is obtained,

$$\hat{H}(t) = \begin{pmatrix} \hat{T} + V_g(\hat{R}) + \omega_g^S(t) & \chi(t)e^{-i\varphi(t)} \\ \chi(t)e^{i\varphi(t)} & \hat{T} + V_e(\hat{R}) + \Delta_{2P} + \omega_e^S(t) \end{pmatrix}, \tag{4}$$

where $\omega_{g/e}^S(t)$ denotes the dynamic Stark shift of the ground and excited state, $\chi(t)$ the two-photon coupling and Δ_{2P} the two-photon detuning, $\Delta_{2P} = \omega_{eg} - 2\omega_L$. The two-photon coupling is given in terms of the one-photon couplings and one-photon detunings of the ground and excited state to the intermediate levels,

$$\chi(t) = -\frac{1}{4}E_0^2 S(t)^2 \sum_m \frac{\mu_{em}\mu_{mg}}{\omega_{mg} - \omega_L}, \tag{5}$$

Similarly, the dynamic Stark shifts are obtained as

$$\omega_i^S(t) = -\frac{1}{2}E_0^2 |S(t)|^2 \sum_m |\mu_{mi}|^2 \frac{\omega_{mi}}{\omega_{mi}^2 - \omega_L^2}, \quad i = g, e. \tag{6}$$

The transition dipole matrix elements, μ_{ij}, and the transition frequencies, ω_{mi}, are in general R-dependent. They tend to a constant value (the atomic dipole moments and transition frequencies) at large internuclear distances that are important for photoassociation. The R-dependence is therefore neglected in the following.

Eqn (5) and (6) contain couplings between the intermediate and the ground and excited states, but couplings between intermediate states are neglected. For very strong fields, these higher order couplings should be included. This type of adiabatic elimination is valid if the Stark shift is small relative to the energy spacing between the ground and excited states. The validity of the conditions for adiabatic elimination and two-photon RWA are easily checked in the atomic case by comparing the dynamics under the effective Hamiltonian, eqn (4), to those under the full Hamiltonian, eqn (1), including all intermediate levels. Possibly the Born–Oppenheimer potential energy surfaces in eqn (3) show crossings or avoided crossings in the range of R that is addressed by the pulse. In such a case, eqn (4) needs to be extended to include all coupled electronic states.

For strong-field control, it is instructive to choose the rotating frame such that the phase terms of the coupling in eqn (4) are expressed as time-dependent frequencies† and to time-dependently shift the origin of energy.[23] The transformed Hamiltonian is written,

---

† This corresponds to the fact that in the RWA a chirp can be expressed either as a phase term of the electric field envelope or as instantaneous laser frequency.

$$\hat{\mathbf{H}}(t) = \begin{pmatrix} \hat{T} + V_g(\hat{R}) & \chi(t) \\ \chi(t) & \hat{T} + V_e(\hat{R}) + \left(\delta_\omega^S(t) + \Delta_{2P} + \varphi(t)\right) \end{pmatrix}, \quad (7)$$

where the differential Stark shift

$$\delta_\omega^S(t) = \omega_e^S(t) - \omega_g^S(t) \quad (8)$$

is introduced. In the atomic case, strong-field control consists of locking the phase corresponding to the term in parenthesis in eqn (7), *i.e.* in keeping

$$\Phi(t) = \int_{-\infty}^{t} \delta_\omega^s(\tau)d\tau + \Delta_{2P}t + \varphi(t) \quad (9)$$

constant. This can be achieved by proper choice of phase of the laser pulse, $\varphi(t)$, and it corresponds to maintaining resonance by instantaneously correcting for the dynamic Stark shift.[25]

III. Two-photon dark pulses

A. Solution in the weak- and intermediate-field regime

At low field intensities, the population transfer is dominated by the resonant two-photon pathways. Therefore control strategies are best understood in the frequency domain.[21] Starting from the Hamiltonian in the two-photon RWA, eqn (4), the condition for a dark pulse at the atomic transition can be derived analytically. The atomic transition corresponds to the asymptotic limit of eqn (4), *i.e.* to a two-level system. Since the two-photon coupling, $\chi(t)$, is proportional to the field amplitude squared, $E_0^2 S(t)^2$, we look for a spectral phase such that the Fourier transform of $S(t)^2$,

$$F(\omega) = \int_{-\infty}^{+\infty} S^2(t)e^{i\omega t}dt = \int_{-\infty}^{+\infty} \tilde{S}(\omega')\tilde{S}(\omega - \omega')d\omega',$$

becomes zero at the two-photon resonance, $F(\omega_{eg}) = 0$ ($\tilde{S}(\omega)$ is the Fourier transform of the envelope function $S(t)$). We assume a two-photon resonant pulse, $\omega_L = \omega_{eg}/2$. Then

$$F(\omega_{eg}) = 0 = F(2\omega_L) = \int_{-\infty}^{+\infty} \tilde{S}(\omega')\tilde{S}(2\omega_L - \omega')d\omega' \\ = \int_{-\infty}^{+\infty} \tilde{S}(\omega_L + \omega)\tilde{S}(\omega_L - \omega)d\omega. \quad (10)$$

The condition in eqn (10) is formally equivalent to that obtained in second order perturbation theory.[19,21] However, here we have derived it by considering the effective two-photon Hamiltonian, eqn (4). The condition of eqn (10) is sufficient for a two-photon dark pulse as long as the states are not strongly Stark-shifted, *i.e.* it holds also at intermediate field intensities. This finding is rationalized by considering the next order contribution. The fourth-order terms are classified into on-resonant and near-resonant terms.[21] The dominant fourth-order term is the one that includes the resonant second-order term and is therefore cancelled by the same condition.

We assume the pulse shape to be Gaussian. Then

$$\tilde{S}(\omega) = A_0 e^{-\frac{(\omega - \omega_L)^2}{2\sigma_\omega^2}} e^{i\phi(\omega)}$$

with σ_ω the spectral bandwidth, and the two-photon Fourier transform becomes

$$F(2\omega_L) = A_0^2 \int_{-\infty}^{+\infty} e^{-\frac{\omega^2}{\sigma_\omega^2}} e^{i(\phi(\omega_L + \omega) + \phi(\omega_L - \omega))} d\omega.$$

We now need to choose the spectral phase $\phi(\omega)$ such as to render $F(2\omega_L)$ zero. This can be achieved by a step of the phase $\phi(\omega)$ by π.[19,20]

We derive here the position of this step, ω_s. Let us assume that $\omega_s > \omega_L$, and denote $\Delta\omega_s = \omega_s - \omega_L$. Then

$$\phi(\omega_L + \omega) + \phi(\omega_L - \omega) = \begin{cases} 0 & \text{if} \quad \omega \geq \Delta\omega_s, \\ 0 & \text{if} \quad \omega \leq -\Delta\omega_s, \\ \pi & \text{if} \quad -\Delta\omega_s < \omega < \Delta\omega_s. \end{cases} \quad (11)$$

The π-step implies a change of sign in the integral, and the dark pulse condition is rewritten accordingly,

$$F(2\omega_L)/A_0^2 = 0 = \int_{-\infty}^{-\Delta\omega_s} e^{-\frac{\omega^2}{\sigma_\omega^2}} d\omega + \int_{\Delta\omega_s}^{+\infty} e^{-\frac{\omega^2}{\sigma_\omega^2}} d\omega - \int_{-\Delta\omega_s}^{\Delta\omega_s} e^{-\frac{\omega^2}{\sigma_\omega^2}} d\omega$$

$$= 2\int_{\Delta\omega_s}^{+\infty} e^{-\frac{\omega^2}{\sigma_\omega^2}} d\omega - 2\int_{0}^{-\Delta\omega_s} e^{-\frac{\omega^2}{\sigma_\omega^2}} d\omega.$$

We recognize the error function up to a factor and obtain

$$\text{Erf}\left(\frac{\Delta\omega_s}{\sigma_\omega}\right) = \frac{1}{2}, \quad (12)$$

which determines $\Delta\omega_s$ and hence the step position, ω_s. That is, if one chooses $\Delta\omega_s$ such that $\text{Erf}(\sigma\Delta\omega_s) = 1/2$, the two-photon spectrum is zero at the two-photon transition frequency, $F(2\omega_L) = 0$, and a dark pulse is obtained. Interestingly we can also combine the conditions of a symmetrical function and a phase step. This is achieved by defining two phase steps symmetrical around ω_L, $\omega_L \pm \omega_s$. Following a derivation along the same lines as outlined above, the identical condition for ω_s is obtained.

B. Solution in the strong-field regime

At high intensities, the effective Hamiltonian varies during the pulse. A time-domain picture is therefore best adapted to develop control strategies. Population inversion in the effective Hamiltonian can be achieved by adjusting the temporal phase of the laser such as to maintain a π-pulse condition.[23,24] Here we turn this strategy upside-down to achieve a dark pulse, *i.e.* we ask for the 2π-pulse condition. We restrict the notation to the atomic case for simplicity. The Hamiltonian, eqn (7), can be transformed such that the diagonal elements are zero,[25]

$$\hat{H}(t) = \begin{pmatrix} 0 & \chi(t)e^{-i\Phi(t)} \\ \chi(t)e^{i\Phi(t)} & 0 \end{pmatrix}. \quad (13)$$

Evaluating Rabi's formula for the final state amplitude leads to the condition

$$\left|\int_0^{t_0} \chi(t)e^{i\Phi(t)} dt\right| = 2k\pi, \quad k = 0,1\ldots \quad (14)$$

for zero population transfer to the excited state. We again assume a two-photon resonant pulse, $\omega_{ge} = 2\omega_L$. The condition, eqn (14), is fulfilled given the phase $\Phi(t)$ obeys

$$\Phi(t) = \begin{cases} \pi & \text{if} \quad t_0 - \Delta t_x < t < t_0 + \Delta t_x, \\ 0 & \text{elsewhere}. \end{cases} \quad (15)$$

This corresponds to the following temporal laser phase

$$\varphi(t) = \begin{cases} -\pi - \int_0^t \delta_\omega^S(\tau)d\tau & \text{if} \quad t_0 - \Delta t_x < t < t_0 + \Delta t_x, \\ -\int_0^t \delta_\omega^S(\tau)d\tau & \text{elsewhere} \end{cases} \quad (16)$$

For a Gaussian pulse, the time-dependence of the differential Stark shift is also Gaussian, and $\varphi(t)$ can be evaluated analytically. Inserting the definition of $\Phi(t)$, eqn (9), into the condition, eqn (14), and evaluating the integrals yields a condition for Δt_x,

$$\text{Erf}\left(\frac{\Delta t_x}{\sigma}\right) = \frac{1}{2}. \quad (17)$$

A pulse with the temporal phase defined by eqn (16) and (17) suppresses population transfer by keeping the two levels out of resonance.

Such a solution can also be obtained by applying local control[26] to the effective Hamiltonian, eqn (13). Local control theory allows the extension of the analysis to the full molecular problem. We can seek conditions which eliminate atomic two-photon transitions while maximizing the molecular population transfer. To this end, we define an operator to be minimized, $\hat{P}_{at} = \int_{R_0}^{+\infty} dR |e\rangle\langle e|$ which projects onto the atomic levels, $|e\rangle$ denotes the excited electronic state. R_0 is a distance which is larger than the bond length of the last bound level of the excited state potential. The objective to be maximized is defined by the projection onto the molecular levels for photoassociation,

$$\hat{P}_{mol} = \sum_i |\varphi_i^e\rangle\langle\varphi_i^e|, \quad (19)$$

where $|\varphi_i^e\rangle$ denote vibrational eigenstates of the electronically excited state. In local control theory, we seek the phase of the field such that

$$\frac{d}{dt}\langle\hat{P}_{at}\rangle = 0, \quad \frac{d}{dt}\langle\hat{P}_{mol}\rangle > 0, \quad (20)$$

i.e. atomic transitions are suppressed while the molecular population is monotonically increasing.[27] Eqn (16) is a specific analytical example of local control. The first part of the pulse up to time $t \leq t_0 - \Delta t_x$ builds up an excited state population and a phase relation between the ground and excited levels. At this instant in time the phase of the field jumps by π so that it becomes perpendicular to the phase of the instantaneous transition dipole, thus eliminating additional population transfer. The last part of the pulse restores the original population.

IV. Two-photon excitation schemes for photoassociation

Our goal is to populate molecular levels in a long-range potential while suppressing excitation of the atomic transition. We will restrict the discussion to homonuclear dimers where potentials with $1/R^5$ behavior occur below the $S + D$ asymptote‡. In heteronuclear dimers, the potentials go as $1/R^6$, and the improvement of two-photon

---

‡ One may also consider transitions into potentials correlating to an $S + P$ asymptote. While this is two-photon forbidden for atoms, it might be two-photon allowed at short internuclear distances. However, in that case, photoassociation will occur only at distances below 20 Bohr. This corresponds to large detunings, and a two-photon control scheme is then not required.

Table 1 Two-photon transitions which are candidates for two-photon photoassociation in the near-infrared for alkali and alkaline earth atoms

Atom	Transition	Wavelength
Cs	$6s \rightarrow 7d$	2×768 nm
Mg	$3s^2 \rightarrow 3s3d$	4×862 nm
Ca	$4s^2 \rightarrow 3d4s$	2×915 nm
Sr	$5s^2 \rightarrow 4d5s$	2×992 nm

over one-photon femtosecond photoassociation is expected to be less significant. Alkali and alkaline earth species, where cooling and trapping of atoms is well established, are considered. Carrier frequencies in the near-infrared are assumed throughout.

A. Two-photon transitions in alkali dimers

A one-photon transition cannot be controlled by interference between different photon pathways. A prerequisite for our scheme is therefore that no one-photon resonances be contained within the pulse spectral bandwidth. This excludes both potassium and rubidium as possible candidates. Of the remaining alkalis, only caesium shows a two-photon resonance between the ground state and a d-level in the near-infrared, cf. Table 1.

B. Two-photon transitions in alkaline earth dimers

Of the alkaline earth atoms, calcium and strontium show two-photon resonances at the desired wavelength, cf. Table 1, and in magnesium a near-IR four-photon transition to a D level is possible. In the following we will study calcium, since potential curves for a large number of electronically excited states are known with sufficient accuracy.[28] §The two-photon excitation scheme is shown in Fig. 2: Two atoms interacting via the $X^1\Sigma_g^+$-state are excited into bound levels of the $(1)^1\Pi_g$-state or the $(2)^1\Sigma_g^+$-state that both correlate to the $^1S + {}^1D$ asymptote. Note that the ^1D term of calcium occurs energetically below the $^1P^0$ term. The dipole coupling is provided by the states with one-photon allowed transitions from the ground state, i.e. the $A^1\Sigma_u^+$, $A'^1\Pi_u$, and $B^1\Sigma_u^+$ states and further electronic states at higher energies. The effective two-photon coupling, cf. eqn (5), is determined by the dipole matrix elements of the one-photon-allowed transitions and by the one-photon detunings. Compared to the atomic two-photon transitions studied in caesium[19] and sodium,[21,22,24,25] the one-photon detunings in the alkaline earths are much larger. Higher laser intensities will therefore be required.

Fig. 3 demonstrates the existence of long-range vibrational wavefunctions in a potential with $1/R^5$ behavior for large internuclear distances R. Vibrational wavefunctions of the $(1)^1\Pi_g$ excited state potential are plotted in Fig. 3a for three different binding energies. The wavefunctions need to be compared to the initial scattering state, shown in Fig. 3b for two ground state atoms colliding with approximately 40 µK: Large free-bound Franck–Condon factors are due to sufficient probability density at large R in the excited state vibrational wavefunctions. In the following, population shall be transferred from the initial ground state atomic density into these vibrational wavefunctions close to the excited-state dissociation limit. The efficiency will be determined by the width of the dark atomic

---

§ We expect that our reasoning can be carried over to strontium with only slight modifications due to the similar electronic structure.

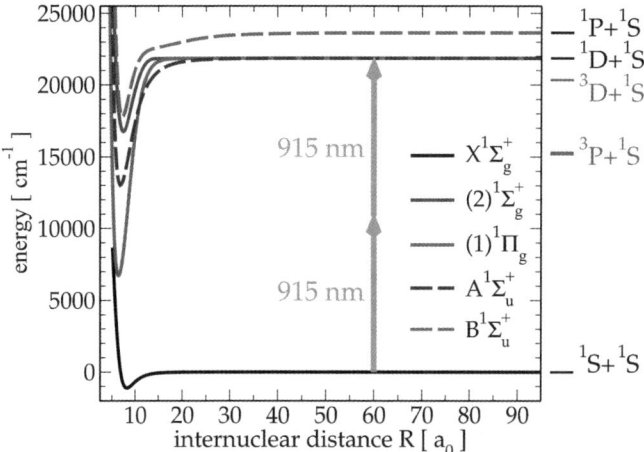

Fig. 2 Two-photon photoassociation of two atoms colliding in the $X^1\Sigma_g^+$ ground state into the $(1)^1\Pi_g$-state or the $(2)^1\Sigma_g^+$-state correlating to the $S + D$ asymptote (solid lines). The dipole coupling is provided by the states with one-photon allowed transitions from the ground state such as the $A^1\Sigma_u^+$-state or the $B^1\Sigma_u^+$-state (dashed lines). The example shown here is Ca$_2$.

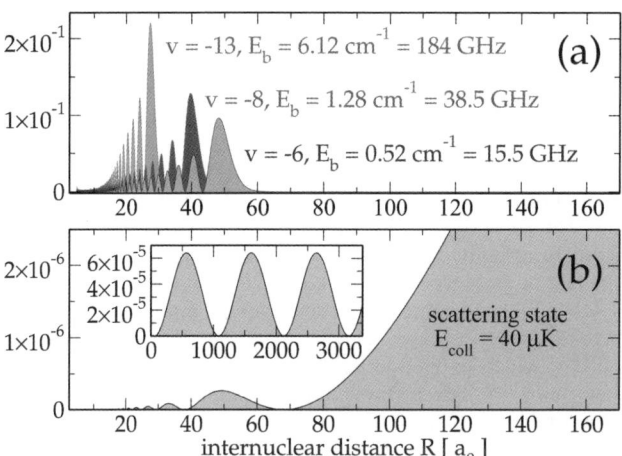

Fig. 3 Weakly bound vibrational wavefunctions of the $(1)^1\Pi_g$ excited state for three different binding energies relevant for photoassociation (a) and scattering wavefunction of two ground state calcium atoms (b).

resonance, *i.e.* by how close to the dissociation limit excitation into molecular levels can occur.

V. Two-photon photoassociation in calcium

Photoassociation of two calcium atoms is studied by numerically solving the time-dependent Schrödinger equation for the effective two-photon Hamiltonian, eqn (4), with a Chebychev propagator.[29] The Hamiltonian is represented on a grid using an adaptive grid step size.[30–32] The two-photon couplings and dynamic Stark shifts are calculated from the atomic transition dipole matrix elements and frequencies found in ref. 33. The potentials are gathered from ref. 28. An initial state with a scattering energy corresponding to 40 μK, *cf.* Fig. 3b, is chosen.

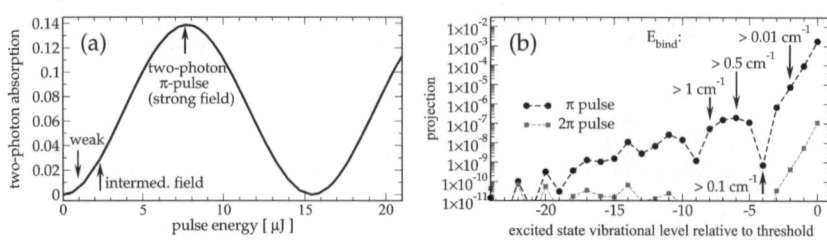

Fig. 4 (a) The excited state population shows Rabi oscillations as the pulse energy is increased. The arrows indicate the pulsed energies which are employed for shaped pulses. (b) The projection of the final excited state wave packet onto the molecular levels (with highest vibrational quantum number set to $v = 0$), i.e. the excited state vibrational distribution after the pulse for a two-photon π-pulse and a two-photon 2π-pulse. The arrows indicate the range of binding energies relevant in the discussion of the results for shaped pulses.

Fig. 4 presents calculations with transform-limited pulses of 100 fs full-width half-maximum and a central wavelength of 915 nm. They serve as a reference point for the shaped pulse calculations discussed below. Fig. 4a shows the final excited state population including both atoms and molecules as a function of pulse energy. The arrows in Fig. 4a indicate pulse energies in the weak, intermediate and strong field regime that are employed in shaped pulse calculations. Rabi oscillations are observed as the pulse energy is increased. The contrast of the oscillations is reduced from 100% to 14%. This is a clear sign of the dynamic Stark shift for strong fields. For strong (intermediate) field, the maximum Stark shift, eqn (6), amounts to 104 cm^{-1} (32 cm^{-1}) for the ground state and to 247 cm^{-1} (76 cm^{-1}) for the excited state. The weak field corresponds to a pulse energy of 0.94 µJ and yields an excited state population of 5.0×10^{-3}, the intermediate field to 2.4 µJ and $P_{exc}(t_{final}) = 2.9 \times 10^{-2}$, and the strong field to 7.6 µJ and $P_{exc}(t_{final}) = 1.4 \times 10^{-1}$.

The projection of the final wave packet onto the molecular levels of the excited state, i.e. the excited state vibrational distribution after the pulse, is shown in Fig. 4b for transform-limited two-photon π and 2π-pulses. For a π-pulse, the population of the last bound level amounts to 1.8×10^{-3} compared to an atomic population of 1.4×10^{-1}. For comparison, one-photon photoassociation with a 10 ps pulse detuned by 4 cm^{-1} achieves a population transfer into molecular levels on the order of 10^{-4} which corresponds, depending on the trap conditions, to 1–10 molecules per pulse.[13,14] The arrows in Fig. 4b indicate the binding energies of a few weakly bound levels. Due to their smaller free-bound Franck–Condon factors, deeper molecular levels are populated significantly less than the last bound level. The minimum of the excited state vibrational distribution at $v = -4$ is due to the node of the scattering wavefunction at $R \sim 68\ a_0$, cf. Fig. 3b.

Frequency-domain control is studied for weak and intermediate fields in Fig. 5. The shaped pulses are obtained by applying a spectral phase function consisting of a π-step to a transform-limited 100 fs pulse, cf. eqn (11). The position of the π-step is varied. Atomic population transfer (black filled circles) can be strongly suppressed, by 10^{-6}, for the weak field, see Fig. 5a. For the intermediate field, Fig. 5b, the dynamic Stark shift moves the levels during the pulse. The dark-pulse condition, taking into account only the static resonance frequency, eqn (10), is then no longer sufficient to strongly suppress atomic population transfer. The maximum suppression, i.e. the minimum of the final excited state population, is therefore reduced to be in the order of 10^{-3}. It is observed at the position of the π-step corresponding to eqn (12), detuned by 42.2 cm^{-1} from the carrier frequency for both weak and intermediate field.

When population transfer into all molecular levels is considered, no difference between atoms and molecules is observed; compare the filled circles and open squares in Fig. 5. In this case the last two levels which have the largest free-bound Franck–Condon factors dominate the molecular population transfer. Since their

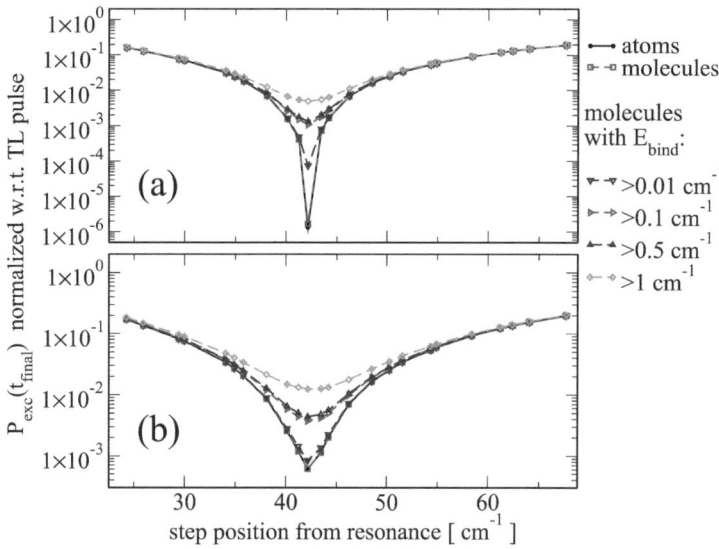

Fig. 5 Frequency-domain control with pulses shaped according to eqn (11): The excited state population of atoms (filled circles) and of molecules (open symbols), normalized with respect to the results obtained for a transform-limited pulse, is shown as a function of the π-step position for (a) weak and (b) intermediate fields. Population transfer can be strongly suppressed, by 10^{-6}, for the weak field (a). For the intermediate field (b), the dynamic Stark shift sets in, yielding a maximum suppression on the order of 10^{-3}. When all molecular levels are considered, the last two levels (that have binding energies of less than 0.01 cm^{-1}) dominate and no difference between atoms and molecules is observed. For increasing binding energy of the molecular levels, and hence increasing detuning from the atomic resonance, the two-photon transition becomes less dark.

binding energy is very small, less than 0.01 cm^{-1}, there is almost no detuning with respect to the atomic two-photon resonance, and the dark-state condition applies to the atomic transition and transitions into the last two bound levels alike. As the binding energy of the molecular levels and hence the detuning from the atomic resonance increases, the two-photon dark state condition, which is defined at the atomic resonance, becomes less applicable. Subsequently, molecular two-photon transitions become less dark, cf. open triangles and diamonds in Fig. 5. Atomic transitions are suppressed by about four orders of magnitude more then transitions into molecular levels with binding energies larger than 1 cm^{-1} for the weak field. This needs to be compared to the ratio of atomic and molecular transition matrix elements. For weakly bound levels, this ratio amounts to four to five orders of magnitude. The suppression of the atomic transition in Fig. 5a implies, therefore, that about the same number of atoms and of molecules bound by more than 1 cm^{-1} will be excited. The excited atoms will be lost. However, this loss is sufficiently small, and the trap is not depleted. Due to the high repetition rate in femtosecond experiments, many photoassociation pulses can be applied before any significant depletion of the trap occurs. At intermediate field strength, the atomic transition relative to transitions into molecular levels with binding energies larger than 1 cm^{-1} is suppressed by only two orders of magnitude. In that case, loss of atoms from the trap becomes significant, and the control strategy of applying a spectral π-step phase function is not sufficient to accumulate molecules by applying many photoassociation pulses.

For intermediate and strong fields, pulse shaping in the time-domain is more appropriate since it allows one to correct for the dynamic Stark shift. Time-domain control is studied in Fig. 6. The shaped pulses are obtained by chirping the pulse according to eqn (16) with the chirp time interval, Δt_x, determined from eqn (17). The

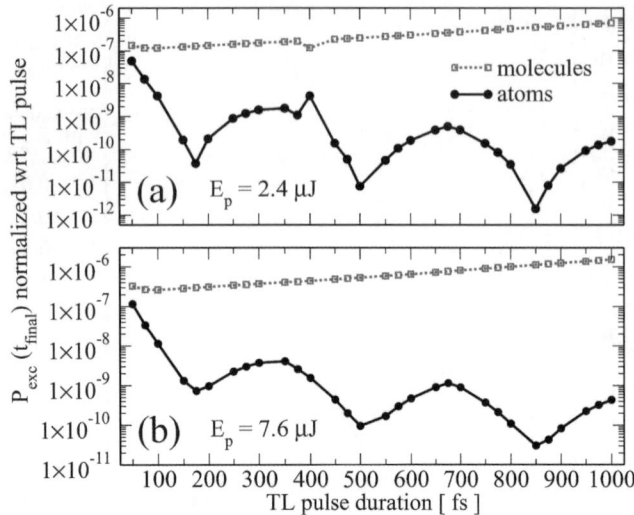

Fig. 6 Time-domain control with pulses shaped according to eqn (16): The excited state population of atoms (filled circles) and of molecules (open symbols), normalized with respect to the results obtained for a transform-limited pulse, is shown as a function of the duration of the corresponding transform-limited pulse for intermediate (a) and strong field (b). Two-photon transitions for atoms are more strongly suppressed than for molecules. However, the suppression of the molecular transitions is too strong to achieve any significant photoassociation yield.

final excited state population, normalized with respect to the transform limited case, is compared for atoms and molecules in Fig. 6 for intermediate (a) and strong (b) fields. It is plotted for increasing duration of the transform-limited pulse from which the chirped pulse is generated, keeping the pulse energy constant. For time-domain control, two-photon transitions for atoms are clearly more strongly suppressed than for molecules. The ratio of suppression of atoms *versus* molecules varies as a function of the transform-limited pulse duration between one and five orders of magnitude demonstrating a large amount of control. The overall increase of the difference between atoms and molecules with increasing pulse duration is rationalized in terms of the control strategy: The dynamic Stark shift together with the two-photon detuning leads to the accumulation of a phase in the atoms and molecules. It is countered by the laser phase $\varphi(t)$, *cf.* eqn (9). The difference between atoms and molecules becomes more perceptible for longer phase accumulation times. The modulations seem to be caused by a periodic accumulated phase of the instantaneous transition dipole. More analysis using local control theory is required. Overall, however, the suppression of the molecular transitions is much too strong to achieve any significant photoassociation yield. It is hence not enough to enforce a minimization of the atomic transition probability. The molecular transition probabilities need to be maximized explicitly, *e.g.* employing local control and eqn (20).

VI. Summary and conclusions

We have studied coherent control of femtosecond photoassociation employing two-photon transitions. The goal of our study was to identify pulses which populate molecular levels close to the atomic transition without exciting the atoms themselves. An effective two-photon Hamiltonian was derived by extending previous work on atoms[23,24] by the vibrational degree of freedom. In the weak-field to intermediate-field regime, frequency-domain control was pursued based on applying a spectral phase function. This leads to constructive or destructive interference between all pathways contributing to the two-photon absorption.[19–22] Time-domain control is more appropriate for the intermediate-field to strong-field regime where the dynamics are dominated by dynamic Stark shift.[23,24] A temporal phase function allows one to lock the atomic phase to that of the laser, keeping the atoms on or out of resonance. For both regimes, we derived analytical conditions to construct pulses which leave the atomic two-photon transition dark. These pulses were tested in a numerical study of photoassociation in calcium. An excited-state potential correlating to the $^1S + {}^1D$ asymptote was chosen. It shows a $1/R^5$ behavior at long range and supports weakly bound molecular levels with large free-bound Franck–Condon factors required for efficient photoassociation.

In the weak-field regime, applying a π-phase step reduces the atomic transition probability by a factor of 10^{-6}. Molecular levels with binding energies of about 1 cm^{-1} are sufficiently detuned from the atomic two-photon resonance that the dark condition does not apply. Transitions into these levels are suppressed by only a factor of 10^{-2}. The resulting four orders of magnitude difference between atomic and molecular transitions is sufficient to counter the different excitation probabilities for atoms and molecular levels. It hence allows for employing femtosecond pulses without depleting the trap due to excitation of atoms. However, weak-field control implies that absolute excitation probabilities are very small.

In the intermediate-field regime, the Stark shift becomes large enough to move the levels during the pulse. The spectral dark pulse condition taking into account only the static two-photon resonance is then not sufficient to suppress atomic transitions. A reduction of the atomic transition probability by merely a factor of 10^{-3} was found. Applying a femtosecond pulse will then excite significantly more atoms than molecules. Since the excited atoms are lost, trap depletion is expected to set in after a few pulse cycles.

Time-domain control based on maintaining a 2π-pulse condition for the atomic transition is applicable for both intermediate and strong fields. The ratio of suppression of atomic *versus* molecular transitions depends on the duration of the transform-limited pulses from which the shaped pulses are generated. It reaches up to five orders of magnitude. This finding is rationalized in terms of accumulation of different phases for atoms and molecules. In absolute numbers, however, the molecular transitions themselves are strongly suppressed. A significant photoassociation yield can therefore not be achieved by the simple control strategy based on the analytical result for the atomic two-photon excitation. Local control should be employed to maximize the molecular transition probabilities while keeping the atomic line dark. This is beyond the scope of the present study.

In conclusion, an impressive amount of control yielding several orders of magnitude difference between atomic and molecular transitions could be demonstrated with both spectral and temporal phase shaping. However, the two-photon photoassociation efficiencies need to be improved in terms of absolute numbers of molecules per pulse to yield a feasible photoassociation scheme. In the strong-field regime, local control offers a means for explicit maximization of molecular transition probabilities. Moreover, our discussion could be extended from two-photon to three-photon transitions. Then, excited-state potentials correlating to $S + P$ asymptotes and showing a $1/R^3$ behavior at long range could be employed. A $1/R^3$ long-range potential supports spatially more extended vibrational wavefunctions than a $1/R^5$ potential and exhibits an overall larger density of vibrational levels. Therefore, the efficiency of a shaped pulse, which allows one to excite weakly bound molecular levels in a three-photon transition while keeping the atomic resonance dark, is expected to far exceed that of the two-photon scheme.

Acknowledgements

We wish to thank Zohar Amitay and Yaron Silberberg for fruitful discussions. Financial support from the Deutsche Forschungsgemeinschaft through the Emmy Noether programme and SFB 450 is gratefully acknowledged.

References

1. P. Brumer and M. Shapiro, *Principles and Applications of the Quantum Control of Molecular Processes*, Wiley Interscience, Hoboken, New Jersey, 2003.
2. S. A. Rice and M. Zhao, *Optical control of molecular dynamics*, John Wiley & Sons, New York, 2000.
3. T. Baumert, B. Bühler, R. Thalweiser and G. Gerber, *Phys. Rev. Lett.*, 1990, **64**(7), 733.
4. V. D. Kleiman, L. Zhu, J. Allen and R. J. Gordon, *J. Chem. Phys.*, 1995, **103**(24), 10800.
5. K. M. Jones, E. Tiesinga, P. D. Lett and P. S. Julienne, *Rev. Mod. Phys.*, 2006, **78**, 483.
6. C. A. Arango, M. Shapiro and P. Brumer, *J. Chem. Phys.*, 2006, **125**, 094315.
7. J. Weiner, V. S. Bagnato, S. Zilio and P. S. Julienne, *Rev. Mod. Phys.*, 1998, **71**(1), 1.
8. F. Masnou-Seeuws and P. Pillet, *Adv. in At., Mol. and Opt. Phys.*, 2001, **47**, 53.
9. J. M. Hutson and P. Soldan, *Int. Rev. Phys. Chem.*, 2006, **25**(4), 497.
10. J. Vala, O. Dulieu, F. Masnou-Seeuws, P. Pillet and R. Kosloff, *Phys. Rev. A*, 2000, **63**, 013412.
11. E. Luc-Koenig, R. Kosloff, F. Masnou-Seeuws and M. Vatasescu, *Phys. Rev. A*, 2004, **70**, 033414.
12. C. P. Koch, E. Luc-Koenig and F. Masnou-Seeuws, *Phys. Rev. A*, 2006, **73**, 033408.
13. C. P. Koch, R. Kosloff and F. Masnou-Seeuws, *Phys. Rev. A*, 2006, **73**, 043409.
14. C. P. Koch, R. Kosloff, E. Luc-Koenig, F. Masnou-Seeuws and A. Crubellier, *J. Phys. B*, 2006, **39**, S1017.
15. C. P. Koch and R. Moszyński, *Phys. Rev. A*, 2008, **78**, 043417.
16. W. Salzmann, U. Poschinger, R. Wester, M. Weidemüller, A. Merli, S. M. Weber, F. Sauer, M. Plewicki, F. Weise, A. Mirabal Esparza, L. Wöste and A. Lindinger, *Phys. Rev. A*, 2006, **73**, 023414.
17. B. L. Brown, A. J. Dicks and I. A. Walmsley, *Phys. Rev. Lett.*, 2006, **96**, 173002.
18. W. Salzmann, T. Mullins, J. Eng, M. Albert, R. Wester, M. Weidemüller, A. Merli, S. M. Weber, F. Sauer, M. Plewicki, F. Weise, L. Wöste and A. Lindinger, *Phys. Rev. Lett.*, 2008, **100**, 233003.
19. D. Meshulach and Y. Silberberg, *Nature*, 1998, **396**, 239.
20. D. Meshulach and Y. Silberberg, *Phys. Rev. A*, 1999, **60**(2), 1287.
21. L. Chuntonov, L. Rybak, A. Gandman and Z. Amitay, *Phys. Rev. A*, 2008, **77**, 021403.
22. L. Chuntonov, L. Rybak, A. Gandman and Z. Amitay, *J. Phys. B*, 2008, **41**(3), 035504.
23. C. Trallero-Herrero, D. Cardoza, T. C. Weinacht and J. L. Cohen, *Phys. Rev. A*, 2005, **71**, 013423.
24. C. Trallero-Herrero, J. L. Cohen and T. Weinacht, *Phys. Rev. Lett.*, 2006, **96**, 063603.
25. C. Trallero-Herrero and T. C. Weinacht, *Phys. Rev. A*, 2007, **75**, 063401.
26. R. Kosloff, A. D. Hammerich and D. Tannor, *Phys. Rev. Lett.*, 1992, **69**(15), 2172.
27. A. Bartana, R. Kosloff and D. J. Tannor, *J. Chem. Phys.*, 1993, **99**(1), 196.
28. B. Bussery-Honvault and R. Moszynski, *Mol. Phys.*, 2006, **104**(13–14), 2387.
29. R. Kosloff, *Annu. Rev. Phys. Chem.*, 1994, **45**, 145.
30. V. Kokoouline, O. Dulieu, R. Kosloff and F. Masnou-Seeuws, *J. Chem. Phys.*, 1999, **110**, 9865.
31. K. Willner, O. Dulieu and F. Masnou-Seeuws, *J. Chem. Phys.*, 2004, **120**, 548.
32. S. Kallush and R. Kosloff, *Chem. Phys. Lett.*, 2006, **433**(1–3), 221.
33. NIST Atomic Spectra Database, http://physics.nist.gov/PhysRefData/ASD.

A pump–probe study of the photoassociation of cold rubidium molecules

Jovana Petrovic,[*a] David McCabe,[a] Duncan England,[a] Hugo Martay,[a] Melissa Friedman,[a] Alexander Dicks,[a] Emiliya Dimova[b] and Ian Walmsley[a]

Received 20th October 2008, Accepted 5th January 2009
First published as an Advance Article on the web 26th May 2009
DOI: 10.1039/b818494a

The dynamics of the excited state during the photoassociation of cold molecules from cold rubidium atoms is studied in a series of pump–probe experiments. Dipole transitions similar to those of the atoms are observed in the molecular signal. While such behaviour is characteristic of the long-range molecules, the photoassociation of bound molecules is confirmed in additional experiments. The pump–probe signal observed on a 250 ps time scale did not, however, reveal wavepacket oscillations predicted by theory. This result is discussed using numerical simulations of photoassociation and a modification to the current experiments that could lead to the detection of wavepacket dynamics is suggested.

1 Introduction

The success of laser cooling of atoms,[1] which culminated in the production of the Bose–Einstein condensate[2,3] and atom laser,[4] has inspired research into more complex ultracold systems. Intensive research into cold molecules over the last decade has opened up new and exciting prospects, such as the development of ultra-high precision spectroscopy,[5] coherent control of chemical reactions[6] and new techniques for quantum computing.[7,8] Atomic cooling techniques, however, cannot be generalized to molecules due to the complex internal energy level structure and subsequent unavailability of a closed-loop cooling cycle; instead, alternative cooling approaches must be taken.

Several competing techniques for cold molecule generation have been demonstrated and can be divided into two groups: the direct cooling of molecules and the association of cold molecules from cold atoms. The former approaches are efficient in the stabilization of molecules into the lowest vibrational levels but are unable to cool them translationally to μK temperatures.[9] On the other hand, methods that use cold atoms as the starting point for the formation of molecules, such as photoassociation (PA),[10–12] association *via* Feshbach resonances[13,14] and three-body collisions,[15,16] produce translationally cold molecules but with substantial vibrational energy. Recently a scheme has been demonstrated that favours spontaneous decay to the lowest vibrational ground singlet state by optically repumping all vibrational levels higher than $v = 0$.[17] In another approach, stimulated Raman adiabatic passage was used to coherently transfer weakly bound molecules held in an optical lattice to the lowest ro-vibrational ground triplet state.[18] The major drawbacks of these two

[a]*Clarendon Laboratory, Parks Road, Oxford, UK OX1 3PU. E-mail: j.petrovic1@physics.ox.ac.uk*
[b]*Institute of Solid State Physics "Acad. G. Nadjakov,", Bulgarian Academy of Science, 72 Tzarigradsko Shoussee, 1784 Sofia, Bulgaria*

methods—the lack of coherence of molecules in the ground state in the former case and the requirement for a detailed knowledge of the spectroscopic structure of molecules in the latter case—could be overcome by coherently controlled photoassociation and stabilization to the ground state. Independence of the coherent control approach from the exact spectroscopy of the molecule makes it a promising technique for the ground-state stabilization of ultracold heteronuclear molecules.

In order to preserve coherence, the excited state molecules must be stabilized during a time period shorter than their spontaneous decay time—of the order of tens of nanoseconds for the alkalis—dictating application of short-pulse lasers. If shaped, the pulses can change the cold molecule formation rate by modifying the wavepacket dynamics in the excited state as predicted by theory.[19–21] Experiments described in refs 22 and 23 have demonstrated coherent control of the formation of ground triplet molecules, resulting in the suppression instead of the expected enhancement of the formation rate. Stabilization of coherent molecules into the lowest vibrational ground states requires an additional step as proposed in ref. 21. In the suggested pump-dump scheme a broadband pump pulse excites a coherent superposition of vibrational states, creating a wavepacket that evolves within the excited electronic potential. When the wavepacket reaches the position at which its Franck–Condon overlap with the wavefunction of a bound ground state is appreciable, a dump pulse resonant with the transition between these two states is applied to transfer molecules to the stable ground state. Learning about the dynamics of the excited state during photoassociation is therefore an important step towards efficient stabilization into the ground state. Recently, a pump-probe technique has been used to study the short-delay dynamics of rubidium atoms and molecules during photoassociation[24] and it showed that both atoms and long-range molecules exhibit coherent transient oscillations during the PA pulse. In this paper, we report on pump–probe experiments performed at probe delays beyond this range. We apply coherent control to differentiate between the behaviour of atoms and molecules in order to identify the favourable PA regime for dumping into the ground state.

In Sections 2 and 3, the experimental and theoretical methods used in this work are described. Section 4 reports on the results of long (hundreds of picoseconds) and short (tens of picoseconds) pump–probe delay scans and back-to-back comparison of the atomic and molecular dynamics as functions of the PA pulse shape. These findings are discussed using the results of numerical simulations of photoassociation.

2 Experimental method

2.1 Pump–probe pulse sequence

A pump–probe experiment was designed to examine the behaviour of the excited state during photoassociation of cold Rb atoms in a magneto-optical trap (MOT). The pulse sequence used in the experiment is shown in Fig. 1a). A pump pulse excites atom pairs from the ground 5s–5s state to long-range molecules with a range of vibrational states just below the D1 line. After an adjustable time delay it is followed by the probe pulse, which ionizes molecules and atoms preparing them for detection using time-of-flight (TOF) mass spectroscopy. To avoid excitation by the MOT lasers and ensure that the atomic pairs are initially in the ground state, the MOT lasers were shut off 1 μs prior to the arrival of the pump–probe pulses and kept off until the ionization was finished. Atomic and molecular ion signals are recorded with up to 250 ps delay, long enough to detect the predicted wavepacket oscillations with estimated periods within a few hundreds of picoseconds.[20,25] Below we give a detailed description of the experimental apparatus.

2.2 Experimental set-up

2.2.1 Magneto-optical trap. ^{85}Rb atoms were cooled and trapped in a MOT generated in an ultrahigh vacuum stainless-steel chamber with multiple windows

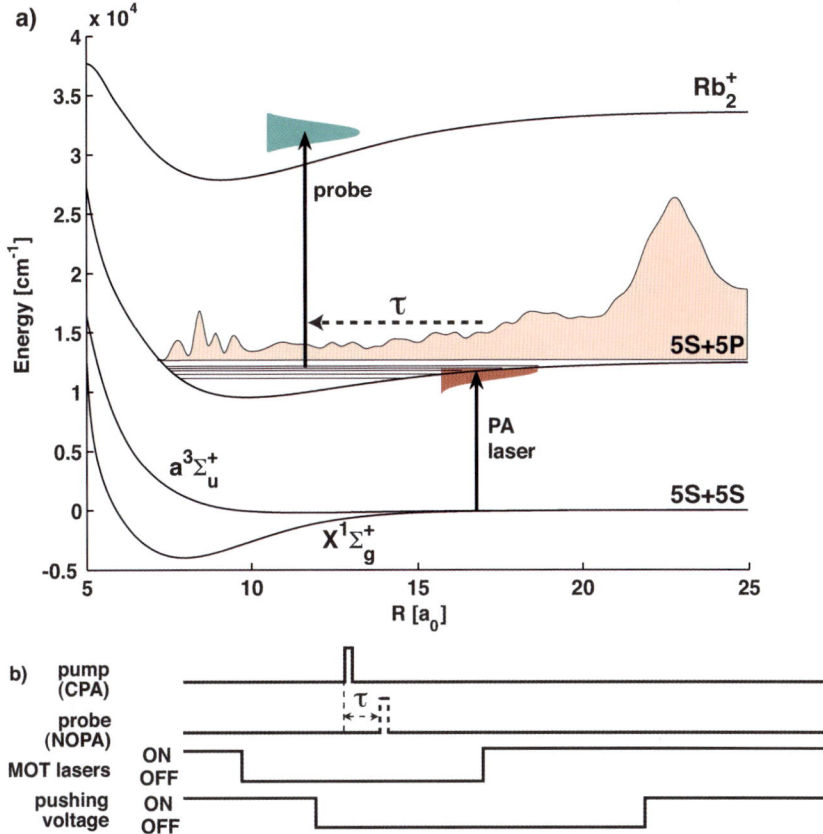

Fig. 1 a) Schematics of the pump–probe experiment. A femtosecond pump pulse photoassociates atoms from the ground state into a range of vibrational states in the excited electronic potential forming a wavepacket. An example of the simulated population density in the excited state is shown by the shaded area. The excited molecules are ionized by the probe pulse to enable their detection. b) Pulse sequence applied during the experiment. The MOT lasers are shut off 1 µs before the arrival of the pump pulse. The pushing electrode is switched to a positive voltage 3.3 µs after the ionization.

for optical access. The trapping region was formed by the intersection of three orthogonal pairs of retro-reflected laser beams of opposite circular polarizations centred in a quadrupole magnetic field generated by a pair of anti-Helmholtz coils. The trapping and repumping lasers were home-built external-cavity diode lasers. Frequency stabilization to the trapping, $F = 3 \rightarrow F' = 4$, and repumping, $F = 2 \rightarrow F' = 3$, hyperfine transitions of ^{85}Rb was realized using Doppler-free saturated absorption spectroscopy and servo-feedback on the current and the grating angle of the external-cavity diode lasers. The frequencies of the lasers were additionally tuned by acousto-optical modulators. The output power of the trapper was enhanced by a slave diode laser and its spatial mode was cleaned by propagating it through 1 m of polarization maintaining fibre, resulting in a total power of 30 mW and a $1/e^2$ beam diameter of 6.8 mm. The power of the repumper was 6 mW. The background magnetic field in the trap was nulled by three pairs of compensation coils. Rubidium was loaded into the trap from a background vapour produced by three commercial dispensers.

Fluorescence from the atoms was monitored by a photodiode and a CCD camera was used to monitor the density profile of the cold atom cloud. 4×10^6 atoms were trapped in the MOT with an estimated maximum density of 1.1×10^9 cm^{-3} and a $1/e^2$ diameter of 1.4 mm. The MOT temperature was measured by the release-and-recapture method[26] to be 110 µK.

2.2.2 Photoassociation and probe pulses.
A chirped-pulse amplified Ti:sapphire laser with an 800 nm central wavelength, 20 nm FWHM, 2kHz repetition rate and 250 µJ pulse energy was used as the master laser in the experiment. Its beam was split into two parts: one beam with pulse energies of a few µJ was used for photoassociation and the rest of the power was sent as a seed into a noncolinear optical parametric amplifier (NOPA)[27] to generate probe pulses. The spectrum of the pump pulse was cut by a razor blade in the Fourier plane of a 4f-line with a resolution of 0.1 nm. The razor blade was mounted on a precision translation stage to enable fine tuning of the cut-off wavelength. The shaped pump pulse was focused onto the MOT by a 20 cm focal length lens. The spectrum of the NOPA was tuned to 500 nm (20 nm FWHM) to correspond to the transitions between the 5s–5p$_{1/2}$ and molecular ionization states. A long-pass filter with a 480 nm cut-off was applied to prevent resonant ionization of atoms. The typical energy of a NOPA pulse was 4 µJ with a fluctuation of less than 5%. The probe pulse was combined with the pump pulse by a dichroic mirror. A variable time delay between the pump and probe pulses was introduced by reflecting the pump pulse from a retroreflector mounted on a computer-controlled translation stage. The pulses were characterized by autocorrelation and spectral phase interferometry for direct electric field reconstruction (SPIDER).[28]

2.2.3 Detection scheme.
Atomic and molecular ions were separated by time-of-flight mass spectroscopy and detected by a multi-channel plate (MCP) detector. The ions were accelerated towards the MCP by a gated pushing electrode. The electrode was switched from 0 V to a positive voltage 3.3 µs after the arrival of the PA pulse, which allowed for the expansion of the ion cloud before it was pushed towards the detector. Thereby, the concentration of ions into the central MCP channels, which causes saturation and limits the signal, was avoided. All the molecular ions arrived at the MCP within a time window narrower than 100 ns. The sensitivity of the detection was changed by varying the pushing electrode voltage from 100 V to 1100 V.

At each pump–probe delay the time-of-flight signal from the MCP was averaged 1000 times by the digital oscilloscope (Tektronix TDS744A) and then integrated numerically. The minimum averaging time for taking one data point at a fixed pump–probe delay was determined by the characteristic time of the fluctuation of the NOPA (a few seconds). The signal-to-noise ratio of the pump–probe signal was further improved by averaging several pump–probe scans.

3 Theory

A model Hamiltonian was used to calculate the expected time-dependent wavefunction of the rubidium dimers after the pump pulse. All accessible electronic states in the 5s–5s and 5s–5p manifolds were modelled using 21 electronic states – 4 ground states and 17 excited states. Hyperfine interaction and rotation of the molecule were neglected since they happen on the timescales longer than the picosecond delays used in the experiment. The model Hamiltonian has the form

$$H_{j,k} = -\frac{\hbar^2}{2\mu}\frac{\partial^2}{\partial R^2} + \delta_{j,k}V_j(R) + \xi(t)D_{j,k} + W_{j,k}(R), \qquad (1)$$

where R is the internuclear separation, μ is the reduced mass of the dimer, V_j is the Born–Oppenheimer potential of the electronic state j, $\xi(t)$ is the time-dependent electric field, $D_{j,k}$ is the transition dipole strength of the transition from the electronic state j to k and $W_{j,k}(R)$ is the spin–orbit coupling between these states. The potentials were taken from refs 29 and 30 for all of the excited states except the $A^1\Sigma_u^+$ and $b^3\Pi_u$ states which were taken from ref. 5. The ground state potentials were described by refs 31 and 32. The spin–orbit couplings between the A and b states were taken from ref. 5. The couplings between the other electronic states and the transition electric dipole strengths were approximated by their respective asymptotic values. The electric field of the pump pulse was inferred from the measured pulse spectrum and temporal profile.

A set of initial wavefunctions of both bound and unbound states was chosen and was propagated using a Chebyshev propagator on an analytic mapped grid to a time of 250 ps. Wavefunctions of stationary states were calculated using a single-channel propagation scheme.

Calculations are presented that make one of two assumptions about the initial state. The first assumption was that the molecular ions observed in the experiment originate from scattering pairs of atoms that were photoassociated by the CPA. The second was that the molecular ions instead come from loosely bound molecules in the electronic ground state photoassociated by the MOT trapping lasers.

Population densities from dimers that were initially in scattering states are compiled by adding the populations from three simulations: one singlet scattering state and two triplet scattering states. Population density calculations from loosely bound molecules are compiled by adding population density from fifteen vibrational states times two triplet spin states, weighted by the expected relative population in each of the thirty states. In this second scenario, different vibrational levels in the molecule are assumed to be populated according to the decay ratio of each of the 0_g^- states multiplied by their overlap with a distribution of ground scattering states at the MOT temperature of 110 μK:

$$\rho_{v''} \propto \sum_{v'} |\langle v''|D|v'\rangle|^2 \frac{\int_0^\infty |\langle v'|D|\varepsilon\rangle|^2 e^{-\frac{\varepsilon}{k_B T}} d\varepsilon}{\int_0^\infty e^{-\frac{\varepsilon}{k_B T}} d\varepsilon}. \quad (2)$$

Here $|v''\rangle$ is a vibrational state in the ground triplet electronic state, $|v'\rangle$ is a vibrational state in the excited 0_g^- state, and $|\varepsilon\rangle$ is a scattering state in the ground triplet state with energy ε. D is the electric dipole operator and is assumed to be constant.

The ionization step has been mimicked by a simple estimate of the detected population density. It was approximated by the expectation value of an operator O, which was calculated as a function of time to illustrate a possible time-varying observable. The observable was chosen to be

$$O_j(R) = \frac{\int_{\frac{1}{\hbar}(V_{ion}(R) - V_j(R))}^\infty \exp\left\{-4\ln 2\frac{(\omega - \omega_0)^2}{\delta^2}\right\} d\omega}{\int_{V_0}^\infty \exp\left\{-4\ln 2\frac{(\omega - \omega_0)^2}{\delta^2}\right\} d\omega}, \quad (3)$$

with ω_0 and δ chosen to correspond to the ionization pulse with FWHM of 15 nm centered at 490 nm.

4 Results and discussion

Taking into account the previous theoretical predictions that the typical periods of the wavepacket oscillations range from several tens to a few hundreds of picoseconds,[20,25] we performed the pump–probe experiment with probe delays of up to 250 ps and resolution of 3.1 ps. In order to avoid excitation to unbound atomic states, all wavelengths shorter than 795.5 nm (below the rubidium D1 line at 794.7 nm) were cut out from the pump spectrum. The pump pulse duration at the output of the shaper was 85 fs and the duration of the probe pulse was 390 fs. The energies of the pump and probe at the MOT window were 400 nJ and 1 µJ, respectively.

The molecular pump–probe signal shown in Fig. 2 comprises a step at zero delay and an oscillatory structure at positive delays. The maximum near zero delay corresponds to the dipole transients between the unbound atoms in the ground state and the long-range molecules in the excited state, as was first shown in ref. 24. The spectrum of the oscillatory structure was spread over all periods detectable at the given resolution and the length of the pump–probe scan but none of the features were dominant.

The experimental data were compared to the results of numerical simulations. When a scattering state was taken as the initial state, the dynamics of the pump–probe signal developed on the time scale of approximately 100 ps. It is attributed to the evolution of the wavepacket in the excited state. The increase in the population of the ungerade state at about 120 ps observed in Fig. 3a) can be compared to the detected increase in the molecular pump–probe signal, as shown in the inset in Fig. 2. However, the other features observed in the experimental signal at long delays are not present in the simulation and additional investigation is needed. On the other hand, when the molecules photoassociated by the MOT lasers into the ground triplet state were taken as the initial condition, the simulated pump–probe

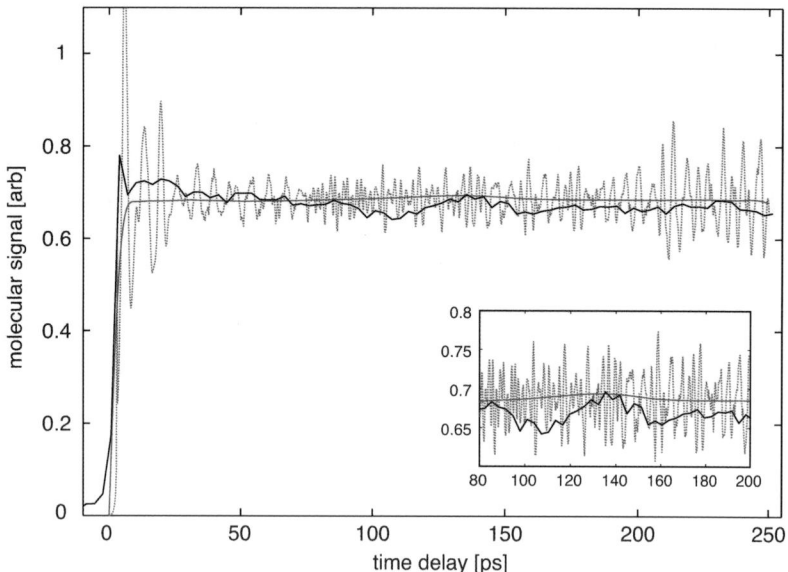

Fig. 2 Molecular pump–probe signal: black line – experimental result, solid blue line – simulation with the scattering state as the initial state, dotted blue line – simulation with the bound triplet states as initial state. The signals are normalized to have the same step at zero delay. Inset: a close-up of the same signals at the point of focusing of the wavepacket in the excited ungerade state.

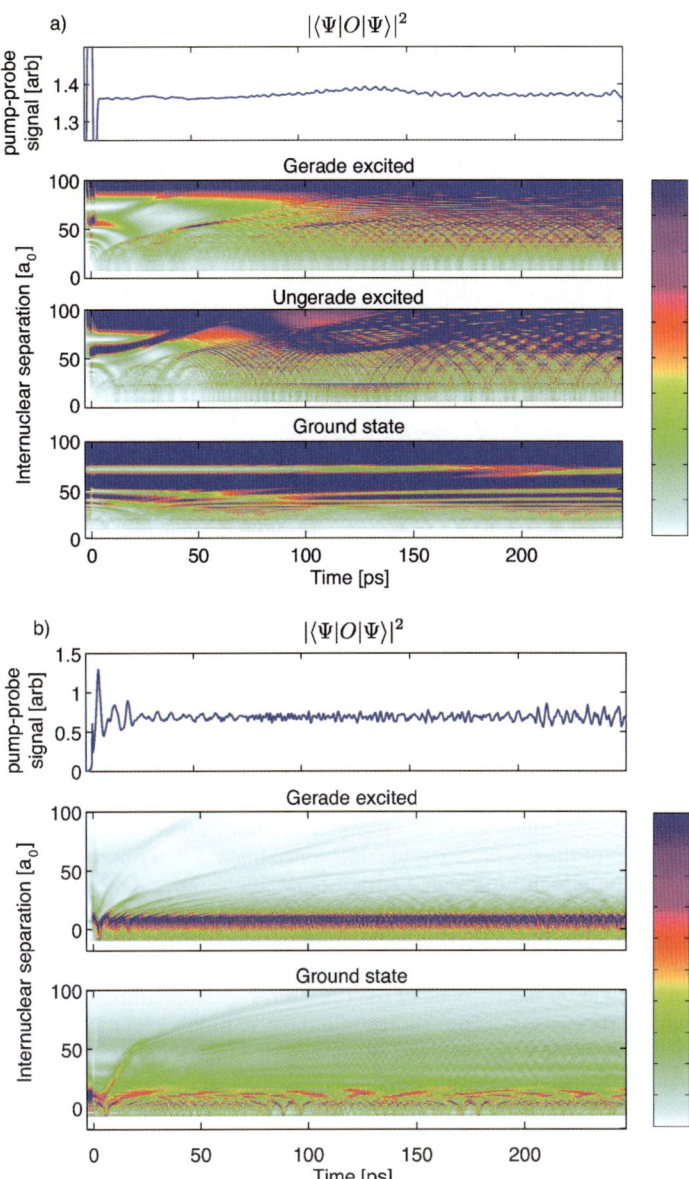

Fig. 3 Population density of the ground and excited states. The pump–probe signal obtained from these states by applying the operator O defined in Section 3. a) Population is initially in the scattering state b) Population is initially in the ground triplet bound states. The excited ungerade states do not couple to the ground triplet and hence are omitted from the simulation in b).

signal had a much richer spectrum dominated by periods of several ps oscillations, Fig. 3b). This result is comparable with the periods found in ref. 33. Although both fast and slow dynamics are observed in the experimental signal, quantitative comparison between theory and experiment can be accurate only if the initial atomic distribution between the scattering ground state and the ground triplet is known.

In order to record the atomic alongside the molecular signal, the off-resonant atomic ionization was reduced by decreasing the energy of the PA pulse to 85 nJ.

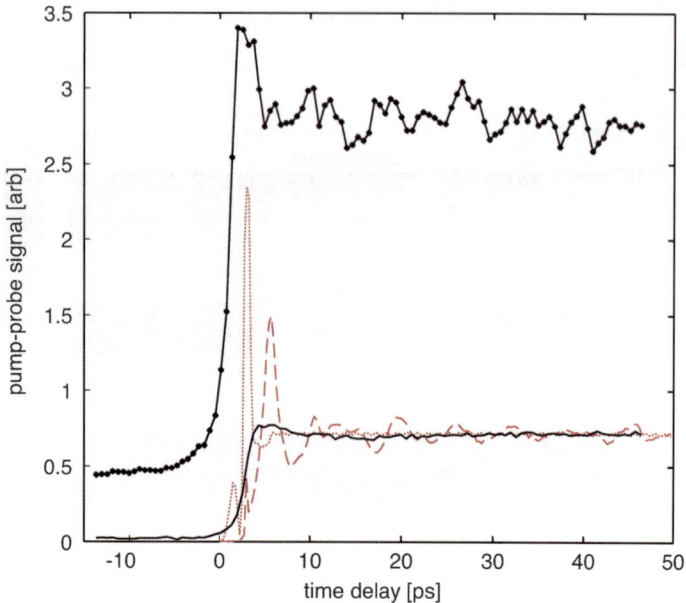

Fig. 4 Atomic (solid line with dots) and molecular signal (solid line) measured in the pump–probe experiment and simulated molecular pump–probe signals with the initial state comprising unbound states (dotted red line) and bound triplet states (dashed red line). Molecular signals are normalized to have the same step at zero delay. Atomic signal is scaled and shifted up.

The resolution of the scans was increased to 0.6 ps. The results in Fig. 4 confirmed the similarity of the behaviour of atoms and photoassociated molecules for the case of the spectrum of the PA pulse cut at 795.3 nm. This was first observed in ref. 24 by comparing the molecular ion signal with atomic fluorescence. Our study of atomic and molecular ion signals as functions of the cut-off wavelength corroborates this conclusion. The pump–probe scans were taken at cut-off wavelengths below, near and above the D1 line, Fig. 5a) and the results were compared to a model of an atom oscillating in the electric field of the PA laser, Fig. 5b). Although the resolution of the pump–probe scan was not sufficient to resolve dipole cycling observed in ref. 24, the dynamics of a free two-level atom are strikingly similar to the measured signal. As predicted, the signal decreases with the detuning of the spectral cut from the atomic resonance.

The molecular dynamics observed in the experiments described above indicate the existence of weakly bound dimers but do not give information on the, more relevant, bound state molecules. As these molecules have comparatively large Franck–Condon overlap with the low vibrational ground state molecules, photoassocation to the lower vibrational levels of the excited state is of importance for the stabilization of cold molecules in a pump-dump scheme. Here, photoassociation into the vibrational states of the excited electronic potential was investigated by scanning the position of the spectral cut. Increase in the detuning from the D1 line causes faster decay in the atomic than in the molecular signal. The shaded area in Fig. 6 shows the cut-off region at which the photoassociation of the bound molecules dominates the photoassociation of atoms. Indeed, transfer of atoms becomes negligible at cut-off wavelengths longer than 796 nm, while a substantial number of molecules are formed. The numerical simulations using the model described in Section 3 showed the same trends in molecular and atomic signals (see inset in Fig. 6). Therefore, we demonstrated that a fraction of the photoassociated molecules is bound in the vibrational states of the excited electronic state.

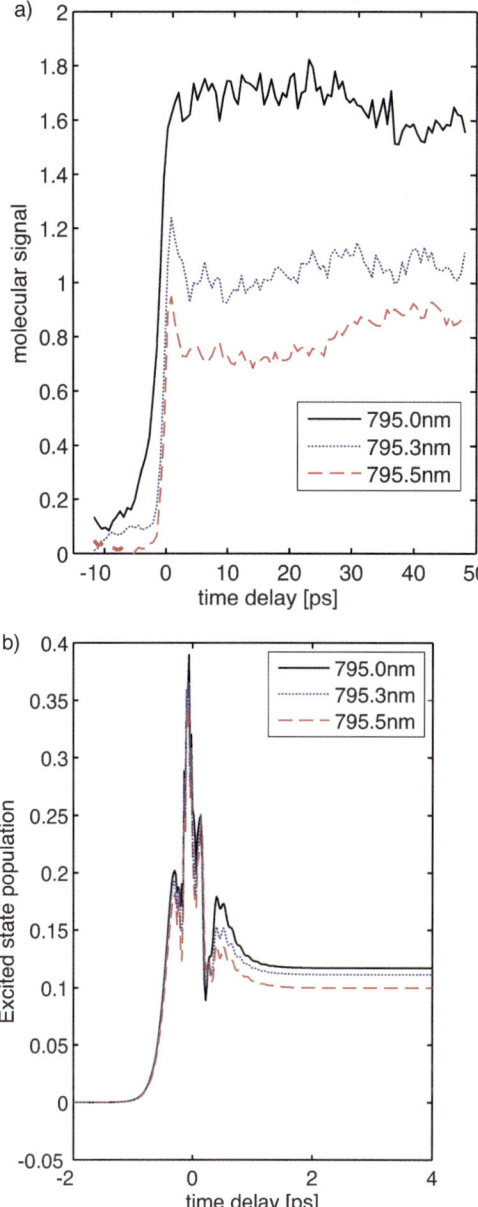

Fig. 5 a) Pump–probe dynamics of molecules recorded for different spectral cuts of the probe pulse. b) Simulation of free rubidium atoms in the laser electric field. Six excited states and two ground states were used to calculate the population transfer between the two electronic states. Spin–orbit coupling is taken into account, as is the coupling caused by a time-varying electric field. The electric field is assumed to have a Gaussian spectrum cut at a given wavelength by a sigmoid function.

The photoassociated bound molecular wavepackets are expected to exhibit dynamics different from the dipole oscillations in the laser electric field. However, spectrally more selective pump and probe pulses may be needed to detect such dynamics. The accordant refinements of the current set-up are in progress.

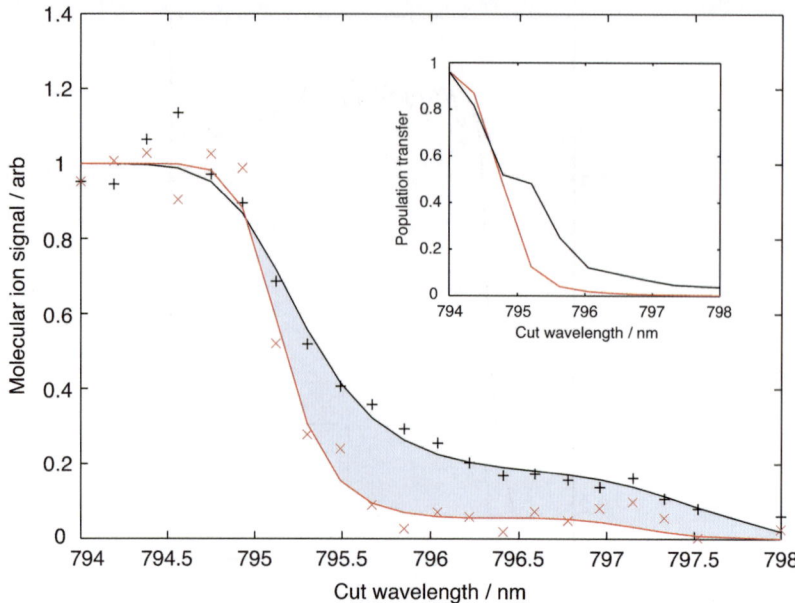

Fig. 6 Atomic (red line) and molecular (black line) signals measured at a fixed pump–probe delay for different spectral cuts of the pump pulse. The shaded area indicates evidence for the formation of bound excited-state dimers. Inset: theoretical results obtained in the weak field regime.

5 Conclusion

The dynamics of the photoassociated molecules in the excited electronic potential were investigated in a series of pump–probe experiments. The observed pump–probe signals confirmed the formation of long-range molecules, the dynamic behaviour of which are similar to the dipole oscillations of free atoms. On the other hand, the molecular dynamics over a few hundreds of ps were observed to qualitatively agree with the numerical simulations of photoassociation. An experiment in which the cut-off wavelength of the pump pulse spectrum was scanned to the red of the D1 line was performed to investigate the photoassociation of bound molecules. A roll-off of the atomic ion signal with an increase in the detuning from the D1 line was much faster than the decrease in the molecular signal, proving the photoassociation of bound molecules in the excited electronic potential. This opens prospects for a relatively simple pump-dump scheme for stabilization of molecules in the singlet ground state.

Acknowledgements

The authors are grateful to Beatrice Chatel and Jordi Mur-Petit for valuable discussions and suggestions. Hugo Martay acknowledges help from Thorsten Köhler. This work was supported by the EPSRC Grant No. EP/D002842/1 and the visit of Emiliya Dimova by the EU projects CAMEL, EMALI and FASTQUAST and the Bulgarian NSF Grant No. WU-301/07.

References

1 W. D. Phillips, *Rev. Mod. Phys.*, 1998, **70**(3), 721–741.
2 M. Anderson, J. Ensher, M. Matthews, C. Wieman and E. Cornell, *Science*, 1995, **269**(5221), 198–201.

3 K. Davis, M.-O. Mewes, M. Andrews, N. van Druten, D. Durfee, D. Kurn and W. Ketterle, *Phys. Rev. Lett.*, 1995, **75**(22), 3969–3973.
4 M. Andrews, C. Townsend, H. Miesner, D. Durfee, D. Kurn and W. Ketterle, *Science*, 1997, **275**(5300), 637.
5 T. Bergeman, J. Qi, D. Wang, Y. Huang, H. K. Pechkis, E. E. Eyler, P. L. Gould, W. C. Stwalley, R. A. Cline, J. D. Miller and D. J. Heinzen, *J. Phys. B*, 2006, **39**, S813.
6 R. V. Krems, *Phys. Chem. Chem. Phys.*, 2008, **10**, 4079.
7 C. Tesch and R. de Vivie-Riedle, *Phys. Rev. Lett.*, 2002, **89**(15), 157901.
8 D. DeMille, *Phys. Rev. Lett.*, 2002, **88**(6), 067901.
9 J. Doyle, B. Friedrich, R. Krems and F. Masnou-Seeuws, *Eur. Phys. J. D*, 2004, **31**, 149–164.
10 H. Thorsheim, J. Weiner and P. Julienne, *Phys. Rev. Lett.*, 1987, **58**(23), 2420–2423.
11 A. Fioretti, D. Comparat, A. Crubellier, O. Dulieu, F. Masnou-Seeuws and P. Pillet, *Phys. Rev. Lett.*, 1998, **80**(20), 4402.
12 C. Gabbanini, A. Fioretti, A. Lucchesini, S. Gozzini and M. Mazzoni, *Phys. Rev. Lett.*, 2000, **84**(13), 2814–2817.
13 M. Theis, G. Thalhammer, K. Winkler, M. Hellwig, G. Ruff, R. Grimm and J. H. Denschlag, *Phys. Rev. Lett.*, 2004, **93**(12), 123001.
14 M. Zwierlein, C. Stan, C. Schunck, S. Raupach, S. Gupta, Z. Hadzibabic and W. Ketterle, *Phys. Rev. Lett.*, 2003, **91**(25), 250401.
15 M. Greiner, C. Regal and D. Jin, *Nature*, 2003, **426**(6966), 537–540.
16 S. Jochim, M. Bartenstein, A. Altmeyer, G. Hendl, S. Riedl, C. Chin, J. Hecker Denschtag and R. Grimm, *Science*, 2003, **302**(5653), 2101–2103.
17 M. Viteau, A. Chotia, M. Allegrini, N. Bouloufa, O. Dulieu, D. Comparat and P. Pillet, *Science*, 2008, **321**, 232–234.
18 F. Lang, K. Winkler, C. Strauss, R. Grimm and J. Hecker Denschlag, *Phys. Rev. Lett.*, 2008, **101**, 133005.
19 J. Vala, O. Dulieu, F. Masnou-Seeuws, P. Pillet and R. Kosloff, *Phys. Rev. A.*, 2001, **63**(1), 013412.
20 E. Luc-Koenig, R. Kosloff, F. Masnou-Seeuws and M. Vatasescu, *Phys. Rev. A*, 2004, **70**(3), 033414.
21 C. P. Koch, J. P. Palao, R. Kosloff and F. Masnou-Seeuws, *Phys. Rev. A*, 2004, **70**(1), 013402.
22 B. Brown, A. Dicks and I. Walmsley, *Phys. Rev. Lett.*, 2006, **96**(17), 173002.
23 W. Salzmann, U. Poschinger, R. Wester, M. Weidemuller, A. Merli, S. M. Weber, F. Sauer, M. Plewicki, F. Weise, A. M. Esparza, L. Woste and A. Lindinger, *Phys. Rev. A*, 2006, **73**(2), 023414.
24 W. Salzmann, T. Mullins, J. Eng, M. Albert, R. Wester, M. Weidemuller, A. Merli, S. Weber, F. Sauer, M. Plewicki, F. Weise, L. Woste and A. Lindinger, *Phys. Rev. Lett.*, 2008, **100**(23), 233003.
25 C. P. Koch, E. Luc-Koenig and F. Masnou-Seeuws, *Phys. Rev. A*, 2006, **73**(3), 033408.
26 P. D. Lett, N. Watts, C. I. Westbrook, W. D. Phillips, P. L. Gould and H. J. Metcalf, *Phys. Rev. Lett.*, 1988, **61**(2), 169.
27 G. Cerullo and S. De Silvestri, *Rev. Sci. Instrum.*, 2003, **74**(1), 2.
28 C. Iaconis and I. A. Walmsley, *Opt. Lett.*, 1998, **23**(10), 792.
29 S. J. Park, S. W. Suh, Y. S. Lee and G. H. Jeung, *J. Mol. Spectrosc.*, 2001, **207**, 129.
30 G. H. Jeung, personal communication.
31 N. N. Klausen, J. L. Bohn and C. H. Greene, *Phys. Rev. A*, 2001, **64**, 053602.
32 C. H. Greene, personal communication.
33 J. Mur-Petit, E. Luc-Koenig and F. Masnou-Seeuws, *Phys. Rev. A*, 2007, **75**, 061404.

Fano profiles in two-photon photoassociation spectra

Maximilien Portier, Michèle Leduc* and Claude Cohen-Tannoudji

Received 3rd November 2008, Accepted 26th January 2009
First published as an Advance Article on the web 10th June 2009
DOI: 10.1039/b819470j

In this article we derive the lineshapes observed in two-photon photoassociation spectrocopy of molecules using an effective Hamiltonian adapted from previous work in atomic physics. The lineshape is decomposed in terms of sums and products of Breit–Wigner and Fano profiles, which we associate with physical absorption and emission processes, and to quantum interferences between transition amplitudes. We emphasize the specific features which do not exist in the atomic case, linked to the dissociation width of the photoassociated molecules in the electronic ground state.

1 Introduction

Photoassociation is a convenient way of preparing ultracold molecules starting from ultracold atoms. Two colliding atoms are brought by one-photon absorption or by two-photon stimulated Raman processes into a bound molecular state. Two-photon photoassociation processes with two coherent laser beams are particularly interesting because they give rise to dark resonances analogous to those observed for atoms. For a certain value of the frequency difference between the two lasers, the two atoms are trapped in a linear combination of two states which cannot absorb light because of destructive interference, a phenomenon called "dark states" or "coherent population trapping". One of the two states which are linearly superposed in the dark state is the initial state of the two colliding atoms, whereas the second state is the lower molecular bound state where one wants to prepare ultracold molecules.

Dark resonances in two-photon photoassociation processes have been observed by several groups,[1–3] including our group.[4,5] The original feature of our experiment is that the two atoms, which are involved in the two-photon photoassociation process, are two helium atoms in a highly excited metastable state (2^3S_1) located at about 20 eV above the ground state. Penning collisions are thus expected to play an important role. Since the two atoms are spin polarized in the magnetic trap that contains them, Penning collisions are partially inhibited[6] and can be ignored in the initial colliding state. In the final molecular bound state, the two atoms remain close to each other and Penning collisions are expected to give a finite lifetime γ_2 to this state. From the observation of dark resonances in our experiment, one can extract two types of information:

• First, from the position of the dark resonance, one can infer the binding energy of the molecular bound state, which in our case is the least bound state in the same electronic molecular potential as the one in which the two colliding atoms are moving. We have obtained in this way a determination of the scattering length of

---

Laboratoire Kastler Brossel, Ecole Normale Supérieure, Université Pierre et Marie Curie, CNRS, IFRAF, 24 rue Lhomond, 75231 PARIS CEDEX 05, FRANCE. E-mail: Leduc@lkb.ens.fr

two metastable helium atoms[4] with an accuracy about 100 times better than all previous measurements of this quantity, which is essential for understanding the properties of the recently observed Bose–Einstein condensates of metastable helium atoms.

• Secondly, from the shape and the width of the dark resonance, one can extract a value of the width γ_2 of the molecular bound state and compare it with the theoretical treatments of Penning collisions in a molecular bound state.

We present in this paper a theoretical treatment of the dark resonances observed in two-photon photoassociation experiments using an effective Hamiltonian method. This method inspired by previous atomic physics work on Λ type level schemes[7] leads to results which are found to be identical to those derived from the MQDT theory.[8,9] Its interest is that it clearly shows that Fano profiles appear in the photoassociation lineshapes. Analytical lineshapes in terms of sums and products of Breit–Wigner and Fano profiles are given. This approach goes beyond the qualitative discussion presented in reference 2 for interpreting the lineshapes observed in two-photon photoassociation of cold Na atoms. It provides quantitative expressions for the interfering scattering amplitudes responsible for the Fano profile. Furthermore it points out the important differences between the atomic and molecular case.

2 Fano profiles in dark resonances observed in atomic physics

Fano profiles are universal lineshapes which are observed in atomic physics. They usually appear when several paths interfere, leading to the same final state lying in a continuum, one of them passing through an intermediate discrete state.[10,11] The experimental study of such profiles provides information on the energy and lifetime of these discrete states.

Fano profiles have been shown to appear in the lineshapes of the so-called dark resonances which can be observed in coherent population trapping experiments involving three-level atomic systems with two ground state sublevels g_1 and g_2 and one excited sublevel e.[7,12] One laser with frequency ω_1 excites the transition $g_1 \rightarrow e$ and another one with frequency ω_2 excites the transition $g_2 \rightarrow e$. If, ω_2 being fixed, one scans ω_1, one observes several resonances in the variations with ω_1 of the fluorescence rate R_F emitted from e, as shown in the inset (c) of Fig. 1:

(1) a broad resonance when ω_1 coincides with the frequency ω_{eg_1} of the $g_1 \rightarrow e$ transition, with a width of the order of the natural width γ_e of the excited state e

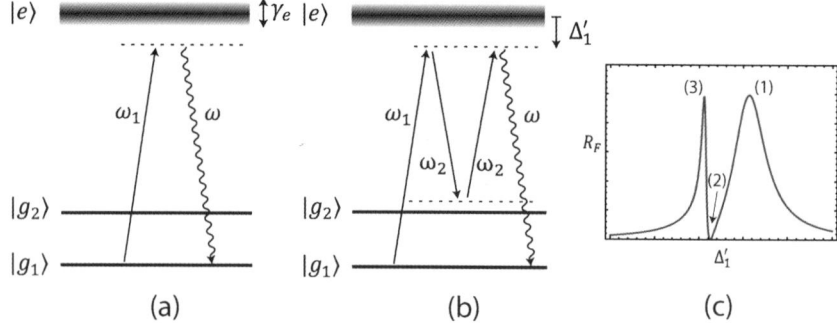

Fig. 1 Processes leading to spontaneous emission of an ω photon for an atom with a Λ level scheme coupled to two lasers ω_1 and ω_2: (a) absorption of a ω_1 photon, followed by spontaneous emission (b) stimulated Raman process bringing the atom from g_1 to g_2 (absorption of ω_1 followed by stimulated emission of ω_2 followed by a spontaneous Raman process bringing the atoms from g_2 to g_1. (c) The inset shows the fluorescence rate as a function of the detuning of the laser ω_1 from the $g_1 \rightarrow e$ transition. The broad resonance (1), the narrow dip (2) and the narrow peak (3) observed in the fluorescence rate R_F are detailed in the text.

(2) a very narrow dark resonance (R_F vanishes) when the resonance Raman condition between the unperturbed states g_1 and g_2 is fulfilled: $\hbar(\omega_1 - \omega_2) = E_{g_1} - E_{g_2}$

(3) a narrow peak associated with resonant Raman processes between perturbed (light shifted) states.

The widths of the narrow dip and of the narrow peak are of the order of the width of g_1 and g_2 and therefore much smaller than γ_e. A Fano profile is observed in the vicinity of the dark resonance and interpreted as being due to the interference between the two paths shown in Fig. 1:[7]

• The atom absorbs a photon ω_1 from g_1, and reaches the state e from which it emits a fluorescence photon. Because of the large value of γ_e compared to the widths of g_1 and g_2, the state e can be considered as a continuum.

• The atom absorbs a photon ω_1 from g_1, reaches the state e, then emits a photon ω_2 in a stimulated way, reaches the state g_2, reabsorbs a photon ω_2, which allows it to again reach the state e from which it finally emits the fluorescence photon.

The first path goes directly to the continuum e. The second path passes through the discrete intermediate state g_2. Their interference gives rise to a Fano profile.

3 Fano profiles observed in molecular photoassociation

Two-photon photoassociation processes involve three-state systems analogous to the one described in section 2 (see Fig. 2) and have been widely probed experimentally:[13] the state g_1 is, in the molecular case, a colliding state $|E, g_1\rangle$ of a pair of atoms in a given electronic ground molecular potential $|g_1\rangle$ with a scattering energy E; a laser ω_1 drives the transition from this state $|E, g_1\rangle$ to a vibrational state $|v_e, e\rangle$ in an electronically excited molecular potential $|e\rangle$; another laser ω_2 drives the transition from this state $|v_e, e\rangle$ to another vibrational state $|v_2, g_2\rangle$ in an electronic ground state molecular potential $|g_2\rangle$. Dark resonances have been observed in these configurations.[1–4]

There are important differences between the molecular and the atomic cases which should be emphasized:

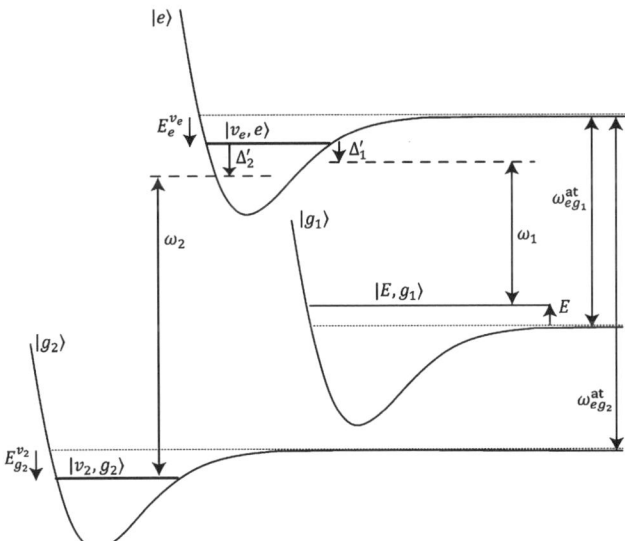

Fig. 2 Level scheme for two-photon photoassociation lineshapes showing Fano profiles. A molecular state $|v_e, e\rangle$ of vibrational number v_e in an electronically excited potential is coupled by a photon ω_1 to a scattering state $|E, g_1\rangle$ in a $|g_1\rangle$ channel and by a photon ω_2 to a molecular state of vibrational number v_2 in a $|g_2\rangle$ channel.

• The two colliding atoms can interact directly and undergo an elastic collision without involving coupling with the lasers.
• The three-level scheme in the molecular case is not closed: radiative decay of a molecule from the electronically excited state will not in general lead only to the two other states (scattering and bound) with which it is coupled by the lasers.
• The colliding state g_1 is not a discrete state. It belongs to a continuum and has an energy dispersion. This should broaden the dark resonance. But, the experiments we have in mind are done with ultracold atoms and this broadening is, in general, negligible.
• In the atomic case, the state g_2 is a ground state and cannot decay. More precisely, the sum of the populations of the three levels is conserved. This is why R_F goes to zero at the center of the dark resonance. In the molecular case, the state g_2 can decay because of dissociation processes and we will see that R_F does not go to zero. This is why Breit–Wigner profiles appear in addition to Fano profiles.

We give in section 4 a brief outline of the calculation of the scattering amplitude of atoms in the context of photoassociation, generalizing the calculation developed for the atomic case. The interested reader can find more details in reference 20, available on-line. A physical interpretation of the photoassociation lineshapes is then given in section 5 where we identify the paths whose interference contributes to a Fano resonance and calculate the amplitude of the corresponding processes as a function of the characteristics of the system.

4 Calculation of the scattering matrix

In this section, we first express the Hamiltonian which describes photoassociation as a function of the relevant parameters. We then separate the effect of the interaction of the lasers from the effect of the interaction between the atoms. This allows us to express the scattering matrix for the photoassociation problem as a function of the wavefunction solutions of the problem of two atoms scattering in the absence of light, and which are supposed to be known. We then give the interpretation of the different factors in the scattering amplitude.

4.1 Presentation of the problem

Photoassociation is described here as light assisted scattering between atoms.[14] We use a basis for a light dressed collisional channel which is a product of a basis for collision channels and a basis for the light radiation.[15,16] A collision channel is characterized by the electronic state of the two atoms, and their relative rotational state.[17] In this problem we consider three collision channels noted g_1, g_2 and e, the first two of which correspond to an electronic ground state (g_1 and g_2 can coincide), and the third one corresponds to an electronically excited state. We also consider two laser modes characterized by their polarizations, their frequencies ω_1 and ω_2, and their numbers of photons N_1 and N_2. The laser fields 1 and 2 couple the dressed channel $|e, N_1 - 1, N_2\rangle$ to the dressed channels $|g_1, N_1, N_2\rangle$ and $|g_2, N_1 - 1, N_2 + 1\rangle$ respectively.

The initial state of the experiment is $|E, g_1, N_1, N_2\rangle$, where E denotes the scattering energy of the two colliding atoms. We will calculate here the scattering matrix element $S_{g_1g_1}$ and the quantity $1 - |S_{g_1g_1}|^2$ which will be interpreted as the probability per collision event that relaxation occurs, and which is proportional to the photoassociation rate. This relaxation may correspond to spontaneous emission from a $|e, N_1 - 1, N_2\rangle$ state or relaxation of a $|g_2, N_1 - 1, N_2 + 1\rangle$ molecule due to dissociation or atom–molecule collision. In the basis $\{|g_1, N_1, N_2\rangle, |e, N_1-1, N_2\rangle, |g_2, N_1 - 1, N_2 + 1\rangle\}$, the Hamiltonian describing light assisted collision is written $H = K + V$, where K is the radial kinetic energy and V describes the effects of interacting potentials and light couplings:

$$K = -\frac{\hbar^2}{2\mu}\frac{d^2}{dr^2} \qquad (1)$$

$$V = \begin{pmatrix} V_{g_1}(r) & V_{rad1}(r) & 0 \\ V_{rad1}(r) & V_e(r) - i\frac{\gamma_e(r)}{2} - \Delta_1 & V_{rad2}(r) \\ 0 & V_{rad2}(r) & V_{g_2}(r) - i\frac{\gamma_{g_2}(r)}{2} - \Delta_1 + \Delta_2 \end{pmatrix} \quad (2)$$

The reduced mass for the atom pair is noted μ. The potentials V_{g_1}, V_{g_2} and V_e describe the radial interaction between the colliding atoms in the g_1, g_2 and e states. The origin of the energy scale is such that $V_{g_1}(\infty) = V_{g_2}(\infty) = V_e(\infty) = 0$. The quantities $\Delta_1 = \omega_1 - \omega^{at}_{eg_1}$ and $\Delta_2 = \omega_2 - \omega^{at}_{eg_2}$ are the detunings of the lasers 1 and 2 from the atomic frequencies between the asymptotic states e and g_1 at $\omega^{at}_{eg_1}$, and e and g_2 at $\omega^{at}_{eg_2}$ (see Fig. 2). The couplings V_{rad1} and V_{rad2} are the matrix elements of the electric-dipole interaction, and depend on the dipole of the molecule which varies with the interatomic distance r. The widths $\gamma_e(r)$ and $\gamma_{g_2}(r)$ account for the decay processes which limit the lifetime of the atoms in the e state and in the g_2 state. These widths depend on the interatomic distance because of the interaction between the atoms. For large interatomic distance $\gamma_e(r)$ tends to a constant proportional to an atomic radiative width, and is modulated for shorter interatomic distance due to the electrostatic interaction between the atoms. Depending on the considered atoms, the width $\gamma_{g_2}(r)$ can be associated for example with autoionization, or spin relaxation, and takes non vanishing values for small interatomic distances where the two atoms interact.

As the Hamiltonian H is not Hermitian owing to the decay widths $\gamma_e(r)$ and $\gamma_{g_2}(r)$, the probability current is not conserved during the collision: if the atoms collide initially in the g_1 channel, there is a non zero probability that they undergo photoassociation leading to a molecule in the e or g_2 potential, followed by decay to another state different from g_1, g_2 or e. As a consequence, the scattering matrix is non unitary. In order to estimate the photoassociation rate K_{PA}, we will calculate the loss of unitarity of the scattering matrix, which is proportional to the photoassociation rate:

$$K_{PA} \propto 1 - |S_{g_1g_1}|^2 \quad (3)$$

4.2 Introduction of the elastic scattering unperturbed by the laser light

The coupling V is separated as $V = V_A + V_B$. V_A describes in an uncoupled way the interaction between atoms in the $|g_1, N_1, N_2\rangle$ state in the one hand, and the interaction between the atoms in the $|g_1, N_1 - 1, N_2 + 1\rangle$ and $|e, N_1 - 1, N_2\rangle$ states perturbed by the second laser on the other hand. The coupling V_B describes the interaction with the laser ω_1.

$$V_A = \begin{pmatrix} V_{g_1}(r) & 0 & 0 \\ 0 & V_e(r) - i\frac{\gamma_e(r)}{2} - \Delta_1 & V_{rad2}(r) \\ 0 & V_{rad2}(r) & V_{g_2}(r) - i\frac{\gamma_{g_2}(r)}{2} - \Delta_1 + \Delta_2 \end{pmatrix} \quad (4)$$

$$V_B = \begin{pmatrix} 0 & V_{rad1}(r) & 0 \\ V_{rad1}(r) & 0 & 0 \\ 0 & 0 & 0 \end{pmatrix} \quad (5)$$

We note $H_0 = K + V_A$, and Q and P are the projectors onto the subspace spanned by the states $|e, N_1 - 1, N_2\rangle$ and $|g_2, N_1 - 1, N_2 + 1\rangle$ on the one hand, and $|g_1, N_1, N_2\rangle$ on the other hand:

$$P = |g_1, N_1, N_2\rangle\langle g_1, N_1, N_2| \qquad (6)$$

$$Q = |e, N_1 - 1, N_2\rangle\langle e, N_1 - 1, N_2| + |g_2, N_1 - 1, N_2 + 1\rangle\langle g_2, N_1 - 1, N_2 + 1| \qquad (7)$$

Notice that PH_0P describes an elastic collision of an atom pair in the g_1 state, whose properties are supposed to be known. We note $S^{(0)}{}_{g_1g_1} = e^{i\eta_{g_1}}$ the scattering matrix element which describes this elastic collision, and which is expressed as a function of a phase shift η_{g_1}.

We calculate the scattering matrix element $S_{g_1g_1}$ which characterizes the collision perturbed by the lasers as a function of $S^{(0)}{}_{g_1g_1}$. The two-potential scattering formula gives a relation between the transfer matrices elements, where the transfer matrix is defined as a function of a scattering matrix as $S = 1 - 2\pi i T$:[18]

$$T_{g_1g_1} = T^{(0)}{}_{g_1g_1} + \langle \phi^{E-}{}_{g_1}, g_1, N_1, N_1 | V_B | \psi^{E+}{}_{g_1}, g_1, N_1, N_2 \rangle \qquad (8)$$

where $\phi^{E-}{}_{g_1}$ (†) and $\psi^{E+}{}_0$ are the radial part of the scattering wavefunctions describing a stationary state of the Hamiltonian PH_0P with ingoing boundary condition, and a stationary state of the Hamiltonian PHP with outgoing boundary condition respectively.

The matrix element on the right hand side of eqn (8). can be expressed using only solutions $\phi^{E\pm}{}_{g_1}$ for the scattering described by the PH_0P Hamiltonian which are supposed to be known, and the resolvent operator $G^+(E) = \lim_{\varepsilon \to 0^+}(E - H + i\varepsilon)^{-1}$:

$$T_{g_1g_1} = T^{(0)}{}_{g_1g_1} + \langle \phi^{E-}{}_{g_1}, g_1, N_1, N_1 | PV_BQG^+(E)QV_BP | \phi^{E+}{}_{g_1}, g_1, N_1, N_2 \rangle \qquad (9)$$

Using (9) and the definition of the T matrix as a function of the S matrix, we get:

$$S_{g_1g_1} = S^{(0)}{}_{g_1g_1} - \langle \phi^{E-}{}_{g_1}, g_1, N_1, N_1 | PV_BQG^+(E)QV_BP | \phi^{E+}{}_{g_1}, g_1, N_1, N_2 \rangle \qquad (10)$$

Expanding $QG^+(E)Q$ as a power series of V_B and using projection operator techniques to make a partial resummation of this perturbation series (see for example chapter III in reference 15), one can write (10) as

$$S_{g_1g_1} = S^{(0)}_{g_1g_1} - 2\pi i \left\langle \phi^{E-}_{g_1}, g_1, N_1, N_1 \left| PV_B \frac{Q}{E - H_{\text{eff}}} V_B P \right| \phi^{E+}_{g_1}, g_1, N_1, N_2 \right\rangle \qquad (11)$$

where H_{eff} is an effective Hamiltonian defined by:

$$H_{\text{eff}} = QH_0Q + QV_BPG_0^+(E)PV_0Q \qquad (12)$$

with $G_0^+(E) = \lim_{\varepsilon \to 0^+}(E - H_0 + i\varepsilon)^{-1}$. The effective Hamiltonian H_{eff} describes the reduced evolution in the subspace (7). Note that it is calculated to the second order in V_B, but $S_{g_1g_1}$ in (11) contains all orders in V_B since H_{eff} appears in the denominator.

---

† The scattering wavefunctions $\phi^{E\pm}{}_{g_1}$ are written as a function of a wavefunction $f^E{}_{g_1}(r)$ as:

$$\phi^{E\pm}_{g_1}(r) = \sqrt{\frac{2\mu}{\pi\hbar^2 k}} e^{\pm i\eta_{g_1}} f^E_{g_1}(r)$$

with $f^E_{g_1}(r \to \infty) = \sin(kr - \frac{l_{g_1}\pi}{2} + \eta_{g_1})$ where $k = \sqrt{2\mu E/\hbar^2}$ is the wavevector, l_{g_1} is the angular momentum for the rotation of the atoms associated with g_1, and η_{g_1} the phase shift of the wave function.

We assume that the frequency of the lasers and the level scheme are such that the only transitions that may be excited involve a bound state in the e potential with a vibrational quantum number v, a binding energy $E^v{}_e$, a wavefunction $\phi^v{}_e$, and a bound state in the g_2 potential with a vibrational quantum number v_2, a binding energy $E^{v_2}{}_{g_2}$ and a wavefunction $\phi^{v_2}{}_{g_2}$. The projector Q given by (7) can then be approximated by $\tilde{Q}$:

$$Q \approx \tilde{Q} = |e,N_1-1,N_2\rangle\langle e,N_1-1,N_2|\otimes|\phi^{b_e}{}_e\rangle\langle\phi^{b_e}{}_e| \\ + |g_2,N_1-1,N_2+1\rangle\langle g_2,N_1-1,N_2+1|\otimes|\phi^{v_2}{}_{g_2}\rangle\langle\phi^{v_2}{}_{g_2}| \quad (13)$$

We thus have the following expression by using (13) and (11):

$$S_{g_1g_1} = S_{g_1g_1}^{(0)} - 2\pi i \langle\phi^{E-}{}_{g_1}|V_{rad1}|\phi^{v_e}{}_e\rangle\langle\phi^{v_e}{}_e,e,N_1-1,N_2| \\ (E-H_{\text{eff}})^{-1}|\phi^{v_e}{}_e,e,N_1-1,N_2\rangle\langle\phi^{b_e}{}_e|V_{rad1}|\phi^{E+}{}_{g_1}\rangle \quad (14)$$

We have expressed the scattering matrix $S_{g_1g_1}$ as a function of the characteristics of the elastic scattering unperturbed by light (the scattering matrix $S^{(0)}{}_{g_1g_1}$, and the wavefunctions $\phi^{E\pm}{}_{g_1}$) on the one hand, and as a function of an operator H_{eff} which contains all the physics associated with the laser interaction on the other hand.

4.3 Expression of the effective Hamiltonian

In order to calculate the matrix element $\langle\phi^{v_e}{}_e,e,N_1-1,N_2|(E-H_{\text{eff}})^{-1}|\phi^{v_e}{}_e,e,N_1-1,N_2\rangle$, which appears in (14), we write $E-H_{\text{eff}}$ in the basis spanned by $|\phi^{v_e}{}_e,e,N_1-1,N_2\rangle$ and $|\phi^{v_2}{}_{g_2},g_2,N_1-1,N_2+1\rangle$ using the definition (12):

$$E - H_{\text{eff}} = \begin{pmatrix} E+\Delta_1-E^{v_e}_e-\delta^{v_e}_e+i\frac{\gamma_1+\Gamma}{2} & -\Omega \\ -\Omega & E+\Delta_1-\Delta_2-E^{v_2}_{g_2}+i\frac{\gamma_2}{2} \end{pmatrix} \quad (15)$$

The quantities appearing in (15) are the following:
- the widths $\gamma_1 = \langle\phi^{v_e}{}_e|\gamma_e(r)|\phi^{v_e}{}_e\rangle$ and $\gamma_2 = \langle\phi^{v_2}{}_{g_2}|\gamma_{g_2}(r)|\phi^{v_2}{}_{g_2}\rangle$ of the bound states in the e and g_2 potentials respectively,
- the Rabi frequency $\Omega = \langle\phi^{v_2}{}_{g_2}|V_{rad2}(r)|\phi^{v_e}{}_e\rangle$ for the coupling induced by the second laser between the two bound states.
- the light induced shift $\delta^{v_e}{}_e$ and broadening Γ of the excited bound state (‡) due to its coupling with the laser ω_1.

Introducing new detunings (see Fig. 3) Δ'_1 and Δ'_2 which include the shift $\Delta^{v_e}{}_e$

---

‡ According to the definition of H_{eff} (12), the light induced shift $\Delta^{be}{}_e$ and broadening Γ are given by

$$\delta^{v_e}_e = i\frac{\Gamma}{2} = \left\langle\phi^{v_e}_e,e,N_1-1,N_2\left|QV_BPG_0^+(E)PV_BQ\right|\phi^{v_e}_e,e,N_1-1,N_2\right\rangle$$

Using the decomposition of the identity operator $\mathbb{1}$ on the eigenstates of $PH_0P = k + V_{g_1}$ which are noted $\phi^{vg_1}{}_{g_1}$ (bound state of vibrational numberb v_{g_1}) and $\phi^{E'+}{}_{g_1}$ (scattering state of energy E'), we obtain the explicit expression for the light induced shift and broadening within this model:

$$\delta^{v_e}_e = \sum_{v_{g_1}} \frac{\left|\left\langle\phi^{v_e}_e|V_{rad1}|\phi^{v_{g_1}}_{g_1}\right\rangle\right|^2}{E-E^{v_{g_1}}_{g_1}} + vp\int dE' \frac{\left|\left\langle\phi^{v_e}_e|V_{rad1}|\phi^{E'+}_{g_1}\right\rangle\right|^2}{E-E'}$$

$\Gamma = 2\pi|\langle\phi^{v_e}{}_e|V_{rad1}|\phi^{E+}{}_{g_1}\rangle|^2.$

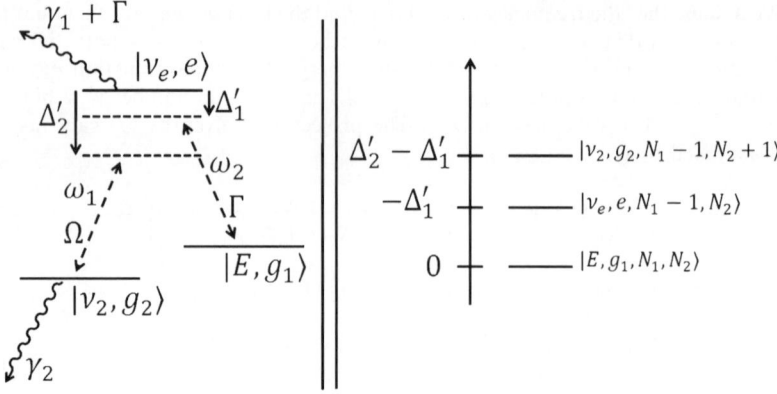

Fig. 3 Equivalent level scheme (left: molecular states, right: states including the photon quantum numbers) for the two-photon photoassociation scheme depicted in Fig. 2.

$$\begin{aligned}\Delta'_1 &= E + \Delta_1 - E^{ve}_e - \delta^{ve}_e \\ \Delta'_2 &= \Delta_2 + E^{ve}_g - E^{ve}_e - \delta^{ve}_e\end{aligned} \quad (16)$$

We rewrite (15)(15)

$$E - H_{\text{eff}} = \begin{pmatrix} \Delta'_1 + i\dfrac{\gamma_1 + \Gamma}{2} & -\Omega \\ -\Omega & \Delta'_1 - \Delta'_2 + i\dfrac{\gamma_2}{2} \end{pmatrix} \quad (17)$$

4.4 Calculation of the scattering matrix element $S_{g_1 g_1}$

We note $\mathcal{D}$ the determinant of (17). The central matrix element in (14) can be expressed as:

$$\langle \phi^{b_e}_e, e, N_1 - 1, N_2 | (E - H_{\text{eff}})^{-1} | \phi^{b_e}_e, e, N_1 - 1, N_2 \rangle = \frac{\Delta'_1 - \Delta'_2 + i\dfrac{\gamma_2}{2}}{\mathcal{D}} \quad (18)$$

Using the expression of $\phi^{E\pm}_{g_1}$ given in footnote (†) and the expression of Γ given in footnote (‡) one can show that:

$$2\pi i \, \langle \phi^{E-}_{g_1} | V_{radI} | \phi^{b_e}_e \rangle \langle \phi^{b_e}_e | V_{radI} | \phi^{E+}_{g_1} \rangle = i \Gamma e^{2i\eta_{g_1}} \quad (19)$$

From (18), (19) and (14), we finally get the scattering matrix element:

$$S_{g_1 g_1} = e^{2i\eta_{g_1}} - i\Gamma e^{2i\eta_{g_1}} \frac{\Delta'_1 - \Delta'_2 + \dfrac{i\gamma_2}{2}}{\mathcal{D}} = e^{2i\eta_{g_1}} \frac{\mathcal{N}}{\mathcal{D}}$$

with $\mathcal{N} = \Delta'_1(\Delta'_1 - \Delta'_2) - \Omega^2 - \dfrac{\gamma_2(\gamma_1 - \Gamma)}{4} + \dfrac{i}{2}\left[(\gamma_1 - \Gamma)(\Delta'_1 - \Delta'_2) + \gamma_2 \Delta'_1\right]$

and $\mathcal{D} = \Delta'_1(\Delta'_1 - \Delta'_2) - \Omega^2 - \dfrac{\gamma_2(\gamma_1 + \Gamma)}{4} + \dfrac{i}{2}\left[(\gamma_1 + \Gamma)(\Delta'_1 - \Delta'_2) + \gamma_2 \Delta'_1\right]$

$$(20)$$

We have recovered the expression given in ref. 9. In the following section, we identify $S_{g_1 g_1}$ as a sum of amplitudes associated with three distinct processes.

4.5 Eigenvalues of the effective Hamiltonian: physical interpretation

If Z_I and Z_{II} are the eigenvalues of H_{eff}, the determinant $\mathcal{D}$ of (17) can be written:

$$\mathcal{D} = (\Delta'_1 - Z_I)(\Delta'_1 - Z_{II}) \tag{21}$$

If the coupling Ω induced by the laser ω_2 is zero, Z_I and Z_{II} are the complex energies of the states $|v_e,e,N_1-1,N_2\rangle$ and $|v_2,g_2,N_1-1,N_2+1\rangle$. In the case $\Omega \neq 0$, Z_I and Z_{II} include a light shift and a broadening due to the perturbation induced by the laser ω_2. Inserting the following expression:

$$\frac{1}{\mathcal{D}} = \frac{1}{Z_I - Z_{II}} \left[\frac{1}{\Delta'_1 - Z_I} - \frac{1}{\Delta'_1 - Z_{II}} \right] \tag{22}$$

into (20) shows that, considered as a function of Δ'_1 (Δ'_2 being fixed), the scattering matrix element $S_{g_1g_1}$ is a sum of three amplitudes: the first one, independent of Δ'_1 corresponds to direct scattering in the potential $|g_1\rangle$; the second one is resonant when the detuning Δ'_1 coincides with the perturbed energy Z_I of $|v_e,e,N_1-1,N_2\rangle$ (resonant excitation in a direct one photon absorption process); the third one is resonant when Δ'_1 coincides with the perturbed energy of $|v_2,g_2,N_1-1,N_2\rangle$ (resonant Raman process between the two perturbed ground states $|E,g_1,N_1,N_2\rangle$ and $|v_e,g_2,N_1-1,N_2+1\rangle$ by absorption of a photon ω_1 and stimulated emission of a photon ω_2).

It thus clearly appears that the scattering amplitude $S_{g_1g_1}$ contains the sum of two contributions corresponding to the two eigenstates of the effective Hamiltonian H_{eff}.

The two eigenvalues Z_I and Z_{II} can be calculated perturbatively when Ω is small compared to the real and imaginary parts of the difference $\Delta'_2 + \frac{i(\Gamma + \gamma_1 - \gamma_2)}{2}$ of the diagonal elements of (17) One finds:

$$Z_I = -s\Delta'_2 - i\left[\frac{(\Gamma + \gamma_1)}{2} - s\frac{(\Gamma + \gamma_1 - \gamma_2)}{2} \right] \tag{23}$$

$$Z_{II} = \Delta'_2 + s\Delta'_2 - i\left[\frac{\gamma_2}{2} - s\frac{(\Gamma + \gamma_1 - \gamma_2)}{2} \right] \tag{24}$$

where s is a saturation parameter ($s \ll 1$ in the perturbative limit), and given by

$$s = \frac{\Omega^2}{{\Delta'_2}^2 + \frac{(\Gamma + \gamma_1 - \gamma_2)^2}{4}} \tag{25}$$

5 Interpretation of the lineshapes

5.1 Photoassociation rate

The calculation of the photoassociation rate K_{PA} which is proportional to the loss rate per collision event gives:

$$K_{\text{PA}} \propto 1 - |S_{g_1g_1}|^2 = \frac{\gamma_1 \Gamma}{|\mathcal{D}|^2}\left[(\Delta'_2 - \Delta'_1)^2 + \frac{\gamma_2^2}{4}\right] + \frac{\Omega^2 \gamma_2 \Gamma}{|\mathcal{D}|^2} \tag{26}$$

In the rate (26). we identify two contributions noted $|S_{g_1e}|^2$ and $|S_{g_1g_2}|^2$:

$$|S_{g_1e}|^2 \equiv \frac{\gamma_1 \Gamma}{|\mathcal{D}|^2}\left[(\Delta'_2 - \Delta'_1)^2 \frac{\gamma_2^2}{4}\right] \tag{27}$$

$$|S_{g_1g_2}|^2 \equiv \frac{\Omega^2 \gamma_2 \Gamma}{|\mathcal{D}|^2} \quad (28)$$

The first contribution (27) vanishes if $g_1 = 0$ and is attributed to the radiative decay of the $|v_e,e,N_1 - 1,N_2\rangle$ state. The second contribution (28) vanishes if $g_2 = 0$ and is attributed to the decay of the $|v_2,g_2,N_1 - 1,N_2 + 1\rangle$ state. By construction, we have $|S_{g_1g_2}|^2 + |S_{g_1e}|^2 + |S_{g_1g_1}|^2 = 1$ which corresponds to the fact that for a collision in the initial channel $|g_1,N_1,N_2\rangle$, the final state can be either in the $|g_1,N_1,N_2\rangle$ channel (elastic collision), or in a channel resulting from the decay of $|e,N_1 - 1,N_2\rangle$ or $|g_2,N_1 - 1,N_2 + 1\rangle$.

5.2 Loss rate associated with the decay width g_1

5.2.1 Case $\Delta'_2 \neq 0$.
Using the factorization of $\mathcal{D}$ (21) and the expression (23) and (24) of Z_I and Z_{II} in the perturbative limit, we get for the loss rate $|S_{g_1e}|^2$

$$|S_{g_1e}|^2 = \frac{\gamma_1 \Gamma}{(\Delta'_1 + s\Delta'_2)^2 + \dfrac{\left(\gamma_1 + \Gamma - s(\Gamma + \gamma_1 - \gamma_2)\right)^2}{4}} \times \left[\frac{(\Delta'_1 - \Delta'_2)^2}{(\Delta'_1 - \Delta'_2 - s\Delta'_2)^2 + \dfrac{\left(\gamma_2 + s(\Gamma + \gamma_1 - \gamma_2)\right)^2}{4}} + \frac{\dfrac{\gamma_2^2}{4}}{(\Delta'_1 - \Delta'_2 - s\Delta'_2)^2 + \dfrac{\left(\gamma_2 + (\Gamma + \gamma_1 - \gamma_2)\right)^2}{4}} \right] \quad (29)$$

which is noted:

$$|S_{g_1e}|^2 = A \times (B_1 + B_2) \quad (30)$$

We now interpret the three factors A, B_1 and B_2 and illustrate their contribution in Fig. 4.

Factor A: direct photoassociation of the $|v_e,e\rangle$ state. The factor A corresponds to the photoassociation of the $|v_e,e\rangle$ state and is a Breit–Wigner profile for which the resonance condition is written $\Delta'_1 = -s\Delta'_2$. This is a one-photon resonance condition between the state $|E,g_1\rangle$ and the state $|v_e,e\rangle$ shifted by the second photon. Taking $s = 0$ in the factor A, this is to say $\Omega = 0$, we recover the one-photon photoassociation lineshape.

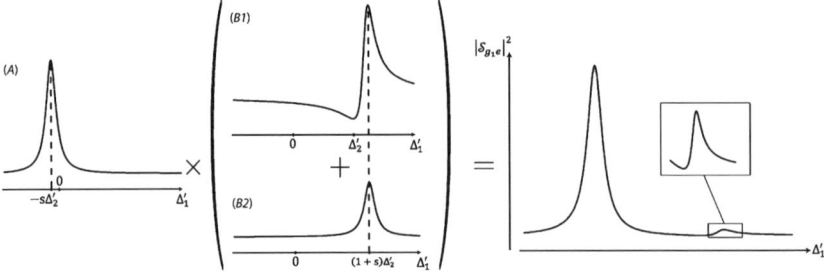

Fig. 4 Decomposition of the loss rate $|S_{g_1e}|^2$ induced by photoassociation and associated with the decay of the molecule $|v_e,e\rangle$ in the case $\Delta'_2 \neq 0$.

Factor B_1: Fano profile. The factor B_1 can be expressed as a Fano profile of parameter q defined by (33) and centered on $\Delta'_1 = (1+s)\Delta'_2$ close to the two-photon resonance with $|v_2, g_2, N_1 - 1, N_2 + 1\rangle$:

$$B_1 = \frac{(\Delta'_1 - \Delta'_2)^2}{\left[(\Delta'_1 - \Delta'_2 - s\Delta'_2)^2 + \frac{(\gamma_2 + s(\Gamma + \gamma_1 - \gamma_2))^2}{4}\right]} \equiv \frac{(\varepsilon + q)^2}{\varepsilon^2 + 1} \quad (31)$$

With:

$$\varepsilon = \frac{\Delta'_1 - \Delta'_2 - s\Delta'_2}{(\gamma_2 + s(\Gamma + \gamma_1 - \gamma_2))/2} \quad (32)$$

$$q = \frac{s\Delta'_2}{(\gamma_2 + s(\Gamma + \gamma_1 - \gamma_2))/2} \quad (33)$$

The general shape of a Fano profile $f_q(\varepsilon)$ for different q parameters are drawn on Fig. 5 as a function of ε. Since $\Delta'_2 \neq 0$, eqn (33) and Fig. 5 show that the corresponding lineshape is assymetric as shown in Fig. 4.

Suppose now that $g_2 = 0$. According to (29), the rate $|S_{g_1 e}|^2$ for decay of the radiatively excited molecule $|v_e, e\rangle$ reduces to $A \times B_1$, and has the same structure as the fluorescence rate R_F calculated in the atomic case. The physical interpretation depicted in Fig. 1 can be generalized in the molecular case (see Fig. 6): the Fano lineshape is associated with the interference between two paths coupling the initial state $|E, g_1, N_1, N_2\rangle$ to the continuum of decay products of $|v_e, e, N_1 - 1, N_2\rangle$. The first path is direct and involves only the first laser. The second path is indirect and involves once the coupling induced by the first laser, and twice the coupling induced by the second laser. For $\varepsilon = -q$, this is to say $\Delta'_1 = \Delta'_2$, there is a destructive interference between the two amplitudes corresponding to the two paths, and $B_1 = 0$.

Factor B_2: specific features associated with the width $\gamma_2 \neq 0$. In the case $\gamma_2 \neq 0$ which is specific to the molecular problem, the inhibition of the loss rate $|S_{g_1 e}|^2$ is never complete for any detuning Δ'_1 due to the factor B_2. This factor is a Breit–Wigner profile for which the resonance condition is $\Delta'_2 - \Delta'_1 = -s\Delta'_2$. This is a Raman resonance condition between the state $|E, g_1\rangle$ and the state $|v_2, g_2\rangle$, the latter being contaminated by the state $|v_e, e\rangle$ due to the coupling Ω induced by the laser ω_2. This is why, close to the resonance with $|v_2, g_2\rangle$, there is a slight enhancement of the decay rate $|S_{g_1 e}|^2$ from the $|v_e, e\rangle$ state.

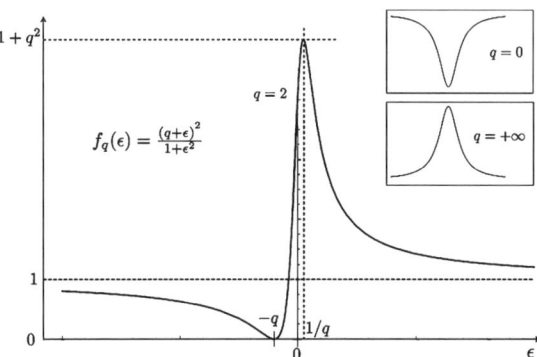

Fig. 5 Shape of a Fano resonance $f_q(\varepsilon)$ as a function of a reduced energy ε for different q parameters.

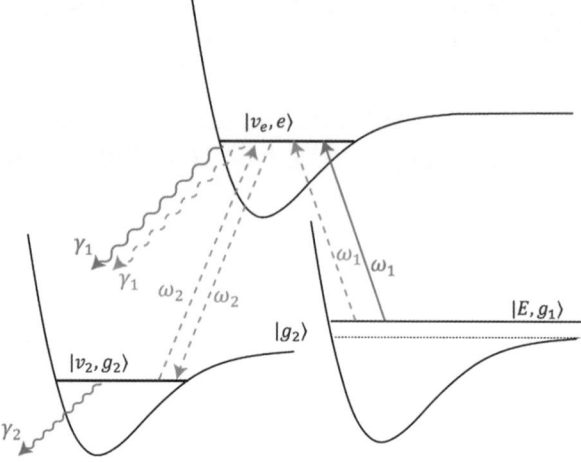

Fig. 6 Paths leading to the decay products of the $|v_e,e\rangle$ molecule of width γ_1. The solid line represents the direct path: absorption of a ω_1 photon followed by the decay of the $|v_e,e\rangle$ molecule. The dashed line represents the indirect path: absorption of a ω_1 photon, stimulated emission and absorption of a ω_2 photon followed by the decay of the $|v_e,e\rangle$ molecule. Specific features of the lineshape are due to the width γ_2 of the $|v_2,g_2\rangle$ molecule.

Finally, we notice in (29) that in the case $s \to 0$ (small intensity of laser ω_2), the width of the resonance associated with $|v_2,g_2\rangle$ in the rate $|S_{g_1e}|^2$ has a characteristic width of g_2. This shows that in an experiment where only the decay from the $|v_e,e\rangle$ can be monitored, it is still possible to measure the lifetime $\frac{1}{\gamma_2}$ of the $|v_2,g_2\rangle$ state.[5]

5.2.2 Case $\Delta'_2 = 0$. We illustrate here loss rate $|S_{g_1e}|^2$ in the case where $\Delta'_2 = 0$ but still $s \ll 0$ (see Fig. 7). The factor B_1 in (31) is a specific case of a Fano lineshape with a parameter $q = 0$ (see eqn (33) and Fig. 5). The lineshape shows even more clearly than in the $\Delta'_2 \neq 0$ case the inhibition of photoassociation when the Raman resonance condition between the unperturbed states $\Delta'_1 = \Delta'_2$ is fulfilled. However the inhibition is not complete due to the contamination of the energy of the electronically excited state $|v_e,e\rangle$ by the width γ_2. The study of such a lineshape is very useful to measure the energy of the $|v_2,g_2\rangle$ molecule as the resonance condition is independent of light shifts.[4]

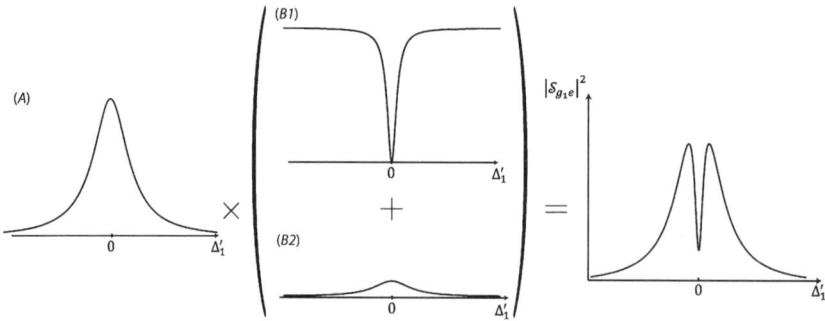

Fig. 7 Decomposition of the loss rate $|S_{g_1e}|^2$ induced by photoassociation and associated with the decay of the molecule $|v_e,e\rangle$ in the case $\Delta'_2 \neq 0$.

5.3 Loss rate associated with the decay width g_2

Using the factorization of $\mathcal{D}$ and the expressions (23) and (24) of Z_I and Z_{II} in the perturbative limit ($s \ll 1$), we get for the loss rate $|S_{g_1g_2}|^2$

$$|S_{g_1g_2}|^2 = \frac{\Omega^2 \gamma_2 \Gamma}{\left[(\Delta'_1 + s\Delta'_2)^2 + \dfrac{\left(\gamma_1 + \Gamma - s(\Gamma + \gamma_1 - \gamma_2)\right)^2}{4}\right]\left[(\Delta'_1 - \Delta'_2 - s\Delta'_2)^2 + \dfrac{\left(\gamma_2 + s(\Gamma + \gamma_1 - \gamma_2)\right)^2}{4}\right]} \quad (34)$$

The rate $|S_{g_1g_2}|^2$ is therefore the product of two Lorentzians centered on $\Delta'_1 = -s\Delta'_2$ (resonance associated with the excitation of the state $|v_e,e,N_1 - 1,N_2\rangle$ perturbed by the second photon) and $\Delta'_1 = (1 + s)\Delta'_2$ (resonance associated with the excitation of the state $|v_2,g_2,N_1 - 1,N_2 + 1\rangle$ perturbed by the second photon). This rate is represented in Fig. 8.

The structure of $|S_{g_1g_2}|^2$ is similar to that of the factor $A \times B_2$ in $|S_{g_1e}|^2$ as the same resonances are involved. However the rate $|S_{g_1g_2}|^2$ shows no Fano lineshape: as can be seen in Fig. 6 there is only one path leading to the dissociation of the $|v_2,g_2\rangle$ molecule which is the stimulated Raman excitation from the $|E,g_1\rangle$ state.

6 Conclusion

In this discussion we proposed the physical interpretation of two-photon photoassociation lineshapes adapted from a previous work in atomic physics.[7] The analytical expression of the scattering matrix was found using an effective Hamiltonian method rather than MQDT theory.[9] This allowed us to decompose the photoassociation lineshape in sums and products of Breit–Wigner and Fano profiles whose resonances are associated with the excitation of the eigenstates of the effective Hamiltonian. The interfering paths leading to a Fano profile were identified. We emphasized the differences from the atomic case. Because of the width associated with the finite lifetime of the molecule in the electronic ground state probed by photoassociation, there is no complete inhibition of the fluorescence rate induced by the radiative decay of the electronically excited molecule.

The analytical expressions derived in this discussion can be used to interpret experimental data and provide information on the molecule photoassociated in the electronic ground state. In reference 1, the production of translationally cold Cs_2 molecules was investigated through the interpretation of two-photon

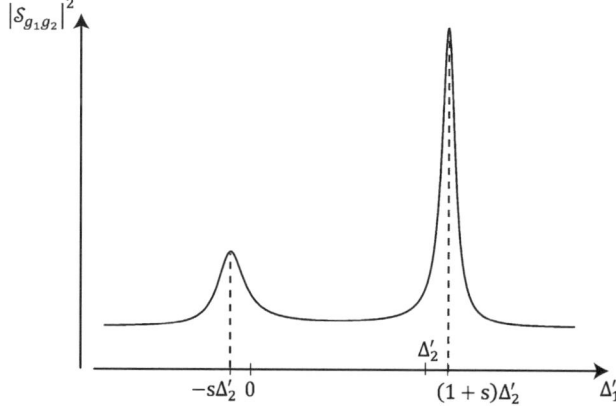

Fig. 8 Loss rate $|S_{g_1g_2}|^2$ associated with the decay of the $|v_2,g_2\rangle$ molecule.

photoassociation data. In references 4 and 5 where metastable helium dimers were investigated, the photoassociation data were remarkably well fitted by the analytical expressions derived in this paper. The measurements gave a lifetime of $\tau = 1.4 \pm 0.3$ μs which is in discrepancy with theoretical expectations[19] and could stimulate new investigations on Penning ionization.

An experimental illustration of the interpretation derived in this paper is emphasized in reference 2. The authors exhibited photoassociation lineshape features narrower than the width of the excited molecular state and attributed their observation to a quantum interference between processes similar to that found in atomic physics. The present paper gives a solid theoretical background to this interpretation and emphasizes the similarities and discrepancies with an atomic lambda system. Interestingly enough, the authors of reference 3 studied a similar problem with a Bose–Einstein condensate, where the initial state is not a thermal distribution of colliding states but a condensate corresponding to a macroscopically occupied atomic state in a trap. It would be interesting to try to extend our theoretical approach to this case in order to exhibit explicitly Fano profiles and the corresponding interfering amplitudes.

References

1 C. Lisdat et al., *Eur. Phys. J. D*, 2002, **21**, 299.
2 R. Dumke et al., *Phys. Rev. A*, 2005, **72**, 041801.
3 K. Winkler et al., *Phys. Rev. Lett.*, 2005, **95**, 063202.
4 S. Moal et al., *Phys. Rev. Lett.*, 2006, **96**, 023203.
5 S. Moal et al., *Phys. Rev. A*, 2007, **75**, 033415.
6 G. V. Shlyapnikov et al., *Phys. Rev. Lett.*, 1994, **73**, 24.
7 B. Lounis and C. Cohen-Tannoudji, *J. Phys. II*, 1992, **2**, 579.
8 J. L. Bohn and P. S. Julienne, *Phys. Rev. A*, 1996, **54**, R4637.
9 J. L. Bohn and P. S. Julienne, *Phys. Rev. A*, 1999, **60**, 414.
10 U. Fano, *Nuovo Cimento.*, 1935, **12**, 154.
11 U. Fano, *Phys. Rev.*, 1961, **124**, 1866.
12 G. Alzetta et al., *Nuovo Cimento.*, 1988, **36B**, 5.
13 K. M. Jones et al., *Rev. Mod. Phys.*, 2006, **78**, 483.
14 H. R. Thorsheim, J. Weiner and P. S. Julienne, *Phys. Rev. Lett.*, 1987, **58**, 2420.
15 C. Cohen-Tannoudji, *Atom-Photon Interactions: Basic Processes and applications*, Wiley, New-York, 1992.
16 A. Simoni et al., *Phys. Rev. A*, 2002, **66**, 063406.
17 F. H. Mies, *Mol. Phys.*, 1980, **41**, 953.
18 C. Joachain, *Quantum Collision Theory*, North Holland, 1983.
19 T. J. Beams, G. Peach and I. B. Whittingham, *Phys. Rev. A*, 2006, **74**, 014702.
20 M. Portier, PhD Thesis, 2007. http://tel.archives-ouvertes.fr/tel-00258383/en/.

General discussion

Dr Aldegunde opened the discussion by commenting: Yesterday we presented a brief description of the Cs_2 hyperfine structure.[1] Hyperfine splittings are also important for experiments involving polar alkali dimers like $^7Li^{133}Cs$[2] and $^{40}K^{87}Rb$,[3] which have been recently produced in their ground rovibrational state. We have calculated these splittings for non-rotating ($N = 0$) $^7Li^{133}Cs$ and $^{40}K^{87}Rb$.[4]

The hyperfine splitting for the ground rotational state in the absence of field is given in Figs 1 and 2. As for the Cs_2 ($N = 0$) molecule,[1] the scalar spin–spin interaction ($c_4 I_1 \cdot I_2$) splits the $^7Li^{133}Cs$ and $^{40}K^{87}Rb$ $N = 0$ levels in several sublevels that can be labelled by means of the total nuclear spin I. The only difference with the Caesium dimer is that, for heteronuclear dimers, the nuclear exchange symmetry does not limit the values of I that can occur when N is equal to zero. The values for the scalar spin–spin coupling constant c_4, 1.6 and -2.0 kHz for $^7Li^{133}Cs$ and $^{40}K^{87}Rb$ respectively, were calculated using DFT calculations with relativistic corrections.[4]

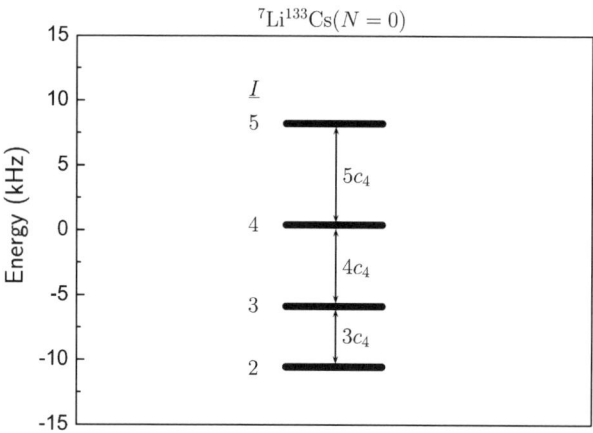

Fig. 1 Zero field splitting for the $N = 0$ state of $^7Li^{133}Cs$.

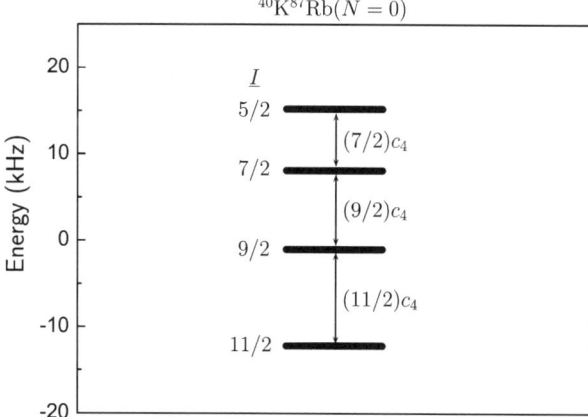

Fig. 2 Zero field splitting for the $N = 0$ state of $^{40}K^{87}Rb$.

When the value of c_4 is positive (negative) the higher energy corresponds to the larger (smaller) value of I.

Each of the zero field levels splits into $2I + 1$ components in the presence of a magnetic field. These splittings are shown in Figs 3 and 4 for ^{7}Li^{133}Cs and ^{40}K^{87}Rb.

The Zeeman patterns show noticeable differences to the Cs$_2$ ($N = 0$) case.[1] The energy levels corresponding to the same value of M_I (the projection of I on the laboratory axis) display avoided crossings as a function of the magnetic field. The crossings are exemplified in Fig. 5, which shows the Zeeman splitting for the $M_I = -7/2$ levels of the ^{40}K^{87}Rb ($N = 0$) molecule. A second difference with the homonuclear case is that the total nuclear spin I only remains a nearly good quantum number for low values of the field. For fields above the avoided crossings, I is destroyed and M_I is the only good quantum number. However, if the value of the field continues increasing and the nuclear Zeeman term becomes dominant in the molecular Hamiltonian, the individual projections of the nuclear spins M_{I_1} and M_{I_2} become well defined as well (see Fig. 3 and right panel of Fig. 4).

The $M_I = -7/2$ energy levels presented in Fig. 5 are relevant for the experiments described by Ye and coworkers attempting to create a ^{40}K^{87}Rb Bose–Einstein condensate.[3]

Those experiments transfer Feshbach molecules in a $M_{\text{tot}} = -7/2$ state into the ground rovibrational state using one STIRAP (stimulated Raman adiabatic passage) cycle. The geometry of the lasers is such that M_{tot} is conserved during the process and the ^{40}K^{87}Rb ($N = 0$) molecules are produced in one of the states of Fig. 5.

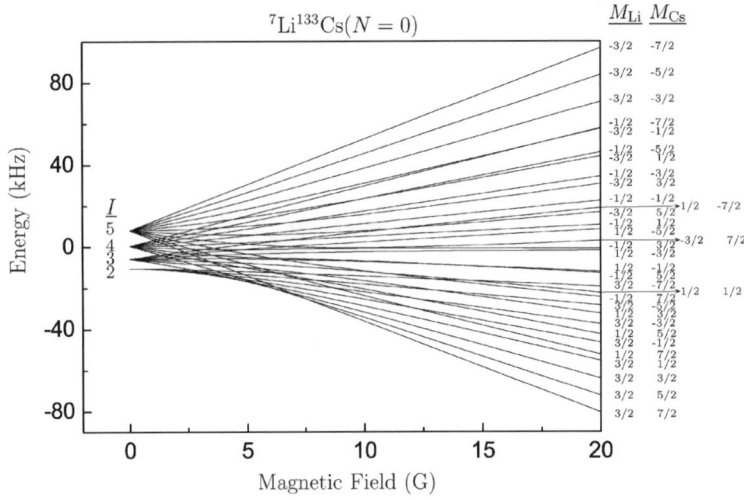

Fig. 3 Zeeman splitting of hyperfine levels for ^{7}Li^{133}Cs ($N = 0$).

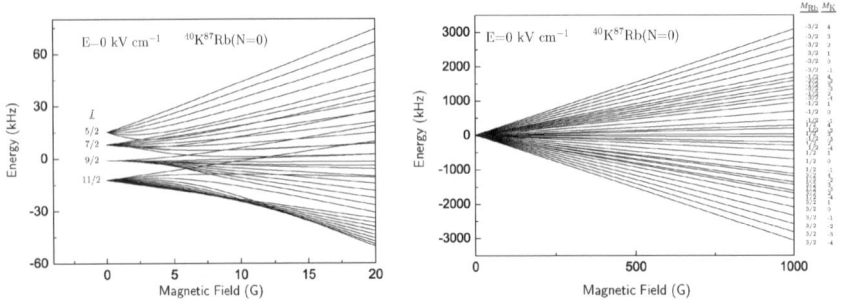

Fig. 4 Zeeman splitting of hyperfine levels for ^{40}K^{87}Rb ($N = 0$).

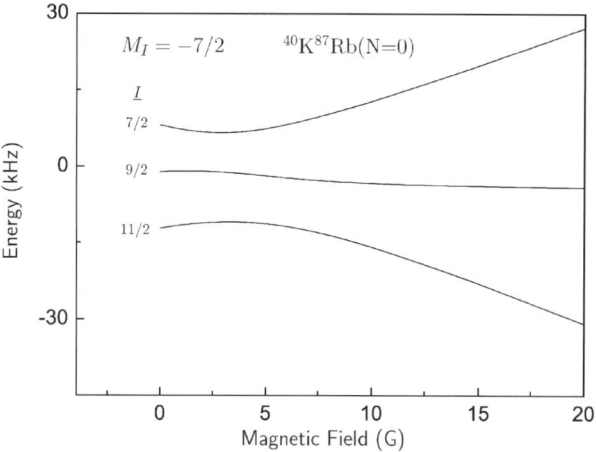

Fig. 5 Zeeman splitting and avoided crossings for the $M_I = -7/2$ levels of ^{40}K^{87}Rb ($N = 0$).

1 J. Aldegunde, in general discussion section, *Faraday Discuss.*, 2009, **142**, DOI: 10.1039/b910121g.
2 J. Deiglmayr, A. Grochola, M. Repp, K. Mörtlbauer, C. Glück, J. Lange, O. Dulieu, R. Wester, and M. Weidemüller, *Phys. Rev. Lett.*, 2008, **101**, 133004.
3 K.-K Ni, S. Ospelkaus, M. H. G. de Miranda, A. Pe'er, B. Neyenhuis, J. J. Zirbel, S. Kotochigova, P. S. Julienne, D. S. Jin and J. Ye, *Science*, 2008, **322**, 231.
4 J. Aldegunde, B. A. Rivington, P. S. Żuchowski and J. M. Hutson, *Phys. Rev. A*, 2008, **78**, 033434.

Professor Kotochigova contributed: We are publishing a paper in a special issue of *New J. Phys.*.[1] This paper describes our theoretical calculation of the KRb molecule. This work has been done in support of the JILA experiments of J. Ye and D. Jin. In the Raman transition, which was used to create the ground state KRb molecules, the

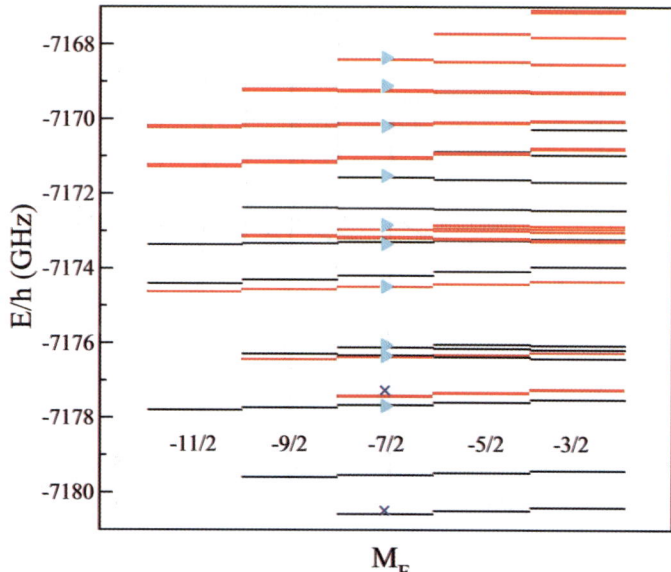

Fig. 6 The hyperfine and Zeeman structure at $B = 545.9$ G of the $\ell = 0$ (black lines) and 2 (red lines) rotational levels of the $v = 0$ vibrational state of the $a^3\Sigma^+$ potential of ^{40}K^{87}Rb. Adapted from ref. 1.

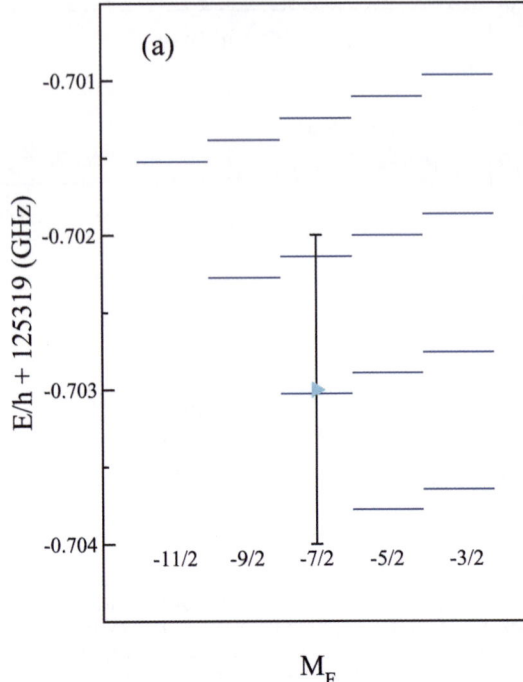

Fig. 7 The hyperfine and Zeeman structure of the $v = 0$ $\ell = 0$ of the $X^1\Sigma^+$ state of ^{40}K^{87}Rb at $B = 545.9$ G. Adapted from ref. 1.

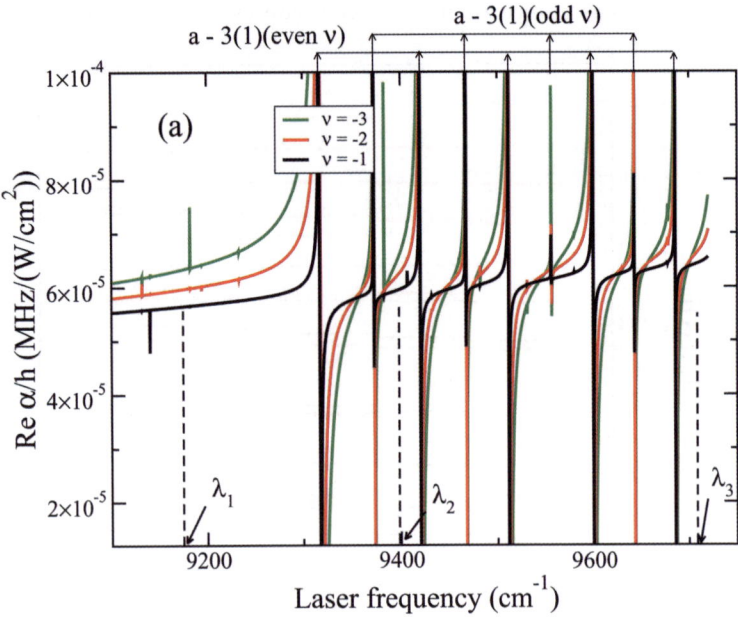

Fig. 8 Real part of the dynamic polarizability of the $v = -1, -2,$ and $-3, J = 0$ ro-vibrational levels of the $a^3\Sigma^+$ ground state of KRb as a function of laser frequency. The most-widely-used trapping laser wavelengths, $\lambda_1 = 1090$ nm, $\lambda_2 = 1064$ nm, and $\lambda_3 = 1030$ nm, are indicated. Adapted from ref. 1.

initial and final bound vibrational levels belong to the ground $X^1\Sigma^+$ and $a^3\Sigma^+$ states. Knowledge of the rovibrational and hyperfine structure of the ground states is important for efficient state preparation and to identify the observed spectra. The two states are coupled *via* hyperfine interactions: the Fermi-contact interaction, which couples its electron spin, here 1/2, to its nuclear spin, and electron and nuclear Zeeman interactions for each of the constituent atoms. The Zeeman interaction is non-zero since an external magnetic field is used to create Feshbach KRb molecules. Each vibrational level has hyperfine, Zeeman and rotational sublevels. In our calculation we only include the usual atomic hyperfine interactions plus magnetic electron–electron spin dipole $1/r^3$ interactions.

The final states of the Raman transitions are $v = 0$ levels of the "X" and "a" potentials. Fig. 6 shows the rotational hyperfine and Zeeman structure of the $v = 0$ level of the "a" triplet potential at $B = 545.9$ G as calculated with the coupled-channel method. The black lines are the $\ell = 0$ bound states and the red lines are the $\ell = 2$ bound states. Here, ℓ is the relative orbital angular momentum of the two atoms and M_F is the projection of the total atomic angular momentum along the magnetic field direction. The JILA experiment[2] has located more than ten sublevels of the $v = 0$ vibrational level of the singlet and triplet potentials, indicated by triangles and crosses in Figs 6 and 7. The agreement between the theory and the experiment is good.

We have also examined the dynamic polarizability of vibrationally cold KRb molecules as a function of laser frequency. Based on this knowledge, laser frequencies can be selected to minimize decoherence from loss of molecules due to spontaneous or laser-induced transitions. Figs 8 and 9 show the real part of the polarizability of the three most weakly bound $J = 0$ ro-vibrational levels ($v = -1$, -2, -3) of the $a^3\Sigma^+$ potential and $v = 0$ of the $X^1\Sigma^+$ potential as a function of laser frequency. Most of the resonances in Fig. 8 are due to vibrational levels of the $2^3\Sigma(1)$ excited state potential. The uncertainty in these vibrational level energies is about 5 cm^{-1}. We can see from Fig. 8 that the wavelength $\lambda_1 = 1090$ nm, a commonly used laser wavelength for trapping, is far away from the resonances. The experimental measurement of the dynamic polarizability of the Feshbach molecule at this wavelength gives $\alpha/h_{\text{Feshbach}} = 5.7 \times 10^{-5}$ MHz/(W/cm^2),[3] whereas our calculation predicts 5.6×10^{-5} MHz/(W/cm^2). The two other trap wavelengths, $\lambda_2 = 1064$ nm

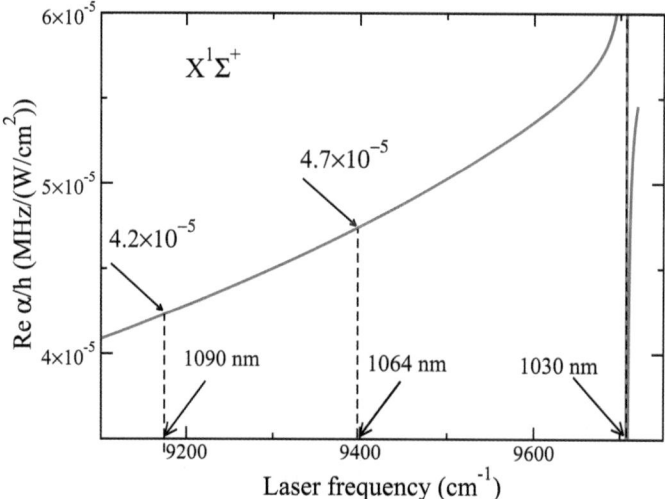

Fig. 9 Real part of the dynamic polarizability of the $v = 0$ $J = 0$ level of the $X^1\Sigma^+$ ground state of KRb as a function of laser frequency. Adapted from ref. 1.

and $\lambda_3 = 1030$ nm, lie around stronger resonances and their loss rate is predicted to be ten to one hundred times larger. One atomic unit of polarizability (1 a_0^3) corresponds to 4.69×10^{-8} MHz/(W/cm^2).

Fig. 9 shows that only one resonance is visible. It is due to the lowest ro-vibrational level of the excited $1^3\Pi(0^+)$ potential. This contribution to the polarizability is due to spin-orbit mixing with the $2^1\Sigma(0^+)$ potential. Comparison with the experimental polarizability at 1090 nm, measured by the JILA group,[3] gives an uncertainty for our calculation of 13%.

1 S. Kotochigova, E. Tiesinga, and P. S. Julienne, *New J. Phys.*, 2009, **11**, 055043.
2 K.-K. Ni, S. Ospelkaus, M. H. G. de Miranda, A. Peer, B. Neyenhuis, J. J. Zirbel, S. Kotochigova, P. S. Julienne, D. S. Jin, and J. Ye, *Science*, 2008, **322**, 231.
3 S. Ospelkaus, K.-K. Ni, M. H. G. de Miranda, B. Neyenhuis, D. Wang, S. Kotochigova, P. S. Julienne, D. S. Jin and J. Ye, *Faraday Discuss.*, 2009, **142**, DOI:10.1039/b821298h.

Professor Hutson asked Professor Kotochigova and Professor Julienne: I am very interested to see that your coupled-channel calculations of singlet KRb molecules in the lowest rovibrational level of the ground singlet state are split into sublevels with different values of the total nuclear spin quantum number *I*. In our calculations of these levels, this splitting comes from the scalar nuclear spin–nuclear spin coupling, which is a "through-bond" (electron-mediated) coupling. However, I do not think that this scalar coupling appears explicitly in your Hamiltonian. Have you investigated where this splitting comes from?

More generally, which hyperfine terms are included in your coupled channel calculations? We calculate nuclear electric quadrupole and tensor spin–spin coupling constants as well as the scalar spin–spin coupling: the additional terms can be quite important for rotationally excited states, though they average to zero for rotationless states. In particular, I think quadrupole couplings are likely to be important for polar molecules in electric fields, where the electric field itself will mix the rotational states.

Professor Kotochigova replied: In our calculation we only include the usual atomic hyperfine interactions plus magnetic electron–electron spin dipole interactions. Two interaction terms that we do not include are the magnetic nuclear–nuclear dipole $1/r^3$ interaction and the electron–nuclear dipole $1/r^3$ interaction. They depend on the rotation of the molecule and for any $J = 0$ state the first-order correction is zero. For a $J = 2$ state it could matter. My estimate of the nuclear–nuclear effect is 1 kHz for the *X* state $v = 0$, $J > 0$.

Professor Julienne answered: It is true that the scalar nuclear spin coupling does not appear explicitly in our Hamiltonian, which is based on atomic spin properties and a frame transformation into the molecular basis. Consequently, we get an effective coupling between nuclear spins in the singlet ground state vibrational levels mediated as a second-order interaction through the (energetically) distant ground triplet state. The triplet state provides unpaired electron spins with which the nuclear spins interact in second order through off-diagonal matrix elements (in a molecular basis), so one gets a splitting that takes on the form of the scalar nuclear spin interaction, with the order of magnitude of the coupling constant given by the atomic hyperfine coupling divided by the energy splitting between the ground singlet and triplet states. We find a splitting in the $v = 0$ $N = 0$ singlet level of 41 kHz between the highest and lowest energy states of total nuclear spin $I= 5/2$ and 11/2 respectively. This compares with the splitting of about 30 kHz from the calculation for the same splitting reported by Aldegunde *et al.* (private communication). Thus, we get the right order of magnitude, but of course, can not get the full molecular effect, which in the language of our reduced-space Hamiltonian, would require summing over an infinite series over the second-order contribution from all excited molecular states with unpaired electron spins. To the extent that the lowest triplet state provides a good approximation, we get a reasonable approximate representation of the splitting.

There is one good test that could be done experimentally or theoretically for this idea about the second order coupling. We have calculated the splitting for all singlet vibrational levels in our coupled channels calculations.

The $v = 73$ singlet level, the last singlet level below the $v = 0$ triplet level, has a zero magnetic field splitting between the 5/2 and 11/2 levels of 725 MHz, an increase of a factor of 18 from the $v = 0$ splitting.

This is because of much smaller energy denominators in the second order sum. It would be helpful to have both experimental and *ab initio* theoretical determinations of the splitting of these two levels, to verify if indeed the second order coupling between nuclear spins in the ground state is indeed primarily mediated through the atomic-like ground triplet state. If this is true, then the predicted 725 MHz splitting from the atomic Hamiltonian should be a good approximation for this level.

We have not looked at the nuclear quadrupole coupling, but it clearly is not explicitly in our coupled channels calculations, and I do not believe that our restricted space ground state Hamiltonian would bring it in in second order. It is evident that it will be necessary to measure, or at least to calculate, all of the coupling constants that are relevant to understanding the nuclear spin structure in polar molecule states. Two papers by Aldegunde *et al.*[1,2] provide a very helpful calculation for several mixed alkali species. Ideally, a measurement of the spin structure of one or more singlet levels would permit the fitting of a spin-Hamiltonian model to the observed structure to deduce accurate values of the various nuclear spin coupling constants. Such measurements will need to be done. The role of nuclear hyperfine structure will clearly be significant on the scale of microkelvin experiments. A

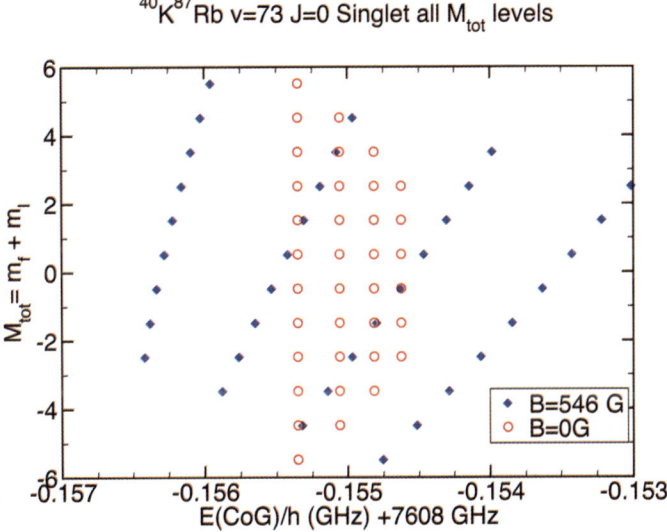

Fig. 10 Bound state energy for all M_{tot} $v = 73$ $J = 0$ singlet state of the ^{40}K^{87}Rb molecule at $B = 0$ (open circles) and 54.6 mT (diamond symbols). Energy is relative to the energy center of gravity of the atomic hyperfine multiplets. These levels lie about 400 GHz below the location of the $v = 0$ triplet state. At $B = 0$ the levels divide into four degenerate groups with total nuclear spin quantum number $I_{tot} =$ 11/2, 9/2, 7/2, and 5/2 in order of increasing energy, with a spread of 725 kHz between the highest and lowest energy. This is to be compared with Figure 16 of my paper, which shows similar structure for the $v = 0$ $J = 0$ state of the same molecule but with much smaller spread of 41 kHz at $B = 0$.

multiplicity of nuclear spin states increases the control possibilities, as well as the possibility of harmful spin-changing inelastic collisions.

1 J. Aldegunde, B. A. Rivington, P. S. Żuchowski and J. M. Hutson, *Phys. Rev. A*, 2008, **78**, 033434.
2 J. Aldegunde and J. M. Hutson, *Phys. Rev. A*, 2009, **79**, 013401.

Mr Deiglmayr said: I want to comment on the issue of R-transfer in the association of pairs of atoms into molecules, which was discussed by William Stwalley yesterday. As presented in our paper, we observe a strong enhancement of the rate for photoassociation into deeply bound levels of the $B^1\Pi$ state of LiCs which allows us to directly form molecules in the rovibrational ground state. In a cooperation with Philippe Pellegrini and Robin Cote from Storrs we were able to identify the mechanism leading to this enhacement:[1] a bound level at roughly 200 mK is coupled to the incoming scattering state by hyperfine interactions. Although very far away in energy, this Feshbach resonance induces a significant "echo" of the triplet-like wavefunction in the singlet component of the scattering state, which leads to a very strong enhancement of the overlap between continuum and excited state wavefunctions at the inner turning point of the lowest triplet potential.

This enhanced R-transfer due to hyperfine-coupling in the ground state is complementary to the spin-orbit mediated couplings in excited molecular states commonly used for the transfer of magneto-associated molecules into deeply bound levels as reported by the groups from Innsbruck and Boulder. We suggest, that the use of hyperfine couplings in the ground state as observed in our experiment might also be helpful for the transfer of Feshbach molecules into the absolute ground state. The possibility to use unperturbed excited singlet states for the STIRAP transfer might even increase the efficiency of such schemes.

1 J. Deiglmayr, P. Pellegrini, A. Grochola, M. Repp, R. Côté, O. Dulieu, R. Wester and M. Weidemüller, *New J. Phys.*, 2009, **11**, 055034.

Professor Julienne responded: Generally, the near-threshold levels associated with Feshbach resonances have mixed singlet-triplet character, with several different options available for which vibrational wave function character to choose from (*i.e.*, the last, next-to-last, *etc.* character), depending on the mass of the system. Heavier systems tend to be more mixed with more options, lighter systems less so. Examples are given in my paper (see Fig. 7, 10, 11 and 14) for the KRb system. The 546 G resonance used in this case takes advantage of the enhanced amplitude of the wave function component of $n = -2$ triplet character to get a good Franck–Condon factor and reach the excited state manifold. Choosing other nearby resonances (see Fig. 7 of my paper) would permit $n = -1$ or $n = -3$ triplet-character levels to be selected. Even these levels have some singlet character. In light systems, where the singlet and triplet manifolds can be more separated and less mixed (depending on the quantum defects of each), there can be good opportunity to use mixed wave functions with significant singlet character. But even in KRb, there are several other possible pathways to making cold molecules than the one selected by the JILA group that could take advantage of different types of up and down Franck–Condon factors. In many mixed species, there may be more than one route to the desired ground state. One should take advantage of the known properties of the species of interest, and use whatever properties of the ground or excited state that are understood to find a successful route. This generally means understanding the level structure and at least qualitative Franck–Condon patterns associated with both ground and excited states. In other words, one needs some good data to start with. Using mixed threshold levels of the ground state with singlet character will certainly be possible in some cases, if the quantum defects are favorable. Such routes should be looked for, especially in light species with Li or Na in them. It will be interesting to see if such routes can be found for heavier species with Rb or Cs in them.

Professor Kotochigova addressed Professor Weidemüller: This is a comment on the lifetime of the $v = 0$, $J = 0$ LiCs molecules. Simple estimates based on the dissociation energy of the three Li_2, Cs_2, and LiCs molecules suggests that there is ~400 cm^{-1} of energy that can be released.

Professor Weidemüller responded: Indeed, there might be exothermic channels in reactive collisions of two LiCs atoms. The densities of molecules produced in our current experiments are far too low to expect any signatures of molecule–molecule collisions, either elastic or inelastic. We are currently preparing experiments on the accumulation of LiCs molecules in an optical dipole trap with the aim of finding signatures of atom–molecule and, possibly, molecule–molecule collisions.

Professor Hutson commented: Many of the methods that are used to produce molecules in their ground electronic state rely on singlet–triplet mixing in the excited state. We have done some calculations on RbCs that illustrate some important general features of this mixing.[1]

When two electronic states cross and mix, there are two limits that may be thought of as "almost uncoupled". If the matrix element between the two states is *small*, then it will produce only localised perturbations when vibrational states in the two wells are near-degenerate and have significant Franck–Condon overlap. Conversely, when the matrix element connecting the two states is *large* near their crossing point, the system will be well described in an adiabatic picture in which rovibrational levels exist on the upper and lower adiabatic curves formed by diagonalising a 2 × 2 (or larger) matrix at each value of the internuclear distance.

For the heteronuclear alkali-metal dimers, there are two excited states that display just such a crossing, as shown for RbCs in Fig. 11. The $A^1\Sigma^+$ and $b^3\Pi$ states are mixed by spin-orbit coupling to form two states of $\Omega = 0^+$ symmetry. At long range the two spin-free states are degenerate, so there is always strong singlet–triplet mixing.

However, the nature of the mixed vibrational states depends crucially on the strength of the spin-orbit coupling near the crossing point.

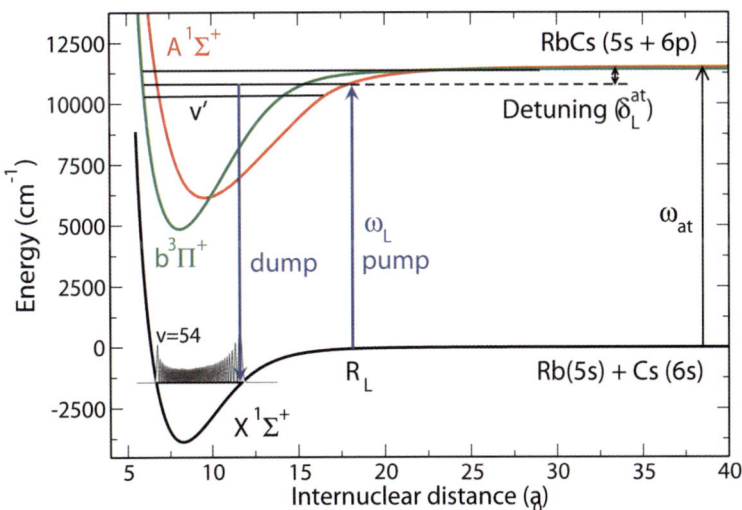

Fig. 11 The potential curves for the ground $X^1\Sigma^+$ and the excited $X^1\Sigma^+$ and $b^3\Pi$ states of RbCs, with a possible pump-dump scheme for transferring population *via* the mixed excited state.

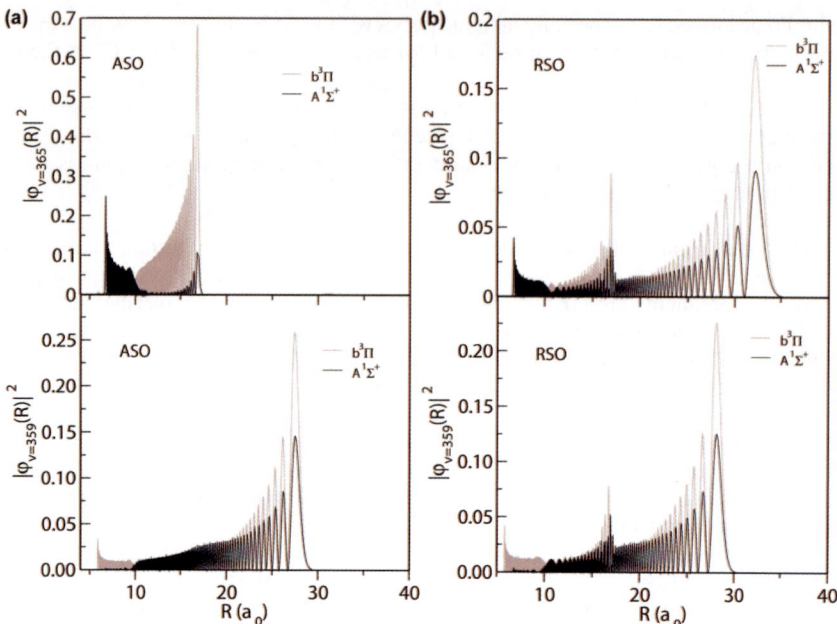

Fig. 12 Eigenfunctions of the coupled excited-state potentials corresponding to $v = 359$ and 365. (a) Results obtained with constant asymptotic spin-orbit coupling (ASO). (b) Results obtained with R-dependent spin-orbit coupling (RSO).

We have calculated the mixed states for two different spin-orbit couplings, and the results are shown in Fig. 12.

The two left-hand panels show what the mixed states would look like if the spin-orbit coupling was held fixed at its asymptotic value (asymptotic spin-orbit, ASO). The resulting coupling function is large near the crossing point and gives near-adiabatic behaviour: the state in the upper panel is essentially confined to the upper adiabatic curve, and the state in the lower panel is confined to the lower curve. Both of them show a sharp transition from singlet to triplet character around the crossing point, in opposite directions.

However, it is noteworthy that the state that has strong singlet character at short range (so gives significant intensity for transitions to low-lying states of $X^1\Sigma^+$) does not penetrate significantly to long range (ASO, upper panel).

The two right-hand panels show the same two states calculated with an R-dependent spin-orbit coupling function (RSO) that is the same asymptotically but much smaller near the crossing point. In this case the vibronic functions show strong *nonadiabatic* mixing. They still show the switchover from singlet to triplet character near the crossing point, but in this case *both* functions have significant amplitude on the lower adiabatic curve at long range. They also both show density peaks in both singlet and triplet components at the outer turning point of the upper adiabatic curve. The mixed state in the upper panel provides good intensities for both photoassociation (long-range character) and for transitions to low-lying states of $X^1\Sigma^+$.

The take-home message here is that it is important in conceptual terms to understand whether two states that cross do so near-diabatically (weak coupling), near-adiabatically (strong coupling) or nonadiabatically (intermediate coupling).

The wavefunctions have quite different character in the three cases and present different opportunities for laser-based experiments.

1 S. Ghosal, R. J. Doyle, C. P. Koch and J. M. Hutson, *New J. Phys.*, 2009, **11**, 055011

Professor Stwalley opened the discussion of Professor Ye's paper: I would like to ask about the harpooning ideas you presented. First of all, we estimated an ionization potential IP(KRb) = 27982 ± 100 cm^{-1} with respect to ground state K and Rb atoms; what value did you use? Secondly, the ion pair K$^+$ + KRb$^-$ may be close in energy to K$^-$ + KRb$^+$. Also KRb$^+$ + KRb$^-$ may be significant in considering the collision of two KRb molecules. However, the distances in your figure are quite small and probably correspond to relatively small cross sections. In addition, once harpooning occurs, the two KRb molecules are no longer identical fermions. Moreover, in the case of Rb + KRb, what is the exit channel? It seems to me that it must be simply a change in hyperfine state, since the reaction itself (to form Rb$_2$) is exoergic by ~215 cm^{-1}.[1]

1 D. Wang *et al.*, *Eur. Phys. J. D*, 2004, **31**, 165.

Professor Ye replied: Experimentally we have observed similar loss rates for the KRb ground state molecules when they collide with either K or Rb atoms, with only one exception, when both KRb molecule and Rb atom are prepared in their respectively lowest hyperfine states. At the time of the discussion, we had not prepared KRb molecules in their lowest hyperfine state, now we have. What we find now is a universal scattering loss behavior. When a KRb molecule and an atom (K or Rb) get close to each other, within the van der Waals length (as defined by their C_6 coefficient), as long as there is a high probability for something to happen in the short range (could be chemical reaction as in the case of KRb + K → K$_2$ + Rb, or simply a scramble of the KRb hyperfine state), we see that the loss rate for KRb is nearly a constant, regardless of the hyperfine state of KRb or K. For Rb atoms in a hyperfine state higher than the lowest, the loss rate is also the same. Only when Rb atoms and KRb molecules are both prepared in their lowest hyperfine states do we see that the KRb loss rate is reduced by at least two orders of magnitude.

Dr Pillet asked Professor Weidemüller: Could you comment more on your strategy to increase the rate of formation of cold molecules in $v = 0$ for a large accumulation?

Professor Weidemüller responded: The most straightforward strategy to increase the formation rate of LiCs molecules by photoassociation is to increase the densities of the Li and Cs atoms, *e.g.* by using an optical dipole trap in combination with appropriate cooling techniques. However, the molecular phase-space densities achievable in this way are limited due to the spontaneous emission which distributes population among other vibrational and rotational states. One may combine this approach with the broad band optical pumping scheme for vibrational population presented by you. However, it turns out that application of this scheme is not as straightforward as in the case of Cs$_2$. In my view, the most promising route towards high phase-space densities of molecules is the combination of magnetoassociation followed by STIRAP, as presented by Jun Ye for heteronuclear molecules, and Johannes Hecker Denschlag and Hanns Christoph Nägerl for homonuclear molecules at this Faraday Discussion.

Dr Lane addressed Professor Ye: The ability to control molecules so that they perform the chemical reactions we wish has been a dream of chemists from the very beginning of reaction dynamics. At the risk of venturing close to the edge of science fiction rather than fact, I was wondering whether Professor Ye or the other "ultracold synthetic chemists" have considered attempting a non-exchange chemical reaction with their freshly prepared molecules? By injecting a few halogen atoms into the chamber, the unique properties of the harpoon reaction[1] can be exploited to produce alkali halide molecules and alkali atoms. These reactions are essentially barrierless, have an enormous reaction cross-section and are very exothermic so definitely belong within the realms of true chemistry. Crucially, the alkali atoms can be produced in electronically excited states that are chemiluminescent and thus the nascent reaction products can be observed. Whereas "room temperature synthetic chemists" are happy just to control the type of molecules used and produced in a chemical reaction the physical chemist wants to go further and control the quantum states of the reagents and products as well. Using Feshbach resonances between ultracold atoms, ultracold alkali metal dimers are formed in atom traps where no molecules exist in the absence of the applied magnetic field and in the exact quantum states required. Consequently, the magnetic field could be used both to initiate a chemical reaction and to control what reagents are used *via* the magnetic Feshbach resonances.[2] Using a mixture of ultracold atoms, say ^{40}K and ^{87}Rb, the chemist can select whether ^{40}K$_2$, ^{87}Rb$_2$ or ^{40}K^{87}Rb molecules are used in the reaction.

1 W. S. Struve, T. Kitgawa and D. R. Hershbach, *J. Chem. Phys.*, 1981, **54**, 2759.
2 E. Tiesinga, B. J. Verhaar and H. T. C. Stoof, *Phys. Rev. A*, 1993, **47**, 4114.

Professor Ye answered: Laser cooling of halogen atoms appears difficult at this time, due to the lack of suitable lasers, but we should keep this in mind as it will open up an incredible range of experimental possibilities.

Dr González-Férez asked Professor Weidemüller: What do you mean by the expression: "the change of coupling between angular momenta" in the following sentence of your manuscript: "The observation of these higher *J* components is probably due to the change of coupling between angular momenta as the molecules become less tightly bound". I would also like to point out that the electronic ground state $X^1\Sigma^+$ of LiCs has a *d*-wave shape resonance around 1 mK.

Professor Weidemüller replied: At high vibrational states, the coupling of angular momenta can give rise to higher rotational states emerging from the association of ultracold atoms.

Dr Dulieu responded: Concerning the second point, we also observe in our calculations such a shape resonance, if we consider scattering in the unperturbed singlet potential extracted from high-resolution molecular spectroscopy. However if we add hyperfine coupling to the triplet scattering potential, this resonance is no longer present in the entrance channel chosen for the experiment.

Dr Koch asked Mr Deiglmayr : I would like to follow up on a question asked by Rosario Gonzalez-Ferez on the higher rotational components in the photoassociation spectra (*cf.* Fig. 4). The observation of these higher rotational components is attributed to angular momentum coupling at long range. However, photoassociation has been explained to occur mainly at short range, near the inner turning point. How does that fit together?

Mr Deiglmayr answered: The higher rotational components ($J' > 2$) of photoassociation (PA) resonances are only present for vibrational levels $v' > \sim26$. The Condon points for these transitions are located above $R = 16$ a$_0$ and we attribute the appearance of the higher rotational components to a recoupling of all angular momenta (rotation, electronic orbital momentum, electronic and nuclear spins) at these larger distances. The PA into deeply bound levels of $B^1\Pi$ in fact occurs around the inner turning point of the lowest triplet state $a^3\Sigma^+$, but in the excited vibrational level the Condon radius is actually closer to the outer turning point. For PA in levels below $v' = 25$ of the $B^1\Pi$ state we only observe rotational components $J' = 1$ and 2, the latter being actually the stronger one. The reason for this is yet unclear, but it indicates the importance of considering higher partial waves in the continuum scattering wavefunction, for instance *via* the possible existence of a *d*-wave resonance; we note however that we have not found such a resonance in the present coupled-channel calculations.

Professor Hutson commented: We have done quite a lot of calculations on the barrierless atom exchange reactions that can occur in collisions between alkali metal atoms and alkali metal dimers.[1–6] Our work was on collisions of molecules in triplet states but the general conclusions apply to singlet states as well. We found generally very large cross sections for both inelastic and reactive collisions whenever they were energetically allowed, though with strong resonance structure.

The ^{7}Li + ^{7}Li$_2$ case[2,3,5] is particularly interesting. We calculated elastic and inelastic/reactive cross sections for initial vibrational quantum numbers $v = 1$ to 3 and explored the effect of small variations in a potential energy scaling factor λ.[5] We observed very strong resonances as a function of λ for $v = 0$ but the resonances died away progressively for higher v as shown in Fig. 13. For $v = 3$ the total inelastic/reactive cross sections vary by only about a factor of 2 above and below a limiting value of 1×10^{-9} cm^2 at a kinetic energy of 0.928 nK, corresponding to a limiting low-energy rate constant of 2×10^{-10} cm^3 s^{-1}. This is remarkably close to the value obtained from the "full absorption" rate predicted by long-range theory (equation 14 of ref. 6), $k_{\text{inel}} = 2h\bar{a}/\mu = 3.7 \times 10^{-10}$ cm^3 s^{-1} for a mean scattering length $\bar{a} = 21.5$ Å.

I would however caution that strong collisions are not necessarily diagnostic of a harpooning mechanism. In the alkali metal dimer + atom systems, the "harpooning radius" calculated from a long-range model is actually fairly small, and at such small distances there are other forces that come into play. In our calculations on systems such as Na + Na$_2$ and K + K$_2$ we saw large inelastic and reactive cross sections, even though the potential energy surfaces we used are dominated by dispersion forces.[8] Even these relatively weak forces give potential wells several hundred cm^{-1} deep, which is quite enough to cause strong inelastic and reactive collisions.

In the case of K + KRb, it is entirely likely that reaction to form Rb + K$_2$ will occur quickly enough to destroy the molecules at a rate close to the "full absorption" rate predicted by long-range theory. For Rb + KRb, where the reaction to form Rb$_2$ is energetically forbidden, it may still be possible for atom exchange collisions to

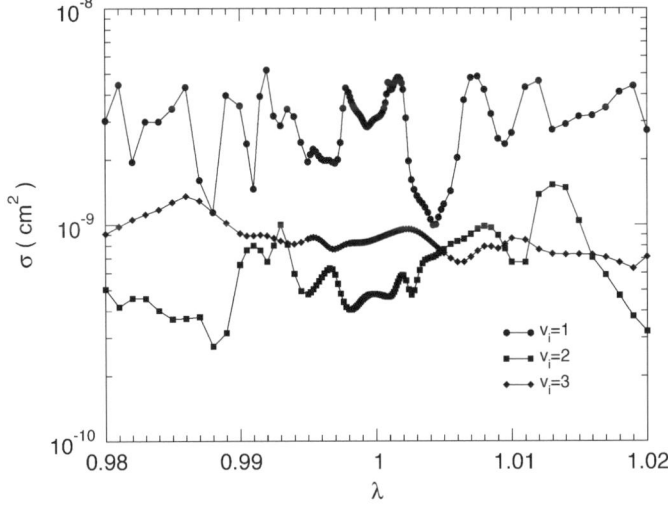

Fig. 13 Dependence of the total inelastic cross sections for ^{7}Li + ^{7}Li$_2$ (v_i, $n_i = 0$) on the scaling factor λ of the nonadditive part of the potential.

scramble the nuclear spins and lead to apparent loss of (detectable) KRb molecules even if the products remain trapped.

1 P. Soldán, M. T. Cvitaš, J. M. Hutson, P. Honvault and J.-M. Launay, *Phys. Rev. Lett.*, 2002, **89**, 153201.
2 M. T. Cvitaš, P. Soldán, J. M. Hutson, P. Honvault and J.-M. Launay, *Phys. Rev. Lett.*, 2005, **94**, 033201.
3 M. T. Cvitaš, P. Soldán, J. M. Hutson, P. Honvault and J.-M. Launay, *Phys. Rev. Lett.*, 2005, **94**, 200402.
4 G. Quéméner, P. Honvault, J.-M. Launay, P. Soldán, D. E. Potter and J. M. Hutson, *Phys. Rev. A*, 2005, **71**, 032722.
5 M. T. Cvitaš, P. Soldán, J. M. Hutson, P. Honvault and J.-M. Launay, *J. Chem. Phys.*, 2007, **127**, 074302.
6 J. M. Hutson and P. Soldán, *Int. Rev. Phys. Chem.*, 2007, **26**, 1.
7 P. S. Julienne, *Faraday Discuss.*, 2009, **142**, DOI: 10.1039/b820917k.
8 P. Soldán, M. T. Cvitaš and J. M. Hutson, *Phys. Rev. A*, 2003, **67**, 054702.

Professor Julienne replied: It is gratifying to find another example of an ultracold collision that is characterized by a rate coefficient determined by the reflection from the long-range part of the potential. The reflection probability R determines the transmission probability $T = 1 - R$ that the colliding atoms with long De Broglie wavelength will pass through the long-range region and reach the short-range region where some strong inelastic transition occurs, thereby "absorbing" all of the incoming flux. This kind of model will not work if the short range probability of an inelastic event is small, since there will be back-reflected amplitude from the inner region that interferes with the amplitude of the incoming wave function, so that the amplitude at short range will then depend on the scattering length of the potential. But when there is perfect "absorption" in the inner region, the inelastic rate coefficient does not depend on the scattering length of the potential, but only on the reflection from the long range part of the potential, since there is no reflection from the inner region. In this case, all that is needed to get a "universal" inelastic rate constant in the E to 0 s-wave limit is the isotropic van der Waals coefficient and the reduced mass of the pair. The model can also be implemented numerically, including higher partial waves, as done by Orzel et al.[1] for Penning ionization. Such a transmission/reflection model with higher partial waves could provide a way to put threshold quantum corrections into the usual semiclassical Langevin model, *e.g.*, eqn (14) of Hutson and Soldán,[2] and will go correctly to the quantum s-wave capture limit when the collision energy E is low enough, see eqn (14) of ref. 3. It will be important to determine what kind of atom–molecule or molecule–molecule collisions are intrinsically strong collisions controlled by capture of the colliding species by the long-range potential. It seems that $v = 0$, $N = 0$ K + KRb or Rb +KRb collisions in at least some spin channels may be such strong collisions. That could be bad news for the prospects of doing sympathetic cooling of $v = 0$ KRb with Rb, for example. However, it may be possible to shield two polar molecules from strong collisions using active field control in reduced dimension, such the "blue shielding" scheme for long-range collisions proposed by Micheli et al.[4]

1 C. Orzel, M. Walhout, U. Sterr, P. S. Julienne, and S. L. Rolston, *Phys. Rev. A*, 1999, **59**, 1926.
2 J. M. Hutson and P. Soldán, *Int. Rev. Phys. Chem.*, 2007, **26**, 1.
3 P. S. Julienne, *Faraday Discuss.*, 2009, **142**, DOI: 10.1039/b820917k.
4 A. Micheli, G. Pupillo, H. P. Büchler, and P. Zoller, *Phys. Rev. A*, 2007, **76**, 043604.

Professor Ye also responded: That's great. Please refer to my earlier reply to Professor Bill Stwalley on our recent experimental observations of atom–molecule collisions. With our recent capability of manipulating the ground state molecules to any specific nuclear spin state, we have made systematic and detailed observations of these atom–molecule collisions and it looks like experiment and theory are coming together.

Dr Ferlaino asked: How is the Wigner Threshold law modified or extended to the case of a $1/R^3$ dipole–dipole interaction?

Professor Hutson replied: There is one special case in which a dipole–dipole interaction will give the same threshold law as for an atom–atom interaction. For molecules in states with well-defined parity, such as freely rotating states in the absence of an electric field, the dipole–dipole R^{-3} term averages to zero.

However, there is still a second-order term involving mixing in excited rotational states that contributes an effective R^{-6} interaction, which will give the same threshold law as an R^{-6} dispersion interaction.

Professor Julienne responded: This is certainly true – in the absence of an electric field, but with or without a magnetic field, two polar molecules interact *via* the van der Waals interaction (plus the quadrupole–quadrupole interaction for higher partial waves). The C_6 coefficient will typically be significantly larger than for atoms (see Hudson *et al.*[1] for a RbCs example). Even a small electric field will mix in dipolar character to the long range interaction and change the threshold laws. Understanding the practical consequences of this transition will be an important early task for the experimental and theoretical studies of cold polar molecule collisions. In particular, the density of near-threshold levels will increase for the dipole–dipole potential so there should be interesting field-tunable Feshbach resonances. Long-range quantum defect theory for the dipole–dipole potential (or a mixed dipole–dipole plus van der Waals potential) may prove to be quite useful for this.

Hudson *et al.*, *Phys. Rev. Lett.*, 2008, **100**, 203201.

Dr Groenenboom remarked: The (R^{-3}) dipole–dipole interaction in the long range is only non-vanishing in the case of a resonant interaction or in the presence of an external field.

Professor Hutson answered: I think there are some semantic differences here. If you think of the "interaction" as a function of the intermolecular angles then it is non-vanishing between any pair of permanent dipoles at most angles. If you average it over the eigenstates then it has no diagonal matrix elements except for a resonant interaction between identical species or in the presence of a field.

Professor Grimm asked Professor Ye: The experiments on KRb ground-state molecules are quite close to quantum degeneracy. Is it straightforward to reach this goal by further evaporative cooling?

Professor Ye replied: We are making good experimental progress in this direction.† We need to understand how these molecules collide. At zero electric field, Fermi molecules do not interact. As we raise the electric field, we have seen preliminary evidence of thermalization of these molecules inside the trap. Of course we need to avoid certain electric field values where inelastic collisions between molecules have been observed.

Professor Hutson asked: Several times you've mentioned properties of your KRb molecules that stem from their fermionic nature, and some of them (though not all) seem to be disadvantages. I realise that the current experiment uses fermionic ^{40}K, but is there anything fundamental to prevent you extending the experiments to bosonic isotopes?

---

† Since the discussion took place, we have now extended the molecular lifetime in an optical trap to be near 1 s.

Professor Ye responded: No, there is no fundamental limit, one could certainly explore the bosonic isotopes. A technical challenge might be to find a reasonably long-lived Feshbach state to enable a successful second step for STIRAP transfer. Fermions need to be cooled to a much lower temperature for the dipolar interaction energy to become dominant. On the other hand, there is an interesting and unique perspective to demonstrate dipolar interactions. At zero electric field, fermionic molecules do not interact at ultralow temperatures. As the electric field increases, the strength of the dipolar interaction increases, which lowers the p-wave barrier, as shown by recent calculations of John Bohn, and molecules can have resonant interactions at certain electric field values. Preliminary experimental data suggest that we are indeed observing this novel behavior.

Professor Weidemüller asked Professor Julienne: One could imagine a Mott insulator state of atomic pairs in a regular lattice which is transformed into a molecular ensemble trapped in the same lattice. Due to the finite efficiency of the association process, one has to expect defects and impurities in the distribution of molecules over the lattice sites. If one then "melts" such an ensemble by ramping down the lattice depth, how do these defects transform? Do they possibly give rise to topological effects like vortices, or do they create elementary excitations?

Professor Julienne answered: These are very good questions. The process of transforming such "pre-formed pairs" of atoms into molecules seems to me to be a very good way to make a molecular lattice. In fact, I am currently engaged with James Freericks of Georgetown University in some calculations of the phases of cold ^{40}K and ^{87}Rb mixtures in a 2D optical lattice. The calculations show that by lowering the temperature it should be possible to realize extended spatial regions where there are exactly one atom of each type at each lattice site. Because of imperfections or inefficiency of STIRAP conversion, there are likely to be some defects in the resulting lattice structure. So your question is quite relevant to possible future experiments. I expect that one of the most exciting prospects for ultracold molecules research in the future will be to explore a variety of condensed matter phenomena with lattices of such molecules. So we will need to understand such things. This is one of the reasons for my questions about the efficiency of STIRAP in a previous session. It relates to the number of defects you might initially create in a molecular lattice.

Dr Petrovic said: Two essential steps are needed to form ultracold molecules by broadband photoassociation: (i) photoassocation of molecules in a superposition of vibrational states in an excited state and (ii) manipulation of the thereby created wavepacket to achieve a good Franck–Condon overlap with the wavefunction of the targeted ground state. A suitable dump pulse applied at the time delay which corresponds to the optimal overlap, efficiently transfers the population to the ground state. We have achieved the first step and have performed a series of pump-probe experiments to investigate dynamics of the wavepacket in the 5s + 5p manifold.

In order to determine the initial state of rubidium we performed the resonant two-photon ionization of the rubidium in the magneto-optical trap and found that, besides atoms, it comprises molecules pre-photoassociated by the trapping lasers.[1]

In the experiment we illuminated the MOT by a focused femtosecond laser pulse with the spectrum above the D1 line blocked to avoid the atomic resonance and followed by an ionization femtosecond pulse which arrived after a controllable delay. Fig. 14 shows molecular and atomic ion signals separated in a TOF mass spectrometer at a fixed delay as functions of the detuning of the cut-off frequency of the photoassociation pulse from the D1 line. The presence of molecules at larger detunings and a quicker roll-off of the atomic than the molecular signal, suggests that the molecules are formed in the bound states below the 5s + 5p asymptote. In order to distinguish between the photoassociation of the free scattering state atoms and

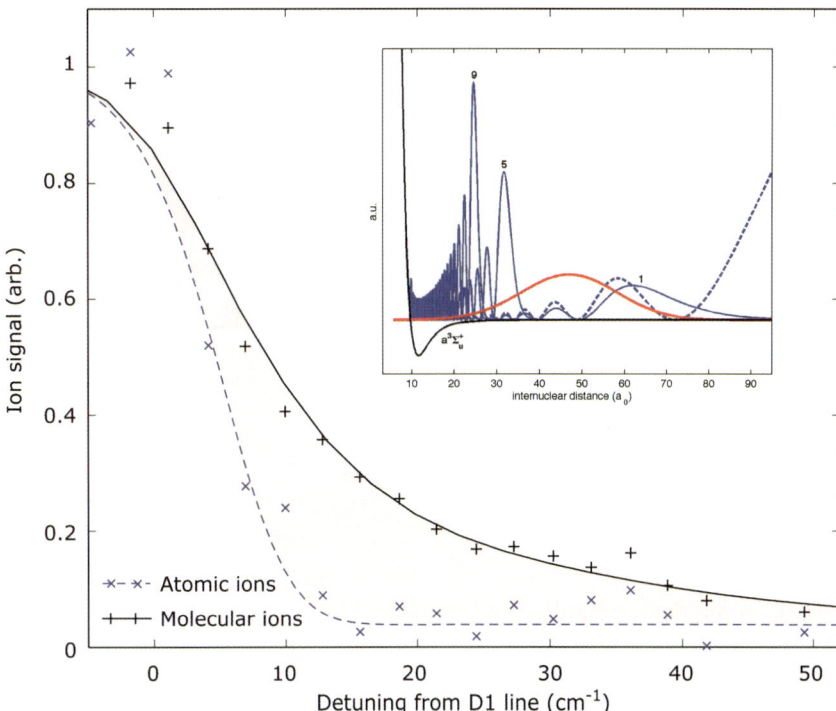

Fig. 14 Molecular and atomic ionic signals *versus* the cut-off wavelength of the photoassociation pulse. Inset: the initial state which best reproduces the experimental molecular signal (red line) and typical wavefunctions of the measured initial state (solid lines – ground triplet states numbered from the highest, dashed line – scattering state).

excitation of pre-formed ground triplet molecules, which both can result in formation of bound molecules, we fitted the molecular yield from the Gaussian distribution of atoms in the initial state to the experimental molecular signal. The red line in the inset in Fig. 14 shows the distribution which gives the best fit. Comparison with the probability of finding the pre-associated triplet molecules and a typical scattering state (blue lines) suggests that the scattering state is a dominant source of molecules, *i.e.* that the photoassociation is indeed taking place. To investigate the dynamical behaviour of the excited state, we performed a series of pump-probe experiments in which we recorded molecular and atomic ionic signals with delays between pump and probe pulses of up to 250 ps. The signals were similar with an apparent maximum during the overlap of the pulses. The expected wavepacket dynamics in the molecular signal were not observed. This may be due to the complicated incoherent initial state which produces molecules in a number of the excited state wavefunctions which do not interfere or do not interfere constructively or due to the poor spectral selectivity of the pump pulse.

D. J. McCabe, D. G. England, H. E. L. Martay, M. E. Friedman, J. Petrovic, E. Dimova, B. Chatel and I. A Walmsley, 2009, arXiv:0904.0244.

Professor Weidemüller opened the discussion of Dr Petrovic's paper by commenting: We have performed conceptually similar investigations in a collaboration between my group and the group of Professor Ludger Wöste from the FU Berlin. In a recent series of experiments we have studied the interaction of ultracold atoms with shaped femtosecond laser pulses and, in particular, coherent transients in the photoassociation of ultracold molecules.[1–7]

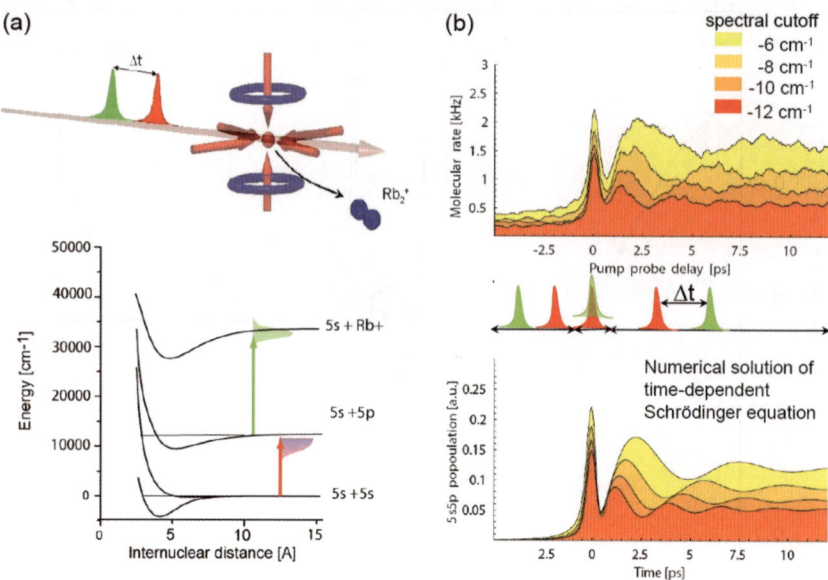

Fig. 15 (a) Sketch of the experiment and relevant molecular energy levels. Ultracold rubidium atoms in a magneto-optical trap interact with two successive ultrashort laser pulses (pump-probe scheme). The pump pulse is tuned below the Rb 5s–5p transition and is spectrally cut to suppress resonant and blue-detuned frequency components. The probe pulse ionizes excited molecular states, the Rb_2^+ ions are detected. (b) Molecular ion signal as a function of the delay between pump and probe pulse. The upper trace shows the results of the experiment, the lower trace depicts the solution of the time-dependent Schrödinger equation. The coherent oscillations stem from the energy exchange of the molecular dipole and the electric field of the laser pulse. The finite molecular ion signal at negative delays (probe before pump) can be attributed to ultracold molecules in the ground state which are ionized by the probe pulse. For further details see ref. 4–6.

Time-resolved pump-probe spectra, as shown in Fig. 15, reveal coherent oscillations of the molecular formation rate, which are due to coherent transient dynamics in the electronic excitation. The oscillation frequency corresponds to the detuning of the spectral cut position to the asymptotic transition frequency of the rubidium D1 or D2 lines, respectively. Measurements of the molecular photoassociation signal as a function of the pulse energy reveal a non-linear dependence and indicate a non-perturbative excitation process. Chirping the association laser pulse allowed us to change the phase of the coherent transients. Furthermore, a signature for molecules in the electronic ground state is found, which is attributed to molecule formation by femtosecond photoassociation followed by spontaneous decay.

1 W. Salzmann et al., Phys. Rev. A, 2006, **73**, 023414.
2 U. Poschinger et al., J. Phys. B, 2006, **39**, S1001.
3 F. Weise et al., Phys. Rev. A, 2007, **76**, 063404.
4 W. Salzmann et al., Phys. Rev. Lett., 2008, **100**, 233003.
5 W. Salzmann et al., submitted (arXiv:0903.4549).
6 A. Merli et al., submitted (arXiv:0903.4401).
7 F. Eimer et al., Eur. Phys. J. D, 2009, in press.

Dr Mudrich commented: This comment addresses the sub-communities working on ultracold molecule formation either by Feshbach association or using photoassociation with cw or ultrashort pulsed lasers. The goal is to draw their attention to the femtosecond (fs) pump-probe technique of doped helium nanodroplets as a spectroscopic method to obtain high-resolution spectra of alkali dimers and trimers in high-spin states. Such spectroscopic data may be valuable for finding new routes to ultracold ground state molecules and for interpreting spectra of ultracold molecules. Furthermore, the characterization of vibrational wave-packet dynamics in the lowest and excited triplet states may provide complementary information to those studies aiming at exciting molecular wavepackets starting with a sample of ultracold atoms.

Our experiment combines a beam of He nanodroplets doped with on average 2 Rb atoms, with fs laser ionization. The experimental setup is described in detail in ref. 1. In short, Rb_2 dimers are excited by a first pump pulse and subsequently desorb from the helium droplet surface on a time scale of about <10 ps.[1] A second identical probe pulse ionizes the free Rb_2 molecules by multiphoton ionization. The photoions are detected mass-selectively using a quadrupole mass spectrometer. Thus, recording pump-probe transients of the vibrational dynamics with delay times between pump and probe pulses t_D ranging between 10 and 1500 ps yields vibrational frequency differences of free Rb_2 with high spectral resolution.

A typical pump-probe transient is shown in Fig. 16(a). The fast wavepacket motion excited at 1006 nm in the first excited triplet state $(1)^3\Sigma_g^+$ is periodically amplitude-modulated due to fractional ($t_D \approx 82$ ps) and full ($t_D \approx 165$ ps) revivals of the vibrational wavepacket in the anharmonic $(1)^3\Sigma_g^+$-potential. Further details are obtained by means of "sliding-window" Fourier analysis. The resulting spectrogram is shown in Fig. 16(b) for the transient recorded at 1025 nm. In this representation one can clearly discern vibrations in the lowest triplet state $a^3\Sigma_u^+$ excited by impulsive Raman scattering from vibrations in the $(1)^3\Sigma_g^+$-state (combs on top of Fig. 16(b)). Clearly, half-period fractional revivals ($t_D \approx n \cdot 82$ ps, $n = 1, 2, 3,...$) are induced by the beating of vibrational levels v separated by $\Delta v = 2$.[2]

Fourier transformation of the entire data set yields vibrational frequency differences with a spectral resolution $\Delta \nu < 0.005$ cm^{-1}. Fig. 17(a) summarizes the Fourier spectra obtained for $\Delta v = 1$-beatings of the two stable isotopes of Rb_2 for various laser wavelengths. Due to the low temperature $T < 0.4$ K of Rb_2 formed on helium nanodroplets only the lowest vibrational state $v'' = 0$ is initially populated. Therefore, the impulsive Raman Fourier spectrum (Fig. 17(b)) is dominated by one frequency component $\nu_{v'' = 1} - \nu_{v'' = 0}$. The measured vibrational frequencies are in excellent agreement with the data obtained from high-resolution cw-spectroscopy using Feshbach molecules.[3]

1 P. Claas, G. Droppelmann, C. P. Schulz, M. Mudrich and F. Stienkemeier, *J. Phys. B*, 2006, **39**, S1151.

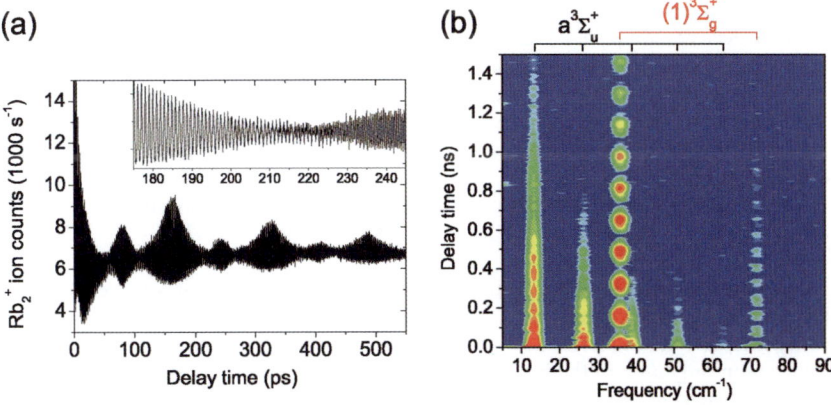

Fig. 16 (a) Experimental yield of Rb_2^+ photo ions as a function of delay time between pump and probe laser pulses recorded at laser wavelength 1006 nm. (b) Spectrogram representation of the vibrational wavepacket dynamics of Rb_2 at 1025 nm.

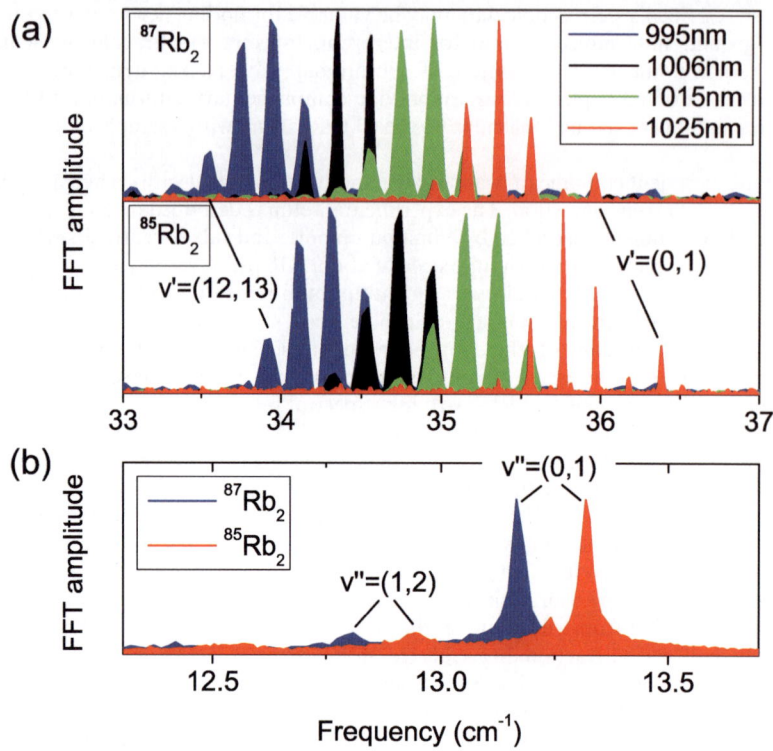

Fig. 17 Detailed views of the Fourier spectra around the fundamental frequencies of wavepacket dynamics in the first excited triplet state $(1)^3\Sigma_g^+$ (a) and in the lowest triplet state $a^3\Sigma_u^+$ (b).

2 M. J. J. Vrakking, D. M. Villeneuve and A. Stolow, *Phys. Rev. A*, 1996, **54**, R37.
3 F. Lang and J. Hecker Denschlag, private communication.

Dr Koch asked Dr Mudrich: Please clarify the description of the femtosecond pump-probe spectra for Rb_2 on a helium droplet shown in your comment.

As we've learned from Dudley Herschbach in this meeting, one should be careful with language since some terms tend to take on a life of their own.

The pump-probe spectra you have shown are decaying oscillations. You attributed the decay to wavepacket dephasing while later on rephasing due to wavepacket revivals were observed. I would like to point out that the term dephasing is generally used in the context of dissipative dynamics. It refers to a true loss of phase information. This is in contrast to dispersion, *i.e.* the spreading and refocussing of a wavepacket. Since the pump-probe experiments are performed on a helium droplet, the droplet could act as a reservoir causing dephasing. I would like to ask whether the shape of the pump-probe spectra is indeed attributed to dephasing in the sense of dissipation? If so, what would be the dissipative mechanisms?

Dr Mudrich replied: While 'dephasing' without decoherence of a wavepacket as a consequence of the anharmonic shape of a potential curve may be more properly called 'dispersion', the term 'dephasing' in the context of 'rephasing' and 'revivals' is quite commonly used in the literature (see, *e.g.*, ref. 1). In our experiments on vibrational wavepacket dynamics using pump-probe spectroscopy of He nanodroplets doped with Rb_2, we observe three effects causing damping of oscillation amplitudes. Besides the mentioned wavepacket dispersion due to anharmonicity, we see approximately exponential damping with time constants 0.1–1 ns depending on the vibrational states excited (Fig. 16(a)). This is probably due to different rotational states contributing to the signal in combination with vibration–rotation coupling. In addition, we observe fast irreversible damping at pump-probe delay times 0–10 ps. Under certain conditions, this fast initial drop in amplitude is accompanied by slightly shifting vibrational frequencies as well as by other frequency components appearing during this delay time interval, similarly to observations with K_2 molecules.[2] We interpret these dynamics as the interaction of the vibrating molecule with the He environment during the desorption process. It can be modelled by adding to the free molecular vibration shifted potential curves, vibrational relaxation, and a statistical desorption probability.

1 M. J. J. Vrakking, D. M. Villeneuve and A. Stolow, *Phys. Rev. A*, 1996, **54**, R37.
2 P. Claas, G. Droppelmann, C. P. Schulz, M. Mudrich and F. Stienkemeier, *J. Phys. B*, 2006, **39**, S1151.

Professor Weidemüller asked: How do the results of the time-resolved spectroscopy of molecules attached to helium nanodroplets, as you presented them, compare to recent results on association spectroscopy with ultracold gases, *e.g.* on rubidium in the triplet ground state as presented by Johannes Hecker Denschlag?

Dr Mudrich responded: The energies of the lowest 15 vibrational states of the first excited triplet state $(1)^3\Sigma_g^+$ of Rb_2 and of the lowest 5 vibrational states of the $a^3\Sigma_u^+$, which we have studied by fs pump-probe spectroscopy of doped He nanodroplets, are in perfect agreement within the experimental errors with the results measured by Johannes Hecker Denschlag and coworkers. This is due to the fact that the Rb_2 desorb off the He droplets upon laser excitation within about 10 ps, such that gas-phase molecules are probed at pump-probe delay times >10 ps. However, in contrast to cw-spectroscopy performed by the group of Johannes, our Fourier spectra do not show any line splitting due to rotation, fine or hyperfine structure despite high spectral resolution. This may be related to fast decoherence induced by the helium droplet environment.

Dr Petrovic said: The major difference between the pump-probe experiment just presented and ours is in the initial conditions. While we start from a mixture of cold atoms and high-lying ground state molecules pre-photoassociated by the trapping lasers, rubidium atoms on He nano-droplets are in a precisely defined state, in this case the $v = 0$ vibrational state of the ground triplet. It would be interesting to repeat our experiment starting from a precisely defined ground vibrational state. I, therefore, invite the group which has demonstrated formation of such a molecular state in ultra-cold rubidium to work with us on studying the wavepacket oscillations in the excited state.

Dr Mudrich answered: The pump-probe experiment on Rb_2 molecules initially attached to helium nanodroplets and the experiment presented by J. Petrovic are indeed complementary as far as starting conditions are concerned. On the one hand, this could offer the opportunity to find out new routes to forming ground state molecules by first optimizing the opposite process – the excitation into highly excited vibronic states or even dissociation. On the other hand, once the 'clean' ultracold experiment is in the position of repeating the pump-probe experiment starting from the vibronic ground state as in the helium droplet experiment, comparing the two will reveal the details of interactions of the surrounding (superfluid?) helium with the rotating and vibrating molecule as it desorbs off the droplet surface.

Professor Kosloff commented on Dr Koch's paper: We would like to point out that current experiments on femtosecond photoassociation[1] could be used for pump-probe spectroscopy of the two-body correlations in ultracold gases[2] with very little modification.

Conceptually the crucial point is that the reduced pair wave function that describes two-body correlations in a many-body system obeys a standard Schrödinger equation if (i) the length scale of the correlations is much smaller than the condensate length scale and (ii) higher than second order terms do not affect the dynamics of the reduced pair wave function.[3] This is true for dilute, weakly interacting ultracold gases in a standard trap.

The reduced pair wave function shows a distinct nodal structure at short and intermediate range where the potential energy of the two-body interaction is equal to or larger than the scattering energy of two atoms. A 'hole' can be cut into the peaks of the reduced pair wave function by exciting population to some electronically excited state with a suitably chosen pump pulse. The 'hole' corresponds to a non-stationary state that is made up of scattering states and bound levels of the ground state. A subsequent probe pulse, applied simultaneously with an ionization pulse, can monitor the ensuing dynamics of the ground state pair wave function. Our simulations predict a distinct time-dependent signal whose spectrum reveals the frequencies of the contributing levels.[1]

In conclusion, we suggest probing two-atom correlations in ultracold gases with time-resolved spectroscopy using pulses with a bandwidth of a few wavenumbers.

1 J. Petrovic, D. McCabe, D. England, H. Martay, M. Friedman, A. Dicks, E. Dimova and I. Walmsley, *Faraday Discuss.*, 2009, **142**, DOI:10.1039/b818494a and comment on this paper by M. Weidemüller.
2 C. P. Koch and R. Kosloff, arXiv:0904.2408.
3 P. Naidon and F. Masnou-Seeuws, *Phys. Rev. A*, 2003, **68**, 033612.

Dr Hecker Denschlag commented: You considered 'cutting' holes into scattering wavefunctions and mentioned that this creates population in bound states. Can you elaborate on this?

Professor Kosloff replied: 'The dynamical hole' is based on the idea of separating time scales between the relative motion of a pair of atoms and the timescale of electronic excitation. In the extreme limit the electronic excitation redistributes the scattering wavefunction between the ground and excited electronic potentials. The total phase space density of the relative R and P coordinates does not change but its projection on the ground and excited states is coordinate and momentum dependent.[1-3] The detuning from the atomic line positions the maximum transfer in coordinate space. A gradient in the difference potential or a chirp in the excitation pulse generates momentum in the ground state projection. Once the pulse is over, a stationary initial scattering state, *i.e.* is an eigenstate of the ground electronic state Hamiltonian, becomes a non stationary state. The generated hole becomes dynamic. The result is that an initial scattering eigenstate becomes redistributed on both bound and scattering states. Typically, only the last few bound eigenstates have significant amplitude. Nevertheless this amplitude is on the same order of magnitude as photoassociated amplitude on the excited state. An extreme case is generated by a short pulse detuned to the blue from the atomic line. Such a pulse generates a dynamical "hole" which includes significant bound amplitude in the ground electronic state.[4]

1 E. Luc-Koenig, R. Kosloff, F. Masnou-Seeuws and M. Vatasescu, *Phys. Rev. A*, 2004, **70**, 033407.
2 S. Kallush and R. Kosloff, *Phys. Rev. A*, 2007, **76**, 053408.
3 E. Luc-Koenig, F. Masnou-Seeuws, and Ronnie Kosloff, *Phys. Rev. A*, 2007, **76**, 053415.
4 S. Kallush and R. Kosloff, *Phys. Rev. A*, 2008, **77**, 023421.

Professor Whitaker said: I have a question for Dr Koch about the differences between phase control between bound–bound states and free–bound states, as in the proposed photoassociation experiment described in her paper, but I would like to preface it with a few remarks.

We have recently investigated phase control of three photon excitation in molecular iodine.[1] In our experiments we investigated the efficiency of various spectral phase functions in promoting the excitation of I_2 from its electronic ground state ($X^1\Sigma_{0+g}$) to the first ion pair state ($D^1\Sigma_{0+u}$) using wavelengths of 515–625 nm. This excitation scheme requires three photons and makes use of quasi resonant enhancement *via* the bound $B^3\Pi_{0+u}$ state and a dissociative state of $^1\Sigma_{0+g}$ symmetry. One of the experiments we reported involved scanning a π-step across the spectral profile of the excitation pulse along the lines proposed in your paper. The spectrum of the laser pulses was an approximate top-hat function, ~50 nm wide, as set by a pulse shaper which also controlled the spectral phase. The results for various central wavelengths (see Fig. 5 of ref. 1) showed a broad enhancement (as measured against a TL pulse) in the transfer of population from the $X^1\Sigma_{0+g}$ to the $D^1\Sigma_{0+u}$ state by a factor of >3 for central wavelengths below 570 nm for phase functions with a π flip close to the central frequency of the pulse. For this range of central wavelengths the combined excitation of the 3 photons is not enough to reach the Franck–Condon window onto the $D^1\Sigma_{0+u}$ state starting from $v = 0$ in the $X^1\Sigma_{0+g}$ state, and we showed in ref. 1 that it was the double pulse nature (in the temporal domain) of an antisymmetric phase function that was responsible for the enhancement. That is, the first pulse excited a wavepacket on the $B^3\Pi_{0+u}$ state which moved to larger internuclear separations to arrive at an optimal position for the second pulse to excite the ensemble to the D state *via* a resonance with the repulsive intermediate $^1\Sigma_{0+g}$ surface on a two photon transition. More interestingly for central wavelengths from 580 to 600 nm we observed very highly structured "phase" spectra as the position of the π-flip was scanned across the spectral profile. For these wavelengths resonantly enhanced transitions to the B state starting from $v = 0$ and hot bands originating from $v = 1$ in the X state are possible. We simulated these profiles with wavepacket calculations, an example of which is shown in Fig. 18. Due to technical limitations of the pulse shaper the experimental phase flip takes place over a narrow but finite spectral range (~0.4 nm) so the structure in the simulations is sharper than the experiment, nonetheless very sharp enhancement by constructive interference (and suppression *via* destructive interference) is visible close to wavelengths where the position of the phase

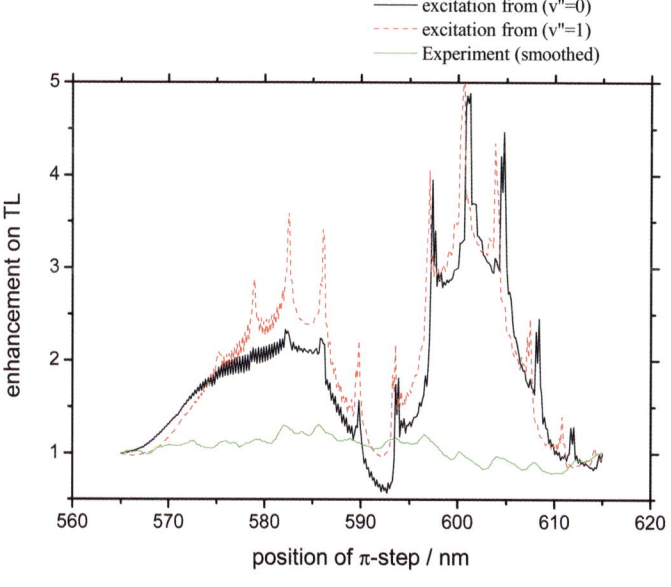

Fig. 18

flip coincides with the energy of resonant transitions. I should also say that the simulated profiles are extremely sensitive to the shapes of the potential curves, some of the details of which we may not have quite right yet, but I think the overall picture is clear.

My question is: might there be any advantage in the proposed photoassociation control scheme of making use of intermediate resonant states to create a stronger, sharper, filter with which to suppress the atomic excitation, and is there the possibility by so doing of constructing an edge filter rather than a notch filter?

1. N. T. Form, B. J. Whitaker and C Meier, *J. Phys. B: At. Mol. Opt. Phys.*, 2008, **41**, 074011.

Dr Koch replied: I would like to thank Professor Whitaker for his question which is very helpful to clarify the difference between bound–bound transitions and free–bound transitions. Bound–bound transitions occur in most applications of coherent control to molecular systems, while free–bound transitions are key in photoassociation. There are two points I would like to stress:

1. Resonance-enhanced vs. non-resonant multi-photon transitions: The experiment on I_2 cited by Professor Whitaker[1] relies on resonance enhancement of the multi-photon absorption due to intermediate electronic states that are accessible by one-photon transitions. These intermediate states are actually populated during the interaction of the molecule with the pulse. In the two-photon photoassociation scheme proposed in our paper,[2] it is absolutely imperative to avoid population transfer in one-photon transitions. One-photon transitions would lead to excitation of the atoms which subsequently undergo spontaneous emission and are lost from the trap. The sample of ultracold atoms would thus be destroyed.[3,4] The central laser frequency in our proposal is therefore far-detuned from any one-photon resonance.

The key idea of our paper[2] is the creation of destructive interference of two photons whose frequencies sum up to the atomic transition energy, *i.e.* the creation of a dark resonance for the atoms. Atomic transitions have an excitation probability that is several orders of magnitude larger than photoassociative transitions into bound molecular levels. Relative to the free–bound transitions, atomic transitions therefore need to be suppressed by these several orders of magnitude. Otherwise most of the atoms would be lost, and only very few molecules photoassociated.

2. Time-domain vs. frequency domain control: The phase step serves a different function in the two control approaches. In Professor Whitaker's experiment, it is a means to create a long pulse with subpulse structure that allows one to take advantage of wavepacket dynamics on the intermediate electronic state.[1] The control mechanism is hence based on a time-domain picture, the spectral structure of the pulse is less important. Since this approach is not sensitive to the exact matching of the multi-photon resonance, it should not depend on laser intensity.

In our paper, we use the phase step for frequency-domain control, where the phase function serves to annihilate $E^2(\omega)$ at the two-photon resonance.[5] This works only for weak fields; for strong fields the resonance condition becomes time-dependent due to the dynamic Stark shifts. The dark condition is then not fulfilled throughout the pulse. Note that we have also introduced a strong-field control scheme.[2] It is, however, not based on spectral phase steps but on time-domain chirps. The laser chirp compensates for the dynamic Stark shift, and the intensity of the pulse is chosen to create a 2π Rabi cycle for the atoms.

1 N. T. Form, B. J. Whitaker, and C. Meier, *J. Phys. B*, 2008, **41**, 074011.
2 C. P. Koch, M. Ndong, and R. Kosloff, *Faraday Discuss.*, 2009, **142**, DOI:10.1039/b919458e.
3 W. Salzmann, U. Poschinger, R. Wester, M. Weidemüller, A. Merli, S. M. Weber, F. Sauer, M. Plewicki, F. Weise, A. Mirabal Esparza, *et al.*, *Phys. Rev. A*, 2006, **73**, 023414.
4 B. L. Brown, A. J. Dicks, and I. A. Walmsley, *Phys. Rev. Lett.*, 2006, **96**, 173002.
5 D. Meshulach and Y. Silberberg, *Nature*, 1998, **396**, 239.

Dr Petrovic said: May I suggest an alternative to your experiment? For the reasons described in the paper presented by Dr Koch, atomic resonance should be avoided during the photoassociation. By using two oppositely chirped pulses and fixing the sum of their instantaneous frequencies to the targeted transition frequency while blocking the intermediate atomic resonances in a 4f-line, the two-photon transition can be realised without exciting the unwanted atomic transitions. Such a scheme would extend the proposed type of coherent control to the alkaline atoms.

Mr McCabe asked Professor Weidemüller: This question relates to your earlier comment on Dr Petrovic's paper, in which you presented your observations of transient oscillations of the excited-state population in rubidium.

In your related paper[1] you describe these oscillations as being attributable to the interaction between the induced molecular dipole and the long temporal tail of the pump pulse induced by the sharp cut in the spectral domain. What evidence is there to support your claim that the interacting dipole is molecular in nature, rather than merely pertaining to an excited free atom? This behaviour can be described by a two-level atom picture, and the detection of these oscillations in the molecular ion signal could be explained by the creation of a molecular ion from a pair of free 5s + 5p atoms through the action of the probe pulse (which is detuned from the transition to the atomic ionization threshold).

1 W. Salzmann *et al.*, *Phys. Rev. Lett.*, 2008, **100**, 233003.

Professor Weidemüller responded: The most important evidence for the photoassociation of ultracold atoms by the femtosecond laser pulse stems from the observation of *molecular* ions in the pump-probe scheme. However, molecular ions might also be created by alternative pathways: (i) ionization of cold ground state molecules produced by the trap light, and (ii) collisional autoionization of colliding Rydberg pairs. We have thoroughly investigated these scenarios (for details see ref. 1) and exclude them on the basis of their systematics with respect to experimental parameters, which do not match the observations. This reasoning is supported by the fact that our observations are well reproduced by the solution of the time-dependent Schrödinger equation (see ref. 2) which explicitly includes the molecular dynamics of the photoassociated atomic pairs. In fact, the results of these simulations show that the molecules produced by the femtosecond pulse are very weakly bound. In this sense, there is no qualitative difference in the dynamics of a molecular dipole as compared to the atomic dipole.

1 W. Salzmann *et al.*, 2009, arXiv:0903.4549
2 W. Salzmann *et al.*, *Phys. Rev. Lett.*, 2008, **100**, 233003; arXiv:0903.4401

Professor Julienne remarked: There are two quite different kinds of approaches to getting some kind of optical control of cold collisions or molecules, one using the frequency domain and the other using the time domain. The former has been highly successful in the level of precision and control that has been demonstrated, either through photoassociation spectroscopy or STIRAP. Its main disadvantage is that it requires very detailed measurement and understanding of often complex excited state level structure. The latter has been more difficult and challenging experimentally, since the very fast time domain seems not to be well-matched to the extraordinarily small energy scale (and corresponding long time scale) of the ultracold domain. However, in principle, time domain experiments offer the advantage that detailed information about excited states may not be needed in order to achieve some level of control over molecular dynamics. I would like to ask experimentalists or theorists working on the time domain to comment upon the merits of working in the time domain relative to the frequency domain, and suggest the most practical ways in which experiments might take advantages of working in the time domain (I count STIRAP as being in the frequency domain, since it operates on a relatively long time scale and takes advantage of highly resolved molecular level structure).

Dr Koch answered: I see two answers to the question of time *vs.* frequency domain approaches.

(1) A time-dependent approach allows us to dynamically *probe* a structure or a slow process. An example was given in Professor Kosloff's earlier comment where pump-probe spectroscopy of the pair correlations in an ultracold gas is suggested.

(2) A time-dependent approach is best suited to analyse processes in strong fields when the energy levels are *e.g.* strongly Stark-shifted and many levels are mixed. A frequency domain analysis on the other hand is best suited for the weak-field regime. This is discussed in detail in our paper.

Mr Lemeshko answered: I would like to bring to your attention the possibility of using a pulsed nonresonant laser field to exert control in the time domain.

Since the frequency of nonresonant light is far removed from any molecular resonance, such light only interacts with the body fixed dipole moment of the molecule, typically with the dipole moment induced by the anisotropic polarizability interaction. As a result, the usual limitations on the energy resolution in the "fast time domain" do not arise for nonresonant radiation, as it is just the strength of the electric field carried by the radiation that matters and not its frequency.

We recently described the possibility of probing weakly bound molecules using nonresonant laser pulses shorter than the vibrational period. Since such pulses are much shorter than the rotational period, the interaction is inherently nonadiabatic: the field endows the molecule with an angular momentum which the molecule keeps even after the pulse has passed. The imparted value of the angular momentum is tunable – for a given molecule and pulse duration it depends on the pulse intensity.

The angular momentum imparted by the laser pulse changes the effective molecular potential, by introducing a centrifugal term. If this term is large enough to push a weakly bound vibrational level out of the potential, the molecule dissociates. The laser intensity needed to impart a preordained value of angular momentum varies characteristically with the internuclear distance. This characteristic dependence can be used to map out the probability density of the vibrational state from which the molecule was forced to dissociate.

Our numerical simulations applied to the Rb_2 and KRb Feshbach molecules indicate that dissociation by feasible laser pulses enables us to accurately recover the square of the vibrational wavefunction and, by inversion, also the long-range part of the molecular potential. This route to an accurate potential, independent of spectroscopy or scattering, complements what can be learned from either.

Please see poster presentation 'Probing halo molecules with nonresonant light', of this Discussion meeting, and the e-print arXiv: 0903.0811.

Professor Weidemüller answered: I see three interesting perspectives for working in the time domain: First, pulses in the picosecond domain, which can be shaped, are much better suited to the actual dynamics of ultracold atoms under the influence of the long-range forces. Technologically, such pulses appear to be realizable. Second, it might be an interesting option to use coherent trains of femtosecond pulses, as was suggested by Moshe Shapiro, Jun Ye and coworkers. With these pulse trains one may continuously bridge the single ultrashort pulse domain and the continuous wave domain. And third, one may consider application of the techniques of femtosecond coherent control to create non-stationary wavepackets in ultracold ground state molecules which are formed by other approaches, and study collision of these molecules in non-stationary states with other atoms or molecules.

Professor De Vivie-Riedle commented: In case it is possible, *i.e. via* a Feshbach resonance starting from an atomic BEC and applying femtosecond light pulses, to generate a molecular wavepacket in an electronically excited state, this wavepacket can be transferred with high efficiency to the electronic ground state or to any selected vibrational eigenstate in the ground state by optimized light fields. This is possible as the wavepacket is a coherent state and optimal control theory can find the appropriate unitary transformation from one coherent state to the other. Moreover I think that the tools and concepts offered by optimal control theory will become extremely helpful in manipulating cold molecules in the sense of state to state chemistry.

Dr Koch remarked: In the current discussion of photoassociation, and formation of ultracold molecules in general, there is some confusion on the role of coherent and incoherent processes, *e.g.* ref. 1. In that context, I would like to stress that a dissipation mechanism is required when the starting point is a thermal ensemble and the goal to reach a coherent state. It is true that using photoassociation one can create a coherent wavepacket in the excited state from a thermal ensemble of ground state atoms, *e.g.* ref. 2, and that in principle a coherent excited state wavepacket can be transferred to any coherent state in the electronic ground state. However, the initial step of creating the excited state wavepacket can never address the full ensemble, or even a significant part of it. In a way, one could view photoassociation as an analogue to evaporative cooling. The real problem becomes apparent when one looks at absolute numbers. Usually a single photoassociation step is not sufficient to create a reasonable number of molecules. An obvious remedy consists of a repetitive application of the photoassociation pulses.[1] Then two effects need to be taken into account: (i) The coherent wavepacket is coupled to the thermal ensemble and can be destroyed by the process that is the inverse of its creation. This is sometimes referred to as back stimulation. It can be suppressed if an incoherent step is included in the molecule formation process.[3] (ii) The modification of the initial thermal ensemble by the preceding pulses cannot be neglected. In other words, the time between pulses needs to be long enough to allow for re-thermalization of the ensemble. In summary, only a dissipative step can assure unidirectional formation of molecules from a thermal ensemble of atoms.

1 E. Kuznetsova, S. F. Yelin, M. Gacesa, P. Pellegrini and R. Côté, *New J. Phys.*, 2009, **11**, 055028.
2 C. P. Koch, R. Kosloff, E. Luc-Koenig, F. Masnou-Seeuws and A. Crubellier, *J. Phys. B*, 2006, **39**, S1017.
3 C. P. Koch and R. Moszynski, *Phys. Rev. A*, 2008, **78**, 043417.

Professor Julienne asked: As a follow up to my previous question on time-domain experiments, let me comment that one of the difficulties of working with excited molecules made from alkali atom species is the complex set of excited state potentials, at least two of which are bright (*i.e.*, have optically allowed transitions) at any given wavelength near the S to P excitation energy. Thus, one necessarily gets rather complex and difficult-to-interpret molecular dynamics for any excited state wavepacket formed. Consequently, there might be some advantage in working with simpler molecules, such as can be formed from alkaline earth atoms, such as Ca or Sr, or similar species such as Yb, where the excited state structure is simpler. My question is whether any of the experimentalists or theorists working in the time domain have any experiments to suggest with alkaline earth species?

Dr Koch replied: I would like to point out a recent theory proposal that makes use of the comparatively simple molecular structure of alkaline earths. In short-pulse photoassociation of calcium dimers, non-adiabatic mixing can be induced by employing a strong infrared laser field that couples two electronically excited states.[1] This mimics resonant coupling known from the alkaline species (as mentioned in some of the earlier comments) and qualitatively changes the Franck–Condon factors governing pump-dump photoassociation.

1 C. P. Koch and R. Moszynski, *Phys. Rev. A*, 2008, **78**, 043417.

Professor Gianturco opened the discussion of Dr Portier's paper by asking: The conventional analysis of Fano profiles for potential scattering resonances relies on the reasonable assumption that the background contributions to the overall scattering are weakly dependent on energy across the resonance interval of energy location. However, when one is dealing with multichannel molecular scattering, as you propose to do in the near future, the background cross sections, especially at ultralow energies, are not weak functions of that energy and therefore can strongly mask the "pure Fano" profile of even an isolated resonance, if it exists. How do you propose to deal with this problem, which is already known for scattering processes at room temperatures, when you move to ultralow energies?

Dr Portier answered: The Fano profiles described in the present paper are obtained in the following experiments: a cloud of atoms is sufficiently cooled so that the scattering energy for the collisions between atoms is well defined. The photoassociation rates are probed using Raman lasers and plotted *versus* the energy of the first photon, the energy of the second photon being fixed. The scattering energy of the atoms is also a fixed parameter. If the initial state of the experiment is a molecule, then the scattering energy of the atom pair has to be replaced by the binding energy of the molecule in the model, and is a fixed parameter as well.

The phenomenon that you describe was studied by Dürr *et al.*[1] where the autodissociation rate of Feshbach molecules was studied *versus* the near-zero scattering energy of the dissociated atoms. As in that case the background scattering rate is not a weak function of the scattering energy, the Fano profiles are indeed expected to be distorted.

1 Dürr *et al.*, *Phys. Rev. A*, 2005, **72**, 052707.

Dr Hecker Denschlag asked: At the end of your talk/paper you mentioned that it will be interesting to extend your theoretical approach to the BEC regime. Can you explain how this should be done and why it should be much different from the thermal regime?

Dr Portier replied: For the BEC regime case, atoms and molecules are described as coherent matter fields. A model Hamiltonian that accounts for statistics for both atoms and molecules was derived for example by Javanainen and Mackie.[1]

Some additional effects which can be taken into account beyond this model, and which may be important for some experimental parameters are the following:

Kostrun *et al.*[2] studied the effect of "rogue" dissociation, this is to say photodissociation of molecules into a pair of atoms, neither of which belongs to the atomic condensate.

Mean-field interactions between atom and molecular species lead to coupled Gross–Pitaevskii equations as derived by Heinzen *et al.*[3]

Effects of correlations were investigated by Naidon and Masnou-Seeuws.[4] The effect on the symmetry of one-photon photoassociation lineshapes was also investigated.

The above phenomena are specific to the BEC case, and it should be checked whether they lead to specific features in the lineshapes of two-photon photoassociation compared to the thermal case.

1 Javanainen and Mackie, *Phys. Rev. A*, 1999, **59**, R3186.
2 Kostrun *et al.*, *Phys. Rev. A*, 2000, **62**, 063616.
3 Heinzen *et al.*, *Phys. Rev. Lett.*, 2000, **84**, 5029.
4 Naidon and Masnou-Seeuws, *Phys. Rev. A*, 2006, **73**, 043611.

Professor Grimm asked: In the quantum-gas community, the creation of degenerate gas of ground-state molecules seems to be in close reach, and it is a big goal on the time scale of one year. Are there any similar short-term big goals for the photoassociation work? And, more generally, what are such goals in our whole field?

Dr Petrovic responded: Building on Dr Vivie-Riedle's comment, the flexibility of photoassociation by broadband pulses stems from the ability to shape the amplitude and phase of these pulses and therefore steer the photoassociation towards the wanted molecular product. This implies not only the formation of the zero vibrational state molecules, but coherent transfer to a variety of distributions over different vibrational states.

Dr Koch answered: Coherent control of photoassociation is one example for coherent control of a binary reaction which has yet to be demonstrated. The difficulty arises from the fact that there is no well-defined phase relation between the initial state and the control field. From a more practical point of view, enhancing the efficiency of photoassociation clearly represents a short term goal that needs to be realized if photoassociation is to compete with other molecule formation methods. A promising approach uses a combination of external fields and resonances in ultracold scattering to enhance the pair density at short range prior to photoassociation.[1] On a more general level, ultracold gases offer the possibility to demonstrate the true potential of coherent control. Control experiments on hot samples are hampered by thermal averaging and often yield rather small improvement factors. Using ultracold samples instead allows us to see transfer probabilities similar to the STIRAP ones[2] beyond the adiabatic regime.

1 P. Pellegrini, M. Gacesa, and R. Côté, *Phys. Rev. Lett.*, 2008, **101**, 053201.
2 F. Lang, C. Strauss, K. Winkler, T.u Takekoshi, R. Grimm and J. Hecker Denschlag, *Faraday Discuss.*, 2009, **142**, DOI:10.1039/b818964a; J. G. Danzl, M. J. Mark, E. Haller, M. Gustavsson, N. Bouloufa, O. Dulieu, H. Ritsch, R. Hart and H.-C. Nägerl, *Faraday Discuss.*, 2009, **142**, DOI:10.1039/b820542f; S. Ospelkaus, K.-K. Ni, M. H. G. de Miranda, B. Neyenhuis, D. Wang, S. Kotochigova, P. S. Julienne, D. S. Jin and J. Ye, *Faraday Discuss.*, 2009, **142**, DOI:10.1039/b821298h.

Professor Hutson addressed all: Since we're nearing the end of the Discussion, I'd like to open the field even wider. For several years we've been talking about the production of a degenerate quantum gas of dipolar molecules as a 5-year goal, a holy grail. But the enormous advances in ground-state molecule production over the last year now make it look likely that this will be achieved within a year or two. So what are our new 5-year goals?

I'm sure different people will have different answers to the question. Coming from a chemistry perspective, my own 5-year goal would be to develop ways to achieve chemical transformations coherently on entire samples of molecules by controlling them with applied electric and magnetic fields and photons. Traditional coherent control with tailored femtosecond laser pulses has been restricted to unimolecular processes, but ultracold molecules open the way to coherent bimolecular processes as well, and the molecule formation processes already achieved are a prototype example of that. But what are the other big promises of our field now?

Dr van de Meerakker replied: In my opinion, a large promise of our field is the detailed experimental study of bimolecular scattering in the 1–10 K temperature range. Although not quite in the (ultra)cold regime, this temperature range is very interesting in its own right as only a few partial waves contribute to the scattering and scattering resonances can be expected. Experimental data in this temperature regime, and the careful experimental mapping of scattering resonances, will provide detailed and unprecedented information on molecular potential energy surfaces. Furthermore, this temperature regime bridges the gap between the low temperature phenomena that are now beginning to be explored by the (ultra)cold molecules community and the scattering phenomena that occur at higher temperatures that are studied in the mature field of molecular reaction dynamics. Experimental methods to obtain full control over the velocity and internal state distribution have now advanced sufficiently to be used in molecular scattering experiments, and I expect that state-to-state scattering experiments between state-selected and velocity controlled neutral molecules will become reality in the next few years.

Professor Weidemüller responded: Besides the perspectives mentioned by Professor Hutson, I see further important mid- and long-term perspectives of our field, which are well distinguished from the spectacular achievements in traditional physical chemistry, in particular involving crossed beams. One perspective is the investigation of resonances in reactive atom–molecule or molecule–molecule scattering due to the finite number of partial waves involved. These resonances would provide an ideal test for potential surfaces *etc.*, and they have long been sought by researchers in physical chemistry. Another perspective is the investigation of phenomena in the dynamics of many-body quantum systems, where the field of ultracold quantum gases offers unique possibilities. In this context, the creation of degenerate gases of dipolar molecules is of particular importance, but also homonuclear molecules, *e.g.* trapped in lattices, might be of interest.

Professor Hinds answered: Because molecules have fundamentally non-spherical structure, they offer important new possibilities for high-precision measurement. One example of this is in the search for the permanent electric dipole moment of elementary particles, which is greatly aided by the polarisation of polar molecules. Another is the search for a variation of fundamental constants, which benefits from the wide range of frequency scales occurring in molecules. For these kinds of experiments the dream is to have, say, ten million molecules, each occupying one site of an optical lattice at the lowest temperature possible and able to interact coherently with an external field for many seconds. It does not seem to me unreasonable to strive for this as a 5-year goal, which would lead to spectacular precision being possible.

Professor Gianturco commented: When one desires to make the chemistry community at large increasingly more aware of the potential of ultracold molecules and of molecular processes at ultralow energies for the better understanding of chemical processes in general, I think that one should not forget that conventional chemical understanding relies quite a bit on the "local" symmetry properties of a given compound or of its interaction with radiation in general. In this vein, therefore, it may be very useful if the experiments realized in MOT devices could bring out more explicitly the role that "collective symmetry" can play and does play in such experiments. Such efforts could therefore indicate novel ways in which symmetry as the chemists know it is appearing under different forms but with the same helpful role for our understanding of processes at a deeper level.

Dr Küpper remarked: Chemistry – including biology – is to a large extent carbon based. What are the prospects of creating degenerate or ultracold samples of ^{12}C atoms? What are the prospects of having ultracold reactions of hydrogen and carbon to form first easy hydrocarbons?

Professor Julienne commented: It is worth noting that the really new aspect of ultracold atom and molecule research is the ability to prepare the atoms or molecules at a very precisely defined energy, on an energy scale on the order of kHz in the "deep ultracold" regime where we have the ability to make atoms and now perform initial experiments with molecules. The scale increases to the MHz range or higher for the relatively hotter decelerated or buffer gas cooled systems. Since during the course of a collision, the typical energy scale of a chemical bond is on the order of 100 to 1000 THz, more than 9 orders of magnitude larger than the energy of the starting species, we see that any control that one gets by going to the ultracold domain is in the initial preparation of the colliding species and in their long range interactions. Relatively weak forces can be used for exquisite and precise control of the motion of the two species and how they are "inserted" into the collision to make some "collision complex" at short range where they are strongly interacting. The energy of the collision complex is very precisely determined by the initial energy of the reactants, and subject to tunable resonance control. Thus, in order to take advantage of the ultracold domain, one must understand the weak long range interactions that determine the spectrum of near-threshold scattering resonances. Threshold laws will be quite different depending on the nature of the long-range forces between the separated species, *i.e.*, van der Waals for atoms, and dipolar for polar molecules in an electric field. One must also learn how one might take advantage of spatially engineered structures like 1-, 2-, or 3-dimensional optical lattices to control the way the two species get together or act collectively. This is where the really new experimental and theoretical opportunities are with ultracold chemistry – the ability to prepare the separated species in specific quantum states at a precise energy, and then understand and control how they come together to make a collision complex that undergoes essentially uncontrolled dynamics of strongly interacting species.

Professor Grimm said: We have only talked about interesting physics with molecules in the rovibrational ground state. I want to point out that there also is a lot of interesting physics for molecules near threshold (Feshbach molecules). One example is few-body physics, like Efimov states, where ultracold molecules have opened up a new window for research; few-body physics with cold atoms and molecules is an emerging field. Another example would be a quantum gas of strongly magnetic Feshbach molecules, such as erbium dimers with a magnetic dipole moment of up to 14 times the Bohr magneton. Erbium can be laser-cooled. Dysprosium would be a laser-coolable candidate with molecules having up to 20 times the Bohr magneton.

Dr Lane remarked: Chemists are not exclusively interested in molecules in the ground rovibronic state. The techniques presented at this meeting could be adapted to produce molecules in any rovibrational state in the ground state electronic state the user desired. Such molecules should demonstrate remarkably different reactivities compared to $v = 0$, $J = 0$. When using a room temperature gas it is virtually impossible to select a single rovibrational state in the ground electronic state of a homonuclear diatomic, even by laser excitation, because there will always be a background of internal states thanks to the Boltzmann distribution. Therefore, a particularly appealing feature of these ultracold molecules in a single rovibrational level is their creation in the absence of any other molecular states. In addition, the radial wavefunctions of the ultracold dimers prepared with magnetic fields are enormous and strongly asymmetric, so "Feshbach chemistry" should explore uncharted parts of the PE surface ignored by rival techniques. Such extreme nuclear motion, often far from the traditional "transition state" is a hot topic in the field of reaction dynamics[1] because it can play a significant role in determining the reaction products.

1 U. Lourderaj and W. L. Hase, *J. Phys. Chem. A*, 2009, **113**, 2236.

Dr Groenenboom asked: What are the prospects in the coming years to break the mK barrier in the direct cooling of 'real molecules'?

Mr Zeppenfeld replied: In a recent paper,[1] we have proposed an opto-electrical cooling scheme which will hopefully allow a broad class of molecules to be cooled below 1mK in the near future. The main feature of this scheme is to use the differing Stark shift of various internal molecular states to implement a Sisyphus-type cooling cycle. Due to the large, ~1 K, amount of kinetic energy which can be removed in a single cycle, the requirements imposed on the spontaneous decay which is used for dissipation is many orders of magnitude weaker than for standard laser cooling.

1 M. Zeppenfeld, M. Motsch, P. W. H. Pinkse and G. Rempe, 2009, arXiv:0904.4144.

Professor Meijer responded: At the Fritz Haber Institute in Berlin we will pursue sympathetic cooling experiments of polar molecules in high-field seeking states in an AC electric trap (in particular deuterated ammonia molecules) by spatially overlapping them with Rb atoms in a magnetic trap.

Professor Hutson said: Sympathetic cooling of molecules with alkali metal atoms is not going to be as easy as we had hoped, because there will often be strong inelastic collisions that cause trap loss. However, I am now really quite optimistic about the prospects for sympathetic cooling of molecules in closed-shell singlet states, and ND_3 is a very good candidate. This is a change; if you had asked me a year ago I would have said the prospects looked poor.

As Piotr Żuchowski described earlier in the Discussion,[1] the Rb–NH_3 interaction potential is very strongly anisotropic[2] and is likely to drive fast inelastic collisions between rotation-inversion states of NH_3 or ND_3. Even those are slower than we expected,[3] but are probably fast enough to prevent sympathetic cooling for molecules in low-field-seeking states. However, there are only very small terms in the collision Hamiltonian that connect the *molecular* degrees of freedom to the *atomic* spins.

It thus seems likely that collisions that change the atomic Zeeman/hyperfine level will be very slow. This suggests that it will be possible to achieve sympathetic cooling for NH_3 or ND_3 in *high-field-seeking* states, which are the absolute ground state in the field, with Rb or other alkali-metal atoms that are magnetically trapped in low-field-seeking states.

A limiting factor here may be relaxation of the *molecular* nuclear spins, but that should not be an issue until well below 1 mK. The situation is rather different for molecules in non-singlet states, where there are likely to be strong spin-exchange transitions that change both atomic and molecular spin states.

Although spin-exchange collisions can be prevented by using spin-stretched states, there is still much stronger spin relaxation than in atomic systems because the potential anisotropy can change L quantum numbers. For non-singlet molecules the best prospect for sympathetic cooling is probably to use closed-shell atoms such as alkaline earths or ground-state rare gases.

1 P. Żuchowski, in *Faraday Discuss.*, 2009, **142**, DOI:10.1039/b910119p.
2 P. S. Żuchowski and J. M. Hutson, *Phys. Rev. A*, 2008, **78**, 022701.
3 P. S. Żuchowski and J. M. Hutson, *Phys. Rev. A*, 2009, **79**, 062708.
4 P. Soldán, P. S. Żuchowski and J. M. Hutson, *Faraday Discuss.*, 2009, **142**, DOI:10.1039/b822769c.

Professor Barker remarked: In response to the question about the general methods for creation of ultracold molecules, we note that the use of rare gas atomic species co-trapped with Stark decelerated species in a deep optical trap should be a very general method for creating cold complex molecular species *via* sympathetic cooling. This scheme is capable of trapping all rovibrational states in the electronic ground state, and avoids chemistry by utilizing ground state rare gas atoms as the ultracold collision partner.

Professor Julienne commented: While there are certainly many new experimental opportunities available with cold or ultracold molecules, there also are many new opportunities for theory as well. Advances in experiment and theory have tended to go hand-in-hand in the development of cooling and ultracold matter studies, and this is likely to continue. *Ab initio* quantum chemistry calculations will be even more necessary for the molecular domain than they were for work with ultracold atoms. Low energy scattering theory and few-body theory will need to continue to be developed for molecules. The hyperfine, Zeeman, and Stark structure of molecules will need to be worked out, and their field-modified response to electromagnetic radiation in various frequency domains (rf, microwave, infrared, optical). It will be important to develop quantum defect theories based on the long range potential to characterize molecular interactions (I say theories since quantum defect theory is a tool that can be implemented in various ways depending on the problems being addressed). Such theories help bridge the gap between the threshold domain heavily influenced by the long range potential, and the strongly interacting domain at chemical bonding distances. The transition between an essentially semiclassical "high energy" domain $\gg$ 1 mK and the quantum threshold law domain also needs to be worked out experimentally and theoretically for different kinds of species and long range forces, with an emphasis on understanding the role of shape or Feshbach resonances as collision energy decreases. Quantum defect theory may give useful ways to classify such resonances, as well as characterize their effects on elastic or inelastic collisions. The ratio between elastic and inelastic collision rates is very important to understand and control if possible. Few-body theory for three polar molecules is virtually nonexistent, and many aspects of many-body theory for polar molecules in quantum gases or lattices still needs to be worked out. The ability to engineer lattice Hamiltonians and interactions between molecules in lattices, with resultant control of many-body properties, is a very rich subject that will need to be developed if the experimental promise of such control is to be realized.

PAPER

Concluding remarks: achievements and challenges in cold and ultracold molecules

F. A. Gianturco*[a] and M. Tacconi*[b]

Received 22nd May 2009, Accepted 27th May 2009
First published as an Advance Article on the web 9th July 2009
DOI: 10.1039/b910178k

Introduction

The breadth and variety of the presentations, and of the additional contributions, which have been analyzed and discussed during this Faraday discussion meeting, the first one in the area of *Cold and Ultracold Molecules*, certainly testify to the very impressive growth and to the broad range of topics that are being considered by so many research groups around the world; they are pushing our knowledge and understanding of chemistry under extreme conditions, that have been taken as unattainable for many previous years, even further. Although the concept of an absolute zero of temperature has been familiar to scientists for a long time, and therefore the challenges related to getting as close as possible to that limit are nothing entirely new, the dramatic advances of the last few years have made that remote possibility even more attainable.

We already know that the lowest temperatures accessible, and known to science, have been generated in our laboratories: around us, the distribution of the frequencies of the cosmic background radiation corresponds to around 2.7 K, so that any temperature below 1 K has only been reached in laboratory experiments. Molecules, the main object of study in our present Discussion, have been observed in space at temperatures above that background value, since they chiefly appear in the cold cores of the Dense Interstellar Clouds (DISCs) which have temperatures of around 10 K. We therefore see that the majority of the studies presented here have gone way beyond such *natural* situations and have explored the behavior of molecular species down to fractions of 1 K (cold species) and further down below mK, all the way to μK and nK: a real occurrence of VERY extreme conditions under which to observe molecular and chemical phenomena.

One important result of plunging down to such low temperatures is that one slows down the species under observation, thereby extending the time during which the atoms and molecules can be analyzed and therefore enhancing the accuracy of the corresponding experiments: the most accurate measurements to date can be attained by working on cold atoms and cold molecules. Furthermore, by exploring the nature of cold matter physics and cold matter chemistry, one quickly discovers that the corresponding matter waves, which are characterized by their associated deBroglie wavelength, take up a new role and importance in affecting the processes at hand: while at room temperatures of about 300 K, in fact, the associated λ of a typical diatomic species like CaH (mass of about 40 g mol^{-1}) is less than 10^{-9} cm, the same molecule at a temperature of 1 mK has an associated λ of about 200, which becomes more than 1000 at temperatures of nK. Hence, under such extreme conditions the molecules are best described as quantum objects that behave like matter

[a]*Department of Chemistry and CNISM, The University of Rome Sapienza, P.le A. Moro 5, 00185 Rome, Italy. E-mail: fa.gianturco@caspur.it; Web: http://www.chem.uniroma1.it/gianturc/*
[b]*Department of Chemistry and CNISM, The University of Rome Sapienza, P.le A. Moro 5, 00185 Rome, Italy. E-mail: tacconi@caspur.it*

waves whose associated λ values far exceed their molecular dimensions. One sees immediately, therefore, that the molecules now exhibit a new collective ability to interact with each other, from which originates a corresponding new variety of phenomena that involve their chemical features and structural characteristics. Such novelties, therefore, are the main object of the 22 presentations at the Discussion and of the 56 posters that were displayed during the meeting.

I. The paths to cooling molecules

Over recent years we have seen how the inability to directly apply laser cooling to molecules has created a challenge for experiments which has been taken up by many groups and has drastically widened the range of cold-matter studies in order to include molecules as much as possible. We clearly sensed during the present Discussion that necessity has indeed been the mother of invention and that the variants followed by the presentations have been remarkable in terms of their achievements.

In what may be called the use of indirect methods, the experimental device seeks to directly slow or select pre-existing molecules and therefore to achieve the translational and/or internal cooling of the examined species. There have been numerous presentations of this type of path that have indeed shown how the initial, explanatory studies have developed into proper avenues to general molecular cooling. There were also several, detailed presentations during the poster session, but the following discussion papers are certainly worth mentioning:

1. *Collision experiments with Stark decelerated beams*, by the Fritz-Haber Institute group in Berlin;[1]

2. *Continuous guided beams of slow and internally cold polar molecules* by the MPI Institute in Garching;[2]

3. *Production of cold ND_3 by kinematic cooling* which involved a very interesting collaboration between the group at the Sandia National Laboratory in the US and again groups at the Fritz-Haber Institute in Berlin.[3]

The use of Stark deceleration confirmed in the above experiments its current role as a very robust method with many possibilities for the control of both internal and external degrees of freedom of polar molecules, control which has been shown to be carried out with very high energy resolution and directly during crossed molecular beam experiments as reported in the example of Fig. 1.

The interesting experiments carried out at the Max-Planck Institute for Quantum Optics in Garching showed a combination of buffer-gas cooling with electrostatic velocity filtering that allowed them to produce a high-flux, continuous guided beam of internally cold and slow polar molecules.[2] Their results, shown for example in Fig. 2, indicate very clearly the versatility of the source in being able to produce guided beams of different molecular species.

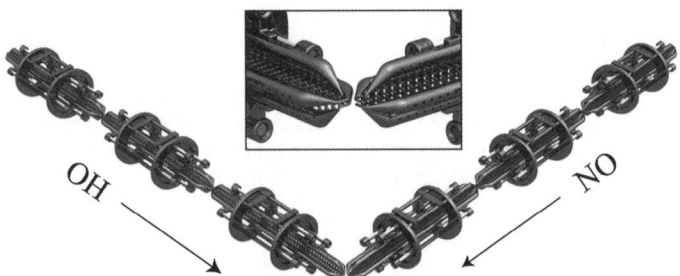

Fig. 1 Schematic representation of the crossed beam machine used in the experiments at the Fritz-Haber Institute in Berlin (adapted from Fig. 4 of van de Meerakker and Meijer).[1]

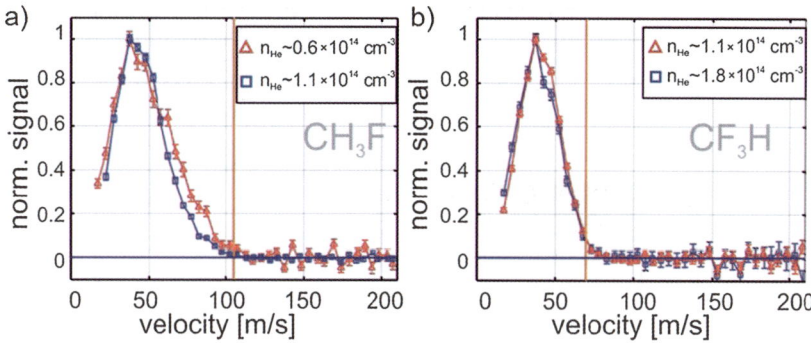

Fig. 2 Velocity distributions of CH$_3$F (a) and CF$_3$H (b) using He densities resulting in maximum flux of molecules (adapted from Fig. 11 of Rempe et al.).[2]

The measurements carried out at the Sandia National Laboratories in the group of D. W. Chandler, and in collaboration with colleagues at the Fritz-Haber Institute, show again an example of experimental ingenuity in producing a versatile technique whereby molecules of a very broad variety are brought to rest by single collisions with atomic partners of similar mass.[3] The velocity cancellation that occurs in these setups causes the molecules of interest to be scattered in such a way that their post-collision velocity in the center-of-mass frame cancels the center-of-mass velocity of the collision pair; thus, molecules of very broad variety can be cooled down to mK temperatures that are comparable with those achieved using Stark deceleration, Zeeman deceleration and buffer-gas cooling.

A sample of their data in Table 1 shows measured and calculated velocities of ND$_3$ in low-lying quantum states and confirms the success of the method.

Another interesting example of the robustness and versatility of the Stark deceleration method was provided by the results illustrated in another contribution from the Fritz-Haber Institute in Berlin.[4] The data were able to show that the method can handle large molecules that have complicated potential energy surfaces with many local minima. By using an electric deflector the authors were able to show how one could disperse quantum states of particles of identical mass to their effective dipole moments and therefore devise experiments which use selected samples of species in their most polar quantum states.

Table 1 Velocities of ND$_3$ in low-lying quantum states (from Chandler et al.)[3]

(J, K)	v_{ND_3}' (meas.)[a]	$\bar{E}_{trans}/k$(meas.)[c]	$\Delta E_{trans}/k$ (meas.)[d]	v_{ND_3}'(calc.)[b]	$\bar{E}_{trans}/k$ (calc.)[c]	$\Delta E_{trans}/k$ (calc.)[d]
(2, 0)	26 ± 20 m s^{-1}	810 mK	460 mK	16 ± 0.6 m s^{-1}	310 mK	360 μK
(2, 1)	27 ± 18 m s^{-1}	880 mK	390 mK	11 ± 0.4 m s^{-1}	140 mK	170 μK
(2, 2)	23 ± 16 m s^{-1}	640 mK	320 mK	8 ± 0.3 m s^{-1}	70 mK	80 μK
(3, 1)	28 ± 16 m s^{-1}	940 mK	320 mK	27 ± 1.0 m s^{-1}	880 mK	1.1 mK
(3, 2)	21 ± 17 m s^{-1}	530 mK	340 mK	24 ± 0.8 m s^{-1}	680 mK	810 μK
(3, 3)	32 ± 16 m s^{-1}	1.2 K	790 mK	23 ± 0.8 m s^{-1}	630 mK	750 μK

[a] Indicates measured center velocity and velocity spread. Uncertainty in center velocity is ±10 m s^{-1}. [b] Indicates calculated center velocity and velocity spread. [c] Ratio of translational kinetic energy to k in the laboratory frame of reference, assuming stated center velocity. [d] Ratio of translational kinetic energy to k in the moving frame of reference, assuming stated velocity spread.

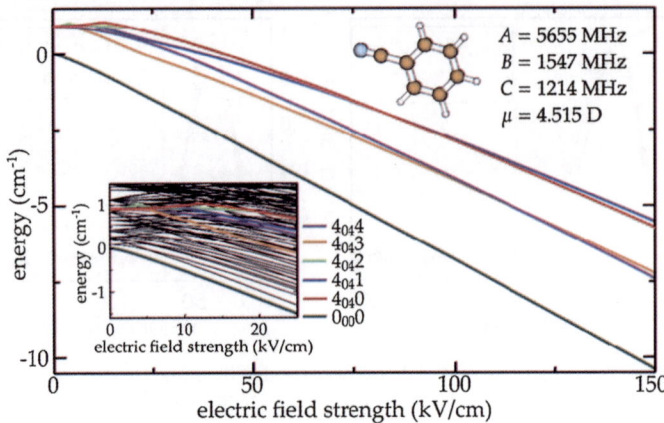

Fig. 3 Energy-selected rotational states of benzonitrile as a function of electric field strength (adapted from Fig. 1 of Küpper et al.).[4]

The results reported by Fig. 3 illustrate this behavior for the molecule of benzonitrile. A particular example of the invention demonstrated by presentations on preparing and manipulating cold molecules was given by the experimental group of the Chemistry Department at the University of Alberta, Canada.[5] These authors presented the rotational spectra of single carbonyl sulfide molecules embedded in superfluid helium nanodroplets. The data showed, very interestingly, that such experiments can indeed measure the *pure* rotational transitions between levels pertaining to the lowest vibrational state of OCS and that the observations reveal a clear distribution of different B-values and clear oscillations of such values in the smaller droplets, thus indicating the presence of several mechanisms which preside over the couplings between the dopant molecule's rotational motion and the solvent helium's translational motion.

As mentioned before, pre-existing molecules can also be sympathetically cooled in a buffer gas of pre-cooled atoms and, once slowed down to a cold regime, they can be subjected to further sympathetic cooling *via* thermal contact with an ultracold gas of atoms or, in the case in which the molecule has been confined in a trap, to further evaporative cooling. Several groups around the world are pursuing the above method chiefly using helium as a pre-cooled buffer gas and therefore several papers at the Discussion have analyzed and reported applications of such a method:

1. *Sympathetic cooling by collision with ultracold rare-gas atoms* was presented by an experiment and theory collaboration at University College London;[6]

2. *Prospects for sympathetic cooling of polar molecules* was analyzed by the computational collaboration between the Durham and Prague theory groups;[7]

3. *Dynamics of OH($^2\Pi$) – He collisions in combined electric and magnetic fields* was presented by a computational collaboration between groups at ITAMP, at Nijmegen and Vancouver.[8]

In the experimental scenario proposed by the UCL collaboration, a general scheme for sympathetic cooling of molecules down to μK on a timescale of seconds was described. The method basically utilizes ultracold, laser-cooled metastable rare-gas atoms quenched to their ground state when employed as collision partners to co-trapped molecular species within a deep optical trap of about 150 mK.[6]

The feasibility of such studies was further confirmed by calculations of the relevant scattering attributes for *s*-wave collisions at ultralow energies, as illustrated by the data in Table 2.

The computational analysis in ref. 7 additionally discusses in detail a very interesting option for the use of a buffer-gas coolant on polar molecules, *i.e.* the use of

Table 2 Scattering length a_s and zero-energy elastic cross section σ calculated for Rg–H_2 complexes (adapted from Barker et al.)[6]

	$|a_s|/\text{Å}$	$\sigma(E=0)/\text{Å}^2$
^{3}He–H_2	67.6–90.6	57500–103000
^{4}He–H_2	22.7–24.7	6500–7800
Ne–H_2	3.30–3.85	140–190
Ar–H_2	8.71–10.1	950–1300
Kr–H_2	5.51–6.96	380–610
Xe–H_2	1.82	42

alkali-earth atoms interacting with NH molecules.[7] One of the systems which could indeed provide a feasible example of weak couplings between the partner atom and the internal energy content of the NH molecule is the use of Mg as a buffer-gas.

The contours of the potential energy surface reporting the dispersion-bound complex of Mg–NH are shown in Fig. 4, where one could hope that external energy quenching processes may turn out to be more efficient than those leading to internal quenching of NH. That the additional presence of external fields could be a viable device for controlling the outcomes of collisional exchanges during sympathetic cooling dynamics was discussed and analyzed in a computational presentation (ref. 8) involving collaboration between three, very productive and active groups in the area of collisional cooling. The work presented there was able to demonstrate by computational models that spin relaxation in ^{3}He–OH collisions at temperatures below 0.01 K can be effectively suppressed by moderate electric fields of the order of 10 K V cm^{-1}.

An example of this behavior is given by the data reported in Fig. 5, where the spin relaxation cross sections as a function of electric field, and at fixed magnetic fields B of fractions of Tesla, are shown.

Another interesting aspect of the study of cold and ultracold molecular species has been provided by the analysis of ionic processes and ionic reactions. The chemistry of charged partners is certainly very important since many chemical reactions involve the presence of ions and, furthermore, the stronger types of interaction forces which are present usually cause much more marked features of the final

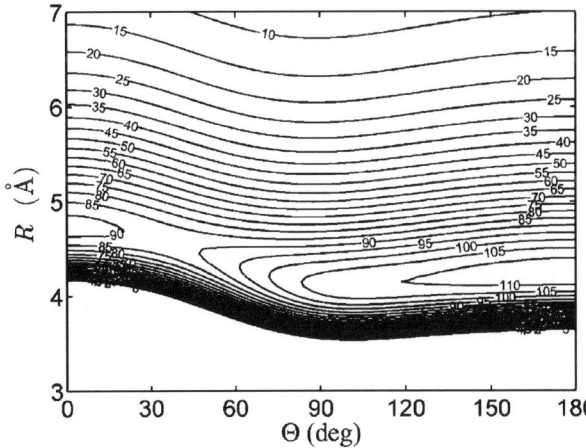

Fig. 4 Computed potential energy surface contour plot for the complex Mg–NH (adapted from Hutson et al.).[7]

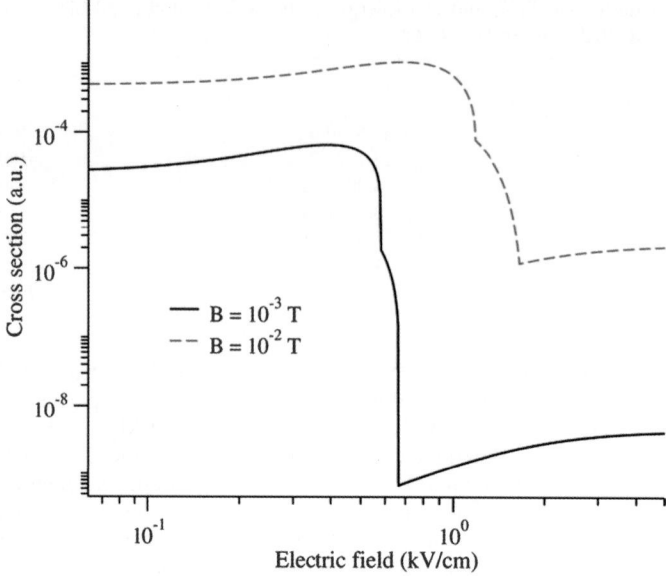

Fig. 5 Computed spin-relaxation cross section for OH–He collisions as a function of electric field and for two values of the magnetic field. The collision energy was 1 mK for both curves (adapted from Groenenboom et al.).[8]

chemical outcomes. One interesting contribution on *buffer-gas cooling of polyatomic ions in rf multi-electrode traps* was presented by the group of D. Gerlich of the University of Chemnitz .[9] The authors describe a new method in which a beam of slow atoms or molecules is used for cooling charged particles that are, in turn, confined in a multi-electrode radiofrequency trap. In order to reach sub-mK temperatures the fast part of a cold, effusive beam is removed with a shutter before the slow, remaining neutrals interact with the ion cloud. The ions are therefore cooled by numerous, multiple collisions with the colder beam of the buffer He gas and can be in turn employed to study chemical reactions in their own environment.

Another example of studying ionic reactions, at ultralow energies this time, was given at the meeting by a very interesting collaboration between the experimental group at the University of Oxford and the group at the University of Basel.[10] In that presentation the authors discussed cold chemical reactions between Coulomb-crystallized ions and velocity-selected neutral molecules. The experiment relies on the combination of a quadrupole-guide velocity selector for the generation of translationally cold neutral molecules with an added facility to produce ordered structures of cold ions (Coulomb crystals) by laser cooling in a linear quadrupole ion trap. The strong localization of the ions in the trap, combined with the high sensitivity of laser-induced fluorescence detection was able to allow for the study of chemical reactions, at $T \sim 1$ K, between laser-cooled Ca^+ ions and velocity selected CH_3F molecules, and for the additional analysis of sympathetically cooled ions down to ~ 10 mK with neutral species, as is the case for OCS^+ with ND_3.

The versatility of the method for an ample spectrum of molecules prepared to interact with the externally cold ions is demonstrated by the data shown in Fig. 6.

II. Great strides in photoassociation studies

As mentioned before, the process of photo-associating an atomic pair into forming a diatomic molecule is, in principle, the epitome of the indirect paths to obtaining cold molecules: to expose two free atoms on a radiative collision course to laser

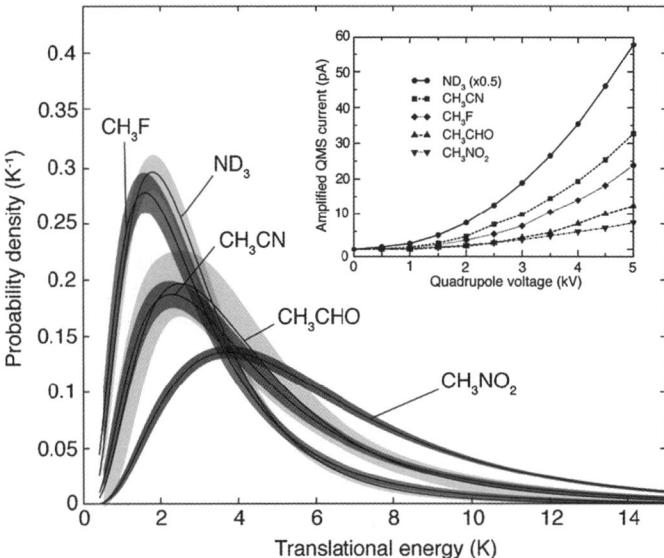

Fig. 6 Kinetic energy distributions for various guided molecules in the quadrupolar guide (adapted from Willitsch et al.).[10]

radiation of a suitable wavelength can excite them to make cold molecules from those cold atoms. However, the internal temperatures of the products are not internally "cold" and therefore a great variety of processes have been explored by several groups in order to finally produce translationally cold molecules which are also into their lowest electronic and rotovibrational states. The formation of an excited state of the molecule is, in fact, not the end of the experiment but rather an intermediate step: the state is only a metastable one and can therefore spontaneously decay (within microseconds) either back to the initial atoms or to a ground electronic state which remains vibrationally excited. The latter, unfortunately, can still collide with the other formed molecules, or residual atoms, and quite efficiently convert its internal energy into external, translational energy that will defeat the purpose of the photoassociation (PA) process and generate molecules which are back to temperatures of a few kelvins or more. Hence the need to additionally stimulate transitions from the electronically excited state using another laser field that can help achieve a population transfer that will prevent the collisional exchanges which will translationally heat the sample molecules. The physical situation of the initially formed photoassociation molecules is also of great theoretical interest because of the very marked proximity of the metastable bound states to the corresponding molecular continuum: the region of interaction potentials which is sampled there becomes very sensitive to the existence of resonant states, of virtual states and to the sign of the scattering length, a very important indicator of the collisional behavior of the system at very low energies and for dominant s-wave contributions. Thus, three papers at the meeting have examined in detail the theoretical implications of such situations:

1. *Ultracold molecules from ultracold atoms*, from P. S. Julienne, of the NIST Joint Quantum Institute in Maryland;[11]

2. *Two-photon coherent control of femtosecond photoassociation*, by the theory group of Christiane Koch at the Free University of Berlin;[12]

3. *Fano profiles in two-photon photoassociation spectra*, from C. Cohen-Tannoudji's group at the ENS in Paris;[13]

The first analysis shows in detail how the use of resonant scattering theory at ultralow energies can be used to great advantage when in combination with the methods of multichannel quantum defect theory.[11] Thus, several features of ultra-cold associative collisions can be revealed and illustrated by using calculations of bound and scattering states at vanishing collision energies and even down to the deeply bound $\nu = 0$ level of the ^{40}K^{87}Rb example.

These calculations show that to analyze collisions of prepared species indeed makes available a large part of the entire bound state spectrum of the collision complex for high-resolution probing and coherent population transfer. The delicate relationship between scattering length behavior and bound state formation is shown by the results reported by Fig. 7, where one clearly sees how moderate changes of the external field can move the system through a series of bound and virtual states.

The properties of studying photoassociation with short laser pulses using broad-band excitations to drive the necessary series of excitation and de-excitation steps have been analyzed by Koch *et al.*, which was mentioned before, where the possibility of two-photon femtosecond photoassociation for alkali and alkaline earth metal dimers (using calcium as an example) has been discussed within the framework of coherent control theory.[12]

The goal was to identify pulses which populate molecular levels close to the atomic transitions without exciting the atoms themselves. An example of the computational achievements is given by the data in Fig. 8 which indeed show that controlling processes by coherent light is making evident progress.

The theoretical analysis of the different line shapes which can be observed in photoassociation spectroscopy, and which can be rationalized using multichannel scattering theory with an effective Hamiltonian, has been explored by the ENS atomic theory group mentioned before.[13] Thus, by discussing the photoassociation process as a light-assisted scattering process between atoms they were able to derive different forms of Fano profiles depending on both the collision channels and the basis employed to represent the radiation. Hence, analytic expressions for the scattering matrix were found using an effective Hamiltonian, and not multichannel quantum defect theory (MQDT), thereby directly producing the relevant line shapes from the characteristics of the scattering matrix.

To demonstrate the variety of experiments which are correctly carried out by several groups in order to be able to attain the *holy grail* of today's photoassociation

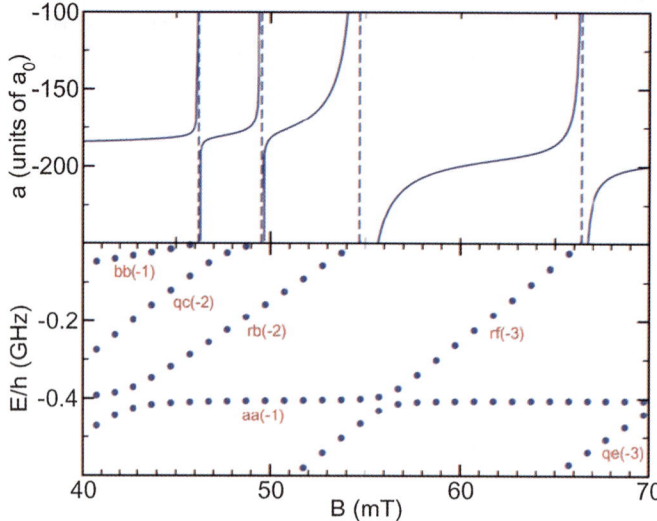

Fig. 7 Scattering length (upper panel) and bound-state energy as a function of the magnetic field B for a specific channel of the ^{40}K^{87}Rb molecule (adapted from Julienne).[11]

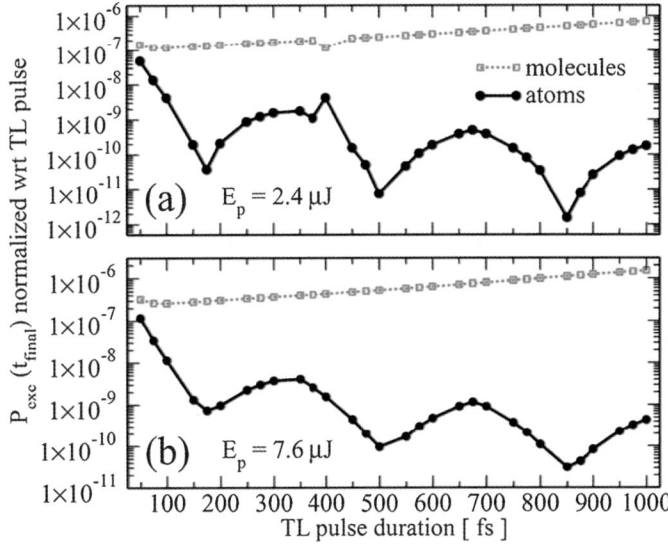

Fig. 8 Time-domain control with shaped pulse at two different values of the applied field E_p (adapted from Koch et al.).[12]

processes, i.e. the production of cold and ultracold molecules in their ground electronic and vibrational states, six different contributions were presented during the meeting:

1. *Broadband lasers to detect and cool the vibrations of cold molecules*, by the experimental group of the Laboratoire Aimé Cotton in Orsay, Paris;[14]
2. *Dark-state experiments with ultracold, deeply-bound triplet molecules*, by the Quantum Optics Center of the University of Innsbruck, Austria;[15]
3. *Precision molecular spectroscopy for ground state transfer of molecular quantum gases*, from H.-C. Nägerl's experimental group of the Quantum Optics Center in Innsbruck;[16]
4. *Formation of ultracold dipolar molecules in the lowest vibrational level*, from M. Weidemüller's group now at the University of Heidelberg;[17]
5. *Ultracold polar molecules near quantum degeneracy*, from a collaboration between JILA, Temple University and NIST, in the US;[18]
6. *A pump–probe study of the photoassociation of cold rubidium molecules*, from the Clarendon Laboratory in Oxford.[19]

The above series of experiments certainly certifies to the current ingenuity of the experimental groups that are attempting to attain the formation of cold, stable molecules by variants of the basic PA mechanism.

In ref. 14, for instance, a broadband femtosecond laser is used to electronically excite the molecules, leading to a redistribution of the vibrational population in the ground state *via* a few absorption-spontaneous emission cycles: a technique that the authors believe could be generalized to laser cooling of molecular rotations. The REMPI scheme they suggest is schematically presented in Fig. 9. The work reported by Hecker Denschlag *et al.* examines instead the formation of dark quantum superposition states of weakly bound Rb_2 Feshbach molecules and of strongly bound Rb_2 molecules in their lowest rovibrational state: a simpler square laser pulse scheme was thus used to replace the more complex STIRAP coherent transfer process, thereby demonstrating control of molecular motion in an optical lattice.[15] The work reported in ref. 16 employs instead the STIRAP procedure by first producing samples of Cs_2 Feshbach molecules, for which the experiment observed several optical transitions to deeply bound rotovibrational levels of the O_u^+ excited

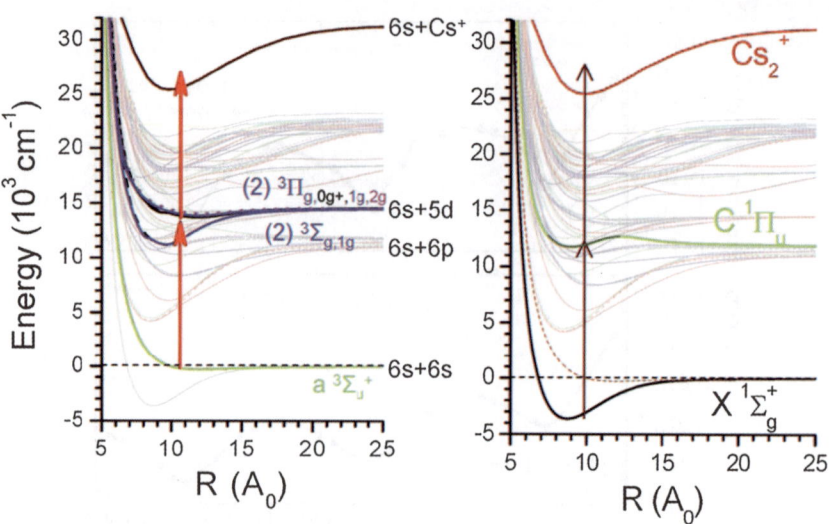

Fig. 9 REMPI scheme of the first triplet state (a) and the deeply bound ground state of Cs$_2$ (adapted from Pillet *et al.*).[14]

molecular state. The rovibrational ground levels were then reached from that state using a maximum of two successive two-photon STIRAP processes or one single four-photon STIRAP step.

A simplified scheme of the molecular levels involved is given in Fig. 10, where one can appreciate the experimental skills involved in making sure that the correct molecular levels are coupled by the chosen laser frequencies.

The analysis of a polar molecule (LiCs) formed in its rovibrational ground state $X^1\Sigma^+(v = 0, j = 0)$ is presented in ref. 17. The ground state formation is then proved by ionizing the molecules by resonantly-enhanced multiphoton ionization and by subsequent detection of the molecular ions.

Molecular samples with densities of the order of 10^{10} cm^{-3} are reached, at temperatures around 20 μK. The richness of the observed species during PA into various levels is shown schematically in Fig. 11.

The experimental observation of molecules near quantum degeneracy has been reported in ref. 18, where ^{40}K^{87}Rb in its absolute rovibrational ground state was obtained, starting from weakly bound KRb Feshbach molecules and by implementing a single-step coherent transfer between those states. One important property of such ground state molecules is their expected long lifetimes which would allow for further cooling of the molecules to quantum degeneracy. As an example, measured lifetimes of the order of 70 ms are seen from the decay curve of Fig. 12.

An additional pump–probe study of cold Rb$_2$ formation was discussed by Petrovic *et al.*, where several experimental setups are analyzed by using numerical simulations *via* wavepacket dynamics.[19] Using the theoretical prediction that the typical periods of the wavepacket oscillations can range from several tens to few hundreds of picoseconds, the experiments were performed and confirmed the formation of molecules bound in the vibrational states of the excited electronic state of Rb$_2$, an example of the measured signals are visible in the data of Fig. 13.

III. Broadening the usage of ultracold samples

Besides providing substantial proof of the variety of methods and devices that are currently used all over the world to form cold and ultracold molecular samples, the present meeting has also shown how ultracold molecular species can indeed serve as a special testing ground for physical phenomena which, in principle, are expected

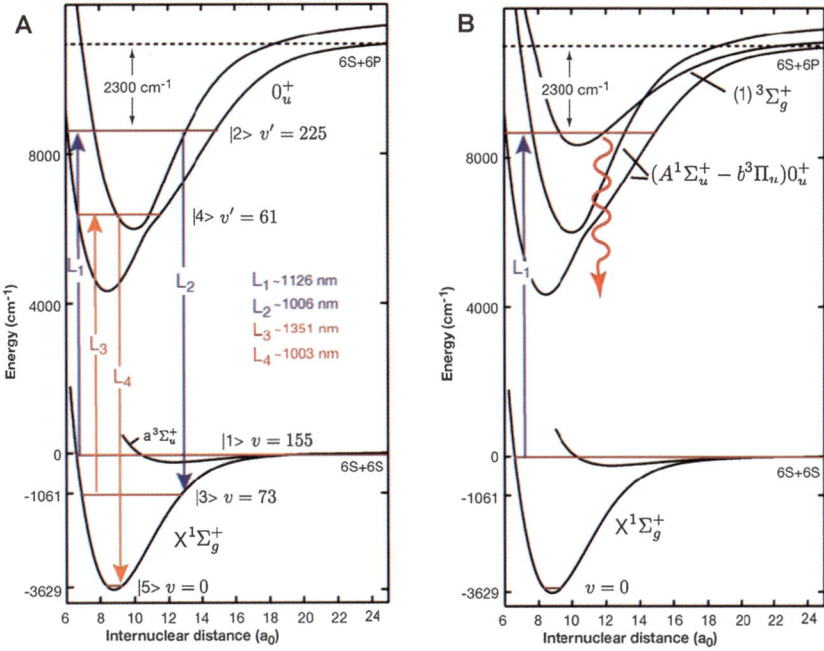

Fig. 10 Simplified molecular level schemes for Cs_2, showing in (A) the levels involved in the transfer. The scheme (B) shows the candidate levels of the first laser excitation (adapted from Nägerl et al.).[16]

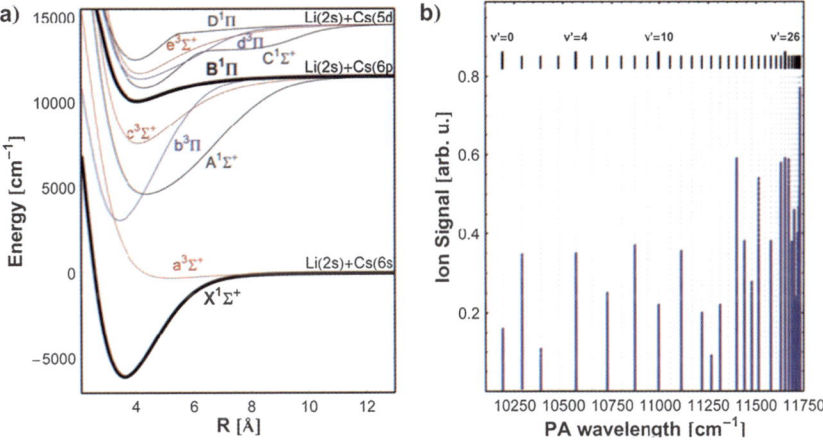

Fig. 11 Potential energy curves (a) and observed vibrational levels (b) of LiCs (adapted from Weidemüller et al.).[17]

to be present in very different fields and in a very different community. One example is: *Testing the time invariance of fundamental constants using microwave spectroscopy on cold diatomic radicals*, by Dr Bethlem of the Laser Center of the Free University, Amsterdam.[20] The paper focused on the fine constant α and on the proton/electron mass ratio μ and proposes that the metastable $^3\Pi$ state of the CO molecule could be employed for detecting μ variations by taking advantage of the near degeneracy between rotational levels within the fine-structure ladder of the metastable state.

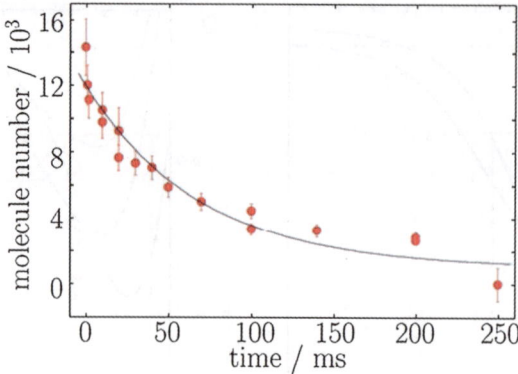

Fig. 12 Lifetime of ground-state polar KRb molecules in the optical dipole trap (adapted from Ye et al.).[18]

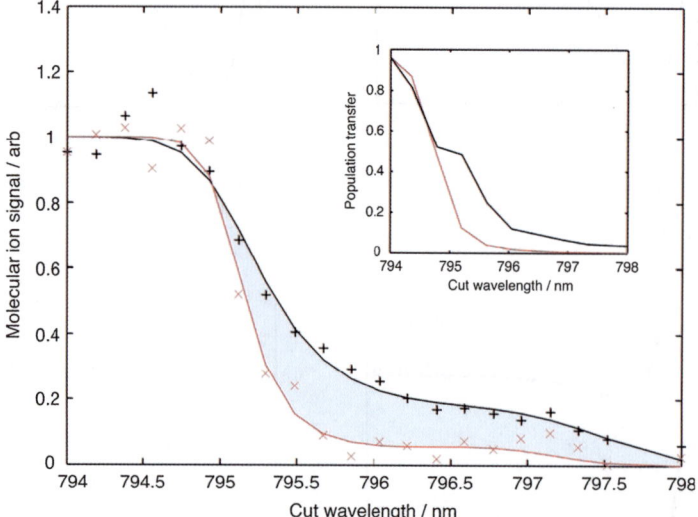

Fig. 13 Atomic (×) and molecular (+) signals for Rb and Rb_2 measured at fixed pump–probe delay and different spectral cuts (adapted from Petrovic et al.).[19]

Thus, the proposed experiment is that of measuring 2-photon microwave transitions in metastable CO, focusing on the $J = 6$, $\Omega = 1$, + → $J = 8$, $\Omega = 0$, + transition in the $\nu = 0$ and $\nu = 1$ levels of $^{12}C^{16}O$. Since astronomical observations suggest that $\Delta\mu\backslash\mu$ might change by 2×10^{-15} g^{-1}, in order to achieve that sensitivity, the 2-photon transitions should be measured with a fractional accuracy of 10^{-12}, which seems demanding but experimentally possible.

Another interesting test of fundamental properties was reported in the work discussed by the paper entitled: *Prospects for measuring the electric dipole moment of the electron using electrically trapped polar molecules*, by the Hinds group at the Center for Cold Matter at Imperial College, London.[21] Many of the proposed extensions of the Standard Model, in fact, contain new sources of CP (charge + parity) violations and result in substantially large values of the permanent electron dipole moment (edm), which bring its measurement very close to the current experimental limit. This proposal is therefore that of measuring the electron edm using electrically trapped molecules of YbF that undergo Stark deceleration. The process allows for a sensitivity gain which brings the cold molecule closer to the expected experimental detection. A further feasibility study, was directed at *Self-organization of large ensemble of particles in optical cavities*, a collaboration between groups at Heriot-Watt University in Edinburgh, UCL in London and the East China Normal University in Shanghai.[22] The study analyzes the dynamics of a neutral cloud of particles set in a cavity, which is pumped by an optical field: the paper addresses the key issue of the pump threshold scaling law that controls the onset of cavity cooling behavior for a large ensemble of particles, using a dense CN molecular cloud as a working example. This work is able to provide useful indications on the pumping conditions which need to be selected for the control of the cloud dynamics.

One of the important issues raised by various talks and comments has also been that of broadening the concept of a *chemical* reaction between ultracold partners, if it exists. For example, the work of the Sims experimental group at Rennes has presented results on an apparently simple reaction of $S(^1D) + H_2 \rightarrow SH + H$, where an open-shell atom reacts with the prototype molecular species of H_2. The results at very low temperatures are very difficult to reproduce by accurate calculations, but the data in Fig. 14 clearly indicate that agreement is very good between a reaction under special conditions and the accurate quantum calculations of its rates. A further, very interesting example has been also given by the experiments carried out by the Innsbruck group on the presence of a marked resonance enhancement associated with what one might call a *reaction* between specific hyperfine structure states of Cs_2 in collision with Cs, whereby a strong enhancement of the exchange process at nK temperatures is observed in the experiment: clearly calculations will be needed to provide convincing nanoscopic explanations for such an intriguing finding. The data reported by Fig. 15 attest to the sensitivity of cold environments for detecting even small "chemical" features such as those of the present example.

IV. Future outlook

The great variety of experimental methods presented by the participating group in order to produce and manipulate cold and ultracold molecules has certainly been the main feature of the present Discussion. However, this research field is very dynamic and rapidly expanding and therefore it is difficult to provide a conclusive description of what is in store for its future. Taking the work on cold and ultracold atomic gases carried out thirty years ago as a possible reference, the present interest

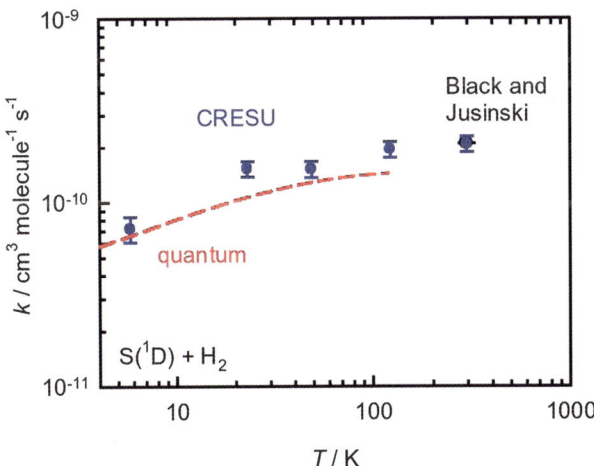

Fig. 14 Reactive measurements and calculated rates of the Rennes experiment (courtesy of Prof. Ian Sims).

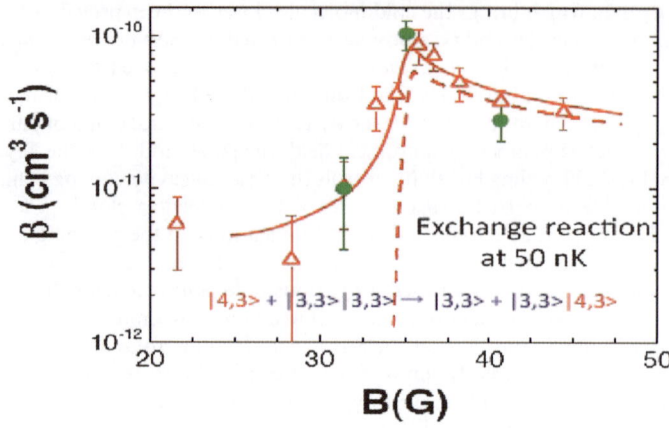

Fig. 15 Observed enhancement of the *exchange* reaction between different hyperfine states of a cesium atom and a cesium dimer molecule at ultralow temperatures (courtesy of Dr F. Ferlaino, Innsbruck).

in cold and ultracold molecules is also stimulated by the promise of new applications and potential for fundamental discoveries. Furthermore, since chemical applications invariably require the possibility of manipulating a dense ensemble of ultracold molecules, one of the main challenges for experiments now is that of producing molecular gases with large phase-space densities. To confine molecules into low-field seeking states requires us to further be able to limit the molecular losses created by collisional relaxation (depolarization) of the confined species; thus, it is important to design additional external fields shaped to confine the absolute ground state: examples were given at the meeting of how to go about this task without limiting the trap depths in such a way that few practical applications would be possible.

PA processes are also a promising way to create very cold molecular ensembles, although they are often produced in rovibrationally excited states which are collisionally unstable. We have seen during this meeting that many avenues for producing these molecules in their ground rovibrational levels have been explored and that several new possibilities have been successfully tested. Hence, the stable and robust realization of such devices will provide sources of cold and ultracold molecules for a wide range of applications which span quantum computing, controllable models of quantum lattices, controllable chemical processes and reactions followed in confined geometries for better control. On the other hand, to expand the field into the realm of *real chemistry*, and therefore win the hearts of those who work in it, one needs to achieve several important goals:

1. to be able to produce dense ensembles of *polyatomic* molecules with reaction properties of interest to mainstream chemistry;

2. to be able to set up devices that are simple enough to be learned to use by occasional practitioners and stable enough to allow for long-term manipulative procedures of the confined, ultracold species;

3. to match the experimental data collection with nearly *real time* computational modeling that can realistically guide experimental choices and further understand from simulations its verifiable outcomes.

The above bullet-points may appear to be a tall order for our community, but they are certainly similar to those met almost forty years ago by the then fledgling community of molecular beam experimentalists whose methods have now become fairly standard tools for the study of a very broad range of molecular processes. We should therefore confidently expect a similar evolution for the field of cold and ultracold chemistry in the forthcoming years.

Acknowledgements

The financial support of the National Research Project PRIN2006 and of the CAS-PUR Supercomputing Consortium is gratefully acknowledged. One of us (M. T.) thanks the University of Rome *La Sapienza* for the award of a postdoctoral fellowship during which the present work was completed. We are grateful to Professor Ian Sims and Dr Francesca Ferlaino for making their data available to us before publication. We also thank all of the participating authors of the 22 contributions to this Discussion for allowing us to present some of their data in the above work.

References

1. S. Y. T. van de Meerakker and G. Meijer, *Faraday Discuss.*, 2009, **142**, DOI: 10.1039/b819721k.
2. C. Sommer, L. D. van Buuren, M. Motsch, S. Pohle, J. Bayerl, P. W. H. Pinkse and G. Rempe, *Faraday Discuss.*, 2009, **142**, DOI: 10.1039/b819726a.
3. J. J. Kay, S. Y. T. van de Meerakker, K. E. Strecker and D. W. Chandler, *Faraday Discuss.*, 2009, **142**, DOI: 10.1039/b819256c.
4. J. Küpper, F. Filsinger and G. Meijer, *Faraday Discuss.*, 2009, **142**, DOI: 10.1039/b820045a.
5. R. Lehnig, P. L. Raston and W. Jäger, *Faraday Discuss.*, 2009, **142**, DOI: 10.1039/b819844f.
6. P. F. Barker, S. M. Purcell, P. Douglas, P. Barletta, N. Coppendale, C. Maher-McWilliams and J. Tennyson, *Faraday Discuss.*, 2009, **142**, DOI: 10.1039/b819079h.
7. P. Soldán, P. S. Żuchowski and J. M. Hutson, *Faraday Discuss.*, 2009, **142**, DOI: 10.1039/b822769c.
8. T. V. Tscherbul, G. C. Groenenboom, R. V. Krems and A. Dalgarno, *Faraday Discuss.*, 2009, **142**, DOI: 10.1039/b819198k.
9. D. Gerlich and G. Borodi, *Faraday Discuss.*, 2009, **142**, DOI: 10.1039/b820977d.
10. M. T. Bell, A. D. Gingell, J. Oldham, T. P. Softley and S. Willitsch, *Faraday Discuss.*, 2009, **142**, DOI: 10.1039/b818733a.
11. P. Julienne, *Faraday Discuss.*, 2009, **142**, DOI: 10.1039/b820917k.
12. C. P. Koch, M. Ndong and R. Kosloff, *Faraday Discuss.*, 2009, **142**, DOI: 10.1039/b818458e.
13. M. Portier, M. Leduc and C. Cohen-Tannoudji, *Faraday Discuss.*, 2009, **142**, DOI: 10.1039/b819470j.
14. M. Viteau, A. Chotia, D. Sofikitis, M. Allegrini, N. Bouloufa, O. Dulieu, D. Comparat and P. Pillet, *Faraday Discuss.*, 2009, **142**, DOI: 10.1039/b819697d.
15. F. Lang, C. Strauss, K. Winkler, T. Takekoshi, R. Grimm and J. Hecker Denschlag, *Faraday Discuss.*, 2009, **142**, DOI: 10.1039/b818964a.
16. J. G. Danzl, M. J. Mark, E. Haller, M. Gustavsson, N. Bouloufa, O. Dulieu, H. Ritsch, R. Hart and H.-C. Nägerl, *Faraday Discuss.*, 2009, **142**, DOI: 10.1039/b818653g.
17. J. Deiglmayr, M. Repp, A. Grochola, K. Mörtlbauer, C. Glück, O. Dulieu, J. Lange, R. Wester and M. Weidemüller, *Faraday Discuss.*, 2009, **142**, DOI: 10.1039/b818391k.
18. S. Ospelkaus, K. K. Ni, M. H. G. de Miranda, B. Neyenhuis, D. Wang, S. Kotochigova, P. S. Julienne, D. S. Jin and J. Ye, *Faraday Discuss.*, 2009, **142**, DOI: 10.1039/b821298h.
19. J. Petrovic, D. McCabe, D. England, H. Martay, M. Friedman, A. Dicks, E. Dimova and I. Walmsley, *Faraday Discuss.*, 2009, **142**, DOI: 10.1039/b818494a.
20. H. L. Bethlem and W. Ubachs, *Faraday Discuss.*, 2009, **142**, DOI: 10.1039/b819099b.
21. M. R. Tarbutt, J. J. Hudson, B. E. Sauer and E. A. Hinds, *Faraday Discuss.*, 2009, **142**, DOI: 10.1039/b810625b.
22. Y. Zhao, W. Lu, P. F. Barker and G. Dong, *Faraday Discuss.*, 2009, **142**, DOI: 10.1039/b818653g.

Poster titles

Probing halo molecules with nonresonant light, **Mikhail Lemeshko and Bretislav Friedrich**, *Fritz-Haber-Institut der Max-Planck Gesellschaft, Germany*

Methods for studying translationally cold ion-neutral reactions, **Martin T. Bell, William G. Doherty, Alexander D. Gingell, James M. Oldham, Timothy P. Softley and Stefan Willitsch**, *University of Oxford, UK*

Metals in superfluid helium nanodroplets, **Adrian Boatwright, Jay Jeffs, Hisham Al-Zubaidi and Anthony J. Stace**, *University of Nottingham, UK*

Application of a general-order 4-component multi-reference coupled-cluster on LiCs and RbYb, **Lasse Kragh Sørensen Timo Fleig, Christel Marian and Jeppe Olsen**, *Heinrich-Heine-Universität Düsseldorf, Germany*

Local control of two-photon femtosecond photoassociation, **Mamadou Ndong, Ruzin Ağanoğlu and Christiane P. Koch**, *Freie Universität Berlin, Germany*

Low-energy dynamics of the H_3-system: a prototype for cold collisions between molecules and negative ions, **Mehdi Ayouz, Romain Guérout, Olivier Dulieu, M. Raoult and Jacques Robert**, *Université Paris-Sud, France*

Calculations of Cs_3 potential energy surfaces, **Romain Guérout, Mirielle Aymar and Olivier Dulieu**, *Université Paris-Sud, France*

Rovibrational dynamics and photoassociation of cold heteronuclear dimers in electric fields, **M. Mayle, R. González-Férez and P. Schmelcher**, *University of Heidelberg, Germany*

Towards improved rovibrational spectroscopy of cold trapped HD^+-ions, **T. Schneider, B. Roth, U. Bressel, I. Ernsting, M. Hansen, H. Duncker, S. Vasilyev, A. Nevsky and S. Schiller**, *Heinrich-Heine-Universität Düsseldorf, Germany*

Cooling and spectroscopy of gas-phase molecules at 100 mK, **T. Schneider, D. Offenberg, B. Roth, Ch. Wellers, and S. Schiller**, *Heinrich-Heine-Universität Düsseldorf, Germany*

Ab initio potential energy surfaces of NH–NH, **Liesbeth M. C. Janssen, Gerrit C. Groenenboom and Ad van der Avoird**, *Radboud University Nijmegen, The Netherlands*

A molecular synchrotron, **Peter Zieger, Cynthia E. Heiner, Andre J. A. Van Roij, Hendrick L. Bethlem, Sebastiaan Y. T. van de Meerakker and Gerard Meijer**, *Fritz-Haber-Institut der Max-Planck-Gesellschaft, Germany*

Production and high-resolution spectroscopy of buffer gas cooled YbF molecules, **S. M. Skoff, R. J. Hendricks, J. J. Hudson, B. E. Sauer, D. M. Segal, M. R. Tarbutt and E. A. Hinds**, *Imperial College London, UK*

Polarization-mixing as a signature for the formation of molecules in an optical lattice, **Hashem Zoubi and Helmut Ritsch**, *University of Innsbruck, Austria*

A novel method for producing state-selected and translationally cold molecular ions, **Xin Tong and Stefan Willitsch**, *University of Basel, Switzerland*

Broadband vibrational cooling of Cs_2 molecules, **M. Pichler, D. Sofikitis, A. Fioretti, S. Weber, R. Horchani, X. Li, M. Allegrini, B. Chatel, D. Comparat and P. Pillet**, *Université Paris-Sud and Université Paul Sabatier Toulouse, France*

Towards ultracold polar KCs molecules, **M. Pichler, D. Barker, M. Garman and O. Dulieu**, *Université Paris-Sud, France*

Ultra cold ion atom and ion molecule collisions: dynamics, chemical reactions and charge exchange, **Enrico Bodo, Franco Gianturco and A. Dalgarno**, *University of Rome "La Sapienza", Italy*

Cold atom-molecule interactions, **L. Paul Parazzoli, Noah Fitch, D. Lobser and Heather Lewandowski**, *University of Colorado, USA*

Impact of electric fields on highly excited rovibrational states of polar dimers, **Rosario González-Férez and Peter Schmelcher**, *University of Granada, Spain*

Cold ion chemistry in a linear Paul trap: MgH^+, Mg^+ with Rb, **M. Tacconi and F. A. Gianturco**, *University of Rome "La Sapienza", Italy*

Retardation effects in the interactions between ultracold atoms, **Wojciech Skomorowski, Krzysztof Pachucki and Robert Moszynski**, *University of Warsaw, Poland*

Sympathetic cooling of molecules by collisions with ultra-cold rare gas atoms in a quasi-electrostatic trap, **C. Maher-McWilliams, P. Douglas and P. F. Barker**, *University College London, UK*

Coherent control of ultracold collisions with nonlinear frequency chirps: experiment and simulations, **J. A. Pechkis, J. L. Carini, C. E. Rogers III and P. L. Gould**, *University of Connecticut, USA*

Resonant coupling in the heteronuclear alkali dimers for direct photoassociative formation of X(0,0) ultracold molecules, **J. Ray Majumder, M. Bellos, R. Carollo, M. Recore, M. Mastroianni, V. Tagliamonti, W. C. Stwalley, E. E. Eyler and P. L. Gould**, *University of Connecticut, USA*

Progress towards forming ultracold $^{85}Rb_2$ in an optical dipole trap, **H. K. Pechkis, M. Bellos, R. Carollo, J. Ray Majumder, E. E. Eyler, P. L. Gould and, W. C. Stwalley**, *University of Connecticut, USA*

Relativistic multi-reference configuration interaction calculations on Rb-Yb and $(Rb-Ba)^+$. Prospects for ultracold molecular formation, **Stefan Knecht, Hans Jørgen Aagaard Jensen, Timo Fleig and Christel M. Marian**, *Heinrich-Heine-Universität Düsseldorf, Germany*

Towards sympathetic cooling of molecules with ultracold atoms, **Adela Marian, Henrik Haak, Sophie Schlunk, Amudha K. Durasaimy, Wieland Schöllkopf and Gerard Meijer**, *Fritz-Haber-Institut der Max-Planck-Gesellschaft, Germany*

Towards a quantum gas of polar RbCs molecules, **D. J. McCarron, D. L. Jenkin and S. L. Cornish**, *Durham University, UK*

Cooling of large (neutral) molecules, **Hendrik Ulbricht, Sarayut Deachapunya, Markus Marksteiner, Philipp Haslinger, Michele Sclafani, Helmut Ritsch and Markus Arndt**, *University of Southampton, UK*

A laser system for chirped optical Stark deceleration, **Nicholas Coppendale, Peter Douglas, Lei Wang and P. F. Barker**, *University College London, UK*

Prospects for sympathetic cooling of optically stark decelerated molecules, **P. Barletta, P. F. Barker and J. Tennyson**, *University College London, UK*

Cavity-enhanced Rayleigh scattering from atoms and molecules, **Pepijn W. H. Pinkse, Michael Motsch, Martin Zeppenfeld and Gerhard Rempe**, *Max-Planck-Institut für Quantenoptik, Germany*

Photoassociation of polar molecules in the rovibrational ground state, **M. Repp, J. Deiglmayr, A. Grochola, P. Pellegrini, C. Glück, K. Mörtlbauer, J. Lange, O. Dulieu, R. Côté, R. Wester and M. Weidemüller**, *Albert-Ludwigs-Universität Freiburg, Germany*

Adiabatic cooling of a large number of strongly-confined particles in a relatively leaky optical cavity, **Almut Beige and Giuseppe Vitiello**, *University of Leeds, UK*

Cold reactive collisions with open shell ^{1}D atoms: the role of the long range interactions, **F. Dayou, M. Lara and J. M. Launay**, *University of Rennes 1, France*

Exploring fermionic Feshbach resonances of ^{6}Li-^{40}K, **Devang Naik, Frederik Spiegelhalder, Andreas Trenkwalder, Erik Wille, Gerhard Hendl, Florian Schreck and Rudolf Grimm**, *Austrian Academy of Sciences, Austria*

Resolved sideband cooling of dipole interacting atoms, **R. N. Palmer and A. Beige**, *University of Leeds, UK*

Cavity-enhanced sideband cooling of molecules to the ground state of a harmonic trap, **Markus Kowalewski, Giovanna Morigi, Pepijn W. H. Pinkse and Regina de Vivie-Riedle**, *Ludwig-Maximilians-Universität München, Germany*

Vibrational wave packet dynamics of Rb_2 formed on helium nanodroplets, **P. Heister, T. Hippler, M. Mudrich and F. Stienkemeier**, *Universität Freiburg, Germany*

A quantum gas of deeply bound ground state molecules, **Johann Danzl, Manfred Mark, Elmar Haller, Mattias Gustavsson, Russell Hart, Nadia Bouloufa, Olivier Dulieu, Houssam Salami, Tom Bergeman, Helmut Ritsch and Hanns-Christoph Nägerl**, *Universität Innsbruck, Austria*

1D systems with tunable interactions, **Elmar Haller, Mattias Gustavsson, Manfred Mark, Johann G. Danzl, Russell Hart and Hanns-Christoph Nägerl**, *Universität Innsbruck, Austria*

Progress towards sympathetic cooling of Stark decelerated lithium hydride molecules by laser cooled atoms, **S. K. Tokunaga, J. O. Stack, J. J. Hudson, B. E. Sauer, E. A. Hinds and M. R. Tarbutt**, *Imperial College, London, UK*

Laser cooling of group 13 atoms, **L. Rutherford, I. C. Lane and J. F. McCann**, *Queen's University Belfast, UK*

Trapping and cooling using micro-mirror arrays, **H. Ohadi, A. Xuereb, R. Murray, M. Himsworth, J. Bateman and T. Freegarde**, *University of Southampton, UK*

Zeeman deceleration and trapping of molecules: using travelling waves, **Manabendra N. Bera, Jacques Robert, Pierre Pillet and Nicholas Vanhaecke**, *Université Paris-Sud, France*

Li_2F: sympathetic cooling and ultracold chemistry, **K. Wright and Ian Lane**, *Queen's University Belfast, UK*

Rotational cooling of translationally and vibrationally cold MgH^+ ions, **A. K. Hansen, K. Højbjerre, P. S. Skyt, P. F. Staanum and M. Drewsen**, *University of Aarhus, Denmark*

Inside nature's smallest blackbody, **Andreas Kurcz, Antonio Capolupo and Almut Beige**, *University of Leeds, UK*

Continous guided beams of slow and internally cold polar molecules, **Laurens D. van Buuren, Michael Motsch, Christian Sommer, M. Zeppenfeld, Sebastian Pohle, M. Schenk, Josef Bayerl, Pepijn W. H. Pinkse and Gerhard Rempe**, *Max-Planck-Institut für Quantenoptik, Germany*

Chemical reactivity at extremely low temperatures: rate coefficients for $S(^1D) + H_2$ down to 5.8 K, **Coralie Berteloite, Manuel Lara, Sébastien D. Le Picard, Fabrice Dayou, Jean-Michel Launay, André Canosa and Ian R. Sims**, *Université de Rennes 1, France*

Rotational study of carbon monoxide solvated with *para*-hydrogen molecules, **Paul Raston and Wolfgang Jäger**, *University of Alberta, Canada*

Conical intersections between molecular states in external fields, **Alisdair O. G. Wallis, S. A. Gardiner and Jeremy Hutson**, *Durham University, UK*

Low-energy collisions of ND_3 with ultracold Rb atoms, **Piotr S. Żuchowski and Jeremy M. Hutson**, *Durham University, UK*

Hyperfine levels of alkali metal dimers, **J. Aldegunde, Hong Ran and Jeremy M Hutson**, *Durham University, UK*

Theoretical treatment of Feshbach resonances for trimers at thermal and ultra-cold temperatures: example of H_3, **Juan Blandon, Françoise Masnou-Seeuws and Viatcheslav Kokoouline**, *University of Central Florida, USA*

The Skinner Prize for the best poster was awarded to Dr Johann Danzl of the University of Innsbruck, Austria, for his poster on a quantum gas of deeply bound ground state molecules.

List of participants

Dr J. Aldegunde, *Durham University, United Kingdom*
Mr J. Allen, *University of Nottingham, United Kingdom*
Mr S. Banerjee, *University of Connecticut, USA*
Professor P. Barker, *University College London, United Kingdom*
Dr P. Barletta, *University College London, United Kingdom*
Dr A. Beige, *University of Leeds, United Kingdom*
Mr M. Bell, *University of Oxford, United Kingdom*
Mr M. Bera, *Laboratoire Aimé Cotton, France*
Dr H. Bethlem, *Laser Centre, Vrije Universiteit, The Netherlands*
Mr T. Blake, *University of Leeds, United Kingdom*
Dr A. Boatwright, *University of Nottingham, United Kingdom*
Dr E. Bodo, *University of Rome, 'La Sapienza', Italy*
Dr N. Bouloufa, *Université Paris-Sud (Paris XI), Orsay, France*
Dr D. Carty, *Durham University, United Kingdom*
Dr D. Chandler, *Sandia National Laboratories, USA*
Dr M. Chapman, *Royal Society of Chemistry, United Kingdom*
Dr S. Chervenkov, *Max Planck Institute of Quantum Optics, Germany*
Professor D. Clary, *University of Oxford, United Kingdom*
Mr N. Coppendale, *University College London, United Kingdom*
Dr S. Cornish, *Durham University, United Kingdom*
Mr J. Croft, *Durham University, United Kingdom*
Dr J. Danzl, *University of Innsbruck, Austria*
Professor R. De Vivie-Riedle, *Ludwig Maximilians University, Munich, Germany*
Mr M. Debatin, *University of Innsbruck, Austria*
Mr J. Deiglmayr, *University of Freiburg, Germany*
Emeritus Professor. A. Dickinson, *Newcastle University, United Kingdom*
Dr J. Dickinson, *Institute of Physics Publishing Ltd, United Kingdom*
Mr W. Doherty, *Oxford University, United Kingdom*
Dr P. Douglas, *University College London, United Kingdom*
Dr O. Dulieu, *Université Paris-Sud XI, France*
Mr D. England, *Oxford University, United Kingdom*
Dr E. Favilla, *Instituto Per I Processi Chimico Fisica (IPCF) - CNR, Italy*
Dr F. Ferlaino, *University of Innsbruck, Austria*
Dr S. Gardiner, *Durham University, United Kingdom*
Professor D. Gerlich, *Technical University, Chemnitz, Germany*
Professor F. Gianturco, *University of Rome, Italy*
Mr A. Gingell, *University of Oxford, United Kingdom*
Mrs S. Godfrey, *Royal Society of Chemistry, United Kingdom*
Dr R. González-Férez, *University of Granada, Spain*
Professor R. Grimm, *University of Innsbruck, Austria*
Dr A. Grochola, *University of Freiburg, Germany*
Dr G. Groenenboom, *Radboud University, Nijmegen, The Netherlands*
Dr R. Guerout, *Université Paris Sud XI, France*
Miss S. Haendel, *Durham University, United Kingdom*
Mr A. Hansen, *University of Aarhus, Denmark*
Mr L. Harper, *Oxford University, United Kingdom*
Dr R. Hart, *University of Innsbruck, Austria*
Dr J. Hecker Denschlag, *University of Innsbruck, Austria*
Dr A. Helman, *European Science Foundation, France*
Dr R. Hendricks, *Imperial College London, United Kingdom*
Professor D. Herschbach, *Harvard University, USA*
Professor E. Hinds, *Imperial College London, United Kingdom*

Dr J. Hudson, *Imperial College London, United Kingdom*
Dr I. Hughes, *Durham University, United Kingdom*
Professor J. Hutson, *Durham University, United Kingdom*
Dr Z. Idziaszek, *University of Warsaw, Poland*
Professor W. Jäger, *University of Alberta, Canada*
Mr P. Jansen, *Laser Centre, Vrije Universiteit, The Netherlands*
Ms L. Janssen, *Radboud University, Nijmegen, The Netherlands*
Mr D. Jenkin, *Durham University, United Kingdom*
Dr M. Jones, *Durham University, United Kingdom*
Professor P. Julienne, *National Institute of Standards and Technology Physics Laboratory, USA*
Dr J. Kay, *Sandia National Laboratories, USA*
Mrs L. Kennedy, *European Science Foundation, United Kingdom*
Mr S. Knecht, *Heinrich-Heine University, Germany*
Dr C. Koch, *Freie Universität Berlin, Germany*
Dr V. Kokoouline, *University of Central Florida, USA*
Professor R. Kosloff, *Hebrew University of Jerusalem, Israel*
Professor S. Kotochigova, *Temple University, USA*
Mr M. Kowalewski, *Ludwig Maximilians University, Munich, Germany*
Dr J. Küpper, *Fritz-Haber-Institut Der Max Planck Gesellschaft, Germany*
Mr A. Kurcz, *University of Leeds, United Kingdom*
Dr I. Lane, *Queens University Belfast, United Kingdom*
Dr M. Lara Garrido, *Université De Rennes 1, France*
Professor J. Launay, *Université De Rennes 1, France*
Dr R. Le Sueur, *Durham University, United Kingdom*
Mr M. Lemeshko, *Fritz Haber Institute of the Max Planck Society, Germany*
Professor H. Lewandowski, *University of Colorado, USA*
Dr W. Lu, *Heriot-Watt University, United Kingdom*
Mr C. Maher-McWilliams, *University College London, United Kingdom*
Miss A. Marchant, *Durham University, United Kingdom*
Dr A. Marian, *Fritz Haber Institute Der Max Planck Gesellschaft, Germany*
Mr D. McCabe, *University of Oxford, United Kingdom*
D. McCarron, *Durham University, United Kingdom*
Professor G. Meijer, *Fritz Haber Institute, Berlin, Germany*
Professor R. Moszynski, *University of Warsaw, Poland*
Dr M. Mudrich, *University of Freiburg, Germany*
Dr H. Nägerl, *University of Innsbruck, Austria*
Dr N. Nahler, *Durham University, United Kingdom*
Mr D. Naik, *Institute of Quantum Optics and Quantum Information, Austria*
Ms F. Nalden, *Royal Society of Chemistry, United Kingdom*
Dr M. Ndong, *Freie Universität, Berlin, Germany*
Miss R. Needham, *Royal Society of Chemistry, United Kingdom*
Dr H. Ohadi, *University of Southampton, United Kingdom*
Mr J. Oldham, *University of Oxford, United Kingdom*
Dr J. Pachos, *University of Leeds, United Kingdom*
Dr R. Palmer, *University of Leeds, United Kingdom*
Ms H. Pechkis, *University of Connecticut, USA*
Mr J. Pechkis, *University of Connecticut, USA*
Mr D. Petrov, *Université Paris Sud, France*
Dr J. Petrovic, *University of Oxford, United Kingdom*
Prof Dr M. Pichler, *Goucher College, USA*
Dr P. Pillet, *Laboratoire Aimé Cotton, CNRS, France*
Dr P. Pinkse, *Max Planck Institute of Quantum Optics, Germany*
Dr M. Portier, *Laboratoire Kastler Brossel, France*
Dr J. Pugh, *Royal Society of Chemistry, United Kingdom*
Miss T. Ran, *Durham University, United Kingdom*

Dr P. Raston, *University of Alberta, Canada*
Ms J. Ray Majumder, *University of Connecticut, USA*
Professor G. Rempe, *Max Planck Institut, Germany*
Mr M. Repp, *University of Heidelberg, Germany*
Mr A. Rowland, *Durham University, United Kingdom*
Miss L. Rutherford, *Queens University Belfast, United Kingdom*
Dr B. Sauer, *Imperial College London, United Kingdom*
Professor Dr P. Schmelcher, *Theoretical Chemistry, Germany*
Dr T. Schneider, *University of Dusseldorf, Germany*
Dr D. Segal, *Imperial College, United Kingdom*
Professor I. Sims, *UMR 6251 CNRS - Université de Rennes 1, France*
Ms S. Skoff, *Imperial College London, United Kingdom*
Mr W. Skomorowski, *University of Warsaw, Poland*
Professor I. Smith, *University of Cambridge, United Kingdom*
Mr L. Soerensen, *Heinrich-Heine-University, Germany*
Professor T. Softley, *University of Oxford, United Kingdom*
Dr P. Soldán, *Charles University, Czech Republic*
Mr C. Sommer, *Max Planck Institute for Quantum Optics, Germany*
Dr P. Staanum, *University of Aarhus, Denmark*
Mr J. Stack, *Imperial College London, United Kingdom*
Dr T. Stoecklin, *Université Bordeaux 1, France*
Professor W. Stwalley, *University of Connecticut, USA*
Dr M. Tacconi, *University of Rome, 'La Sapienza', Italy*
Dr M. Tarbutt, *Imperial College London, United Kingdom*
Professor J. Tennyson, *Imperial College London, United Kingdom*
Mr S. Tokunaga, *Imperial College London, United Kingdom*
Dr X. Tong, *Universität Basel, Switzerland*
Professor M. Trippenbach, *University of Warsaw, Poland*
Dr H. Ulbricht, *University of Southampton, United Kingdom*
Dr L. Van Buuren, *Max Planck Institute of Quantum Optics, Germany*
Dr S. Van De Meerakker, *Fritz Haber Institute der Max Planck Gesellschaft, Germany*
Professor Dr A. van der Avoird, *Radboud University, Nijmegen, The Netherlands*
Dr N. Vanhaecke, *Laboratoire Aimé Cotton, France*
Mr T. Wall, *Imperial College London, United Kingdom*
Mr A. Wallis, *Durham University, United Kingdom*
Professor M. Weidemüller, *University of Heidelberg, Germany*
Dr R. Wester, *Universität Freiburg, Germany*
Professor B. Whitaker, *University of Leeds, United Kingdom*
Mr S. Will, *Johannes Gutenberg University, Mainz, Germany*
Professor Dr S. Willitsch, *University of Basel, Switzerland*
Dr E. Wrede, *Durham University, United Kingdom*
Professor J. Ye, *University of Colorado, USA*
Dr M. Zaccanti, *Lens, University of Florence, Italy*
Mr M. Zeppenfeld, *Max Planck Institute for Quantum Optics, Germany*
Dr H. Zoubi, *Institute for Theoretical Physics, Austria*
Dr P. Żuchowski, *Durham University, United Kingdom*

Index of contributors*

Aldegunde, J., 319, 429
Allegrini, M., **257**
Barker, P. F., **175**, 221, **311**, 319, 429
Barletta, P., **175**
Bayerl, J., **203**
Bell, M. T., **73**
Berteloite, C., 221
Bethlem, H. L., **25**, 93
Borodi, G., **57**
Bouloufa, N., **257, 283**
Canosa, A., 221
Chandler, D. W., 93, **143**, 221
Chotia, A., **257**
Cohen-Tannoudji, C, **415**
Comparat, D., **257**
Coppendale, N., **175**
Dalgarno, A., **127**
Danzl, J. G., **283**, 319
Dayou, F., 221
Deiglmayr, J., 319, **335**, 429
de Miranda, M. H. G., **351**
de Vivie-Riedle, R., 429
Dicks, A., **403**
Dimova, E., **403**
Dong, G., **311**
Douglas, P., **175**
Dulieu, O., 221, **257, 283**, 319, **335**, 429
England, D., **403**
Ferlaino, F., 429
Filsinger, F., **155**
Friedman, M., **403**
Gerlich, D., 93, **57**
Gianturco, F. A., 93, 221, 429, **463**
Gingell, A. D., **73**
Glück, C., **335**
González-Féres, R., 429
Grimm, R., 221, **271**, 319, 429
Grochola, A., **335**
Groenenboom, G. C., **127**, 221, 429
Gustavsson, M., **271**
Haller, E., **283**
Hart, R., **283**
Hecker Denschlag, J., **271**, 319, 429
Herschbach, D., **9**, 93
Hinds, E. A., **37**, 221, 319, 429
Hudson, J. J., **37**, 93, 221
Hutson, J. M., 93, **191**, 221, 319, 429
Jäger, W., 93, **297**
Jin, D. S., **351**

Julienne, P. S., 221, 319, **351, 361**, 429
Kay, J. J., **143**
Koch, C. P., 319, **389**, 429
Kosloff, R., 319, **389**, 429
Kotochigova, S., 221, **351**, 429
Krems, R. V., **127**
Küpper, J., 93, **155**, 221, 429
Lane, I., 93, 221, 429
Lang, F., **271**
Lange, J., **335**
Lara, M., 221
Launay, J.-M., 221
Leduc, M., **415**
Lehnig, R., **297**
Lemeshko, M., 93, 319, 429
Le Picard, S. D., 221
Lu, W., **311**, 319
Maher-McWilliams, C., **175**
Marian, A., 221
Mark, M. J., **283**
Martay, H., **403**
McCabe, D., **403**, 429
Meijer, G., 93, **113, 155**, 221
Mörtlbauer, K., **335**
Motsch, M., **203**
Mudrich, M., 93, 429
Nägerl, H. C., **283**
Ndong, M., **389**
Neyenhuis, B., **351**
Ni, K.-K., **351**
Oldham, J., **73**
Ospelkaus, S., **351**
Petrovic, J., **403**, 429
Pichler, M., 319
Pillet, P., **257**, 319, 429
Pinkse, P. W. H., 93, **203**, 221
Pohle, S., **203**
Portier, M., 221, **415**, 429
Purcell, S. M., **175**
Raston, P. L., 221, **297**
Rempe, G., **203**, 221
Repp, M., **335**
Ritsch, H., **283**
Sauer, B. E., **37**
Segal, D., 93
Sims, I., 93, 221
Sofikitis, D., **257**
Softley, T. P., **73**, 93, 221
Soldán, P., **191**
Sommer, C., **203**

Stoecklin, T., 93, 221
Strauss, C., **271**
Strecker, K. E., **143**
Stwalley, W. C., 93, 319, 429
Tacconi, M., **463**
Takekoshi, T., **271**
Tarbutt, M. R., **37**, 93, 221
Tennyson, J., 93, **175**
Tscherbul, T. V., **127**
Ubachs, W., **25**
van Buuren, L. D., **203**
van de Meerakker, S. T., **113**, **143**, 221, 429
Vanhaecke, N., 93, 221
Viteau, M., **257**
Wallis, A., 221
Walmsley, I., **403**
Weidemüller, M., 221, **335**, 429
Wester, R., 93, 221, **335**
Whitaker, B., 221, 429
Wang, D., **351**
Willitsch, S., **73**, 93
Winkler, K., **271**
Wrede, E., 221
Ye, J., 93, 319, **351**, 429
Zeppenfeld, M., 93, 429
Zhao, Y., **311**
Żuchowski, P. S., **191**, 221

* The page numbers in **bold** type indicate papers submitted for discussions.